Flavia Smarrazzo and Alberto Tesei

Measure Theory and Nonlinear Evolution Equations

De Gruyter Studies in Mathematics

Volume 86

Flavia Smarrazzo and Alberto Tesei

Measure Theory and Nonlinear Evolution Equations

DE GRUYTER

Mathematics Subject Classification 2010
Primary: 28-01, 28Axx, 28Bxx, 31A15; Secondary: 35D55, 35K55, 35L65, 35R25

Authors
Dr. Flavia Smarrazzo
Università Campus Bio-Medico di Roma
Via Alvaro del Portillo 21
00128 Roma
Italy
flavia.smarrazzo@gmail.com

Prof. Dr. Alberto Tesei
Università Sapienza di Roma
Dipartimento di Matematica
Piazzale Aldo Moro 5
00185 Roma RM
Italy
albertotesei@gmail.com

ISBN 978-3-11-055600-1
e-ISBN (PDF) 978-3-11-055690-2
e-ISBN (EPUB) 978-3-11-055604-9
ISSN 0179-0986

Library of Congress Control Number: 2022930253

Bibliographic information published by the Deutsche Nationalbibliothek
The Deutsche Nationalbibliothek lists this publication in the Deutsche Nationalbibliografie; detailed bibliographic data are available on the Internet at http://dnb.dnb.de.

Typesetting: VTeX UAB, Lithuania
Printing and binding: CPI books GmbH, Leck

www.degruyter.com

Contents

Preface

The study of measure-valued solutions of partial differential equations (PDEs in the sequel) combines two distant mathematical areas, measure theory and theory of PDEs. Since measure-valued solutions describe singularities of solutions of PDEs, a subject related to the concept of capacity, fundamental results of potential theory also appear in the discussion.

As a consequence, gathering from the literature what is needed for the study is often difficult. In books devoted to PDEs, often the results of measure theory (e. g., on Young measures) that are strictly necessary for applications are presented, to the detriment of the understanding of the theory as a whole. On the other hand, books on measure theory and probability frequently use terminology and arguments unfamiliar to the PDE scholar, while aspects related to the analytic functional framework remain in the background. To some extent, similar remarks apply to treatises on potential theory.

This book is aimed at presenting the topics mentioned above in a unified framework; analytical methods of proof are mostly used, and general aspects of functional analysis are highlighted. It is written for a wide range of possible interested parties, including the students and advanced mathematicians. Being self-contained, it is also intended both for self-study and as a reference book for well-known and less well-known things. The reader is expected to have a background in real analysis, topology, and functional analysis at the level of textbooks like [90]. Anyway, necessary preliminaries on topology are recalled in Appendix A at the end of the book.

A detailed description of the contents of the chapters is given at the beginning of each part. We do not consider it useful to suggest specific paths to the reader, who will proceed for himself according to his own taste and interests.

https://doi.org/10.1515/9783110556902-201

Introduction

Measure-valued solutions have an important role in problems of calculus of variations and nonlinear partial differential equations (PDEs) suggested by physics, chemistry, biology, and engineering.

Speaking of “measure-valued solutions” is actually ambiguous, since this term refers both to *Radon measure-valued* and to *Young measure-valued* solutions, two notions that remarkably fit in the same mathematical framework. It is informative to highlight some areas of research in which one or the other notion, or both, have played an important role. This is the content of the following section, which, though, is not intended to cover the whole spectrum of evolution problems where measure-valued solutions arise (in particular, we refer the reader to [6] for the theory of gradient flows in the Wasserstein space of probability measures and its applications to the transport equation).

Motivations

1 Radon measure-valued solutions of elliptic and parabolic PDE's

In the elliptic case, we are concerned with equations of the form

$$Lu = \mu \quad \text{in } \Omega, \tag{E_0}$$

where $\Omega \subseteq \mathbb{R}^N$ is open, $u = u(x)$ $(x \in \Omega)$, L is a linear or nonlinear elliptic operator, and μ is a Radon measure on Ω. If $\Omega \subset \mathbb{R}^N$ properly, boundary conditions at the boundary $\partial\Omega$ must be satisfied:

$$u = \nu \quad \text{on } \partial\Omega, \tag{BC}$$

the boundary data ν possibly being a Radon measure.

In the parabolic case, we think of initial value problems of the form

$$\begin{cases} \partial_t u + Lu = \mu & \text{in } \Omega \times (0,T) =: Q, \\ u = u_0 & \text{in } \Omega \times \{0\}, \end{cases} \tag{P_0}$$

where Ω is as above, $T > 0$, $u = u(x,t)$ $((x,t) \in Q)$, and either μ or u_0, or both, are Radon measures. Again, if $\Omega \subset \mathbb{R}^N$ properly, suitable boundary conditions at the lateral boundary $\partial\Omega \times (0,T)$, possibly involving Radon measures, must be satisfied.

A specimen of equation (E_0) and problem (P_0) are the Poisson equation and the Cauchy problem for the heat equation, respectively. These cases (which correspond to the choice $\Omega = \mathbb{R}^N$, $L = -\Delta$, $\mu = \delta_{x_0}$ in (E_0) and $\mu = 0$, $u_0 = \delta_{x_0}$ in (P_0), δ_{x_0} being the Dirac mass located at some point $x_0 \in \Omega$) played a central role in the linear theory of

https://doi.org/10.1515/9783110556902-202

PDEs, leading to the investigation of *fundamental solutions*, whose superposition by linearity allowed representing more general solutions.

On the strength of the linear theory, it seemed natural to study analogous problems in the nonlinear case. However, it soon became apparent that even in simple semilinear cases, there were obstacles to the very existence of solutions in the usual L^1-framework.

1.1 The elliptic case

A first case of nonexistence is described in vivid terms in [14] concerning the elliptic semilinear equation

$$-\Delta u + |u|^{p-1}u = \mu \quad \text{in } \mathbb{R}^N \quad (p > 0) \tag{0.1}$$

with $N \geq 3$. Equation (0.1) is related to the equation

$$\lambda v - \Delta(|v|^{\alpha-1}v) = \mu \quad \text{in } \mathbb{R}^N \quad (\lambda, \alpha > 0) \tag{0.2}$$

(setting $u = |v|^{\alpha-1}v$, $p = \frac{1}{\alpha}$, and scaling out λ), which is the resolvent equation for the porous medium equation

$$\partial_t v = \Delta(|v|^{\alpha-1}v) \quad \text{in } \mathbb{R}^N \times (0,T) \quad (\alpha, T > 0). \tag{0.3}$$

Since the nonlinear semigroup theory suggests to study equation (0.3) in the function space $L^1(\mathbb{R}^N)$, it is natural to assume that $\mu \in L^1(\mathbb{R}^N)$ in (0.1)–(0.2). A different motivation to study (0.1) came from the Thomas–Fermi model for the electron density of large atoms, in which case μ is a finite superposition of Dirac masses [65].

Let $\Omega \subseteq \mathbb{R}^N$ be open and bounded with smooth boundary $\partial\Omega$. It was proven in [28] that for any $\mu \in L^1(\Omega)$ and $p > 0$, there exists a unique solution $u \in W_0^{1,1}(\Omega)$, with $|u|^p \in L^1(\Omega)$, of the homogeneous Dirichlet problem

$$\begin{cases} -\Delta u + |u|^{p-1}u = \mu & \text{in } \Omega, \\ u = 0 & \text{in } \partial\Omega. \end{cases}$$

Well-posedness results for equation (0.1) were subsequently proven in [15].

Concerning the case where the right-hand side of (0.1) is a Radon measure, it was soon realized that the condition

$$p < \frac{N}{N-2} \tag{0.4}$$

plays a central role. In fact, let $\mathfrak{R}_f(\Omega)$ denote the Banach space of finite (signed) Radon measures on $\Omega \subseteq \mathbb{R}^N$, endowed with the norm $\|\mu\|_{\mathfrak{R}_f(\Omega)} := |\mu|(\Omega)$ ($|\mu|(\Omega) < \infty$ being the total variation of the measure μ; see Subsection 1.8.2). If (0.4) holds, then for every

$\mu \in \mathfrak{R}_f(\mathbb{R}^N)$, there exists a unique function u in the Marcinkiewicz space $M^{\frac{N}{N-2}}(\mathbb{R}^N)$ that solves equation (0.1) in the following sense: we have $v := \Delta u + \mu \in L^1(\mathbb{R}^N)$ and $v = |u|^{p-1}u$ a. e. in $\mathbb{R}^N$ (see [14, Appendix A]). On the other hand, no function exists that satisfies (0.1) with $\mu = \delta_{x_0}$ ($x_0 \in \mathbb{R}^N$) if condition (0.4) is not satisfied (this follows by a simple direct argument; see [14, Remark A.4]).

The above results are now well understood. In fact, the method used to prove the existence of solutions of (0.1) is natural and widely used (possibly with some variant; e. g., see [27]). Consider the family of *approximating problems*

$$-\Delta u_n + |u_n|^{p-1}u_n = \mu_n \quad \text{in } \mathbb{R}^N, \tag{0.5}$$

where $\{\mu_n\} \subseteq L^1(\mathbb{R}^N) \cap C^\infty(\mathbb{R}^N)$ is an approximating sequence of μ such that

$$\|\mu_n\|_{L^1(\mathbb{R}^N)} \le \|\mu\|_{\mathfrak{R}_f(\mathbb{R}^N)}, \tag{0.6}$$

$$\lim_{n\to\infty} \int_{\mathbb{R}^N} \rho\,\mu_n\,dx = \int_{\mathbb{R}^N} \rho\,d\mu \quad \text{for any } \rho \in C_0(\mathbb{R}^N) \tag{0.7}$$

(namely, $\lim_{n\to\infty} \mu_n = \mu$ weakly* in $\mathfrak{R}_f(\mathbb{R}^N)$; see Section 5.1). By the results in [15], for each $n \in \mathbb{N}$, there exists a unique solution $u_n \in M^{\frac{N}{N-2}}(\mathbb{R}^N)$ of (0.5). A priori estimates of u_n and the uniform bound (0.6) ensure that (*i*) the sequence $\{u_n\}$ is relatively compact in $L^1_{\mathrm{loc}}(\mathbb{R}^N)$ and (*ii*) the sequence $\{v_n\}$, $v_n := \Delta u_n + \mu_n$, is bounded in $L^1(\mathbb{R}^N)$. Moreover, if (0.4) holds, then the sequence $\{v_n\}$ is *uniformly integrable* on the bounded subsets of $\mathbb{R}^N$ and hence relatively weakly compact in $L^1_{\mathrm{loc}}(\mathbb{R}^N)$ by the Dunford–Pettis theorem (see [14, Lemma A.1] and Theorem 2.8.18). Then letting $n \to \infty$ in (0.5) and using (0.7) the stated existence result follows.

If (0.4) does not hold, then the uniform integrability of $\{v_n\}$ fails. In view of the Dunford–Pettis theorem, in this case, $\{v_n\}$ is not relatively weakly compact in $L^1_{\mathrm{loc}}(\mathbb{R}^N)$, in agreement with the above nonexistence statement. However, even in this case the sequence $\{v_n\}$ is bounded in $L^1(\mathbb{R}^N) \subseteq \mathfrak{R}_f(\mathbb{R}^N)$ and thus relatively compact in the weak* topology of $\mathfrak{R}_f(\mathbb{R}^N)$ (see Theorem 5.1.7). Hence there exists a *finite Radon measure* v such that

$$\lim_{n\to\infty} \int_{\mathbb{R}^N} \rho\,v_n\,dx = \int_{\mathbb{R}^N} \rho\,dv \quad \text{for any } \rho \in C_0(\mathbb{R}^N).$$

This suggests to seek the existence of solutions of (a suitably modified version of) equation (0.1) in a larger class of *Radon measure-valued solutions* (see [97]).

On the other hand, the measures $\mu \in \mathfrak{R}_f(\mathbb{R}^N)$ for which equation (0.1) has a solution $u \in L^p(\mathbb{R}^N)$ have been characterized in [10] (see also [52]): for any $p \in (1, \infty)$, this happens if and only if $\mu \in L^1(\mathbb{R}^N) + W^{-2,p}(\mathbb{R}^N)$. By Theorem 3.4.15 an equivalent statement is that (0.1) has a solution $u \in L^p(\mathbb{R}^N)$ if and only if μ is *diffuse* with respect to the *Sobolev capacity* $C_{2,q}$, where $q := \frac{p}{p-1}$ is the conjugate exponent of p, that is, if and only if

$$\mu(E) = 0 \quad \text{for any Borel set } E \subseteq \mathbb{R}^N \text{ such that } C_{2,q}(E) = 0 \tag{0.8}$$

(see Subsections 1.8.4 and 3.3.3).

By Remark 3.3.2(ii) there are no Borel sets $E \subseteq \mathbb{R}^N$ such that $C_{2,q}(E) = 0$ if $p \in (1, \infty)$ and $2q > N \Leftrightarrow p < \frac{N}{N-2}$, and hence condition (0.8) is void if (0.4) holds (see Proposition 3.4.11). On the other hand, when $p \geq \frac{N}{N-2}$, we have $C_{2,q}(\{x_0\}) = 0$ for any $x_0 \in \mathbb{R}^N$, thus the Dirac mass δ_{x_0} is *concentrated* with respect to $C_{2,q}$, and condition (0.8) is not satisfied (see Definition 1.8.10 and Proposition 3.3.5). In other words, if $p \geq \frac{N}{N-2}$, then the Dirac mass is not a *removable singularity* for (0.1) (see [105] and references therein). This explains the above existence and nonexistence results for equation (0.1).

Remark 0.1.1. With reference to equation (0.2), it follows from the above remarks that no function solves

$$\lambda v - \Delta(|v|^{\alpha-1}v) = \mu \quad \text{in } \mathbb{R}^N \quad (\lambda, \alpha > 0)$$

if $N \geq 3$, μ is *concentrated* with respect to the *Sobolev capacity* $C_{2,\frac{1}{1-\alpha}}$, and $\alpha \leq \frac{N-2}{N}$. This is in agreement with the parabolic results in [26] and [80, Theorems 1, 2] (see below).

Let us mention that the above existence and nonexistence results concerning equation (0.1) stimulated subsequent work in various directions, in particular, concerning different nonlinear elliptic operators in problem (E_0) and/or the case where $\partial\Omega \neq \emptyset$ and $g \in \mathfrak{R}_f(\partial\Omega)$ in (BC) (see [14, 13, 37, 55, 69, 106] and references therein).

1.2 The parabolic case

Consider the Cauchy problem for the *fast diffusion* porous medium equation

$$\begin{cases} \partial_t u = \Delta u^\alpha & \text{in } \mathbb{R}^N \times (0, T) =: S, \\ u = u_0 & \text{in } \mathbb{R}^N \times \{0\}, \end{cases} \tag{C}$$

where u_0 belongs to the cone $\mathfrak{R}_f^+(\mathbb{R}^N)$ of nonnegative finite Radon measures on $\mathbb{R}^N$ $(N \geq 3)$, and $\alpha \in (0, \frac{N-2}{N}]$. In [80] a solution of problem (C) is, by definition, any $u \in L^1(S)$ that satisfies the first equation in distributional sense and the second in the sense of the *narrow convergence*, namely,

$$\operatorname*{ess\,lim}_{t \to 0^+} \int_{\mathbb{R}^N} \rho\, u(\cdot, t)\, dx = \int_{\mathbb{R}^N} \rho\, du_0 \quad \text{for any } \rho \in C_b(\mathbb{R}^N)$$

(see Definition 5.1.3; here $C_b(\mathbb{R}^N)$ denotes the space of continuous bounded real functions on $\mathbb{R}^N$). It is proven that such a solution – more precisely, a solution $u \in C(0,T;L^1(\mathbb{R}^N))$ – exists if and only if u_0 is diffuse with respect to the Sobolev capacity $C_{2,\frac{1}{1-\alpha}}(\mathbb{R}^N)$ (see [80, Theorems 1, 2]). This is in agreement with [26], where it was observed that such a solution of (*C*) with $u_0 = \delta_0$ exists if and only if $\alpha \in (\frac{N-2}{N},1)$; in fact, for a point $x_0 \in \mathbb{R}^N$, we have $C_{2,\frac{1}{1-\alpha}}(\{x_0\}) = 0$, and thus the Dirac mass δ_{x_0} is concentrated with respect to $C_{2,\frac{1}{1-\alpha}}$, if and only if $\alpha \in (0,\frac{N-2}{N}]$ (see Proposition 3.4.11). The result also agrees with Theorem 6.7.1 (see equality (6.135)). The connection with the above elliptic results is apparent (see Remark 0.1.1).

Following (*C*), initial value problems for quasilinear parabolic equations whose initial datum is a Radon measure have been widely investigated (in particular, see [16, 26, 35, 36, 81, 104] and references therein), always seeking *function-valued solutions*. To highlight this point, consider the initial-boundary value problem

$$\begin{cases} \partial_t u = \Delta\phi(u) & \text{in } \Omega\times(0,T) =: Q, \\ u = 0 & \text{on } \partial\Omega\times(0,T), \\ u = u_0 & \text{in } \Omega\times\{0\}, \end{cases} \tag{P}$$

which will be discussed at length in Chapter 6; here $\Omega \subseteq \mathbb{R}^N$ is an open bounded set with smooth boundary $\partial\Omega$, $T > 0$, $u_0 \in \mathfrak{R}_f(\Omega)$, and ϕ is assumed to be continuous and nondecreasing in $\mathbb{R}$. In our parlance, a solution u of (*P*) is *function-valued* if for any $t > 0$, $u(\cdot,t) \in L^1(\Omega)$. Such a solution can be viewed as a particular *Radon measure-valued solution*, namely, a solution u of (*P*) such that $u(\cdot,t) \in \mathfrak{R}_f(\Omega)$ for positive times (see [85, 86, 87] and references therein). Whether or not such a Radon measure-valued solution becomes function-valued for some $t > 0$ is a *regularity* problem, which can be also regarded as a problem of *removal of singularities*. Not surprisingly, as we will see below, this issue is related to some characteristic *capacity* pertaining to the evolution equation under consideration.

Defining Radon measure-valued solutions of (*P*) raises the question of how to define nonlinear functions of such measures. In [11] a natural definition was proposed for a similar problem by a heuristic argument, in agreement with the notion of a nonlinear function of measures given in [38, 39]. Accordingly, we think of a Radon measure-valued solution of (*P*) as a map from $(0,T)$ to $\mathfrak{R}_f(\Omega)$ that satisfies (*P*) in the following weak sense:

$$\int_0^T \langle u(\cdot,t), \partial_t\zeta(\cdot,t)\rangle dt = \iint_Q \nabla[\phi(u_r)]\cdot\nabla\zeta\, dxdt - \langle u_0, \zeta(\cdot,0)\rangle$$

for a suitable class of test functions ζ (see Chapter 6 for technical details). Specifically, the measure $u \in \mathfrak{R}_f(Q)$ is required to belong to the space $L^\infty_{w*}(0,T;\mathfrak{R}_f(\Omega))$, thus the Radon measure $u(\cdot,t)$ on Ω is defined for a. e. $t \in (0,T)$, and the function $\phi(u_r)$ is required to belong to the space $L^1(0,T;W_0^{1,1}(\Omega))$. Here $u_r \in L^1(Q)$ is the density of the absolutely continuous part u_{ac} of u with respect to the Lebesgue measure, and $\langle\cdot,\cdot\rangle$ denotes the duality map between $C_0(\overline{\Omega})$ and $\mathfrak{R}_f(\Omega)$ (see Chapters 2 and 4; by abuse of notation, we sometimes further identify u_r with u_{ac}).

Radon measure-valued solutions of (P) can be constructed as outlined before in the elliptic case, approximating the initial data u_0 by a suitable sequence $\{u_{0n}\} \subseteq L^1(\Omega)$ and studying the convergence in $L^\infty_{w*}(0,T;\mathfrak{R}_f(\Omega))$ of the corresponding sequence of solutions of the approximating problems (see Chapter 6). As for regularity, if the function $|\phi(z)|$ diverges more than $|z|^{\frac{N-2}{N}}$ as $|z| \to \infty$, then every constructed solution of (P) belongs to $L^\infty(0,T;L^1(\Omega))$, that is, an *instantaneous* $\mathfrak{R}_f$-L^1 *regularization* occurs (see Proposition 6.7.9). On the other hand, if for some $M > 0$ and $\alpha \in (0, \frac{N-2}{N}]$, $|\phi(z)| \le M(1+|z|)^\alpha$ ($z \in \mathbb{R}$), then the $C_{2,\frac{1}{1-\alpha}}$-concentrated part of any weak solution of (P) is constant in time, in agreement with [80] (see Theorem 6.7.1). Therefore, in this case, every such solution is Radon measure-valued for any positive time.

2 Young and Radon measure-valued solutions of hyperbolic conservation laws

Consider the Cauchy problem

$$\begin{cases} \partial_t u + \operatorname{div}[\phi(u)] = 0 & \text{in } S, \\ u = u_0 & \text{in } \mathbb{R}^N \times \{0\}, \end{cases} \tag{CL}$$

with $\phi \equiv (\phi_1,\ldots,\phi_N) \in (C^1(\mathbb{R}))^N$, $\phi_j : \mathbb{R} \mapsto \mathbb{R}$ nonlinear ($j = 1,\ldots,N$).

2.1 Young measure-valued solutions

Let $u_0 \in L^\infty(\mathbb{R}^N)$. A classical way to prove the existence of weak solutions of (CL) is the *vanishing viscosity method*, which relies on the companion parabolic problem

$$\begin{cases} \partial_t u_\epsilon + \operatorname{div}[\phi(u_\epsilon)] = \epsilon\,\Delta u_\epsilon & \text{in } S \\ u_\epsilon = u_{0\epsilon} & \text{in } \mathbb{R}^N \times \{0\} \end{cases} \quad (\epsilon > 0), \tag{CL_ϵ}$$

where $u_{0\epsilon} \in C_b(\mathbb{R}^N)$ and $\|u_{0\epsilon}\|_{L^\infty(\mathbb{R}^N)} \le \|u_0\|_{L^\infty(\mathbb{R}^N)}$. For every $\epsilon > 0$, there exists a unique classical solution u_ϵ of (CL_ϵ), which by the maximum principle satisfies the inequality

$$\|u_\epsilon\|_{L^\infty(S)} \le \|u_0\|_{L^\infty(\mathbb{R}^N)}. \tag{0.9}$$

By inequality (0.9) and the Banach–Alaoglu theorem there exist a sequence $\{u_k\} \equiv \{u_{\epsilon_k}\} \subseteq L^\infty(S)$ with $\epsilon_k \to 0^+$ as $k \to \infty$ and $u \in L^\infty(S)$ such that $\lim_{k\to\infty} u_k = u$ weakly* in $L^\infty(S)$. Then u is an obvious candidate to be a weak solution of (CL). However, letting $k \to \infty$ in the right-hand side of the weak formulation of (CL_{ϵ_k}),

$$\iint_S \{u_k\, \partial_t\zeta + \epsilon_k u_k \Delta\zeta\}\, dxdt + \int_{\mathbb{R}} u_0\, \zeta(x,0)\, dx = -\iint_S [\phi(u_k)] \cdot \nabla\zeta\, dxdt, \tag{0.10}$$

is cumbersome, since the weak* convergence in $L^\infty(S)$ of u_k to u does not imply that of $\phi_j(u_k)$ to $\phi_j(u)$, $j = 1, \dots, N$ (see Example 2.8.1(i); here ζ is a suitable test function).

To overcome this difficulty, we need further information, for instance, the convergence $u_k \to u$ a. e. in S (of some subsequence of $\{u_k\}$, not relabeled for simplicity). In fact, by inequality (0.9), the continuity of ϕ, and the dominated convergence theorem, this convergence would imply that

$$\lim_{k\to\infty} \iint_S [\phi(u_k)] \cdot \nabla\zeta\, dxdt = \iint_S [\phi(u)] \cdot \nabla\zeta\, dxdt.$$

Then letting $k \to \infty$ in (0.10), we would obtain, as anticipated,

$$\iint_S \{u\, \partial_t\zeta + [\phi(u)] \cdot \nabla\zeta\}\, dxdt + \int_{\mathbb{R}} u_0\, \zeta(x,0)\, dx = 0. \tag{0.11}$$

To establish the a. e. convergence of $\{u_k\}$ in S, a typical approach is setting $\mathbb{R}^N = \bigcup_{n=1}^\infty K_n$ with each K_n compact and then proving by the Fréchet–Kolmogorov theorem that for any fixed $n \in \mathbb{N}$, the sequence of restrictions $\{u_k|_{K_n\times(0,T)}\}$ are relatively compact in $L^1(K_n \times (0,T))$. Then by classical results and a diagonal argument the result follows (see the proof of [68, Theorem 4.62 of Chapter 2]).

Let us outline a different method, which makes use of Young measures (e. g., see [68] and references therein). Let λ_N denote the Lebesgue measure on $\mathbb{R}^N$, let $U \subseteq S \subseteq \mathbb{R}^{N+1}$ be open and bounded, and let $\mathfrak{R}_f^+(U \times \mathbb{R})$ be the cone of finite (positive) Radon measures on $U \times \mathbb{R}$. By definition a Young measure on $U \times \mathbb{R}$ is any $\nu \in \mathfrak{R}_f^+(U \times \mathbb{R})$ such that

$$\nu(E \times \mathbb{R}) = \lambda_{N+1}(E) \quad \text{for any Borel set } E \subseteq U.$$

If $f : U \mapsto \mathbb{R}$ is measurable, then the Young measure ν_f such that

$$\nu_f(E \times F) = \lambda_{N+1}(E \cap f^{-1}(F)) \quad \text{for all Borel sets } E \subseteq U \text{ and } F \subseteq \mathbb{R}$$

is called the *Young measure associated with* the function f (see Definition 2.5.3).

Let us denote by $\mathfrak{Y}(U;\mathbb{R})$ the set of Young measures on $U \times \mathbb{R}$ $(U \subseteq S)$ and by $\mathscr{C}_b(U \times \mathbb{R})$ that of *bounded Carathéodory integrands* on $U \times \mathbb{R}$ (see Definition 5.3.1). The following results are particular consequences of Proposition 5.3.1, Theorem 5.4.5, and Proposition 5.4.10.

Proposition 0.2.1. *Let $\nu \in \mathfrak{Y}(U;\mathbb{R})$. Then for a.e. $(x,t) \in U$, there exists a probability measure $\nu_{(x,t)}$ on $\mathbb{R}$ such that for any $g \in \mathscr{C}_b(U;\mathbb{R})$, the map $(x,t) \to \int_{\mathbb{R}} g(x,t,y)\,d\nu_{(x,t)}(y)$ is integrable on U, and*

$$\int_{U\times\mathbb{R}} g\,d\nu = \int_U dxdt \int_{\mathbb{R}} g(x,t,y)\,d\nu_{(x,t)}(y).$$

The family $\{\nu_{(x,t)}\} \equiv \{\nu_{(x,t)}\}_{(x,t)\in U}$ is a *parameterized measure* on $\mathbb{R}$, called *disintegration* of ν (see Definitions 5.2.1 and 5.2.2). The function u^* defined as

$$u^*(x,t) := \int_{\mathbb{R}} y\,d\nu_{(x,t)}(y) \quad ((x,t) \in U) \tag{0.12}$$

is called the *barycenter* of the disintegration $\{\nu_{(x,t)}\}$.

Theorem 0.2.2. *Let $\{u_n\} \subseteq L^1(U)$ be bounded, and let $\{\nu_n\} \equiv \{\nu_{u_n}\}$ be the sequence of associated Young measures. Then:*

(i) *there exist a subsequence $\{u_{n_k}\} \subseteq \{u_n\}$ and a Young measure ν on $U \times \mathbb{R}$ such that $\nu_{n_k} \to \nu$ narrowly;*

(ii) *for any $f \in C(\mathbb{R})$ such that the sequence $\{f(u_{n_k})\} \subseteq L^1(U)$ is bounded and uniformly integrable, we have*

$$\int_U dxdt \int_{\mathbb{R}} |f(y)|\,d\nu_{(x,t)}(y) < \infty, \quad f(u_{n_k}) \rightharpoonup f^* \quad \text{in } L^1(U),$$

where

$$f^*(x,t) := \int_{\mathbb{R}} f(y)\,d\nu_{(x,t)}(y) \quad \text{for a.e. } (x,t) \in U; \tag{0.13}$$

(iii) *for any Carathéodory function $g : U \times \mathbb{R} \to \mathbb{R}$ such that the sequence $\{g(\cdot,u_{n_k})\} \subseteq L^1(U)$ is bounded and uniformly integrable, we have*

$$\lim_{k\to\infty} \int_U g(x,t,u_{n_k}(x,t))\,dxdt = \int_{U\times\mathbb{R}} g\,d\nu.$$

(iv) [Biting lemma] *there exist a subsequence $\{u_{n_{k_j}}\} \equiv \{u_{n_j}\} \subseteq \{u_{n_k}\}$ and a sequence of measurable sets $\{U_j\}$ such that $U_{j+1} \subseteq U_j \subseteq U$ for any $j \in \mathbb{N}$, $\lambda_{N+1}(U_j) \to 0$ as $j \to \infty$, and the sequence $\{u_{n_j}\chi_{U\setminus U_j}\}$ is uniformly integrable. Moreover, $u^* \in L^1(U)$, and $u_{n_j}\chi_{U\setminus U_j} \rightharpoonup u^*$ in $L^1(U)$.*

Let us go back to the problem of letting $k \to \infty$ in (0.10). Set again $\mathbb{R}^N = \bigcup_{n=1}^{\infty} K_n$ with each K_n compact. For any fixed $n \in \mathbb{N}$, consider the sequence of restrictions $\{u_k|_{K_n\times(0,T)}\}$, which is bounded in $L^1(K_n \times (0,T))$ and uniformly integrable by Lemma 2.8.12. Clearly, the same holds for the sequence $\{\phi(u_k)|_{K_n\times(0,T)}\}$. Then by Theorem 0.2.2(ii) (see also Remark 5.4.2(ii)) and a diagonal argument there exist a subsequence of $\{u_k\}$ (not relabeled for simplicity) and a Young measure $\nu \in \mathfrak{Y}(S;\mathbb{R})$ such that

$$u_k \rightharpoonup u^*, \quad \phi_j(u_k) \rightharpoonup \phi_j^* \quad \text{in } L^1(U)$$

for any bounded open $U \subseteq S$, where

$$\phi_j^*(x,t) := \int_{\mathbb{R}} \phi_j(y)\,d\nu_{(x,t)}(y) \quad \text{for a. e. } (x,t) \in U \quad (j = 1,\dots,N). \tag{0.14}$$

Since $\lim_{k\to\infty} u_k = u$ weakly* in $L^\infty(S)$, it follows that $u = u^*$. We say that the couple (u,ν) is a *weak Young measure-valued solution* of (CL), meaning that

$$u(x,t) = \int_{\mathbb{R}} y\,d\nu_{(x,t)}(y) \quad \text{for a. e. } (x,t) \in S, \tag{0.15a}$$

and for any $\zeta \in C^1([0,T];C_c^1(\mathbb{R}))$ with $\zeta(\cdot,T) = 0$ in $\mathbb{R}$, we have

$$\iint_S \{u\,\partial_t\zeta + \phi^*\cdot\nabla\zeta\}\,dxdt + \int_{\mathbb{R}} u_0\,\zeta(x,0)\,dx = 0 \tag{0.15b}$$

with $\phi^* \equiv (\phi_1^*,\dots,\phi_N^*)$ given by (0.14).

So far only inequality (0.9) has been used. More can be said if each component ϕ_j of the flux is *genuinely nonlinear* (see assumption (A_2) in Chapter 7). In this case, it can be proven that ν is in fact the Young measure associated with u, that is, $\nu = \nu_u$. Then by Proposition 5.3.2 $\nu_{(x,t)} = \delta_{u(x,t)}$ for a. e. $(x,t) \in S$ (see (5.53)), whence by (0.14) $\phi^* = \phi(u)$ a. e. in S, and equality (0.15b) reduces to (0.11). Equivalently, knowing that $\nu = \nu_u$ implies that, up to a subsequence, $u_k \to u$ a. e. in S (see Proposition 5.4.1(ii)), and thus we can argue as before to get (0.11).

The proof of the key equality $\nu = \nu_u$ relies on the so-called Murat–Tartar equality, which in turn makes use of the div–curl lemma (see [46, Corollary 1.3.1 and Theorem 5.2.1], [68, Section 3.3]; we will use a similar argument to prove Proposition 7.5.6).

2.2 Radon measure-valued solutions

Problem (CL) in one space dimension,

$$\begin{cases} \partial_t u + \partial_x[\phi(u)] = 0 & \text{in } \mathbb{R}\times(0,\infty), \\ u = u_0 & \text{in } \mathbb{R}\times\{0\}, \end{cases} \tag{CL_1}$$

was studied in [67] with $u_0 \in \mathfrak{R}_f^+(\mathbb{R})$ for a class of fluxes ϕ superlinear at infinity (the model case being $\phi(u) = u^m$, $m > 1$). Entropy solutions of (CL_1) were meant in the following sense:

- for positive times, u is a *function*: $u \in L^\infty(0,T;L^1(\mathbb{R})) \cap L^\infty((\tau,T)\times\mathbb{R})$ for every $\tau \in (0,T)$;
- for all $k \in \mathbb{R}$ and $\zeta \in C_c^1(S)$, $\zeta \geq 0$, we have the *entropy inequality*

$$\iint_S \{|u-k|\,\partial_t\zeta + \operatorname{sgn}(u-k)[\phi(u)-\phi(k)]\partial_x\zeta\}\,dxdt \geq 0;$$

- the initial condition is satisfied in the sense of the narrow convergence:

$$\operatorname*{ess\,lim}_{t\to0^+} \int_{\mathbb{R}} u(\cdot,t)\rho\,dx = \int_{\mathbb{R}} \rho\,du_0 \quad \text{for any } \rho \in C_b(\mathbb{R}).$$

The following result was proven.

Proposition 0.2.3. *Let $\phi : [0,\infty) \mapsto [0,\infty)$, $\phi(0) = 0$ be increasing, and let $z \mapsto [\phi(z)]^{1-\alpha}$ be convex on $(0,\infty)$ for some $\alpha \in (0,1)$. Then there exists a unique entropy solution $u \geq 0$ of (CL_1).*

The existence part of Proposition 0.2.3 follows by a constructive procedure, in which the initial Radon measure $u_0 \geq 0$ is approximated by a sequence $\{u_{0n}\} \subseteq L^\infty(\mathbb{R}) \cap L^1(\mathbb{R})$, $u_{0n} \to u_0$ narrowly in $\mathbb{R}$. If $\phi(u) = u^m$ with $m > 1$ and $u_0 = \delta_0$, then this procedure gives the fundamental solution of the Burgers equation, called *N-wave*:

$$u(x,t) = \left(\frac{x}{mt}\right)^{1/(m-1)} \chi_A(x,t), \tag{0.16a}$$

$$A := \left\{(x,t) \mid 0 \leq x \leq \left(\frac{m}{m-1}\right)^{(m-1)/m} t^{1/m},\ 0 < t \leq T\right\}. \tag{0.16b}$$

Hence the solution of (CL_1) in this case is function valued (plainly, the same holds for $m \in (0,1)$). Instead, in the linear case $m = 1$ the initial singularity δ_0 obviously persists, and thus the solution is Radon measure valued for all times.

A novel situation prevails when studying (CL_1) with *bounded* ϕ, a problem (motivated by a mathematical model of ion etching [88]; see also [50]) that is addressed in Chapter 7. To illustrate its main features, consider the model

$$\begin{cases} \partial_t u + \partial_x[1 - \frac{1}{(1+u)^m}] = 0 & \text{in } \mathbb{R}\times(0,\infty), \quad m > 0, \\ u = \delta_0 & \text{in } \mathbb{R}\times\{0\}, \end{cases} \tag{CL_2}$$

where $\phi(z) = 1 - \frac{1}{(1+z)^m}$ is increasing and concave and belongs to a class for which the constructed entropy solution of problem (CL_2) is unique (see Definition 7.3.1 and Theorems 7.3.3–7.3.4). As in [67], we construct entropy solutions of (CL_2) as limiting points (in a suitable topology) of the sequence of entropy solutions of the *approximating Riemann problems*:

$$\begin{cases} \partial_t u_n + \partial_x[1 - \frac{1}{(1+u_n)^m}] = 0 & \text{in } S, \\ u_n = n\chi_{(-\frac{1}{2n},\frac{1}{2n})} & \text{in } \mathbb{R} \times \{0\}. \end{cases} \qquad (R_n)$$

An elementary analysis of (R_n) shows that a shock starts from the point $(-\frac{1}{2n}, 0)$, a rarefaction fan from $(\frac{1}{2n}, 0)$. They meet at a point (x_n, t_n),

$$x_n := \frac{t_n}{2}\left[\frac{\phi(n)}{n} + \phi'(n)\right], \quad t_n := \frac{1}{\phi(n) - n\phi'(n)} \qquad (n \in \mathbb{N}),$$

where a new shock starts. It is easily seen that $\lim_{n\to\infty} t_n = 1$ and $\lim_{n\to\infty} x_n = 0$. This explains the following result (see [18]).

Proposition 0.2.4. *The unique (constructed) entropy solution of problem (CL_2) is $u = u_r + u_s$, where*

$$u_r(x,t) := \left[\left(\frac{mt}{x}\right)^{\frac{1}{1+m}} - 1\right]\chi_A(x,t), \quad u_s(\cdot,t) := [1-t]_+\delta_0,$$
$$A := \{(x,t) \mid 0 < x \le mt, 0 \le t \le 1\} \cup \{(x,t) \mid \xi(t) \le x \le mt, 1 < t \le T\}, \qquad (0.17)$$

and ξ solves the problem

$$\begin{cases} \xi' = -\frac{(mt\xi^{-1})^{-\frac{m}{1+m}} - 1}{(mt\xi^{-1})^{\frac{1}{1+m}} - 1} & \text{in } (1,T), \\ \xi(1,1) = 0. \end{cases}$$

In view of (0.17), in this case the singular part of the solution of (CL_1) is nonzero until the *waiting time* $t = 1$ and vanishes for $t \ge 1$, and hence we have *deferred regularization*. Another important remark is that the singular part $u_s(\cdot,t)$ is concentrated at $x_0 = 0$ for $0 < t < \tau_1 = 1$ (see (0.17)). When approaching $\operatorname{supp} u_s(\cdot,t) = \{0\}$, for any $t \in (0,\tau_1)$, the density $u_r(\cdot,t)$ satisfies the *compatibility conditions*

$$\lim_{x\to 0^-} \phi(u_r(x,t)) \equiv \phi(u_r(0^-,t)) = 0, \quad \lim_{x\to 0^+} \phi(u_r(x,t)) \equiv \phi(u_r(0^+,t)) = 1. \qquad (0.18)$$

We refer the reader to Chapter 7 to highlight the essential role played by the compatibility conditions in the uniqueness proof for problem (CL_1) with bounded ϕ (see Definition 7.2.1 and Theorem 7.2.4).

3 Young and Radon measure-valued solutions of ill-posed problems

Let us now consider the *forward-backward* parabolic equation

$$\partial_t u = \Delta\phi(u) \quad \text{in } \Omega \times (0,T) =: Q, \tag{FB}$$

where $\Omega \subseteq \mathbb{R}^N$ ($N \geq 1$) is open, and $\phi \in C^1(\mathbb{R})$ is *nonmonotonic*. Equation (*FB*) can be seen as the gradient flow (in the H^{-1} topology) of the functional

$$\mathcal{E}[u] := \int_\Omega \Phi(u)\,dx, \tag{0.19}$$

where $\Phi' = \phi$. Since ϕ is nonmonotonic, the functional $\mathcal{E}$ is nonconvex, and initial-boundary value problems for (*FB*) are ill-posed. The dynamics associated with (*FB*) is relevant whenever nonconvex functionals arise, e. g., in phase transitions, nonlinear elasticity, and image processing (see [29, 73, 79]). Ill-posedness is revealed by the lack of uniqueness of solutions (see [59], [108] for ϕ satisfying (A_1), respectively, (A_2) below).

In the Landau theory of phase transitions the function $\Phi(u) = (1-u^2)^2$ is the double-well potential, and thus ϕ is a cubic:

$$\begin{cases} \phi'(u) > 0 \ \text{ if } \ u \in (-\infty, b) \cup (a, \infty), \ \phi'(u) < 0 \ \text{ if } \ u \in (b,a); \\ \phi(u) \to \pm\infty \text{ as } u \to \pm\infty. \end{cases} \tag{A_1}$$

The three monotone branches of the graph of $v = \phi(u)$ are denoted by

$$\begin{aligned} u &:= s_1(v),\ v \in (-\infty, B) &\Leftrightarrow&\quad v = \phi(u),\ u \in (-\infty, b), \\ u &:= s_0(v),\ v \in (A, B) &\Leftrightarrow&\quad v = \phi(u),\ u \in (b, a), \\ u &:= s_2(v),\ v \in (A, \infty) &\Leftrightarrow&\quad v = \phi(u),\ u \in (a, \infty) \end{aligned} \tag{0.20}$$

(here $A := \phi(a)$, $B := \phi(b)$; see Fig. 1). Heuristically, the branches s_1 and s_2 correspond to stable phases, s_0 corresponds to an unstable phase, and equation (*FB*) describes the dynamics of transition between stable phases.

Ill-posedness calls for several refinements of the Landau theory, including in particular *nonlocal spatial* effects and/or *time delay* effects. In the first case, starting from the so-called Allen–Cahn functional

$$\mathcal{E}[u] := \int_\Omega \left\{\Phi(u) + \frac{\kappa}{2}|\nabla u|^2\right\} dx,$$

we obtain the Cahn–Hilliard equation

$$\partial_t u = \Delta[\phi(u) - \kappa\Delta u] \quad (\kappa > 0), \tag{0.21}$$

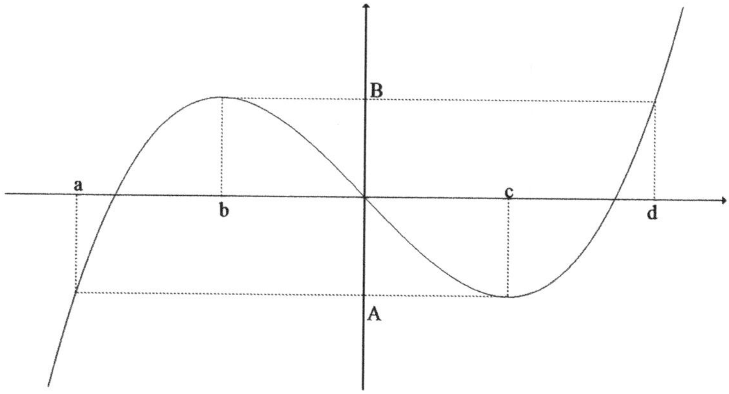

Figure 1: Cubic ϕ.

namely a *fourth-order regularization* of (*FB*). Instead, taking time delay effects into account leads to the *Sobolev regularization* of (*FB*):

$$\partial_t u = \Delta[\phi(u) + \epsilon\, \partial_t u] \quad (\epsilon > 0).$$

If both effects are included, then we obtain the viscous Cahn–Hilliard equation

$$\partial_t u = \Delta[\phi(u) - \kappa\Delta u + \epsilon\, \partial_t u].$$

Other regularizations of (*FB*), based on time or space discretization, have been used by several authors [12, 43, 57, 61].

Beside phase transitions, a motivation for studying equation (*FB*) comes from population dynamics, where it arises as a quasi-continuous approximation of a discrete model for aggregating populations (see [76, 77] and references therein). In this framework, $u \geq 0$ has the meaning of population density, and $\phi \in C^2([0,\infty))$ is a nonmonotonic function with the following properties:

$$\begin{cases} \phi(u) > 0 & \text{if } u > 0,\ \phi(0) = 0; \\ \phi'(u) > 0 & \text{if } 0 \leq u < \bar{u},\ \phi'(u) < 0 \text{ if } u > \bar{u}; \\ \phi'(\bar{u}) = 0,\ \phi''(\bar{u}) \neq 0,\ \phi(u) \to 0 & \text{as } u \to \infty; \\ \phi \in L^p(0,\infty) & \text{for some } p \in [1,\infty) \end{cases} \qquad (A_2)$$

(see Fig. 2). Functions satisfying (A_2) are often called "of Perona–Malik type", since for $N = 1$, the Perona–Malik equation reads

$$\partial_t z = \partial_x[\phi(\partial_x z)] \quad \text{with} \quad \phi(u) = \frac{u}{1+u^2} \quad (u \geq 0), \qquad (0.22)$$

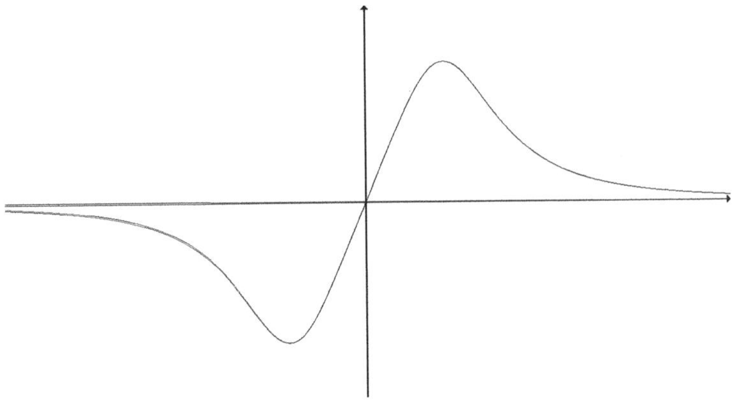

Figure 2: ϕ of Perona–Malik type.

and ϕ above satisfies (A_2) (see [79]). Equation (0.22) also models the formation of layers of constant temperature (or salinity) in the ocean [11]. Setting $u := \partial_x z$ in (0.22) gives

$$\partial_t u = \partial_{xx}\left(\frac{u}{1+u^2}\right) \quad (u \geq 0), \tag{0.23}$$

a particular case of (*FB*) (let us mention that equation (*FB*) with $N = 1$ and cubic ϕ was studied in [77]). Observe that the first equation in (0.22) is the gradient flow of (0.19) with the Perona–Malik potential $\Phi(u) = \log(1+u^2)$.

Sound modeling arguments were produced in [11] to motivate the *pseudoparabolic regularization* of (0.22):

$$\partial_t z = \partial_x[\phi(\partial_x z)] + \epsilon\, \partial_{tx}[\psi(\partial_x z)],$$

where the regularization term with

$$\psi(u) := -\phi(u) + \int_0^u \frac{\phi(z)}{z}\, dz \quad (u \geq 0) \tag{0.24}$$

is obtained introducing in (*FB*) time delay effects. For the function ϕ in (0.22), we obtain $\psi(u) = -\frac{u}{1+u^2} + \operatorname{arctg} u$, and hence

$$\psi'(u) > 0 \quad \text{for all } u \geq 0, \quad \psi(u) \to \frac{\pi}{2} \quad \text{as } u \to \infty.$$

In general, if ϕ satisfies (A_2), then the function $\psi : [0,\infty) \to [0,\infty)$ satisfies the following requirements:

$$\psi'(u) > 0 \quad \text{for all } u \geq 0, \quad \psi(u) \to \psi_\infty \in (0,\infty) \quad \text{as } u \to \infty. \tag{A_3}$$

Clearly, the pseudoparabolic regularization is weaker than the Sobolev regularization, which formally corresponds to the case $\psi(u) = u$. In agreement with the correspondence between (0.22) and (0.23), the pseudoparabolic regularization of (*FB*) reads

$$\partial_t u = \Delta\phi(u) + \epsilon\,\partial_t\Delta[\psi(u)] \quad (u \geq 0).$$

3.1 Young measure-valued solutions

Consider the initial-boundary value problem for (*FB*) with Sobolev regularization:

$$\begin{cases} \partial_t u = \Delta v & \text{in } Q, \\ \partial_\nu v = 0 & \text{in } \partial\Omega \times (0, T), \\ u = u_0 & \text{in } \Omega \times \{0\}, \end{cases} \tag{S_ϵ}$$

where $v := \phi(u) + \epsilon\,\partial_t u$, $\Omega \subseteq \mathbb{R}^N$ $(N \geq 1)$ is open and bounded with smooth boundary $\partial\Omega$, and ∂_ν denotes the outer normal derivative. The following results were proven in [74] (see also [76]).

Lemma 0.3.1. *Let $\phi \in C^1(\mathbb{R})$, and let $u_0 \in C(\overline{\Omega})$. Then for every $\epsilon > 0$, there exists $T_\epsilon > 0$ such that problem (S_ϵ) has a unique strong solution (u^ϵ, v^ϵ) in $Q_{T_\epsilon} := \Omega \times (0, T_\epsilon)$. Moreover:*

(i) *let $g \in C^1(\mathbb{R})$, $g' \geq 0$, and $G(u) := \int_0^u g(\phi(s))ds + c$ $(c \in \mathbb{R})$. Then*

$$\int_\Omega G(u^\epsilon(x,t))\,dx \leq \int_\Omega G(u_0(x))\,dx \quad \text{for any } t \in (0, T_\epsilon); \tag{0.25}$$

(ii) *let $u_1, u_2 \in \mathbb{R}$, $u_1 < u_2$, and*

$$\phi(u_1) \leq \phi(u) \leq \phi(u_2) \quad \text{for any } u \in [u_1, u_2]. \tag{0.26}$$

Let $u_0(x) \in [u_1, u_2]$ for every $x \in \Omega$. Then $u^\epsilon(x,t) \in [u_1, u_2]$ for every $(x,t) \in Q_{T_\epsilon}$.

Proof. Claim (ii) above follows from (i) by a proper choice of the function g. To prove (i), observe that

$$\begin{aligned} \partial_t[G(u^\epsilon)] &= g(\phi(u^\epsilon))\partial_t u^\epsilon = g(v^\epsilon)\Delta v^\epsilon + [g(\phi(u^\epsilon)) - g(v^\epsilon)]\Delta v^\epsilon \\ &= \operatorname{div}[g(v^\epsilon)\nabla v^\epsilon] - g'(v^\epsilon)|\nabla v^\epsilon|^2 + \underbrace{[g(\phi(u^\epsilon)) - g(v^\epsilon)]\frac{v^\epsilon - \phi(u^\epsilon)}{\epsilon}}_{\leq 0}. \end{aligned}$$

Integrating on Ω the above inequality gives (0.25). □

If (A_1) holds, then the function ϕ satisfies (0.26) for a suitable choice of $u_1 < 0 < u_2$ with $|u_1|$, $|u_2|$ sufficiently large. Hence we get a uniform L^∞-estimate of $\{u^\epsilon\}$. The other estimates in (0.27) below follow similarly from the proof of inequality (0.25), and thus we have the following result.

Proposition 0.3.2. *Let (A_1) be satisfied, and let $u_0 \in C(\overline{\Omega})$. Then:*
(i) *for every $\epsilon > 0$, there exists a unique strong solution (u^ϵ, v^ϵ), with $v^\epsilon = \phi(u^\epsilon) + \epsilon\, \partial_t u^\epsilon$, of the regularized problem (S_ϵ);*
(ii) *there exists $M > 0$ such that for any $\epsilon > 0$,*

$$\max\{\|u^\epsilon\|_{L^\infty(Q)}, \sqrt{\epsilon}\,\|\partial_t u^\epsilon\|_{L^2(Q)}, \|v^\epsilon\|_{L^\infty(Q)}, \|v^\epsilon\|_{L^2(0,T;H^1(\Omega))}\} \le M. \tag{0.27}$$

By inequality (0.27) there exist sequences $\{\epsilon_k\}$ with $\epsilon_k \to 0^+$ as $k \to \infty$, $\{u_k\} \equiv \{u^{\epsilon_k}\}$, and $\{v_k\} \equiv \{v^{\epsilon_k}\}$, and $u \in L^\infty(Q)$ and $v \in L^\infty(Q) \cap L^2(0,T;H^1(\Omega))$ such that

$$\begin{cases} u_k \overset{*}{\rightharpoonup} u \text{ in } L^\infty(Q), & \phi(u_k) \overset{*}{\rightharpoonup} v \text{ in } L^\infty(Q), \\ v_k \overset{*}{\rightharpoonup} v \text{ in } L^\infty(Q), & v_k \rightharpoonup v \text{ in } L^2(0,T;H^1(\Omega)) \end{cases} \tag{0.28}$$

(observe that the sequences $\{v_k\}$ and $\{\phi(u_k)\}$ admit the same weak* limit in $L^\infty(Q)$). By (0.28), letting $k \to \infty$ in the weak formulation of (S_{ϵ_k}), we easily obtain

$$\iint_Q \{u\,\partial_t\zeta - \nabla v \cdot \nabla\zeta\}\,dxdt + \int_\Omega u_0\,\zeta(x,0)\,dx = 0 \tag{0.29}$$

for any $\zeta \in C^1([0,T]; C^1(\overline{\Omega}))$, $\zeta(\cdot,T) = 0$ in Ω.

Since the convergences in (0.28) do not imply the equality $v = \phi(u)$, u need not be a weak solution of the problem

$$\begin{cases} \partial_t u = \Delta[\phi(u)] & \text{in } Q, \\ \partial_\nu[\phi(u)] = 0 & \text{in } \partial\Omega \times (0,T), \\ u = u_0 & \text{in } \Omega \times \{0\}. \end{cases} \tag{S}$$

However, a weak Young measure-valued solution of (S) is easily obtained. Since $\|u_k\|_{L^\infty(Q)} \le M$ (see (0.27)) and Q is bounded, the sequence $\{u_k\}$ is bounded in $L^1(Q)$ and uniformly integrable (see Lemma 2.8.12), and the same holds for the sequence $\{\phi(u_k)\}$. Then there exist a subsequence of $\{u_k\}$ (not relabeled) and a Young measure $\nu \in \mathfrak{Y}(Q;\mathbb{R})$ such that

$$u_k \rightharpoonup u^*, \quad \phi(u_k) \rightharpoonup \phi^* \quad \text{in } L^1(Q),$$

where u^* and ϕ^* are given by (0.12) and (0.13), respectively, with $U = Q$ and $f = \phi$. From (0.28) and the above convergences we get $u = u^*$, $v = \phi^*$, whence by (0.29)

$$\iint_Q \{u\,\partial_t\zeta - \nabla\phi^* \cdot \nabla\zeta\}\,dxdt + \int_{\mathbb{R}} u_0\,\zeta(x,0)\,dx = 0. \tag{0.30}$$

Therefore the couple (u, ν) is a weak Young measure-valued solution of (S).

If ϕ satisfies assumption (A_1), then we have the following characterization of the Young measure ν ([83]; see also [82, 84]):

$$\nu_{(x,t)} = \sum_{i=0}^{2} \lambda_i(x,t)\delta_{s_i(v(x,t))} \quad \text{for a. e. } (x,t) \in Q. \tag{0.31}$$

Here the coefficients λ_i take values in the interval $[0,1]$, $\sum_{i=0}^{2} \lambda_i = 1$, and the functions $s_i(v)$ are defined by (0.20). By equalities (0.12) and (0.31) we have

$$u(x,t) = \sum_{i=0}^{2} \lambda_i(x,t)s_i(v(x,t)) \quad \text{for a. e. } (x,t) \in Q. \tag{0.32}$$

Remark 0.3.1. In view of (0.32), u is a superposition of different phases. By definition a *two-phase solution* of (S) is any solution as above such that $\lambda_0 = 0$ a. e. in Q and $\lambda_i = 1$ a. e. in V_i $(i = 1,2)$, where $\overline{Q} = \overline{V}_1 \cup \overline{V}_2$, $V_1 \cap V_2 = \emptyset$ with smooth common boundary $\overline{V}_1 \cap \overline{V}_2$.

When $N = 1$, a major qualitative feature of two-phase solutions is that they display *hysteresis effects*: the only admissible phase changes are those that take place from the extremum points (b, B) and (c, A) of the cubic (see Fig. 1). The proof is analogous to that of the Oleinik entropy condition for piecewise smooth solutions of hyperbolic conservation laws (which to some extent are the counterpart of two-phase solutions of (S); see [48, 54, 70, 71, 82, 83, 84, 98]).

Among others, this feature points out that solutions of (S) obtained by Sobolev regularization are definitely different from those obtained by Cahn–Hilliard regularization. In fact, it was shown in [12] that the limiting dynamics as $\kappa \to 0^+$ of solutions of (0.21) on the one-dimensional torus is governed by the *Maxwell equal area law*:

$$\int_{-1}^{1} \phi^{**}(u)\,du = 0 = \int_{-1}^{1} \phi(u)\,du,$$

which is incompatible with hysteresis phenomena (here $\phi^{**} := (\Phi^{**})'$, Φ^{**} being the convex envelope of the double-well potential Φ). Similar results were obtained by an *implicit variational scheme* in [43], concerning an ill-posed problem for a gradient equation. Problems of the same kind were also investigated by Sobolev and Cahn–Hilliard regularizations in [30, 95] (see Chapter 8).

3.2 Radon measure-valued solutions

Let ϕ and ψ satisfy (A_2) and (A_3), respectively. Since $\psi'(u) \to 0$ as $u \to \infty$, the initial-boundary value problem

$$\begin{cases} \partial_t u = \partial_{xx}[\phi(u)] + \epsilon\,\partial_{txx}[\psi(u)] & \text{in } Q, \\ \phi(u) + \epsilon\,\partial_t[\psi(u)] = 0 & \text{on } \partial\Omega \times (0,T), \quad u \geq 0, \\ u = u_0 & \text{in } \Omega \times \{0\}, \end{cases} \qquad (PM_\epsilon)$$

can be guessed to have for $\epsilon > 0$ the same qualitative features displayed as $\epsilon \to 0^+$ by the problem

$$\begin{cases} \partial_t u = \partial_{xx}[\phi(u)] + \epsilon\,\partial_{txx} u & \text{in } Q, \\ \phi(u) + \epsilon\,\partial_t u = 0 & \text{on } \partial\Omega \times (0,T), \quad u \geq 0, \\ u = u_0 & \text{in } \Omega \times \{0\}, \end{cases} \qquad (\Sigma_\epsilon)$$

with Sobolev regularization, as far as the behavior of singularities is concerned.

Problem (PM_ϵ) with $u_0 \in \mathfrak{R}_f^+(\Omega)$ was studied in [99, 100]. Its most striking feature, clearly connected to the weakness of the pseudoparabolic regularization, is the *spontaneous appearance of singularities* for positive times when u_0 is smooth (see [22, Theorem 4.2], Definition 1.8.9 and Theorem 3.1.8):

Proposition 0.3.3. *Let (A_2)–(A_3) be satisfied. There exist solutions of (PM_ϵ) with $u_0 \in C(\overline{\Omega})$, $u_0 \geq 0$, such that for some $t \in (0,T)$, either $u_s(\cdot,t)$ is purely atomic and contains countably many Dirac masses, or $u_s(\cdot,t)$ is singular continuous.*

This result also depends on the negativity of $\phi'(u)$ for large values of u, which gives rise to *concentration* phenomena. It is in formal agreement with the result in [11] that solutions of the problem

$$\begin{cases} \partial_t z = \partial_x[\phi(\partial_x z)] + \epsilon\,\partial_{tx}[\psi(\partial_x z)] & \text{in } Q, \\ \phi(\partial_x z) + \epsilon\,\partial_t[\psi(\partial_x z)] = 0 & \text{in } \partial\Omega \times (0,T), \\ z = z_0 \in C^1(\overline{\Omega}) & \text{in } \Omega \times \{0\}, \end{cases}$$

may become *discontinuous with respect to x* for positive times.

Finally, it is interesting to compare the effect of (A_1) versus (A_2) when letting $\epsilon \to 0^+$ in problem (Σ_ϵ). If ϕ is a cubic, then every interval $[u_1, u_2]$ sufficiently large satisfies condition (0.26), and thus Lemma 0.3.1(ii) gives a uniform L^∞-estimate of $\{u^\epsilon\}$. Therefore, as $\epsilon \to 0^+$, we obtain equality (0.30) with $u \in L^\infty(Q)$, and hence no singularities arise in this case. On the other hand, if ϕ is of Perona–Malik type, then only those intervals $[u_1, u_2] \subseteq [0, \infty)$ where ϕ' is positive satisfy (0.26). Hence a uniform L^∞-estimate of $\{u^\epsilon\}$ only holds if $\{u^\epsilon\}$ takes values in the stable phase, which is trivial.

However, in the latter case, it can be proven that the half-line $[0,\infty)$ is positively invariant for solutions of (PM_ϵ) (see [96]). As a consequence, if $u_0 \in L^1(\Omega)$, $u_0 \geq 0$, then the *conservation of mass*,

$$\int_\Omega u^\epsilon(x,t)\,dx = \int_\Omega u_0(x)\,dx \quad (t \in (0,T)),$$

gives a uniform estimate of $\{u^\epsilon\}$ in $L^\infty(0,T;L^1(\Omega))$:

$$\|u^\epsilon(\cdot,t)\|_{L^1(\Omega)} = \int_\Omega u^\epsilon(x,t)\,dx = \int_\Omega u_0(x)\,dx = \|u_0\|_{L^1(\Omega)}.$$

It follows that

$$\|u^\epsilon\|_{L^1(Q)} \leq T\,\|u_0\|_{L^1(\Omega)} \quad \text{for any } \epsilon > 0.$$

Then by the biting lemma (see Theorem 0.2.2(iv) and Theorem 5.4.12) there exist a sequence $\{u_k\} \equiv \{u^{\epsilon_k}\} \subseteq \{u^\epsilon\}$, a Young measure $\nu \in \mathfrak{Y}(Q;\mathbb{R})$, a Radon measure $\sigma \in \mathfrak{R}_f^+(Q)$, and a sequence $Q_{k+1} \subseteq Q_k \subseteq Q$, with Lebesgue measure $\lambda_{N+1}(Q_k)$ vanishing as $\epsilon_k \to 0^+$, such that (possibly up to a subsequence, not relabeled)

$$u_k\chi_{Q\setminus Q_k} \rightharpoonup u^* := \int_{[0,\infty)} y\,d\nu_{(x,t)}(y) \quad \text{in } L^1(Q)$$

and

$$u_k\chi_{Q_k} \overset{*}{\rightharpoonup} \sigma \quad \text{in } \mathfrak{R}_f(Q). \tag{0.33}$$

As in Proposition 0.3.2, it is also easily checked that there exists $v \in L^\infty(Q) \cap L^2(0,T;H_0^1(\Omega))$ such that

$$v_k \overset{*}{\rightharpoonup} v \quad \text{in } L^\infty(Q), \quad v_k \rightharpoonup v \quad \text{in } L^2(0,T;H_0^1(\Omega)),$$

where $v_k := \phi(u_k) + \epsilon_k\,\partial_t u_k$. Then letting $k \to \infty$ in the weak formulation of (Σ_{ϵ_k}), we easily obtain

$$\iint_Q (u^*\,\partial_t\zeta - \nabla v \cdot \nabla\zeta)\,dxdt + \int_\Omega u_0\,\zeta(x,0)\,dx = -\langle\sigma,\partial_t\zeta\rangle$$

for any $\zeta \in C^1([0,T];C_c^1(\Omega))$, $\zeta(\cdot,T) = 0$ in Ω.

The right-hand side of the above equality, which arises because of the convergence in (0.33), accounts for possible concentration phenomena in (Σ_ϵ) as $\epsilon \to 0^+$ (already observed in (PM_ϵ) for $\epsilon > 0$ because of the weaker regularization). On the other hand, as in the case of cubic ϕ, it can be checked that $v = \phi^*$ with $\phi^*(x,t) := \int_{\mathbb{R}} \phi(y)\,d\nu_{(x,t)}(y)$

for a. e. $(x,t) \in Q$. Therefore both Radon and Young measures are needed to describe the vanishing viscosity limit of (Σ_ϵ) in the Perona–Malik case.

The above discussion suggests the following picture:

(i) the behavior of ϕ at infinity determines the behavior of the singular part of the solution with respect to the Lebesgue measure, whereas the changes of monotonicity affect that of the absolutely continuous part;

(ii) the behavior of the singular part is described by Radon measures and is related to disappearance, persistence, or appearance of singularities (*concentration* or *regularization* phenomena);

(iii) the behavior of the absolutely continuous part is described by Young measures and is related to *oscillation* phenomena.

Mathematical tools

Let us draw some conclusions from the above discussion.

Addressing nonlinear differential problems that involve measures always makes use of some approximation procedure. The solutions of the approximating problems are a family in some function space, which is naturally embedded in the Banach space of finite Radon measures. The limiting points of the family in some suitable topology, as some regularization parameter goes to zero, are natural candidates as solutions of the original problem.

Therefore the main tool to prove the existence of solutions is provided by results concerning convergence and compactness of sequences of finite Radon measures in various topologies. These results fit in (and their proofs make use of) the general framework of Lebesgue spaces of measure-valued functions defined on some given Banach space. Indeed, the clarity and elegance of the proofs greatly benefit from a clear understanding of the functional analytic context.

Applying the general theory to the sequence of solutions of the approximating problems gives convergence results, which allow us to take the limit in the weak formulation of the approximating problems. Depending on the nature of the problem, such results account both for concentration and for oscillation phenomena, if any (see Section 5.4).

In this general framework a major issue is that of regularity of Radon measure-valued solutions and thus of the possible presence, location, and evolution of singularities. As outlined before, these features are related to suitable capacities which are characteristic of the problem. Hence the theory of capacities is also an important tool to study problems of this kind.

Contents and brief user's guide

Part I of the book is devoted to the general theory. Chapters 1 and 2 have a preliminary character, presenting general results of abstract measure theory, respectively, of measurability and integration of scalar functions. Chapters 3, 4, and 5 are the core of the book, dealing respectively with capacity theory, vector integration, and Lebesgue spaces of vector functions, and convergence of sequences of finite Radon measures.

In Part II (Chapters 6, 7, and 8), we describe three applications of the general theory. Each application fits in one of the three research areas (quasilinear parabolic problems, hyperbolic conservation laws, ill-posed evolution problems) mentioned in this Introduction.

A description of the content of the chapters is given at the beginning of each part. Necessary preliminaries on topology are recalled in Appendix A at the end of the book. Although far from being exhaustive, the bibliography contains ample references to the different topics treated in the book.

Let us add some practical remarks:

1. By a "scalar function" we mean any real-valued function, as opposite to "vector function", a map taking values in any Banach space other than $\mathbb{R}$.
2. All vector spaces considered are real.
3. Within each section theorems, propositions, lemmata and corollaries are labeled sequentially (e. g., Theorem 1.2.1 is followed by Proposition 1.2.2, etc.), whereas definitions, remarks, and examples are numbered separately. Equations are numbered sequentially within each chapter.

Part I: **General theory**

Outline of Part I

The first chapter provides general results of abstract measure theory, including topics like outer measures and abstract capacity, Carathéodory construction, Hausdorff measure and capacity, general results of decomposition and vector measures.

In the second chapter, after recalling classical results on measurability and integration of scalar functions, we introduce two subjects which are fundamental in the development of the book, Young measures and Riesz representation theorems (both for positive and for bounded functionals). Various notions of convergence in Lebesgue spaces are then discussed, pointing out the important role of uniform integrability (Vitali Theorem, Dunford–Pettis Theorem). The final section deals with a major result of differentiation, the Radon–Nikodým Theorem, and with differentiation of Radon measures on the Euclidean space $\mathbb{R}^N$.

In the third chapter first we present some material concerning Sobolev and Bessel potential spaces, then we recall results concerning functions of bounded variations and Sobolev functions. Building on these notions we discuss in detail various concepts of capacity on subsets of $\mathbb{R}^N$ (Bessel, Riesz, Sobolev and Hausdorff capacities) and their mutual relationships. The results concerning Bessel and Riesz capacities are derived from the Meyer's theory of capacities associated with a kernel. Measures concentrated or diffuse with respect to the Sobolev capacity are also discussed and characterized.

Chapter 4 deals with vector integration. After introducing several concepts of measurability of vector functions and discussing their mutual relationships (in particular, the Pettis Theorem), we present results concerning both the Bochner integral and weaker notions of integral. The subsequent step is introducing the vector Lebesgue spaces $L^p(X;Y)$, $L^p_w(X;Y)$ and $L^p_{w^*}(X;Y^*)$ (where $p \in [1,\infty]$, and X, Y Banach spaces), proving completeness and separability results. Then we discuss the duality theory of such spaces (in particular, of $L^p_{w^*}(X;Y^*)$ when Y^* or Y is separable), pointing out the central role of the so-called Radon–Nikodým property, or equivalently of the Riesz representability of linear continuous operators from $L^1(X)$ to Y. The application we have in mind is the case where $Y = C_0(Z)$ and $Y^* = \mathfrak{R}_f(Z)$, the Banach space of finite Radon measures on some metric space Z, since the space $L^\infty_{w^*}(0,T;\mathfrak{R}_f(\Omega))$ ($T > 0$, $\Omega \subseteq \mathbb{R}^N$) is the natural framework for the evolution problems dealt with in Part II of the book. The last two sections of the chapter deal with vector Lebesgue spaces where Y is a space of real functions, and with vector Sobolev spaces, respectively.

Chapter 5 is devoted to convergence of sequences of finite Radon measures. Whereas boundedness in the Banach space $\mathfrak{R}_f(X)$ ensures compactness in the weak* topology (by the Riesz and Banach–Alaoglu Theorems), compactness in the stronger narrow topology requires both boundedness and tightness, a concept which plays a central role in this context. Criteria both for narrow convergence (given by the Portmanteau Theorem, or using the Prokhorov distance) and for tightness are provided.

https://doi.org/10.1515/9783110556902-001

The subsequent step is to prove the Disintegration Theorem for finite Radon measures on the product of Hausdorff spaces. Applying the Disintegration Theorem to some sequence $\{\nu_{u_j}\}$ of Young measures associated with measurable functions $u_j : X \mapsto \mathbb{R}$ $(j \in \mathbb{N})$, it is possible to provide a relation between the boundedness of the sequence $\{u_j\}$ in Lebesgue spaces and the (relative) compactness of $\{\nu_{u_j}\}$ with respect to the narrow topology. In particular, if $\{u_j\}$ is bounded in $L^1(X)$, the sequence $\{\nu_{u_j}\}$ of the associated Young measures converges in the narrow sense (possibly up to a subsequence) to some Young measure ν. On the other hand, by the Banach–Alaoglu Theorem there exist $\{u_{j_k}\} \subseteq \{u_j\}$ and $\sigma \in \mathfrak{R}_f(X)$ such that $u_{j_k} \overset{*}{\rightharpoonup} \sigma$ in $\mathfrak{R}_f(X)$. Then the Biting Lemma (which is also proven) provides an accurate description of the concentration phenomena connected with such convergence, as well as a relationship between the limiting measure σ and the aforementioned Young measure ν.

1 Measure theory

1.1 Preliminaries

We will use the usual notations of set theory. As usual, $\mathbb{N}$, $\mathbb{Z}$, $\mathbb{Q}$, $\mathbb{R}$, and $\mathbb{R}^N$ ($N \in \mathbb{N}$; $\mathbb{R}^1 \equiv \mathbb{R}$) denote the natural numbers, integers, rational numbers, real numbers, and N-tuples of real numbers, respectively. The familiar notations of intervals, (a,b), $[a,b]$, $(a,b]$, $[a,b]$ will be used also for subintervals of the extended real line $\overline{\mathbb{R}} := \mathbb{R} \cup \{\pm\infty\}$. Occasionally, we use the symbol $[-\infty,+\infty]$ for $\overline{\mathbb{R}}$; we also write ∞ instead of $+\infty$.

Let X be a set, and let $\mathcal{P}(X) := \{E \mid E \subseteq X\}$. For any $E \subseteq X$, we denote by $E^c := \{x \in X \mid x \notin E\}$ the *complementary set* of E in X, by $\emptyset = X^c$ the *empty set*, and by $\chi_E : X \mapsto [0,\infty)$ the *characteristic function* of E,

$$\chi_E(x) := \begin{cases} 1 & \text{if } x \in E, \\ 0 & \text{otherwise.} \end{cases}$$

For any $E, F \subseteq X$, we denote by $E \setminus F := E \cap F^c$ the *difference* and by $E \triangle F := (E\setminus F) \cup (F \setminus E)$ the *symmetric difference* of E and F. If $\mathcal{F}_1 \subseteq \mathcal{F}_2 \subseteq \mathcal{P}(X)$ and $f : \mathcal{F}_2 \mapsto [-\infty,\infty]$ is any *set function*, then the restriction of f to $\mathcal{F}_1$ is denoted by $f|_{\mathcal{F}_1}$. For any $x \in X$, the set $\{x\} \in \mathcal{P}(X)$ is called a *singleton*. A *cover* of $E \subseteq X$ is any nonempty family $\{F_i\}_{i \in I} \subseteq \mathcal{P}(X)$ such that $E \subseteq \bigcup_{i\in I} F_i$. A family $\{F_i\}_{i\in I} \subseteq \mathcal{P}(X)$ is *disjoint* if $F_i \cap F_j = \emptyset$ for all $i,j \in I$, $i \neq j$. A finite disjoint family $\{F_1,\dots,F_n\} \subseteq \mathcal{P}(X)$ such that $X = \bigcup_{i=1}^n F_i$ is called a *partition* of X. A sequence $\{E_n\} \subseteq \mathcal{P}(X)$ is *nondecreasing* (respectively, *nonincreasing*) if $E_n \subseteq E_{n+1}$ ($E_n \supseteq E_{n+1}$, respectively) for all $n \in \mathbb{N}$. Nondecreasing or nonincreasing sequences are called *monotone*.

1.2 Families of sets

1.2.1 Measurable spaces

Let us recall the following definitions.

Definition 1.2.1. A nonempty family $\mathcal{S} \subseteq \mathcal{P}(X)$ is called a semialgebra if:
(i) $E, F \in \mathcal{S} \Rightarrow E \cap F \in \mathcal{S}$;
(ii) for any $E \in \mathcal{S}$, there exists a partition of E^c.

A family $\mathcal{A} \subseteq \mathcal{P}(X)$ is called an algebra if:
(i) $\emptyset \in \mathcal{A}$;
(ii) $E \in \mathcal{A} \Rightarrow E^c \in \mathcal{A}$;
(iii) $E, F \in \mathcal{A} \Rightarrow E \cup F \in \mathcal{A}$.

https://doi.org/10.1515/9783110556902-002

For any family $\mathcal{F} \subseteq \mathcal{P}(X)$, the intersection $\mathcal{A}_0(\mathcal{F}) := \bigcap\{\mathcal{A} \mid \mathcal{A} \text{ algebra}, \mathcal{A} \supseteq \mathcal{F}\}$ is an algebra, which contains $\mathcal{F}$ and is contained in all algebras $\mathcal{A} \supseteq \mathcal{F}$. It is called the *algebra generated by* $\mathcal{F}$ or the *minimal algebra* containing $\mathcal{F}$.

Example 1.2.1. Let $a \equiv (a_1, \dots, a_n)$ and $b \equiv (b_1, \dots, b_n)$ with $a_i, b_i \in [-\infty, \infty]$ and $a_i < b_i$ for every $i = 1, \dots, n$ ($n \in \mathbb{N}$). Consider the *N-cell*

$$(a, b] := \{x \equiv (x_1, \dots, x_n) \mid x_i \in [-\infty, \infty],\ a_i < x_i \le b_i\ \forall i = 1, \dots, n\}$$

with *volume* $\operatorname{vol}((a, b]) := \prod_{i=1}^{n}(b_i - a_i) \le \infty$. It is easily seen that the family $\mathcal{I}_n$ of the N-cells is a semialgebra and the algebra $\mathcal{A}_0(\mathcal{I}_n)$ generated by $\mathcal{I}_n$ consists of finite disjoint unions of N-cells, that is, for every $E \in \mathcal{A}_0(\mathcal{I}_n)$, there exist disjoint $(a^1, b^1], \dots, (a^p, b^p]$ such that $E = \bigcup_{k=1}^{p}(a^k, b^k]$.

Definition 1.2.2. A family $\mathcal{A} \subseteq \mathcal{P}(X)$ is called a σ-algebra if:
(i) $\emptyset \in \mathcal{A}$;
(ii) $E \in \mathcal{A} \Rightarrow E^c \in \mathcal{A}$;
(iii) for any sequence $\{E_k\} \subseteq \mathcal{A}$, we have $\bigcup_{k=1}^{\infty} E_k \in \mathcal{A}$.

Definition 1.2.3. Let $\mathcal{A} \subseteq \mathcal{P}(X)$ be a σ-algebra. The couple $(X, \mathcal{A})$ is called a measurable space, and the elements of $\mathcal{A}$ are called measurable sets.

Remark 1.2.1. Let $F \subseteq X$. It is easily seen that if $\mathcal{A}$ is a σ-algebra, then the family $\mathcal{A} \cap F := \{E \cap F \mid E \in \mathcal{A}\} \subseteq \mathcal{P}(F)$ is also a σ-algebra, called the *trace* of $\mathcal{A}$ on F. The measurable space $(F, \mathcal{A} \cap F)$ is called a *measurable subspace* of $(X, \mathcal{A})$.

Definition 1.2.4. A nonempty family $\mathcal{M} \subseteq \mathcal{P}(X)$ is called a monotone class if for any nondecreasing (respectively, nonincreasing) sequence $\{E_k\} \subseteq \mathcal{M}$, we have $\bigcup_{k=1}^{\infty} E_k \in \mathcal{M}$ (respectively, $\bigcap_{k=1}^{\infty} E_k \in \mathcal{M}$).

Clearly, every σ-algebra is a monotone class. As in the case of algebras, for any family $\mathcal{F} \subseteq \mathcal{P}(X)$, we can consider the *minimal σ-algebra*

$$\sigma_0(\mathcal{F}) := \bigcap\{\mathcal{A} \mid \mathcal{A}\ \sigma\text{-algebra}, \mathcal{A} \supseteq \mathcal{F}\}$$

and the *minimal monotone class*

$$\mathcal{M}_0(\mathcal{F}) := \bigcap\{\mathcal{M} \mid \mathcal{M} \text{ monotone class}, \mathcal{M} \supseteq \mathcal{F}\}$$

generated by $\mathcal{F}$.

Definition 1.2.5. A measurable space $(X, \mathcal{A})$ is called *separable* if there exists a countable family $\mathcal{S} \subseteq \mathcal{P}(X)$ such that $\mathcal{A} = \sigma_0(\mathcal{S})$.

Theorem 1.2.1. *Let $\mathcal{A} \subseteq \mathcal{P}(X)$ be an algebra. Then the minimal monotone class $\mathcal{M}_0(\mathcal{A})$ and the minimal σ-algebra $\sigma_0(\mathcal{A})$ coincide.*

Proof. Clearly, $\emptyset \in \mathcal{M}_0(\mathcal{A})$. Let us first show that the monotone class $\mathcal{M}_0(\mathcal{A})$ is an algebra. Set $\tilde{\mathcal{A}} := \{E \in \mathcal{M}_0(\mathcal{A}) \mid E^c \in \mathcal{M}_0(\mathcal{A})\}$. Since $\mathcal{A}$ is an algebra and $\mathcal{A} \subseteq \mathcal{M}_0(\mathcal{A})$, we have $\mathcal{A} \subseteq \tilde{\mathcal{A}}$. On the other hand, it is easily seen that $\tilde{\mathcal{A}}$ is a monotone class, and hence $\mathcal{M}_0(\mathcal{A}) \subseteq \mathcal{M}_0(\tilde{\mathcal{A}}) \subseteq \tilde{\mathcal{A}}$. Since by definition $\tilde{\mathcal{A}} \subseteq \mathcal{M}_0(\mathcal{A})$, it follows that $\tilde{\mathcal{A}} = \mathcal{M}_0(\mathcal{A})$. Therefore, for any $E \in \mathcal{M}_0(\mathcal{A})$, we have $E^c \in \mathcal{M}_0(\mathcal{A})$. It is similarly seen that for any $E, F \in \mathcal{M}_0(\mathcal{A})$, we have $E \cup F \in \mathcal{M}_0(\mathcal{A})$. Hence the claim follows.

Let $\{E_k\} \subseteq \mathcal{M}_0(\mathcal{A})$ and set $F_j := \bigcup_{k=1}^{j} E_k$ $(j, k \in \mathbb{N})$. Then the sequence $\{F_j\}$ is nondecreasing, and $\{F_j\} \subseteq \mathcal{M}_0(\mathcal{A})$ since $\mathcal{M}_0(\mathcal{A})$ is an algebra. Since $\mathcal{M}_0(\mathcal{A})$ is a monotone class, we obtain that $\bigcup_{k=1}^{\infty} E_k = \bigcup_{j=1}^{\infty} F_j \in \mathcal{M}_0(\mathcal{A})$. Therefore $\mathcal{M}_0(\mathcal{A})$ is a σ-algebra.

Since $\mathcal{M}_0(\mathcal{A})$ is a σ-algebra containing $\mathcal{A}$, we have $\mathcal{M}_0(\mathcal{A}) \supseteq \sigma_0(\mathcal{A})$. On the other hand, the σ-algebra $\sigma_0(\mathcal{A})$ is a monotone class containing $\mathcal{A}$, and thus $\sigma_0(\mathcal{A}) \supseteq \mathcal{M}_0(\mathcal{A})$. Hence the conclusion follows. □

Let us define the product of two measurable spaces $(X_1, \mathcal{A}_1)$ and $(X_2, \mathcal{A}_2)$. Consider the family of *measurable rectangles* $\mathcal{R} \subseteq \mathcal{P}(X_1 \times X_2)$,

$$\mathcal{R} := \{E_1 \times E_2 \mid E_1 \in \mathcal{A}_1,\ E_2 \in \mathcal{A}_2\}.$$

Definition 1.2.6. Let $(X_1, \mathcal{A}_1)$ and $(X_2, \mathcal{A}_2)$ be measurable spaces. The minimal σ-algebra $\sigma_0(\mathcal{R})$, denoted $\mathcal{A}_1 \times \mathcal{A}_2$, is called a *product σ-algebra*. The measurable space $(X_1 \times X_2, \mathcal{A}_1 \times \mathcal{A}_2)$ is called a *product measurable space*.

Let $E \subseteq X_1 \times X_2$. The set

$$E_{x_1} := \{x_2 \in X_2 \mid (x_1, x_2) \in E\} \quad (x_1 \in X_1) \tag{1.1}$$

is called the x_1-*section* of E. The x_2-*section* of E with $x_2 \in X_2$ is similarly defined. Observe that

$$(E^c)_{x_i} = (E_{x_i})^c, \quad \left(\bigcup_{n=1}^{\infty} E_n\right)_{x_i} = \bigcup_{n=1}^{\infty} (E_n)_{x_i} \qquad (i = 1, 2).$$

Proposition 1.2.2. *Let $(X_1, \mathcal{A}_1)$ and $(X_2, \mathcal{A}_2)$ be measurable spaces, and let $E \in \mathcal{A}_1 \times \mathcal{A}_2$. Then $E_{x_1} \in \mathcal{A}_2$ for any $x_1 \in X_1$, and $E_{x_2} \in \mathcal{A}_1$ for any $x_2 \in X_2$.*

Proof. Set $\Omega := \{E \in \mathcal{A}_1 \times \mathcal{A}_2 \mid E_{x_1} \in \mathcal{A}_2 \ \ \forall x_1 \in X_1\}$. It is easily checked that Ω is a σ-algebra containing the family $\mathcal{R}$ of measurable rectangles, and hence $\Omega = \mathcal{A}_1 \times \mathcal{A}_2$. It is similarly seen that $E_{x_2} \in \mathcal{A}_1$ for any $x_2 \in X_2$, and thus the result follows. □

1.2.2 Borel σ-algebras

Definition 1.2.7. Let $(X, \mathcal{T})$ be a topological space. The σ-algebra $\sigma_0(\mathcal{T})$ generated by the topology $\mathcal{T}$ (denoted $\mathcal{B} \equiv \mathcal{B}(X) \equiv \mathcal{B}(X, \mathcal{T})$) is called the Borel σ-algebra. Every set $E \in \mathcal{B}$ is called a Borel set.

Example 1.2.2. The family $\mathcal{G}_\delta$ of countable intersections of open sets and the family $\mathcal{F}_\sigma$ of countable unions of closed sets are contained in $\mathcal{B}$.

Definition 1.2.8. Let $(X, \mathcal{T})$ be a compact topological space, and let $\mathscr{G}_\delta \subseteq \mathcal{G}_\delta$ be the family of compact sets $E \in \mathcal{G}_\delta$. The σ-algebra generated by the family $\mathscr{G}_\delta$ (denoted $\mathcal{B}_a(X)$) is called the Baire σ-algebra. Every set $E \in \mathcal{B}$ is called a Baire set.

Remark 1.2.2. Let $X = \mathbb{R}$, and let $\mathcal{T} = \mathcal{T}(\mathbb{R})$ be the real topology. It is easily seen that the Borel σ-algebra $\mathcal{B}(\mathbb{R})$ coincides with the σ-algebra generated by the family of open bounded intervals. It is generated also by other families of intervals, e. g., by the family of half-lines $\{(a, \infty) \mid a \in \mathbb{R}\}$. Similar remarks hold for the Borel σ-algebra $\mathcal{B}(\mathbb{R}^N)$ $(N \geq 2)$.

Let us mention the following result, concerning the product of Borel σ-algebras.

Theorem 1.2.3. *Let $(X_1, \mathcal{T}_1)$ and $(X_2, \mathcal{T}_2)$ be topological spaces with countable bases, and let $(X_1, \times X_2, \mathcal{T}_1 \times \mathcal{T}_2)$ be the product topological space. Then*

$$\mathcal{B}(X_1, \mathcal{T}_1) \times \mathcal{B}(X_2, \mathcal{T}_2) = \mathcal{B}(X_1 \times X_2, \mathcal{T}_1 \times \mathcal{T}_2).$$

In particular, $\mathcal{B}(\mathbb{R}^M) \times \mathcal{B}(\mathbb{R}^N) = \mathcal{B}(\mathbb{R}^{M+N})$ $(M, N \in \mathbb{N})$.

1.3 Measures

1.3.1 General properties

Definition 1.3.1. Let $\emptyset \in \mathcal{F} \subseteq \mathcal{P}(X)$. A set function $\varphi : \mathcal{F} \to [0, \infty]$ is called:

(i) monotone if $\varphi(E) \leq \varphi(F)$ for any $E, F \in \mathcal{F}$ such that $E \subseteq F$;

(ii) additive if $\varphi(\bigcup_{k=1}^n E_k) = \sum_{k=1}^n \varphi(E_k)$ for any finite disjoint family $\{E_1, \ldots, E_n\} \subseteq \mathcal{F}$ such that $\bigcup_{k=1}^n E_k \in \mathcal{F}$;

(iii) σ-subadditive if $\varphi(\bigcup_{k=1}^\infty E_k) \leq \sum_{k=1}^\infty \varphi(E_k)$ for any sequence $\{E_k\} \subseteq \mathcal{F}$ such that $\bigcup_{k=1}^\infty E_k \in \mathcal{F}$;

(iv) σ-additive if $\varphi(\bigcup_{k=1}^\infty E_k) = \sum_{k=1}^\infty \varphi(E_k)$ for any disjoint sequence $\{E_k\} \subseteq \mathcal{F}$ such that $\bigcup_{k=1}^\infty E_k \in \mathcal{F}$.

Definition 1.3.2. Let $(X, \mathcal{A})$ be a measurable space. A set function $\mu : \mathcal{A} \to [0, \infty]$ is called a (positive) measure on $\mathcal{A}$ (or on X) if $\mu(\emptyset) = 0$ and μ is σ-additive. The triple $(X, \mathcal{A}, \mu)$ is called a measure space.

A measure μ is called finite if $\mu(X) < \infty$; it is called σ-finite if there exists a sequence $\{E_k\} \subseteq \mathcal{A}$ such that $X = \bigcup_{k=1}^\infty E_k$ and $\mu(E_k) < \infty$ for all $k \in \mathbb{N}$. A measure space $(X, \mathcal{A}, \mu)$ is called finite (respectively, σ-finite) if μ is finite (σ-finite, respectively).

If $\mu(X) = 1$, then the measure μ is called a probability measure, and the space $(X, \mathcal{A}, \mu)$ is called a probability space. The set of probability measures on X will be denoted by $\mathfrak{P}(X)$.

If $\mathcal{F} \subseteq \mathcal{P}(X)$ is any family of subsets, then the above definition can be generalized as follows. A map $\mu : \mathcal{F} \to [0, \infty]$ is a *measure on* $\mathcal{F}$ if $\mu(\emptyset) = 0$ when $\emptyset \in \mathcal{F}$ and for any disjoint sequence $\{E_k\} \subseteq \mathcal{F}$ such that $\bigcup_{k=1}^{\infty} E_k \in \mathcal{F}$, we have $\mu(\bigcup_{k=1}^{\infty} E_k) = \sum_{k=1}^{\infty} \mu(E_k)$. A measure μ on $\mathcal{F}$ is σ-finite if there exists a sequence $\{E_k\} \subseteq \mathcal{F}$ such that $X = \bigcup_{k=1}^{\infty} E_k$ and $\mu(E_k) < \infty$ for every $k \in \mathbb{N}$.

Remark 1.3.1. Let $(X, \mathcal{A}, \mu)$ be a measure space, and let $F \in \mathcal{A}$. The restriction $\mu|_{\mathcal{A} \cap F}$ of μ to the trace σ-algebra $\mathcal{A} \cap F$ is a measure *induced* by μ on $\mathcal{A} \cap F$. The measure space $(F, \mathcal{A} \cap F, \mu|_{\mathcal{A} \cap F})$ is called a *measure subspace* of $(X, \mathcal{A}, \mu)$.

The following properties of measures are easily proven.

Proposition 1.3.1. *Let $(X, \mathcal{A}, \mu)$ be a measure space. Then:*
(i) $\mu(\bigcup_{k=1}^{n} E_k) = \sum_{k=1}^{n} \mu(E_k)$ *for any finite disjoint family* $\{E_1, \dots, E_n\} \subseteq \mathcal{A}$;
(ii) $E \subseteq F \Rightarrow \mu(E) \leq \mu(F)$ *for any* $E, F \in \mathcal{A}$;
(iii) $\mu(\bigcup_{k=1}^{\infty} E_k) \leq \sum_{k=1}^{\infty} \mu(E_k)$ *for any sequence* $\{E_k\} \subseteq \mathcal{A}$;
(iv) $\mu(\bigcup_{k=1}^{\infty} E_k) = \lim_{k\to\infty} \mu(E_k)$ *for any nondecreasing sequence* $\{E_k\} \subseteq \mathcal{A}$;
(v) $\mu(\bigcap_{k=1}^{\infty} E_k) = \lim_{k\to\infty} \mu(E_k)$ *for any nonincreasing sequence* $\{E_k\} \subseteq \mathcal{A}$ *such that* $\mu(E_1) < \infty$.

Example 1.3.1. (i) Let X be a set. The map $\mu^{\#} : \mathcal{P}(X) \to [0, \infty]$ defined as

$$\mu^{\#}(E) := \begin{cases} \text{number of elements of } E & \text{if } E \text{ is finite,} \\ \infty & \text{otherwise} \end{cases}$$

is a measure called the *counting measure*. Clearly, $\mu^{\#}$ is σ-finite if X is countable.
(ii) Let $X \neq \emptyset$, and let $x \in X$. The map $\delta_x : \mathcal{P}(X) \to [0, \infty)$ defined as

$$\delta_x(E) := \begin{cases} 1 & \text{if } x \in E, \\ 0 & \text{otherwise} \end{cases}$$

is a measure called the *Dirac measure concentrated in* $\{x\}$.

1.3.2 Borel and Radon measures

Let $(X, \mathcal{T})$ be a Hausdorff space, and let $\mathcal{B} = \mathcal{B}(X)$ be the Borel σ-algebra generated by the topology $\mathcal{T}$. Let $\mathcal{K} \subseteq \mathcal{P}(X)$ denote the family of compact subsets.

Definition 1.3.3. Let $\mathcal{A} \supseteq \mathcal{B}$ be a σ-algebra, and let $\mu : \mathcal{A} \to [0, \infty]$ be a measure.

(a) μ is called locally finite if for any $x \in X$, there exists a neighborhood U of x such that $\mu(U) < \infty$. A locally finite measure $\mu : \mathcal{A} \to [0, \infty]$ is called a Borel measure on X.

(b) A set $E \in \mathcal{A}$ is called *μ-outer regular* (or *outer regular*) if

$$\mu(E) = \inf\{\mu(A) \mid A \supseteq E,\, A \in \mathcal{T}\}.$$

The measure μ is called *outer regular* if every $E \in \mathcal{A}$ is μ-outer regular.

(c) A set $E \in \mathcal{A}$ is called *μ-inner regular* (or *inner regular*) if

$$\mu(E) = \sup\{\mu(K) \mid K \subseteq E,\, K \in \mathcal{K}\}.$$

The measure μ is called *inner regular* if every $E \in \mathcal{A}$ is μ-inner regular.

(d) A set $E \in \mathcal{A}$ is called *μ-regular* (or *regular*) if it is both μ-outer regular and μ-inner regular. The measure μ is called *regular* if every $E \in \mathcal{A}$ is μ-regular.

(e) An inner regular Borel measure is called a *Radon measure*.

(f) A Borel measure is called *moderate* if there exists $\{A_n\} \subseteq \mathcal{T}$ such that $X = \bigcup_{n=1}^{\infty} A_n$ and $\mu(A_n) < \infty$ for all $n \in \mathbb{N}$.

(g) The support of a Borel measure μ (denoted by $\operatorname{supp}\mu$) is the closed set of points $x \in X$ such that $\mu(U) > 0$ for any neighborhood U of x.

The collections of Borel and Radon measures on X will be denoted by $\mathfrak{B}^+(X)$ and $\mathfrak{R}^+(X)$, respectively.

Remark 1.3.2. Let X, $\mathcal{A}$, $\mathcal{B}$, and μ be as in Definition 1.3.3.

(i) A set $E \in \mathcal{A}$ is μ-outer regular if and only if for any $\epsilon > 0$, there exists $A \in \mathcal{T}$ such that $A \supseteq E$ and $\mu(A \setminus E) < \epsilon$. A set $E \in \mathcal{A}$ is μ-inner regular if and only if for any $\epsilon > 0$, there exists $K \in \mathcal{K}$ such that $K \subseteq E$ and $\mu(E \setminus K) < \epsilon$. If $E \in \mathcal{A}$ is μ-outer regular, then there exists $G \in \mathcal{B}$ such that $G \supseteq E$ and $\mu(G \setminus E) = 0$.

(ii) A σ-compact set $E \subseteq X$ is μ-inner regular. In fact, by definition there exists a nondecreasing sequence $\{K_n\} \subseteq \mathcal{K}$ such that $E = \bigcup_{n=1}^{\infty} K_n$, and thus $E \in \mathcal{B}$. Then by Proposition 1.3.1(iv) we have

$$\mu(E) = \lim_{n\to\infty} \mu(K_n) = \sup\{\mu(K) \mid K \subseteq E,\, K \in \mathcal{K}\}.$$

(iii) If μ is locally finite (in particular, if μ is a Radon measure), then $\mu(K) < \infty$ for any $K \in \mathcal{K}$. In fact, for any $K \in \mathcal{K}$, there exists $A \in \mathcal{T}$ such that $A \supseteq K$ and $\mu(A) < \infty$, and thus $\mu(K) \le \mu(A) < \infty$.

(iv) If the Hausdorff space X is locally compact, then μ is locally finite if and only if $\mu(K) < \infty$ for any $K \in \mathcal{K}$. In fact, the "if" part of the claim follows from the very definition of locally compact space, whereas the "only if" part follows from (iii). In particular, if X is a locally compact Hausdorff space, then $\mu \in \mathfrak{B}^+(X)$ if and only if $\mu(K) < \infty$ for any $K \in \mathcal{K}$.

(v) Every moderate Borel measure is σ-finite. Conversely, every σ-finite outer regular Borel measure is moderate.

(vi) If the Hausdorff space X is σ-compact, then every Borel measure μ on X is moderate. In fact, let $X = \bigcup_{n=1}^{\infty} K_n$ with $\{K_n\} \subseteq \mathcal{K}$. Since μ is locally finite, by (iii) for every $n \in \mathbb{N}$, there exists $A_n \in \mathcal{T}$ such that $A_n \supseteq K_n$ and $\mu(A_n) < \infty$, and thus the claim follows.

Lemma 1.3.2. *Let $(X, \mathcal{T})$ be a Hausdorff space, and let $\mu \in \mathfrak{R}^+(X)$. Then:*

(i) *if $\mu(X) < \infty$, then μ is regular;*

(ii) *every $K \in \mathcal{K}$ is μ-outer regular.*

Proof. (i) Let $E \in \mathcal{A} \supseteq \mathcal{B}$. Since μ is inner regular, the set $E^c \in \mathcal{A}$ is μ-inner regular,

$$\mu(E^c) = \sup\{\mu(K) \mid K \subseteq E^c,\ K \in \mathcal{K}\}.$$

Therefore

$$\begin{aligned}\mu(E) &= \mu(X) - \sup\{\mu(K) \mid K \subseteq E^c,\ K \in \mathcal{K}\} \\ &= \inf\{\mu(X) - \mu(K) \mid K \subseteq E^c,\ K \in \mathcal{K}\} = \inf\{\mu(K^c) \mid K^c \supseteq E,\ K \in \mathcal{K}\}.\end{aligned}$$

Since

$$\mu(E) \le \inf\{\mu(A) \mid A \supseteq E,\ A \in \mathcal{T}\} \le \inf\{\mu(K^c) \mid K^c \supseteq E,\ K \in \mathcal{K}\},$$

the claim follows.

(ii) Let $K \in \mathcal{K}$ be fixed. Since μ is locally finite, as in Remark 1.3.2(iii), there exists $A \in \mathcal{T}$ such that $A \supseteq K$ and $\mu(A) < \infty$. Since μ is inner regular, the set $A \setminus K \in \mathcal{B}$ is μ-inner regular. Then for any $\epsilon > 0$, there exists $L \in \mathcal{K}$, $L \subseteq A \setminus K$, such that $\mu(L) > \mu(A \setminus K) - \epsilon$. Hence $M := A \setminus L \in \mathcal{T}$, $M \supseteq K$, and $\mu(K) \le \mu(M) = \mu(A) - \mu(L) < \mu(A) - \mu(A \setminus K) + \epsilon = \mu(K) + \epsilon$. Then by the arbitrariness of ϵ the result follows. □

Proposition 1.3.3. *Let $(X, \mathcal{T})$ be a locally compact Hausdorff space with countable basis, and let $\mu \in \mathfrak{B}^+(X)$. Then μ is moderate and regular.*

Proof. By Remark A.1 X is σ-compact, and hence by Remark 1.3.2(vi) μ is moderate.

To prove that μ is outer regular, let $\{A_n\} \subseteq \mathcal{T}$, $\mu(A_n) < \infty$ for all $n \in \mathbb{N}$, and $X = \bigcup_{n=1}^{\infty} A_n$. For every $E \in \mathcal{A} \supseteq \mathcal{B}$, set $E_n := E \cap A_n$, and thus $E = \bigcup_{n=1}^{\infty} E_n$ and $\mu(E_n) \le \mu(A_n) < \infty$ for all $n \in \mathbb{N}$. Then $\mu_n := \mu|_{\mathcal{B}(A_n)}$ is finite; moreover, by Remark A.1 every open subset of A_n is σ-compact and thus is μ_n-inner regular (see Remark 1.3.2(ii)). Arguing as in the proof of Lemma 1.3.2(i) shows that μ_n is regular and thus in particular outer regular (see [45, Satz VIII.1.5] for details). Therefore, for any $\epsilon > 0$ and $n \in \mathbb{N}$, there exists $A_n' \in \mathcal{T}$ such that $E_n \subseteq A_n' \subseteq A_n$ and $\mu_n(A_n' \setminus E_n) = \mu(A_n' \setminus E_n) < \frac{\epsilon}{2^n}$. Then $A' := \bigcup_{n=1}^{\infty} A_n' \in \mathcal{T}$, $A' \supseteq E$, and $\mu(A' \setminus E) \le \sum_{n=1}^{\infty} \mu(A_n' \setminus E_n) < \epsilon$. Hence μ is outer regular (see Remark 1.3.2(i)).

Let us now prove that μ is inner regular. Let $E \in \mathcal{A}$, $E_n := E \cap A_n$, and μ_n as above. For any $\alpha < \mu(E)$, let $p \in \mathbb{N}$ be so large that $\mu(\bigcup_{n=1}^p E_n) > \alpha$; then set $\epsilon := \mu(\bigcup_{n=1}^p E_n) - \alpha$. For every $n = 1, \ldots, p$, the measure μ_n is regular and thus inner regular; hence there exists a compact $K_n \subseteq E_n$ such that $\mu_n(E_n \setminus K_n) = \mu(E_n \setminus K_n) < \frac{\epsilon}{p}$ $(n = 1, \ldots, p)$. Then $K := \bigcup_{n=1}^p K_n \in \mathcal{K}$, $K \subseteq \bigcup_{n=1}^p E_n$, and

$$\mu\left(\bigcup_{n=1}^p E_n \setminus K\right) \leq \sum_{n=1}^p \mu(E_n \setminus K_n) < \epsilon = \mu\left(\bigcup_{n=1}^p E_n\right) - \alpha.$$

From the above inequality we get $\mu(K) > \alpha$, and thus the conclusion follows. □

Remark 1.3.3. In view of Proposition 1.3.3, in a locally compact Hausdorff space with countable basis (e. g., in $\mathbb{R}^N$), Borel and Radon measures coincide and are regular.

1.3.3 Null sets

Definition 1.3.4. Let $(X, \mathcal{A}, \mu)$ be a measure space.
(i) A set $N \subseteq X$ is called μ-null (or *null*) if $N \in \mathcal{A}$ and $\mu(N) = 0$. A set $E \subseteq X$ is called μ-negligible (or negligible) if there exists a μ-null set N such that $E \subseteq N$.
(ii) The space $(X, \mathcal{A}, \mu)$ is called complete if all negligible sets are measurable and hence null. In such a case, μ is a complete measure, and the σ-algebra $\mathcal{A}$ is complete for μ.

We denote the family of null sets by $\mathcal{N}_\mu$ and that of negligible sets by $\mathcal{U}_\mu$. Both families $\mathcal{N}_\mu$ and $\mathcal{U}_\mu$ are stable with respect to the countable union. Moreover, $\mathcal{U}_\mu$ is *hereditary*, that is, $F \subseteq E$, $E \in \mathcal{U}_\mu \Rightarrow F \in \mathcal{U}_\mu$. Clearly, $N_\mu = \mathcal{U}_\mu$ if and only if $(X, \mathcal{A}, \mu)$ is complete.

Theorem 1.3.4. *Let $(X, \mathcal{A}, \mu)$ be a measure space. Then:*
(i) *the family*

$$\bar{\mathcal{A}} := \{E \subseteq X \mid \exists F, G \in \mathcal{A} \text{ such that } F \subseteq E \subseteq G,\ \mu(G \setminus F) = 0\} \tag{1.2}$$

is a σ-algebra containing $\mathcal{A}$;
(ii) *there exists a complete measure $\bar{\mu} : \bar{\mathcal{A}} \to [0, \infty]$ such that $\bar{\mu}|_{\mathcal{A}} = \mu$.*

Moreover, $(X, \bar{\mathcal{A}}, \bar{\mu})$ is the smallest complete measure space containing $(X, \mathcal{A}, \mu)$.

Definition 1.3.5. The space $(X, \bar{\mathcal{A}}, \bar{\mu})$ is called the *Lebesgue completion* of $(X, \mathcal{A}, \mu)$.

Proof of Theorem 1.3.4. (i) Clearly, $\bar{\mathcal{A}} \supseteq \mathcal{A}$; in particular, $\emptyset \in \bar{\mathcal{A}}$. It follows immediately from (1.2) that $E \in \bar{\mathcal{A}} \Rightarrow E^c \in \bar{\mathcal{A}}$. Let, moreover, $\{E_n\} \subseteq \bar{\mathcal{A}}$. Then there exist $\{F_n\} \subseteq \mathcal{A}$ and $\{G_n\} \subseteq \mathcal{A}$ such that $F_n \subseteq E_n \subseteq G_n$, $\mu(G_n \setminus F_n) = 0$ for any $n \in \mathbb{N}$. Hence $F :=$

$\bigcup_{n=1}^{\infty} F_n \subseteq \bigcup_{n=1}^{\infty} E_n \subseteq \bigcup_{n=1}^{\infty} G_n =: G$, whence by the σ-subadditivity of μ we have $\mu(G \setminus F) \leq \sum_{n=1}^{\infty} \mu(G_n \setminus F_n) = 0$. Hence $\bigcup_{n=1}^{\infty} E_n \in \bar{\mathcal{A}}$, and thus the claim follows.

(ii) For any $E \in \bar{\mathcal{A}}$, set $\bar{\mu}(E) := \mu(F) = \mu(G)$ with F, G as in (1.2). Observe that the definition is well posed, since it does not depend on the choice of F, G. In fact, let $F_1, G_1 \in \mathcal{A}$ be such that $F_1 \subseteq E \subseteq G_1$ and $\mu(G_1 \setminus F_1) = 0$. Then $F_1 \subseteq E \subseteq G$ and $F \subseteq E \subseteq G_1$, whence $G \setminus F_1 = (G \setminus E) \cup (E \setminus F_1) \subseteq (G \setminus F) \cup (G_1 \setminus F_1)$. It follows that $\mu(G \setminus F_1) = 0$; we similarly see that $\mu(G_1 \setminus F) = 0$. The equalities

$$\mu(G \setminus F) = \mu(G_1 \setminus F_1) = \mu(G \setminus F_1) = \mu(G_1 \setminus F) = 0$$

imply that $\mu(F) = \mu(F_1) = \mu(G_1) = \mu(G)$.

Let us show that $\bar{\mu}$ is a measure on $\bar{\mathcal{A}}$. Clearly, $\bar{\mu}(\emptyset) = 0$. To prove the σ-additivity of $\bar{\mu}$, let $\{E_n\} \subseteq \bar{\mathcal{A}}$ be a disjoint sequence. Thus every sequence $\{F_n\} \subseteq \mathcal{A}$ such that $F_n \subseteq E_n$ for all $n \in \mathbb{N}$ is also disjoint. Hence by the σ-additivity of μ

$$\bar{\mu}\left(\bigcup_{n=1}^{\infty} E_n\right) = \mu\left(\bigcup_{n=1}^{\infty} F_n\right) = \sum_{n=1}^{\infty} \mu(F_n) = \sum_{n=1}^{\infty} \bar{\mu}(E_n),$$

and thus the claim follows. To prove that $\bar{\mu}$ is complete, fix any $N \in \bar{\mathcal{A}}$ such that $\bar{\mu}(N) = 0$, and let $F, G \in \mathcal{A}$ satisfy $F \subseteq N \subseteq G$, $\mu(G \setminus F) = 0$. By the definition of $\bar{\mu}$ we have $\mu(G) = \bar{\mu}(N) = 0$. Then for any $E \subseteq N$, we have $\emptyset \subseteq E \subseteq G$, $\mu(G \setminus \emptyset) = \mu(G) = 0$. This shows that $E \in \bar{\mathcal{A}}$, and thus $\bar{\mu}$ is complete.

By its very definition we have $\bar{\mu}|_{\mathcal{A}} = \mu$. It remains to prove that if $(X, \mathcal{A}_1, \mu_1)$ is a complete measure space such that $\mathcal{A} \subseteq \mathcal{A}_1$ and $\mu_1|_{\mathcal{A}} = \mu$, then $\bar{\mathcal{A}} \subseteq \mathcal{A}_1$ and $\mu_1|_{\bar{\mathcal{A}}} = \bar{\mu}$. To this purpose, let $E \in \bar{\mathcal{A}}$, and let $F, G \in \mathcal{A}$ be such that $F \subseteq E \subseteq G$ and $\mu(G \setminus F) = 0$. Since μ_1 is complete, $\mu_1(G \setminus F) = \mu(G \setminus F) = 0$, and $G \setminus E \subseteq G \setminus F$, we obtain that $G \setminus E \in \mathcal{A}_1$. Then $E = G \setminus (G \setminus E) \in \mathcal{A}_1$, whence $\bar{\mathcal{A}} \subseteq \mathcal{A}_1$. The remaining claims are clear, and thus the result follows. □

Remark 1.3.4. It is worth observing the following characterization of the σ-algebra $\bar{\mathcal{A}}$ defined in (1.2):

$$\bar{\mathcal{A}} = \{F \cup E_0 \mid F \in \mathcal{A},\ E_0 \in \mathcal{U}_\mu\}.$$

Indeed, let $E \in \bar{\mathcal{A}}$. Then by (1.2) we have $E_0 := E \setminus F \subseteq G \setminus F$ with $G \setminus F \in \mathcal{N}_\mu$; hence $E_0 \in \mathcal{U}_\mu$ and $E = F \cup E_0$. Conversely, let $E = F \cup E_0$ with $F \in \mathcal{A}$, $E_0 \in \mathcal{U}_\mu$. Then by the definition of $\mathcal{U}_\mu$ there exists $N_0 \in \mathcal{N}_\mu$ such that $E_0 \subseteq N_0$. Set $G := F \cup (N_0 \setminus F)$. Then we have $F \subseteq E \subseteq G$, $\mu(G \setminus F) = \mu(N_0 \setminus F) = 0$, and thus $E \in \bar{\mathcal{A}}$.

Definition 1.3.6. Let $(X, \mathcal{A}, \mu)$ be a measure space. A property P holds μ-almost everywhere (written μ-a. e. or a. e.) if $\{x \in X \mid P(x) \text{ false}\} \in \mathcal{N}_\mu$.

As particular cases of the above definition, (a) two functions $f, g : X \to \mathbb{R}$ are *equal μ-a. e. in X* if $\{x \in X \mid f(x) \neq g(x)\} \in \mathcal{N}_\mu$; ($b$) a sequence $\{f_n\}$ with $f_n : X \to \mathbb{R}$ *converges μ-a. e. in X* if $\{x \in X \mid \{f_n(x)\}$ does not converge$\} \in \mathcal{N}_\mu$.

Clearly, the equality μ-a. e. is a relation of equivalence:

$$g \sim f \quad \overset{\text{def}}{\Longleftrightarrow} \quad g = f \quad \mu\text{-a. e. in } X. \tag{1.3}$$

Every class of equivalence with respect to (1.3) is uniquely determined by anyone of its elements, which is called a *representative* of the class. It is thus natural to regard the whole class as a unique map f defined μ-a. e. in X in the following sense.

Definition 1.3.7. Let $(X, \mathcal{A}, \mu)$ be a measure space, and let Y be a set. A function $f : E \to Y$ $(E \in \mathcal{A})$ is *defined μ-a. e. in X* if $E^c \in \mathcal{N}_\mu$.

Whenever appropriate, to stress the difference between Definition 1.3.7 and the classical pointwise definition we will say that:

(i) a function $f : X \to \mathbb{R}$ is *classical* if it is defined at every point $x \in X$;
(ii) a classical function g is equal to a function f defined μ-a. e. if g is a classical representative of f.

In this connection, it is worth observing that the limit μ-a. e. of a sequence is unique in the sense of Definition 1.3.7 (yet not in the pointwise sense): in fact, $f_n \to f$ μ-a. e. and $f_n \to g$ μ-a. e. imply $f = g$ μ-a. e.

Let us prove for further reference the following technical lemma.

Lemma 1.3.5 (Exhaustion lemma). *Let $(X, \mathcal{A}, \mu)$ be a finite measure space. Let $P = P(x)$ $(x \in X)$ be a property such that:*

(i) *P holds on every μ-null subset of X;*
(ii) *for any $E \in \mathcal{A}$ with $\mu(E) > 0$, there exists $F \in \mathcal{A}$, $F \subseteq E$, with $\mu(F) > 0$ such that P holds on F.*

Then there exists a disjoint sequence $\{E_k\} \subseteq \mathcal{A}$ such that $\bigcup_{k=1}^\infty E_k = X$ and P holds on every set E_k.

Proof. Set $\Sigma := \{E \in \mathcal{A} \mid P \text{ holds on } E\}$ and $c := \sup_{E \in \Sigma} \mu(E)$. Let $\{F_n\} \subseteq \Sigma$ be such that $\lim_{n\to\infty} \mu(F_n) = c$. Set $H_k := \bigcup_{n=1}^k F_n$. Then for any $k \in \mathbb{N}$, the property P holds on H_k, and $H_k \subseteq H_{k+1}$; moreover, we have $\lim_{k\to\infty} \mu(H_k) = c$.

Let us prove that the set $E_1 := X \setminus (\bigcup_{k=1}^\infty H_k) = \bigcap_{k=1}^\infty H_k^c$ is μ-null. By absurd let $\mu(E_1) > 0$. Then by assumption (ii) there exists $F \in \mathcal{A}$, $F \subseteq E_1$, with $\mu(F) > 0$ such that the property P holds on F, and thus $F \in \Sigma$; moreover, we have $F \subseteq H_k^c$ for all $k \in \mathbb{N}$. Set $G_k := H_k \cup F$ $(k \in \mathbb{N})$. Then $\{G_k\} \subseteq \Sigma$ and $\lim_{k\to\infty} \mu(G_k) = \lim_{k\to\infty}(\mu(H_k) + \mu(F)) = c + \mu(F) > c$, a contradiction.

Now set $E_2 := H_1$ and $E_k := H_{k-1} \setminus H_{k-2}$ for any $k \geq 3$. The sequence $\{E_k\}$ has the stated properties, and thus the conclusion follows. ☐

Let us finally prove the following characterization of the support of a Radon measure.

Lemma 1.3.6. *Let $(X, \mathcal{T})$ be a Hausdorff space, and let $\mu \in \mathfrak{R}^+(X)$. Let $A \in \mathcal{T}$ be the largest open μ-null set. Then* $\operatorname{supp}\mu = A^c$.

Proof. Let us first show that the largest open μ-null set does exist. Set $\mathcal{N}_\mu \cap \mathcal{T} \equiv \{A_j\}_{j\in J}$ and $A := \bigcup_{j\in J} A_j$. Clearly, $A \in \mathcal{T}$ and $A \supseteq A_j$ for all $j \in J$. If $\mathcal{N}_\mu \cap \mathcal{T} = \{\emptyset\}$, then the claim is obvious. Otherwise, let $K \subseteq A$ be compact. Thus there exist $A_1, \dots, A_n \in \mathcal{N}_\mu \cap \mathcal{T}$ such that $K \subseteq \bigcup_{j=1}^n A_j$. It follows that $\mu(K) = 0$, whence $\mu(A) = 0$ by the arbitrariness of K and the inner regularity of μ.

Let $x \in [\operatorname{supp}\mu]^c$. Then by Definition 1.3.3(g) there exists an open neighborhood $\tilde{A} \ni x$ such that $\mu(\tilde{A}) = 0$; thus, in particular, $\tilde{A} \in \mathcal{N}_\mu \cap \mathcal{T}$. Hence we have $x \in \tilde{A} \subseteq A$. Conversely, let $x \in A$. Then A is an open μ-null neighborhood of x, and thus $x \in [\operatorname{supp}\mu]^c$. Then we have $[\operatorname{supp}\mu]^c = A$, and the result follows. □

1.4 Measures and outer measures

Definition 1.4.1. A map $\mu^* : \mathcal{P}(X) \to [0, \infty]$ is called an outer measure if
(i) $\mu^*(\emptyset) = 0$;
(ii) $\mu^*(E_1) \le \mu^*(E_2)$ for any $E_1 \subseteq E_2$;
(iii) $\mu^*(\bigcup_{n=1}^\infty E_n) \le \sum_{n=1}^\infty \mu^*(E_n)$ for any sequence $\{E_n\} \subseteq \mathcal{P}(X)$.

As shown further, outer measures are easily constructed, and complete measures are obtained from them by a general restriction procedure.

1.4.1 Carathéodory construction

Definition 1.4.2. Let $\mu^* : \mathcal{P}(X) \to [0, \infty]$ be an outer measure. A set $E \subseteq X$ is called μ^*-measurable if

$$\mu^*(Z) = \mu^*(Z \cap E) + \mu^*(Z \cap E^c) \quad \text{for any } Z \subseteq X. \tag{1.4}$$

Remark 1.4.1. (i) By the subadditivity of μ^*, a set $E \subseteq X$ is μ^*-measurable if and only if

$$\mu^*(Z) \ge \mu^*(Z \cap E) + \mu^*(Z \cap E^c) \quad \text{for any } Z \subseteq X. \tag{1.5}$$

(ii) Every set E such that $\mu^*(E) = 0$ (in particular, the empty set) is μ^*-measurable. Indeed, by the monotonicity of μ^*, for any $Z \subseteq X$,

$$\mu^*(Z \cap E) + \mu^*(Z \cap E^c) \le \mu^*(E) + \mu^*(Z) = \mu^*(Z).$$

Given an outer measure μ^*, there exists a σ-algebra $\mathcal{L} \subseteq \mathcal{P}(X)$ such that the restriction $\mu^*|_{\mathcal{L}}$ is a measure. This is the content of the following theorem.

Theorem 1.4.1 (Carathéodory). *Let μ^* be an outer measure. Then:*
(i) *the family*

$$\mathcal{L} := \{E \subseteq X \mid E \ \mu^*\text{-measurable}\} \tag{1.6}$$

is a σ-algebra;
(ii) *the restriction $\mu^*|_{\mathcal{L}}$ is a complete measure on $\mathcal{L}$.*

Proof. (i) Let us first prove that $\mathcal{L}$ is an algebra. It was already observed that the empty set is μ^*-measurable. Moreover, it is apparent from (1.4) and (1.6) that for any $E \in \mathcal{L}$, we have $E^c \in \mathcal{L}$. Let E and F be μ^*-measurable; then for any $Z \subseteq X$, we have

$$\mu^*(Z) = \mu^*(Z \cap E) + \mu^*(Z \cap E^c), \quad \mu^*(Z \cap E^c) = \mu^*(Z \cap E^c \cap F) + \mu^*(Z \cap E^c \cap F^c).$$

Therefore

$$\mu^*(Z) = \mu^*(Z \cap E) + \mu^*(Z \cap E^c \cap F) + \mu^*(Z \cap E^c \cap F^c). \tag{1.7}$$

Observe that

$$(Z \cap E) \cup (Z \cap E^c \cap F) = Z \cap [E \cup (E^c \cap F)] = Z \cap [E \cup (F \setminus E)] = Z \cap (E \cup F),$$

and thus by subadditivity

$$\mu^*(Z \cap (E \cup F)) \leq \mu^*(Z \cap E) + \mu^*(Z \cap E^c \cap F). \tag{1.8}$$

Moreover, since $Z \cap E^c \cap F^c = Z \cap (E \cup F)^c$, by (1.7)–(1.8) for any $Z \subseteq X$,

$$\mu^*(Z) \geq \mu^*(Z \cap (E \cup F)) + \mu^*(Z \cap (E \cup F)^c).$$

It follows that $E \cup F$ is μ^*-measurable, and thus $\mathcal{L}$ is an algebra.

Let us now prove that if $\{E_k\}$ is a sequence of μ^*-measurable disjoint sets and $S := \bigcup_{k=1}^{\infty} E_k$, then

$$\mu^*(Z \cap S) = \sum_{k=1}^{\infty} \mu^*(Z \cap E_k) \quad \text{for any } Z \subseteq X. \tag{1.9}$$

Set $S_n := \bigcup_{k=1}^{n} E_k$ ($n \in \mathbb{N}$). Let us first show that for any $n \in \mathbb{N}$,

$$\mu^*(Z \cap S_n) = \sum_{k=1}^{n} \mu^*(Z \cap E_k) \quad \text{for any } Z \subseteq X. \tag{1.10}$$

Since $\mathcal{L}$ is an algebra, the set S_n is μ^*-measurable for any $n \in \mathbb{N}$, and

$$\mu^*(Z \cap S_{n+1}) = \mu^*(Z \cap S_{n+1} \cap S_n) + \mu^*(Z \cap S_{n+1} \cap S_n^c) = \mu^*(Z \cap S_n) + \mu^*(Z \cap E_{n+1}).$$

If (1.10) holds for some n, then by the above equality we have

$$\mu^*(Z \cap S_{n+1}) = \sum_{k=1}^{n} \mu^*(Z \cap E_k) + \mu^*(Z \cap E_{n+1}),$$

and thus equality (1.10) holds for $n + 1$. Clearly, (1.10) holds for $n = 1$, and hence by induction the claim follows.

Let us now prove equality (1.9). Since $S_n \subseteq S$, by the monotonicity of μ^* and (1.10) we have

$$\mu^*(Z \cap S) \geq \mu^*(Z \cap S_n) = \sum_{k=1}^{n} \mu^*(Z \cap E_k) \quad (n \in \mathbb{N}),$$

whence, letting $n \to \infty$, we obtain that $\mu^*(Z \cap S) \geq \sum_{k=1}^{\infty} \mu^*(Z \cap E_k)$. The inverse inequality follows from the σ-subadditivity of μ^*, since $Z \cap S = \bigcup_{k=1}^{\infty}(Z \cap E_k)$. Then (1.9) follows.

Now we can prove that $\mathcal{L}$ is a σ-algebra. Since for any $n \in \mathbb{N}$, the set S_n is μ^*-measurable and $S_n^c \supseteq S^c$, by (1.10) and the monotonicity of μ^* we have

$$\mu^*(Z) = \mu^*(Z \cap S_n) + \mu^*(Z \cap S_n^c) \geq \sum_{k=1}^{n} \mu^*(Z \cap E_k) + \mu^*(Z \cap S^c) \quad (n \in \mathbb{N})$$

for any $Z \subseteq X$. Letting $n \to \infty$ in the above inequality and using (1.9), we get

$$\mu^*(Z) \geq \sum_{k=1}^{\infty} \mu^*(Z \cap E_k) + \mu^*(Z \cap S^c) = \mu^*(Z \cap S) + \mu^*(Z \cap S^c).$$

Hence the set S is μ^*-measurable. The same holds for any countable union of μ^*-measurable sets, and hence the claim follows.

(ii) Clearly, $\mu^*|_{\mathcal{L}}(\emptyset) = \mu^*(\emptyset) = 0$. Let $\{E_k\}$ be a sequence of μ^*-measurable disjoint sets. Choosing in equality (1.9) $Z = S := \bigcup_{k=1}^{\infty} E_k$, we obtain that $\mu^*(\bigcup_{k=1}^{\infty} E_k) = \sum_{k=1}^{\infty} \mu^*(E_k)$. Since $E_k, S \in \mathcal{L}$, it follows that $\mu^*|_{\mathcal{L}}$ is σ-additive and thus is a measure. To prove that it is complete, let $N \in \mathcal{L}$ satisfy $\mu^*|_{\mathcal{L}}(N) = \mu^*(N) = 0$, and let $E \subseteq N$. By the monotonicity of μ^* we have $\mu^*(E) \leq \mu^*(N) = 0$; hence $E \in \mathcal{L}$ by Remark 1.4.1(ii). This completes the proof. □

A general procedure to construct outer measures is as follows. Let $\mathcal{C} \subseteq \mathcal{P}(X)$ with $\emptyset \in \mathcal{C}$, and let $\zeta : \mathcal{C} \to [0, \infty]$ satisfy $\zeta(\emptyset) = 0$. Define the map $\mu^* : \mathcal{P}(X) \to [0, \infty]$ as follows:

$$\mu^*(E) := \inf\left\{\sum_{n=1}^{\infty} \zeta(E_n) \mid E \subseteq \bigcup_{n=1}^{\infty} E_n,\ \{E_n\} \subseteq \mathcal{C}\right\} \tag{1.11a}$$

if E can be covered by a countable union of sets $E_n \in \mathcal{C}$, or

$$\mu^*(E) := \infty \quad \text{otherwise.} \tag{1.11b}$$

Then the following holds.

Theorem 1.4.2. *Let $\mathcal{C} \subseteq \mathcal{P}(X)$ with $\emptyset \in \mathcal{C}$, and let $\zeta : \mathcal{C} \to [0, \infty]$ satisfy $\zeta(\emptyset) = 0$. Then the map μ^* defined in* (1.11a)–(1.11b) *is an outer measure on X.*

Proof. By (1.11a) we have $\mu^*(\emptyset) \le \zeta(\emptyset) = 0$. If $E_1 \subseteq E_2$, then the inequality $\mu^*(E_1) \le \mu^*(E_2)$ follows from (1.11b) if E_2 cannot be covered by a countable union of sets $E_n \in \mathcal{C}$ and from (1.11a) otherwise, since every countable cover of E_2 is also a countable cover of E_1. Let us prove that for any sequence $\{E_n\} \subseteq \mathcal{P}(X)$, we have $\mu^*(\bigcup_{n=1}^{\infty} E_n) \le \sum_{n=1}^{\infty} \mu^*(E_n)$. This is obvious if the right-hand side is infinite. Otherwise, we have $\mu^*(E_n) < \infty$ for any $n \in \mathbb{N}$, and thus for any $\epsilon > 0$ and $n \in \mathbb{N}$, there exists a sequence $\{E_{n,k}\} \subseteq \mathcal{C}$ such that

$$E_n \subseteq \bigcup_{k=1}^{\infty} E_{n,k}, \quad \mu^*(E_n) + \frac{\epsilon}{2^n} > \sum_{k=1}^{\infty} \zeta(E_{n,k}).$$

Since $\bigcup_{n=1}^{\infty} E_n \subseteq \bigcup_{n,k=1}^{\infty} E_{n,k}$ and $\{E_{n,k}\} \subseteq \mathcal{C}$ $(n, k \in \mathbb{N})$, we get

$$\mu^*\left(\bigcup_{n=1}^{\infty} E_n\right) \le \sum_{n,k=1}^{\infty} \zeta(E_{n,k}) < \sum_{n=1}^{\infty}\left[\mu^*(E_n) + \frac{\epsilon}{2^n}\right] = \sum_{n=1}^{\infty} \mu^*(E_n) + \epsilon.$$

Then by the arbitrariness of ϵ the conclusion follows. □

Definition 1.4.3. The outer measure μ^* given by Theorem 1.4.2 is said to be *generated by the couple* $(\mathcal{C}, \zeta)$.

1.4.2 Extension of measures

Much more can be said about the outer measure μ^* generated by a couple $(\mathcal{C}, \zeta)$ if the latter has additional properties. Let us first prove the following result.

Proposition 1.4.3 (Coincidence criterion). *Let $\mathcal{A} \subseteq \mathcal{P}(X)$ be an algebra, and let ν be a σ-finite measure on $\mathcal{A}$. Let μ_1 and μ_2 be two measures on the minimal σ-algebra $\sigma_0(\mathcal{A})$ such that $\mu_1|_{\mathcal{A}} = \mu_2|_{\mathcal{A}} = \nu$. Then $\mu_1 = \mu_2$.*

Proof. Since ν is a σ-finite measure on $\mathcal{A}$, there exists a nondecreasing sequence $\{E_n\} \subseteq \mathcal{A}$ such that $X = \bigcup_{n=1}^{\infty} E_n$, $\nu(E_n) < \infty$ for every $n \in \mathbb{N}$. Set $\mathcal{E}_n := \{E_n \cap F \mid F \in \sigma_0(\mathcal{A})\}$ $(n \in \mathbb{N})$. It is easily seen that the equality

$$\mu_1|_{\mathcal{E}_n} = \mu_2|_{\mathcal{E}_n} \quad \text{for any } n \in \mathbb{N} \tag{1.12}$$

implies that $\mu_1 = \mu_2$. In fact, if (1.12) holds, for any $F \in \sigma_0(\mathcal{A})$, we have

$$\mu_1(F) = \lim_{n\to\infty} \mu_1(E_n \cap F) = \lim_{n\to\infty} \mu_2(E_n \cap F) = \mu_2(F),$$

since $F = \bigcup_{n=1}^{\infty}(E_n \cap F)$ and the sequence $\{E_n \cap F\}$ is nondecreasing (see Proposition 1.3.1(iv)).

Therefore it suffices to prove equality (1.12). Observe that for any $n \in \mathbb{N}$, we have $\mu_1(E_n \cap F) \le \mu_1(E_n) = \nu(E_n) < \infty$; similarly, $\mu_2(E_n \cap F) < \infty$, and hence $\mu_1|_{\mathcal{E}_n}$ and $\mu_2|_{\mathcal{E}_n}$ are finite. Let us show that

$$\mathcal{Z}_n := \{F \in \sigma_0(\mathcal{A}) \mid \mu_1(E_n \cap F) = \mu_2(E_n \cap F)\} \quad (n \in \mathbb{N})$$

is a monotone class. Indeed, for any nondecreasing sequence $\{F_k\} \subseteq \mathcal{Z}_n$, we have

$$\mu_1\left(E_n \cap \left(\bigcup_{k=1}^{\infty} F_k\right)\right) = \lim_{k\to\infty} \mu_1(E_n \cap F_k) = \lim_{k\to\infty} \mu_2(E_n \cap F_k) = \mu_2\left(E_n \cap \left(\bigcup_{k=1}^{\infty} F_k\right)\right),$$

whereas for any nonincreasing sequence $\{F_k'\} \subseteq \mathcal{Z}_n$, since $\mu_1|_{\mathcal{E}_n}$ and $\mu_2|_{\mathcal{E}_n}$ are finite measures,

$$\mu_1\left(E_n \cap \left(\bigcap_{k=1}^{\infty} F_k'\right)\right) = \lim_{n\to\infty} \mu_1(E_n \cap F_k') = \lim_{n\to\infty} \mu_2(E_n \cap F_k') = \mu_2\left(E_n \cap \left(\bigcap_{k=1}^{\infty} F_k'\right)\right)$$

(see Proposition 1.3.1(v)). Hence the claim follows.

We also have that $\mathcal{A} \subseteq \mathcal{Z}_n$. indeed, if $F \in \mathcal{A}$, then we have $\mu_1(E_n \cap F) = \mu_2(E_n \cap F)$, since by assumption $\mu_1|_{\mathcal{A}} = \mu_2|_{\mathcal{A}}$. Then by Theorem 1.2.1 we have $\sigma_0(\mathcal{A}) \subseteq \mathcal{Z}_n$. Since $\mathcal{Z}_n \subseteq \sigma_0(\mathcal{A})$ by definition, we obtain that $\mathcal{Z}_n = \sigma_0(\mathcal{A})$, that is, for any $F \in \sigma_0(\mathcal{A})$ and $n \in \mathbb{N}$, we have $\mu_1(E_n \cap F) = \mu_2(E_n \cap F)$. This proves (1.12), and thus the conclusion follows. □

Relying on the coincidence criterion given by Proposition 1.4.3, we can prove the following refinement of Theorem 1.4.1.

Theorem 1.4.4. *Let $\mathcal{A} \subseteq \mathcal{P}(X)$ be an algebra, and let μ be a measure on $\mathcal{A}$. Let $\mu^* : \mathcal{P}(X) \to [0, \infty]$ be the outer measure generated by the couple $(\mathcal{A}, \mu)$. Then:*
(i) $\mu^*|_{\mathcal{A}} = \mu$;
(ii) $\sigma_0(\mathcal{A}) \subseteq \mathcal{L}$, *where $\mathcal{L}$ is defined by (1.6);*
(iii) $\hat{\mu} := \mu^*|_{\sigma_0(\mathcal{A})}$ *is a measure that extends μ to the minimal σ-algebra $\sigma_0(\mathcal{A})$.*

Moreover, let μ be σ-finite. Then:
(iv) *$\hat{\mu}$ is the unique extension of μ to a measure on $\sigma_0(\mathcal{A})$;*
(v) *the measure space $(X, \mathcal{L}, \mu^*|_{\mathcal{L}})$ is the completion of $(X, \sigma_0(\mathcal{A}), \hat{\mu})$.*

Proof. (i) By (1.11a) we have $\mu^*(E) \leq \mu(E)$ for any $E \in \mathcal{A}$. To prove the inverse inequality, let $E \in \mathcal{A}$, and let the sequence $\{E_n\} \subseteq \mathcal{A}$ be a cover of E. Set

$$F_1 := E \cap E_1, \quad F_n := E_n \cap \left[E \setminus \left(\bigcup_{k=1}^{n-1} E_k \right) \right] \quad (n \in \mathbb{N},\ n \geq 2).$$

Then $\{F_n\} \subseteq \mathcal{A}$ since $\mathcal{A}$ is an algebra. Moreover, the sets F_n are disjoint, $F_n \subseteq E_n$ for any $n \in \mathbb{N}$, and $E = \bigcup_{n=1}^{\infty} F_n$. Then by the monotonicity and σ-additivity of μ on $\mathcal{A}$ we get $\mu(E) = \sum_{n=1}^{\infty} \mu(F_n) \leq \sum_{n=1}^{\infty} \mu(E_n)$, whence $\mu(E) \leq \mu^*(E)$. Hence claim (i) follows.

(ii) Since $\mathcal{L}$ is a σ-algebra (see Theorem 1.4.1(i)), it suffices to show that $\mathcal{A} \subseteq \mathcal{L}$, that is, that for any $E \in \mathcal{A}$ and $Z \subseteq X$, inequality (1.5) holds. Let $\mu^*(Z) < \infty$ (otherwise, the conclusion is obvious); then for any $\epsilon > 0$, there exists a sequence $\{E_n\} \subseteq \mathcal{A}$ such that $Z \subseteq \bigcup_{n=1}^{\infty} E_n$ and $\sum_{n=1}^{\infty} \mu(E_n) < \mu^*(Z) + \epsilon$. Let $E \in \mathcal{A}$. By the monotonicity and σ-subadditivity of μ^*, in view of (i), we get

$$\mu^*(Z \cap E) \leq \sum_{n=1}^{\infty} \mu(E_n \cap E), \quad \mu^*(Z \cap E^c) \leq \sum_{n=1}^{\infty} \mu(E_n \cap E^c).$$

Therefore by the additivity of μ we have

$$\mu^*(Z \cap E) + \mu^*(Z \cap E^c) \leq \sum_{n=1}^{\infty} \mu(E_n \cap E) + \sum_{n=1}^{\infty} \mu(E_n \cap E^c) = \sum_{n=1}^{\infty} \mu(E_n) < \mu^*(Z) + \epsilon.$$

By the arbitrariness of ϵ we obtain that E is μ^*-measurable, and thus the claim follows.

(iii)–(iv) By Theorem 1.4.1(ii) $\mu^*|_{\mathcal{L}}$ is a measure on $\mathcal{L}$. Then, since $\sigma_0(\mathcal{A}) \subseteq \mathcal{L}$, $\hat{\mu}$ is a measure. Moreover, by (i) we have $\hat{\mu}|_{\mathcal{A}} = \mu^*|_{\mathcal{A}} = \mu$, whence claim (iii) follows. Claim (iv) is a consequence of Proposition 1.4.3.

(v) Let $\overline{\sigma_0(\mathcal{A})}$ be the Lebesgue completion of $\sigma_0(\mathcal{A})$. Since $\sigma_0(\mathcal{A}) \subseteq \mathcal{L}$, $\mu^*|_{\sigma_0(\mathcal{A})} = \hat{\mu}$, and by Theorem 1.4.1 $(X, \mathcal{L}, \mu^*|_{\mathcal{L}})$ is a complete measure space, we have $\overline{\sigma_0(\mathcal{A})} \subseteq \mathcal{L}$ (see Theorem 1.3.4). Conversely, let us prove that $\mathcal{L} \subseteq \overline{\sigma_0(\mathcal{A})}$. To this purpose, consider a sequence $\{E_n\} \subseteq \mathcal{A}$ such that $X = \bigcup_{n=1}^{\infty} E_n$ and $\mu(E_n) < \infty$ for all $n \in \mathbb{N}$. Let $F \in \mathcal{L}$; it suffices to prove that $E_n \cap F \in \overline{\sigma_0(\mathcal{A})}$ for any $n \in \mathbb{N}$, since this implies that $F = \bigcup_{n=1}^{\infty} (E_n \cap F) \in \overline{\sigma_0(\mathcal{A})}$.

We will prove the following:

Claim. For every $n \in \mathbb{N}$, there exists $F_n' \in \sigma_0(\mathcal{A})$ such that

$$E_n \cap F \subseteq F_n' \subseteq E_n, \quad \mu^*(E_n \cap F) = \hat{\mu}(F_n'). \tag{1.13a}$$

Using the claim we can conclude the proof. Indeed, since $F^c \in \mathcal{L}$, there also exists $F_n'' \in \sigma_0(\mathcal{A})$ such that

$$E_n \setminus F \subseteq F_n'' \subseteq E_n, \quad \mu^*(E_n \setminus F) = \hat{\mu}(F_n''). \tag{1.13b}$$

From (1.13) we get

$$E_n = F_n' \cup F_n'', \quad F_n' \setminus (E_n \cap F) \subseteq F_n' \cap F_n''. \tag{1.14}$$

Since $E_n \in \mathcal{A}$, we have $\mu(E_n) = \mu^*(E_n)$ (see (i)). Since $E_n \cap F \in \mathcal{L}$, $E_n \setminus F \in \mathcal{L}$, and $\mu^*|_{\mathcal{L}}$ is a measure, using the equalities in (1.13)–(1.14), we obtain

$$\begin{aligned}\mu(E_n) &= \mu^*(E_n) = \mu^*(E_n \cap F) + \mu^*(E_n \setminus F) = \hat{\mu}(F_n') + \hat{\mu}(F_n'') \\ &= \hat{\mu}(F_n' \setminus F_n'') + \hat{\mu}(F_n' \cap F_n'') + \hat{\mu}(F_n'' \setminus F_n') + \hat{\mu}(F_n'' \cap F_n') \\ &= \hat{\mu}(F_n' \triangle F_n'') + 2\hat{\mu}(F_n' \cap F_n'') = \hat{\mu}(F_n' \cup F_n'') + \hat{\mu}(F_n' \cap F_n'') \\ &= \mu(E_n) + \hat{\mu}(F_n' \cap F_n'')\end{aligned}$$

(recall that $F_n' \triangle F_n'' := (F_n' \setminus F_n'') \cup (F_n'' \setminus F_n') = F_n' \cup F_n'' \setminus (F_n' \cap F_n''))$. Since $\mu(E_n) < \infty$, it follows that $\hat{\mu}(F_n' \cap F_n'') = 0$.

Since $F_n' \setminus (E_n \cap F) \subseteq F_n' \cap F_n''$ (see (1.14)), we obtain that $\hat{\mu}(F_n' \setminus (E_n \cap F)) = 0$. On the other hand, by (1.13) we have

$$F_n' \setminus F_n'' \subseteq F_n' \setminus (E_n \cap F) \subseteq F_n', \quad \hat{\mu}(F_n' \setminus (F_n' \setminus F_n'')) = \hat{\mu}(F_n' \cap F_n'') = 0.$$

Since $F_n', F_n'' \in \sigma_0(\mathcal{A})$, we obtain that $F_n' \setminus (E_n \cap F) \in \overline{\sigma_0(\mathcal{A})}$ (see (1.2)). Then for any $n \in \mathbb{N}$, we have $E_n \cap F = F_n' \setminus (F_n' \setminus (E_n \cap F)) \in \overline{\sigma_0(\mathcal{A})}$, and hence the conclusion follows.

It remains to prove the claim. For simplicity, we omit the dependence on n, and thus we have that $F \in \mathcal{L}, E \equiv E_n \in \mathcal{A}, G := F \cap E \in \mathcal{L}$ (since $\mathcal{A} \subseteq \mathcal{L}$ by (ii)), and $\mu(E) < \infty$. We must show that there exists $F' \in \sigma_0(\mathcal{A})$ such that

$$G \subseteq F' \subseteq E, \quad \mu^*(G) = \hat{\mu}(F').$$

Since $G \subseteq E$, we have $\mu^*(G) \le \mu(E) < \infty$, and thus for any fixed $k \in \mathbb{N}$, there exists a sequence $\{G_n^k\} \subseteq \mathcal{A}$ such that

$$G \subseteq \bigcup_{n=1}^{\infty} G_n^k, \quad \sum_{n=1}^{\infty} \mu(G_n^k) < \mu^*(G) + \frac{1}{k}.$$

Without loss of generality, we can suppose that $G_n^k \subseteq E$ for all $k, n \in \mathbb{N}$.

Set $F' := \bigcap_{k=1}^{\infty}[\bigcup_{n=1}^{\infty} G_n^k]$. Then $F' \in \sigma_0(\mathcal{A})$, and for any fixed $\tilde{k} \in \mathbb{N}$, we have $G \subseteq F' \subseteq \bigcup_{n=1}^{\infty} G_n^{\tilde{k}}$. It follows that

$$\hat{\mu}(F') \le \hat{\mu}\left(\bigcup_{n=1}^{\infty} G_n^{\tilde{k}}\right) \le \sum_{n=1}^{\infty} \hat{\mu}(G_n^{\tilde{k}}) < \mu^*(G) + \frac{1}{\tilde{k}} \le \hat{\mu}(F') + \frac{1}{\tilde{k}}.$$

Since $\tilde{k}$ is arbitrary, we obtain that $\hat{\mu}(F') = \mu^*(G)$, and thus the claim follows. This completes the proof. □

Definition 1.4.4. The measure $\hat{\mu}$ given by Theorem 1.4.4 is called the Carathéodory extension of the measure μ on $\mathcal{A}$.

The following result shows that the conclusions of Theorem 1.4.4 can be refined by extending a measure μ defined on a semialgebra $\mathcal{S} \subseteq \mathcal{P}(X)$.

Proposition 1.4.5. *Let $\mathcal{S} \subseteq \mathcal{P}(X)$ be a semialgebra containing $\emptyset$, and let ν be a measure on $\mathcal{S}$. Then there exists a unique measure μ on the minimal algebra $\mathcal{A}_0(\mathcal{S})$ such that $\mu|_{\mathcal{S}} = \nu$.*

Proof. For any $E \in \mathcal{A}_0(\mathcal{S})$, there exist disjoint sets $E_1, \dots, E_n \in \mathcal{S}$ such that $E = \bigcup_{k=1}^{n} E_k$. Define $\mu : \mathcal{A}_0(\mathcal{S}) \mapsto [0, \infty]$ by setting $\mu(E) := \sum_{k=1}^{n} \nu(E_k)$. It is easily seen that the definition does not depend on the choice of the sets E_k, thus is well posed, and μ is a measure on $\mathcal{A}_0(\mathcal{S})$.

Let $\tilde{\mu}$ be a measure on $\mathcal{A}_0(\mathcal{S})$ such that $\tilde{\mu}|_{\mathcal{S}} = \nu$. For any n-tuple of disjoint sets $E_1, \dots, E_n \in \mathcal{S}$, we have

$$\tilde{\mu}\left(\bigcup_{k=1}^{n} E_k\right) = \sum_{k=1}^{n} \tilde{\mu}(E_k) = \sum_{k=1}^{n} \nu(E_k) = \mu\left(\bigcup_{k=1}^{n} E_k\right).$$

Every $E \in \mathcal{A}_0(\mathcal{S})$ is the union of such an n-tuple, and thus $\tilde{\mu} = \mu$ on $\mathcal{A}_0(\mathcal{S})$. This proves the uniqueness claim and the result. □

1.5 Lebesgue and Lebesgue–Stieltjes measures

1.5.1 Lebesgue measure in $\mathbb{R}^N$

Let $a \equiv (a_1, \dots, a_N)$ and $b \equiv (b_1, \dots, b_N)$ with $a_i, b_i \in [-\infty, \infty]$ and $a_i \le b_i$ for every $i = 1, \dots, N$ ($N \in \mathbb{N}$). Consider the *N-cell*

$$(a, b] := \{x \equiv (x_1, \dots, x_N) \mid x_i \in [-\infty, \infty],\ a_i < x_i \le b_i \ \forall\, i = 1, \dots, N\}$$

with *volume* $\mathrm{vol}\,((a, b]) := \prod_{i=1}^{N}(b_i - a_i) \le \infty$ (the subsets of $\mathbb{R}^N$ (a, b), $[a, b]$, and $[a, b)$ are similarly defined). Recall that the family $\mathcal{I}_N$ of the N-cells is a semialgebra and that for every $E \in \mathcal{A}_0(\mathcal{I}_N)$, there exist disjoint $(a^1, b^1], \ldots, (a^p, b^p]$ such that $E = \bigcup_{k=1}^p (a^k, b^k]$ (see Example 1.2.1). Set

$$\lambda_N(E) := \sum_{k=1}^{p} \mathrm{vol}((a^k, b^k]). \tag{1.15}$$

It is easily checked that λ_N is a σ-finite measure on $\mathcal{A}_0(\mathcal{I}_N)$. By the results in Subsection 1.4 (in particular, Theorem 1.4.1) we can state the following definition.

Definition 1.5.1. The outer measure λ_N^* generated by the couple $(\mathcal{A}_0(\mathcal{I}_N), \lambda_N)$ with λ_N given by (1.15) is called the *Lebesgue outer measure*. The elements of the σ-algebra

$$\mathcal{L}^N \equiv \mathcal{L}(\mathbb{R}^N) := \{E \subseteq \mathbb{R}^N \mid E\ \lambda_N^*\text{-measurable}\}$$

are called *Lebesgue sets*. The measure $\lambda_N := \lambda_N^*|_{\mathcal{L}^N}$ is called the *Lebesgue measure* on $\mathbb{R}^N$. If $N = 1$, then we set $\lambda_1 \equiv \lambda$ and $\mathcal{L}^1 \equiv \mathcal{L} \equiv \mathcal{L}(\mathbb{R})$.

Remark 1.5.1. (i) Since the σ-algebra $\mathcal{B}^N = \mathcal{B}(\mathbb{R}^N)$ of Borel sets in $\mathbb{R}^N$ is generated by the semialgebra $\mathcal{I}_N$, by Theorem 1.4.4 we have $\mathcal{B}^N \subseteq \mathcal{L}^N$, and thus λ_N is a Borel measure on $\mathbb{R}^N$, and the measure space $(\mathbb{R}^N, \mathcal{L}^N, \lambda_N)$ is the completion of the space $(\mathbb{R}^N, \mathcal{B}^N, \lambda_N|_{\mathcal{B}^N})$.

(ii) Since $\mathbb{R}^N$ is a locally compact Hausdorff space with countable basis, λ_N is a Radon measure and is regular (see Remark 1.3.3). By the regularity of λ_N it is easily seen that for any $E \subseteq \mathbb{R}^N$, there exist $F, G \in \mathcal{B}^N$ such that $F \subseteq E \subseteq G$ and $\lambda_N(F) = \lambda_N(G) = \lambda_N^*(E)$.

Some interesting properties of λ_N are related to the following result, whose elementary proof is omitted.

Proposition 1.5.1. *Let $T : \mathbb{R}^N \mapsto \mathbb{R}^N$ be affine,*

$$Tx := Ax + b \quad \text{with } A \equiv (a_{ij}) \text{ and } b \in \mathbb{R}^N \quad (x \in \mathbb{R}^N;\ i,j = 1, \ldots, N).$$

Let the determinant $\det(A)$ *be nonzero. Then:*
(i) *for any $E \subseteq \mathbb{R}^N$, we have $\lambda_N^*(T(E)) = |\det(A)|\, \lambda_N^*(E)$;*
(ii) *$T(E) := \{Tx \mid x \in E\} \in \mathcal{L}^N$ if and only if $E \in \mathcal{L}^N$.*

Remark 1.5.2. If $E \in \mathcal{L}^N$ and T is homothetic (i. e., $Tx = Ax$ with $A = a(\delta_{ij})$, $a > 0$), then from Proposition 1.5.1(i) we obtain that $\lambda_N(T(E)) = a^N \lambda_N(E)$. In particular, if $E = B(x_0, r)$, then $\lambda_N(B(x_0, r)) = \kappa_N r^N$, where

$$\kappa_N := \frac{\pi^{\frac{N}{2}}}{\Gamma\left(\frac{N}{2}+1\right)}, \quad \Gamma(t) := \int_0^\infty x^{t-1} \exp(-x)\, dx \quad (t > 0), \tag{1.16}$$

is the volume of the unitary ball $B(0,1) \subseteq \mathbb{R}^N$. In this connection, observe that λ_N is a measure of dimension $s = N$ and is doubling with constant $c_D = 2^N$ (see Definition 1.6.4).

An interesting geometric inequality is given by the following result, whose proof will be given in Subsection 2.4.2.

Proposition 1.5.2 (Isodiametric inequality). *For any $E \subseteq \mathbb{R}^N$, we have*

$$\lambda_N^*(E) \le \frac{\kappa_N}{2^N}[\operatorname{diam}(E)]^N \tag{1.17}$$

with κ_N given by (1.16).

Proposition 1.5.1 implies the invariance of the Lebesgue measure under rotations and translations. If coupled with a suitable normalization condition, translation invariance is a characteristic feature of λ_N. This is the content of the following proposition.

Proposition 1.5.3. *Let $\mu : \mathcal{B}^N \mapsto [0,\infty]$ be translation invariant and such that $\mu((0,1]^N) = 1$. Then $\mu = \lambda_N|_{\mathcal{B}^N}$.*

Proof. Let $(K_1,\dots,K_N) \in \mathbb{N}^N$, and consider the points $(\frac{k_1}{K_1},\dots,\frac{k_N}{K_N})$ with $(k_1,\dots,k_N) \in \mathbb{N}^N$, $0 \le k_i \le K_i$ for $i = 1,\dots,N$. These points are the vertices of $P := K_1K_2\dots K_N$ disjoint N-cells I_k such that $(0,1]^N = \bigcup_{k=1}^P I_k$. Since μ is translation invariant, we have $\mu(I_k) = \mu(I_l)$ for every $k,l = 1,\dots,P$. Therefore, since $\mu((0,1]^N) = 1$, we have $\mu(\prod_{i=1}^N(0,\frac{1}{K_i}]) = \frac{1}{P} = \lambda_N(\prod_{i=1}^N(0,\frac{1}{K_i}])$. By the translation invariance of μ and λ_N the same equality holds on the family $\mathcal{I}_{N,\mathbb{Q}}$ of the N-cells with vertices in $\mathbb{Q}^N$ and thus on the family $\mathcal{I}_N$ of the N-cells. Then by Theorem 1.4.4 and Proposition 1.4.5 the conclusion follows. □

The proof of the following covering result is connected with the above remarks.

Definition 1.5.2. Let $E \subseteq \mathbb{R}^N$. A family $\mathcal{F}$ of closed balls of $\mathbb{R}^N$ is a *fine cover* of E if for any $x \in E$ and any $\epsilon > 0$, there exists a ball $B \in \mathcal{F}$ such that $x \in B$ and $\operatorname{diam}(B) < \epsilon$.

Theorem 1.5.4 (Vitali covering lemma). *Let $E \subseteq \mathbb{R}^N$ be such that $\lambda_N^*(E) < \infty$, and let $\mathcal{F}$ be a fine cover of E. Then there exists a disjoint sequence $\{B_k\} \subseteq \mathcal{F}$ such that $\lambda_N^*(E \setminus (\bigcup_{k=1}^\infty B_k)) = 0$.*

Proof. It suffices to prove that for any $\epsilon > 0$, there exist $B_1,\dots,B_K \in \mathcal{F}$ such that $B_j \cap B_k = \emptyset$ for $j,k = 1,\dots,K$, $j \ne k$, and $\lambda_N^*(E \setminus (\bigcup_{k=1}^K B_k)) < \epsilon$.

Since $\lambda_N^*(E) < \infty$, there exists an open set $U \supseteq E$ with $\lambda_N(U) < \infty$, and by Definition 1.5.2 we can assume that $B \subseteq U$ for all $B \in \mathcal{F}$. We construct inductively a disjoint sequence $\{B_k\} \subseteq \mathcal{F}$ as follows. Suppose that p disjoint closed balls $B_1,\dots,B_p \in \mathcal{F}$ have already been chosen and that $E \setminus (\bigcup_{k=1}^{p} B_k) \neq \emptyset$. Set

$$k_p := \sup\left\{\operatorname{diam}(B) \mid B \in \mathcal{F},\, B \subseteq E \setminus \left(\bigcup_{k=1}^{p} B_k\right)\right\} \leq \sup\{\operatorname{diam}(B) \mid B \in \mathcal{F}\} < \infty.$$

Then we can choose B_{p+1} such that

$$\operatorname{diam}(B_{p+1}) \geq \frac{k_p}{2}, \quad B_{p+1} \cap B_k = \emptyset \quad \text{for } k = 1, \dots, p. \tag{1.18}$$

Since the sequence $\{B_k\}$ is disjoint and $\bigcup_{k=1}^{\infty} B_k \subseteq U$, we have

$$\frac{\kappa_N}{2^N} \sum_{k=1}^{\infty} [\operatorname{diam}(B_k)]^N = \sum_{k=1}^{\infty} \lambda_N(B_k)) = \lambda_N\left(\bigcup_{k=1}^{\infty} B_k\right) \leq \lambda_N(U) < \infty \tag{1.19}$$

with κ_N given by (1.16), whence

$$\lim_{k\to\infty} \operatorname{diam}(B_k) = 0. \tag{1.20}$$

Let $\epsilon > 0$ be fixed. By (1.19) there exists $K \in \mathbb{N}$ such that

$$\sum_{k=K+1}^{\infty} [\operatorname{diam}(B_k)]^N < \frac{2^N \epsilon}{5^N \kappa_N}. \tag{1.21}$$

Set $R := E \setminus (\bigcup_{k=1}^{K} B_k)$. The conclusion will follow if we prove that $\lambda_N^*(R) < \epsilon$.

Let $x \in R$. Since $\bigcup_{k=1}^{K} B_k$ is closed and $x \notin \bigcup_{k=1}^{K} B_k$, by Definition 1.5.2 there exists $B \in \mathcal{F}$ such that $x \in B$ and $B \cap B_k = \emptyset$ for $k = 1, \dots, K$. In general, if $B \cap B_k = \emptyset$ for $k = 1, \dots, M$, then for some $M \in \mathbb{N}$, we have $B \subseteq E \setminus (\bigcup_{k=1}^{M} B_k)$, and hence by (1.18)

$$\operatorname{diam}(B) \leq k_M \leq 2 \operatorname{diam}(B_{M+1}).$$

If there were $B \cap B_k = \emptyset$ for all $k \in \mathbb{N}$, then we would have a contradiction with (1.20). Therefore there is a smallest integer $m \in \mathbb{N}$ such that $B \cap B_m \neq \emptyset$, and thus $m > k$. Since $B \subseteq E \setminus (\bigcup_{k=1}^{m-1} B_k)$, by the above remarks we have

$$\operatorname{diam}(B) \leq k_{m-1} \leq 2 \operatorname{diam}(B_m). \tag{1.22}$$

Let $B_m \equiv B(x_m, r_m)$. Since $x \in B$ and $B \cap B_m \neq \emptyset$, we have that

$$|x - x_m| \leq \operatorname{diam}(B) + \tfrac{1}{2} \operatorname{diam}(B_m) \leq \tfrac{5}{2} \operatorname{diam}(B_m),$$

and thus $x \in B(x_m, 5r_m)$. Therefore we have $R \subseteq \bigcup_{m=K+1}^{\infty} B(x_m, 5r_m)$, which implies that

$$\lambda_N^*(R) \leq \sum_{m=K+1}^{\infty} \lambda_N(B(x_m, 5r_m)) = \frac{5^N \kappa_N}{2^N} \sum_{m=K+1}^{\infty} [\operatorname{diam}(B_m)]^N < \epsilon.$$

This completes the proof. □

Remark 1.5.3. Plainly, the proof of Theorem 1.5.4 also provides the following statement:

Let $E \in \mathcal{B}^N$ with $\lambda_N(E) < \infty$, and let $\mathcal{F}$ be a fine cover of E. Let $\delta > 0$. Then there exists a disjoint sequence $\{B_k\} \subseteq \mathcal{F}$ with $\operatorname{diam} B_k < \delta$ *for all $k \in \mathbb{N}$ such that $\lambda_N(E \setminus (\bigcup_{k=1}^{\infty} B_k)) = 0$.*

A crucial role in the proof of Theorem 1.5.4 is played by Remark 1.5.2. For a general Radon measure on $\mathbb{R}^N$, more involved arguments must be used, which lead to the following result (e. g., see [5] for the proof).

Theorem 1.5.5 (Vitali–Besicovitch). *Let $U \in \mathcal{B}^N$ be bounded, and let $\mathcal{F} \equiv \{F_i\}_{i\in I}$ be a fine cover of U. Let $\mu \in \mathfrak{R}^+(\mathbb{R}^N)$. Then there exists a disjoint sequence $\{B_k\} \subseteq \mathcal{F}$ such that $\mu(U \setminus (\bigcup_{k=1}^{\infty} B_k)) = 0$.*

1.5.2 Lebesgue–Stieltjes measure

The above construction can be generalized as follows (we only consider the case $N = 1$, referring to [45] for the general case). Let $\phi : \mathbb{R} \to \mathbb{R}$ be nondecreasing and right-continuous. Consider the semialgebra $\mathcal{I}_1$ of intervals $(a, b]$ with $a, b \in [-\infty, \infty]$ and the algebra $\mathcal{A}_0(\mathcal{I}_1)$, consisting of finite disjoint unions of elements of $\mathcal{I}_1$, that is, for every $E \in \mathcal{A}_0(\mathcal{I}_1)$, there exist disjoint $(a^1, b^1], \dots, (a^p, b^p]$ with $b^k \leq a^{k+1}$ $(k = 1, \dots, p-1)$ such that $E = \bigcup_{k=1}^{p}(a^k, b^k]$. Set

$$\lambda^{\phi}(E) := \sum_{k=1}^{p}[\phi(b^k) - \phi(a^k)], \tag{1.23}$$

where $\phi(b^p) := \lim_{x\to\infty}\phi(x)$ if $b^p = \infty$ and $\phi(a^1) := \lim_{x\to-\infty}\phi(x)$ if $a^1 = -\infty$. Since ϕ is nondecreasing and right-continuous, it is easily seen that λ^{ϕ} is a σ-finite measure on the algebra $\mathcal{A}_0(\mathcal{I}_1)$. Then by the results of Section 1.4 the following definition is well posed.

Definition 1.5.3. The measure $\lambda^{\phi} := (\lambda^{\phi})^*|_{\mathcal{L}_{\phi}}$, where $(\lambda^{\phi})^*$ is the outer measure generated by the couple $(\mathcal{A}_0(\mathcal{I}_1), \lambda^{\phi})$ with λ^{ϕ} given by (1.23), and $\mathcal{L}_{\phi}$ is the σ-algebra of the $(\lambda^{\phi})^*$-measurable sets, is called the Lebesgue–Stieltjes measure on $\mathbb{R}$.

Analogous results hold in $\mathbb{R}^N$ for $N \geq 2$ (see [45, Korollar II.3.10, Beispiel II.4.7]). The collection of Lebesgue–Stieltjes measures on $\mathbb{R}$, corresponding to nondecreasing right-continuous functions $\phi : \mathbb{R} \to \mathbb{R}$, is denoted by $\mathfrak{L}_{\phi}(\mathbb{R})$. These measures coincide with the Borel measures and thus with the Radon measures on $\mathbb{R}$ (see Remark 1.3.3), as the following proposition shows.

Proposition 1.5.6. *Let $\mathfrak{L}_\phi(\mathbb{R})$ be the family of Lebesgue–Stieltjes measures on $\mathbb{R}$ with nondecreasing and right-continuous $\phi : \mathbb{R} \to \mathbb{R}$. Then $\mathfrak{L}_\phi(\mathbb{R}) = \mathfrak{B}^+(\mathbb{R}) = \mathfrak{R}^+(\mathbb{R})$.*

Proof. Let ϕ be fixed. Since the σ-algebra $\mathcal{B}(\mathbb{R})$ is generated by the semialgebra of intervals $\mathcal{I}_1$, by Theorem 1.4.4 we have $\mathcal{B} \subseteq \mathcal{L}_\phi$, and thus λ^ϕ is a Borel measure on $\mathbb{R}$. Conversely, let us show that $\mathfrak{B}^+(\mathbb{R}) \subseteq \mathfrak{L}_\phi(\mathbb{R})$. Let μ be a Radon measure on $\mathbb{R}$. Fix $c \in \mathbb{R}$, and define $\phi_\mu : \mathbb{R} \to \mathbb{R}$ as follows:

$$\phi_\mu(x) := \begin{cases} \mu((c,x]) & \text{if } x > c, \\ 0 & \text{if } x = c, \\ -\mu((x,c]) & \text{if } x < c. \end{cases} \tag{1.24}$$

Then ϕ_μ is nondecreasing and right-continuous, and we have

$$\phi_\mu(b) - \phi_\mu(a) = \mu((a,b]) \quad (-\infty < a < b \leq \infty).$$

By Proposition 1.4.5 μ coincides with the Lebesgue–Stieltjes measure associated with ϕ_μ. Hence the claim follows. □

Remark 1.5.4. In view of Proposition 1.5.6, there is one-to-one correspondence between Radon measures on $\mathbb{R}$ on one side and classes of equivalence of nondecreasing right-continuous functions $\phi : \mathbb{R} \to \mathbb{R}$ on the other, the equivalence relation being

$$\psi \sim \phi \quad \overset{\text{def}}{\Longleftrightarrow} \quad \psi = \phi + c \quad \text{for some } c \in \mathbb{R}.$$

If $\mu \in \mathfrak{R}^+(\mathbb{R})$, then a representative of its class of equivalence is called the *distribution function* of μ.

If $\phi : \mathbb{R} \mapsto \mathbb{R}$ is nondecreasing and continuous, then for each interval $I = (a,b]$, we have $\phi(I) := \{\phi(x) \mid x \in I\} = (\phi(a), \phi(b)]$, and thus

$$\lambda^\phi(I) = \phi(b) - \phi(a) = \lambda(\phi(I)). \tag{1.25}$$

For every nondecreasing and right-continuous $\phi : \mathbb{R} \to \mathbb{R}$, set

$$\varphi : \mathcal{P}(\mathbb{R}) \to \mathcal{P}(\mathbb{R}), \qquad \begin{cases} \varphi(\emptyset) := \emptyset, \\ \varphi(E) := \bigcup_{x \in E} [\phi(x^-), \phi(x)] \ \forall E \subseteq \mathbb{R},\ E \neq \emptyset. \end{cases} \tag{1.26}$$

Then equality (1.25) can be extended as follows.

Proposition 1.5.7. *Let $\phi : \mathbb{R} \to \mathbb{R}$ be nondecreasing and right-continuous, and let φ be defined by* (1.26). *Then:*

(i) $\varphi(E) \in \mathcal{L}(\mathbb{R})$ *for any* $E \in \mathcal{B}(\mathbb{R})$;

(ii) *we have*

$$\lambda^\phi(E) = \lambda(\varphi(E)) \quad \text{for any } E \in \mathcal{B}(\mathbb{R}). \tag{1.27}$$

Proof. It is easily seen that the family $\Sigma := \{E \in \mathcal{B}(\mathbb{R}) \mid \varphi(E) \in \mathcal{L}(\mathbb{R})\}$ is a σ-algebra containing the semialgebra of intervals $\mathcal{I}_1$. Since the Borel σ-algebra $\mathcal{B}(\mathbb{R})$ is generated by $\mathcal{I}_1$, claim (i) follows by Theorem 1.4.4 and Proposition 1.4.5. As for (ii), it suffices to prove that the map

$$m : \mathcal{B}(\mathbb{R}) \to [0, \infty], \quad m(E) := \lambda(\varphi(E)),$$

is a measure on $\mathcal{B}(\mathbb{R})$ such that

$$\begin{cases} m((a,b]) = \phi(b) - \phi(a) & \text{if } -\infty \le a < b < \infty, \\ m((a,\infty)) = \phi(\infty) - \phi(a) & \text{if } a \in \mathbb{R}. \end{cases} \tag{1.28}$$

Then the conclusion follows by Proposition 1.4.5.

By definition we have $m(\emptyset) = 0$. To prove that m is σ-additive consider a disjoint sequence $\{E_k\} \subseteq \mathcal{B}(\mathbb{R})$ and denote by $F \subseteq \phi(\mathbb{R})$ the countable set of points y such that $\phi^{-1}(\{y\})$ is not a singleton. Then the sets $\varphi(E_k) \setminus F$ are disjoint, and

$$\bigcup_{k=1}^{\infty} (\varphi(E_k) \setminus F) = \left[\varphi\left(\bigcup_{k=1}^{\infty} E_k\right)\right] \setminus F.$$

It follows that

$$\lambda\left(\varphi\left(\bigcup_{k=1}^{\infty} E_k\right)\right) = \lambda\left(\left[\varphi\left(\bigcup_{k=1}^{\infty} E_k\right)\right] \setminus F\right) = \sum_{k=1}^{\infty} \lambda(\varphi(E_k) \setminus F) = \sum_{k=1}^{\infty} \lambda(\varphi(E_k)).$$

Let us finally prove equalities (1.28). From the equality $\varphi((a,b]) = (\phi(a), \phi(b)]$ we obtain the first one, whence the second follows by the monotonicity of ϕ and Proposition 1.3.1(iv). This completes the proof. □

Corollary 1.5.8. *Let $\phi : \mathbb{R} \to \mathbb{R}$ be nondecreasing and right-continuous. Then ϕ is continuous at $x_0 \in \mathbb{R}$ if and only if $\lambda^\phi(\{x_0\}) = 0$.*

In connection with Proposition 1.5.7, let us state the following definition.

Definition 1.5.4. A function $\phi : \mathbb{R} \mapsto \mathbb{R}$ satisfies the Lusin condition if for any λ-null set $E \subseteq \mathbb{R}$, the value $\varphi(E)$ of the map (1.26) is λ-null.

Let us discuss two significant cases where the Lusin condition is *not* satisfied.

(α) If ϕ is the *Heaviside map* $H := \chi_{[0,\infty)}$, then the map (1.26) reads

$$\varphi(\{x\}) = \begin{cases} \{0\} & \text{if } x < 0, \\ [0,1] & \text{if } x = 0, \\ \{1\} & \text{if } x < 0. \end{cases}$$

Then by equality (1.27) for any Borel set $E \neq \emptyset$, we have

$$\lambda^H(E) = \begin{cases} 1 & \text{if } 0 \in E, \\ 0 & \text{if } 0 \notin E, \end{cases}$$

that is, λ^H is the Dirac mass δ_0. More generally, for the nondecreasing step function

$$\phi := c_0\chi_{(-\infty,x_1)} + \sum_{p=1}^{n-1} c_p\chi_{[x_p,x_{p+1})} + c_n\chi_{[x_n,\infty)} \quad (x_1 < x_2 < \cdots < x_n)$$

with $c_p \neq c_{p+1}$ ($p = 0,\ldots,n-1$), we have $\lambda^\phi = \sum_{p=1}^n [c_p - c_{p-1}]\delta_{x_p}$, and thus the Lusin condition is not satisfied.

(β) Let $\{m_n\} \subseteq \mathbb{N} \cup \{0\}$ be a decreasing sequence such that $m_0 = 1$ and $0 < 2m_{n+1} < m_n$ for all $n = 0,1,\ldots$. Set $K_0 \equiv [0,1]$, and let $K_1 \subseteq K_0$ be obtained by removing from K_0 an open interval of length $m_0 - 2m_1$ in the middle (namely, with center $\frac{1}{2}$). Hence K_1 is the disjoint union of two closed intervals $J_{1,1}$ and $J_{1,2}$ of length m_1. Then we remove an open interval of length $m_1 - 2m_2$ in the middle of both $J_{1,1}$ and $J_{1,2}$. Iterating the procedure, after n steps, we obtain a disjoint union K_n of 2^n closed intervals $J_{n,j}$ ($j = 1,\ldots,2^n$) with

$$\lambda(J_{n,j}) = m_n, \quad \lambda(K_n^c) = \sum_{k=0}^{n} 2^k(m_k - 2m_{k+1}) \quad (n \in \mathbb{N}). \tag{1.29}$$

The intersection $K := \bigcap_{n=1}^\infty K_n$ is the *Cantor set* corresponding to the sequence $\{m_n\}$. A possible choice is $m_n = a^n$ with $a \in (0,\frac{1}{2})$ (which for $a = \frac{1}{3}$ gives the standard *Cantor middle third set*). In this case, from (1.29) we plainly get $\lambda(K^c) = 1$, and thus $\lambda(K) = 0$.

Let $m_n = a^n$ with $a \in (0,1/2)$, and let $V \in C([0,1])$ be the uniform limit in $[0,1]$ of the sequence of the piecewise linear functions $V_n \in C([0,1])$, where $V_n(0) := 0$ and

$$V_n \text{ has slope } \begin{cases} (2a)^{-n} & \text{in } J_{n,j} \text{ for each } j = 1,\ldots,2^n, \\ 0 & \text{otherwise.} \end{cases} \quad (n \in \mathbb{N}).$$

If $n \in \mathbb{N}$ is fixed, then for every $p > n$, we have $V_p = V_n$ in the set K_n^c. It follows that

$$\sup_{x\in[0,1]} |V_{n+1}(x) - V_n(x)| = \sup_{x\in[0,a^n]} \left| \frac{x}{(2a)^{n+1}} - \frac{x}{(2a)^n} \right| = \frac{1-2a}{a}\frac{1}{2^{n+1}},$$

and thus for any $p > n$,

$$\sup_{x\in[0,1]} |V_p(x) - V_n(x)| \leq \frac{1-2a}{a} \sum_{k=n}^{p-1} \frac{1}{2^{k+1}} \leq \frac{1-2a}{a} \frac{1}{2^n}.$$

By this inequality the sequence $\{V_n\}$ converges uniformly in $[0,1]$, and hence the function V is well defined. Moreover, V is nondecreasing, $V(0) = 0$, $V(1) = 1$, and $V' = 0$ a. e. in $(0,1)$. We extend V to all of $\mathbb{R}$ by setting $V = 0$ in $(-\infty, 0)$ and $V = 1$ in $(1, \infty)$. The function V is called the *Cantor–Vitali function.*

By (1.25) we have that $\lambda^V(K) = 1$, and thus the Lusin condition is not satisfied. Further remarks concerning λ^V will be made in Section 2.9.2.

Remark 1.5.5. (i) For any $n \in \mathbb{N}$ and $j = 1, \ldots, 2^n$, by the above construction we have $\lambda^V(J_{n,j}) = 2^{-n}$. Since the 2^n intervals $J_{n,j}$ are disjoint, by additivity we get $\lambda^V(K_n) = 1$ for each n, and hence again $\lambda^V(K) = 1$ as $n \to \infty$ by Proposition 1.3.1(v).

(ii) Denote by K_n^N the product of N copies of the set K_n $(n, N \in \mathbb{N})$. The set $K^N := \bigcap_{n=1}^{\infty} K_n^N$ is called the *N-dimensional Cantor set* corresponding to the sequence $\{m_n\}$ (we set $K^1 \equiv K$).

1.6 Metric outer measures and capacities

1.6.1 Capacities

Definition 1.6.1. Let (X, d) be a metric space, and let $\mathcal{F} \subseteq \mathcal{P}(X)$ have the following properties: (a) the family $\mathcal{K}$ of compact subsets is contained in $\mathcal{F}$; (b) for every sequence $\{E_n\} \subseteq \mathcal{F}$, we have $\bigcup_{n=1}^{\infty} E_n \in \mathcal{F}$.

A map $C : \mathcal{F} \to [0, \infty]$ is called a capacity on $\mathcal{F}$ if:

(i) $C(\emptyset) = 0$;

(ii) $C(E_1) \leq C(E_2)$ for any $E_1, E_2 \in \mathcal{F}$ such that $E_1 \subseteq E_2$;

(iii) $C(\bigcup_{n=1}^{\infty} E_n) \leq \sum_{n=1}^{\infty} C(E_n)$ for any sequence $\{E_n\} \subseteq \mathcal{F}$.

A capacity on $\mathcal{P}(X)$ is called a capacity on X.

Remark 1.6.1. If $\mathcal{F}$ satisfies the above properties (a)–(b), then by Remark A.1 the family $\mathcal{T}$ of open subsets is contained in $\mathcal{F}$. Moreover, by Definitions 1.4.1 and 1.6.1 a capacity on X is an outer measure, and an outer measure on a metric space X is a capacity.

Definition 1.6.2. A capacity is called an outer capacity if for any $E \in \mathcal{F}$,

$$C(E) = \inf\{C(A) \mid A \supseteq E,\, A \in \mathcal{T}\},$$

and an inner capacity if for any $E \in \mathcal{F}$,

$$C(E) = \sup\{C(K) \mid K \subseteq E,\ K \in \mathcal{K}\}.$$

A set $E \in \mathcal{F}$ is called C-capacitable if

$$C(E) = \inf\{C(A) \mid A \supseteq E,\ A \in \mathcal{T}\} = \sup\{C(K) \mid K \subseteq E,\ K \in \mathcal{K}\}. \tag{1.30}$$

Remark 1.6.2. Let X and $\mathcal{F} \subseteq \mathcal{P}(X)$ be as in Definition 1.6.1, and let C be a capacity on $\mathcal{F}$. For any family $\mathcal{F}_1 \subseteq \mathcal{F}$, define $C_{\mathcal{F}_1} : \mathcal{P}(X) \mapsto [0,\infty]$ setting

$$C_{\mathcal{F}_1}(E) := \inf\{C(F) \mid F \supseteq E,\ F \in \mathcal{F}_1\} \quad (E \subseteq X)$$

if there exists $F \in \mathcal{F}_1$ such that $F \supseteq E$ and $C_{\mathcal{F}_1}(E) := \infty$ otherwise. By Theorem 1.4.2 $C_{\mathcal{F}_1}$ is an outer measure and thus a capacity on X (see Remark 1.6.1). In particular, if $\mathcal{F}_1 = \mathcal{T}$ is the topology on X, then $C_{\mathcal{T}}$ is by definition an outer capacity, and C is an outer capacity if and only if $C = C_{\mathcal{T}}$.

The concepts in Definition 1.6.2 are analogous to those in Definition 1.3.3(b–d). In Chapter 3, we will use the following result (see [2, Theorem 2.3.11], [32], [109], and related references therein).

Theorem 1.6.1 (Choquet). *Let C be a capacity on $\mathbb{R}^N$ such that:*
(i) *for any nonincreasing sequence $\{K_k\}$ of compact subsets of $\mathbb{R}^N$,*

$$C\left(\bigcap_{k=1}^{\infty} K_k\right) = \lim_{k\to\infty} C(K_k); \tag{1.31}$$

(ii) *for any nondecreasing sequence $\{E_k\} \subseteq \mathcal{P}(\mathbb{R}^N)$,*

$$C\left(\bigcup_{k=1}^{\infty} E_k\right) = \lim_{k\to\infty} C(E_k). \tag{1.32}$$

Then every $E \in \mathcal{B}(\mathbb{R}^N)$ is C-capacitable.

Remark 1.6.3. The conclusion of Theorem 1.6.1 actually holds for the larger class of *Suslin sets*.

In connection with Theorem 1.6.1, let us note the following result.

Proposition 1.6.2. *Let X and $\mathcal{F} \subseteq \mathcal{P}(X)$ be as in Definition 1.6.1, and let $C : \mathcal{F} \to [0,\infty]$ be an outer capacity. Then for any nonincreasing sequence $\{K_n\} \subseteq \mathcal{K}$, equality (1.31) is satisfied.*

Proof. Set $K := \bigcap_{n=1}^{\infty} K_n$, and consider any $A \supseteq K$, $A \in \mathcal{T}$. Since $\{K_n\}$ is nonincreasing, there exists $\bar{n} \in \mathbb{N}$ such that $K_n \subseteq A$ for all $n > \bar{n}$, and thus by the monotonicity of C we plainly have

$$C(K) \le \lim_{n\to\infty} C(K_n) \le \inf_{A\supseteq K, A\in\mathcal{T}} C(A).$$

On the other hand, by assumption we have $\inf_{A\supseteq K, A\in\mathcal{T}} C(A) = C(K)$. Then the result follows. □

Let X and $\mathcal{F} \subseteq \mathcal{P}(X)$ be as in Definition 1.6.1, and let $C : \mathcal{F} \to [0,\infty]$ be a capacity. The following notions are analogous to those of Subsection 1.3.3:

(a) a set $E \in \mathcal{F}$ such that $C(E) = 0$ is called *C-null*;

(b) a property P *holds C-quasi-everywhere* (written C-q. e. or q. e.) if the set $\{x \in U \mid P(x) \text{ false}\}$ is C-null.

By Definition 1.6.1 the family $\mathcal{N}_C$ of C-null sets is stable with respect to the countable union and inheritance, that is, $F \subseteq E$, $E \in \mathcal{N}_C \Rightarrow F \in \mathcal{N}_C$. Particular cases of (b) are the following:

- let $f, f_n : X \mapsto \mathbb{R}$ $(n \in \mathbb{N})$. We say that f_n *converges to* f *C-quasi-everywhere in* X if there exists $N \in \mathcal{N}_C$ such that $f_n(x)$ converges to $f(x)$ for any $x \in N^c$;
- two functions $f, g : X \to \mathbb{R}$ are *equal C-q. e. in X* if $\{x \in X \mid f(x) \neq g(x)\} \in \mathcal{N}_C$.

Clearly, the equality C-q. e. is a relation of equivalence:

$$g \sim f \overset{\text{def}}{\iff} g = f \quad C\text{-q. e. in } X, \tag{1.33}$$

and each class of equivalence with respect to (1.33) is uniquely determined by anyone of its elements, a *representative* of the class. As for functions equal μ-a. e., we identify functions that are equal C-q.e, thus regarding the whole class as a unique map f defined C-q. e. in X.

1.6.2 Metric outer measures

Definition 1.6.3. Let (X, d) be a metric space. An outer measure μ^* is a metric outer measure if for any nonempty sets $E, F \subseteq X$ with $d(E, F) > 0$,

$$\mu^*(E \cup F) = \mu^*(E) + \mu^*(F). \tag{1.34}$$

The existence of metric outer measures can be proven as in Subsection 1.4.1 by a slight modification of definition (1.11). Let (X, d) be a metric space, and let $\mathcal{C} \subseteq \mathcal{P}(X)$

with $\emptyset \in \mathcal{C}$ and $\eta : \mathcal{C} \to [0,\infty]$ with $\eta(\emptyset) = 0$. For any $\delta \in (0,\infty]$, set

$$\eta^*_\delta(E) := \inf\left\{\sum_{n=1}^{\infty} \eta(E_n) \mid E \subseteq \bigcup_{n=1}^{\infty} E_n,\ \{E_n\} \subseteq \mathcal{C},\ \operatorname{diam}(E_n) \le \delta\ \forall n \in \mathbb{N}\right\} \tag{1.35}$$

if a countable cover $\{E_n\}$ of E as in (1.35) exists and $\eta^*_\delta(E) := \infty$ otherwise. Arguing as in the proof of Theorem 1.4.2 shows that the map $\eta^*_\delta : \mathcal{P}(X) \mapsto [0,\infty]$ is an outer measure (in particular, η^*_∞ is the outer measure μ^* defined in (1.11)). It is easily seen that the map

$$\eta^* : \mathcal{P}(X) \mapsto [0,\infty], \quad \eta^*(E) := \lim_{\delta\to 0^+} \eta^*_\delta(E) = \sup_{\delta>0} \eta^*_\delta(E) \quad (E \subseteq X) \tag{1.36}$$

also is an outer measure. In fact, for any $E \subseteq X$, the map $\delta \mapsto \eta^*_\delta(E)$ is nonincreasing; thus the definition is well posed, and

$$\eta^*_\infty(E) \le \eta^*_\delta(E) \le \eta^*(E) \quad \text{for any } \delta \in (0,\infty). \tag{1.37}$$

Hence for any $\{E_n\} \subseteq \mathcal{P}(X)$ and $\delta > 0$,

$$\eta^*_\delta\left(\bigcup_{n=1}^{\infty} E_n\right) \le \sum_{n=1}^{\infty} \eta^*_\delta(E_n) \le \sum_{n=1}^{\infty} \eta^*(E_n) \quad \Rightarrow \quad \eta^*\left(\bigcup_{n=1}^{\infty} E_n\right) \le \sum_{n=1}^{\infty} \eta^*(E_n).$$

This proves that η^* is σ-subadditive. Clearly, it is monotone, $\eta^*(\emptyset) = 0$, and thus η^* is an outer measure. The following lemma completes the claim.

Lemma 1.6.3. *Let (X,d) be a metric space. Then the outer measure η^* defined in* (1.36) *is a metric outer measure.*

Proof. Let $E,F \subseteq X$ be nonempty, and let $d(E,F) > 0$. The claim follows if we prove that $\eta^*(E \cup F) \ge \eta^*(E) + \eta^*(F)$. Assume that $\eta^*(E \cup F) < \infty$, since otherwise the conclusion is obvious. Let $\delta \in (0, d(E,F))$, and let $\{G_n\} \subseteq \mathcal{C}$ with $\operatorname{diam}(G_n) \le \delta$ for all $n \in \mathbb{N}$ be such that $E \cup F \subseteq \bigcup_{n=1}^{\infty} G_n$. Clearly, by the choice of δ no $\bar{n} \in \mathbb{N}$ exists such that $G_{\bar{n}}$ intersects both E and F. Hence for all $n \in \mathbb{N}$, there exist G'_n, $G''_n \subseteq X$, such that $G'_n \cup G''_n = G_n$; thus $\max\{\operatorname{diam}(G'_n), \operatorname{diam}(G''_n)\} \le \delta$, and $E \subseteq \bigcup_{n=1}^{\infty} G'_n$, $F \subseteq \bigcup_{n=1}^{\infty} G''_n$. Plainly, this implies that $\eta^*_\delta(E) + \eta^*_\delta(F) \le \eta^*_\delta(E \cup F)$, whence letting $\delta \to 0^+$ the result follows. □

Remark 1.6.4. By Remark 1.6.1 the outer measures η^*_δ ($\delta \in (0,\infty]$) and η^* defined in (1.35)–(1.36) are capacities on X.

Metric outer measures can be characterized as follows.

Proposition 1.6.4. *Let (X,d) be a metric space, and let $\mu^* : \mathcal{P}(X) \mapsto [0,\infty]$ be an outer measure. Then the following statements are equivalent:*

(i) *every Borel set is μ^*-measurable;*

(ii) *μ^* is a metric outer measure.*

Proof. (i)⇒(ii). By assumption, for any open set $A \subseteq X$, we have

$$\mu^*(Z) = \mu^*(Z \cap A) + \mu^*(Z \cap A^c) \quad \text{for any } Z \subseteq X \tag{1.38}$$

(see (1.4)). Let $E, F \subseteq X$ be nonempty with $0 < \delta < d(E, F)$. Then $G := \{x \in X \mid d(x, E) < \delta\}$ is open, and $E \subseteq G$, $F \subseteq G^c$. Therefore equality (1.38) with $Z = E \cup F$ and $A = G$ reduces to (1.34). Hence the claim follows.

(ii)⇒(i). It suffices to prove that every closed set $C \subseteq X$ is μ^*-measurable. For any set $D \subseteq C^c$, set $D_n := \{x \in D \mid d(x, C) \geq \frac{1}{n}\} \subseteq D$ $(n \in \mathbb{N})$. Let us prove that

$$\lim_{n\to\infty} \mu^*(D_n) = \mu^*(D). \tag{1.39}$$

Indeed, for all $n \in \mathbb{N}$, we have $\mu^*(D_n) \leq \mu^*(D)$, since $D_n \subseteq D_{n+1} \subseteq D$. It remains to prove that

$$\lim_{n\to\infty} \mu^*(D_n) \geq \mu^*(D). \tag{1.40}$$

Let $\lim_{n\to\infty} \mu^*(D_n) < \infty$ (otherwise, the claim is obvious), and set $P_n := D_{n+1} \setminus D_n$ $(n \in \mathbb{N})$. If $P_{2k} \neq \emptyset$ $(k = 1, \ldots, n)$, it is easily seen that $d(\bigcup_{k=1}^n P_{2k}, P_{2n+2}) > 0$, and thus

$$\mu^*\left(\bigcup_{k=1}^{n+1} P_{2k}\right) = \mu^*\left(\bigcup_{k=1}^{n} P_{2k}\right) + \mu^*(P_{2n+2})$$

since μ^* is a metric outer measure. Arguing inductively, we obtain that

$$\mu^*\left(\bigcup_{k=1}^{n} P_{2k}\right) = \sum_{k=1}^{n} \mu^*(P_{2k}) \quad \text{for all } n \in \mathbb{N}$$

(notice that the above equality also holds if $P_{2k} = \emptyset$ for some $k = 1, \ldots, n$).

It is similarly shown that $\mu^*(\bigcup_{k=0}^n P_{2k+1}) = \sum_{k=0}^n \mu^*(P_{2k+1})$. Since $\bigcup_{k=1}^n P_k \subseteq D_n$ for all $n \in \mathbb{N}$, we obtain that

$$\begin{aligned} \sum_{k=1}^{\infty} \mu^*(P_k) &= \lim_{n\to\infty}\left(\sum_{k=0}^{n} \mu^*(P_{2k+1}) + \sum_{k=1}^{n} \mu^*(P_{2k})\right) \\ &= \lim_{n\to\infty}\left(\mu^*\left(\bigcup_{k=1}^{n} P_{2k}\right) + \mu^*\left(\bigcup_{k=0}^{n} P_{2k+1}\right)\right) \leq 2 \lim_{n\to\infty} \mu^*(D_n) < \infty. \end{aligned} \tag{1.41}$$

Finally, since C is closed, for all $n \in \mathbb{N}$, we have $D = D_n \cup (\bigcup_{k=n}^{\infty} P_k)$, whence

$$\mu^*(D) \leq \mu^*(D_n) + \sum_{k=n}^{\infty} \mu^*(P_k) \quad (n \in \mathbb{N}). \tag{1.42}$$

By (1.41) we have $\lim_{n\to\infty}\sum_{k=n}^{\infty}\mu^*(P_k) = 0$, and thus letting $n \to \infty$ in (1.42), we obtain (1.40). Hence (1.39) follows.

Now let $Z \subseteq X$ be fixed, and choose $D = Z \cap C^c$. For any $n \in \mathbb{N}$, we have $Z \supseteq (Z \cap C) \cup D_n$, whence by the monotonicity of μ^*

$$\mu^*(Z) \geq \mu^*((Z \cap C) \cup D_n) = \mu^*(Z \cap C) + \mu^*(D_n) \quad (n \in \mathbb{N}),$$

since $d(Z\cap C, D_n) \geq \frac{1}{n} > 0$ and μ^* is a metric outer measure. On the other hand, by (1.39) we have $\lim_{n\to\infty}\mu^*(D_n) = \mu^*(D) = \mu^*(Z \cap C^c)$, and thus letting $n \to \infty$ in the above inequality the result follows. □

Let us finally add two notions relative to Borel measures on metric spaces.

Definition 1.6.4. Let (X, d) be a metric space. A regular Borel measure μ on X is called:
(i) *of dimension* $s > 0$ if there exists $c \geq 1$ such that $\frac{1}{c}r^s \leq \mu(B(x,r)) \leq cr^s$ for all $x \in X$ and $r > 0$;
(ii) *doubling* if there exists $c_D \geq 1$ such that $\mu(B(x, 2r)) \leq c_D\mu(B(x,r))$ for all $x \in X$ and $r > 0$.

Remark 1.6.5. (i) If μ is of dimension s, then it is doubling with constant $c_D = c^2 2^s$.

(ii) If μ is doubling, then it is easily seen that for all $x \in X$ and $r > 0$, we have $\mu(B(x,r)) < \infty$, and if $\mu \not\equiv 0$, then $\mu(B(x,r)) > 0$.

(iii) If μ is doubling and $0 < r < R$, then

$$\mu(B(x,R)) \leq c_D\left(\frac{R}{r}\right)^s \mu(B(x,r)) \tag{1.43}$$

with $s = \log_2 c_D = \frac{\log c_D}{\log 2}$. To prove (1.43) (if $\mu \not\equiv 0$; otherwise, (1.43) is obvious), let $k := [\log_2(\frac{R}{r})]$ be the integer part of $\log_2(\frac{R}{r})$. Since $2^k \leq \frac{R}{r} \leq 2^{k+1}$, using the monotonicity of μ and iterating the doubling condition give

$$\begin{aligned}\mu(B(x,R)) &\leq \mu(B(x, 2\cdot 2^k r)) \leq c_D\,\mu(B(x, 2^k r))\\ &\leq c_D^2\,\mu(B(x, 2^{k-1}r)) \leq \ldots \leq c_D^{k+1}\,\mu(B(x,r)).\end{aligned}$$

If $c_D = 1$, then the above inequality gives (1.43), since $s = 0$ in this case. If $c_D > 1$, then since $\frac{\mu(B(x,R))}{\mu(B(x,r))} \leq c_D^{k+1}$, we get

$$\log_{c_D}\left(\frac{\mu(B(x,R))}{c_D\,\mu(B(x,r))}\right) \leq k \leq \log_2\left(\frac{R}{r}\right) = \log_{c_D}\left(\frac{R}{r}\right)\log_2 c_D = \log_{c_D}\left(\frac{R}{r}\right)^s,$$

and thus (1.43) also follows in this case. This proves the claim.

1.7 Hausdorff measure and capacities

Let (X, d) be a metric space. Let $h : (0, \infty) \mapsto (0, \infty]$ be nondecreasing with $\lim_{r \to 0^+} h(r) = 0$. Arguing as in Subsection 1.6.2, for any $\delta \in (0, \infty]$ and $E \subseteq X$, set

$$\mathcal{H}^*_{h,\delta}(E) := \inf\left\{ \sum_{k=1}^{\infty} h(\operatorname{diam}(E_k)) \mid E \subseteq \bigcup_{k=1}^{\infty} E_k, \operatorname{diam}(E_k) \leq \delta \ \forall k \in \mathbb{N} \right\} \tag{1.44}$$

if a cover $\{E_k\}$ of E as in (1.44) exists and $\mathcal{H}^*_{h,\delta}(E) := \infty$ otherwise (this corresponds to choosing $\eta = h \circ \operatorname{diam}$ in (1.35)). Then for any $\delta \in (0, \infty]$, the map $\mathcal{H}^*_{h,\delta} : \mathcal{P}(X) \mapsto [0, \infty]$ is an outer measure and thus, in particular, a capacity on X (see Remark 1.6.4). The same holds for the map

$$\mathcal{H}^*_h : \mathcal{P}(X) \mapsto [0, \infty], \quad \mathcal{H}^*_h(E) := \lim_{\delta \to 0^+} \mathcal{H}^*_{h,\delta}(E) = \sup_{\delta > 0} \mathcal{H}^*_{h,\delta}(E) \quad (E \subseteq X), \tag{1.45}$$

which, moreover, is a metric outer measure (see Lemma 1.6.3).

Definition 1.7.1. Let (X, d) be a metric space.

(i) Each capacity $\mathcal{H}^*_{h,\delta}$ ($\delta \in (0, \infty]$) on X, as well as $\mathcal{H}^*_h$, is called a *Hausdorff capacity*. The measure obtained by restriction of $\mathcal{H}^*_h$ to the σ-algebra of the $\mathcal{H}^*_h$-measurable sets is called a *Hausdorff measure* and denoted by $\mathcal{H}_h$. The function h is called a *gauge function* of $\mathcal{H}^*_h$.

(ii) If $h(r) = r^s$ ($s \in (0, \infty)$), then the Hausdorff capacities and the Hausdorff measure are denoted by $\mathcal{H}^*_{s,\delta}$, $\mathcal{H}^*_s$, and $\mathcal{H}_s$, respectively, and $\mathcal{H}_s$ is called an *s-dimensional Hausdorff measure*. The 0*-dimensional Hausdorff measure* is by definition the counting measure $\mu^{\#}$.

Remark 1.7.1. (i) Since $\mathcal{H}^*_h$ is a metric outer measure, by Theorem 1.4.1 the definition of $\mathcal{H}_h$ is well posed, and $\mathcal{H}_h$ is a complete measure. Observe that by Proposition 1.6.4 all Borel sets are $\mathcal{H}^*_h$-measurable.

(ii) By its very definition the Hausdorff measure is invariant under every map $\phi : X \mapsto X$ such that $d(\phi(x), \phi(y)) = d(x, y)$ $(x, y \in X)$, in particular, under rotations and translations.

The following link between Hausdorff capacities will be used (see Theorem 3.4.4).

Proposition 1.7.1. *Let (X, d) be a metric space, and let $E \subseteq X$. Then for any $\delta \in (0, \infty]$, we have $\mathcal{H}^*_{h,\delta}(E) = 0$ if and only if $\mathcal{H}^*_h(E) = 0$.*

Proof. By inequality (1.37) for any $E \subseteq X$ we have

$$\mathcal{H}^*_{h,\infty}(E) \leq \mathcal{H}^*_{h,\delta}(E) \leq \mathcal{H}^*_h(E) \quad (\delta \in (0, \infty)), \tag{1.46}$$

and thus the "if" part of the claim immediately follows. To prove the "only if" part, let $\mathcal{H}_h^*(E) > 0$, let $c \in (0, \mathcal{H}_h^*(E))$ be fixed, and let $\delta > 0$ be so small that $\mathcal{H}_{h,\delta}^*(E) > c$. Then by definition (1.44) we have $\sum_{k=1}^\infty h(\mathrm{diam}(E_k)) > c$ for all coverings of E by a countable union of sets E_k with $\mathrm{diam}(E_k) \le \delta$. For every other covering of the same kind, we have $\mathrm{diam}(E_{\bar{k}}) > \delta$ for some $\bar{k} \in \mathbb{N}$, and thus $\sum_{k=1}^\infty h(\mathrm{diam}(E_k)) \ge h(\delta) > 0$. This implies that $\mathcal{H}_{h,\infty}^*(E) \ge \min\{c, h(\delta)\} > 0$, and hence the conclusion follows. □

In the same spirit of Proposition 1.7.1, when $X = \mathbb{R}^N$, we have the following result of local equivalence between Hausdorff capacities (see [102, Lemma 3.4.1 and Proposition 3.4.15] for the proof).

Proposition 1.7.2. *Let $\alpha \in [0, N)$, and let $E \subseteq B(x_0, \rho) \subseteq \mathbb{R}^N$ ($x_0 \in \mathbb{R}^N$, $\rho > 0$). Then there exists $C_1 = C_1(\alpha, N) > 0$ such that for any $\delta \in (0, \infty)$,*

$$\mathcal{H}_{N-\alpha,\infty}^*(E) \le \mathcal{H}_{N-\alpha,\delta}^*(E) \le C\, \mathcal{H}_{N-\alpha,\infty}^*(E), \tag{1.47}$$

where

$$C := C_1 \frac{\mathcal{H}_{N-\alpha,\delta}^*(B(x_0, \rho))}{\mathcal{H}_{N-\alpha,\infty}^*(B(x_0, \rho))} \max\left\{1, \left(\frac{\rho}{\delta}\right)^\alpha\right\}.$$

Lemma 1.7.3. *Let (X, d) be a metric space. Then for any $E \subseteq X$ and $s, t \in (0, \infty)$:*
(i) *if $s < t$, then $\mathcal{H}_s^*(E) < \infty \Rightarrow \mathcal{H}_t^*(E) = 0$;*
(ii) *if $t < s$, then $\mathcal{H}_s^*(E) > 0 \Rightarrow \mathcal{H}_t^*(E) = \infty$.*

Proof. Let $s < t$. By (1.44) for any $\delta \in (0, \infty)$, we have $\mathcal{H}_{t,\delta}^*(E) \le \delta^{t-s} \mathcal{H}_{s,\delta}^*(E)$. Letting $\delta \to 0^+$, we obtain that $\mathcal{H}_t^*(E) = 0$ if $\mathcal{H}_s^*(E) < \infty$. This proves claim (i), whence (exchanging s with t) claim (ii) follows. □

The lemma motivates the following definition.

Definition 1.7.2. Let (X, d) be a metric space. The *Hausdorff dimension* of a set $E \subseteq X$ is the number

$$\dim_H(E) := \inf\{s > 0 \mid \mathcal{H}_s^*(E) = 0\} = \sup\{s > 0 \mid \mathcal{H}_s^*(E) = \infty\}.$$

Remark 1.7.2. If there exists a measure μ on X of dimension s, then X has Hausdorff dimension $\dim_H(X) = s$, and there exists $c > 0$ such that

$$\frac{1}{c}\mathcal{H}_s(E) \le \mu(E) \le c\, \mathcal{H}_s(E) \quad \text{for all } E \in \mathcal{B}(X)$$

(e. g., see [49, 2.10], [62, 4.11]). In particular, $\dim_H(\mathbb{R}^N) = N$.

Concerning the Hausdorff dimension of Cantor sets, we have the following:

Proposition 1.7.4. *Let K^N be the N-dimensional Cantor set corresponding to the sequence $\{a^n\}$ with $a \in (0, \frac{1}{2})$. Then $\dim_H(K^N) = N \frac{\log 2}{\log(1/a)}$ ($N \in \mathbb{N}$).*

In particular, the Hausdorff dimension of the one-dimensional Cantor middle third set is $\frac{\log 2}{\log 3}$. Proposition 1.7.4 follows immediately from the following result (see [2, Theorem 5.3.1] for the proof).

Theorem 1.7.5. *Let $h : [0,\infty) \mapsto [0,\infty)$ be nondecreasing with $\lim_{r\to 0^+} h(r) = h(0) = 0$. Let K^N be the N-dimensional Cantor set corresponding to the sequence $\{m_n\}$. Then there exists $C > 0$ such that*

$$\frac{1}{C}\liminf_{n\to\infty}(2^{nN} h(m_n)) \le \mathcal{H}_h(K^N) \le C \liminf_{n\to\infty}(2^{nN} h(m_n)). \tag{1.48}$$

Proof of Proposition 1.7.4. Choosing in (1.48) $h(r) = r^s$ and $m_n = a^n$ with $a \in (0, \frac{1}{2})$ gives $\liminf_{n\to\infty}(2^N a^s)^n = \mathcal{H}_s(K^N) = 0$ for all $s > N \frac{\log 2}{\log(1/a)}$. Hence by Definition 1.7.2 the result follows. □

We finish this section by proving the following link between the Lebesgue measure λ_N and the N-dimensional Hausdorff measure $\mathcal{H}_N$.

Theorem 1.7.6. *We have $\lambda_N = \frac{\kappa_N}{2^N}\mathcal{H}_N$ with κ_N given by (1.16) ($N \in \mathbb{N}$).*

Remark 1.7.3. The relationship established by Theorem 1.7.6 in fact holds between the Lebesgue outer measure and the Hausdorff outer measure, that is, $\lambda_N^* = k_N\mathcal{H}_N^*$ (see [45, Satz III.2.9]). Therefore $E \in \mathcal{L}^N$ if and only if E is $\mathcal{H}_N^*$-measurable. Also, observe that $\frac{\kappa_1}{2} = \frac{\sqrt{\pi}}{2\Gamma(\frac{3}{2})} = 1$, and thus for $N = 1$, the Lebesgue and the Hausdorff outer measures coincide.

Proof of Theorem 1.7.6. Set $W := (0,1]^N$, $k_N := [\mathcal{H}_N^*(W)]^{-1}$ (observe that $0 < \mathcal{H}_N^*(W) < \infty$). Since $k_N\mathcal{H}_N^*$ is a metric outer measure, by Proposition 1.6.4 all Borel sets are $\mathcal{H}_N^*$-measurable. Moreover, by Remark 1.7.2 (with $\mu = \lambda_N$) the restriction $k_N\mathcal{H}_N^*|_{\mathcal{B}^N}$ is a locally finite measure and thus a Borel measure, which clearly satisfies the assumptions of Proposition 1.5.3. It follows that $k_N\mathcal{H}_N^*|_{\mathcal{B}^N} = \lambda_N|_{\mathcal{B}^N}$, and hence $\lambda_N = k_N\mathcal{H}_N$ by Theorem 1.4.4.

It remains to prove that $k_N = \frac{\kappa_N}{2^n}$. By the Vitali covering lemma (see Remark 1.5.3), for any $\delta > 0$, there exists a disjoint sequence $\{B_k\}$ of closed balls with $B_k \subseteq W$ and $\operatorname{diam} B_k < \delta$ for all $k \in \mathbb{N}$ such that $\lambda_N(W \setminus (\bigcup_{k=1}^\infty B_k)) = 0$. Then we also have that $\mathcal{H}_N(W \setminus (\bigcup_{k=1}^\infty B_k)) = 0$. Moreover, by (1.44) for any $\delta > 0$, we have

$$\begin{aligned}\mathcal{H}_{N,\delta}^*\left(\bigcup_{k=1}^\infty B_k\right) &\le \sum_{k=1}^\infty [\operatorname{diam}(B_k)]^N = \frac{2^N}{\kappa_N}\sum_{k=1}^\infty \lambda_N(B_k)\\ &= \frac{2^N}{\kappa_N}\lambda_N\left(\bigcup_{k=1}^\infty B_k\right) \le \frac{2^N}{\kappa_N}\lambda_N(W) = \frac{2^N}{\kappa_N}.\end{aligned}$$

Letting $\delta \to 0^+$ in this inequality, we obtain

$$\mathcal{H}_N(W) = \mathcal{H}_N\left(\bigcup_{k=1}^{\infty} B_k\right) = \mathcal{H}_N^*\left(\bigcup_{k=1}^{\infty} B_k\right) \le \frac{2^N}{\kappa_N}. \tag{1.49}$$

Let $W \subseteq \bigcup_{k=1}^{\infty} E_k$ with $\operatorname{diam}(E_k) \le \delta$ for all $k \in \mathbb{N}$. Then by inequality (1.17)

$$1 = \lambda_N(W) \le \sum_{k=1}^{\infty} \lambda_N^*(E_k) \le \frac{\kappa_N}{2^N} \sum_{k=1}^{\infty} [\operatorname{diam}(E_k)]^N,$$

whence $\mathcal{H}_N(W) \ge \frac{2^N}{\kappa_N}$. This inequality and (1.49) prove that $k_N = [\mathcal{H}_N^*(W)]^{-1} = \frac{\kappa_N}{2^N}$, and thus the result follows. □

1.8 Signed measures

Definition 1.8.1. Let $(X, \mathcal{A})$ be a measurable space. A map $\mu : \mathcal{A} \mapsto [-\infty, \infty]$ is called a signed measure on $\mathcal{A}$ if:

(i) $\mu(\emptyset) = 0$;

(ii) $\mu(\mathcal{A})$ is contained either in $[-\infty, \infty)$ or in $(-\infty, \infty]$;

(iii) for any disjoint sequence $\{E_n\} \subseteq \mathcal{A}$, we have $\mu(\bigcup_{n=1}^{\infty} E_n) = \sum_{n=1}^{\infty} \mu(E_n)$.

A signed measure μ is called finite if $\mu(\mathcal{A}) \subseteq \mathbb{R}$ and σ-finite if there exists a sequence $\{E_k\} \subseteq \mathcal{A}$ such that $X = \bigcup_{k=1}^{\infty} E_k$ with $\mu(E_k)| < \infty$ for any $k \in \mathbb{N}$.

By the above request (ii) the series $\sum_{n=1}^{\infty} \mu(E_n)$ in Definition 1.8.1 is well defined. Moreover, by (iii) its convergence is absolute: in fact, its sum cannot depend on the order of its terms since $\bigcup_{n=1}^{\infty} E_n$ does not.

Remark 1.8.1. Let $(X, \mathcal{A})$ be a measurable space, and let μ be a signed measure on $\mathcal{A}$.

(i) For any $E, F \in \mathcal{A}$, $E \subseteq F$, then $|\mu(F)| < \infty \Rightarrow |\mu(E)| < \infty$.

(ii) It is easily seen that:
   - $\mu(\bigcup_{n=1}^{\infty} E_n) = \lim_{n\to\infty} \mu(E_n)$ for any nondecreasing sequence $\{E_n\} \subseteq \mathcal{A}$;
   - $\mu(\bigcap_{n=1}^{\infty} E_n) = \lim_{n\to\infty} \mu(E_n)$ for any nonincreasing sequence $\{E_n\} \subseteq \mathcal{A}$ such that $|\mu(E_1)| < \infty$.

The set of signed measures on X is a real vector space, denoted by $\mathfrak{M} = \mathfrak{M}(X)$. It is an *ordered* vector space if the *order* "$\le$" is defined as follows:

$$\nu \le \mu \quad \overset{\text{def}}{\Longleftrightarrow} \quad \nu(E) \le \mu(E) \quad \text{for any } E \in \mathcal{A} \quad (\mu, \nu \in \mathfrak{M}).$$

The subspace of finite signed measures will be denoted by $\mathfrak{M}_f(X) \subset \mathfrak{M}(X)$, whereas $\mathfrak{M}_f^+(X)$ denotes the cone of finite (positive) measures.

1.8.1 Hahn and Jordan decompositions

Definition 1.8.2. Let $(X, \mathcal{A})$ be a measurable space, and let μ be a signed measure on $\mathcal{A}$.
(i) A set $P \in \mathcal{A}$ is called *μ-positive* if $\mu(E) \geq 0$ for any $E \in \mathcal{A}$, $E \subseteq P$;
(ii) A set $N \in \mathcal{A}$ is called *μ-negative* if $\mu(E) \leq 0$ for any $E \in \mathcal{A}$, $E \subseteq N$;
(iii) A set $F \in \mathcal{A}$ is called *μ-null* if $\mu(E) = 0$ for any $E \in \mathcal{A}$, $E \subseteq F$.

Lemma 1.8.1. *Let $(X, \mathcal{A})$ be a measurable space, and let μ be a signed measure on $\mathcal{A}$ such that $\mu(\mathcal{A}) \subseteq [-\infty, \infty)$. Then for any set $E \in \mathcal{A}$ such that $\mu(E) > -\infty$, there exists a μ-*positive *set $P \subseteq E$ such that $\mu(P) \geq \mu(E)$.*

Proof. It the set E is μ-positive, then the claim is obviously satisfied. Otherwise, the conclusion is implied by the following:

Claim. For any $\epsilon > 0$, there exists $E_\epsilon \in \mathcal{A}$, $E_\epsilon \subseteq E$, with $\mu(E_\epsilon) \geq \mu(E)$ such that $\mu(F) \geq -\epsilon$ for any $F \in \mathcal{A}$, $F \subseteq E_\epsilon$.

In fact, if the claim holds, then the set $P := \bigcap_{n=1}^{\infty} E_{1/n}$ has the stated properties.

By contradiction, let the claim be false for some $\bar{\epsilon} > 0$. Since by assumption E is not μ-positive, there exists $G \in \mathcal{A}$ such that $G \subseteq E$ and $\mu(G) < 0$. This implies that $E \setminus G \in \mathcal{A}$ is a subset of E with $\mu(E \setminus G) = \mu(E) - \mu(G) > \mu(E)$, whence there exists $F_1 \in \mathcal{A}$ such that $F_1 \subseteq E \setminus G$ and $\mu(F_1) < -\bar{\epsilon}$. Next, we observe that the set $E \setminus F_1 \in \mathcal{A}$ satisfies the following properties: $E \setminus F_1 \subseteq E$ and $\mu(E \setminus F_1) = \mu(E) - \mu(F_1) > \mu(E) + \bar{\epsilon} > \mu(E)$. Then there exists $F_2 \in \mathcal{A}$ such that $F_2 \subseteq E \setminus F_1$ and $\mu(F_2) < -\bar{\epsilon}$. Thus, by induction, there exists a disjoint sequence $\{F_n\}$ with $F_1 \subseteq E$, $F_n \subseteq E \setminus (\bigcup_{k=1}^{n-1} F_k)$ $(n \geq 2)$ such that $\mu(F_n) < -\bar{\epsilon}$ for all n. It follows that $\mu(\bigcup_{n=1}^{\infty} F_n) = -\infty$. However, this is a contradiction, since $\bigcup_{n=1}^{\infty} F_n \subseteq E$ and $\mu(E) > -\infty$ (see Remark 1.8.1-(i)). □

Theorem 1.8.2 (Hahn). *Let $(X, \mathcal{A})$ be a measurable space, and let μ be a signed measure on $\mathcal{A}$. Then there exist a μ-positive set P and a μ-negative set N such that $P \bigcup N = X$, $P \cap N = \emptyset$. Moreover, the couple (P, N) is uniquely determined up to μ-null sets.*

Proof. Without loss of generality, let $\mu(\mathcal{A}) \subseteq [-\infty, \infty)$. Set $\alpha := \sup_{E \in \mathcal{A}} \mu(E)$. Then by Lemma 1.8.1 there exists a sequence $\{P_n\}$ of μ-positive sets P_n such that $\mu(P_n) \to \alpha$ as $n \to \infty$. Then $P := \bigcup_{n=1}^{\infty} P_n$ is μ-positive and satisfies $\mu(P) \geq \mu(P_n)$ for all n, and thus $\mu(P) = \alpha$ (in particular, $\alpha \in \mathbb{R}$). It follows easily that $N := P^c$ is μ-negative: in fact, should $G \in \mathcal{A}$, $G \subseteq N$, with $\mu(G) > 0$ exist, we would have $\mu(P \bigcup G) = \alpha + \mu(G) > \alpha$, a contradiction.

To prove the last statement, let a couple (P', N') with μ-positive P' and μ-negative N' satisfy $P' \bigcup N' = X$, $P' \cap N' = \emptyset$. For any $E \in \mathcal{A}$, $E \subseteq P \setminus P' = P \cap N'$, we have both $\mu(E) \geq 0$ and $\mu(E) \leq 0$, and thus E is μ-null. The same holds for every measurable subset of $P' \setminus P$, and thus $\mu(P \triangle P') = \mu(N \triangle N') = 0$ (here $P \triangle P' := (P \setminus P') \cup (P' \setminus P)$, and similarly for N, N'). This proves the result. □

Definition 1.8.3. Let $(X, \mathcal{A})$ be a measurable space, let μ be a signed measure on $\mathcal{A}$, and let $P, N \in \mathcal{A}$ be given by Theorem 1.8.2.

(i) The couple (P, N) is called the *Hahn decomposition* of X.

(ii) The measure μ^+, $\mu^+(E) := \mu(E \cap P)$ for any $E \in \mathcal{A}$, is called the *positive part* (or *positive variation*) of μ.

(iii) The measure μ^-, $\mu^-(E) := -\mu(E \cap N)$ for any $E \in \mathcal{A}$, is called the *negative part* (or *negative variation*) of μ.

(iv) The measure $|\mu|$, $|\mu|(E) := \mu^+(E) + \mu^-(E)$ $(E \in \mathcal{A})$, is called the *variation* of μ. The quantity $|\mu|(X)$ is called *total variation* of μ.

In view of Theorem 1.8.2, the above definitions do not depend on the choice of the couple (P, N) and thus are well posed. In particular, we get

$$\mu = \mu^+ - \mu^-, \quad |\mu| = \mu^+ + \mu^-, \tag{1.50}$$

whence, in particular,

$$|\mu(E)| \le |\mu|(E) \quad \text{for all } E \in \mathcal{A}. \tag{1.51}$$

Definition 1.8.4. The first equality in (1.50) is called the *Jordan decomposition* of the signed measure μ.

Remark 1.8.2. (i) The Jordan decomposition of μ is *minimal* in the following sense: if there exist (positive) measures ρ and σ such that $\mu = \rho - \sigma$, then $\mu^+(E) \le \rho(E)$ and $\mu^-(E) \le \sigma(E)$ for all $E \in \mathcal{A}$. Indeed,

$$\mu^+(E) := \mu(E \cap P) = \rho(E \cap P) - \sigma(E \cap P) \le \rho(E \cap P) \le \rho(E),$$

and similarly for the other inequality.

(ii) It is easily seen that for any $E \in \mathcal{A}$,

$$\mu^+(E) = \infty \Rightarrow \mu(E) = \infty, \quad \mu^-(E) = \infty \Rightarrow \mu(E) = -\infty.$$

By definition a signed measure μ cannot attain both values $\pm\infty$, and thus at least one of $\mu^\pm$ is finite.

(iii) A signed measure μ is finite if and only if its total variation is finite. In fact, if $|\mu|(X) < \infty$, then $|\mu(E)| \le |\mu|(E) \le |\mu|(X) < \infty$ for all $E \in \mathcal{A}$, and thus μ is finite. Conversely, let $|\mu(E)| < \infty$ for $E \in \mathcal{A}$. Then by (ii) we have $|\mu|(E) < \infty$ for all $E \in \mathcal{A}$, and thus, in particular, $|\mu|(X) < \infty$.

Proposition 1.8.3. *Let $(X, \mathcal{A})$ be a measurable space, and let μ be a signed measure on $\mathcal{A}$. Then for any $E \in \mathcal{A}$,*

$$\mu^+(E) = \sup\{\mu(F) \mid F \in \mathcal{A},\, F \subseteq E\}, \quad \mu^-(E) = -\inf\{\mu(F) \mid F \in \mathcal{A},\, F \subseteq E\},$$

and thus $\mu^- = (-\mu)^+$. *Moreover,*

$$|\mu|(E) = \sup\left\{\sum_{i=1}^{n}|\mu(E_i)| \mid E_1,\dots,E_n \in \mathcal{A} \text{ disjoint}, E = \bigcup_{i=1}^{n} E_i,\ n \in \mathbb{N}\right\}$$
$$= \sup\left\{\sum_{i=1}^{\infty}|\mu(E_i)| \mid E_i \in \mathcal{A} \text{ disjoint}, E = \bigcup_{i=1}^{\infty} E_i\right\}. \tag{1.52}$$

Proof. We only prove (1.52). Denote by S_f and S_i the first and second suprema in the right-hand side. Then

$$|\mu|(E) = |\mu(E \cap P)| + |\mu(E \cap N)| \le S_f \le S_i.$$

On the other hand, for any finite disjoint family $\{E_1,\dots,E_n\} \subseteq \mathcal{A}$ with $\bigcup_{i=1}^{n} E_i \subseteq E$,

$$\sum_{i=1}^{n}|\mu(E_i)| \le \sum_{i=1}^{n}[\mu^+(E_i) + \mu^-(E_i)] \le \mu^+(E) + \mu^-(E) = |\mu|(E),$$

whence we get $S_f \le S_i \le |\mu|(E)$. Hence (1.52) follows. □

The following definition extends Definition 1.3.3.

Definition 1.8.5. Let $(X, \mathcal{T})$ be a Hausdorff space, and let $\mathcal{B} = \mathcal{B}(X) = \sigma_0(\mathcal{T})$ be the Borel σ-algebra. A finite signed measure μ on $\mathcal{B}$ is a *finite signed Borel measure* (respectively, a *finite signed Radon measure* or a *finite signed regular measure*) if it is the difference of two finite Borel measures (respectively, two finite Radon measures or two finite regular measures).

The vector space of finite signed Radon measures on X will be denoted by $\mathfrak{R}_f(X)$, whereas $\mathfrak{R}_f^+(X)$ denotes the cone of finite (positive) Radon measures.

Remark 1.8.3. (i) It is easily seen that the following statements are equivalent:
(a) μ is a finite signed regular measure;
(b) $\mu^\pm$ are finite and regular;
(c) $|\mu|$ is finite and regular.

Indeed, let $\mu = \mu_1 - \mu_2$ with μ_1, μ_2 finite, positive, and regular. Let $E \in \mathcal{B}$. Then for any $\epsilon > 0$, there exist an open set A and a compact set K such that $A \supseteq E \supseteq K$ and $\mu_i(A \setminus K) < \epsilon$, and thus $\mu^\pm(A \setminus K) < \epsilon$ ($i = 1, 2$; see Remark 1.3.2(i) and Remark 1.8.2). Hence (a)⇒(b); the other implications are clear.

(ii) A finite signed measure is regular if and only if it is a finite signed Radon measure. Indeed, if $\mu = \mu_1 - \mu_2$ with μ_1, μ_2 finite and regular, then by definition μ_1, μ_2 are finite Radon measures, and thus μ is a finite signed Radon measure. On the other hand, if $\mu = \mu_1 - \mu_2$ and μ_1, μ_2 are finite Radon measures on X, then they are regular by Lemma 1.3.2(i), and thus μ is regular.

(iii) In a locally compact Hausdorff space with countable bases, signed Borel and signed Radon measures coincide and are regular (see Remark 1.3.3).

1.8.2 The Banach space of finite signed measures

Let $(X, \mathcal{A})$ be a measurable space. It is easily seen that the map $\|\cdot\| : \mathfrak{M}_f(X) \mapsto \mathbb{R}_+$, $\|\mu\| := |\mu|(X)$ for any $\mu \in \mathfrak{M}(X)$, is a norm on the vector space $\mathfrak{M}_f(X)$ of finite signed measures on X. A subset $\mathcal{M} \subseteq \mathfrak{M}_f(X)$ is bounded if $\sup_{\mu\in\mathcal{M}} \|\mu\| < \infty$. Moreover, we have the following:

Proposition 1.8.4. *Let $(X, \mathcal{A})$ be a measurable space. Then the vector space $\mathfrak{M}_f(X)$ endowed with the norm $\|\cdot\|$ is a Banach space.*

Proof. To prove the completeness, let $\{\mu_k\} \subseteq \mathfrak{M}_f(X)$ be a Cauchy sequence. Then for any $\epsilon > 0$, there exists $\bar{k} \in \mathbb{N}$ such that for all $k, l > \bar{k}$ and all $E \in \mathcal{A}$,

$$|\mu_k(E) - \mu_l(E)| = |(\mu_k - \mu_l)(E)| \le |\mu_k - \mu_l|\,(E) \le \|\mu_k - \mu_l\| < \epsilon.$$

Hence there exists $\mu : \mathcal{A} \mapsto \mathbb{R}$ such that $\|\mu_k - \mu\| \to 0$ as $k \to \infty$. Clearly, $\mu(\emptyset) = 0$, and a standard $\epsilon/3$ argument shows that μ is σ-additive, and thus $\mu \in \mathfrak{M}_f(X)$. Hence the result follows. □

1.8.3 Absolutely continuous and singular measures

Definition 1.8.6. Let $(X, \mathcal{A})$ be a measurable space, and let μ be a signed measure on $\mathcal{A}$. The restriction $\mu \llcorner E$ of μ to a set $E \in \mathcal{A}$ is defined as follows:

$$(\mu \llcorner E)(F) := \mu(E \cap F) \quad \text{for every } F \in \mathcal{A}.$$

We say that μ is concentrated on a set $E \in \mathcal{A}$ if $(\mu \llcorner E)(F) = \mu(F)$ for all $F \in \mathcal{A}$.

Remark 1.8.4. (i) It is easily seen that there does not exist a unique set on which μ is concentrated. For instance, if $E, F \in \mathcal{A}$ and $\mu(E \triangle F) = 0$, then $\mu \llcorner E = \mu \llcorner F$.

(ii) Let $(X, \mathcal{T})$ be a Hausdorff space, and let $\mu \in \mathfrak{R}^+(X)$. By Lemma 1.3.6 μ is concentrated on $\operatorname{supp}\mu$, and $\operatorname{supp}\mu$ is the smallest closed set on which μ is concentrated.

Definition 1.8.7. Let $(X, \mathcal{A})$ be a measurable space, and let μ, ν be two signed measures on $\mathcal{A}$. We say that:

(i) μ and ν are mutually singular (written $\mu \perp \nu$) if there exists $E \in \mathcal{A}$ such that $\mu = \mu \llcorner E$ and $\nu = \nu \llcorner E^c$;

(ii) ν is *absolutely continuous* with respect to μ (written $\nu \ll \mu$) if for any $|\mu|$-null set $E \in \mathcal{A}$, we have $\nu(E) = 0$.

Remark 1.8.5. (i) It is easily seen that μ, ν are mutually singular if and only if there exists $E \in \mathcal{A}$ such that $|\mu|(E^c) = |\nu|(E) = 0$.

(ii) The positive part μ^+ and the negative part μ^- of any signed measure μ are mutually singular. In fact, let (P, N) be any Hahn decomposition of X. Then $P = N^c$ by Theorem 1.8.2, and by definition we have $\mu^+ = \mu(E \llcorner P)$, $\mu^- = \mu(E \llcorner N)$.

Lemma 1.8.5. *Let $(X, \mathcal{A}, \mu)$ be a measure space, and let ν be a signed measure on $\mathcal{A}$. Then $\nu \ll \mu$ if and only if $|\nu| \ll \mu$.*

Proof. Let $\nu \ll \mu$, and let (P, N) be the Hahn decomposition of X associated with the measure ν (see Theorem 1.8.2). Let $E \in \mathcal{A}$ satisfy $\mu(E) = 0$, thus $\mu(E \cap P) = \mu(E \cap N) = 0$. Then by Definition 1.8.7(ii) we have that $\nu(E \cap P) = \nu(E \cap N) = 0$, that is, $\nu_+(E) = \nu_-(E) = 0$. It follows that $|\nu|(E) = 0$, and thus $|\nu| \ll \mu$. The converse is obvious from inequality (1.51). □

Lemma 1.8.6. *Let $(X, \mathcal{A})$ be a measurable space, let μ, ν be two signed measures on $\mathcal{A}$, and let ν be finite. Then $\nu \ll \mu$ if and only if $\lim_{|\mu|(G)\to 0} |\nu|(G) = 0$ $(G \in \mathcal{A})$.*

Proof. We only prove the necessity, since the converse is transparent. Let $\nu \ll \mu$, and by contradiction let $\limsup_{|\mu|(G)\to 0} |\nu|(G) > 0$. Then there exist $\epsilon > 0$ and a sequence $\{E_k\} \subset \mathcal{A}$ such that $|\nu|(E_k) \geq \epsilon$ and $|\mu|(E_k) \leq 2^{-k}$ for all $k \in \mathbb{N}$. Set $F_n := \bigcup_{k=n}^{\infty} E_k$ $(n \in \mathbb{N})$, and thus $|\nu|(F_n) \geq |\nu|(E_n) \geq \epsilon$ for all $n \in \mathbb{N}$. On the other hand, since the sequence $\{F_n\}$ is nonincreasing and $|\mu|(F_1) \leq \sum_{k=1}^{\infty} |\mu|(E_k) \leq \sum_{k=1}^{\infty} 2^{-k} < \infty$, we have $|\mu|(\bigcap_{n=1}^{\infty} F_n) = \lim_{n\to\infty} |\mu|(F_n) = 0$. Since $\nu \ll \mu$ by assumption, it follows that $|\nu|(\bigcap_{n=1}^{\infty} F_n)) = 0$.

Now set $G_1 := F_1^c$ and $G_n := F_{n-1} \setminus F_n$ for every $n \geq 2$. The sequence $\{G_n\}$ is disjoint, and for any $m \geq 2$, we have $\bigcup_{n=m}^{\infty} G_n = F_{m-1} \setminus (\bigcap_{n=1}^{\infty} F_n))$. Moreover, since ν is finite and $\bigcup_{n=1}^{\infty} G_n = X$, we have that

$$|\nu| \left(\bigcup_{n=1}^{\infty} G_n \right) = \sum_{n=1}^{\infty} |\nu| (G_n) = |\nu| (X) < \infty.$$

Since $|\nu|(\bigcap_{n=1}^{\infty} F_n)) = 0$, it follows that

$$\lim_{m\to\infty} |\nu| (F_{m-1}) = \lim_{m\to\infty} |\nu| \left(\bigcup_{n=m}^{\infty} G_n \right) = \lim_{m\to\infty} \sum_{n=m}^{\infty} |\nu| (G_n) = 0.$$

However, this contradicts the fact that $|\nu|(F_n) \geq \epsilon$ for all $n \in \mathbb{N}$. Then the result follows. □

Lemma 1.8.7. *Let $(X, \mathcal{A}, \mu)$ be a measure space, and let ν be a signed measure on $\mathcal{A}$. If $\nu \ll \mu$ and $\nu \perp \mu$, then $\nu = 0$.*

Proof. Since $\nu \perp \mu$, by Remark 1.8.5 there exists $G \in \mathcal{A}$ such that $\mu(G) = |\nu|(G^c) = 0$, and thus, in particular, $|\nu|(E \cap G^c) = 0$ for any $E \in \mathcal{A}$. On the other hand, since $\nu \ll \mu$, by Lemma 1.8.5 $|\nu|(G) = 0$, and thus $|\nu|(E \cap G) = 0$ for all $E \in \mathcal{A}$. Hence $|\nu|(E) = 0$ for all $E \in \mathcal{A}$, and thus the result follows. □

Remark 1.8.6. Let $(X, \mathcal{A})$ be a measurable space, and let $\Phi : \mathcal{A} \mapsto [0, \infty]$ be monotone. Let ν be a signed measure on $\mathcal{A}$ such that (i) $\nu(E) = 0$ for any $E \in \mathcal{A}$ with $\Phi(E) = 0$ and (ii) ν is concentrated on a set $G \in \mathcal{A}$ such that $\Phi(G) = 0$. Then the same argument used to prove Lemma 1.8.7 shows that $\nu = 0$.

We will use the following decomposition result ([51]; see Chapter 3).

Proposition 1.8.8 (Fukushima–Sato–Taniguchi). *Let $(X, \mathcal{A})$ be a measurable space, and let ν be a σ-finite signed measure on $\mathcal{A}$. Let $\Phi : \mathcal{A} \mapsto [0, \infty]$ be σ-subadditive and monotone. Then there exists a unique couple (ν_1, ν_2) of σ-finite signed measures on $\mathcal{A}$ such that:*

(a) $\nu = \nu_1 + \nu_2$;

(b) $\nu_1(E) = 0$ *for any* $E \in \mathcal{A}$ *such that* $\Phi(E) = 0$;

(c) ν_2 *is concentrated on a set* $N \in \mathcal{A}$ *such that* $\Phi(N) = 0$.

Proof. We only prove the result when ν is finite and positive, since the extension to the general case is standard. Let us first prove the uniqueness claim. Let (ν_1, ν_2) and (ν_1', ν_2') have the stated properties. Then by equality (a) for any $E \in \mathcal{A}$, we get $\nu_1(E) - \nu_1'(E) = \nu_2'(E) - \nu_2(E)$. This implies that $\nu_1 - \nu_1'$ satisfies the assumptions of Remark 1.8.6, and thus $\nu_1 = \nu_1'$, whence $\nu_2 = \nu_2'$. Then the claim follows.

To prove the existence, set $\alpha := \sup\{\nu(E) \mid E \in \mathcal{A},\ \Phi(E) = 0\}$ (observe that $\alpha < \infty$, since $|\nu|(X) < \infty$ by Remark 1.8.2(iii)). Let $\{E_n\} \subseteq \mathcal{A}$ be a nondecreasing sequence such that $\Phi(E_n) = 0$ for each $n \in \mathbb{N}$ and $\lim_{n\to\infty} \nu(E_n) = \alpha$. Set $E_\infty := \bigcup_{n=1}^{\infty} E_n$. Then $E_\infty \in \mathcal{A}$, $\nu(E_\infty) = \lim_{n\to\infty} \nu(E_n) = \alpha$, and by the σ-subadditivity of Φ we have $\Phi(E_\infty) \le \sum_{n=1}^{\infty} \Phi(E_n) = 0$. Therefore, for any $E \in \mathcal{A}$ such that $\Phi(E) = 0$, we have $0 \le \nu(E \setminus E_\infty) = \nu(E) - \alpha \le 0$, and thus $\nu(E \setminus E_\infty) = (\nu \llcorner E_\infty^c)(E) = 0$. On the other hand, the measure $\nu \llcorner E_\infty$ is concentrated on the set E_∞, and $\Phi(E_\infty) = 0$. Then by setting $N := E_\infty$, $\nu_1 := \nu \llcorner E_\infty^c$, and $\nu_2 := \nu \llcorner E_\infty$ the conclusion follows. □

Proposition 1.8.8 immediately gives the following result.

Theorem 1.8.9 (Lebesgue). *Let $(X, \mathcal{A}, \mu)$ be a measure space, and let ν be a σ-finite signed measure on $\mathcal{A}$. Then there exists a unique couple (ν_{ac}, ν_s) of σ-finite signed measures on $\mathcal{A}$ such that $\nu_{ac} \ll \mu$, $\nu_s \perp \mu$, and*

$$\nu = \nu_{ac} + \nu_s. \tag{1.53}$$

Definition 1.8.8. Equality (1.53) is called the *Lebesgue decomposition* of ν with respect to μ.

Let us mention for future reference the following result.

Lemma 1.8.10. *Let the assumptions of Theorem* 1.8.9 *be satisfied. Then*

$$[\nu_{ac}]^{\pm} = [\nu^{\pm}]_{ac}, \quad [\nu_s]^{\pm} = [\nu^{\pm}]_s. \tag{1.54}$$

Proof. By Definitions 1.8.3 and 1.8.6 we have $\nu^+ = \nu \llcorner P$ and $\nu^- = \nu \llcorner N$ with P, N given by Theorem 1.8.2. Then by (1.53) we get

$$\nu^+ = \nu_{ac} \llcorner P + \nu_s \llcorner P, \quad \nu^- = \nu_{ac} \llcorner N + \nu_s \llcorner N, \tag{1.55}$$

whence

$$[\nu^+]_{ac} = \nu_{ac} \llcorner P, \quad [\nu^-]_{ac} = \nu_{ac} \llcorner N \tag{1.56}$$

and

$$[\nu^+]_s = \nu_s \llcorner P, \quad [\nu^-]_s = \nu_s \llcorner N. \tag{1.57}$$

On the other hand, by (1.55) we have

$$\nu = \nu^+ - \nu^- = (\nu_{ac} \llcorner P - \nu_{ac} \llcorner N) + (\nu_s \llcorner P - \nu_s \llcorner N),$$

and thus

$$\nu_{ac} = \nu_{ac} \llcorner P - \nu_{ac} \llcorner N, \quad \nu_s = \nu_s \llcorner P - \nu_s \llcorner N. \tag{1.58}$$

By (1.56) and the first equality in (1.58) the set P is ν_{ac}-positive, whereas N is ν_{ac}-negative. Hence by Definition 1.8.3

$$[\nu_{ac}]^+ = \nu_{ac} \llcorner P, \quad [\nu_{ac}]^- = \nu_{ac} \llcorner N. \tag{1.59}$$

From (1.56) and (1.59) we obtain the first equality in (1.54). The proof of the second is similar using (1.57) and the second equality in (1.58). □

Definition 1.8.9. Let $(X, \mathcal{A})$ be a measurable space, and let μ be a signed measure on $\mathcal{A}$.
(i) A set $E \in \mathcal{A}$ is called a *μ-atom* if $|\mu|(E) > 0$ and for all $F \in \mathcal{A}$, $F \subseteq E$, either $|\mu|(F) = 0$, or $|\mu|(E \setminus F) = 0$.
(ii) The measure μ is called nonatomic (or continuous) if no μ-atoms exist.
(iii) The measure μ is called purely atomic if there exists a sequence $\{E_n\} \subseteq \mathcal{A}$ of μ-atoms such that $\mu = \mu \llcorner (\bigcup_{n=1}^{\infty} E_n)$.

Remark 1.8.7. (i) A signed measure μ is nonatomic if and only if for any $E \in \mathcal{A}$ with $|\mu|(E) > 0$, there exists $F \in \mathcal{A}$, $F \subseteq E$, such that $0 < |\mu|(F) < |\mu|(E)$. The set of μ-atoms is at most finite if μ is finite and at most countable if μ is σ-finite.

(ii) If μ is both purely atomic and nonatomic, then $\mu = 0$.

Proposition 1.8.11. *Let $(X, \mathcal{A})$ be a measurable space, and let μ be a σ-finite signed measure on $\mathcal{A}$. Then there exists a sequence (possibly empty or finite) $\{E_n\} \subseteq \mathcal{A}$ of μ-atoms such that:*

(i) *$\mu_{pa} := \mu \llcorner (\bigcup_{n=1}^{\infty} E_n)$ is purely atomic, and $\mu_{na} := \mu \llcorner ((\bigcup_{n=1}^{\infty} E_n)^c)$ is nonatomic;*
(ii) *μ is uniquely represented by the sum*

$$\mu = \mu_{pa} + \mu_{na}. \tag{1.60}$$

Proof. Claim (i) and equality (1.60) are obvious from the definition of μ_{pa} and μ_{na}. To prove the uniqueness, let $\mu = \mu'_{pa} + \mu'_{na}$ with μ'_{pa} purely atomic and μ'_{na} nonatomic. It follows that $\mu_{na} - \mu'_{na} = \mu'_{pa} - \mu_{pa}$, and hence $\mu_{na} - \mu'_{na}$ and $\mu_{pa} - \mu'_{pa}$ are both discrete and nonatomic. Thus by Remark 1.8.7(ii) the claim follows. □

1.8.4 Concentrated and diffuse measures

The following definition is the counterpart for capacities of Definition 1.8.7.

Definition 1.8.10. Let (X, d) be a metric space. Let ν be a signed measure on $\mathcal{B}(X)$, and let $C : \mathcal{B}(X) \to [0, \infty]$ be a capacity. We say that:

(i) ν is *concentrated* with respect to C if there exists a C-null set $E \in \mathcal{B}(X)$ such that $\nu = \nu \llcorner E$;
(ii) ν is *diffuse* with respect to C if for any C-null set $E \in \mathcal{B}(X)$, we have $|\nu|(E) = 0$.

We denote by $\mathfrak{R}_{C,c}(X)$ (respectively, $\mathfrak{R}_{C,d}(X)$) the set of finite signed Radon measures on U that are concentrated (diffuse, respectively) with respect to C. Clearly, $\mathfrak{R}_{C,c}(X) \bigcap \mathfrak{R}_{C,d}(X) = \{0\}$. In view of Proposition 1.8.8, we have the following:

Proposition 1.8.12. *Let (X, d) be a metric space. Let ν be a signed measure on $\mathcal{B}(X)$, and let $C : \mathcal{B}(X) \to [0, \infty]$ be a capacity. Then there exists a unique couple $(\nu_{C,c}, \nu_{C,d})$ such that $\nu_{C,c} \in \mathfrak{R}_{C,c}(X)$, $\nu_{C,d} \in \mathfrak{R}_{C,d}(X)$, and*

$$\nu = \nu_{C,d} + \nu_{C,c}. \tag{1.61}$$

Definition 1.8.11. The measures $\nu_{C,c}$ and $\nu_{C,d}$ given by Proposition 1.8.12 are called the *concentrated* and *diffuse parts* of ν with respect to C.

1.9 Vector measures

All the measures so far considered are *scalar* measures, as opposed to the *vector* measures to be now introduced. Throughout this section, $(X, \mathcal{A})$ is a measurable space, and Y is a Banach space with norm $\|\cdot\|_Y$.

1.9.1 Definitions and general results

Definition 1.9.1. A map $\mu : \mathcal{A} \to Y$ is called *σ-additive* if for any disjoint sequence $\{E_k\} \subseteq \mathcal{A}$,

$$\lim_{n\to\infty} \left\| \sum_{k=1}^{n} \mu(E_k) - \mu\left(\bigcup_{k=1}^{\infty} E_k\right) \right\|_Y = 0.$$

Remark 1.9.1. A σ-additive map $\mu : \mathcal{A} \to Y$ is *additive*, that is, for any finite family $\{E_1, \dots, E_n\} \subseteq \mathcal{A}$ of disjoint elements, we have $\mu(\bigcup_{k=1}^{n} E_k) = \sum_{k=1}^{n} \mu(E_k)$.

Definition 1.9.2. (i) A σ-additive map $\mu : \mathcal{A} \to Y$ such that $\mu(\emptyset) = 0$ is called a *vector measure* on $\mathcal{A}$.

(ii) The function $|\mu| : \mathcal{A} \mapsto [0, \infty]$ defined as

$$|\mu|(E) := \sup\left\{ \sum_{i=1}^{n} \|\mu(E_i)\|_Y \mid E_1, \dots, E_n \in \mathcal{A} \text{ disjoint}, E = \bigcup_{i=1}^{n} E_i,\ n \in \mathbb{N} \right\} \quad (E \in \mathcal{A}) \tag{1.62}$$

is called the *variation* of μ. The quantity $|\mu|(X)$ is called the *total variation* of μ.

(iii) A vector measure $\mu : \mathcal{A} \to Y$ is said to be *of bounded variation* if $|\mu|(X) < \infty$.

Example 1.9.1. Let Y be a Banach space, and let $T : L^1(0,1) \mapsto Y$ be a bounded linear operator. For any Borel set $E \in \mathcal{B} \cap (0,1)$, set $\mu(E) := T(\chi_E)$. Clearly, $\|\mu(E)\|_Y \le \|T\|\,\lambda(E)$, and thus, in particular, $|\mu|(0,1) \le \|T\|$ (here $\|T\|$ denotes the operator norm of T). Then for any disjoint sequence $\{E_k\} \subseteq \mathcal{A}$,

$$\lim_{n\to\infty} \left\| \sum_{k=1}^{n} \mu(E_k) - \mu\left(\bigcup_{k=1}^{\infty} E_k\right) \right\|_Y = \lim_{n\to\infty} \left\| \mu\left(\bigcup_{k=n+1}^{\infty} E_k \right) \right\|_Y \le \|T\| \lim_{n\to\infty} \lambda\left(\bigcup_{k=n+1}^{\infty} E_k \right) = 0.$$

Hence μ is a vector measure of bounded variation.

Remark 1.9.2. If $Y = \mathbb{R}$, then the notion of vector measure coincides with that of *finite* signed measure.

Proposition 1.9.1. *Let $\mu : \mathcal{A} \to Y$ be a vector measure. Then its variation $|\mu|$ is a measure.*

Proof. By definition $|\mu|(\emptyset) = 0$. For any disjoint sequence $\{E_k\} \subseteq \mathcal{A}$, set $E := \bigcup_{k=1}^{\infty} E_k$, and let $F_1, \dots, F_n \in \mathcal{A}$ be disjoint such that $E = \bigcup_{i=1}^{n} F_i$. Then by Remark 1.9.1 we get

$$\|\mu(E)\|_Y = \left\| \sum_{i=1}^{n} \mu(F_i) \right\|_Y \le \sum_{i=1}^{n} \|\mu(F_i)\|_Y = \sum_{i=1}^{n} \|\mu(F_i \cap E)\|_Y$$
$$\le \sum_{i=1}^{n} \sum_{k=1}^{\infty} \|\mu(F_i \cap E_k)\|_Y \le \sum_{k=1}^{\infty} |\mu|(E_k),$$

since by (1.62)

$$E_k = E_k \cap E = \bigcup_{i=1}^{n} (F_i \cap E_k) \quad \Rightarrow \quad \sum_{i=1}^{n} \|\mu(F_i \cap E_k)\|_Y \le |\mu|\,(E_k) \quad (k \in \mathbb{N}).$$

It follows that

$$|\mu|\left(\bigcup_{k=1}^{\infty} E_k\right) = |\mu|\,(E) \le \sum_{k=1}^{\infty} |\mu|\,(E_k). \tag{1.63}$$

On the other hand, it is easily seen that the map $|\mu| : \mathcal{A} \mapsto [0,\infty]$ is additive and monotone. Then for any $n \in \mathbb{N}$,

$$\sum_{k=1}^{n} |\mu|\,(E_k) = |\mu|\left(\bigcup_{k=1}^{n} E_k\right) \le |\mu|\left(\bigcup_{k=1}^{\infty} E_k\right),$$

whence

$$\sum_{k=1}^{\infty} |\mu|\,(E_k) \le |\mu|\left(\bigcup_{k=1}^{\infty} E_k\right). \tag{1.64}$$

By (1.63)–(1.64) $|\mu|$ is σ-additive, and thus the claim follows. □

Lemma 1.9.2. *Let $\mu : \mathcal{A} \to Y$ be a vector measure. Then for any monotone sequence $\{E_k\} \subset \mathcal{A}$, the sequence $\{\mu(E_k)\} \subseteq Y$ is convergent in Y.*

Proof. If $\{E_k\} \subseteq \mathcal{A}$ is nondecreasing, then for every $k \in \mathbb{N}$, we have $E_k = \bigcup_{i=1}^{k} F_i$, where $F_i := E_i \setminus E_{i-1}$ ($i \in \mathbb{N}$, $E_0 := \emptyset$). Since μ is σ-additive and $\{F_i\}$ is a disjoint sequence, we have

$$\lim_{k\to\infty} \mu(E_k) = \lim_{k\to\infty} \sum_{i=1}^{k} \mu(F_i) = \mu\left(\bigcup_{i=1}^{\infty} F_i\right) = \mu\left(\bigcup_{k=1}^{\infty} E_k\right) \quad \text{in } Y,$$

and thus the conclusion follows in this case. If $\{E_k\} \subseteq \mathcal{A}$ is nonincreasing, then the sequence $\{E_k^c\} \subseteq \mathcal{A}$ is nondecreasing, and $\mu(E_k) = \mu(X) - \mu(E_k^c)$ for all $k \in \mathbb{N}$. Then the sequence $\{\mu(E_k)\}$ is convergent in Y by the first part of the proof, and thus the result follows. □

Denote by $\langle \cdot, \cdot \rangle_{Y^*,Y} : Y^* \times Y \mapsto \mathbb{R}$ the duality map between the Banach space Y and its dual space Y^*.

Proposition 1.9.3. *Let $\mu : \mathcal{A} \to Y$ be a vector measure, and let $y^* \in Y^*$. Then the map*

$$\langle y^*, \mu \rangle : \mathcal{A} \mapsto \mathbb{R}, \quad \langle y^*, \mu \rangle(E) := \langle y^*, \mu(E) \rangle_{Y^*,Y} \quad \text{for } E \in \mathcal{A} \tag{1.65}$$

is a finite signed measure.

Proof. By definition we have $\langle y^*,\mu\rangle(\emptyset) = 0$ and $\langle y^*,\mu\rangle(\mathcal{A}) \subseteq \mathbb{R}$. By Remark 1.9.1 the map $\langle y^*,\mu\rangle$ is additive, i. e., $\langle y^*,\mu\rangle(\bigcup_{k=1}^n E_k) = \sum_{k=1}^n \langle y^*,\mu\rangle(E_k)$. Then for any disjoint sequence $\{E_k\} \subseteq \mathcal{A}$,

$$\begin{aligned}\lim_{n\to\infty}\left|\langle y^*,\mu\rangle\left(\bigcup_{k=1}^{\infty} E_k\right) - \sum_{k=1}^{n}\langle y^*,\mu\rangle(E_k)\right| &= \lim_{n\to\infty}\left|\langle y^*,\mu\rangle\left(\bigcup_{k=n+1}^{\infty} E_k\right)\right| \\ = \lim_{n\to\infty}\left|\left\langle y^*,\mu\left(\bigcup_{k=n+1}^{\infty} E_k\right)\right\rangle_{Y^*,Y}\right| &\le \|y^*\|_{Y^*}\lim_{n\to\infty}\left\|\mu\left(\bigcup_{k=n+1}^{\infty} E_k\right)\right\|_Y = 0,\end{aligned}$$

and hence the result follows. □

Let $|\langle y^*,\mu\rangle|$ denote the variation of the signed measure $\langle y^*,\mu\rangle$.

Definition 1.9.3. The function $|\mu|_w : \mathcal{A} \mapsto [0,\infty)$,

$$|\mu|_w(E) := \sup\{|\langle y^*,\mu\rangle|(E) \mid y^* \in Y^*,\ \|y^*\|_{Y^*} \le 1\} \quad (E \in \mathcal{A}), \tag{1.66}$$

is called the *semivariation* of the vector measure $\mu : \mathcal{A} \to Y$.

Remark 1.9.3. By (1.62) and (1.66) we have $|\mu|_w(E) \le |\mu|(E)$ for all $E \in \mathcal{A}$. If $Y = \mathbb{R}$, then $|\mu|_w = |\mu|$ is the variation of the finite signed measure μ (see Remark 1.9.2).

Lemma 1.9.4. *Let $\mu : \mathcal{A} \to Y$ be a vector measure. Then for all $E \in \mathcal{A}$,*

$$\sup_{F\in\mathcal{A},F\subseteq E}\|\mu(F)\|_Y \le |\mu|_w(E) \le 2\sup_{F\in\mathcal{A},F\subseteq E}\|\mu(F)\|_Y. \tag{1.67}$$

Proof. For any $E \in \mathcal{A}$ and $F \in \mathcal{A}$, $F \subseteq E$, we have

$$\begin{aligned}\|\mu(F)\|_Y &= \sup_{y^*\in Y^*,\|y^*\|_{Y^*}\le 1}|\langle y^*,\mu(F)\rangle_{Y^*,Y}| = \sup_{y^*\in Y^*,\|y^*\|_{Y^*}\le 1}|\langle y^*,\mu\rangle(F)| \\ &\le \sup_{y^*\in Y^*,\|y^*\|_{Y^*}\le 1}|\langle y^*,\mu\rangle|(F) = |\mu|_w(E),\end{aligned}$$

and thus the first inequality in (1.67) follows.

Let $E = \bigcup_{i=1}^n F_i$ with disjoint $F_1,\dots,F_n \in \mathcal{A}$, and let $y^* \in Y^*$, $\|y^*\|_{Y^*} \le 1$. Let $\pi_\pm$ be subsets of $\{1,\dots,n\}$ such that $\operatorname{sgn}(\langle y^*,\mu\rangle(F_i)) = \pm 1 \Leftrightarrow i \in \pi_\pm$. Then

$$\begin{aligned}\sum_{i=1}^{n}|\langle y^*,\mu\rangle(F_i)| &= \sum_{i\in\pi_+}\langle y^*,\mu\rangle(F_i) - \sum_{i\in\pi_-}\langle y^*,\mu\rangle(F_i) \\ &= \left\langle y^*,\sum_{i\in\pi_+}\mu(F_i) - \sum_{i\in\pi_-}\mu(F_i)\right\rangle_{Y^*,Y} \\ &= \left\langle y^*,\mu\left(\bigcup_{i\in\pi_+}F_i\right)\right\rangle_{Y^*,Y} - \left\langle y^*,\mu\left(\bigcup_{i\in\pi_-}F_i\right)\right\rangle_{Y^*,Y}\end{aligned}$$

$$\le 2 \sup_{F \in \mathcal{A}, F \subseteq E} \|\mu(F)\|_Y.$$

Then by (1.52) and (1.66) the second inequality in (1.67) follows. □

Proposition 1.9.5. *Let $\mu : \mathcal{A} \to Y$ be a vector measure. Then $|\mu|_w(X) < \infty$.*

Proof. By contradiction let $|\mu|_w(X) = \infty$. Then by the second inequality in (1.67) $\sup_{F \in \mathcal{A}} \|\mu(F)\|_Y = \infty$. Hence there exists $F_1 \in \mathcal{A}$ such that

$$\|\mu(F_1)\|_Y \ge 1 + \|\mu(X)\|_Y,$$

whence

$$\|\mu(F_1^c)\|_Y \ge \|\mu(F_1)\|_Y - \|\mu(X)\|_Y \ge 1.$$

It follows that

$$\min\{\|\mu(F_1)\|_Y, \|\mu(F_1^c)\|_Y\} \ge 1. \tag{1.68}$$

On the other hand, since $\infty = |\mu|_w(X) \le |\mu|_w(F_1) + |\mu|_w(F_1^c)$, we also have

$$\max\{|\mu|_w(F_1), |\mu|_w(F_1^c)\} = \infty. \tag{1.69}$$

Set either $E_1 := F_1$ or $E_1 := F_1^c$. Then $E_1 \in \mathcal{A}$, and by (1.69)–(1.68)

$$E_1 \subseteq X, \quad |\mu|_w(E_1) = \infty, \quad \|\mu(E_1)\|_Y \ge 1.$$

Iterating the argument gives a sequence $\{E_k\} \subset \mathcal{A}$ such that for all $k \ge 2$,

$$E_k \subseteq E_{k-1}, \quad |\mu|_w(E_k) = \infty, \quad \|\mu(E_k)\|_Y \ge k.$$

The sequence $\{E_k\}$ is nonincreasing, and $\lim_{k\to\infty} \|\mu(E_k)\|_Y = \infty$, which contradicts Lemma 1.9.2. Hence the result follows. □

The following definition generalizes Definition 1.8.7(ii).

Definition 1.9.4. Let $\nu : \mathcal{A} \to Y$ be a vector measure, and let $\mu : \mathcal{A} \to [-\infty, \infty]$ be a signed measure. We say that ν is *absolutely continuous* with respect to μ (written $\nu \ll \mu$) if for any $E \in \mathcal{A}$,

$$|\mu|(E) = 0 \quad \Rightarrow \quad \nu(E) = 0.$$

Remark 1.9.4. Let μ, ν be as in Definition 1.9.4, and let $\lim_{|\mu|(E)\to 0} \nu(E) = 0$, that is, for any $\epsilon > 0$, there exists $\delta > 0$ such that for any $E \in \mathcal{A}$ with $|\mu|(E) < \delta$, we have $\|\nu(E)\|_Y < \epsilon$. Then, clearly, $\nu \ll \mu$. The converse is true if μ is finite (see [41, Theorem 1.2.1]).

The following result generalizes the Lebesgue decomposition theorem to vector measures (see [41, Theorem 1.5.9] for the proof).

Theorem 1.9.6 (Lebesgue decomposition). *Let $\nu : \mathcal{A} \to Y$ be a vector measure, and let $\mu : \mathcal{A} \to \mathbb{R}$ be a finite signed measure. Then there exists a unique couple of vector measures ν_{ac}, $\nu_s : \mathcal{A} \mapsto Y$ such that:*
(i) $\nu_{ac} \ll \mu$;
(ii) $\langle y^*, \nu_s \rangle \perp \mu$ *for all* $y^* \in Y^*$;
(iii) $\nu = \nu_{ac} + \nu_s$.

If ν is of bounded variation, then ν_{ac} and ν_s are of bounded variation, and $|\nu| = |\nu_{ac}| + |\nu_s|$, $|\nu_s| \perp \mu$.

2 Scalar integration and differentiation

2.1 Measurable functions

2.1.1 Definition and general properties

Definition 2.1.1. Let $(X, \mathcal{A})$ and $(X', \mathcal{A}')$ be measurable spaces. A function $f : X \to X'$ is called $(\mathcal{A}, \mathcal{A}')$-*measurable* (or simply *measurable*) if for any $E' \in \mathcal{A}'$, we have $f^{-1}(E') \in \mathcal{A}$.

Remark 2.1.1. If X' is a topological space, then the σ-algebra $\mathcal{A}'$ is the Borel σ-algebra $\mathcal{B}(X')$. In particular, we say that $f : X \to \mathbb{R}^N$ is $\mathcal{A}$-*measurable* (or simply *measurable*) if it is $(\mathcal{A}, \mathcal{B}(\mathbb{R}^N))$-measurable.

Proposition 2.1.1. *Let $(X, \mathcal{A})$, $(X', \mathcal{A}')$, and $(X'', \mathcal{A}'')$ be measurable spaces. Let $f : X \to X'$ be $(\mathcal{A}, \mathcal{A}')$-measurable, and let $g : X' \to X''$ be $(\mathcal{A}', \mathcal{A}'')$-measurable. Then $g \circ f : X \to X''$ is $(\mathcal{A}, \mathcal{A}'')$-measurable.*

Proof. By assumption, for any $E' \in \mathcal{A}'$, we have $f^{-1}(E') \in \mathcal{A}$, and for any $E'' \in \mathcal{A}''$, we have $g^{-1}(E'') \in \mathcal{A}'$. Therefore $(g \circ f)^{-1}(E'') = f^{-1}(g^{-1}(E'')) \in \mathcal{A}$ for any $E'' \in \mathcal{A}''$. □

Lemma 2.1.2. *Let $(X, \mathcal{A})$ and $(X', \mathcal{A}')$ be measurable spaces. Let $\mathcal{C}' \subseteq \mathcal{P}(X')$ be a family of sets such that the minimal σ-algebra $\sigma_0(\mathcal{C}') = \mathcal{A}'$. Then $f : X \to X'$ is $(\mathcal{A}, \mathcal{A}')$-measurable if and only if*

$$f^{-1}(E') \in \mathcal{A} \quad \text{for any } E' \in \mathcal{C}'. \tag{2.1}$$

Proof. Let f be $(\mathcal{A}, \mathcal{A}')$-measurable, thus $f^{-1}(E') \in \mathcal{A}$ for any $E' \in \mathcal{A}'$. Since $\mathcal{C}' \subseteq \sigma_0(\mathcal{C}') = \mathcal{A}'$, the "only if" part of the claim follows. To prove the "if" part, set

$$\Sigma := \{E' \subseteq X' \mid f^{-1}(E') \in \mathcal{A}\}.$$

Using the equalities

$$f^{-1}(X') = X, \quad f^{-1}(X' \setminus E') = X \setminus f^{-1}(E'), \quad f^{-1}\left(\bigcup_{k=1}^{\infty} E'_k\right) = \bigcup_{k=1}^{\infty} f^{-1}(E'_k), \tag{2.2}$$

it is easily seen that Σ is a σ-algebra. On the other hand, if (2.1) holds, then $\mathcal{C}' \subseteq \Sigma$. It follows that $\mathcal{A}' = \sigma_0(\mathcal{C}') \subseteq \Sigma$. □

Proposition 2.1.3. *Let $(X, \mathcal{A})$ and $(X', \mathcal{A}')$ be measurable spaces, and let $X = \bigcup_{k=1}^{\infty} E_k$ with $\{E_k\} \subseteq \mathcal{A}$. Then $f : X \to X'$ is $(\mathcal{A}, \mathcal{A}')$-measurable if and only if all restrictions $f|_{E_k} : E_k \to X'$ are $(\mathcal{A} \cap E_k, \mathcal{A}')$-measurable.*

https://doi.org/10.1515/9783110556902-003

Proof. Let $f|_{E_k}$ be $(\mathcal{A} \cap E_k, \mathcal{A}')$-measurable for every $k \in \mathbb{N}$. For any $F' \in \mathcal{A}'$, we have $(f|_{E_k})^{-1}(F') \in \mathcal{A} \cap E_k \subseteq \mathcal{A}$, since $E_k \in \mathcal{A}$ $(k \in \mathbb{N})$. Therefore

$$f^{-1}(F') = \left(\bigcup_{k=1}^{\infty} E_k\right) \bigcap f^{-1}(F') = \bigcup_{k=1}^{\infty} (E_k \cap f^{-1}(F')) = \bigcup_{k=1}^{\infty} (f|_{E_k})^{-1}(F') \in \mathcal{A},$$

and thus f is $(\mathcal{A}, \mathcal{A}')$-measurable. To prove the converse, observe that $f|_{E_k} = f \circ j_{E_k}$, where j_{E_k} denotes the injection of E_k in X. For any $E \in \mathcal{A}$, we have $j_{E_k}^{-1}(E) := \{x \in E_k \mid j_{E_k}(x) \in E\} = E \cap E_k$, and thus j_{E_k} is $(\mathcal{A} \cap E_k, \mathcal{A})$-measurable. Then by Proposition 2.1.1 $f|_{E_k}$ is $(\mathcal{A} \cap E_k, \mathcal{A}')$-measurable for any $k \in \mathbb{N}$, and thus the result follows. □

For any $f : X \mapsto \mathbb{R}$ and $\alpha \in \mathbb{R}$, we set $\{f > \alpha\} := \{x \in X \mid f(x) > \alpha\} = f^{-1}((\alpha, \infty))$; the sets $\{f \geq \alpha\}$, $\{f < \alpha\}$, $\{f \leq \alpha\}$, and $\{f = \alpha\}$ are similarly defined. For real-valued functions, we have the following measurability criterion, which easily follows from Lemma 2.1.2 and Remark 1.2.2.

Proposition 2.1.4. *Let $(X, \mathcal{A})$ be a measurable space, and let $f : X \to \mathbb{R}$. Then the following statements are equivalent:* (i) *f is $\mathcal{A}$-measurable;* (ii) *$\{f > \alpha\} \in \mathcal{A}$ for any $\alpha \in \mathbb{R}$;* (iii) *$\{f \geq \alpha\} \in \mathcal{A}$ for any $\alpha \in \mathbb{R}$;* (iv) *$\{f < \alpha\} \in \mathcal{A}$ for any $\alpha \in \mathbb{R}$; and* (v) *$\{f \leq \alpha\} \in \mathcal{A}$ for any $\alpha \in \mathbb{R}$.*

Using the equality $\{\sup f_n > \alpha\} = \bigcup_{n=1}^{\infty} \{f_n > \alpha\}$, from Proposition 2.1.4 we clearly get the following:

Corollary 2.1.5. *Let $(X, \mathcal{A})$ be a measurable space.*
(i) *If $f, g : X \to \mathbb{R}$ are $\mathcal{A}$-measurable, then the functions $f+g, fg, |f|, \min\{f, g\}, \max\{f, g\}$, and f/g (the latter if $g \neq 0$ in X) are $\mathcal{A}$-measurable.*
(ii) *If the functions $f_n : X \to [-\infty, \infty]$ are $\mathcal{A}$-measurable for every $n \in \mathbb{N}$, then the functions $\sup_{n\in\mathbb{N}} f_n$, $\inf_{n\in\mathbb{N}} f_n$, $\limsup_{n\to\infty} f_n$, and $\liminf_{n\to\infty} f_n$ are $\mathcal{A}$-measurable. In particular, if the pointwise limit $\lim_{n\to\infty} f_n$ exists, then it is $\mathcal{A}$-measurable.*

Proposition 2.1.6. *Let $(X, \mathcal{A}, \mu)$ be a complete measure space, and let $(X', \mathcal{A}')$ be a measurable space. Let $f, g : X \to X'$ be equal μ-a. e. in X. Then f is $(\mathcal{A}, \mathcal{A}')$-measurable if and only if g is $(\mathcal{A}, \mathcal{A}')$-measurable.*

Proof. By assumption there exists a null set $N \in \mathcal{A}$ such that $g(x) = f(x)$ for every $x \in N^c$. It is easily seen that for all $E' \in \mathcal{A}'$,

$$g^{-1}(E') = (g^{-1}(E') \cap N) \cup (f^{-1}(E') \cap N^c).$$

If f is $(\mathcal{A}, \mathcal{A}')$-measurable, then we have $f^{-1}(E') \cap N^c \in \mathcal{A}$. Moreover, since μ is complete and $g^{-1}(E') \cap N \subseteq N$, we have $g^{-1}(E') \cap N \in \mathcal{A}$, and thus g is $(\mathcal{A}, \mathcal{A}')$-measurable. By inverting the roles of f and g the result follows. □

If μ is not complete, then the above conclusion need not be true. To avoid this difficulty, it is convenient to extend the definition of measurability to functions defined almost everywhere.

Definition 2.1.2. Let $(X, \mathcal{A}, \mu)$ be a measure space, and let $(X', \mathcal{A}')$ be a measurable space. Let $N \subseteq X$ be a null set, and let $f : N^c \to X'$. We say that f is $(\mathcal{A}, \mathcal{A}')$-measurable if it is $(\mathcal{A} \cap N^c, \mathcal{A}')$-measurable.

If $N = \emptyset$, then the above definition reduces to Definition 2.1.1. If $N^c \subseteq X$ properly and Definition 2.1.2 holds, then it is easy to extend f to X to obtain an $(\mathcal{A}, \mathcal{A}')$-measurable function (in the sense of Definition 2.1.1). Clearly, the conclusion of Proposition 2.1.6 is true in the sense of Definition 2.1.2.

The following remark will be tacitly used hereafter.

Remark 2.1.2. Let $(X, \mathcal{A}, \mu)$ be a complete measure space. If $\{f_n\}$ is a sequence of $\mathcal{A}$-measurable functions and $f_n \to g$ μ-a. e. in X, then by Corollary 2.1.5 and Proposition 2.1.6 g is $\mathcal{A}$-measurable. If μ is not complete, then the same is true in the sense of Definition 2.1.2, since $g := \lim_{n\to\infty} f_n$ is defined on some set $E \in \mathcal{A}$ with $\mu(E^c) = 0$ and is $(\mathcal{A} \cap E, \mathcal{A}')$-measurable by Corollary 2.1.5.

Let us state the following definition.

Definition 2.1.3. Let $(X, \mathcal{T})$ and $(X', \mathcal{T}')$ be topological spaces, and let $\mathcal{B} := \sigma_0(\mathcal{T})$ and $\mathcal{B}' := \sigma_0(\mathcal{T}')$ be the relative Borel σ-algebras. A function $f : X \to X'$ is called *Borel measurable* (or *Borel function*) if it is $(\mathcal{B}, \mathcal{B}')$-measurable.

Remark 2.1.3. (i) If $(X, \mathcal{T})$ and $(X', \mathcal{T}')$ are topological spaces, then from Lemma 2.1.2 it follows that every continuous function $f : X \mapsto X'$ is Borel measurable.

(ii) Let $(X, \mathcal{T})$ be a topological space. A function $f : X \mapsto (-\infty, \infty]$ is lower semicontinuous if and only if for any $\alpha \in \mathbb{R}$, the set $\{f > \alpha\}$ is open. Similarly, a function $f : X \mapsto [-\infty, \infty)$ is upper semicontinuous if and only if for any $\alpha \in \mathbb{R}$, the set $\{f < \alpha\}$ is open. Then by Proposition 2.1.4 lower and upper semicontinuous functions are Borel measurable.

Definition 2.1.4. Let $(X, \mathcal{A})$ be a measurable space. A function $s : X \to \mathbb{R}$ is called *simple* if there exist $c_1, \dots, c_n \in \mathbb{R}$ and a partition $\{E_1, \dots, E_n\} \subseteq \mathcal{A}$ of X such that $s = \sum_{i=1}^n c_i \chi_{E_i}$.

Clearly, simple functions are measurable. For any measurable space $(X, \mathcal{A})$, we will denote by $\mathscr{S}(X)$ the set of real-valued simple functions and by $\mathscr{S}_+(X)$ that of nonnegative simple functions $s : X \to [0, \infty)$.

The proof of the following result is well known.

Theorem 2.1.7. *Let $(X, \mathcal{A})$ be a measurable space, and let $f : X \to \mathbb{R}$ be $\mathcal{A}$-measurable. Then there exists a sequence $\{s_n\}$ of simple functions such that $s_n \to f$ pointwise in X. Moreover:*

(i) *if* $f \geq 0$, *then the sequence* $\{s_n\}$ *is nondecreasing, and we have* $0 \leq s_n \leq f$ *for any* $n \in \mathbb{N}$;
(ii) *if* f *is bounded, then* $s_n \to f$ *uniformly in* X.

It is convenient for future purposes to address the measurability of functions defined on or taking values in the product of measurable spaces (see Section 1.2).

Definition 2.1.5. Let the sets X, X_1, and X_2 be given.
(i) The map $p_k : X_1 \times X_2 \to X_k$, $p_k(x_1, x_2) := x_k$, is called the *projection* of $X_1 \times X_2$ onto X_k $(k = 1, 2)$.
(ii) Let $f : X \mapsto X_1 \times X_2$. The map $f_k : X \to X_k$, $f_k := p_k \circ f$, is called the *kth component* of f $(k = 1, 2)$.
(iii) Let $f : X_1 \times X_2 \to X$. For any $x_1 \in X_1$, the map $f_{x_1} : X_2 \to X$, $f_{x_1}(x_2) := f(x_1, x_2)$, is called the x_1-*trace* of f. Similarly, for any $x_2 \in X_2$, the x_2-*trace* of f is the map $f_{x_2} : X_1 \to X$, $f_{x_2}(x_1) := f(x_1, x_2)$.

Remark 2.1.4. Let χ_E be the characteristic function of $E \subseteq X_1 \times X_2$. Clearly, for any $x_1 \in X_1$, the x_1-trace $(\chi_E)_{x_1}$ of χ_E coincides with the characteristic function $\chi_{E_{x_1}}$ of the x_1-*section* of E, $E_{x_1} := \{x_2 \in X_2 \mid (x_1, x_2) \in E\}$. Similarly, for any $x_2 \in X_2$, the x_2-trace $(\chi_E)_{x_2}$ of χ_E coincides with the characteristic function $\chi_{E_{x_2}}$ of the x_2-*section* of E, $E_{x_2} := \{x_1 \in X_1 \mid (x_1, x_2) \in E\}$.

Lemma 2.1.8. *Let* $(X_1 \times X_2, \mathcal{A}_1 \times \mathcal{A}_2)$ *be the product of two measurable spaces* $(X_1, \mathcal{A}_1)$ *and* $(X_2, \mathcal{A}_2)$. *Then for each* $k = 1, 2$, *the projection* p_k *is* $(\mathcal{A}_1 \times \mathcal{A}_2, \mathcal{A}_k)$-*measurable.*

Proof. For any $E_1 \in \mathcal{A}_1$, we have $p_1^{-1}(E_1) := \{(x_1, x_2) \in X_1 \times X_2 \mid x_1 \in E_1\} = E_1 \times X_2$, a measurable rectangle. Hence $p_1^{-1}(E_1) \in \mathcal{A}_1 \times \mathcal{A}_2$. Similarly for p_2, and thus the claim follows. □

Proposition 2.1.9. *Let* $(X, \mathcal{A})$, $(X_1, \mathcal{A}_1)$, *and* $(X_2, \mathcal{A}_2)$ *be measurable spaces, and let* $(X_1 \times X_2, \mathcal{A}_1 \times \mathcal{A}_2)$ *be the product of* $(X_1, \mathcal{A}_1)$ *and* $(X_2, \mathcal{A}_2)$. *Let* $f : X \to X_1 \times X_2$. *Then* f *is* $(\mathcal{A}, \mathcal{A}_1 \times \mathcal{A}_2)$-*measurable if and only if each component* f_k *is* $(\mathcal{A}, \mathcal{A}_k)$-*measurable* $(k = 1, 2)$.

Proof. If f is $(\mathcal{A}, \mathcal{A}_1 \times \mathcal{A}_2)$-measurable, then by Proposition 2.1.1 and Lemma 2.1.8 f_k is $(\mathcal{A}, \mathcal{A}_k)$-measurable for each $k = 1, 2$. Conversely, let f_k be $(\mathcal{A}, \mathcal{A}_k)$-measurable, and thus $f_k^{-1}(E_k) \in \mathcal{A}$ for any $E_k \in \mathcal{A}_k$ $(i = 1, 2)$. Let us prove that $f^{-1}(E) \in \mathcal{A}$ for any $E \in \mathcal{R}$, where $\mathcal{R}$ denotes the set of measurable rectangles; if so, then the result will follow from Lemma 2.1.2, since $\mathcal{A}_1 \times \mathcal{A}_2 := \sigma_0(\mathcal{R})$. In fact, let $E = E_1 \times E_2$, with $E_k \in \mathcal{A}_k$ $(i = 1, 2)$. Then

$$
\begin{aligned}
f^{-1}(E) &= \{x \in X \mid (f_1(x), f_2(x)) \in E\} \\
&= \{x \in X \mid f_1(x) \in E_1\} \cap \{x \in X \mid f_2(x) \in E_2\} = f_1^{-1}(E_1) \cap f_2^{-1}(E_2) \in \mathcal{A}.
\end{aligned}
$$

Hence the conclusion follows. □

Proposition 2.1.10. *Let $(X_1, \mathcal{A}_1)$, $(X_2, \mathcal{A}_2)$, and $(X, \mathcal{A})$ be measurable spaces, and let $f : X_1 \times X_2 \to X$ be $(\mathcal{A}_1 \times \mathcal{A}_2, \mathcal{A})$-measurable. Then:*
(i) *for any $x_1 \in X_1$, the x_1-trace of f is $(\mathcal{A}_2, \mathcal{A})$-measurable;*
(ii) *for any $x_2 \in X_2$, the x_2-trace of f is $(\mathcal{A}_1, \mathcal{A})$-measurable.*

Proof. For any $\overline{x}_1 \in X_1$, we have $f_{\overline{x}_1} = f \circ g$, where $g : X_2 \to X_1 \times X_2$ and $g(x_2) \equiv (g_1(x_2), g_2(x_2)) := (\overline{x}_1, x_2)$. Clearly, each component g_k is $(\mathcal{A}_2, \mathcal{A}_k)$-measurable ($k = 1, 2$), and thus by Proposition 2.1.9 g is $(\mathcal{A}_2, \mathcal{A}_1 \times \mathcal{A}_2)$-measurable. Then by Proposition 2.1.1 $f_{\overline{x}_1}$ is $(\mathcal{A}_2, \mathcal{A})$-measurable. The proof of (ii) is analogous. □

Remark 2.1.5. Let χ_E be the characteristic function of $E \subseteq X_1 \times X_2$. If $E \in \mathcal{A}_1 \times \mathcal{A}_2$, then χ_E is $\mathcal{A}_1 \times \mathcal{A}_2$-measurable, and by Proposition 2.1.10 the x_1-trace $(\chi_E)_{x_1} = \chi_{E_{x_1}}$ is $\mathcal{A}_2$-measurable (in this connection, observe that $E_{x_1} \in \mathcal{A}_2$ by Proposition 1.2.2). Similarly, for any $x_2 \in X_2$, the x_2-trace $(\chi_E)_{x_2} = \chi_{E_{x_2}}$ is $\mathcal{A}_1$-measurable.

2.1.2 Convergence results

Let $(X, \mathcal{A}, \mu)$ be a measure space, and let $f, f_n : X \to \mathbb{R}$ be $\mathcal{A}$-measurable functions ($n \in \mathbb{N}$). The following notions of convergence (for which we write $f_n \to f$) are well known:
- f_n converges to f *in measure* if $\lim_{n\to\infty} \mu(\{x \in X \mid |f_n(x) - f(x)| > \epsilon\}) = 0$ for all $\epsilon > 0$;
- f_n converges to f *almost uniformly* if for any $\delta > 0$, there exists $E \in \mathcal{A}$ with $\mu(E^c) < \delta$ such that

$$\lim_{n\to\infty} \sup_{x\in E} |f_n(x) - f(x)| = 0. \tag{2.3}$$

Let us recall the relationships between the above concepts.

Proposition 2.1.11. *Let $(X, \mathcal{A}, \mu)$ be a measure space, and let $f, f_n : X \to \mathbb{R}$ be $\mathcal{A}$-measurable.*
(i) *Let $\mu(X) < \infty$. If $f_n \to f$ μ-a. e., then $f_n \to f$ in measure.*
(ii) *If $f_n \to f$ in measure, then there exists $\{f_{n_k}\} \subseteq \{f_n\}$ such that $f_{n_k} \to f$ μ-a. e.*
(iii) *If $f_n \to f$ almost uniformly, then $f_n \to f$ μ-a. e. and in measure.*

The following result partially inverts Proposition 2.1.11(iii).

Theorem 2.1.12 (Egorov). *Let $(X, \mathcal{A}, \mu)$ be a finite measure space, and let $f, f_n : X \to \mathbb{R}$ be $\mathcal{A}$-measurable. If $f_n \to f$ μ-a. e., then $f_n \to f$ almost uniformly.*

Proof. By assumption there exists a μ-null subset $N \subseteq X$ such that for every $x \in N^c$, we have $\lim_{n\to\infty} |f_n(x) - f(x)| = 0$. For all $k, n, p \in \mathbb{N}$, set

$$E_{n,k} := \left\{x \in X \mid |f_n(x) - f(x)| \geq \frac{1}{k}\right\} \in \mathcal{A}, \quad F_{p,k} := \bigcup_{n=p}^{\infty} E_{n,k}$$

(see Remark 2.1.2). It is easily seen that for every $k \in \mathbb{N}$, we have $G_k := \bigcap_{p=1}^{\infty} F_{p,k} \subseteq N$. In fact, let $k \in \mathbb{N}$ and $x \in G_k$ be fixed. Then, by the definition of $F_{p,k}$ and $E_{n,k}$, for every $p \in \mathbb{N}$, there exists $n_p \geq p$ such that $|f_{n_p}(x) - f(x)| \geq \frac{1}{k}$. Clearly, this implies that $|f_{n_p}(x) - f(x)| \nrightarrow 0$, thus $x \notin N^c$, and the claim follows.

Since for any $k \in \mathbb{N}$, the set G_k is measurable and $G_k \subseteq N$, it follows that $\mu(G_k) = 0$. Moreover, since for every fixed $k \in \mathbb{N}$, the sequence $\{F_{p,k}\}$ is nonincreasing and $\mu(F_{1,k}) \leq \mu(X) < \infty$, we have that $\lim_{p\to\infty} \mu(F_{p,k}) = \mu(G_k) = 0$. Then for any $\delta > 0$ and $k \in \mathbb{N}$, there exists $p_k \in \mathbb{N}$ such that $\mu(F_{p_k,k}) < \frac{\delta}{2^{k+1}}$. Set $F_\delta := \bigcup_{k=1}^{\infty} F_{p_k,k}$. Then we have $\mu(F_\delta) < \delta$ and

$$F_\delta^c = \bigcap_{k=1}^{\infty} F_{p_k,k}^c = \bigcap_{k=1}^{\infty} \left(\bigcap_{n=p_k}^{\infty} E_{n,k}^c \right).$$

Fix $x \in F_\delta^c$. Then for every $k \in \mathbb{N}$, there exists $p_k \in \mathbb{N}$ such that $x \in F_{p_k,k}^c$, whence $x \notin E_{n,k}$ for all $n \geq p_k$, that is, $|f_n(x) - f(x)| < \frac{1}{k}$. Since $\mu(F_\delta) < \delta$ with $E = F_\delta^c$, the conclusion follows. □

Similar concepts and results hold if we consider capacities instead of measures.

Definition 2.1.6. Let (X, d) be a metric space, let $\mathcal{F} \subseteq \mathcal{P}(X)$ be as in Definition 1.6.1, and let $C : \mathcal{F} \to [0, \infty]$ be a capacity. Let $f, f_n : X \to \mathbb{R}$ $(n \in \mathbb{N})$. We say that:

(i) f_n *converges to* f *C-quasi-everywhere in* X if $f_n(x)$ converges to $f(x)$ for C-q. e. $x \in X$;

(ii) f_n converges to f *in capacity* if

$$\lim_{n\to\infty} C(\{|f_n - f| > \epsilon\}) = 0 \quad \text{for all } \epsilon > 0;$$

(iii) f_n converges to f *C-quasi-uniformly* if for any $\delta > 0$, there exists $E \in \mathcal{F}$ with $C(E^c) < \delta$ such that (2.3) holds.

Remark 2.1.6. Clearly, $f_n \to f$ in capacity if and only if for any $\delta > 0$ and $\epsilon > 0$, there exists $\bar{n} \in \mathbb{N}$ such that for all $n \geq \bar{n}$, we have $C(\{|f_n - f| \geq \epsilon\}) < \delta$.

The following proposition is similar to Proposition 2.1.11; we give its proof for completeness.

Proposition 2.1.13. *Let* X *and* $f, f_n : X \to \mathbb{R}$ $(n \in \mathbb{N})$ *be as in Definition* 2.1.6.

(i) *If* $f_n \to f$ *C-quasi uniformly, then* $f_n \to f$ *in capacity and C-q. e.*

(ii) *If $f_n \to f$ in capacity, then there exists $\{f_{n_k}\} \subseteq \{f_n\}$ such that $f_{n_k} \to f$ C-quasi-uniformly.*

Proof. (i) Let us prove that $f_n \to f$ in capacity. By assumption, for any $\delta > 0$ there exists $E \in \mathcal{F}$ with $C(E^c) < \delta$ such that $\lim_{n\to\infty} \sup_{x\in E} |f_n(x) - f(x)| = 0$ – namely, for any $\epsilon > 0$ there exists $\bar{n} \in \mathbb{N}$ such that for all $n \geq \bar{n}$ we have $E \subseteq \{|f_n - f| < \epsilon\}$. Hence we have $E^c \supseteq \{|f_n - f| \geq \epsilon\}$, and thus

$$C(\{|f_n - f| \geq \epsilon\}) \leq C(E^c) < \delta.$$

To sum up, for any $\epsilon > 0$ and $\delta > 0$, there exists $\bar{n} \in \mathbb{N}$ such that for all $n \geq \bar{n}$, we have $C(\{|f_n - f| \geq \epsilon\}) < \delta$. Hence the claim follows.

Let us now prove that $f_n \to f$ C-q. e. By assumption, for any $k \in \mathbb{N}$, there exists $E_k \in \mathcal{A}$ such that $C(E_k^c) < \frac{1}{k}$ and $\lim_{n\to\infty} \sup_{x\in E_k} |f_n(x) - f(x)| = 0$. Set $E := \bigcup_{k=1}^{\infty} E_k$. Then $f_n(x) \to f(x)$ for any $x \in E$. Moreover, since $E^c \subseteq (E_k)^c$ for any $k \in \mathbb{N}$, we have $C(E^c) \leq \lim_{k\to\infty} C(E_k^c) = 0$. Hence claim (i) follows.

(ii) By Remark 2.1.6, for any $k \in \mathbb{N}$, there exists $n_k \in \mathbb{N}$ such that $C(\{|f_n - f| \geq \frac{1}{2^k}\}) < \frac{1}{2^k}$ for all $n \geq n_k$. In particular, choosing $n = n_k$, we get $C(B_k) < \frac{1}{2^k}$ for all $k \in \mathbb{N}$, where $B_k := \{|f_{n_k} - f| \geq \frac{1}{2^k}\}$. Clearly, it is not restrictive to assume that $n_{k+1} > n_k$, so that $\{f_{n_k}\}$ is a subsequence of $\{f_n\}$. Set $A_m := \bigcup_{k=m}^{\infty} B_k$ $(m \in \mathbb{N})$. Then for every $m \in \mathbb{N}$, we have $A_m \subseteq A_{m-1}$ and $C(A_m) \leq \sum_{k=m}^{\infty} C(B_k) < \sum_{k=m}^{\infty} \frac{1}{2^k}$, and thus $\lim_{m\to\infty} C(A_m) = 0$.

Therefore, for any $\delta > 0$, there exists $\bar{m} \in \mathbb{N}$ such that $C(A_m) < \delta$ for all $m \geq \bar{m}$. Set $F_m := (A_m)^c$ $(m \geq \bar{m})$. Then $F_m \subseteq (B_k)^c$ for all $k \geq m \geq \bar{m}$, whence

$$\sup_{x\in F_{\bar{m}}} |f_{n_k}(x) - f(x)| \leq \sup_{x\in (B_k)^c} |f_{n_k}(x) - f(x)| \leq \frac{1}{2^k} \leq \frac{1}{2^m}.$$

It follows that $f_{n_k} \to f$ uniformly in $F_{\bar{m}}$. This completes the proof. □

2.1.3 Quasi-continuous functions

Theorem 2.1.14 (Lusin). *Let X be a Hausdorff space, and let μ be a regular measure on $\mathcal{B}(X)$. Let $f : X \to \mathbb{R}$ be $\mathcal{B}(X)$-measurable. Then for any $E \in \mathcal{B}(X)$ with $\mu(E) < \infty$ and any $\delta > 0$, there exists a compact subset $K \subseteq E$ such that $\mu(E \setminus K) < \delta$, and the restriction $f|_K$ is continuous.*

Proof. It suffices to consider the case $E = X$ with $\mu(X) < \infty$. Let $\{B_n\}$ be a countable basis of open intervals in $\mathbb{R}$. Thus $f^{-1}(B_n) \in \mathcal{B}(X)$ $(n \in \mathbb{N})$. Fix $\delta > 0$. Since by assumption the measure μ is regular, for every $n \in \mathbb{N}$, there exist a compact set K_n and an open set A_n such that $K_n \subseteq f^{-1}(B_n) \subseteq A_n$ and $\mu(A_n \setminus K_n) < \frac{\delta}{2^{n+1}}$. Set $A := \bigcup_{n=1}^{\infty}(A_n \setminus K_n)$ and $C := A^c$; clearly, A is open, thus C is closed, and $\mu(A) = \mu(C^c) \leq \frac{\delta}{2}$. Moreover, for any

$n \in \mathbb{N}$,

$$A_n \cap C = K_n \cap C \subseteq f^{-1}(B_n) \cap C \subseteq A_n \cap C,$$

whence

$$f^{-1}(B_n) \cap C = (f|_C)^{-1}(B_n) = A_n \cap C.$$

Since for any n, the set $A_n \cap C$ is open in the relative topology of C, the restriction $f|_C$ is continuous.

Using again the regularity of μ, we can choose a compact subset $K \subseteq C$ such that $\mu(C \setminus K) < \frac{\delta}{2}$. Therefore $\mu(K^c) = \mu(C^c) + \mu(C \setminus K) < \delta$, and $f|_K$ is continuous, since $f|_C$ is continuous and $K \subseteq C$. Then the conclusion follows. □

Definition 2.1.7. Let X be a Hausdorff space, and let μ be a measure on $\mathcal{B}(X)$. A function $f : X \to \mathbb{R}$ is called *μ-quasi-continuous* (or simply *quasi-continuous*) in $E \in \mathcal{B}(X)$ if for any $\delta > 0$, there exists a closed subset $C \subseteq E$ such that $\mu(E \setminus C) < \delta$ and the restriction $f|_C$ is continuous.

Using this definition, the Lusin theorem can be rephrased by saying that if the measure μ is finite and regular, then every $\mathcal{B}(X)$-measurable function is μ-quasi-continuous. The following result states that the converse is true up to null sets.

Proposition 2.1.15. *Let X be a Hausdorff space, and let μ be a finite regular measure on $\mathcal{B}(X)$. Let $f : X \to \mathbb{R}$ be μ-quasi-continuous. Then there exists a $\mathcal{B}(X)$-measurable function $g : X \to \mathbb{R}$ such that $g = f$ μ-a. e. in X.*

Proof. By Definition 2.1.7, for every $n \in \mathbb{N}$, there exists a closed set $C_n \subseteq X$ such that $\mu(C_n^c) < 1/n$ and $f|_{C_n}$ is continuous. Set $C := \bigcup_{n=1}^{\infty} C_n$ and $g := f\chi_C$. It is not restrictive to suppose that $\{C_n\}$ is nondecreasing, and thus $\mu(C^c) = \lim_{n\to\infty} \mu(C_n^c) = 0$. Moreover, observe that for every n, the function $f|_{C_n} = f\chi_{C_n}$ is continuous, thus $(\mathcal{B}(X) \cap C_n)$-measurable, and $f\chi_{C_n} \to f\chi_C = g$ pointwise in X. Then by Proposition 2.1.3 the function g is $(\mathcal{B}(X) \cap C)$-measurable and hence $\mathcal{B}(X)$-measurable. Since $f = g$ on C and $\mu(C^c) = 0$, the result follows. □

Proposition 2.1.16. *Let X be a locally compact Hausdorff space, and let μ be a regular measure on $\mathcal{B}(X)$. Let $f : X \to \mathbb{R}$ be μ-quasi-continuous. Then for any open set $A \subseteq X$ with $\mu(A) < \infty$ and for any $\delta > 0$, there exist a compact subset $K \subseteq A$ and a function $\zeta \in C_c(X)$ with $\operatorname{supp}\zeta \subseteq A$ such that $\mu(\{f \neq \zeta\}) < \delta$ and $\|\zeta\|_\infty = \|f\|_K := \sup_{x\in K} |f(x)|$.*

Proof. Let $A \subseteq X$ be open with $\mu(A) < \infty$, and let $\delta > 0$ be fixed. Then by the Lusin theorem there exists a compact subset $K \subseteq E$ such that $f|_K$ is continuous. By a standard compactification argument and the Tietze extension theorem there exists $h \in C(X)$ such that $h|_K = f|_K$ (e. g., see [45, Korollar VIII.1.19] for details). On the other hand, by Lemma A.9 there exists $g \in C_c(X)$ such that $g(X) \subseteq [0,1]$, $g|_K = 1$, and $\operatorname{supp} g \subseteq A$. Then

$\eta := g\,h|_K \in C_c(X)$, $\operatorname{supp}\eta \subseteq A$, and $\eta|_K = h|_K = f|_K$. It is easily seen that the function

$$\zeta(x) := \begin{cases} \eta(x) & \text{if } |\eta(x)| \le \|f\|_K, \\ \|f\|_K \operatorname{sgn}\eta(x) & \text{otherwise} \end{cases}$$

has the stated properties. Hence the result follows. □

Remark 2.1.7. By the Lusin theorem Proposition 2.1.16 also holds for $\mathcal{B}(X)$-measurable functions.

Similar notions and results hold for capacities.

Definition 2.1.8. Let (X, d) be a metric space, let $\mathcal{F} \subseteq \mathcal{P}(X)$ be as in Definition 1.6.1, and let $C : \mathcal{F} \to [0, \infty]$ be a capacity. A function $f : X \to \mathbb{R}$ is called *C-quasi-continuous* in X if for any $\delta > 0$, there exists a subset $E \subseteq X$ with $C(E^c) < \delta$ such that the restriction $f|_E$ is continuous.

Remark 2.1.8. If C is an outer capacity, then it is not restrictive to assume E^c to be open in Definition 2.1.8.

Proposition 2.1.17. *Let X and $f, f_n : X \to \mathbb{R}$ $(n \in \mathbb{N})$ be as in Definition 2.1.8. Let f_n be C-quasi-continuous for each n, and let $f_n \to f$ in capacity. Then f is C-quasi-continuous.*

Proof. By assumption, for every $\delta > 0$ and for every $n \in \mathbb{N}$, there exists a subset $E_n \subseteq X$ with $C(E_n^c) < \frac{\delta}{2^{n+1}}$ such that the restriction $f_n|_{E_n}$ is continuous. Set $E := \bigcap_{n=1}^{\infty} E_n$. Then $C(E^c) \le \sum_{n=1}^{\infty} C(E_n^c) \le \frac{\delta}{2}$, and $f_n|_E$ is continuous for each n. Moreover, by Proposition 2.1.13(ii) there exist $\{f_{n_k}\} \subseteq \{f_n\}$ and $F \subseteq X$ such that $C(F^c) < \frac{\delta}{2}$ and $f_{n_k} \to f$ uniformly in F. Set $G := E \bigcap F$; thus $C(G^c) \le C(E^c) + C(E^c) < \delta$. Moreover, $f_{n_k}|_G$ is continuous for each n_k and $f_{n_k} \to f$ uniformly in G, and hence $f|_G$ is continuous. Then the result follows. □

2.2 Integration

2.2.1 Definition of integral

Let $(X, \mathcal{A}, \mu)$ be a measure space, and let $s \in \mathscr{S}_+(X)$, that is, $s = \sum_{i=1}^{n} c_i \chi_{E_i}$ with $c_1, \dots, c_n \ge 0$ and measurable $E_1, \dots, E_n$, $E_i \cap E_j = \emptyset$ for $i \ne j$, and $\bigcup_{i=1}^{n} E_i = X$.

Definition 2.2.1. Let $s \in \mathscr{S}_+(X)$. The quantity $\int_X s\,d\mu := \sum_{i=1}^{N} c_i\,\mu(E_i)$ is called the *integral of s* on X.

This definition is well posed, since the integral $\int_X s\,d\mu$ does not depend on the choice of the partition $\{E_1, \dots, E_N\}$. Moreover, the integral $\int_X s\,d\mu$ has all the usual properties: linearity, additivity, monotonicity, and so on.

Definition 2.2.2. Let $f : X \to [0, \infty]$ be $\mathcal{A}$-measurable. The quantity

$$\int_X f\, d\mu := \sup_{s \in \mathscr{S}_+(X),\, s \le f} \int_X s\, d\mu \tag{2.4}$$

is called the *integral of f on X*. For every measurable set $E \subseteq X$, we set $\int_E f\, d\mu := \int_X f\chi_E\, d\mu$.

Remark 2.2.1. It is well known that for any $\mathcal{A}$-measurable $f : X \to [0, \infty]$, the set $\{s \in \mathscr{S}_+(X) \mid s \le f\}$ is nonempty; in fact, there exists a sequence $\{s_n\} \subseteq \mathscr{S}_+(X)$ such that

$$0 \le s_n \le s_{n+1} \le f \quad \text{in } X \text{ for all } n \in \mathbb{N}, \quad s_n \to f \text{ pointwise in } X. \tag{2.5}$$

Clearly, this implies that

$$\int_X f\, d\mu = \lim_{n\to\infty} \int_X s_n\, d\mu. \tag{2.6}$$

For every $f : X \mapsto [-\infty, \infty]$, set $f^{\pm} := \max\{\pm f, 0\}$.

Definition 2.2.3. An $\mathcal{A}$-measurable function $f : X \to [-\infty, \infty]$ is called *integrable* if $\int_X f^{\pm}\, d\mu < \infty$. It is called *quasi-integrable* if at least one of the two integrals $\int_X f^{\pm}\, d\mu$ is finite.

For every quasi-integrable function $f : X \to [-\infty, \infty]$, the (possibly, infinite) quantity $\int_X f\, d\mu := \int_X f^+\, d\mu - \int_X f^-\, d\mu$ is called the *integral of f on X*. For every measurable $E \subseteq X$, we set $\int_E f\, d\mu := \int_X f\chi_E\, d\mu$.

We denote the set of integrable functions $f : X \mapsto [-\infty, \infty]$ by $L^1(X) \equiv L^1(X, \mathcal{A}, \mu)$. It is easily seen that $f \in L^1(X)$ if and only if

$$\|f\|_1 := \int_X |f|\, d\mu < \infty.$$

Linearity, additivity, monotonicity, and other usual properties of the integral $\int_X f\, d\mu$ are easily proven. By the linearity of the integral, the inequality

$$\left|\int_X f\, d\mu\right| \le \int_X |f|\, d\mu,$$

and the triangular inequality, $L^1(X)$ is a vector space.

Remark 2.2.2. Any *Riemann-integrable* $f : D \subseteq \mathbb{R}^N \to \mathbb{R}$ is integrable in the sense of Definition 2.2.3. By abuse of notation, the usual symbol $\int_D f dx$ instead of $\int_D f d\lambda_N$ will often be used.

Remark 2.2.3. Let $(X, \mathcal{A}, \mu)$ be a measure space, and let $f \geq 0$ be $\mathcal{A}$-measurable. It is immediately seen that the map

$$\nu : \mathcal{A} \mapsto [0, \infty], \quad \nu(E) := \int_E f \, d\mu \quad (E \in \mathcal{A}) \tag{2.7}$$

is a measure. Similarly, the map

$$\nu : \mathcal{A} \mapsto [-\infty, \infty], \quad \nu(E) := \int_E f \, d\mu = \int_E f^+ \, d\mu - \int_E f^- \, d\mu \quad (E \in \mathcal{A}) \tag{2.8}$$

is a signed measure if f is quasi-integrable and a finite signed measure if $f \in L^1(X, \mathcal{A}, \mu)$. Moreover, we have

$$\nu^{\pm}(E) = \int_E f^{\pm} \, d\mu, \tag{2.9}$$

whence

$$|\nu|(E) = \int_E |f| \, d\mu < \infty \quad \text{if } f \in L^1(X, \mathcal{A}, \mu).$$

Indeed, set $\varphi^{\pm}(E) := \int_E f^{\pm} \, d\mu$. Then for any $F \subseteq E$, $F \in \mathcal{A}$,

$$\nu(F) = \varphi^+(F) - \varphi^-(F) \leq \varphi^+(F) \leq \varphi^+(E),$$

and thus $\nu^+(E) \leq \varphi^+(E)$ by Proposition 1.8.3. To prove the inverse inequality, set $H := \{f \geq 0\}$. Clearly, we have $\varphi^+(E \cap H^c) = 0 = \varphi^-(E \cap H)$ for any $E \in \mathcal{A}$, and hence

$$\varphi^+(E) = \varphi^+(E \cap H) = \nu(E \cap H) \leq \nu^+(E \cap H) \leq \nu^+(E).$$

It follows that $\nu^+ = \varphi^+$, whence $\nu^- = (-\nu)^+ = \varphi^-$. This proves the claim.

Remark 2.2.4. It follows from Remark 2.2.1 (see (2.6)) that the set $\mathscr{S}(X)$ of simple functions is dense in $L^1(X)$.

The integral with respect to a signed measure is defined as follows (see Subsection 4.2.3 for a generalization of this notion).

Definition 2.2.4. Let $(X, \mathcal{A})$ be a measurable space, and let μ be a signed measure on $\mathcal{A}$. We say that $f \in L^1(X, \mathcal{A}, \mu)$ if $f \in L^1(X, \mathcal{A}, \mu^{\pm})$, and we set

$$\int_X f \, d\mu := \int_X f \, d\mu^+ - \int_X f \, d\mu^-. \tag{2.10}$$

From (2.10) it easily follows that

$$\left|\int_X f\,d\mu\right| \le \int_X |f|\,d|\mu|. \tag{2.11}$$

Let us recall the following classical results.

Theorem 2.2.1 (Beppo Levi). *Let $(X, \mathcal{A}, \mu)$ be a measure space. Let $f, f_n : X \to [0, \infty]$ be $\mathcal{A}$-measurable. Suppose that $f_n \le f_{n+1}$ μ-a.e. in X for any $n \in \mathbb{N}$, and let f be the μ-pointwise limit in X of the sequence $\{f_n\}$. Then*

$$\lim_{n\to\infty} \int_X f_n\,d\mu = \int_X f\,d\mu.$$

Theorem 2.2.2 (Fatou). *Let $(X, \mathcal{A}, \mu)$ be a measure space. Let $f_n : X \mapsto [0, \infty]$ be $\mathcal{A}$-measurable. Then*

$$\int_X \left(\liminf_{n\to\infty} f_n\right) d\mu \le \liminf_{n\to\infty} \int_X f_n\,d\mu.$$

Theorem 2.2.3 (Lebesgue). *Let $(X, \mathcal{A}, \mu)$ be a measure space. Let f, f_n be $\mathcal{A}$-measurable, and let $f_n \to f$ μ-a. e. in X. Suppose that*

$$\text{there exists } g \in L^1(X),\ g \ge 0 \text{ such that } |f_n| \le g\ \mu\text{-a. e. in } X \text{ for all } n \in \mathbb{N}. \tag{2.12}$$

Then $f, f_n \in L^1(X)$, and

$$\lim_{n\to\infty} \int_X |f_n - f|\,d\mu = 0.$$

Theorems 2.2.1 and 2.2.3 are usually referred to as the monotone convergence theorem and dominated convergence theorem, respectively. Using Theorem 2.2.1, we can prove the following result.

Proposition 2.2.4. *Let $(X, \mathcal{A}, \mu)$ be a measure space.*

(i) *Let $f : X \to [0, \infty)$ be $\mathcal{A}$-measurable, and let ν be the measure defined by (2.7). Then for any $\mathcal{A}$-measurable $g : X \mapsto [0, \infty)$, we have*

$$\int_X g\,d\nu = \int_X fg\,d\mu. \tag{2.13}$$

(ii) *Let $f \in L^1(X, \mathcal{A}, \mu)$, and let ν be the finite signed measure defined by (2.8). Then for any $g \in L^1(X, \mathcal{A}, \nu)$, we have $fg \in L^1(X, \mathcal{A}, \mu)$, and (2.13) holds.*

Proof. (i) Equality (2.13) holds by definition when $g = \chi_E$ and thus also when $g = s \in \mathscr{S}_+(X)$. For any $\mathcal{A}$-measurable $g : X \mapsto [0, \infty)$, there exists a sequence $\{s_n\} \subseteq \mathscr{S}_+(X)$ as in (2.5), and hence by Theorem 2.2.1 the claim follows.

(ii) By (2.9)–(2.10) we have

$$\int_X g\,d\nu = \int_X (g^+ - g^-)\,d\nu^+ - \int_X (g^+ - g^-)\,d\nu^-$$
$$= \int_X g^+\,d\nu^+ - \int_X g^-\,d\nu^+ - \int_X g^+\,d\nu^- + \int_X g^-\,d\nu^-. \tag{2.14}$$

From (2.13) we have that $\int_X g^\pm\,d\nu^\pm = \int_X g^\pm f^\pm\,d\mu$; moreover, $\int_X g^\pm\,d\nu^\pm < \infty$ since by assumption $g \in L^1(X, \mathcal{A}, \nu^\pm)$ (see Definition 2.2.4). Then from (2.14) we obtain

$$\int_X g\,d\nu = \int_X (f^+ - f^-)(g^+ - g^-)\,d\mu = \int_X fg\,d\mu,$$

and thus the result follows. □

2.3 Product measures

2.3.1 Tonelli and Fubini theorems

Let $(X_1, \mathcal{A}_1, \mu_1)$ and $(X_2, \mathcal{A}_2, \mu_2)$ be measure spaces. Let $E \in \mathcal{A}_1 \times \mathcal{A}_2$. Thus $E_{x_1} \in \mathcal{A}_2$ for any $x_1 \in X_1$ and $E_{x_2} \in \mathcal{A}_1$ for any $x_2 \in X_2$ (see Proposition 1.2.2). Then the maps from X_1 to $[0, \infty]$, $x_1 \mapsto \mu_2(E_{x_1})$ $(x_1 \in X_1)$, and from X_2 to $[0, \infty]$, $x_2 \mapsto \mu_1(E_{x_2})$ $(x_2 \in X_2)$, are well defined. We have the following:

Proposition 2.3.1. *Let $(X_1, \mathcal{A}_1, \mu_1)$ and $(X_2, \mathcal{A}_2, \mu_2)$ be σ-finite measure spaces. Let $E \in \mathcal{A}_1 \times \mathcal{A}_2$. Then:*

(i) *the map $X_1 \to [0, \infty]$, $x_1 \mapsto \mu_2(E_{x_1})$ $(x_1 \in X_1)$, is $\mathcal{A}_1$-measurable, and the map $X_2 \to [0, \infty]$, $x_2 \mapsto \mu_1(E_{x_2})$ $(x_2 \in X_2)$, is $\mathcal{A}_2$-measurable;*

(ii) *we have*

$$\int_{X_1} \mu_2(E_{x_1})\,d\mu_1 = \int_{X_2} \mu_1(E_{x_2})\,d\mu_2; \tag{2.15a}$$

(iii) *the map $\mu_1 \times \mu_2 : \mathcal{A}_1 \times \mathcal{A}_2 \to [0, \infty]$ defined by*

$$(\mu_1 \times \mu_2)(E) := \int_{X_1} \mu_2(E_{x_1})\,d\mu_1 = \int_{X_2} \mu_1(E_{x_2})\,d\mu_2 \quad (E \in \mathcal{A}_1 \times \mathcal{A}_2) \tag{2.15b}$$

is a σ-finite measure.

Remark 2.3.1. It is easy to produce examples where equality (2.15a) is false if μ_1 and/or μ_2 are not σ-finite.

By Proposition 2.3.1 the following definition is well posed.

Definition 2.3.1. Let $(X_1, \mathcal{A}_1, \mu_1)$ and $(X_2, \mathcal{A}_2, \mu_2)$ be σ-finite measure spaces. The measure $\mu_1 \times \mu_2$ defined by (2.15b) is called a product measure, and $(X_1 \times X_2, \mathcal{A}_1 \times \mathcal{A}_2, \mu_1 \times \mu_2)$ is called a product measure space.

Remark 2.3.2. Let $(X_1 \times X_2, \mathcal{A}_1 \times \mathcal{A}_2, \mu_1 \times \mu_2)$ be a product measure space, and let $P = P(x_1, x_2)$ be some property. Using (2.15b), it is easily seen that the following statements are equivalent: (i) P is true $\mu_1 \times \mu_2$-a. e. in $X_1 \times X_2$; (ii) $P(x_1, \cdot)$ is true μ_2-a. e. in X_2 for μ_1-a. e. $x_1 \in X_1$; (iii) $P(\cdot, x_2)$ is true μ_1-a. e. in X_1 for μ_2-a. e. $x_2 \in X_2$.

Remark 2.3.3. Simple examples show that the product of complete measure spaces need not be complete. However, it can be shown that the Lebesgue space $(\mathbb{R}^{M+N}, \mathcal{L}^{M+N}, \lambda_{M+N})$ is the completion of the product $(\mathbb{R}^{M+N}, \mathcal{L}^M \times \mathcal{L}^N, \lambda_M \times \lambda_N)$ $(M, N \in \mathbb{N}$; see Definition 1.5.1).

Let $(X_1, \mathcal{A}_1, \mu_1)$ and $(X_2, \mathcal{A}_2, \mu_2)$ be measure spaces, and let $f : X_1 \times X_2 \to [0, \infty)$ be $\mathcal{A}_1 \times \mathcal{A}_2$-measurable. By Proposition 2.1.10 the x_1-trace f_{x_1} is $\mathcal{A}_2$-measurable, and thus the function

$$X_1 \to [0, \infty], \quad x_1 \mapsto \int_{X_2} f_{x_1} d\mu_2 \tag{2.16a}$$

is well defined; similarly for the map

$$X_2 \to [0, \infty], \quad x_2 \mapsto \int_{X_1} f_{x_2} d\mu_1. \tag{2.16b}$$

Theorem 2.3.2 (Tonelli). *Let $(X_1, \mathcal{A}_1, \mu_1)$ and $(X_2, \mathcal{A}_2, \mu_2)$ be σ-finite measure spaces. Let $f : X_1 \times X_2 \to [0, \infty)$ be $\mathcal{A}_1 \times \mathcal{A}_2$-measurable. Then:*
(i) *the functions defined in* (2.16) *are $\mathcal{A}_1$- and $\mathcal{A}_2$-measurable, respectively;*
(ii) *we have*

$$\int_{X_1 \times X_2} f \, d(\mu_1 \times \mu_2) = \int_{X_1} d\mu_1 \int_{X_2} f_{x_1} d\mu_2 = \int_{X_2} d\mu_2 \int_{X_1} f_{x_2} d\mu_1. \tag{2.17}$$

Remark 2.3.4. Observe that equality (2.15b) is a particular case of (2.17) with $f = \chi_E$, $E \in \mathcal{A}_1 \times \mathcal{A}_2$. Therefore the assumption in Theorem 2.3.2 that μ_1 and μ_2 are σ-finite cannot be relaxed (see Remark 2.3.1).

For integrable real-valued functions, we have the following:

Theorem 2.3.3 (Fubini). *Let $(X_1, \mathcal{A}_1, \mu_1)$ and $(X_2, \mathcal{A}_2, \mu_2)$ be σ-finite measure spaces. Let $f \in L^1(X_1 \times X_2, \mathcal{A}_1 \times \mathcal{A}_2, \mu_1 \times \mu_2)$. Then:*

(i) $f_{x_1} \in L^1(X_2, \mathcal{A}_2, \mu_2)$ *for* μ_1*-a. e.* $x_1 \in X_1$, *and* $f_{x_2} \in L^1(X_1, \mathcal{A}_1, \mu_1)$ *for* μ_2*-a. e.* $x_2 \in X_2$;

(ii) *the functions from* X_1 *to* $\mathbb{R}$, $x_1 \mapsto \int_{X_2} f_{x_1} d\mu_2$, *and from* X_2 *to* $\mathbb{R}$, $x_2 \mapsto \int_{X_1} f_{x_2} d\mu_1$, *belong to* $L^1(X_1, \mathcal{A}_1, \mu_1)$ *and* $L^1(X_2, \mathcal{A}_2, \mu_2)$, *respectively;*

(iii) *we have*

$$\int_{X_1 \times X_2} f\, d(\mu_1 \times \mu_2) = \int_{X_1} d\mu_1 \int_{X_2} f_{x_1} d\mu_2 = \int_{X_2} d\mu_2 \int_{X_1} f_{x_2} d\mu_1. \tag{2.18}$$

The assumption of integrability of f in Theorem 2.3.3 cannot be relaxed, as simple examples show. Therefore, to apply the Fubini theorem, the following consequence of the Tonelli theorem is expedient.

Theorem 2.3.4. *Let $(X_1, \mathcal{A}_1, \mu_1)$ and $(X_2, \mathcal{A}_2, \mu_2)$ be σ-finite measure spaces. Let $f : X_1 \times X_2 \to \mathbb{R}$ be $\mathcal{A}_1 \times \mathcal{A}_2$-measurable. Then the following statements are equivalent:*

(i) $f \in L^1(X_1 \times X_2, \mathcal{A}_1 \times \mathcal{A}_2, \mu_1 \times \mu_2)$;

(ii) $\int_{X_1} d\mu_1 \int_{X_2} |f_{x_1}|\, d\mu_2 < \infty$;

(iii) $\int_{X_2} d\mu_2 \int_{X_1} |f_{x_2}|\, d\mu_1 < \infty$.

2.4 Applications

2.4.1 A useful equality

Let $(X, \mathcal{A}, \mu)$ be a measure space, and let $f : X \to [0, \infty]$ be $\mathcal{A}$-measurable. Then the function from $[0, \infty)$ to $[0, \infty]$, $t \mapsto \mu(\{f > t\})$, is nonincreasing, and by the Chebychev inequality we have $\lim_{t\to\infty} \mu(\{f > t\}) = 0$ if $f \in L^1(X, \mathcal{A}, \mu)$.

Proposition 2.4.1. *Let $(X, \mathcal{A}, \mu)$ be a σ-finite measure space, and let $f : X \to [0, \infty]$ be $\mathcal{A}$-measurable. Then for any $\alpha > 0$, we have*

$$\int_X f^\alpha\, d\mu = \alpha \int_{[0,\infty)} \mu(\{f > t\})\, t^{\alpha-1} dt. \tag{2.19}$$

Proof. Let us use the Tonelli theorem with $(X_1, \mathcal{A}_1, \mu_1) = (X, \mathcal{A}, \mu)$ and $(X_2, \mathcal{A}_2, \mu_2) = ([0, \infty), \mathcal{B}\cap[0, \infty), \lambda^\phi)$, where λ^ϕ is the Lebesgue–Stieltjes measure on $\mathcal{B}\cap[0, \infty)$ associated with $\phi(t) = t^\alpha$ $(t \geq 0)$. It is easily seen that the set $E := \{(x, t) \in X \times [0, \infty) \mid f(x) > t\}$ is $\mathcal{A}_1 \times \mathcal{A}_2$-measurable. Then from (2.17) with $f = \chi_E$ we get

$$\alpha \int_{[0,\infty)} \mu(\{f > t\})\, t^{\alpha-1} dt = \int_{[0,\infty)} \mu(\{f > t\})\, d\lambda^\phi(t)$$

$$= \int_{[0,\infty)} d\lambda^{\phi}(t) \int_X \chi_{\{f(x)>t\}} \, d\mu(x)$$

$$= \int_X d\mu(x) \int_{[0,\infty)} \chi_{\{f(x)>t\}} \, d\lambda^{\phi}(t)$$

$$= \int_X d\mu(x) \lambda^{\phi}(\{f(x) > 0\}) = \int_X f^{\alpha}(x) \, d\mu(x). \qquad \square$$

2.4.2 Steiner symmetrization

For any $x \in \mathbb{R}^N$, set $x \equiv (x_1, \hat{x})$ with $x_1 \in \mathbb{R}$ and $\hat{x} \in \mathbb{R}^{N-1}$. Let E be a Borel subset of $\mathbb{R}^N$, and for any $\hat{x} \in \mathbb{R}^{N-1}$, let $E_{\hat{x}} := \{x_1 \in \mathbb{R} \mid (x_1, \hat{x}) \in E\}$ be its $\hat{x}$-section. By Proposition 1.2.2 the function

$$f : \mathbb{R}^{N-1} \to [0, \infty], \quad f(\hat{x}) := \lambda(E_{\hat{x}}), \tag{2.20}$$

is well defined. Consider the symmetric interval $I_{\hat{x}} := (-\frac{f(\hat{x})}{2}, \frac{f(\hat{x})}{2})$ if $f(\hat{x}) \neq 0$ and $I_{\hat{x}} := \emptyset$ otherwise, and define

$$S_1(E) := \bigcup_{\hat{x} \in \mathbb{R}^{N-1}} (I_{\hat{x}} \times \{\hat{x}\}) = \left\{(x_1, \hat{x}) \in \mathbb{R}^N \mid |x_1| \leq \frac{f(\hat{x})}{2}\right\}. \tag{2.21}$$

Definition 2.4.1. The set $S_1(E)$ defined in (2.21) is called the *Steiner symmetrization of E with respect to the plane* $x_1 = 0$. The Steiner symmetrization of E with respect to the plane $x_j = 0$ ($j = 2, \dots, N$) is similarly defined. The *Steiner symmetrization of E with respect to the origin* is

$$S(E) := S_N(S_{N-1}(\dots S_1(E) \dots)). \tag{2.22}$$

Some relevant properties of the set $S_1(E)$ are proven in the following lemma.

Lemma 2.4.2. *Let $E \in \mathcal{B}^N$, and let $S_1(E)$ be defined by* (2.21). *Then:*
(i) $(x_1, \hat{x}) \in S_1(E) \Rightarrow (-x_1, \hat{x}) \in S_1(E)$;
(ii) $S_1(E) \in \mathcal{B}^N$;
(iii) $\lambda_N(S_1(E)) = \lambda_N(E)$;
(iv) $\operatorname{diam}(S_1(E)) \leq \operatorname{diam}(E)$.

Proof. (i) is clear from (2.21). To prove (ii), let $\{s_k\} \subseteq \mathcal{S}_+(\mathbb{R}^{N-1})$ be a nondecreasing sequence of simple functions such that $s_k \to f$ pointwise in $\mathbb{R}^{N-1}$ as $k \to \infty$, f being defined by (2.20). For any $k \in \mathbb{N}$, define

$$g_k : \mathbb{R}^N \mapsto \mathbb{R}, \quad g_k(x_1, \hat{x}) := \frac{s_k(\hat{x})}{2} - |x_1|.$$

Clearly, g_k is $\mathcal{B}^N$-measurable, and hence $F_k := \{g_k > 0\} \in \mathcal{B}^N$ ($k \in \mathbb{N}$). Since $s_k \le s_{k+1}$ for all $k \in \mathbb{N}$, it is easily checked that the sequence $\{F_k\}$ is nondecreasing and $S_1(E) = \bigcup_{k=1}^{\infty} F_k$. Then (ii) follows. Concerning (iii), it suffices to observe that by (2.15b) (see also Remark 2.3.3) we have

$$\lambda_N(E) = \int_{\mathbb{R}^{N-1}} \lambda(E_{\hat{x}})\, d\lambda_{N-1}(\hat{x}) = \int_{\mathbb{R}^{N-1}} f(\hat{x})\, d\lambda_{N-1}(\hat{x}) = \lambda_N(S_1(E)).$$

Let us finally prove (iv). Suppose $\operatorname{diam}(E) < \infty$; otherwise, the conclusion is obvious. Let $E_{\hat{x}} \neq \emptyset$, and define $K_{\hat{x}} := [\inf E_{\hat{x}}, \sup E_{\hat{x}}]$. Clearly, we have

$$\lambda(K_{\hat{x}}) = \sup E_{\hat{x}} - \inf E_{\hat{x}} \ge \lambda(E_{\hat{x}}) = f(\hat{x}). \tag{2.23}$$

Let $(x_1, \hat{x}), (y_1, \hat{y}) \in S_1(E)$ be such that

$$\operatorname{diam}(S_1(E)) \le \sqrt{|x_1 - y_1|^2 + |\hat{x} - \hat{y}|^2} + \epsilon. \tag{2.24}$$

It is not restrictive to assume that

$$\sup E_{\hat{y}} - \inf E_{\hat{x}} \ge \sup E_{\hat{x}} - \inf E_{\hat{y}}. \tag{2.25}$$

Since $|x_1| \le \frac{f(\hat{x})}{2}$ and $|y_1| \le \frac{f(\hat{y})}{2}$, by (2.23) and (2.25) we clearly get

$$\begin{aligned} |x_1 - y_1| &\le \frac{f(\hat{x}) + f(\hat{y})}{2} \le \frac{\sup E_{\hat{x}} - \inf E_{\hat{x}}}{2} + \frac{\sup E_{\hat{y}} - \inf E_{\hat{y}}}{2} \\ &= \frac{\sup E_{\hat{y}} - \inf E_{\hat{x}}}{2} + \frac{\sup E_{\hat{x}} - \inf E_{\hat{y}}}{2} \le \sup E_{\hat{y}} - \inf E_{\hat{x}}. \end{aligned} \tag{2.26}$$

Then by (2.24) and (2.26) we have

$$[\operatorname{diam}(S_1(E)) - \epsilon]^2 \le |x_1 - y_1|^2 + |\hat{x}_1 - \hat{y}_1|^2 \le |\hat{x}_1 - \hat{y}_1|^2 + [\sup E_{\hat{y}} - \inf E_{\hat{x}}]^2 \le [\operatorname{diam}(E)]^2,$$

whence by the arbitrariness of ϵ claim (iv) follows. This completes the proof. □

By Lemma 2.4.2 we have the following result.

Theorem 2.4.3 (Steiner symmetrization). *Let $E \in \mathcal{B}^N$, and let $S(E)$ be its Steiner symmetrization with respect to the origin. Then:*
(i) $x \in S(E) \Rightarrow -x \in S(E)$;
(ii) $S(E) \in \mathcal{B}^N$;
(iii) $\lambda_N(S(E)) = \lambda_N(E)$;
(iv) $\operatorname{diam}(S(E)) \le \operatorname{diam}(E)$.

As an application of Theorem 2.4.3, we can now prove the isodiametric inequality (see Proposition 1.5.2).

Proof of Proposition 1.5.2. If $\operatorname{diam}(E) = \infty$, then the conclusion is obvious, thus suppose $E \subseteq \mathbb{R}^N$ to be bounded. Moreover, by Remark 1.5.1(ii) it is not restrictive to assume that $E \in \mathcal{B}^N$. Let $S(E)$ be defined by (2.22). By Theorem 2.4.3(i, iv) $S(E)$ is symmetric with respect to the origin, and $\operatorname{diam}(S(E)) \leq \operatorname{diam}(E)$, and hence $S(E) \subseteq B(0, \frac{\operatorname{diam}(E)}{2})$. Then by Theorem 2.4.3(iii) we have

$$\lambda_N(E) = \lambda_N(S(E)) \leq \kappa_N \left(\frac{\operatorname{diam}(E)}{2} \right)^N$$

with κ_N given by (1.16). Then the conclusion follows. □

2.5 Young measure

Let $(X, \mathcal{A}, \mu)$ be a measure space, and let $(X', \mathcal{A}')$ be a measurable space. Let $f : X \to X'$ be $(\mathcal{A}, \mathcal{A}')$-measurable. Using (2.2), we immediately see that the map

$$\mu_f : \mathcal{A}' \to [0, \infty], \quad \mu_f(E') := \mu(f^{-1}(E')) \quad \text{for } E' \in \mathcal{A}', \tag{2.27}$$

is a measure.

Definition 2.5.1. The measure μ_f defined by (2.27) is called the *image* of μ under f.

Let $(X_1 \times X_2, \mathcal{A}_1 \times \mathcal{A}_2)$ be the product of two measurable spaces $(X_1, \mathcal{A}_1)$ and $(X_2, \mathcal{A}_2)$, and let $\mu : \mathcal{A}_1 \times \mathcal{A}_2 \mapsto [0, \infty]$ be a measure. Then by Lemma 2.1.8 the image measures

$$\mu_{p_1} : \mathcal{A}_1 \to [0, \infty], \quad \mu_{p_1}(E_1) := \mu(p_1^{-1}(E_1)) = \mu(E_1 \times X_2) \quad \text{for } E_1 \in \mathcal{A}_1, \tag{2.28a}$$

$$\mu_{p_2} : \mathcal{A}_2 \to [0, \infty], \quad \mu_{p_2}(E_2) := \mu(p_2^{-1}(E_2)) = \mu(X_1 \times E_2) \quad \text{for } E_2 \in \mathcal{A}_2, \tag{2.28b}$$

are well defined.

Definition 2.5.2. The measure μ_{p_i} defined by (2.28) is called the *projection* of μ onto X_i $(i = 1, 2)$.

Definition 2.5.3. Let X and Y be locally compact Hausdorff spaces, and let $\mu \in \mathfrak{R}_f^+(X)$.
(i) Any measure $\nu \in \mathfrak{R}_f^+(X \times Y)$ such that $\nu(E \times Y) = \mu(E)$ for all $E \in \mathcal{B}(X)$ is called the *Young measure*.
(ii) Let $f : X \to Y$ be $(\mathcal{B}(X), \mathcal{B}(Y))$-measurable. The Young measure ν such that

$$\nu(E \times F) = \mu(E \cap f^{-1}(F)) \quad \text{for all } E \in \mathcal{B}(X), F \in \mathcal{B}(Y) \tag{2.29}$$

is called the *Young measure associated with* f.

By definition we have $\nu_{p_X} = \mu$, p_X denoting the projection of $X \times Y$ on X. We denote the set of Young measures by $\mathfrak{Y}^+(X, \mathcal{B}(X), \mu; Y) \equiv \mathfrak{Y}^+(X; Y)$.

Remark 2.5.1. It is easily checked that $\mathfrak{Y}^+(X;Y)$ is a bounded subset of $\mathfrak{R}_f^+(X \times Y)$. In fact, for any $\nu \in \mathfrak{Y}^+(X;Y)$, we have

$$\|\nu\|_{\mathfrak{R}_f(X\times Y)} = \nu(X \times Y) = \mu(X).$$

Hence $\mathfrak{Y}^+(X;Y)$ is contained in the sphere $\partial B(0, \|\mu\|_{\mathfrak{R}_f(X)}) \subseteq \mathfrak{R}_f^+(X \times Y)$.

Remark 2.5.2. According to (2.29), the Young measure associated with f is concentrated on the graph of f, i. e., on the set $\{(x,f(x)) \mid x \in X\}$ (see Definition 1.8.6). It is also easily seen that if $f_1, f_2 : X \to Y$ are $(\mathcal{B}(X), \mathcal{B}(Y))$-measurable, and ν_1, ν_2 are the associated Young measures, we have $\nu_1 = \nu_2$ if and only if $f_1 = f_2$ μ-a. e. in X. In fact, if $\nu_1 = \nu_2$, then we have

$$\mu(X) = \nu_2(\{(x,f_2(x)) \mid x \in X\}) = \nu_1(\{(x,f_2(x)) \mid x \in X\}) = \mu(\{x \in X \mid f_1(x) = f_2(x)\}),$$

and thus $f_1 = f_2$ μ-a. e. in X. The opposite implication immediately follows from (2.29).

Let us finally mention the following result.

Lemma 2.5.1. *Let $(X, \mathcal{A}, \mu)$ be a measure space, and let $(X', \mathcal{A}')$ be a measurable space. Let $f : X \to X'$ be $(\mathcal{A}, \mathcal{A}')$-measurable, and let μ_f be the image of μ under f. Then for any $g \in L^1(X', \mathcal{A}', \mu_f)$, we have $g \circ f \in L^1(X, \mathcal{A}, \mu)$, and*

$$\int_F g \, d\mu_f = \int_{f^{-1}(F)} g \circ f \, d\mu \quad \textit{for any } F \in \mathcal{A}'. \tag{2.30}$$

Proof. Let $s \in \mathscr{S}(X')$, that is, $s = \sum_{i=1}^n c_i \chi_{E_i'}$ with $c_1, \dots, c_n \in \mathbb{R}$ and partition $\{E_1', \dots, E_n'\} \subseteq \mathcal{A}'$ of X'. Since $s \circ f = \sum_{i=1}^n c_i \chi_{f^{-1}(E_i')}$, equality (2.30) with $g = s$ immediately follows from (2.27). Then by the denseness of $\mathscr{S}(X')$ in $L^1(X', \mathcal{A}', \mu_f)$ (see Remark 2.2.4) the result follows. □

2.6 Riesz representation theorem: positive functionals

Let X, Y be normed vector spaces. By X^* we denote the dual space of X with norm

$$X^* \ni x^* \mapsto \|x^*\|_{X^*} := \sup_{\|x\|_X \le 1} |\langle x^*, x\rangle_{X^*,X}|,$$

where $\langle \cdot, \cdot \rangle_{X^*,X}$ denotes the duality map between X and X^*,

$$\langle \cdot, \cdot \rangle_{X^*,X} : X^* \times X \to \mathbb{R}, \quad (x^*, x) \mapsto \langle x^*, x\rangle_{X^*,X} .$$

Let $(X, \mathcal{T})$ be a Hausdorff space, and let μ be a Borel measure on X. For any $f \in C_c(X)$, we have $|f| \leq \|f\|_\infty \chi_{\operatorname{supp} f}$, and thus $C_c(X) \subseteq L^1_{\mathrm{loc}}(X, \mathcal{B}, \mu)$. Hence the map

$$F : C_c(X) \to \mathbb{R}, \quad \langle F, f \rangle := \int_X f \, d\mu,$$

is a linear functional on $C_c(X)$, which is positive in the following sense.

Definition 2.6.1. A linear functional $F : C_c(X) \mapsto \mathbb{R}$ is called *positive* if $\langle F, f \rangle \geq 0$ for any $f \in C_c(X), f \geq 0$.

Conversely, the following theorem proves that every positive linear functional $F : C_c(X) \to \mathbb{R}$, X locally compact, can be represented as $\langle F, f \rangle = \int_X f \, d\mu$, where μ is a suitable *Radon* measure.

Theorem 2.6.1 (Riesz). *Let X be a locally compact Hausdorff space, and let $F : C_c(X) \to \mathbb{R}$ be a positive linear functional. Then there exists a unique Radon measure μ on X such that*

$$\langle F, f \rangle = \int_X f \, d\mu \quad \text{for any } f \in C_c(X). \tag{2.31}$$

Moreover, μ has the following properties:

(i) *we have*

$$\mu(A) = \sup\{\langle F, f \rangle \mid f \in C_c(X), f(X) \subseteq [0,1], \operatorname{supp} f \subseteq A\} \quad \text{for any open } A \subseteq X, A \neq \emptyset, \tag{2.32}$$

$$\mu(K) = \inf\{\langle F, f \rangle \mid f \in C_c(X), f \geq \chi_K\} \quad \text{for any compact } K \subseteq X; \tag{2.33}$$

(ii) *μ is finite if and only if F is bounded. In such case, we have $\mu(X) = \|F\|$, where*

$$\|F\| := \sup\{|\langle F, f \rangle| \mid f \in C_c(X), \|f\|_\infty = 1\}.$$

For simplicity, we will prove Theorem 2.6.1 under the additional assumption that X has a countable basis (see [45] for the general case). The existence statement relies on the following lemma.

Lemma 2.6.2. *Let X be a locally compact Hausdorff space with countable basis, and let $F : C_c(X) \to \mathbb{R}$ be a positive linear functional. Then there exists a Radon measure μ on X such that* (2.32)–(2.33) *hold.*

Proof. Set $\mu_0(\emptyset) := 0$ and

$$\mu_0(A) := \sup\{\langle F, f \rangle \mid f \in C_c(X), f(X) \subseteq [0,1], \operatorname{supp} f \subseteq A\} \tag{2.34}$$

for any nonempty $A \in \mathcal{T}$. For any $E \subseteq X$, define

$$\mu^*(E) := \inf\left\{\sum_{n=1}^{\infty} \mu_0(A_n) \mid E \subseteq \bigcup_{n=1}^{\infty} A_n, \{A_n\} \subseteq \mathcal{T}\right\} \tag{2.35}$$

(observe that a countable cover of any $E \subseteq X$ with open sets exists, since by assumption X has a countable basis). By Theorem 1.4.2 the map μ^* is an outer measure on X. Then by Theorem 1.4.1 the family of μ^*-measurable sets is a σ-algebra, and the restriction μ of μ^* on this σ-algebra is a complete measure on X.

Let us prove that every $A \in \mathcal{T}$ is μ^*-measurable, that is,

$$\mu^*(Z) \geq \mu^*(Z \cap A) + \mu^*(Z \setminus A) \quad \text{for any } Z \subseteq X \tag{2.36}$$

(see Remark 1.4.1). Since $\mathcal{B} = \sigma_0(\mathcal{T})$, from this it will follow that every Borel set is μ^*-measurable, and thus μ is defined on $\mathcal{B}$. Moreover, since $\mu^*(A) = \mu_0(A)$ for every $A \in \mathcal{T}$ (see (2.35)), equality (2.32) will follow.

Let $\mu^*(Z) < \infty$ (otherwise, (2.36) is obvious), and suppose first that $Z \in \mathcal{T}$. Then $Z \cap A \in \mathcal{T}$, and thus $\mu^*(Z \cap A) = \mu_0(Z \cap A)$. Hence by the definition of μ_0, for any $\epsilon > 0$, there exists $f \in C_c(X)$ with $f(X) \subseteq [0, 1]$ and $\operatorname{supp} f \subseteq Z \cap A$ such that

$$\langle F, f\rangle > \mu_0(Z \cap A) - \epsilon. \tag{2.37}$$

Similarly, since $Z \setminus \operatorname{supp} f \in \mathcal{T}$, we have $\mu^*(Z \setminus \operatorname{supp} f) = \mu_0(Z \setminus \operatorname{supp} f)$, and thus for any $\epsilon > 0$, there exists $g \in C_c(X)$ with $g(X) \subseteq [0, 1]$ and $\operatorname{supp} g \subseteq Z \setminus \operatorname{supp} f$ such that

$$\langle F, g\rangle > \mu_0(Z \setminus \operatorname{supp} f) - \epsilon. \tag{2.38}$$

In particular, we have $f + g \in C_c(X)$ with $(f + g)(X) \subseteq [0, 1]$ and $\operatorname{supp}(f + g) \subseteq Z$. Then by (2.37)–(2.38) we get

$$\begin{aligned}
\mu^*(Z) = \mu_0(Z) &\geq \langle F, f + g\rangle = \langle F, f\rangle + \langle F, g\rangle \\
&\geq \mu_0(Z \cap A) + \mu_0(Z \setminus \operatorname{supp} f) - 2\epsilon = \mu^*(Z \cap A) + \mu^*(Z \setminus \operatorname{supp} f) - 2\epsilon \\
&\geq \mu^*(Z \cap A) + \mu^*(Z \setminus A) - 2\epsilon,
\end{aligned}$$

and thus by the arbitrariness of ϵ we get (2.36) when Z is open. In the general case, by (2.35) for any $\epsilon > 0$, there exists $W \in \mathcal{T}$ such that $Z \subseteq W$ and $\mu_0(W) < \mu^*(Z) + \epsilon$. Then by the above considerations and the monotonicity of μ^* we have

$$\begin{aligned}
\mu^*(Z) &> \mu_0(W) - \epsilon = \mu^*(W) - \epsilon \\
&\geq \mu^*(W \cap A) + \mu^*(W \setminus A) - \epsilon \geq \mu^*(Z \cap A) + \mu^*(Z \setminus A) - \epsilon,
\end{aligned}$$

and thus the claim follows.

Let us now prove (2.33). Fix any $K \in \mathcal{K}$, and let $f \in C_c(X), f \geq \chi_K$. For any $\epsilon \in (0,1)$, set $A_\epsilon := \{f > 1-\epsilon\}$. Thus A_ϵ is open, and $K \subseteq A_\epsilon$. Let $g \in C_c(X)$, $g(X) \subseteq [0,1]$, and $\operatorname{supp} g \subseteq A_\epsilon$. Hence we have $(1-\epsilon)g(x) \leq f(x)$ for any $x \in X$, and thus $\langle F, g\rangle \leq \frac{1}{1-\epsilon}\langle F, f\rangle$. Moreover, since every $K \in \mathcal{K}$ is μ^*-measurable, we have that $\mu^*(K) = \mu(K)$. Then by the monotonicity of μ^* and (2.34) we get

$$\begin{aligned}\mu(K) &= \mu^*(K) \leq \mu^*(A_\epsilon) = \mu_0(A_\epsilon) \\ &= \sup\{\langle F, f\rangle \mid g \in C_c(X), g(X) \subseteq [0,1], \operatorname{supp} g \subseteq A_\epsilon\} \leq \frac{1}{1-\epsilon}\langle F, f\rangle, \end{aligned} \tag{2.39}$$

whence by the arbitrariness of f and $\epsilon \in (0,1)$ we get

$$\mu(K) \leq \inf\{\langle F, f\rangle \mid f \in C_c(X), f \geq \chi_K\}. \tag{2.40}$$

Now observe that from (2.39) with $\epsilon = \frac{1}{2}$ we get $\mu(K) \leq \langle F, 2f\rangle < \infty$ for any $K \in \mathcal{K}$, and thus clearly μ is locally finite (see Remark 1.3.2(iv); recall that by assumption X is locally compact). Since μ is defined on a σ-algebra containing $\mathcal{B}$, it follows that $\mu \in \mathfrak{B}^+(X)$, and thus by Proposition 1.3.3 it is regular. In particular, μ is inner regular and thus is a Radon measure.

Now we can prove the inverse inequality of (2.40). Since μ is outer regular, for any $K \in \mathcal{K}$ and $\epsilon > 0$, there exists $A \in \mathcal{T}$ such that $K \subseteq A$ and $\mu(A) < \mu(K) + \epsilon$. Then by Lemma A.9 there exists $\hat{f} \in C_c(X)$ such that $\hat{f}(X) \subseteq [0,1]$, $\hat{f}|_K = 1$, thus $\hat{f} \geq \chi_K$, and $\operatorname{supp}\hat{f} \subseteq A$. By (2.32) it follows that

$$\inf\{\langle F, f\rangle \mid f \in C_c(X), f \geq \chi_K\} \leq \langle F, \hat{f}\rangle \leq \mu(A) < \mu(K) + \epsilon,$$

whence by the arbitrariness of ϵ

$$\inf\{\langle F, f\rangle \mid f \in C_c(X), f \geq \chi_K\} \leq \mu(K). \tag{2.41}$$

By (2.40)–(2.41) we obtain (2.33), and the result follows. □

Proof of Theorem 2.6.1. (i) Let $f \in C_c(X)$; it is not restrictive to suppose $f(X) \subseteq [0,1]$. Set $K_0 := \operatorname{supp} f$, $K_{j,n} := \{f \geq \frac{j}{n}\}$ for any $n \in \mathbb{N}$ and $j = 1, \dots, n$. Clearly, every set $K_{j,n}$ is compact, and we have $K_{n,n} = \{f = 1\} \subseteq \cdots \subseteq K_{j,n} \subseteq K_{j-1,n} \subseteq \cdots \subseteq K_0$. For any $x \in X$, set

$$f_{j,n}(x) := \begin{cases} \frac{1}{n} & \text{if } x \in K_{j,n}, \\ f(x) - \frac{j-1}{n} & \text{if } x \in K_{j-1,n} \setminus K_{j,n}, \\ 0 & \text{if } x \in X \setminus K_{j-1,n}. \end{cases}$$

Then for all j, n, as above, we have $f_{j,n} \in C_c(X)$, $f_{j,n}(X) \subseteq [0,1]$, $\sum_{j=1}^n f_{j,n} = f$, and

$$\frac{\chi_{K_{j,n}}}{n} \leq f_{j,n} \leq \frac{\chi_{K_{j-1,n}}}{n} \quad \Rightarrow \quad \frac{1}{n}\sum_{j=1}^n \chi_{K_{j,n}} \leq f \leq \frac{1}{n}\sum_{j=1}^n \chi_{K_{j-1,n}}. \tag{2.42}$$

Let μ be the Radon measure given by Lemma 2.6.2. Then from (2.42), for every $n \in \mathbb{N}$, we obtain

$$\frac{1}{n}\sum_{j=1}^{n}\mu(K_{j,n}) \leq \int_X f d\mu \leq \frac{1}{n}\sum_{j=1}^{n}\mu(K_{j-1,n}). \tag{2.43}$$

Let us prove that also

$$\frac{1}{n}\sum_{j=1}^{n}\mu(K_{j,n}) \leq \langle F,f\rangle \leq \frac{1}{n}\sum_{j=1}^{n}\mu(K_{j-1,n}) \quad (n \in \mathbb{N}). \tag{2.44}$$

Indeed, from the first inequality in (2.42) we get $nf_{j,n} \geq \chi_{K_{j,n}}$, and hence by (2.33)

$$\mu(K_{j,n}) \leq \langle F, nf_{j,n}\rangle = n\langle F,f_{j,n}\rangle \quad \Rightarrow \quad \frac{\mu(K_{j,n})}{n} \leq \langle F,f_{j,n}\rangle.$$

Moreover, we have $\operatorname{supp} f_{j,n} \subseteq K_{j-1,n}$ by definition and $nf_{j,n} \leq \chi_{K_{j-1,n}}$ by (2.42). On the other hand, since by Lemma 1.3.2(ii) the compact set $K_{j-1,n}$ is μ-outer regular, there exists a decreasing sequence $\{A_m\} \subseteq \mathcal{T}$ such that $K_{j-1,n} \subseteq A_m$ for any $m \in \mathbb{N}$ and $\mu(K_{j-1,n}) = \lim_{m\to\infty}\mu(A_m)$. Then by (2.32) we obtain that

$$\langle F, nf_{j,n}\rangle = n\langle F,f_{j,n}\rangle \leq \mu(A_m) \quad \forall m \in \mathbb{N} \quad \Rightarrow \quad \langle F,f_{j,n}\rangle \leq \frac{\mu(K_{j-1,n})}{n}.$$

Therefore, for any $n \in \mathbb{N}$ and $j = 1, \dots, n$, we have

$$\frac{\mu(K_{j,n})}{n} \leq \langle F,f_{j,n}\rangle \leq \frac{1}{n}\mu(K_{j-1,n}),$$

whence (2.44) follows.

From (2.43)–(2.44) we obtain

$$\left|\langle F,f\rangle - \int_X f d\mu\right| \leq \frac{1}{n}\sum_{j=1}^{n}[\mu(K_{j-1,n}) - \mu(K_{j,n})] = \frac{\mu(K_0) - \mu(K_{n,n})}{n} \leq \frac{\mu(K_0)}{n}.$$

Letting $n \to \infty$ in this inequality gives (2.31).

Let us prove the uniqueness statement. To this purpose, it suffices to show that every Radon measure μ such that (2.31) holds also satisfies (2.33) (in fact, by inner regularity two Radon measures that coincide on $\mathcal{K}$ coincide on X). Let (2.31) hold. Then for any $K \in \mathcal{K}$ and any $f \in C_c(X)$, $f \geq \chi_K$ we have

$$\langle F,f\rangle = \int_X f\, d\mu \geq \mu(K) \quad \Rightarrow \quad \mu(K) \leq \inf\{\langle F,f\rangle \mid f \in C_c(X), f \geq \chi_K\}.$$

Conversely, by Lemma 1.3.2(ii) every $K \in \mathcal{K}$ is μ-outer regular, and thus for any $\epsilon > 0$, there exists $A \in \mathcal{T}$ such that $K \subseteq A$ and $\mu(A) < \mu(K) + \epsilon$. Then by Lemma A.9 there exists $\bar{f} \in C_c(X)$ such that $\chi_K \leq \bar{f} \leq \chi_A$. Therefore, for any $\epsilon > 0$, we have

$$\langle F, \bar{f} \rangle = \int_X \bar{f}\, d\mu \leq \mu(A) < \mu(K) + \epsilon,$$

whence by the arbitrariness of ϵ

$$\mu(K) \geq \inf\{\langle F, f \rangle \mid f \in C_c(X), f \geq \chi_K\}.$$

Hence the claim follows.

(ii) If μ is finite, then from (2.31) we get

$$|\langle F, f \rangle| = \left|\int_X f\, d\mu\right| \leq \|f\|_\infty \, \mu(X) \quad \text{for any } f \in C_c(X),$$

and thus F is bounded, and $\|F\| \leq \mu(X)$. Conversely, if F is bounded, then by (2.31) we have

$$\left|\int_X f\, d\mu\right| = |\langle F, f \rangle| \leq \|F\| \|f\|_\infty \quad \text{for any } f \in C_c(X).$$

Then for any $K \in \mathcal{K}$ and $f \in C_c(X)$ such that $f(X) \subseteq [0,1]$ and $f \geq \chi_K$, we have $\mu(K) \leq \|F\|$, whence by the inner regularity of μ

$$\mu(X) = \sup\{\mu(K) \mid K \in \mathcal{K}\} \leq \|F\|.$$

This completes the proof. □

Let us prove a companion result of Theorem 2.6.1 for the space $C_0(X)$.

Theorem 2.6.3. *Let X be a locally compact Hausdorff space, and let $F : C_0(X) \to \mathbb{R}$ be a positive linear functional. Then there exists a unique Radon measure μ on X such that $C_0(X) \subseteq L^1(X, \mathcal{B}, \mu)$ and*

$$\langle F, f \rangle := \int_X f\, d\mu \quad \text{for all } f \in C_0(X). \tag{2.45}$$

Moreover, μ is finite and satisfies (2.32)–(2.33).

For the proof, we need the following lemma of independent interest.

Lemma 2.6.4. *Let X be a locally compact Hausdorff space. Then every positive linear functional $F : C_0(X) \to \mathbb{R}$ is bounded.*

Proof. By contradiction let F be unbounded. Then there exists a sequence $\{f_n\} \subseteq C_0(X)$ such that $\|f_n\|_\infty = 1$ and $|\langle F, f_n\rangle| \geq n^3$ for all $n \in \mathbb{N}$. It is not restrictive to assume that $f_n \geq 0$, since $|\langle F, f_n\rangle| \leq \langle F, |f_n|\rangle$ $(n \in \mathbb{N})$. Since $\sum_{n=1}^\infty \frac{\|f_n\|_\infty}{n^2} < \infty$, the series $\sum_{n=1}^\infty \frac{f_n}{n^2}$ converges in $C_0(X)$. Set $f := \sum_{n=1}^\infty \frac{f_n}{n^2}$. Then $\frac{f_n}{n^2} \leq f$ for all $n \in \mathbb{N}$, and thus $n \leq \frac{\langle F, f_n\rangle}{n^2} \leq \langle F, f\rangle < \infty$, a contradiction. □

Proof of Theorem 2.6.3. By Theorem 2.6.1(i) there exists a unique Radon measure μ such that the restriction $\hat{F} := F|_{C_c(X)}$ can be represented as in (2.31), and (2.32)–(2.33) hold. By Lemma 2.6.4 F is bounded, and thus $\hat{F}$ is also bounded. Hence by Theorem 2.6.1(ii) the measure μ is finite, and thus the linear functional

$$G : C_c(X) \to \mathbb{R}, \quad \langle G, f\rangle := \int_X f \, d\mu,$$

is well defined and bounded. Since $C_c(X)$ is dense in $C_0(X)$ and $F = G$ on $C_c(X)$, the result follows. □

It is useful to extend the above considerations to the space $C_b(X)$. The following analogue of Lemma 2.6.2 can be proven (see [45, Lemma VIII.2.11]).

Lemma 2.6.5. *Let X be a locally compact Hausdorff space. Then for any positive linear functional $F : C_b(X) \to \mathbb{R}$, there exists $\mu \in \mathfrak{R}_f^+(X)$ such that*

$$\int_X f \, d\mu \leq \langle F, f\rangle \quad \textit{for all } f \in C_b(X), f \geq 0. \tag{2.46}$$

In particular, we have $C_b(X) \subseteq L^1(X, \mathcal{B}(X), \mu)$.

The proof of Lemma 2.6.5 requires a suitable modification of that of Lemma 2.6.2. Let us only mention that

$$\mu(A) = \sup\{\mu_0(K) \mid K \text{ compact}, \ K \subseteq A\} \quad \text{for any open } A \subseteq X, A \neq \emptyset, \tag{2.47}$$

where

$$\mu_0(K) := \inf\{\langle F, f\rangle \mid f \in C_c(X), f \geq \chi_K\} \quad \text{for any compact } K \subseteq X. \tag{2.48}$$

Now we can prove the following result.

Proposition 2.6.6. *Let X be a locally compact Hausdorff space. Let $F : C_b(X) \to \mathbb{R}$ be linear and positive, and let $\mu \in \mathfrak{R}_f^+(X)$ be given by Lemma 2.6.5. Then the following statements are equivalent:*

(i) *we have*

$$\langle F, f\rangle = \int_X f \, d\mu \quad \textit{for any } f \in C_b(X); \tag{2.49}$$

(ii) *for any* $\epsilon > 0$, *there exists a compact subset* $K \subseteq X$ *such that* $\langle F, f\rangle < \epsilon$ *for all* $f \in C_b(X)$ *such that* $f(X) \subseteq [0,1]$ *and* $f|_K = 0$.

If any of these statements holds, then μ *is the unique element of* $\mathfrak{R}_f^+(X)$ *such that* (2.49) *is satisfied.*

Proof. The uniqueness statement follows from (2.49) and Theorem 2.6.3.

(i)⟹(ii). Choosing $f = 1$ in (2.49) gives $\langle F, 1\rangle = \mu(X)$. Let $\epsilon > 0$ be fixed. By Proposition A.2 X is σ-compact, and thus there exists a compact subset $K \subseteq X$ such that $\mu(K) > \mu(X) - \epsilon$. Then using inequality (2.46), for all $f \in C_b(X)$ with $f(X) \subseteq [0,1]$ and $f|_K = 0$, we get

$$\mu(X) - \langle F, f\rangle = \langle F, 1\rangle - \langle F, f\rangle = \langle F, 1 - f\rangle \geq \mu(K) > \mu(X) - \epsilon,$$

whence the claim follows.

(ii)⟹(i). by (2.46) we have $\mu(X) \leq \langle F, 1\rangle$. Let us prove the inverse inequality. Let $\epsilon > 0$ be fixed. Then by assumption there exists a compact subset $K \subseteq X$ such that $\langle F, f\rangle < \epsilon$ for all $f \in C_b(X)$ such that $f(X) \subseteq [0,1]$ and $f|_K = 0$. By (2.47)–(2.48) and Lemma 1.3.2(ii) there exists $g \in C_b(X)$ such that $g(X) \subseteq [0,1]$, $g|_K = 1$, and $\langle F, g\rangle < \mu(K) + \epsilon$. Then we have $\langle F, 1 - g\rangle < \epsilon$, whence

$$\langle F, 1\rangle = \langle F, g\rangle + \langle F, 1 - g\rangle < \mu(K) + 2\epsilon \leq \mu(X) + 2\epsilon.$$

By the arbitrariness of ϵ we obtain that $\langle F, 1\rangle \leq \mu(X)$. It follows that $\langle F, 1\rangle = \mu(X)$.

We can now prove equality (2.49) for any $f \in C_b(X)$ such that $f(X) \subseteq [0,1]$ (whence the general case plainly follows). Applying inequality (2.46) to the function $1 - f$, we get

$$\mu(X) - \int_X f\, d\mu = \int_X (1 - f)\, d\mu \leq \langle F, 1 - f\rangle = \langle F, 1\rangle - \langle F, f\rangle = \mu(X) - \langle F, f\rangle,$$

whence $\int_X f\, d\mu \geq \langle F, f\rangle$. Combining the above inequality with (2.46) proves the result. □

2.7 Riesz representation theorem: bounded functionals

Let X be a locally compact Hausdorff space. Consider the Banach space $C_0(X)$ endowed with the supremum norm and the Banach space $\mathfrak{R}_f(X)$ of finite signed Radon measures on X endowed with the norm

$$\mu \mapsto \|\mu\|_{\mathfrak{R}_f(X)} := |\mu|(X).$$

We want to prove that the dual space $(C_0(X))^*$ can be identified with $\mathfrak{R}_f(X)$. More exactly, consider the map

$$T : \mathfrak{R}_f(X) \to (C_0(X))^*, \quad \langle T(\mu), f \rangle := \int_X f \, d\mu, \tag{2.50}$$

for $\mu \in \mathfrak{R}_f(X)$ and $f \in C_0(X)$. Then the following holds.

Theorem 2.7.1. *Let X be a locally compact Hausdorff space. Then the map T defined in* (2.50) *is an isometric isomorphism of $\mathfrak{R}_f(X)$ onto $(C_0(X))^*$.*

The proof of Theorem 2.7.1 relies on the following lemma, whose proof is given at the end of this subsection.

Lemma 2.7.2. *Let X be a locally compact Hausdorff space, and let $F : C_0(X) \to \mathbb{R}$ be a bounded linear functional. Then there exist positive linear functionals $F^\pm : C_0(X) \to \mathbb{R}$ such that*

$$F = F^+ - F^-. \tag{2.51}$$

Moreover:

(i) *we have*

$$\langle F^+, f \rangle = \sup\{\langle F, g \rangle \mid g \in C_0(X), \, 0 \le g \le f\}; \tag{2.52}$$

(ii) *decomposition* (2.51) *is minimal in the following sense: if we have $F = F^1 - F^2$ with positive linear functionals F^1, F^2 on $C_0(X)$, then the linear functionals $F^1 - F^+ = F^2 - F^-$ are positive.*

Remark 2.7.1. Decomposition (2.51) is called the *Jordan decomposition* of F. Observe that the minimality statement (ii) is the counterpart of that in Remark 1.8.2(i).

Proof of Theorem 2.7.1. Let us first prove that T is surjective. Fix $F \in (C_0(X))^*$, and let $F^\pm : C_0(X) \to [0, \infty)$ be the positive linear functionals given by Lemma 2.7.2 (recall that by Lemma 2.6.4 $F^\pm$ are bounded). Then by Theorem 2.6.3 there exist two finite Radon measures $\mu^\pm$ on X such that $\langle F^\pm, f \rangle := \int_X f \, d\mu^\pm$ for any $f \in C_0(X)$. Then $\mu := \mu^+ - \mu^-$ belongs to $\mathfrak{R}_f(X)$ (see Definition 1.8.5(ii)), and by equalities (2.10), (2.50), and (2.51) we obtain

$$\langle F, f \rangle = \int_X f \, d\mu = \langle T(\mu), f \rangle \quad \text{for all } f \in C_0(X).$$

Hence $F = T(\mu)$ in $(C_0(X))^*$, and thus the claim follows.

Let us now prove that T is isometric and thus injective. Clearly, by its very definition we have

$$\|T(\mu)\|_{(C_0(X))^*} \le |\mu|(X) = \|\mu\|_{\mathfrak{R}_f(X)} \quad \text{for all } \mu \in \mathfrak{R}_f(X).$$

To prove the inverse inequality, fix $\epsilon > 0$. By the first equality in (1.52) there exist disjoint $E_1, \dots, E_n \in \mathcal{B}$ such that $X = \bigcup_{i=1}^n E_i$ and $\sum_{i=1}^n |\mu(E_i)| > |\mu|(X) - \epsilon$. Then by the regularity of μ there exist $K_1, \dots, K_n \in \mathcal{K}$ and disjoint $A_1, \dots, A_n \in \mathcal{T}$ such that:

(a) $\sum_{i=1}^n |\mu(K_i)| > |\mu|(X) - 2\epsilon$;

(b) $K_i \subseteq E_i \subseteq A_i$, $|\mu|(A_i \setminus K_i)| < \frac{\epsilon}{n}$ for every $i = 1, \dots, n$.

Moreover, by Lemma A.9, for every $i = 1, \dots, n$, there exists $f_i \in C_c(X)$ such that $f_i(X) \subseteq [0,1]$, $f_i|_{K_i} = 1$, and $\operatorname{supp} f_i \subseteq A_i$.

Set $g := \sum_{i=1}^n [\operatorname{sgn} \mu(K_i)] f_i$. Then $\|g\|_\infty \le 1$, and we have

$$\begin{aligned}\left|\int_X g\, d\mu\right| &\ge \int_X g\, d\mu = \sum_{i=1}^n \int_{A_i} g\, d\mu = \sum_{i=1}^n \Big(\int_{A_i \setminus K_i} g\, d\mu + [\operatorname{sgn} \mu(K_i)] \int_{K_i} f_i\, d\mu \Big) \\ &\ge \sum_{i=1}^n \big(-|\mu|(A_i \setminus K_i) + |\mu(K_i)|\big) > |\mu|(X) - 3\epsilon.\end{aligned}$$

By the arbitrariness of ϵ, from the above inequality we get

$$\|T(\mu)\|_{(C_0(X))^*} \ge |\langle T(\mu), g\rangle| = \left|\int_X g\, d\mu\right| \ge |\mu|(X) = \|\mu\|_{\mathfrak{R}_f(X)}.$$

Then the conclusion follows. □

Proof of Lemma 2.7.2. Set

$$\langle F^+, f\rangle := \sup_{g \in S_f} \langle F, g\rangle, \quad S_f := \{g \in C_0(X) \mid 0 \le g \le f\} \quad \text{for all } f \in C_0(X), f \ge 0,$$

$$\langle F^+, f\rangle := \langle F^+, f^+\rangle - \langle F^+, f^-\rangle \quad \text{for all } f \in C_0(X).$$

Clearly, we have $\langle F^+, f\rangle \ge 0$ for $f \ge 0$ and

$$|\langle F^+, f\rangle| \le |\langle F^+, f^+\rangle| + |\langle F^+, f^-\rangle| \le \|F\|_{(C_0(X))^*} \|f\|_\infty$$

for all $f \in C_0(X)$, and thus F^+ is positive and bounded with $\|F^+\|_{(C_0(X))^*} \le \|F\|_{(C_0(X))^*}$. Moreover, by definition we have $\langle F^+, f\rangle \ge \langle F, f\rangle$ for all $f \in C_0(X)$, $f \ge 0$.

Let us prove that F^+ is linear. It is immediately seen that $\langle F^+, cf\rangle = c\,\langle F^+, f\rangle$ for all $c \ge 0$ and $f \ge 0$. Then for all $f \in C_0(X)$:

(a) if $c \geq 0$, then

$$\begin{aligned}\langle F^+, cf\rangle &= \langle F^+, (cf)^+\rangle - \langle F^+, (cf)^-\rangle = \langle F^+, cf^+\rangle - F^+(cf^-)\\ &= c\langle F^+, f^+\rangle - c\langle F^+, f^-\rangle = c\langle F^+, f\rangle;\end{aligned}$$

(b) if $c < 0$, then

$$\begin{aligned}\langle F^+, cf\rangle &= \langle F^+, (cf)^+\rangle - \langle F^+, (cf)^-\rangle = \langle F^+, |c|f^-\rangle - \langle F^+, |c|f^+\rangle\\ &= |c|[\langle F^+, f^-\rangle - \langle F^+, f^+\rangle] = c[\langle F^+, f^+\rangle - \langle F^+, f^-\rangle] = c\langle F^+, f\rangle.\end{aligned}$$

Hence we have $\langle F^+, cf\rangle = c\,\langle F^+, f\rangle$ for all $c \in \mathbb{R}$ and $f \in C_0(X)$. The conclusion will follow if we prove that

$$\langle F^+, f_1 + f_2\rangle = \langle F^+, f_1\rangle + \langle F^+, f_2\rangle \quad \text{for all } f_1, f_2 \in C_0(X),\, f_1, f_2 \geq 0; \tag{2.53}$$

in fact, by the definition of F^+ this entails that the same equality holds for all $f_1, f_2 \in C_0(X)$.

Let $f_1, f_2 \in C_0(X)$, $f_1, f_2 \geq 0$, be fixed. For any $g_1 \in S_{f_1}$ and $g_2 \in S_{f_2}$ we have

$$\langle F, g_1\rangle + \langle F, g_2\rangle = \langle F, g_1 + g_2\rangle \leq \sup_{g \in S_{f_1+f_2}} \langle F, g\rangle = \langle F^+, f_1 + f_2\rangle,$$

and thus $\langle F^+, f_1 + f_2\rangle \geq \langle F^+, f_1\rangle + \langle F^+, f_2\rangle$. To prove the reverse inequality, let $0 \leq g \leq f_1 + f_2$. Set $g_1 := \min\{g, f_1\}$. Thus $g_1 \in S_{f_1}$, and $g_2 := g - g_1$, whence plainly $g_2 = [g - f_1]_+$. Since

$$g_2 - f_2 = [g - f_1]_+ - f_2 = \begin{cases} -f_2 & \text{if } g \leq f_1,\\ g - (f_1 + f_2) & \text{if } g > f_1,\end{cases}$$

we obtain that $0 \leq g_2 \leq f_2$, that is, $g_2 \in S_{f_2}$. It follows that

$$\langle F, g\rangle = \langle F, g_1 + g_2\rangle = \langle F, g_1\rangle + \langle F, g_2\rangle \leq \langle F^+, f_1\rangle + \langle F^+, f_2\rangle$$

for all $g \in S_f$, whence $\langle F^+, f_1 + f_2\rangle \leq \langle F^+, f_1\rangle + \langle F^+, f_2\rangle$.

Therefore F^+ is a positive linear functional on $C_0(X)$. Set $F^- := F^+ - F$; then $F^- : C_0(X) \mapsto \mathbb{R}$ is linear and positive, since $\langle F^-, f\rangle = \langle F^+, f\rangle - \langle F, f\rangle \geq 0$ for all $f \geq 0$. Hence equality (2.51) and claim (i) follow. To prove claim (ii), let $f \in C_0(X)$, $f \geq 0$. Since F^1 and F^2 are positive, for any $g \in S_f$, we have $\langle F^1, f\rangle \geq \langle F^1, g\rangle \geq \langle F^1, g\rangle - \langle F^2, g\rangle = \langle F, g\rangle$, and thus $\langle F^1, f\rangle \geq \langle F^+, f\rangle$. This completes the proof. □

Let us briefly discuss how the above situation changes when $C_0(X)$ is replaced by $C_c(X)$. Let μ_1, μ_2 be (positive) Radon measures. For any $f \in C_c(X)$, set

$$\langle F, f\rangle := \int_X f\, d\mu_1 - \int_X f\, d\mu_2; \tag{2.54}$$

the definition is well posed since both integrals in the right-hand side are finite. Then for any compact set $K \subseteq X$ and for any $f \in C_c(X)$ with $\operatorname{supp} f \subseteq K$, we have

$$|\langle F, f\rangle| \leq [\mu_1(K) + \mu_2(K)]\|f\|_\infty.$$

By this inequality F is *locally bounded* in the following sense.

Definition 2.7.1. A linear functional $F : C_c(X) \to \mathbb{R}$ is *locally bounded*, if for any $K \in \mathcal{K}$, there exists $c_K > 0$ such that

$$|\langle F, f\rangle| \leq c_K\|f\|_\infty \quad \text{for all } f \in C_c(X) \text{ with } \operatorname{supp} f \subseteq K.$$

Lemma 2.7.3. *Let X be a locally compact Hausdorff space.*
(i) *Every positive linear functional on $C_c(X)$ is locally bounded.*
(ii) *Let $F : C_c(X) \to \mathbb{R}$ be a locally bounded linear functional. Then there exist positive linear functionals $F^\pm : C_c(X) \to \mathbb{R}$ such that equality* (2.51) *holds, and claims* (i)–(ii) *of Lemma 2.7.2 are satisfied.*

Proof. We only prove (i), since the proof of (ii) is the same as that of Lemma 2.7.2 with $C_0(X)$ replaced by $C_c(X)$. Let $K \in \mathcal{K}$, and let $\varphi_K \in C_c(X)$, $\varphi_K(X) \subseteq [0,1]$, $\varphi_K = 1$ in K (see Lemma A.9). For any $f \in C_c(X)$ with $\operatorname{supp} f \subseteq K$, we have

$$\|f\|_\infty \varphi_K \pm f = (\|f\|_\infty \pm f)\,\varphi_K \geq 0 \quad \text{in } X.$$

Since $\|f\|_\infty \varphi_K \pm f \in C_c(X)$ and F is positive and linear, it follows that

$$\langle F, \|f\|_\infty \varphi_K \pm f\rangle = \langle F, \varphi_K\rangle\|f\|_\infty \pm \langle F, f\rangle \geq 0,$$

and hence $|\langle F, f\rangle| \leqslant c_K\|f\|_\infty$ with $c_K := \langle F, \varphi_K\rangle$. □

From Lemma 2.7.3 and Theorem 2.6.1 we get the following result.

Theorem 2.7.4. *Let X be a locally compact Hausdorff space, and let $F : C_c(X) \to \mathbb{R}$ be a locally bounded linear functional. Then there exists a unique couple μ_1, μ_2 of mutually singular Radon measures on X such that equality* (2.54) *holds.*

Proof. Since the measures μ_1 and μ_2 in equality (2.54) are mutually singular, the uniqueness claim is easily checked.

Let us address the existence part. Let $F^\pm : C_c(X) \to \mathbb{R}$ be the positive linear functionals associated with F by Lemma 2.7.3. Thus equality (2.51) holds, and claims (i)–(ii) of Lemma 2.7.3 are satisfied. Moreover, by Theorem 2.6.1 there exists a unique pair of Radon measures μ_1 and μ_2 such that

$$\langle F^+, f\rangle = \int_X f\, d\mu_1, \quad \langle F^-, f\rangle = \int_X f\, d\mu_2 \tag{2.55}$$

for all $f \in C_c(X)$. By equalities (2.51) and (2.55) the measures μ_1 and μ_2 satisfy (2.54). Since, by inner regularity, two Radon measures that coincide on $\mathcal{K}$ are the same measure, the conclusion will follow if we prove the following:

Claim. For every $K \in \mathcal{K}$, the measures $\mu_1 \llcorner K$ and $\mu_2 \llcorner K$ are mutually singular.

To prove the claim, fix $K \in \mathcal{K}$ and set $\mu := \mu_1 \llcorner K - \mu_2 \llcorner K$. Clearly, $|\mu|(X) \le \mu_1(K) + \mu_2(K) < \infty$, and thus $\mu \in \mathfrak{R}_f(X)$. The claim will follow if we prove that $\mu_1 \llcorner K = \mu^+$ and $\mu_2 \llcorner K = \mu^-$, where $\mu = \mu^+ - \mu^-$ is the Jordan decomposition of μ. In fact, it suffices to show that

$$\mu_1 \llcorner K \le \mu^+, \quad \mu_2 \llcorner K \le \mu^-, \tag{2.56}$$

since the inverse inequalities follow from the very definition of $\mu^\pm$ (see Remark 1.8.2).

To prove (2.56), let $F_i : C_c(X) \to \mathbb{R}$ $(i = 1, 2)$ be the positive linear functionals

$$\langle F_1, f\rangle := \int_X f\, d\mu^+, \quad \langle F_2, f\rangle := \int_X f\, d\mu^- \quad (f \in C_c(X)).$$

Set

$$G^\pm := F^\pm|_{C_c(K)}, \quad G := F|_{C_c(K)} = G^+ - G^-.$$

Then we have $G^+ - G^- = G = F_1 - F_2$ on $C_c(K)$, and by Lemma 2.7.3(ii) the linear functionals $F_1 - G^+$ and $F_2 - G^-$ are positive on $C_c(K)$. This implies that

$$\int_K f\, d\mu_1 \le \int_K f\, d\mu^+, \quad \int_K f\, d\mu_2 \le \int_K f\, d\mu^- \quad \text{for all } f \in C_c(K), f \ge 0,$$

whence (2.56) follows. This completes the proof. □

Following [24], we state the following definition.

Definition 2.7.2. Let $(X, \mathcal{T})$ be a Hausdorff space. By a *signed Radon measure* on X we mean a locally bounded linear functional on $C_c(X)$.

Remark 2.7.2. In view of Theorem 2.7.4, there is one-to-one correspondence between signed Radon measures and ordered couples (μ_1, μ_2) of (positive) mutually singular Radon measures. Carefully observe that the measures μ_1, μ_2 given by Theorem 2.7.4 need not be finite, and thus in general their difference is not defined (also observe that a signed Radon measure need not be a signed measure; see Definition 1.8.1(ii)). However,

a) Definition 2.7.2 reduces to Definition 1.8.5, that is, namely, a signed Radon measure is a finite signed Radon measure if both μ_1 and μ_2 are finite;
b) a signed Radon measure is a signed measure if at least one of μ_1, μ_2 is finite.

The vector space of signed Radon measures on X will be denoted by $\mathfrak{R}(X)$. It is a locally convex space with topology generated by seminorms

$$p_K(\mu) := (\mu_1 + \mu_2)(K) \quad \text{for all } K \in \mathcal{K}, \mu \equiv (\mu_1, \mu_2) \in \mathfrak{R}(X).$$

2.8 Convergence in Lebesgue spaces

2.8.1 Preliminary remarks

Let $(X, \mathcal{A}, \mu)$ be a measure space. We denote by the usual symbol $L^p(X) \equiv L^p(X, \mathcal{A}, \mu)$ the Lebesgue space of order $p \in [1, \infty]$:

$$L^p(X) := \{f : X \to \mathbb{R} \ \mathcal{A}\text{-measurable} \mid \|f\|_p < \infty\},$$

where

$$\|f\|_p := \Big(\int_X |f|^p \, d\mu \Big)^{\frac{1}{p}} \quad \text{if } p \in [1, \infty),$$

$$\|f\|_\infty := \operatorname{ess\,sup} |f| = \inf\{C > 0 \mid |f(x)| \leq C \text{ for } \mu\text{-a. e. } x \in X\}.$$

The map $\|\cdot\|_p : L^p(X) \to [0, \infty), f \mapsto \|f\|_p$, is a seminorm. As usual, we identify functions that are equal μ-a. e. in X. More exactly:

(i) we replace $L^p(X)$ by its quotient $L^p(X)/\sim$ with respect to the relation of equivalence "$f \sim g \Leftrightarrow f = g$ μ-a.e.";

(ii) by abuse of notation we denote again $L^p(X)/\sim$ by $L^p(X)$.

Then $L^p(X)$ is a normed vector space (in fact, a Banach space) with norm $f \mapsto \|f\|_{L^p(X)} := \|f\|_p$ ($p \in [1, \infty]$).

Remark 2.8.1. For any open set $U \subseteq \mathbb{R}^N$, we set $L^p(U) \equiv L^p(U, \mathcal{B}^N \cap U, \lambda_N|_{\mathcal{B}^N \cap U})$, where $\mathcal{B}^N \cap U$ is the trace of the Borel σ-algebra $\mathcal{B}^N$ on U, and $\lambda_N|_{\mathcal{B}^N \cap U}$ is the restriction of the Lebesgue measure λ_N. We will say that $f \in L^p_{\text{loc}}(U)$ if $f \in L^p(V, \mathcal{B}^N \cap V, \lambda_N|_{\mathcal{B}^N \cap V})$ for every open subset $V \subset\subset X$ (i. e., the closure $\overline{V}$ is compact, and $\overline{V} \subset U$).

An easy application of the dominated convergence theorem gives the following result.

Lemma 2.8.1. *Let $(X, \mathcal{A}, \mu)$ be a measure space. Then for any $p \in [1, \infty)$, the set $\mathscr{S}_\mathbb{Q}(X)$ of simple functions with rational coefficients is dense in $L^p(X)$.*

Theorem 2.8.2. *Let $(X, \mathcal{A})$ be a separable measurable space, and let $\mu : \mathcal{A} \mapsto [0, \infty]$ be σ-finite. Then for any $p \in [1, \infty)$ the space $L^p(X) \equiv L^p(X, \mathcal{A}, \mu)$ is separable.*

Proof. Fix $f \in L^p(X)$ and $\epsilon > 0$. Then there exists $t \in \mathscr{S}_{\mathbb{Q}}$ such that $\mu(\{t \neq 0\}) < \infty$ and $\|f - t\|_p < \epsilon$. Indeed, let $\{E_n\} \subseteq \mathcal{A}$ be a nondecreasing sequence such that $X = \bigcup_{n=1}^{\infty} E_n$, $\mu(E_n) < \infty$ for all $n \in \mathbb{N}$. By Lemma 2.8.1 there exists $s \in \mathscr{S}_{\mathbb{Q}}$ such that $\|f - s\|_p < \frac{\epsilon}{2}$. Set $t_n := s\chi_{E_n}$ $(n \in \mathbb{N})$. Then $\{t_n\} \subseteq \mathscr{S}_{\mathbb{Q}}$, and $\|t_n - s\|_p^p = \int_{E_n^c} |s|^p \, d\mu \to 0$ as $n \to \infty$. Hence the claim follows.

Therefore it is not restrictive to assume that $\mu(X) < \infty$. Since by assumption $(X, \mathcal{A})$ is separable, there exists a countable family $\mathcal{C} \subseteq \mathcal{P}(X)$ such that $\sigma_0(\mathcal{C}) = \mathcal{A}$. Clearly, both the algebra $\mathcal{A}_0 \equiv \mathcal{A}_0(\mathcal{C})$ generated by $\mathcal{C}$ and the set

$$\mathscr{S}_{\mathbb{Q},\mathcal{A}_0} := \left\{ \sum_{k=1}^{n} c_k \chi_{G_k} \mid G_k \in \mathcal{A}_0,\ c_k \in \mathbb{Q},\ n \in \mathbb{N} \right\}$$

are countable. It is a routine matter to show that $\mathscr{S}_{\mathbb{Q},\mathcal{A}_0}$ is dense in $\mathscr{S}_{\mathbb{Q}}$, and hence the result follows. □

Let us also prove the following result.

Lemma 2.8.3. *Let X be a locally compact Hausdorff space, and let $\mu \in \mathfrak{R}^+(X)$ be σ-finite. Then for any $p \in [1, \infty)$, the space $C_c(X)$ is dense in $L^p(X)$.*

Proof. It is not restrictive to suppose $\mu(X) < \infty$, and thus by Lemma 1.3.2(i) μ is regular. Fix $E \in \mathcal{B}(X)$. By Lemma 2.8.1 it suffices to prove that there exists $\{g_n\} \subseteq C_c(X)$ such that $g_n \to g := \chi_E$ in $L^p(X)$. To this purpose, observe that by the regularity of μ, for any $\delta > 0$, there exist a compact set K and open set A such that $K \subseteq E \subseteq A$ and $\mu(A \backslash K) < \delta$. Arguing as in the proof of Proposition 2.1.16, we obtain a function $\zeta \in C_c(X)$ with $\operatorname{supp} \zeta \subseteq A$ such that $\mu(\{g \neq \zeta\}) < \delta$ and $\|\zeta\|_\infty = 1$. Set $\delta = \frac{1}{n}$ $(n \in \mathbb{N})$. Then there exists a sequence $\{g_n\} \subseteq C_c(X)$ such that $g_n \to g$ in measure and thus in $L^p(X)$ for any $p \in [1, \infty)$ (see Proposition 2.8.8(i)). Hence the result follows. □

2.8.2 Uniform integrability

Let us first state the following definition.

Definition 2.8.1. Let $p \in [1, \infty)$. A subset $\mathscr{F} \subseteq L^p(X)$ is called *p-uniformly integrable* if for any $\epsilon > 0$:

(i) there exists $E \in \mathcal{A}$ with $\mu(E) < \infty$ such that $\sup_{f \in \mathscr{F}} \int_{E^c} |f|^p \, d\mu < \epsilon$;
(ii) there exists $\delta > 0$ such that $\sup_{f \in \mathscr{F}} \int_F |f|^p \, d\mu < \epsilon$ for all $F \in \mathcal{A}$ with $\mu(F) < \delta$.

We say that $\mathscr{F} \subseteq L^1(X)$ is *uniformly integrable* if it is 1-uniformly integrable.

Remark 2.8.2. (i) Let $f \in L^p(X)$. The map $\sigma : \mathcal{A} \to [0, \infty)$, $\sigma(E) := \int_E |f|^p \, d\mu$ for $E \in \mathcal{A}$, is a finite measure on $(X, \mathcal{A})$; moreover, $\sigma \ll \mu$. Hence every singleton $\mathscr{F} = \{f\}$,

$f \in L^p(X)$ (more generally, every finite set $\mathscr{F} = \{f_1, \dots, f_n\} \subseteq L^p(X)$) is p-uniformly integrable.

(ii) Condition (i) of Definition 2.8.1 is trivially satisfied with $E = X$ if μ is finite.

If $p = 1$, then uniformly integrable subsets are characterized as follows.

Proposition 2.8.4. *Let $(X, \mathcal{A}, \mu)$ be a measure space. Then the following statements are equivalent:*

(i) *$\mathscr{F} \subseteq L^1(X)$ is uniformly integrable;*

(ii) *for any nonincreasing sequence $\{E_n\} \subseteq \mathcal{A}$ with $\bigcap_{n=1}^{\infty} E_n = \emptyset$, we have*

$$\lim_{n\to\infty} \sup_{f\in\mathscr{F}} \int_{E_n} |f|\, d\mu = 0.$$

Proof. (i)⇒(ii). Fix $\epsilon > 0$, and let $\delta > 0$ be as in Definition 2.8.1(ii). Since $\lim_{n\to\infty} \mu(E_n) = 0$, there exists $\bar{n} \in \mathbb{N}$ such that $\mu(E_n) < \delta$ for all $n \in \mathbb{N}$, and thus $\sup_{f\in\mathscr{F}} \int_{E_n} |f|\, d\mu < \epsilon$. Hence the claim follows.

(ii)⇒(i). Let us show that both requirements of Definition 2.8.1 are satisfied.

By contradiction, let Definition 2.8.1(i) not hold. Thus there exists $k > 0$ such that for any $F \in \mathcal{A}$ with $\mu(F) < \infty$, we have $\int_{F^c} |f|\, d\mu \geq k$ for some $f \in \mathscr{F}$. Let $F_1 \in \mathcal{A}$ with $\mu(F_1) < \infty$, let $f_1 \in \mathscr{F}$ be such that $\int_{F_1^c} |f_1|\, d\mu \geq k$, and let $F_2 \in \mathcal{A}$, $F_2 \supseteq F_1$, with $\mu(F_2) < \infty$ be such that $\int_{F_2^c} |f_1|\, d\mu < \frac{k}{2}$. Then we have

$$\int_{F_2\setminus F_1} |f_1|\, d\mu = \int_{F_1^c} |f_1|\, d\mu - \int_{F_2^c} |f_1|\, d\mu \geq \frac{k}{2}.$$

Arguing recursively gives a nondecreasing sequence $\{F_n\} \subseteq \mathcal{A}$ and a sequence $\{f_n\} \subseteq \mathscr{F}$ such that $\int_{F_{n+1}\setminus F_n} |f_n|\, d\mu \geq \frac{k}{2}$ for any n. Set $F := \bigcup_{n=1}^{\infty} F_n$, $E_n := F \setminus F_n$ $(n \in \mathbb{N})$. Then the sequence $\{E_n\}$ is nonincreasing, and $\bigcap_{n=1}^{\infty} E_n = \emptyset$, and for all n, we have

$$\sup_{f\in\mathscr{F}} \int_{E_n} |f|\, d\mu \geq \int_{E_n} |f_n|\, d\mu \geq \int_{F_{n+1}\setminus F_n} |f_n|\, d\mu \geq \frac{k}{2},$$

a contradiction.

We also argue by absurd to prove that Definition 2.8.1(ii) holds. Indeed, let there exist $k > 0$ such that for any $n \in \mathbb{N}$, there exist $F_n \in \mathcal{A}$ with $\mu(F_n) < \frac{1}{2^n}$ and $f_n \in \mathscr{F}$ such that $\int_{F_n} |f_n|\, d\mu \geq k$. Set $F_\infty := \bigcap_{m=1}^{\infty} \bigcup_{n=m}^{\infty} F_n$, $E_n := \bigcup_{m=n}^{\infty} (F_m \setminus F_\infty)$ $(n \in \mathbb{N})$. It is easily seen that $\mu(F_\infty) \leq \lim_{m\to\infty} \sum_{n=m}^{\infty} \frac{1}{2^n} = 0$. Then the sequence $\{E_n\}$ is nonincreasing and $\bigcap_{n=1}^{\infty} E_n = \emptyset$, and for all n, we have

$$\sup_{f\in\mathscr{F}} \int_{E_n} |f|\, d\mu \geq \sup_{f\in\mathscr{F}} \int_{F_n} |f|\, d\mu \geq \int_{F_n} |f_n|\, d\mu \geq k,$$

again a contradiction. Hence the result follows. □

Remark 2.8.3. If $\mathscr{F} \subseteq L^1(X)$, then set $\mathscr{F}_\pm := \{f_\pm \mid f \in \mathscr{F}\}$ and $\mathscr{F}_h := \{f - h \mid f \in \mathscr{F}\}$ for $h \in L^1(X)$. It is easily seen that $\mathscr{F}$ is uniformly integrable if and only if the same holds for $\mathscr{F}_\pm$ and $\mathscr{F}_h$.

We will use the following lemma.

Lemma 2.8.5. *Let $(X, \mathcal{A}, \mu)$ be a measure space, and let $\mathscr{F} \subseteq L^1(X)$ be not uniformly integrable. Then there exist a sequence $\{E_n\} \subseteq \mathcal{A}$ with $\bigcap_{n=1}^\infty E_n = \emptyset$ and $\lim_{p\to\infty} \mu(\bigcup_{n=p}^\infty E_n) = 0$ and a sequence $\{f_n\} \subseteq \mathscr{F}$ with the following property: for any $h \in L^1(X)$, there exists $k = k_h > 0$ such that for all $n \in \mathbb{N}$, we have*

$$\left| \int_{E_n} (f_n - h)\, d\mu \right| \geq k. \tag{2.58}$$

Proof. Let us first show that if $\mathscr{F}$ is not uniformly integrable, then there exist sequences $\{E_n\} \subseteq \mathcal{A}$ with $\bigcap_{n=1}^\infty E_n = \emptyset$ and $\lim_{p\to\infty} \mu(\bigcup_{n=p}^\infty E_n) = 0$ and $\{f_n\} \subseteq \mathscr{F}$ and a number $m > 0$ such that

$$\left| \int_{E_n} f_n\, d\mu \right| \geq 2m \quad \text{for all } n \in \mathbb{N}. \tag{2.59}$$

In fact, by Remark 2.8.3 at least one of the sets $\mathscr{F}_\pm$ is not uniformly integrable. Let $\mathscr{F}_+$ be not uniformly integrable (for $\mathscr{F}_-$, the argument is analogous). Then by Proposition 2.8.4 we can find $m > 0$, $\{f_n\} \subseteq \mathscr{F}$, and a nonincreasing sequence $\{E_n'\} \subseteq \mathcal{A}$ with $\bigcap_{n=1}^\infty E_n' = \emptyset$ such that

$$\int_{E_n'} f_n^+\, d\mu \geq 2m \quad \text{for all } n \in \mathbb{N}.$$

Set $E_n := E_n' \cap \{f_n \geq 0\}$. Then we have $\bigcap_{n=1}^\infty E_n = \emptyset$ and $\lim_{p\to\infty} \mu(\bigcup_{n=p}^\infty E_n) = \lim_{p\to\infty} \mu(E_p') = 0$ since $\bigcup_{n=p}^\infty E_n \subseteq \bigcup_{n=p}^\infty E_n' = E_p'$. Moreover,

$$\left| \int_{E_n} f_n\, d\mu \right| = \int_{E_n'} f_n^+\, d\mu \geq 2m \quad \text{for all } n \in \mathbb{N},$$

and thus (2.59) follows.

Now observe that for every $h \in L^1(X)$, we have

$$\left| \int_{E_n} f_n\, d\mu \right| \leq \left| \int_{E_n} (f_n - h)\, d\mu \right| + \int_{E_n} |h|\, d\mu.$$

Since $\lim_{n\to\infty}\mu(E_n) = \lim_{n\to\infty}\mu(\bigcup_{j=n}^{\infty} E_j) = 0$, there exists $\bar{n} \in \mathbb{N}$ such that $\int_{E_n} |h|\,d\mu < m$ for all $n > \bar{n}$. Then by the above inequality and (2.59)

$$\left|\int_{E_n} (f_n - h)\,d\mu\right| \geq m \quad \text{for all } n > \bar{n}.$$

Then setting

$$k := \min\left\{m, \min_{j=1,\dots,\bar{n}}\left\{\left|\int_{E_j} (f_j - h)\,d\mu\right|\right\}\right\},$$

we obtain (2.58). This proves the result. □

Proposition 2.8.6. *Let $(X, \mathcal{A}, \mu)$ be a finite measure space, and let $\mathscr{F} \subseteq L^1(X)$. Then the following statements are equivalent:*

(i) *$\mathscr{F}$ is bounded and uniformly integrable;*

(ii) $\lim_{t\to\infty} \sup_{f\in\mathscr{F}} \int_{\{|f|\geq t\}} |f|\,d\mu = 0$;

(iii) (de la Vallée-Poussin criterion) *there exists a continuous increasing function ψ : $[0,\infty) \to [0,\infty]$ such that $\frac{\psi(t)}{t} \to \infty$ as $t \to \infty$ and $\sup_{f\in\mathscr{F}} \int_X \psi(|f|)\,d\mu < \infty$.*

Proof. (i)⇒(ii). Set $C := \sup_{f\in\mathscr{F}} \|f\|_1 < \infty$. By the Chebyshev inequality we have

$$\sup_{f\in\mathscr{F}} \mu(\{|f| \geq t\}) \leq \frac{1}{t} \sup_{f\in\mathscr{F}} \int_{\{|f|\geq t\}} |f|\,d\mu \leq \frac{C}{t} \quad (t > 0). \tag{2.60}$$

Fix $\epsilon > 0$, and set $t_0 := \frac{C}{\delta}$ with δ given by Definition 2.8.1(ii). Then for any $t > t_0$, from (2.60) we get $\sup_{f\in\mathscr{F}} \mu(\{|f| \geq t\}) < \delta$, whence $\sup_{f\in\mathscr{F}} \int_{\{|f|\geq t\}} |f|\,d\mu \leq \epsilon$ by the uniform integrability of $\mathscr{F}$. Hence the claim follows.

(ii)⇒(i). For any $\epsilon > 0$, there exists $t_0 > 0$ such that $\sup_{f\in\mathscr{F}} \int_{\{|f|\geq t\}} |f|\,d\mu < \frac{\epsilon}{2}$ t for all $t \geq t_0$. Then for all $F \in \mathcal{A}$,

$$\sup_{f\in\mathscr{F}} \int_F |f|\,d\mu \leq \sup_{f\in\mathscr{F}} \int_{F\cap\{|f|\leq t_0\}} |f|\,d\mu + \sup_{f\in\mathscr{F}} \int_{F\cap\{|f|>t_0\}} |f|\,d\mu \leq t_0\mu(F) + \frac{\epsilon}{2}. \tag{2.61}$$

If $F = X$, then from (2.61) we obtain

$$\sup_{f\in\mathscr{F}} \|f\|_1 \leq t_0\mu(X) + \frac{\epsilon}{2} < \infty,$$

and thus $\mathscr{F}$ is bounded. Moreover, set $\delta := \frac{\epsilon}{2t_0}$. Then for any $F \in \mathcal{A}$ with $\mu(F) < \delta$, inequality (2.61) gives

$$\sup_{f\in\mathscr{F}} \int_F |f|\,d\mu \leq t_0\mu(F) + \frac{\epsilon}{2} = \epsilon.$$

Hence the claim follows.

(iii)⇒(ii). Set $D := \sup_{f\in\mathscr{F}} \int_X \psi(|f|)\,d\mu$. Since $\frac{\psi(t)}{t} \to \infty$ as $t \to \infty$, for any $\epsilon > 0$, there exists $t_0 > 0$ such that $\frac{t}{\epsilon} \leq \psi(t)$ for all $t > t_0$, and thus

$$\sup_{f\in\mathscr{F}} \int_{\{|f|\geq t\}} |f|\,d\mu \leq \epsilon \sup_{f\in\mathscr{F}} \int_{\{|f|\geq t\}} \psi(|f|)\,d\mu \leq D\epsilon.$$

Hence the claim follows.

(ii)⇒(iii). By assumption there exists a diverging sequence $\{t_k\} \subseteq \mathbb{N}$ such that

$$\sup_{f\in\mathscr{F}} \int_{\{|f|\geq t_k\}} |f|\,d\mu \leq \frac{1}{2^{k+1}}.$$

For every $t > 0$, set $E_t := \{n \in \{t_k\} \mid n < t\}$ and

$$\psi(t) := t \operatorname{card} E_t = t \int_{\mathbb{N}} \chi_{E_t}(n)\,d\mu^{\#}(n) \quad (t > 0),$$

where $\mu^{\#} : \mathcal{P}(\mathbb{N}) \to [0,\infty]$ is the counting measure.

Clearly, $\psi : [0,\infty) \mapsto [0,\infty]$ is increasing, and we have $\frac{\psi(t)}{t} \to \infty$ as $t \to \infty$. In addition, for any $f \in \mathscr{F}$, by the Tonelli theorem we have

$$\begin{aligned}\int_X \psi(|f|)\,d\mu &= \int_X d\mu(t)\,|f(t)| \int_{\mathbb{N}} \chi_{E_{|f(t)|}}(n)\,d\mu^{\#}(n) \\ &= \int_{\mathbb{N}} d\mu^{\#}(k) \int_{\{|f|\geq t_k\}} |f|\,d\mu = \sum_{k=1}^{\infty} \int_{\{|f|\geq t_k\}} |f|\,d\mu \leq \sum_{k=1}^{\infty} \frac{1}{2^{k+1}} \leq 1.\end{aligned}$$

Hence $\sup_{f\in\mathscr{F}} \int_X \psi(|f|)\,d\mu < \infty$, and thus the result follows by a standard regularization of the function ψ. □

Definition 2.8.2. Let $\mathscr{F} \subseteq L^1(X)$ be bounded. The quantity

$$\eta(\mathscr{F}) := \limsup_{t\to\infty} \sup_{f\in\mathscr{F}} \int_{\{|f|\geq t\}} |f|\,d\mu \tag{2.62}$$

is called the *modulus of uniform integrability* of $\mathscr{F}$.

Remark 2.8.4. By Proposition 2.8.6 a bounded subset $\mathscr{F} \subseteq L^1(X)$ is uniformly integrable if and only if $\eta(\mathscr{F}) = 0$.

Let us prove the following result for future reference.

Lemma 2.8.7. *Let $\mathscr{F} \subseteq L^1(X)$ be bounded. Then*

$$\eta(\mathscr{F}) = \lim_{\delta \to 0^+} \sup\left\{\int_E |f|\, d\mu \mid f \in \mathscr{F},\, E \in \mathcal{B} \text{ such that } \mu(E) < \delta\right\}. \tag{2.63}$$

Proof. Set $\eta \equiv \eta(\mathscr{F})$, and denote by $\eta_1 \equiv \eta_1(\mathscr{F})$ the right-hand side of equality (2.63). The result will follow if we prove both inequalities $\eta_1 \geq \eta$ and $\eta_1 \leq \eta$.

Set $C := \sup_{f \in \mathscr{F}} \|f\|_1 < \infty$ and

$$H_\delta := \sup\left\{\int_E |f|\, d\mu \mid f \in \mathscr{F},\, E \in \mathcal{B} \text{ such that } \mu(E) < \delta\right\},$$

and fix arbitrary $\delta > 0$. By (2.62) there exists $t_0 > 0$ such that for all $t > t_0$, there exists $\bar{f} \in \mathscr{F}$ satisfying

$$\int_{\{|\bar{f}| \geq t\}} |\bar{f}|\, d\mu \geq \eta - \delta. \tag{2.64}$$

Set $t_1 := \frac{C}{\delta}$. Then for all $t > t_1$, by the Chebyshev inequality we have

$$\mu\{|\bar{f}| \geq t\} \leq \frac{1}{t} \int_X |\bar{f}|\, d\mu \leq \frac{C}{t} < \delta. \tag{2.65}$$

By (2.64)–(2.65) and the definition of H_δ, for any $t > \max\{t_0, t_1\}$, we have

$$H_\delta \geq \int_{\{|\bar{f}| \geq t\}} |\bar{f}|\, d\mu \geq \eta - \delta \quad \text{for all } \delta > 0,$$

whence letting $\delta \to 0^+$, we obtain $\eta_1 \geq \eta$.

To prove the reverse inequality, fix arbitrary $\epsilon > 0$ and $t > 0$. By the definition of η_1 there exists $\delta_0 \in (0, \frac{\epsilon}{t})$ with the following property: for all $\delta \in (0, \delta_0)$, there exist $\bar{f} \in \mathscr{F}$ and $E \in \mathcal{B}$ with $\mu(E) < \delta$ such that

$$\int_E |\bar{f}|\, d\mu \geq \eta_1 - \epsilon. \tag{2.66}$$

Since $t\mu(E) < t\delta < t\delta_0 < \epsilon$, from (2.66) we get

$$\begin{aligned}\int_{\{|\bar{f}| \geq t\}} |\bar{f}|\, d\mu \geq \int_{E \cap \{|\bar{f}| \geq t\}} |\bar{f}|\, d\mu &= \int_E |\bar{f}|\, d\mu - \int_{E \cap \{|\bar{f}| < t\}} |\bar{f}|\, d\mu \\ &\geq \eta_1 - \epsilon - t\mu(E) > \eta_1 - 2\epsilon.\end{aligned}$$

In view of definition (2.62), from this inequality we obtain $\eta \geq \eta_1 - 2\epsilon$, whence $\eta \geq \eta_1$ by the arbitrariness of ϵ. This completes the proof. □

2.8.3 Strong convergence

If $f, f_n \in L^p(X)$ ($n \in \mathbb{N}$, $p \in [1, \infty]$), we say that f_n converges to f *strongly in $L^p(X)$* (written $f_n \to f$) if $\lim_{n\to\infty} \|f_n - f\|_p = 0$.

Proposition 2.8.8. *Let $(X, \mathcal{A}, \mu)$ be a measure space, and let $p \in [1, \infty)$.*
(i) *Let $f_n \to f$ in measure, and let condition* (2.12) *be satisfied. Then $f_n \to f$ strongly in $L^p(X)$.*
(ii) *If $f_n \to f$ strongly in $L^p(X)$, then there exists $\{f_{n_k}\} \subseteq \{f_n\}$ such that $f_{n_k} \to f$ μ-a. e.*
(iii) *If $f_n \to f$ strongly in $L^p(X)$, then $f_n \to f$ in measure.*
(iv) *If $f, f_n \in L^p(X)$, $f_n \to f$ μ-a. e., and $\|f_n\|_p \to \|f\|_p$, then $f_n \to f$ strongly in $L^p(X)$.*

Proof. We only prove claim (iv). By the Fatou lemma we have

$$2^p \|f\|_p^p \leq \liminf_{n\to\infty} \int_X [2^{p-1}(|f|^p + |f_n|^p) - |f - f_n|^p] d\mu = 2^p \|f\|_p^p - \limsup_{n\to\infty} \|f - f_n\|_p^p,$$

and thus the claim follows. □

Concerning strong convergence in $L^\infty(X)$, we have the following:

Proposition 2.8.9. *Let $f_n \to f$ strongly in $L^\infty(X)$. Then:*
(i) *$f_n \to f$ μ-a. e., in measure and almost uniformly;*
(ii) *if $\mu(X) < \infty$, then there exists $\{f_{n_k}\} \subseteq \{f_n\}$ such that $f_{n_k} \to f$ strongly in $L^p(X)$ for all $p \in [1, \infty)$;*
(iii) *if condition* (2.12) *is satisfied, then there exists $\{f_{n_k}\} \subseteq \{f_n\}$ such that $f_{n_k} \to f$ strongly in $L^1(X)$.*

The following result relies on the concept of p-uniform integrability.

Theorem 2.8.10 (Vitali). *Let $p \in [1, \infty)$. Let $\{f_n\} \subseteq L^p(X)$, and let $f_n \to f$ μ-a. e. in X. Then the following statements are equivalent:*
(i) *$f \in L^p(X)$, and $f_n \to f$ in $L^p(X)$;*
(ii) *the sequence $\{f_n\}$ is p-uniformly integrable.*

Proof. (i)⇒(ii). Fix $\epsilon > 0$. Since $f_n \to f$ in $L^p(X)$, there exists $\bar{n} \in \mathbb{N}$ such that $\|f_n - f\|_p^p < \frac{\epsilon}{2^p}$ for all $n > \bar{n}$. Then for every $F \in \mathcal{A}$, we have

$$\sup_{n>\bar{n}} \left(\int_F |f_n|^p \, d\mu \right)^{\frac{1}{p}} \leq \sup_{n>\bar{n}} \|f_n - f\|_p + \left(\int_F |f|^p \, d\mu \right)^{\frac{1}{p}} \leq \frac{\epsilon^{\frac{1}{p}}}{2} + \left(\int_F |f|^p \, d\mu \right)^{\frac{1}{p}}. \tag{2.67}$$

Moreover (see Remark 2.8.2(i)):
(a) there exists $E \in \mathcal{A}$ with $\mu(E) < \infty$ such that $\int_{E^c} |f|^p \, d\mu < \frac{\epsilon}{2^p}$;
(b) there exists $\delta > 0$ such that $\int_F |f|^p \, d\mu < \frac{\epsilon}{2^p}$ for all $F \in \mathcal{A}$ with $\mu(F) < \delta$.

Then from (2.67) the claim follows.

(ii)⇒(i). By Remark 2.1.2 the function f is $\mathcal{A}$-measurable. Fix arbitrary $\epsilon > 0$, and let $E \in \mathcal{A}$ and $\delta > 0$ be as in Definition 2.8.1. Let us first prove that $f \in L^p(X)$. Since by assumption $\mu(E) < \infty$, by the Egorov theorem there exists a measurable set $F_\delta \subseteq E$ such that $\mu(F_\delta) < \delta$ and

$$\lim_{n\to\infty} \sup_{x\in E\setminus F_\delta} |f_n(x) - f(x)| = 0. \tag{2.68}$$

On the other hand,
(a) for both $G = E^c$ and $G = F_\delta$, by the Fatou lemma

$$\int_G |f|^p \, d\mu \leq \liminf_{n\to\infty} \int_G |f_n|^p \, d\mu \leq \epsilon;$$

(b) the sequence $\{\int_{x\in E\setminus F_\delta} |f_n(x) - f(x)|^p \, d\mu\}$ is bounded, since by (2.68) we have

$$\lim_{n\to\infty} \int_{E\setminus F_\delta} |f_n(x) - f(x)|^p \, d\mu \leq \mu(E) \lim_{n\to\infty} \sup_{x\in E\setminus F_\delta} |f_n(x) - f(x)| = 0,$$

and thus $\int_{E\setminus F_\delta} |f|^p \, d\mu < \infty$.

By (a)–(b) we have

$$\int_X |f|^p \, d\mu = \int_{E^c} |f|^p \, d\mu + \int_{E\setminus F_\delta} |f|^p \, d\mu + \int_{F_\delta} |f|^p \, d\mu < \infty,$$

and hence $f \in L^p(X)$. Now write

$$\begin{aligned} \|f_n - f\|_p^p &= \int_{E^c} |f_n - f|^p \, d\mu + \int_{E\setminus F_\delta} |f_n - f|^p \, d\mu + \int_{F_\delta} |f_n - f|^p \\ &=: I_{n,1} + I_{n,2} + I_{n,3}. \end{aligned} \tag{2.69}$$

Arguing as in (a), by the Fatou lemma we get $I_{n,k} \leq 2^p\epsilon$ $(k = 1, 3)$. Since by (2.68) we have $\lim_{n\to\infty} I_{n,2} = 0$, we obtain that $\limsup_{n\to\infty} \|f_n - f\|_p^p \leq 2^{p+1}\epsilon$. Then by the arbitrariness of ϵ the claim follows. This completes the proof. □

From the Vitali theorem we get the following generalization of the Lebesgue theorem (see Theorem 2.2.3).

Theorem 2.8.11. *Let $p \in [1,\infty)$. Let $f, f_n \in L^p(X)$ $(n \in \mathbb{N})$, and let $f_n \to f$ μ-a. e. in X. Suppose that*

$$\text{there exists } g \in L^p(X),\, g \geq 0 \text{ such that } \sup_{n\in\mathbb{N}} |f_n| \leq g \ \mu\text{-a. e. in } X. \tag{2.70}$$

Then $f \in L^p(X)$ and $f_n \to f$ in $L^p(X)$.

Proof. By Remark 2.8.2(i) and condition (2.70) the sequence $\{f_n\}$ is p-uniformly integrable. Then by Theorem 2.8.10 the conclusion follows. □

Let us mention the following result.

Lemma 2.8.12. *Let $\mu(X) < \infty$, and let $1 \leq p < q \leq \infty$. Then every bounded subset $\mathscr{F} \subseteq L^q(X)$ is p-uniformly integrable.*

Proof. Property (i) of Definition 2.8.1 is obviously satisfied, since by assumption $\mu(X) < \infty$. Concerning (ii), let $C_q := \sup_{f\in\mathscr{F}} \|f\|_q < \infty$. For any $F \in \mathcal{A}$ with $\mu(F) < \delta$, we have

$$\sup_{f\in\mathscr{F}} \int_F |f|^p \, d\mu \leq \mu(F)^{\frac{q-p}{q}} \sup_{f\in\mathscr{F}} \|f\|_q^p \leq C_q^p \, \delta^{\frac{q-p}{q}} \quad \text{if } q < \infty$$

by the Hölder inequality and

$$\sup_{f\in\mathscr{F}} \int_F |f|^p \, d\mu \leq \mu(F) \sup_{f\in\mathscr{F}} \|f\|_\infty^p \leq C_\infty^p \, \delta \quad \text{if } q = \infty.$$

By these inequalities property (ii) of Definition 2.8.1 holds with $\delta := (\frac{\epsilon}{C_q^p})^{\frac{q}{q-p}}$ if $q < \infty$ and $\delta := \frac{\epsilon}{C_\infty^p}$ if $q = \infty$. Hence the result follows. □

The following result is an immediate consequence of the Vitali theorem and Lemma 2.8.12.

Proposition 2.8.13. *Let $\mu(X) < \infty$, and let $1 \leq p < q \leq \infty$. Let $\{f_n\} \subseteq L^q(X)$ be a bounded sequence converging μ-a. e. in X to some function $f : X \to \mathbb{R}$. Then $f_n \to f$ strongly in $L^p(X)$.*

2.8.4 Weak and weak* convergence

Let q be the Hölder conjugate of $p \in [1,\infty]$:

$$q := \begin{cases} 1 & \text{if } p = \infty, \\ \frac{p}{p-1} & \text{if } p \in (1,\infty), \\ \infty & \text{if } p = 1. \end{cases} \tag{2.71}$$

Concerning duality of Lebesgue spaces, the following result is well known.

Theorem 2.8.14 (Riesz). *Let $p \in [1,\infty)$ and q be as in (2.71); if $p = 1$, let μ be σ-finite. Then the map*

$$\Theta : L^q(X) \to (L^p(X))^*, \quad \Theta g := T_g \quad (g \in L^q(X)),$$

where

$$\langle T_g, f\rangle := \int_X fg\, d\mu \quad (f \in L^p(X)),$$

is an isometric isomorphism.

Accordingly, the weak and weak* convergence in $L^p(X)$ are as follows.

Definition 2.8.3. (i) For any $p \in [1,\infty)$, f_n converges *weakly* to f in $L^p(X)$ (written $f_n \rightharpoonup f$) if

$$\lim_{n\to\infty} \int_X f_n g\, d\mu = \int_X fg\, d\mu \quad \text{for all } g \in L^q(X).$$

(ii) For any $p \in (1,\infty]$, f_n converges *weakly** to f in $L^p(X)$ (written $f_n \overset{*}{\rightharpoonup} f$) if

$$\lim_{n\to\infty} \int_X f_n g\, d\mu = \int_X fg\, d\mu \quad \text{for all } g \in L^q(X).$$

Weak and weak* convergence coincide for $p \in (1,\infty)$, since in this range of values of p the space $L^p(X)$ is uniformly convex and thus reflexive (e. g., see [25, 58]).

Example 2.8.1. Let $(X, \mathcal{A}, \mu) = (I, \mathcal{B} \cap I, \lambda|_{\mathcal{B}\cap I})$, where $I \equiv (0,1)$.

(i) Consider the sequence $f_n(x) := \sin(2n\pi x)$ $(n \in \mathbb{N},\ x \in I)$. By the Riemann–Lebesgue lemma, for all $p \in [1,\infty)$, we have $f_n \rightharpoonup 0$ in $L^p(I)$; however, $f_n \not\to 0$ in $L^p(I)$ since $\|f_n\|_{L^p(I)} = c_p > 0$ $(n \in \mathbb{N},\ p \in [1,\infty))$. It is often said in such cases that *mass concentration* occurs (this loose parlance is used also for other phenomena; see Example 5.1.1(ii)). Observe that in this case the inequality given by the lower semicontinuity of the norm,

$$\|f\|_{L^p(I)} \leq \liminf_{n\to\infty} \|f_n\|_{L^p(I)},$$

is strict. This is expressed by saying that a *loss of mass* occurs in the weak limit.

The members of the above sequence are oscillating functions, the frequency of the oscillations growing to infinity as $n \to \infty$. Sequences of oscillating functions often converge weakly, but not strongly, a feature often called the *oscillation phenomenon*.

Alternatively, observe that f_n weakly converges to 0, but the sequence $\{h \circ f_n\}$ with *nonlinear* $h(s) = |s|^p$ does not. It is often said in this connection that *nonlinearity destroys weak convergence*.

(ii) Consider the sequence $f_n := n^{\frac{1}{q}} \chi_{(0,\frac{1}{n})}$ ($n \in \mathbb{N}$, $q \in [1,\infty)$). we have $f_n \to 0$ λ-a. e. in I; however, $\|f_n\|_{L^q(I)} = 1$, thus $f_n \nrightarrow 0$ strongly in $L^q(I)$. Observe that the sequence $\{f_n\}$ is not q-uniformly integrable (otherwise, we would have a contradiction with the Vitali Theorem). Indeed, if $q = 1$, then a direct calculation shows that

$$\sup\left\{n \int_E \chi_{(0,\frac{1}{n})}\, d\lambda \mid n \in \mathbb{N},\, E \in \mathcal{B} \cap I \text{ such that } \lambda(E) < \delta\right\}$$

$$= \sup\left(n\lambda\left(E \cap \left(0, \frac{1}{n}\right)\right)\right) \leq \sup_{n\in\mathbb{N}} \min\{n\delta, 1\} = 1.$$

On the other hand, choosing $E = E_n = (0, \frac{1}{n})$ with $n > \frac{1}{\delta}$ plainly gives $\lambda(E) < \delta$ and $n \int_E \chi_{(0,\frac{1}{n})}\, d\lambda = 1$, whence

$$\sup\left\{n \int_E \chi_{(0,\frac{1}{n})}\, d\lambda \mid n \in \mathbb{N},\, E \in \mathcal{B} \cap I \text{ such that } \lambda(E) < \delta\right\} = 1.$$

Then by (2.63) we have $\eta = 1$. Observe that the positivity of the modulus of uniform integrability is associated with concentration phenomenon (in this connection, see Remark 5.4.6(iii)). Also observe that for $q = 1$, $f_n \nrightharpoonup 0$ in $L^1(I)$, since for all $g \in C(\bar{I})$, as $n \to \infty$,

$$\int_I f_n(x) g(x)\, dx = n \int_{(0,\frac{1}{n})} g(x)\, dx = \int_I g\left(\frac{\xi}{n}\right) d\xi \to g(0)$$

(a similar situation will be discussed in Example 5.1.1).

Let us finally notice that if $1 \leq p < q$, then $\|f_n\|_{L^p(I)} = n^{\frac{p}{q}-1}$ for all $n \in \mathbb{N}$ (i. e., there is no mass concentration in $L^p(I)$ with $p < q$). Hence $f_n \to 0$ strongly in $L^p(I)$, in agreement with Proposition 2.8.13.

The following results concern the relationship between weak convergence and other types of convergence.

Proposition 2.8.15. *Let $p \in (1,\infty)$. Let $\{f_n\} \subseteq L^p(X)$ be such that $\{\|f_n\|_p\}$ is bounded and $f_n \to f$ μ-a. e. in X. Then $f \in L^p(X)$ and $f_n \rightharpoonup f$ in $L^p(X)$.*

Proof. The fact that $f \in L^p(X)$ follows by the Fatou lemma, since the sequence $\{\|f_n\|_p\}$ is bounded. Moreover, for any fixed $\epsilon > 0$ and $g \in L^q(X)$, there exist $\delta > 0$ such that $\int_E |g|^q\, d\mu < \epsilon^q$ if $\mu(E) < \delta$ and $F \in \mathcal{A}$ with $\mu(F) < \infty$ such that $\int_{F^c} |g|^q\, d\mu < \epsilon^q$. Also, by the Egorov theorem there exists $F_\delta \subseteq F$ such that $\mu(F \setminus F_\delta) < \delta$ and $\lim_{n\to\infty} \sup_{x\in F_\delta} |f_n(x) - f(x)| = 0$. Writing $X = F^c \cup (F \setminus F_\delta) \cup F_\delta$, by the additivity of the integral and the Hölder inequality we obtain

$$\limsup_{n\to\infty}\left|\int_X (f_n - f)g\,d\mu\right| \le 2C\left\{\left(\int_{F^c} |g|^q\,d\mu\right)^{\frac{1}{q}} + \left(\int_{F\setminus F_\delta} |g|^q\,d\mu\right)^{\frac{1}{q}}\right\} < 4C\epsilon,$$

where $C := \sup_n \|f_n\|_p < \infty$. Then by the arbitrariness of ϵ the result follows. □

Remark 2.8.5. It is easily seen that the same conclusion holds if the convergence μ-a. e. in X is replaced by the convergence in measure.

Proposition 2.8.16 (Radon). *Let $p \in (1,\infty)$. Let $f_n \rightharpoonup f$ in $L^p(X)$, and $\|f_n\|_p \to \|f\|_p$. Then $f_n \to f$ in $L^p(X)$.*

Proof. If $p = 2$, then the conclusion immediately follows from the identity

$$\|f_n - f\|_2^2 = \|f_n\|_2^2 + \|f\|_2^2 - 2\int_X f_n f\,d\mu.$$

For the general case, see, e. g., [40, Proposition V.11.1]. □

Concerning weak compactness of $L^p(X)$, for any $p \in (1,\infty)$ the following holds.

Theorem 2.8.17. *Let $(X,\mathcal{A},\mu)$ be a separable measure space, and let μ be σ-finite. Let $p \in (1,\infty)$. Then the following statements are equivalent:*
(i) *$\mathscr{F} \subseteq L^p(X)$ is bounded;*
(ii) *$\mathscr{F}$ is relatively sequentially compact in the weak topology.*

Proof. (i)⇒(ii). Since $L^p(X)$ is separable and reflexive, the claim follows by the Banach–Alaoglu theorem.

(ii)⇒(i). Were $\mathscr{F}$ unbounded, there would exist $\{f_n\} \subseteq L^p(X)$ such that $\|f_n\|_p \to \infty$, and hence $\|f_{n_k}\|_p \to \infty$ for every $\{f_{n_k}\} \subseteq \{f_n\}$. On the other hand, by assumption for any $\{f_n\} \subseteq \mathscr{F}$, there exists $\{f_{n_k}\} \subseteq \{f_n\}$ weakly convergent and hence bounded. From the contradiction the result follows. □

In the case $p = 1$ the situation is as follows.

Theorem 2.8.18 (Dunford–Pettis). *Let X be a σ-compact Hausdorff space, and let $\mu \in \mathfrak{R}_f^+(X)$. Then the following statements are equivalent:*
(i) *$\mathscr{F}$ is bounded in $L^1(X)$ and uniformly integrable;*
(ii) *$\mathscr{F} \subseteq L^1(X)$ is relatively sequentially compact in the weak topology.*

To prove Theorem 2.8.18, we need the following lemma.

Lemma 2.8.19. *Let $(X,\mathcal{A},\mu)$ be a measure space. Then the following statements are equivalent:*
(i) *$\{f_n\} \subseteq L^1(X)$ is weakly convergent;*

(ii) $\{f_n\}$ *is bounded in* $L^1(X)$, *and for any* $E \in \mathcal{A}$, *the sequence* $\{\int_E f_n \, d\mu\} \subseteq \mathbb{R}$ *is convergent.*

Proof. If Y is a metric space and $\{y_n\} \subseteq Y$, then y_n is weakly convergent if and only if it is bounded, and the sequence $\langle y^*, y_n \rangle_{Y^*,Y}$ converges for any y^* in a subset $Y' \subseteq Y^*$, the linear span of Y' being dense in Y^* (e. g., see [66, Theorem IV.15.1]). Since $\int_E f_n \, d\mu = \int_X f_n \chi_E \, d\mu$ and $\mathscr{S}(X)$ is dense in $L^\infty(X)$, the result follows. □

Proof of Theorem 2.8.18. (i)⇒(ii). Since $L^1(X)$ is canonically injected in the Banach space $\mathfrak{R}_f(X)$, by assumption $\mathscr{F}$ is a bounded subset of $\mathfrak{R}_f(X)$ (see Remark 2.2.3 and Section 5.1). Also, observe that $C_0(X)$ is separable. Then by the Banach–Alaoglu theorem and Theorem 2.7.1 for any $\{f_n\} \subseteq \mathscr{F}$, there exist $\{f_{n_k}\} \subseteq \{f_n\}$ and $\nu \in \mathfrak{R}_f(X)$ such that $f_{n_k} \overset{*}{\rightharpoonup} \nu$ in $\mathfrak{R}_f(X)$. In view of Lemma 2.8.19, ν can be identified with a function $f \in L^1(X)$ (i. e., there exists $f \in L^1(X)$ such that $\nu(X) = \int_X f \, d\mu$) if we prove that for any $E \in \mathcal{A}$, the sequence $\{\int_E f_{n_k} \, d\mu\} \subseteq \mathbb{R}$ is a Cauchy sequence.

To this purpose, let $E \in \mathcal{A}$ be fixed. By Proposition 2.1.16 and Remark 2.1.7 there exists a sequence $\{\zeta_m\} \subseteq C_c(X)$ such that $\zeta_m \to \chi_E$ μ-a. e. in X and $\|\zeta_m\|_\infty = 1$ for all $m \in \mathbb{N}$. Fix $\epsilon > 0$, and let $\delta = \delta(\epsilon) > 0$ be given by Definition 2.8.1(ii). Since by assumption μ is finite, by the Egorov theorem there exists $E_\delta \in \mathcal{A}$ such that $\mu(E_\delta^c) < \epsilon$ and $\lim_{m\to\infty} \sup_{x \in E_\delta} |\zeta_m(x) - \chi_E(x)| = 0$. Therefore there exists $m_0 \in \mathbb{N}$ such that $\sup_{x\in E_\delta} |\zeta_m(x) - \chi_E(x)| < \epsilon$ for all $m \geq m_0$.

Now observe that for any $k, l \in \mathbb{N}$,

$$
\begin{aligned}
&\left| \int_E (f_{n_k} - f_{n_l}) \, d\mu \right| \\
&\quad = \left| \int_X (f_{n_k} - f_{n_l}) \chi_E \, d\mu \right| \\
&\quad \leq \left| \int_{E_\delta^c} (f_{n_k} - f_{n_l})(\chi_E - \zeta_{m_0}) \, d\mu \right| + \left| \int_{E_\delta} (f_{n_k} - f_{n_l})(\chi_E - \zeta_{m_0}) \, d\mu \right| + \left| \int_X (f_{n_k} - f_{n_l}) \zeta_{m_0} \, d\mu \right| \\
&\quad =: I_1 + I_2 + I_3.
\end{aligned}
$$

By the above considerations we have

$$
I_1 \leq 2 \int_{E_\delta^c} (|f_{n_k}| + |f_{n_l}|) \, d\mu < 4\epsilon, \quad \text{since } \mu(E_\delta^c) < \epsilon \text{ and } \mathscr{F} \text{ is uniformly integrable;}
$$

$$
I_2 \leq \sup_{x \in E_\delta} |\zeta_{m_0}(x) - \chi_E(x)| \int_{E_\delta} (|f_{n_k}| + |f_{n_l}|) \, d\mu < 2\epsilon M, \quad \text{where } M := \sup_{f \in \mathscr{F}} \|f\|_1.
$$

Moreover, we have $I_3 < \epsilon$ for any $k, l \in \mathbb{N}$ sufficiently large because $\{f_{n_k}\}$ is weakly* convergent in $\mathfrak{R}_f(X)$, and thus the sequence $\{\int_X f_{n_k} \zeta_{m_0} \, d\mu\} \subseteq \mathbb{R}$ is convergent since $\zeta_{m_0} \in C_0(X)$.

To summarize, we proved that for any $E \in \mathcal{A}$ and $\epsilon > 0$, there exists $k_0 \in \mathbb{N}$ such that for all $k, l \geq k_0$, we have

$$\left| \int_E (f_{n_k} - f_{n_l})\, d\mu \right| < (2M + 5)\epsilon,$$

that is, $\{\int_E f_{n_k}\, d\mu\} \subseteq \mathbb{R}$ is a Cauchy sequence for any $E \in \mathcal{A}$. Hence the claim follows.

(ii)⇒(i). The boundedness of $\mathscr{F}$ follows as in the proof of Theorem 2.8.17. Let us prove that $\mathscr{F}$ is uniformly integrable. By contradiction let the opposite hold. Then there exist a sequence $\{E_n\} \subseteq \mathcal{A}$ and a sequence $\{f_n\} \subseteq \mathscr{F}$ with the properties stated in Lemma 2.8.5. On the other hand, since by assumption $\mathscr{F}$ is relatively weakly sequentially compact, there exist a subsequence of $\{f_n\}$ (not relabeled) and $f \in L^1(X)$ such that $f_n \rightharpoonup f$ in $L^1(X)$. Choosing $h = f$ in Lemma 2.8.5, we have that there exists $k = k_f > 0$ such that for all $n \in \mathbb{N}$, we have

$$\left| \int_{E_n} (f_n - h)\, d\mu \right| \geq k. \tag{2.72}$$

Using (2.72), we can find $g \in L^\infty(X)$ such that

$$\lim_{n\to\infty} \int_X f_n g\, d\mu \neq \int_X fg\, d\mu, \tag{2.73}$$

a contradiction, whence the claim follows.

In fact, set $F_1 := E_1$ and $n_1 := 1$, and let $\epsilon_1 > 0$ be such that

$$\int_E |f_{n_1} - f|\, d\mu < \frac{k}{3} \quad \text{for all } E \in \mathcal{A} \text{ such that } \mu(E) < \epsilon_1.$$

Let $n_2 \in \mathbb{N}$ be the smallest integer greater than n_1 such that

$$\mu\left(\bigcup_{n=n_2}^{\infty} E_n \right) < \epsilon_1, \quad \left| \int_{F_1} (f_{n_2} - f)\, d\mu \right| < \frac{k}{3}.$$

Observe that such n_2 exists, since $\lim_{p\to\infty} \mu(\bigcup_{n=p}^{\infty} E_n) = 0$, and

$$\lim_{n\to\infty} \int_{F_1} (f_n - f)\, d\mu = \lim_{n\to\infty} \int_X (f_n - f)\chi_{F_1}\, d\mu = 0,$$

since $\chi_{F_1} \in L^\infty(X)$ and $f_n \rightharpoonup f$ in $L^1(X)$.

Set $F_2 := E_{n_2}$, and observe that

$$\int_{\bigcup_{n=n_2}^{\infty} E_n} |f_{n_1} - f|\, d\mu < k/3,$$

since $\mu(\bigcup_{n=n_2}^{\infty} E_n) < \epsilon_1$. Further, fix $\epsilon_2 > 0$ such that

$$\int_E |f_{n_2} - f| d\mu < \frac{k}{3} \quad \text{for all } E \in \mathcal{A} \text{ such that } \mu(E) < \epsilon_2,$$

and let $n_3 \in \mathbb{N}$ be the smallest integer greater that n_2 such that

$$\mu\left(\bigcup_{n=n_3}^{\infty} E_n\right) < \epsilon_2, \quad \left|\int_{F_1 \cup F_2} (f_{n_3} - f)\, d\mu\right| < \frac{k}{3}$$

(as before, it is easily seen that such n_3 exists).

Set $F_3 := E_{n_3}$, and observe that

$$\int_{\bigcup_{n=n_3}^{\infty} E_n} |f_{n_2} - f|\, d\mu < k/3,$$

since $\mu(\bigcup_{n=n_3}^{\infty} E_n) < \epsilon_2$. Iterating the argument, we can construct two subsequences $\{F_p\} \equiv \{E_{n_p}\} \subseteq \{E_n\}$ and $\{f_p\} \equiv \{f_{n_p}\} \subseteq \{f_n\}$ and a sequence $\{\epsilon_p\} \subseteq (0, \infty)$ such that

$$\mu\left(\bigcup_{q \geq p+1} F_q\right) \leq \mu\left(\bigcup_{n=n_{p+1}}^{\infty} E_n\right) < \epsilon_p \quad \Rightarrow \quad \left|\int_{\bigcup_{q>p} F_q} (f_p - f)\, d\mu\right| < \frac{k}{3} \tag{2.74a}$$

(this follows by construction, since $\bigcup_{q \geq p+1} F_q \subseteq \bigcup_{n=n_{p+1}}^{\infty} E_n$) and

$$\left|\int_{\bigcup_{q<p} F_q} (f_p - f)\, d\mu\right| < \frac{k}{3}, \tag{2.74b}$$

$$\left|\int_{F_p} (f_p - f)\, d\mu\right| \geq k \quad \text{for all } p \in \mathbb{N}. \tag{2.74c}$$

Now set $g := \chi_{\bigcup_{q=1}^{\infty} F_q}$. By (2.74a)–(2.74c) we have

$$\left|\int_X (f_p - f)\, g\, d\mu\right| = \left|\int_{\bigcup_{q<p} F_q} (f_p - f)\, d\mu + \int_{F_p} (f_p - f)\, d\mu + \int_{\bigcup_{q>p} F_q} (f_p - f)\, d\mu\right| \geq \frac{k}{3} > 0.$$

This proves (2.73), and hence the conclusion follows. □

2.9 Differentiation

2.9.1 Radon–Nikodým derivative

Theorem 2.9.1 (Radon–Nikodým). *Let $(X, \mathcal{A}, \mu)$ be a σ-finite measure space, and let $\nu \ll \mu$ be a signed measure on $\mathcal{A}$. Then there exists a unique quasi-integrable function $\nu_r : X \to [-\infty, \infty]$ such that*

$$\nu(E) = \int_E \nu_r \, d\mu \quad \textit{for all } E \in \mathcal{A}. \tag{2.75}$$

Moreover: (i) $\nu_r \geq 0$ in X if and only if ν is positive; (ii) $\nu_r \in L^1(X, \mathcal{A}, \mu)$ if and only if ν is finite; and (iii) $\nu_r : X \mapsto \mathbb{R}$ if and only if ν is σ-finite.

Definition 2.9.1. The function $\nu_r \equiv \frac{d\nu}{d\mu}$ in equality (2.75) is called the *Radon–Nikodým derivative* (or the *density*) of ν with respect to μ.

To prove Theorem 2.9.1, we need the following result.

Lemma 2.9.2 (von Neumann). *Let $(X, \mathcal{A}, \mu)$ be a finite measure space, and let $\nu \leq \mu$ be a finite measure on $\mathcal{A}$. Then there exists an $\mathcal{A}$-measurable function $h : X \to [0, 1]$ such that*

$$\nu(E) = \int_E h \, d\mu \quad \textit{for all } E \in \mathcal{A}. \tag{2.76}$$

Proof. The map from $L^2(X, \mathcal{A}, \mu)$ to $\mathbb{R}$, $f \mapsto \int_{\mathbb{R}^N} f \, d\nu$, is linear and continuous, since

$$\left|\int_{\mathbb{R}^N} f \, d\nu\right| \leq \int_{\mathbb{R}^N} |f| \, d\nu \leq [\nu(X)]^{\frac{1}{2}} \left(\int_{\mathbb{R}^N} |f|^2 \, d\nu \right)^{\frac{1}{2}} \leq [\mu(X)]^{\frac{1}{2}} \left(\int_{\mathbb{R}^N} |f|^2 \, d\mu \right)^{\frac{1}{2}}.$$

Hence there exists $h \in L^2(X, \mathcal{A}, \mu)$ such that

$$\int_{\mathbb{R}^N} f \, d\nu = \int_{\mathbb{R}^N} fh \, d\mu \quad \text{for all } f \in L^2(X, \mathcal{A}, \mu). \tag{2.77}$$

Choosing $f = \chi_E$ $(E \in \mathcal{A})$ in the above equality gives (2.76). Clearly, $h \geq 0$ since μ, ν are positive. Were $\mu(\{h > 1\}) > 0$, from (2.76) we would get $\nu(\{h > 1\}) > \mu(\{h > 1\})$, a contradiction. Hence the result follows. □

Remark 2.9.1. (i) Let the assumptions of Lemma 2.9.2 be satisfied. Then for every $\mathcal{A}$-measurable function $f : X \to [0, \infty]$ and $E \in \mathcal{A}$, we have

$$\int_E f \, d\nu = \int_E fh \, d\mu, \tag{2.78}$$

where $h : X \to [0,1]$ is the $\mathcal{A}$-measurable function associated with ν in equality (2.76).

(ii) Under the assumptions of Lemma 2.9.2, we have $h \in L^1(X)$. Conversely, by abuse of language, we can say that $\nu \in L^1(X)$ if there exists $h \in L^1(X)$ such that (2.76) holds.

Proof of Theorem 2.9.1. For the uniqueness, it suffices to recall that if μ is σ-finite and two quasi-integrable functions h, h' satisfy $\int_E h\,d\mu = \int_E h'\,d\mu$ for all $E \in \mathcal{A}$, then $h' = h$ μ-a. e. in X (e. g., see [45, Satz IV.4.5]). Concerning the existence, it suffices to prove the result when ν is positive, since we have $\nu \ll \mu$ if and only if $\nu^{\pm} \ll \mu$ (see Lemma 1.8.5). We will give the proof in three steps: *a*) both μ and ν are finite; *b*) μ is finite; and *c*) μ is σ-finite.

a) Let μ, ν be positive and finite, with $\nu \ll \mu$. Set $\rho := \mu + \nu$. Then ρ is finite, $\mu \le \rho$, and $\nu \le \rho$. Then by Lemma 2.9.2 there exist two $\mathcal{A}$-measurable functions $g, h : X \to [0,1]$ such that $\mu(E) = \int_E g\,d\rho$ and $\nu(E) = \int_E h\,d\rho$ $(E \in \mathcal{A})$. Set $N := \{g = 0\}$; then we have $\mu(N) = 0$, and thus $\nu(N) = 0$ since $\nu \ll \mu$. Define

$$f(x) := \begin{cases} 0 & \text{if } x \in N, \\ \frac{h(x)}{g(x)} & \text{if } x \in N^c. \end{cases}$$

Then f is $\mathcal{A}$-measurable nonnegative, and

$$\nu(E) = \nu\left(E \bigcap N^c\right) = \int\limits_{E \bigcap N^c} h\,d\rho = \int\limits_E fg\,d\rho = \int\limits_E f\,d\mu$$

(here equality (2.78) was used). Then equality (2.75) follows in this case with $\nu_r := f$. Moreover, since ν is positive and finite, by (2.75) ν_r is nonnegative, and $\nu_r \in L^1(X, \mathcal{A}, \mu)$.

b) Let μ, ν be positive, μ finite, and $\nu \ll \mu$. Arguing as in the proof of Proposition 1.8.8, set $\alpha := \sup\{\mu(E) \mid E \in \mathcal{A},\ \nu(E) < \infty\}$ (clearly, $\alpha \le \mu(X) < \infty$). Let $\{E_n\} \subseteq \mathcal{A}$ be a nondecreasing sequence such that $\nu(E_n) < \infty$ for each $n \in \mathbb{N}$ and $\lim_{n\to\infty} \mu(E_n) = \alpha$. Set $E_\infty := \bigcup_{n=1}^\infty E_n$. Then $E_\infty \in \mathcal{A}$ and $\mu(E_\infty) = \lim_{n\to\infty} \mu(E_n) = \alpha$. Therefore, for any $E \in \mathcal{A}$, $E \subseteq E_\infty^c$, such that $\nu(E) < \infty$, we have

$$\alpha + \mu(E) = \lim_{n\to\infty} \mu(E_n) + \mu(E) = \lim_{n\to\infty} \mu\left(E_n \bigcup E\right) \le \alpha,$$

since $E_n \bigcap E = \emptyset$ for each $n \in \mathbb{N}$, and $\nu(E_n \bigcup E) \le \nu(E_n) + \nu(E) < \infty$. Then we have $\mu(E) = 0$, and thus $\nu(E) = 0$ since $\nu \ll \mu$. It follows that for any $E \in \mathcal{A}$, $E \subseteq E_\infty^c$, we have either $\mu(E) = \nu(E) = 0$, or $\mu(E) > 0$ and $\nu(E) = \infty$.

Now set $F_n := E_n \setminus E_{n-1}$ $(n \in \mathbb{N},\ F_1 := \emptyset)$. Then $\{F_n\} \subseteq \mathcal{A}$ is a disjoint sequence such that $E_\infty = \bigcup_{n=1}^\infty F_n$ and $\nu(F_n) = \nu(E_n) - \nu(E_{n-1}) < \infty$ for all $n \in \mathbb{N}$. Define $\nu_n := \nu \llcorner F_n$ $(n \in \mathbb{N})$, $\nu_\infty := \nu \llcorner E_\infty^c$, and observe that

$$\sum_{n=1}^\infty \nu_n(E) = \sum_{n=1}^\infty \nu\left(F_n \bigcap E\right) = \nu\left(E_\infty \bigcap E\right) \quad \text{for all } E \in \mathcal{A}. \tag{2.79}$$

Also, for each $n \in \mathbb{N}$, we have $\nu_n(X) = \nu(F_n) < \infty$ and $\nu_n \ll \nu \ll \mu$. Then by step a) there exists a nonnegative function $f_n \in L^1(X, \mathcal{A}, \mu)$ such that

$$\nu_n(E) = \int_E f_n \, d\mu \quad \text{for all } E \in \mathcal{A} \tag{2.80}$$

and $f_n = 0$ in F_n^c ($n \in \mathbb{N}$). Set also

$$g(x) := \begin{cases} 0 & \text{if } x \in E_\infty, \\ \infty & \text{if } x \in E_\infty^c. \end{cases}$$

Then g is $\mathcal{A}$-measurable and nonnegative, and by (2.79)–(2.80) and the monotone convergence theorem we have

$$\begin{aligned} \nu(E) &= \nu\left(E_\infty \bigcap E\right) + \nu\left(E_\infty^c \bigcap E\right) = \sum_{n=1}^{\infty} \nu_n(E) + \nu\left(E_\infty^c \bigcap E\right) \\ &= \sum_{n=1}^{\infty} \int_E f_n \, d\mu + \nu\left(E_\infty^c \bigcap E\right) = \int_E \left(\sum_{n=1}^{\infty} f_n\right) d\mu + \nu\left(E_\infty^c \bigcap E\right) = \int_E \left(\sum_{n=1}^{\infty} f_n + g\right) d\mu. \end{aligned} \tag{2.81}$$

Hence (2.75) follows also in this case with $\nu_r := \sum_{n=1}^{\infty} f_n + g$.

c) Let μ, ν be positive, μ σ-finite, and $\nu \ll \mu$. Let $\{X_n\} \subseteq \mathcal{A}$ be a disjoint sequence such that $\bigcup_{n=1}^{\infty} X_n = X$ and $\mu(X_n) < \infty$ for all $n \in \mathbb{N}$. Clearly, for any $n \in \mathbb{N}$, the measures $\mu_n := \mu \llcorner X_n$ and $\nu_n := \nu \llcorner X_n$ satisfy the assumptions of step b), and hence there exists an $\mathcal{A}$-measurable function $\hat{h}_n : X \to [0, \infty]$ such that

$$\nu_n(E) = \int_{E \bigcap X_n} \hat{h}_n \, d\mu \quad \text{for all } E \in \mathcal{A}. \tag{2.82}$$

For every $n \in \mathbb{N}$, define

$$h_n(x) := \begin{cases} \hat{h}_n(x) & \text{if } x \in X_n, \\ 0 & \text{if } x \in X_n^c. \end{cases}$$

Then for all $n \in \mathbb{N}$ and $E \in \mathcal{A}$, we have $\int_E h_n \, d\mu = \int_{E \bigcap X_n} \hat{h}_n \, d\mu$, and thus from (2.82) we get

$$\nu(E) = \sum_{n=1}^{\infty} \nu(E \cap X_n) = \sum_{n=1}^{\infty} \nu_n(E) = \sum_{n=1}^{\infty} \int_E h_n \, d\mu = \int_E \left(\sum_{n=1}^{\infty} h_n\right) d\mu.$$

This completes the proof of equality (2.75) with $\nu_r := \sum_{n=1}^{\infty} h_n$.

It remains to prove the final claims (i)–(iii) of the statement. Claims (i)–(ii) are immediate from (2.75). Concerning (iii), let $\nu_r : X \to \mathbb{R}$, and let $\{E_n\} \subseteq \mathcal{A}$ be a nondecreasing sequence such that $\bigcup_{n=1}^{\infty} E_n = X$ and $\mu(E_n) < \infty$ for each $n \in \mathbb{N}$. Then the sequence $\{F_n\} \subseteq \mathcal{A}$, where $F_n := E_n \cap \{|\nu_r| \leq n\}$, is nondecreasing, with $\bigcup_{n=1}^{\infty} F_n = X$; moreover, we have $|\nu(F_n)| \leq n\mu(E_n) < \infty$ for each $n \in \mathbb{N}$. Hence ν is σ-finite.

Conversely, let ν be σ-finite. Let $\{X_n\} \subseteq \mathcal{A}$ be a disjoint sequence such that $\bigcup_{n=1}^{\infty} X_n = X$ and $|\nu(X_n)| < \infty$ for all $n \in \mathbb{N}$, whence $|\nu|(X_n) < \infty$ (see Remark 1.8.2(iii)). Set $\nu_n := \nu \llcorner X_n$. Then plainly ν_n is a finite signed measure, and we have $\nu_n \ll \mu$ $(n \in \mathbb{N})$. Hence by claim (ii) ν_n has a density $\nu_{nr} \in L^1(X_n, \mathcal{A}, \mu)$, and thus $|\nu_{nr}(x)| < \infty$ for μ-a. e. $x \in X_n$. Define $\nu_{nr} := 0$ in X_n^c. Since $\nu = \sum_{n=1}^{\infty} \nu_n$, by uniqueness we have $\nu_r = \sum_{n=1}^{\infty} \nu_{nr}$, and thus (possibly, changing its definition on a μ-null set) ν_r is a quasi-integrable real-valued function on X. This completes the proof. □

Remark 2.9.2. Let $(X, \mathcal{A}, \mu)$ be a σ-finite measure space, and let ν and ρ be finite signed measures on $\mathcal{A}$. Suppose that $\rho \ll \nu$ and $\nu \ll \mu$. Then $\rho \ll \mu$, and from (2.13) we plainly obtain that $\frac{d\rho}{d\mu} = \frac{d\rho}{d\nu}\frac{d\nu}{d\mu}$.

2.9.2 Differentiation of Radon measures on $\mathbb{R}^N$

In this section, we deal with signed Radon measures, which can be identified with the ordered couples $\mu = (\mu_1, \mu_2)$ of (positive) mutually singular Radon measures (see Definition 2.7.2 and Remark 2.7.2). As already remarked, a signed Radon measure need not be a signed measure in the sense of Definition 1.8.1, since μ_1 and μ_2 need not be finite, and thus in general their difference is not defined (however, for every Borel set $E \subset\subset U$, $\mu \llcorner E := \mu_1 \llcorner E - \mu_2 \llcorner E$ is a finite signed measure). Clearly, these two notions coincide if at least one of μ_1 and μ_2 is finite.

Let μ and ν be (positive) Radon measures on an open subset $U \subseteq \mathbb{R}^N$. For any $x_0 \in U$, set

$$\overline{D}_\mu \nu(x_0) := \begin{cases} \limsup_{r \to 0^+} \frac{\nu(B_r(x_0))}{\mu(B_r(x_0))} & \text{if } \mu(B_r(x_0)) > 0 \text{ for all } r > 0, \\ \infty & \text{if } \mu(B_r(x_0)) = 0 \text{ for some } r > 0, \end{cases} \tag{2.83}$$

$$\underline{D}_\mu \nu(x_0) := \begin{cases} \liminf_{r \to 0^+} \frac{\nu(B_r(x_0))}{\mu(B_r(x_0))} & \text{if } \mu(B_r(x_0)) > 0 \text{ for all } r > 0, \\ \infty & \text{if } \mu(B_r(x_0)) = 0 \text{ for some } r > 0; \end{cases} \tag{2.84}$$

here $B_r(x_0)$ denotes the open ball with center at x_0 and radius r. It is easily checked that these definitions do not change if we replace open balls by closed balls.

Definition 2.9.2. (i) Let μ and ν be positive Radon measures on U. If $\overline{D}_\mu \nu(x_0) = \underline{D}_\mu \nu(x_0) < \infty$, then we say that ν is differentiable with respect to μ at x_0 with *derivative* at x_0

$$D_\mu \nu(x_0) := \lim_{r\to 0^+} \frac{\nu(B_r(x_0))}{\mu(B_r(x_0))}. \tag{2.85}$$

(ii) Let μ be a positive and $\nu \equiv (\nu_1, \nu_2)$ a signed Radon measure on U. We say that ν is differentiable with respect to μ at x_0 if so are ν_1 and ν_2, and we set

$$D_\mu \nu(x_0) := D_\mu \nu_1(x_0) - D_\mu \nu_2(x_0). \tag{2.86}$$

The measure ν is differentiable with respect to μ in a subset $V \subseteq U$ if it is differentiable at each $x_0 \in V$. The function $D_\mu \nu : V \mapsto \mathbb{R}$ (nonnegative in case (i)) is called the *derivative* (or the *density*) of ν with respect to μ.

In the next theorem, as always in this section, a $(\mathcal{B}^d \cap U)$-measurable function is shortly called measurable.

Theorem 2.9.3. *Let μ be a positive and ν a signed Radon measure on an open subset $U \subseteq \mathbb{R}^N$. Then:*
(i) *the derivative $D_\mu \nu(x_0)$ exists for μ-a. e. $x_0 \in U$ and is measurable;*
(ii) *if $\nu \llcorner U' \ll \mu \llcorner U'$ for every open subset $U' \Subset U$, then $D_\mu \nu \in L^1_{\mathrm{loc}}(U, \mathcal{B}^d \bigcap U, \mu)$, and*

$$\nu(E) = \int_E D_\mu \nu \, d\mu \quad \textit{for every Borel set } E \subseteq U', \tag{2.87}$$

$$D_\mu \nu = \frac{d(\nu \llcorner U')}{d\mu} \quad \mu\textit{-a. e. in } U'. \tag{2.88}$$

Remark 2.9.3. If μ and ν are (positive) Radon measures on an open subset $U \subseteq \mathbb{R}^N$, such that $\nu \ll \mu$, it follows that $D_\mu \nu : U \to [0, \infty]$ and equality (2.87) is satisfied for any Borel set $E \subseteq U$. The same holds for every Borel set $E \subseteq U$ if ν is a finite signed measure, in which case $D_\mu \nu \in L^1(U, \mathcal{B}^d \bigcap U, \mu)$.

The proof of Theorem 2.9.3 relies on the following lemma, which is a consequence of Theorem 1.5.5.

Lemma 2.9.4. *Let μ and ν be (positive) Radon measures on an open subset $U \subseteq \mathbb{R}^N$. Then:*
(i) *the sets $\{x_0 \in U \mid \underline{D}_\mu \nu(x_0) < \overline{D}_\mu \nu(x_0)\}$ and $\{x_0 \in U \mid \overline{D}_\mu \nu(x_0) = \infty\}$ are μ-null;*
(ii) *the set $\{x_0 \in U \mid \underline{D}_\mu \nu(x_0) = 0\}$ is ν-null.*

Proof. It is easily seen that for any $r > 0$, the maps from U to $\mathbb{R}_+$, $x \mapsto \mu(B(x,r))$ and $x \mapsto \nu(B(x,r))$, are upper semicontinuous and hence measurable (see Remark 2.1.3(ii)). Then by (2.83)–(2.84) and Corollary 2.1.5 the functions $\overline{D}_\mu \nu$, $\underline{D}_\mu \nu$, and $D_\mu \nu$ are measurable, and thus the sets under considerations are also measurable.

Fix any compact subset $K \subset U$ with $\mu(K) > 0$. Observe that

$$\{x_0 \in K \mid \underline{D}_\mu \nu(x_0) < \overline{D}_\mu \nu(x_0)\} = \bigcup_{\alpha,\beta \in \mathbb{Q}_+, \alpha<\beta} \left(A_\beta \bigcap B_\alpha\right), \tag{2.89}$$

$$\{x_0 \in K \mid \overline{D}_\mu \nu(x_0) = \infty\} \subseteq A_\beta \quad \text{for all } \beta \in \mathbb{Q}_+, \tag{2.90}$$

$$\{x_0 \in K \mid \underline{D}_\mu \nu(x_0) = 0\} \subseteq B_\alpha \quad \text{for all } \alpha \in \mathbb{Q}_+, \tag{2.91}$$

where $\mathbb{Q}_+ := \{q \in \mathbb{Q} \mid q > 0\}$, and

$$A_\beta := \{x_0 \in K \mid \overline{D}_\mu \nu(x_0) \geq \beta\}, \quad B_\alpha := \{x_0 \in K \mid \underline{D}_\mu \nu(x_0) \leq \alpha\}.$$

We will prove the following statements:

$$\begin{cases} \text{(i)} & \text{for any Borel set } E \subseteq A_\beta, \text{ we have } \nu(E) \geq \beta\mu(E), \\ \text{(ii)} & \text{for any Borel set } E \subseteq B_\alpha, \text{ we have } \nu(E) \leq \alpha\mu(E) \end{cases} \tag{2.92}$$

$(\alpha, \beta \in (0, \infty))$. Relying on (2.92), the result is easily proven. In fact, we obtain that

$$\beta\mu\left(A_\beta \bigcap B_\alpha\right) \leq \nu\left(A_\beta \bigcap B_\alpha\right) \leq \alpha\mu\left(A_\beta \bigcap B_\alpha\right) \quad \text{for all } \alpha, \beta \in \mathbb{Q}_+, \alpha < \beta,$$

whence $\mu(A_\beta \bigcap B_\alpha) = 0$. Then by (2.89) we obtain

$$\mu(\{x_0 \in K \mid \underline{D}_\mu \nu(x_0) < \overline{D}_\mu \nu(x_0)\}) = 0.$$

Similarly, we have

$$\mu(A_\beta) \leq \frac{\nu(A_\beta)}{\beta} \leq \frac{\nu(K)}{\beta} \quad \text{for all } \beta \in \mathbb{Q}_+, \tag{2.93}$$

$$\nu(B_\alpha) \leq \alpha\mu(B_\alpha) \leq \alpha\mu(K) \quad \text{for all } \alpha \in \mathbb{Q}_+. \tag{2.94}$$

Letting $\beta \to \infty$ in (2.93) and using (2.90), we get

$$\mu(\{x_0 \in K \mid \overline{D}_\mu \nu(x_0) = \infty\}) = 0,$$

whereas letting $\alpha \to 0^+$ in (2.94) and using (2.91) give

$$\nu(\{x_0 \in K \mid \underline{D}_\mu \nu(x_0) = 0\}) = 0.$$

Then by the arbitrariness of K the result follows.

It remains to prove (2.92). We only prove part (i), since the proof of (ii) is analogous. Fix any open bounded $A \supseteq E$ and $\epsilon \in (0, \beta)$. Set

$$\mathcal{F} := \{\overline{B}_r(x_0) \subset A \mid x_0 \in E,\ \nu(\overline{B}_r(x_0)) \geq (\beta - \epsilon)\,\mu(\overline{B}_r(x_0))\}.$$

Since $x_0 \in E \subseteq A_\beta$, two cases are possible (see (2.83)): (i) there exists $r_0 > 0$ such that $\mu(B_{r_0}(x_0)) = 0$, and thus $\nu(\overline{B}_r(x_0)) \geq 0 = (\beta-\epsilon)\mu(\overline{B}_r(x_0))$ for all $r \in (0, r_0)$; (ii) $\mu(B_r(x_0)) > 0$ for all $r > 0$, and $\lim_{r\to 0^+} \sup_{\rho\in(0,r)} \frac{\nu(\overline{B}_\rho(x_0))}{\mu(\overline{B}_\rho(x_0))} \geq \beta$ (recall that in (2.83)–(2.84), we can replace open balls by closed balls). This proves that $\mathcal{F}$ is a fine cover of E. Then by Theorem 1.5.5 there exists a disjoint family $\mathcal{F}' \subseteq \mathcal{F}$ that is a μ-a. e. cover of E. Therefore

$$\begin{aligned}(\beta-\epsilon)\,\mu(E) &\leq (\beta-\epsilon)\,\mu\Big(\bigcup_{\overline{B}_\rho(x_0)\in\mathcal{F}'} \mu(\overline{B}_\rho(x_0))\Big) \leq (\beta-\epsilon) \sum_{\overline{B}_\rho(x_0)\in\mathcal{F}'} \mu(\overline{B}_\rho(x_0)) \\ &\leq \sum_{\overline{B}_\rho(x_0)\in\mathcal{F}'} \nu(\overline{B}_\rho(x_0)) \leq \nu(A).\end{aligned}$$

Then by the arbitrariness of ϵ and the outer regularity of ν the claim follows. □

Proof of Theorem 2.9.3. Suppose first ν to be positive. Then claim (i) follows from Lemma 2.9.4(i) and its proof. Concerning (ii), let $\{K_k\}$ be an increasing sequence of compact sets such that $\bigcup_{k=1}^\infty K_k = \mathbb{R}^N$. For any $k \in \mathbb{N}$, set

$$F_k := \Big\{x_0 \in E \bigcap K_k \mid \exists\, D_\mu \nu(x_0)\Big\}.$$

In view of claim (i), F_k is measurable, and we have $\mu(F_k) = \mu(E \bigcap K_k)$, and thus $\nu(F_k) = \nu(E \bigcap K_k)$ since $\nu \ll \mu$. Set also

$$F_k^0 := \{x_0 \in F_k \mid D_\mu \nu(x_0) = 0\},$$

and for any $m \in \mathbb{Z}$ and fixed $t > 1$,

$$F_{k,m} := \{x_0 \in F_k \mid t^m \leq D_\mu \nu(x_0) < t^{m+1}\}.$$

Then

$$F_k = F_k^0 \bigcup \Big(\bigcup_{m\in\mathbb{Z}} F_{k,m}\Big) \quad \text{for every } k \in \mathbb{N}.$$

By Lemma 2.9.4(ii) we have $\nu(F_k^0) = 0$ for all $k \in \mathbb{N}$, and thus

$$\nu(F_k) = \nu\Big(E \bigcap K_k\Big) = \sum_{m\in\mathbb{Z}} \nu(F_{k,m}). \tag{2.95}$$

On the other hand, by (2.92) and the very definition of $F_{k,m}$ we get

$$\frac{1}{t} \int_{F_{k,m}} D_\mu \nu \, d\mu \leq t^m \mu(F_{k,m}) \leq \nu(F_{k,m}) \leq t^{m+1} \mu(F_{k,m}) \leq t \int_{F_{k,m}} D_\mu \nu \, d\mu.$$

Letting $t \to 1^+$ in the above inequality gives

$$\nu(F_{k,m}) = \int_{F_{k,m}} D_\mu \nu \, d\mu \quad \text{for all } k \in \mathbb{N} \text{ and } m \in \mathbb{Z},$$

whence by (2.95)

$$\nu\left(E \bigcap K_k\right) = \int_{F_k} D_\mu \nu \, d\mu \quad \text{for all } k \in \mathbb{N}.$$

Letting $k \to \infty$ in the above equality proves (2.87) in the present case.

If $\nu \equiv (\nu_1, \nu_2)$ is a signed Radon measure, claims (i)–(ii) hold for the derivatives $D_\mu \nu_1$ and $D_\mu \nu_2$. Then by the very definition of $D_\mu \nu$ (see (2.86)) claim (i) also holds in this case, and, given any open subset $U' \subset\subset U$, the equality in (2.87) is satisfied for every Borel set $E \subseteq U'$. Finally, from equalities (2.75) and (2.87), for any measurable $E \subseteq U'$, we get

$$\int_E \left(D_\mu \nu - \frac{d(\nu \llcorner U')}{d\mu} \right) d\mu = 0.$$

Then by the arbitrariness of E we obtain (2.88), and thus the conclusion follows. □

Corollary 2.9.5. *Let μ be a Radon measure on $U \subseteq \mathbb{R}^N$. Then for μ-a. e. $x_0 \in U$:*
(i) *if $f \in L^1_{\mathrm{loc}}(U)$, then*

$$\lim_{r \to 0^+} \frac{1}{\mu(B_r(x_0))} \int_{B_r(x_0)} f \, d\mu = f(x_0); \tag{2.96}$$

(ii) *if $f \in L^p_{\mathrm{loc}}(U)$ ($p \in [1, \infty)$), then*

$$\lim_{r \to 0^+} \frac{1}{\mu(B_r(x_0))} \int_{B_r(x_0)} |f - f(x_0)|^p \, d\mu = 0. \tag{2.97}$$

Proof. Equality (2.96) coincides with (2.88) when $\nu(E) = \int_E f \, d\mu$ for any measurable E, and hence claim (i) follows. To prove (ii), fix $p \in [1, \infty)$. By (2.96), for any $q \in \mathbb{Q}$, there exists a μ-null set N_q such that for every $x_0 \in N_q^c$,

$$\lim_{r \to 0^+} \frac{1}{\mu(B_r(x_0))} \int_{B_r(x_0)} |f - q|^p \, d\mu = |f(x_0) - q|^p.$$

Set $N := \bigcup_{q\in\mathbb{Q}} N_q$, which is μ-null. Then for every $x_0 \in N^c$, we have

$$\limsup_{r\to 0^+} \frac{1}{[\mu(B_r(x_0))]^{\frac{1}{p}}} \Big(\int_{B_r(x_0)} |f - f(x_0)|^p \, d\mu \Big)^{\frac{1}{p}}$$

$$\le \lim_{r\to 0^+} \frac{1}{[\mu(B_r(x_0))]^{\frac{1}{p}}} \Big(\int_{B_r(x_0)} |f - q|^p \, d\mu \Big)^{\frac{1}{p}} + |f(x_0) - q| = 2|f(x_0) - q|.$$

Then by the denseness of $\mathbb{Q}$ equality (2.97) follows. □

Definition 2.9.3. A point $x_0 \in U$ such that (2.97) holds is called a Lebesgue point of f with respect to μ.

Corollary 2.9.6. *Let μ and ν be (positive) Radon measures on an open subset $U \subseteq \mathbb{R}^N$. Let $\nu_{ac} \ll \mu$ and $\nu_s \perp \mu$ be the absolutely continuous and singular parts of ν with respect to μ, respectively. Then*

$$D_\mu \nu = D_\mu \nu_{ac}, \quad D_\mu \nu_s = 0 \quad \mu\text{-a. e. in } U, \tag{2.98}$$

and for every Borel set $E \subseteq U$,

$$\nu(E) = \int_E D_\mu \nu \, d\mu + \nu_s(E) = \int_E D_\mu \nu_{ac} \, d\mu + \nu_s(E). \tag{2.99}$$

Proof. It suffices to prove that $D_\mu \nu_s = 0$ μ-a. e. in U, whence the first equality in (2.98) and (2.99) (applying (2.87) to ν_{ac} and using Remark 2.9.3) easily follow. Let $N \subseteq U$ be μ-null (thus $\nu_{ac}(N) = 0$) such that $\nu_s(U \setminus N) = 0$. Set $C_\beta := \{x \in U \setminus N \mid \exists D_\mu \nu_s(x) \ge \beta\}$ ($\beta > 0$). Since $D_\mu \nu_s(x)$ exists for μ-a. e. $x \in U$ and $\nu_s(C_\beta) = \nu_s(U \setminus N) = 0$, by (2.92)(i) we have that $\beta\mu(C_\beta) \le \nu_s(C_\beta) = 0$ for all $\beta > 0$. By the arbitrariness of β it follows that $\mu(\{x \in U \setminus N \mid D_\mu \nu_s(x) > 0\}) = 0$, whence $D_\mu \nu_s(x) = 0$ μ-a. e. in U. □

Remark 2.9.4. If $\nu \equiv (\nu_1, \nu_2)$ is a signed Radon measure on $U \subseteq \mathbb{R}^N$, applying Corollary 2.9.6 to both ν_1 and ν_2 proves that (2.98) holds also in this case. Similarly, given any open subset $U' \Subset U$, (2.99) is satisfied for every Borel set $E \subset U'$.

3 Function spaces and capacity

3.1 Function spaces

In this section, we recall some basic notions of scalar distribution theory and Sobolev spaces to fix notations and to make the presentation reasonably self-contained. We refer the reader, e. g., to [1, 64] for lucid and exhaustive accounts of the subjects.

3.1.1 Distributional derivative

Let $U \subseteq \mathbb{R}^N$ be open. Set

$$C^m(U) := \{f : U \mapsto \mathbb{R} \mid \exists D^\alpha f, \text{ and } D^\alpha f \in C(U) \ \ \forall |\alpha| \le m\} \quad (m \in \overline{\mathbb{N}}),$$

where $\overline{\mathbb{N}} := \mathbb{N} \cup \{0\}$, $\alpha \equiv (\alpha_1, \dots, \alpha_N)$ with $\alpha_k \in \overline{\mathbb{N}}$ for $k = 1, \dots, N$, $|\alpha| := \sum_{k=1}^N \alpha_k$, and $D^\alpha \equiv \frac{\partial^{|\alpha|}}{\partial^{\alpha_1} x_1 \dots \partial^{\alpha_N} x_N}$ (if $|\alpha| = 0$, then we set $D^\alpha f \equiv f$). We also write $x^\alpha \equiv x_1^{\alpha_1} \cdots x_N^{\alpha_N}$. Also, set $C^0(U) \equiv C(U)$, $C^\infty(U) := \bigcap_{m=0}^\infty C^m(U)$, and $C_c^\infty(U) := C^\infty(U) \bigcap C_c(U)$.

Definition 3.1.1. Let $U \subseteq \mathbb{R}^N$ be open.

(i) The space $\mathcal{D}(U)$ of *test functions* is the space $C_c^\infty(U)$ with the following notion of convergence: a sequence $\{\zeta_k\} \subseteq C_c^\infty(U)$ *converges to* ζ *in* $\mathcal{D}(U)$ if there exists a compact set $K \subset U$ that contains $\operatorname{supp}\zeta$ and $\operatorname{supp}\zeta_k$ $(k \in \mathbb{N})$, and $D^\alpha \zeta_k \to D^\alpha \zeta$ uniformly in K for all $\alpha \equiv (\alpha_1, \dots, \alpha_N) \in \overline{\mathbb{N}}^N$.

(ii) A function $f \in C^\infty(U)$ belongs to the Schwartz class $\mathcal{S}(U)$ if $\sup_{x \in U} |x^\alpha D^\beta f(x)| < \infty$ for all $\alpha, \beta \in \overline{\mathbb{N}}^N$.

(iii) The space of *distributions* (denoted $\mathcal{D}^*(U)$) is the dual space of continuous linear functionals on $\mathcal{D}(U)$. Namely, $T \in \mathcal{D}^*(U)$ if $T : \mathcal{D}(U) \mapsto \mathbb{R}$ is linear and for any sequence $\zeta_k \to \zeta$ in $\mathcal{D}(U)$, we have $|\langle T, \zeta_k \rangle - \langle T, \zeta \rangle| \to 0$.

In (iii) and in the following, the duality map between $\mathcal{D}(U)$ and $\mathcal{D}^*(U)$ is denoted $(T, \zeta) \mapsto \langle T, \zeta \rangle \in \mathbb{R}$ $(T \in \mathcal{D}^*(U), \zeta \in \mathcal{D}(U))$.

Definition 3.1.2. A sequence of distributions $\{T_k\} \subseteq \mathcal{D}^*(U)$ *converges in* $\mathcal{D}^*(U)$ to $T \in \mathcal{D}^*(U)$ (written $T_k \to T$ in $\mathcal{D}^*(U)$) if $|\langle T_k, \zeta \rangle - \langle T, \zeta \rangle| \to 0$ for all $\zeta \in C_c^\infty(U)$.

Definition 3.1.3. Let $T \in \mathcal{D}^*(U)$. For any $\alpha \in \overline{\mathbb{N}}^N$, by the *αth distributional derivative* of T (denoted $D^\alpha T$) we mean the distribution

$$\langle D^\alpha T, \zeta \rangle := (-1)^{|\alpha|} \langle T, D^\alpha \zeta \rangle \qquad \text{for all } \zeta \in C_c^\infty(U).$$

If $|\alpha| = 0$, then $D^\alpha T = T$. If $|\alpha| = 1$, $\alpha_k = 1$, and $\alpha_l = 0$ for all $l \ne k$ $(k, l = 1, \dots, N)$, then we write $\frac{\partial T}{\partial x_k}$ (or $\frac{dT}{dx}$ if $N = 1$) instead of $D^\alpha T$. We denote by $\nabla T \equiv (\frac{\partial T}{\partial x_1}, \dots, \frac{\partial T}{\partial x_N})$

https://doi.org/10.1515/9783110556902-004

the *distributional gradient* and by div $\underline{T} := \sum_{i=1}^N \frac{\partial T_i}{\partial x_i}$ the *distributional divergence* of $\underline{T} \equiv (T_1, \ldots, T_N) \in (\mathcal{D}^*(U))^N$.

Definition 3.1.4. (i) The *product* of $T \in \mathcal{D}^*(U)$ with a function $f \in C^\infty(U)$ is defined by $\langle fT, \zeta \rangle := \langle T, f\zeta \rangle$ for $\zeta \in C_c^\infty(U)$.
(ii) The *convolution* of $T \in \mathcal{D}^*(\mathbb{R}^N)$ with $\rho \in C_c^\infty(\mathbb{R}^N)$ is defined by $\langle T * \rho, \zeta \rangle := \langle T, \zeta * \hat{\rho} \rangle$ for $\zeta \in C_c^\infty(\mathbb{R}^N)$, where $\hat{\rho}(x) := \rho(-x)$ $(x \in \mathbb{R}^N)$.

If $T = f \in L^1_{loc}(\mathbb{R}^N)$, then it is easily seen that $(T * \rho) = \int_{\mathbb{R}} \rho(\cdot - y) f(y)\, dy$ (that is, Definition 3.1.4(ii) reduces to the definition for locally integrable functions, which is well posed by the Fubini theorem).

Let ρ_ϵ be a standard *mollifier*, that is, $\rho_\epsilon(x) := \epsilon^{-N} \rho(\frac{x}{\epsilon})$ $(\epsilon > 0)$ with $\rho \in C_c^\infty(\mathbb{R}^N)$ such that $\rho \geq 0$ and $\int_{\mathbb{R}^N} \rho\, dx = 1$.

Proposition 3.1.1. (i) *Let $T \in \mathcal{D}^*(\mathbb{R}^N)$. Then $T * \rho_\epsilon \in C_c^\infty(\mathbb{R}^N)$, and $(D^\alpha T) * \rho_\epsilon \to D^\alpha T$ in $\mathcal{D}^*(\mathbb{R}^N)$ as $\epsilon \to 0^+$ for all $\alpha \in \overline{\mathbb{N}}^N$.*
(ii) *Let $f \in L^p(\mathbb{R}^N)$ $(p \in [1, \infty))$. Then $\{\rho_\epsilon * f\} \subset L^p(\mathbb{R}^N) \cap C^\infty(\mathbb{R}^N)$, $\|\rho_\epsilon * f\|_p \leq \|f\|_p$ for all $\epsilon > 0$, and $\rho_\epsilon * f \to f$ in $L^p(\mathbb{R}^N)$ as $\epsilon \to 0^+$.*
(iii) *Let $f \in C(\mathbb{R}^N)$. Then $\rho_\epsilon * f \to f$ as $\epsilon \to 0^+$ uniformly on the compact subsets of $\mathbb{R}^N$.*

Remark 3.1.1. (i) Every Radon measure on U defines a positive distribution, namely, to every $\mu \in \mathfrak{R}^+(U)$ there corresponds $T_\mu \in \mathcal{D}^*(U)$ such that $\langle T_\mu, \zeta \rangle \geq 0$ for any non-negative $\zeta \in \mathcal{D}(U)$, with

$$\langle T_\mu, \zeta \rangle \equiv \langle \mu, \zeta \rangle = \int_U \zeta\, d\mu \quad (\zeta \in C_c^\infty(U)). \tag{3.1}$$

Similarly, every signed Radon measure on U defines a distribution via (3.1), that is, $\mathfrak{R}(U) \subseteq \mathcal{D}^*(U)$.

(ii) Conversely, a slight refinement of Theorem 2.6.1 (e. g., see [64] for details) proves that to every positive $T \in \mathcal{D}^*(U)$ there corresponds $\mu_T \in \mathfrak{R}^+(U)$ such that

$$\langle T, \zeta \rangle = \langle \mu_T, \zeta \rangle = \int_U \zeta\, d\mu_T \quad (\zeta \in \mathcal{D}(U)).$$

Proposition 3.1.2. *Let μ be a Radon measure on $\mathbb{R}$, and let $\phi_\mu : \mathbb{R} \to \mathbb{R}$ be a distribution function of μ. Then the distributional derivative of ϕ_μ is equal to μ.*

Proof. Let $\zeta \in C_c^\infty(\mathbb{R})$ with supp $\zeta \equiv [a, b]$ $(-\infty < a < b < \infty)$. Then by (1.24) and the Fubini theorem we have

$$\left\langle \frac{d\phi_\mu}{dx}, \zeta \right\rangle = -\int_a^b \phi_\mu(x)\, \zeta'(x)\, dx = -\int_a^b \mu((a, x])\, \zeta'(x)\, dx$$

$$= -\int_a^b dx\,\zeta'(x)\int_a^x d\mu(y) = -\int_a^b d\mu(y)\int_y^b \zeta'(x)\,dx = \int_a^b \zeta(y)\,d\mu(y) = \langle\mu,\zeta\rangle.$$

Hence the result follows. □

Remark 3.1.2. By Proposition 3.1.2 the distributional derivative of the step function

$$\phi := c_0\chi_{(-\infty,x_1)} + \sum_{p=1}^{n-1} c_p\chi_{[x_p,x_{p+1})} + c_n\chi_{[x_n,\infty)} \quad (c_p \in \mathbb{R};\ p = 0,\dots,n)$$

is the finite signed Radon measure $\sum_{p=1}^{n}[c_p - c_{p-1}]\delta_{x_p}$. Similarly, the distributional derivative of the Cantor–Vitali function V is the Radon measure λ^V (see Subsection 1.5.1).

3.1.2 Sobolev spaces

Definition 3.1.5. Let $U \subseteq \mathbb{R}^N$ be open, and let $m \in \overline{\mathbb{N}}$ and $p \in [1,\infty]$. The function $f \in L^p(U)$ belongs to the *Sobolev space* $W^{m,p}(U)$ if $D^\alpha f \in L^p(U)$ for all $\alpha \in \overline{\mathbb{N}}^N$ such that $|\alpha| \le m$. The space $W^{m,p}_{\mathrm{loc}}(U)$ is defined by replacing $L^p(U)$ by $L^p_{\mathrm{loc}}(U)$. Elements of $W^{m,p}_{\mathrm{loc}}(U)$ are called *Sobolev functions*.

Clearly, $W^{0,p}(U) \equiv L^p(U)$, and $W^{0,p}_{\mathrm{loc}}(U) \equiv L^p_{\mathrm{loc}}(U)$ $(p \in [1,\infty])$. The space $W^{m,p}(U)$ is in fact a vector space, and the map from $W^{m,p}(U)$ to $[0,\infty)$, $f \mapsto \|f\|_{m,p}$, with

$$\|f\|_{m,p} := \Big(\sum_{|\alpha|\le m} \|D^\alpha f\|_p^p\Big)^{1/p} \quad \text{if } p \in [1,\infty), \tag{3.2a}$$

$$\|f\|_{m,p} := \max_{|\alpha|\le m} \|D^\alpha f\|_\infty \quad \text{if } p = \infty \tag{3.2b}$$

is a norm. The space $W^{m,p}(U)$ endowed with this norm is a Banach space.

Definition 3.1.6. Let $U \subseteq \mathbb{R}^N$ be open, and let $m \in \overline{\mathbb{N}}$ and $p \in [1,\infty]$. By $W^{m,p}_0(U)$ we mean the closure of $C^\infty_c(U)$ with respect to the norm $\|\cdot\|_{m,p}$.

For $p = 2$, the norm $\|\cdot\|_{m,p}$ is induced by the scalar product

$$(f,g) \mapsto (f,g)_{m,2} := \Big(\sum_{|\alpha|\le m}\int_U D^\alpha f Dg\,dx\Big),$$

and hence $H^m(U) \equiv W^{m,2}(U)$ is a Hilbert space. We also denote $H^m_{\mathrm{loc}}(U) \equiv W^{m,2}_{\mathrm{loc}}(U)$ and $H^m_0(U) \equiv W^{m,2}_0(U)$. As in the case $m = 0$, we have the following:

Theorem 3.1.3. *$W^{m,p}(U)$ is separable if $p \in [1,\infty)$ and reflexive and uniformly convex if $p \in (1,\infty)$.*

For all $m \in \overline{\mathbb{N}}$ and $p \in [1,\infty)$, we denote by $W^{-m,q}(U)$, with the Hölder conjugate q of p, the dual space of $W_0^{m,p}(U)$ (that is, $W^{-m,q}(U) := (W_0^{m,p}(U))^*$; when $p = 2$, we set $H^{-m}(U) := (H_0^m(U))^*$). Let us prove the following characterization of $W^{-1,q}(U)$ (a similar result holds for $W^{-m,q}(U)$ with any $m \in \mathbb{N}$; e. g., see [1, Theorem 3.8]).

Proposition 3.1.4. *Let $q \in (1,\infty]$. The following statements are equivalent:*
(i) $\omega \in W^{-1,q}(U)$;
(ii) *there exists* $\underline{f} \equiv (f_0, f_1, \dots, f_N) \in [L^q(U)]^{N+1}$ *such that* $\omega = f_0 + \sum_{k=1}^N \frac{\partial f_k}{\partial x_k}$ *in* $\mathcal{D}^*(U)$.

Proof. We only prove that (i) implies (ii), since the inverse is immediate. Let $X := [L^p(U)]^{N+1}$ be endowed with the norm $\underline{f} \mapsto \|\underline{f}\|_X := (\sum_{k=0}^N \|f_k\|_p^p)^{1/p}$ $(p \in [1,\infty))$. Clearly, the map $T : W_0^{1,p}(U) \mapsto [L^p(U)]^{N+1}$, $T(f) := (f, \frac{\partial f}{\partial x_1}, \dots, \frac{\partial f}{\partial x_N})$, is isometric.

Let $\omega \in W^{-1,q}(U)$. Since $W_0^{1,p}(U)$ can be identified with a closed subspace of X, by the Hahn–Banach theorem there exists $U \in X^*$ that extends ω, and $\|\omega\| = \|U\|$. On the other hand, by the Riesz theorem $X^* = [L^q(U)]^{N+1}$ with Hölder conjugate q of p, and hence there exists $(f_0, -f_1, \dots, -f_N) \in [L^q(U)]^{N+1}$ such that

$$\langle U, \zeta \rangle_{X^*,X} = \int_U \left(f_0 \zeta - \sum_{k=1}^N f_k \frac{\partial \zeta_k}{\partial x_k} \right) dx \quad \text{for all } \zeta \in W_0^{1,p}(U).$$

Hence the claim follows. □

Let us finally recall the following definition.

Definition 3.1.7. We say that $f : U \subseteq \mathbb{R}^N \mapsto \mathbb{R}$ is *Lipschitz continuous* and write $f \in \mathrm{Lip}(U)$ if

$$L_U(f) := \sup_{\substack{x,y\in U\\ x\neq y}} \frac{|f(x) - f(y)|}{|x - y|} < \infty.$$

We say that f is *locally Lipschitz continuous* and write $f \in \mathrm{Lip}_{\mathrm{loc}}(U)$ if for any compact $K \subset U$,

$$L_K(f) := \sup_{\substack{x,y\in K\\ x\neq y}} \frac{|f(x) - f(y)|}{|x - y|} < \infty.$$

Observe that by definition every $f \in \mathrm{Lip}_{\mathrm{loc}}(U)$ is defined in the classical sense. Moreover, every $f \in \mathrm{Lip}(U)$ is uniformly continuous and thus bounded in U. The vector space $\mathrm{Lip}(U)$ is a Banach space with norm

$$f \mapsto \|f\|_{\mathrm{Lip}(U)} := \|f\|_\infty + L_U(f).$$

3.1.3 Bessel potential spaces

For any $r > 0$, consider the *Bessel kernel*

$$g_r(x) := \frac{1}{2^{\frac{N+r-2}{2}} \pi^{\frac{N}{2}} \Gamma(\frac{r}{2})} |x|^{\frac{r-N}{2}} K_{\frac{N-r}{2}}(|x|) \quad (x \in \mathbb{R}^N), \tag{3.3}$$

where Γ denotes the gamma function, and K_ν is the modified Bessel function of the third kind of order ν. Let us mention the following properties of g_r:

(i) $g_r > 0$, $g_r(x) = g_r(|x|)$ is a decreasing function of $|x|$;

(ii) $g_r \in L^1(\mathbb{R}^N)$ with $\|g_r\|_1 = 1$, and its Fourier transform

$$\hat{g}_r(\xi) := \frac{1}{(2\pi)^{\frac{N}{2}}} \int\limits_{\mathbb{R}^N} g_r(x)\, e^{-i\xi\cdot x}\, dx = \frac{1}{(2\pi)^{\frac{N}{2}} (1 + |\xi|^2)^{\frac{r}{2}}}; \tag{3.4}$$

(iii) equality (3.4) yields $g_r * g_s = g_{r+s}$;

(iv) by convolution g_r maps the Schwartz class $\mathcal{S}(\mathbb{R}^N)$ onto itself in a one-to-one manner;

(v) as $|x| \to 0^+$, up to a constant, we have

$$g_r(x) \sim \begin{cases} |x|^{r-N} & \text{if } r \in (0, N), \\ -\log|x| & \text{if } r = N, \\ \text{continuous in } x = 0 & \text{if } r > N; \end{cases} \tag{3.5}$$

(vi) as $|x| \to \infty$, up to a constant, we have

$$g_r(x) \sim |x|^{\frac{r-N-1}{2}} e^{-|x|}. \tag{3.6}$$

Let $f : \mathbb{R}^N \mapsto [0, \infty)$ be Borel measurable. The *Bessel potential* (with density f) is

$$(G_r f)(x) := (g_r * f)(x) = \int\limits_{\mathbb{R}^N} g_r(x - y) f(y)\, dy \quad (x \in \mathbb{R}^N). \tag{3.7}$$

Let $r > 0$ and $p \in [1, \infty]$. The *Bessel potential space* is

$$L^{r,p}(\mathbb{R}^N) := \{g_r * f \mid f \in L^p(\mathbb{R}^N)\}$$

with norm $\|g_r * f\|_{L^{r,p}(\mathbb{R}^N)} := \|f\|_p$. We refer the reader to [1, Theorem 7.63] for the main properties of these spaces. In particular, we have the following;

Theorem 3.1.5 (Calderón). *Let $m \in \mathbb{N}$ and $p \in (1, \infty)$. Then $L^{m,p}(\mathbb{R}^N) = W^{m,p}(\mathbb{R}^N)$, and there exists $M > 0$ (depending on m, p, N) such that*

$$M^{-1} \|h\|_{L^{m,p}(\mathbb{R}^N)} \le \|h\|_{m,p} \le M \|h\|_{L^{m,p}(\mathbb{R}^N)} \quad \textit{for all } h \in W^{m,p}(\mathbb{R}^N). \tag{3.8}$$

Remark 3.1.3. Closely related to the Bessel kernel g_r is the *Riesz kernel*

$$\gamma_r(x) := \frac{\Gamma(\frac{N-r}{2})}{2^r \pi^{\frac{N}{2}} \Gamma(\frac{r}{2})} |x|^{r-N} \equiv \Gamma_r |x|^{r-N} \quad (r \in (0, N),\ x \in \mathbb{R}^N). \tag{3.9}$$

For any Borel-measurable $f \geq 0$ on $\mathbb{R}^N$, the *Riesz potential* (with density f) is the convolution $\gamma_r * f$. Integral representations of the Riesz and Bessel kernels (see [2, Section V1.2.4]) show that for any $r \in (0, N)$,

$$0 < g_r(x) < \gamma_r(x) \quad (x \in \mathbb{R}^N). \tag{3.10}$$

We will also deal with the *inhomogeneous Riesz kernels* $\gamma_{r,\rho}$ $(\rho \in (0, \infty))$

$$\gamma_{r,\rho}(x) := \begin{cases} \Gamma_r |x|^{r-N} & \text{if } 0 < |x| < \rho, \\ 0 & \text{otherwise} \end{cases} \quad (r \in (0, N),\ x \in \mathbb{R}^N). \tag{3.11}$$

By (3.5), for any $r \in (0, N)$ and $\rho \in (0, \infty)$, there exists $M > 0$ (depending on r, ρ, and N) such that

$$0 < M\gamma_{r,\rho}(x) < g_r(x) \quad (x \in \mathbb{R}^N). \tag{3.12}$$

3.1.4 Functions of bounded variation

Definition 3.1.8. Let $U \subseteq \mathbb{R}^N$ be open, and let $f \in L^1(U)$. We say that f is *a function of bounded variation in U* and write $f \in BV(U)$ if for every $k = 1, \ldots, N$, the distributional derivative $\frac{\partial f}{\partial x_k}$ is a finite signed Radon measure on U. The quantity

$$V(f; U) := \sum_{k=1}^{N} \left| \frac{\partial f}{\partial x_k} \right| (U) < \infty \tag{3.13}$$

(where $|\frac{\partial f}{\partial x_k}|(U)$ denotes the total variation of the measure $\frac{\partial f}{\partial x_k}$ in U) is called the *total variation* of f in U.

We say that $f \in L^1_{\text{loc}}(U)$ is *of local bounded variation in U* and write $f \in BV_{\text{loc}}(U)$ if f is of bounded variation in W for any open $W \subset\subset U$.

We denote by $BV(U)$ the Banach space of functions of bounded variation in U with norm

$$\|f\|_{BV(U)} := \|f\|_{L^1(U)} + V(f; U) = \|f\|_{L^1(U)} + \sum_{k=1}^{N} \left| \frac{\partial f}{\partial x_k} \right| (U).$$

Remark 3.1.4. (i) Let $f \in L^1(U)$. Arguing as in [5, Section 3.1] shows that $f \in BV(U)$ if and only if

$$\Gamma(f;U) := \sup\left\{ \int_U f \operatorname{div}(\varphi)\,dx \mid \varphi \in C_c^\infty(U;\mathbb{R}^N),\ |\varphi| \le 1 \right\} < \infty;$$

moreover, $V(f;U) = \Gamma(f;U)$.

(ii) The mapping $BV(U) \ni f \mapsto \Gamma(f;U)$ is lower semicontinuous with respect to the $L^1(U)$-topology. Indeed, let $\{f_k\} \subseteq BV(U)$ be any sequence such that $f_k \to f$ in $L^1(U)$. Then by (i) for any $\epsilon > 0$, there exists $\varphi_\epsilon \in C_c^\infty(U;\mathbb{R}^N)$ such that $|\varphi| \le 1$ and

$$\Gamma(f;U) - \epsilon \le \int_U f \operatorname{div}(\varphi_\epsilon)\,dx,$$

whence

$$\Gamma(f;U) - \epsilon \le \int_U f \operatorname{div}(\varphi_\epsilon)\,dx = \lim_{k\to\infty} \int_U f_k \operatorname{div}(\varphi_\epsilon)\,dx \le \liminf_{k\to\infty} \Gamma(f_k;U).$$

Since ϵ is arbitrary, the claim follows.

If $N = 1$, then the following definition gives a different concept of variation for classical functions.

Definition 3.1.9. (i) Let $I \equiv [a,b]$ with $-\infty < a < b < \infty$, and let $f : I \mapsto \mathbb{R}$. The quantity

$$V^J(f;I) := \sup\left\{ \sum_{k=1}^m |f(x_k) - f(x_{k-1})| \mid m \ge 2,\ a \equiv x_0 < x_1 < \cdots < x_m \equiv b \right\} \tag{3.14}$$

is called the *Jordan variation* of f in I. If $V^J(f;I) < \infty$, then we say that f is *a function of bounded Jordan variation in I* and write $f \in BV^J(I)$.

(ii) Let $U \subseteq \mathbb{R}$ be open, and let $f : U \mapsto \mathbb{R}$. We say that f is *of locally bounded Jordan variation in U* (written $f \in BV^J_{\mathrm{loc}}(U)$) if $f \in BV^J(I)$ for any $I \equiv [a,b] \subset U$.

Remark 3.1.5. (i) Every signed Radon measure $\nu \equiv (\nu_1,\nu_2)$ on $\mathbb{R}$ is an ordered couple of two Stieltjes measures $\lambda^{\phi_{\nu_i}}$ $(i = 1,2)$, where

$$\phi_{\nu_i}(x) := \begin{cases} \nu_i((c,x]) & \text{if } x > c, \\ 0 & \text{if } x = c, \\ -\nu_i((x,c]) & \text{if } x < c \end{cases}$$

with $c \in \mathbb{R}$ arbitrarily fixed (see Proposition 1.5.6). Since the functions ϕ_{ν_i} are nondecreasing, it is well known that $\phi_\nu := \phi_{\nu_1} - \phi_{\nu_2}$ has *locally bounded Jordan variation* in $\mathbb{R}$.

(ii) If $f \in BV^J(I)$, then $\sup_{x\in I} |f(x)| \le |f(a)| + V^J(f;I) < \infty$. Hence $BV^J_{\mathrm{loc}}(U) \subseteq L^\infty_{\mathrm{loc}}(U)$, and $BV^J(U) \subseteq L^\infty(U)$.

The following result is well known.

Proposition 3.1.6. *Let $I \equiv [a,b]$ with $-\infty < a < b < \infty$.*
(i) *Let $f \in BV^J(I)$. Then the map $x \mapsto V^J(f;[a,x])$ $(x \in [a,b])$ is nondecreasing and right- or left-continuous at any point where f is.*
(ii) *$f \in BV^J(I)$ if and only if $f = f_1 - f_2$ with nondecreasing $f_1, f_2 : I \to \mathbb{R}$.*
(iii) *Let $f \in BV^J(I)$. Then f is differentiable λ-a. e. in I, that is, for λ-a. e. $x_0 \in I$, the limit $\lim_{r\to 0} \frac{f(x_0+r)-f(x_0)}{r}$ exists and is finite.*

Remark 3.1.6. By Remark 2.2.3 to every function $f = f_1 - f_2 \in BV^J_{\mathrm{loc}}(\mathbb{R})$, we associate a signed Radon measure $\nu^f := (\lambda^{f_1}, \lambda^{f_2})$ with Stieltjes measures λ^{f_i} $(i = 1,2)$ on $\mathbb{R}$. Then by Theorem 2.9.3(i) for λ-a. e. $x_0 \in \mathbb{R}$, there exists

$$
\begin{aligned}
D_\lambda \nu^f(x_0) &= \lim_{r\to 0^+} \left(\frac{f_1(x_0+r) - f_1(x_0-r)}{2r} - \frac{f_2(x_0+r) - f_2(x_0-r)}{2r} \right) \\
&= (f_1)'(x_0) - (f_2)'(x_0) = f'(x_0).
\end{aligned} \tag{3.15}
$$

Conversely, for any signed Radon measure $\nu \equiv (\nu_1, \nu_2)$, equality (3.15) holds with ν^f replaced by ν, and f replaced by $\phi_\nu := \phi_{\nu_1} - \phi_{\nu_2}$, where ϕ_{ν_i} is any distribution function of ν_i $(i = 1,2)$.

By Theorems 1.8.9, 2.9.1, and 2.9.3, Corollary 2.9.6, and equality (3.15) we have the following:

Theorem 3.1.7. *Let ν be a signed Radon measure on $\mathbb{R}$, and let ϕ_ν be any distribution function of ν. Then*

$$
\phi_\nu' = D_\lambda \nu = D_\lambda \nu_{ac} \quad \lambda\text{-a. e. in } \mathbb{R}, \tag{3.16}
$$

and for every bounded Borel set $E \subseteq \mathbb{R}$, we have

$$
\nu(E) = \int_E \phi_\nu' \, d\lambda + \nu_s(E), \quad \nu_{ac}(E) = \int_E \phi_\nu' \, d\lambda, \tag{3.17}
$$

where $\nu_{ac} \ll \lambda$ and $\nu_s \perp \lambda$ are the absolutely continuous and singular parts of ν, respectively, with respect to the Lebesgue measure λ.

Remark 3.1.7. Let ν be a signed Radon measure on $\mathbb{R}$. Since ν coincides with the distributional derivative $\frac{d\phi_\nu}{dx}$ of the function ϕ_ν (see Proposition 3.1.2), by (3.16) $D_\lambda \nu = \phi_\nu'$

is the Radon–Nikodým derivative of $\frac{d\phi_\nu}{dx}$ with respect to λ. In view of (3.17), we get

$$\left\langle \frac{d\phi_\nu}{dx}, \zeta \right\rangle = \int_{\mathbb{R}} \phi_\nu'(x)\,\zeta(x)\,dx + \left\langle \left(\frac{d\phi_\nu}{dx}\right)_s, \zeta \right\rangle.$$

Let us mention the following refinement of Theorem 3.1.7, where the singular part ν_s of a (positive) Radon measure ν is split into the sum of a *discrete* measure ν_N (that is, ν_N is concentrated on a countable subset $S \subset \mathbb{R}$, and thus $\nu_N \perp \lambda$) and a *singular continuous* measure ν_{sc} (that is, $\nu_{sc}(\{x\}) = 0$ for all $x \in \mathbb{R}$) with respect to the Lebesgue measure λ.

Theorem 3.1.8. *Let ν be a (positive) Radon measure on $\mathbb{R}$. Then there exists a unique triple $(\nu_{ac}, \nu_{sc}, \nu_N)$ of (positive) Radon measures on $\mathbb{R}$ such that:*
(i) $\nu_{ac} \ll \lambda$, $\nu_{sc} \perp \lambda$, $\nu_N \perp \lambda$, $\nu_{sc} \perp \nu_N$;
(ii) *ν_N is discrete, and ν_{sc} is singular continuous;*
(iii) $\nu = \nu_{ac} + \nu_{sc} + \nu_N$.

Analogous statements for signed Radon measure $\nu \equiv (\nu_1, \nu_2)$ follow applying Theorem 3.1.8 to both ν_1 and ν_2.

It is informative for further purposes to compare the concepts of variation given by (3.13) and (3.14). This is the content of the following proposition.

Proposition 3.1.9. *Let $U \subseteq \mathbb{R}$ be open. Then for any $f \in L^1_{\mathrm{loc}}(U)$ and for any $I \equiv [a,b] \subset U$, we have*

$$\Gamma(f;I) = \inf\{V^J(g;I) \mid g : I \to \mathbb{R} \text{ classical representative of } f\}. \tag{3.18}$$

Proof. (i) Let g be defined at each point of I, and let $g = f$ λ-a. e. in I; thus $\Gamma(f;I) = \Gamma(g;I)$. Let us first show that $\Gamma(f;I) \le V^J(g;I)$. Let $V^J(g;I) < \infty$, since otherwise the claim is obviously satisfied. Let $\Delta \equiv \{a \equiv x_0 < x_1 < \cdots < x_m \equiv b\}$, and let $|\Delta| := \max_{i=1,\dots,m} |x_i - x_{i-1}|$. Set

$$g_\Delta(x) := \sum_{i=0}^{m-1} g(x_i)\chi_{[x_i,x_{i+1})}(x)$$

with distributional derivative $\frac{dg_\Delta}{dx} = \sum_{i=1}^m [g(x_i) - g(x_{i-1})]\delta_{x_i}$. Then by (3.13)

$$V(g_\Delta;I) = \sum_{i=1}^m |g(x_i) - g(x_{i-1})| \le V^J(g;I) \tag{3.19}$$

for every partition Δ of I. Now observe that $g_\Delta \to g$ pointwise in I and thus in $L^1(I)$ as $|\Delta| \to 0$, since the family $\{g_\Delta\}$ is bounded in $L^\infty(I)$ (see Remark 3.1.5(ii)). Then by

Remark 3.1.4 and inequality (3.19) we get that $g \in BV(I)$ (thus $f \in BV(I)$) and

$$V(f;I) = V(g;I) \leq \liminf_{|\Delta|\to 0} V(g_\Delta;I) \leq V^J(g;I).$$

By the arbitrariness of g it follows that

$$\Gamma(f;I) \leq \inf\{V^J(g;I) \mid g : I \to \mathbb{R} \text{ classical representative of } f\}.$$

To complete the proof it suffices to exhibit a classical representative g of f such that $\Gamma(f;I) \geq V^J(g;I)$. Clearly, it suffices to assume that $\Gamma(f;I) < \infty$, and thus by Remark 3.1.4 $f \in BV(I)$ and $\Gamma(f;I) = V(f;I) := \|\frac{df}{dx}\|_{\mathfrak{R}_f(I)}$. Consider the distribution function $F : I \to \mathbb{R}$ associated with $\frac{df}{dx}$ defined as $F(x) := \frac{df}{dx}([a,x))$ $(x \in I)$. Then by Proposition 3.1.2 and Remark 1.5.4 there exists $c \in \mathbb{R}$ such that $f = F + c$ λ-a. e. in I, and thus $g := F + c$ is a classical representative of f in I. Moreover, for every partition $\{a \equiv x_0 < x_1 < \cdots < x_m \equiv b\}$ of I, we have

$$\sum_{k=1}^{m} |g(x_k) - g(x_{k-1})| = \sum_{k=1}^{m} |F(x_k) - F(x_{k-1})| = \sum_{k=1}^{m} \left|\frac{df}{dx}([x_{k-1}, x_k))\right| \leq V(f;I).$$

Taking the supremum of the left-hand side over all partitions of I proves that $V(f;I) \geq V^J(g;I)$. Hence the conclusion follows. □

Remark 3.1.8. The proof of equality (3.18) shows that for every $f \in BV_{\mathrm{loc}}(U)$ and $I = [a,b] \subset U$, the infimum in the right-hand side is achieved, that is, there exists a classical representative $\tilde{f}$ of f such that $V(f;I) = V^J(\tilde{f};I)$. Any representative of f with this property is called a *good representative*. Clearly, $f \in BV_{\mathrm{loc}}(U)$ if and only if $\tilde{f} \in BV^J_{\mathrm{loc}}(U)$. Moreover, the Radon–Nikodým derivative of the measure $\frac{df}{dx}$ with respect to the Lebesgue measure λ is equal to $\tilde{f}'$.

Let us recall the following definition.

Definition 3.1.10. (i) Let $I \equiv [a,b]$ with $-\infty < a < b < \infty$. We say that $f : I \to \mathbb{R}$ is *absolutely continuous* and write $f \in AC(I)$ if for any $\epsilon > 0$, there exists $\delta > 0$ such that for any finite disjoint family of subintervals $(a_k, b_k) \subseteq I$ $(k = 1, \ldots, m)$,

$$\sum_{k=1}^{m} (b_k - a_k) < \delta \quad \Rightarrow \quad \sum_{k=1}^{m} |f(b_k) - f(a_k)| < \epsilon.$$

(ii) Let $U \subseteq \mathbb{R}$ be open, and let $f : U \mapsto \mathbb{R}$. We say that $f \in AC_{\mathrm{loc}}(U)$ if $f \in AC(I)$ for any $I \equiv [a,b] \subset U$.

It is well known that equality (3.17) holds with $\nu_s = 0$ (that is, $\nu = \nu_{ac}$) if and only if $\phi_\nu \in AC_{\mathrm{loc}}(\mathbb{R})$ (e. g., see [45]). On the other hand, it is apparent from (3.17) that $\nu = \nu_s$ if and only if $\phi'_\nu = 0$ λ-a. e. in $\mathbb{R}$, that is, if and only if the distribution function ϕ_ν is *singular*. Prototypes of singular functions are the Heaviside function H

and the Cantor–Vitali function V, and hence the associated Stieltjes measures λ^H and λ^V are singular with respect to λ (see Subsection 1.4); moreover, λ^H is discrete, whereas λ^V is singular continuous (see Corollary 1.5.8).

Bearing in mind that neither H nor V satisfies the Lusin condition, further light is shed by the following characterization of the absolutely continuous functions (see [7, Theorem 3.9]).

Theorem 3.1.10 (Vitali–Banach–Zaretskii). *Let $I \equiv [a,b]$ with $-\infty < a < b < \infty$. Then the following statements are equivalent:*
(i) *$f \in AC(I)$;*
(ii) *$f \in C(I) \bigcap BV^J(I)$ and satisfies the Lusin condition.*

Remark 3.1.9. Clearly, $AC(I) \subset BV^J(I)$ for any $I \equiv [a,b]$ $(-\infty < a < b < \infty)$ (see Definitions 3.1.9 and 3.1.10). Hence every $f \in AC(I)$ is differentiable λ-a. e. in I, and $f(d) - f(c) = \int_c^d f'd\lambda$. for all $a \leq c \leq d \leq b$.

3.1.5 Sobolev functions

It is interesting to study the differentiability a. e. in the sense of the following definition, of Sobolev functions (in this connection, see [47]).

Definition 3.1.11. Let $f : U \subseteq \mathbb{R}^N \mapsto \mathbb{R}$ be defined a. e. in U. We say that f is *differentiable* at $x_0 \in U$ if there exists $Df(x_0) \in \mathbb{R}^N$, called the *differential* of f at x_0, such that

$$\operatorname{ess}\lim_{x\to x_0} \frac{|f(x) - f(x_0) - Df(x_0)\cdot(x - x_0)|}{|x - x_0|} = 0.$$

If $N = 1$, then $Df(x_0) \equiv f'(x_0)$ is the *derivative* of f at x_0.

Proposition 3.1.11. *Let $f \in W^{1,p}_{\mathrm{loc}}(U)$ with $N < p \leq \infty$. Then f is differentiable a. e. in U, and $Df = \nabla f$ a. e.*

Proof. Since $W^{1,\infty}_{\mathrm{loc}}(U) \subseteq W^{1,p}_{\mathrm{loc}}(U)$, it is not restrictive to assume that $N < p < \infty$. Then by Morrey's inequality there exists $C = C(m,p) > 0$ such that for a. e. $x_0, x \in U$,

$$\frac{|f(x) - f(x_0) - \nabla f(x_0)\cdot(x - x_0)|}{|x - x_0|} \leq \left(\frac{C}{\lambda_N(B_{|x-x_0|}(x_0))} \int\limits_{B_{|x-x_0|}(x_0)} |\nabla f(y) - (\nabla f)(x_0)|^p \, dy\right)^{\frac{1}{p}}.$$

Since $\nabla f \in (L^p_{\mathrm{loc}}(U))$, letting $|x - x_0| \to 0$ and using (2.97), the result follows. □

Theorem 3.1.12. *Let $U \subseteq \mathbb{R}^N$ be open. Then $f \in W^{1,\infty}_{\mathrm{loc}}(U)$ if and only if f has a classical representative $g \in \mathrm{Lip}_{\mathrm{loc}}(U)$.*

Proof. (i) Let $g \in \mathrm{Lip}_{\mathrm{loc}}(U)$, and let V, W open bounded subsets such that $V \Subset W \Subset U$. Clearly, $g \in L^\infty(V)$. Let $\underline{e}_k$ be the kth unit vector ($k = 1, \dots, N$), and for any $h \in (0, \mathrm{dist}(V, \partial W))$, define

$$g_h^{(k)} : V \mapsto \mathbb{R}, \quad g_h^{(k)}(x) := \frac{g(x + h\,\underline{e}_k) - g(x)}{h}.$$

Observe that for any $\zeta \in C_c^1(V)$,

$$\int_U g_h^{(k)}(x)\,\zeta(x + h\,\underline{e}_k)\,dx = -\int_U g(x)\frac{\zeta(x + h\,\underline{e}_k) - \zeta(x)}{h}\,dx. \tag{3.20}$$

On the other hand, for any $p \in (1, \infty)$,

$$\begin{aligned}\sup_{h \in (0, \mathrm{dist}(V, \partial W))} \|g_h^{(k)}\|_{L^p(V)} &\le [\lambda_N(V)]^{\frac{1}{p}} \sup_{h \in (0, \mathrm{dist}(V, \partial W))} \|g_h^{(k)}\|_{L^\infty(V)} \\ &\le [\lambda_N(V)]^{\frac{1}{p}} L_W(g) < \infty,\end{aligned}$$

where we have denoted by $L_W(g)$ the Lipschitz constant of g in W (see Definition 3.1.7). Then by the Banach–Alaoglu theorem and the reflexivity of $L^p(V)$, there exist a sequence $\{h_m\}$ and a function $g^{(k)} \in L^\infty(V)$ such that $g_{h_m}^{(k)} \rightharpoonup g^{(k)}$ in $L^p(V)$ as $m \to \infty$. Then writing (3.20) with $h = h_m$ and letting $m \to \infty$ gives $g^{(k)} = \frac{\partial g}{\partial x_k}$, and thus $\frac{\partial g}{\partial x_k} \in L^\infty(V)$. Hence the claim follows.

(ii) Let $f \in W^{1,\infty}_{\mathrm{loc}}(U)$, and let $B \Subset U$ be any ball. Thus $\|\nabla f\|_{L^\infty(B)} < \infty$. For any $x \in U$ and $\epsilon \in (0, \mathrm{dist}(x, \partial U))$, set $f_\epsilon(x) := \int_U f(y)\rho_\epsilon(x - y)\,dy$, where ρ_ϵ is a standard mollifier. Then for any $x, x + h \in B$ and ϵ sufficiently small, we have

$$|f_\epsilon(x + h) - f_\epsilon(x)| = \left|\int_0^1 \nabla f_\epsilon(x + th)\cdot h\,dt\right| \le \|\nabla f\|_{L^\infty(B)}\,|h| \quad \text{for } \lambda\text{-a.e. } x \in B.$$

Letting $\epsilon \to 0$ in this inequality, we obtain that

$$|f(x + h) - f(x)| \le \|\nabla f\|_{L^\infty(B)}\,|h|$$

for a. e. $x \in B$. Then the conclusion easily follows. □

Remark 3.1.10. It is easily seen that every $f \in \mathrm{Lip}(U) \bigcap L^\infty(U)$ belongs to $W^{1,\infty}(U)$. However, the inverse implication is in general false (see [5, Section 2.3]).

From Theorem 3.1.12 and Proposition 3.1.11 we immediately obtain the following:

Corollary 3.1.13 (Rademacher). *Let $f \in \mathrm{Lip}_{\mathrm{loc}}(U)$. Then f is differentiable a. e. in U.*

Proposition 3.1.14. *Let $U \subseteq \mathbb{R}$ be open. Then $f \in W^{1,1}_{\mathrm{loc}}(U)$ if and only if f has a classical representative $g \in AC_{\mathrm{loc}}(U)$.*

Proof. Let $I = [a, b] \subset U$ be fixed. If $g \in AC(I)$, then $\frac{dg}{dx} \in L^1(I)$, g is differentiable λ-a. e., and

$$\frac{dg}{dx} = g' \quad \lambda\text{-a. e. in } I, \qquad \int_I |g'|\, d\lambda \le V^J(g; I) < \infty$$

(see Remark 3.1.9). Moreover, $g \in L^\infty(I) \subseteq L^1(I)$, and thus $g \in W^{1,1}(I)$.

Conversely, let $f \in W^{1,1}(I)$. Consider the distribution function $F : I \to \mathbb{R}$ associated with $\frac{df}{dx}$, $F(x) := \frac{df}{dx}([a, x)) = \int_a^x \frac{df}{dx}\, d\lambda$ $(x \in I)$. As in the proof of Proposition 3.1.9, there exists $c \in \mathbb{R}$ such that $f = F + c$ λ-a. e. in I, and thus $g := F + c$ is a classical representative of f in I.

Now let $E := \bigcup_{k=1}^N [x_{k-1}, x_k]$ with $\{a \equiv x_0 < x_1 < \cdots < x_m \equiv b\}$. Then $\lambda(E) = \sum_{k=1}^N (x_k - x_{k-1})$, and

$$\sum_{k=1}^N |g(x_k) - g(x_{k-1})| = \sum_{k=1}^N |F(x_k) - F(x_{k-1})| \le \sum_{k=1}^N \int_{x_{k-1}}^{x_k} \left|\frac{df}{dx}\right| d\lambda = \int_E \left|\frac{df}{dx}\right| d\lambda.$$

Then by elementary results the conclusion follows. □

Theorem 3.1.12 and Proposition 3.1.14 are usually stated by saying, respectively, that $f \in \mathrm{Lip}_{\mathrm{loc}}(U)$ if and only if $f \in W^{1,\infty}_{\mathrm{loc}}(U)$ and that $f \in W^{1,1}_{\mathrm{loc}}(U)$ if and only if $f \in AC_{\mathrm{loc}}(U)$.

3.2 Capacities associated with a kernel

In this section, we present for later use some results from the theory of capacities developed in [72].

3.2.1 Preliminaries and definitions

Hereafter, by a *kernel* on $\mathbb{R}^N \times \mathbb{R}^N$ we mean a lower semicontinuous function $g : \mathbb{R}^N \times \mathbb{R}^N \to [0, \infty]$. By Remark 2.1.3(ii) g is $\mathcal{B}(\mathbb{R}^N) \times \mathcal{B}(\mathbb{R}^N)$-measurable, and hence $g(x, \cdot)$ and $g(\cdot, y)$ are $\mathcal{B}(\mathbb{R}^N)$-measurable for any fixed $x, y \in \mathbb{R}^N$, respectively (see Proposition 2.1.10). Then for any $\mu, \nu \in \mathfrak{R}^+(\mathbb{R}^N)$, the following quantities are well-defined:

$$g(x,\mu) := \int_{\mathbb{R}^N} g(x,y)\,d\mu(y) \quad (x \in \mathbb{R}^N), \tag{3.21a}$$

$$g(\nu,y) := \int_{\mathbb{R}^N} g(x,y)\,d\nu(x) \quad (y \in \mathbb{R}^N), \tag{3.21b}$$

$$g(\nu,\mu) := \int_{\mathbb{R}^N\times\mathbb{R}^N} g\,d(\mu\times\nu) = \int_{\mathbb{R}^N} g(x,\mu)(x)\,d\nu(x) = \int_{\mathbb{R}^N} g(\nu,y)\,d\mu(y). \tag{3.21c}$$

Observe that $g(x,\mu) = g(\delta_x,\mu)$ and $g(\nu,y) = g(\nu,\delta_y)$. If $\tilde{\mu}(E) = \int_E f\,d\mu$ ($E \in \mathcal{B}(\mathbb{R}^N)$) with f nonnegative and $\mathcal{B}(\mathbb{R}^N)$-measurable, then we write $g(\nu,f\mu) \equiv g(\nu,\tilde{\mu})$; we also set

$$(G_\mu f)(x) := g(x,f\mu) = \int_{\mathbb{R}^N} g(x,y)f(y)\,d\mu(y) \quad (x \in \mathbb{R}^N). \tag{3.22}$$

The quantities $g(f\nu,\mu)$ and $(G_\nu f)(y)$ ($y \in \mathbb{R}^N$) are similarly defined.

The functions $g(\cdot,\mu)$, $g(\nu,\cdot)$, and g in (3.21) are called the *potentials* and the *mutual energy* of μ and ν, respectively.

In connection with the following result, observe that the weak* topology on $\mathfrak{R}_f^+(\mathbb{R}^N)$ is characterized by Theorem 2.7.1 (see Definition 5.1.2).

Lemma 3.2.1. *Let g be a kernel, let f be nonnegative and $\mathcal{B}(\mathbb{R}^N)$-measurable, and let $\nu \in \mathfrak{R}_f^+(\mathbb{R}^N)$. Then for any $\mu \in \mathfrak{R}_f^+(\mathbb{R}^N)$:*
(i) *the map $G_\mu f$ is lower semicontinuous;*
(ii) *the map $\nu \mapsto g(\nu,f\mu)$ is lower semicontinuous in the* weak* *topology on $\mathfrak{R}_f^+(\mathbb{R}^N)$.*

The same holds for the maps $G_\nu f$ and $\mu \mapsto g(f\nu,\mu)$.

Proof. (i) Let $\bar{x} \in \mathbb{R}^N$ be fixed, and let $\{x_n\} \subseteq \mathbb{R}^N$, $x_n \to \bar{x}$, be such that $(G_\mu f)(x_n) \to \liminf_{x\to\bar{x}}(G_\mu f)(x)$. Since $g(\cdot,y)$ is lower semicontinuous at $\bar{x}$ for any $y \in \mathbb{R}^N$, by Fatou's lemma we have

$$\begin{aligned}(G_\mu f)(\bar{x}) &= \int_X g(\bar{x},y)f(y)\,d\mu(y) \le \int_X \liminf_{n\to\infty} g(x_n,y)f(y)\,d\mu(y)\\ &\le \liminf_{n\to\infty}\int_X g(x_n,y)f(y)\,d\mu(y) = \lim_{n\to\infty}(G_\mu f)(x_n) = \liminf_{x\to\bar{x}}(G_\mu f)(x).\end{aligned}$$

Hence the claim follows.

(ii) Since by (i) $G_\mu f$ is lower semicontinuous, by Remark 2.1.3(ii) and standard regularization arguments there exists a sequence $\{\varphi_k\} \subseteq C_c(\mathbb{R}^N)$ such that $\varphi_k \le \varphi_{k+1} \le G_\mu f$

for all $k \in \mathbb{N}$ and $\lim_{k\to\infty} \varphi_k = G_\mu f$ in $\mathbb{R}^N$. Then by the monotone convergence theorem

$$g(\nu, f\mu) = \int_{\mathbb{R}^N} G_\mu f \, d\nu = \lim_{k\to\infty} \int_{\mathbb{R}^N} \varphi_k \, d\nu. \tag{3.23}$$

Let $\{\nu_n\} \subseteq \mathfrak{R}_f^+(\mathbb{R}^N)$ converge weakly* to ν. Then for any $k \in \mathbb{N}$,

$$\int_{\mathbb{R}^N} \varphi_k \, d\nu = \lim_{n\to\infty} \int_{\mathbb{R}^N} \varphi_k \, d\nu_n \le \liminf_{n\to\infty} \int_{\mathbb{R}^N} G_\mu f \, d\nu_n = \liminf_{n\to\infty} g(\nu_n, f\mu). \tag{3.24}$$

From (3.23)–(3.24) the conclusion follows. □

3.2.2 The capacities $C_{g,p}$

Fix $\mu \in \mathfrak{R}^+(\mathbb{R}^N)$. Set for simplicity $Gf \equiv G_\mu f$ and $L^p(\mathbb{R}^N) \equiv L^p(\mathbb{R}^N, \mathcal{B}(\mathbb{R}^N), \mu)$ ($p \in [1,\infty)$).

Definition 3.2.1. Let g be a kernel, and let $p \in (1,\infty)$. For any $E \subseteq \mathbb{R}^N$, the *p-capacity of E associated with g* is

$$C_{\mu,g,p}(E) \equiv C_{g,p}(E) := \inf_{f\in\mathscr{V}_{E,p}} \|f\|_p^p, \tag{3.25a}$$

where

$$\mathscr{V}_{E,p} \equiv \mathscr{V}_{E,g,p} := \{f \in L^p(\mathbb{R}^N) \mid f \ge 0,\ (Gf)(x) \ge 1\ \forall x \in E\}, \tag{3.25b}$$

If $\mathscr{V}_{E,p} = \emptyset$, then $C_{g,p}(E) := \infty$.

Definition 3.2.2. Let g be a kernel. For any $E \subseteq \mathbb{R}^N$, we set

$$C_{g,1}(E) := \inf_{\sigma\in\mathscr{V}_{E,1}} \|\sigma\|, \tag{3.26a}$$

$$\mathscr{V}_{E,1} \equiv \mathscr{V}_{E,g,1} := \{\sigma \in \mathfrak{R}_f^+(\mathbb{R}^N) \mid g(x,\sigma) \ge 1\ \forall x \in E\}. \tag{3.26b}$$

If $\mathscr{V}_{E,1} = \emptyset$, then $C_{g,1}(E) := \infty$.

Remark 3.2.1. The restriction $f \ge 0$ in (3.25b) can be removed. Indeed, for all $p \in (1,\infty)$, we clearly have

$$C_{g,p}(E) \ge \inf_{f\in L^p(\mathbb{R}^N),\ Gf\ge 1 \text{ in } E} \|f\|_p^p.$$

On the other hand, for any $f \in L^p(\mathbb{R}^N)$ such that $Gf \ge 1$ in E, we have $Gf^+ \ge Gf^- + 1 \ge 1$ in E, and thus $f^+ \in \mathscr{V}_{E,p}$; moreover, $\|f^+\|_p \le \|f\|_p$. Hence the inverse inequality and the claim follow. It is similarly seen that the requirement of nonnegativity in (3.26b) can be avoided, namely,

$$C_{g,1}(E) = \inf_{\sigma\in\mathfrak{R}_f(\mathbb{R}^N),\, g(x,\sigma)\geq 1 \forall x\in E} \|\sigma\|.$$

Theorem 3.2.2. *Let g be a kernel, and let $p \in [1,\infty)$. Then $C_{g,p}$ is an outer capacity on $\mathbb{R}^N$.*

Proof. Consider first the case $p > 1$. Since $\mathscr{V}_{\emptyset,p} = \{f \in L^p(\mathbb{R}^N) \mid f \geq 0\}$, we have $C_{g,p}(\emptyset) = 0$. If $E_1 \subseteq E_2$ then $\mathscr{V}_{E_2,p} \subseteq \mathscr{V}_{E_1,p}$, and hence $C_{g,p}(E_1) \leq C_{g,p}(E_2)$. Let $E_n \subseteq \mathbb{R}^N$ $(n \in \mathbb{N})$, and set $E := \bigcup_{n=1}^{\infty} E_n$. Let us prove that

$$C_{g,p}(E) \leq \sum_{n=1}^{\infty} C_{g,p}(E_n).$$

If $\sum_{n=1}^{\infty} C_{g,p}(E_n) = \infty$, then there is nothing to prove. Otherwise, let $\epsilon > 0$. Then for any $n \in \mathbb{N}$, there exists $f_n \in \mathscr{V}_{E_n,p}$ such that $\|f_n\|_p^p < C_{g,p}(E_n) + \frac{\epsilon}{2^n}$. Set $E := \bigcup_{n=1}^{\infty} E_n$ and $f := \sup_{n\in\mathbb{N}} f_n$. Since $Gf_n \geq 1$ in E_n $(n \in \mathbb{N})$, we have $(Gf)(x) \geq 1$ for all $x \in E$. Moreover, we have that $\|f\|_p^p \leq \sum_{n=1}^{\infty} \|f_n\|_p^p \leq \sum_{n=1}^{\infty} C_{g,p}(E_n) + \epsilon$. It follows that $f \in \mathscr{V}_{E,p}$ and $C_{g,p}(E) \leq \sum_{n=1}^{\infty} C_{g,p}(E_n) + \epsilon$. Then by the arbitrariness of ϵ the claim follows, and thus $C_{g,p}$ is a capacity.

To complete the proof, we must show that for any $E \subseteq \mathbb{R}^N$,

$$C_{g,p}(E) = \inf\{C_{g,p}(A) \mid A \supseteq E,\ A \text{ open}\}. \tag{3.27}$$

If $\mathscr{V}_{E,p} = \emptyset$, then $\mathscr{V}_{A,p} = \emptyset$, and thus the equality is satisfied. Let $\mathscr{V}_{E,p} \neq \emptyset$. Then $C_{g,p}(E) < \infty$. Let $\epsilon \in (0,1)$, and let $f \in \mathscr{V}_{E,p}$ satisfy $\|f\|_p^p < C_{g,p}(E)+\epsilon$. By Lemma 3.2.1(i) Gf is lower semicontinuous, and thus the set $A_\epsilon := \{Gf > 1-\epsilon\}$ is open. Moreover, since $f \in \mathscr{V}_{E,p}$, we have $(Gf)(x) \geq 1 > 1-\epsilon$ for all $x \in E$, and thus $E \subseteq A_\epsilon$. Obviously, we also have $f_\epsilon := \frac{f}{1-\epsilon} \in \mathscr{V}_{A_\epsilon,p}$ since by definition $Gf_\epsilon > 1$ in A_ϵ. It follows that

$$\inf\{C_{g,p}(A) \mid A \supseteq E,\ A \text{ open}\} \leq C_{g,p}(A_\epsilon) \leq \|f_\epsilon\|_p^p \leq \frac{1}{(1-\epsilon)^p}[C_{g,p}(E)+\epsilon],$$

whence by the arbitrariness of ϵ

$$\inf\{C_{g,p}(A) \mid A \supseteq E,\ A \text{ open}\} \leq C_{g,p}(E).$$

The inverse inequality clearly follows by the monotonicity of $C_{g,p}$. Hence equality (3.27) and the result follow in this case.

Now let $p = 1$. It is easily seen that $C_{g,1}(\emptyset) = 0$ and $C_{g,1}(E_1) \leq C_{g,1}(E_2)$ if $E_1 \subseteq E_2$. To prove that $C_{g,1}(\bigcup_{n=1}^{\infty} E_n) \leq \sum_{n=1}^{\infty} C_{g,1}(E_n)$, we can argue as above, with f replaced by $\sigma := \sum_{n=1}^{\infty} \sigma_n$. Similarly, equality (3.27) with $p = 1$ follows as before by the lower semicontinuity of the map $x \mapsto g(x,\sigma)$ for all $\sigma \in \mathfrak{R}_f^+(\mathbb{R}^N)$ (see Lemma 3.2.1(i) with $\mu = \sigma$ and $f = 1$). This completes the proof. □

In view of Theorem 3.2.2, the terminology and results concerning capacities in Chapters 1–2 can be used for $C_{g,p}$. For shortness, we say that a set $E \subseteq \mathbb{R}^N$ is (g,p)-null if it is $C_{g,p}$-null, and similarly for other notions. In particular, we say that a function $f : \mathbb{R}^N \to \mathbb{R}$ is (g,p)-quasi-continuous if it is $C_{g,p}$-quasi-continuous (see Definition 2.1.8). Then by Proposition 2.1.17 we have the following:

Proposition 3.2.3. *Let g be a kernel, and let $p \in (1,\infty)$. Let $f, f_n : \mathbb{R}^N \to \mathbb{R}$ $(n \in \mathbb{N})$. Let f_n be (g,p)-quasi-continuous for each n, and let $f_n \to f$ in $C_{g,p}$-capacity. Then f is (g,p)-quasi-continuous.*

The following lemma characterizes the (g,p)-null sets.

Lemma 3.2.4. *Let g be a kernel, and let $p \in (1,\infty)$. Then the following statements are equivalent:*
(i) *$E \subseteq \mathbb{R}^N$ is (g,p)-null;*
(ii) *there exists $f \in L^p(\mathbb{R}^N), f \geq 0$, such that $E \subseteq \{Gf = \infty\}$.*

Proof. (i)⇒(ii). if $C_{g,p}(E) = 0$, then by (3.25) there exists $\{f_n\} \subseteq L^p(\mathbb{R}^N)$ such that for each $n \in \mathbb{N}$, we have $f_n \geq 0$, $Gf_n \geq 1$ in E, and $\|f_n\|_p \leq \frac{1}{2^n}$. Then $f := \sum_{n=1}^{\infty} f_n$ satisfies (ii).
(ii)⇒(i). By (3.25), for any $f \in L^p(\mathbb{R}^N)$ and $t > 0$, we have

$$t^p\, C_{g,p}(\{Gf \geq t\}) \leq \int_{\mathbb{R}^N} f^p \, d\mu. \tag{3.28}$$

Letting $t \to \infty$ in this inequality, we get $C_{g,p}(\{Gf = \infty\}) = 0$. Hence the result follows. □

Proposition 3.2.5. *Let g be a kernel, and let $p \in (1,\infty)$. Let $\{f_n\} \subseteq L^p(\mathbb{R}^N)$ be a Cauchy sequence. Then there exists a subsequence $\{f_{n_k}\} \subseteq \{f_n\}$ such that $\{Gf_{n_k}\}$ converges (g,p)-q. e. in $\mathbb{R}^N$, in $C_{g,p}$-capacity, and (g,p)-quasi-uniformly.*

Proof. Since $\{f_n\}$ is a Cauchy sequence in $L^p(\mathbb{R}^N)$, for every $k \in \mathbb{N}$, there exists $n_k \in \mathbb{N}$ such that $\|f_m - f_n\|_p < 2^{-k}k^{-2}$ for all $m, n \geq n_k$. Without loss of generality, we may assume that $n_{k+1} > n_k$, an thus there exists a subsequence $\{f_{n_k}\} \subseteq \{f_n\}$ such that

$$\sum_{k=1}^{\infty} (2^k \|f_{n_k} - f_{n_{k+1}}\|_p)^p < \infty.$$

Set $E_k := \{G\,|f_{n_k} - f_{n_{k+1}}| > \frac{1}{2^k}\}$ $(k \in \mathbb{N})$. Then $2^k |f_{n_k} - f_{n_{k+1}}| \in \mathscr{V}_{E_k,p}$ for all $k \in \mathbb{N}$, and thus

$$C_{g,p}(E_k) \leq 2^{kp} \|f_{n_k} - f_{n_{k+1}}\|_p^p.$$

Set $F_j := \bigcup_{k=j}^{\infty} E_k$. Then by the σ-subadditivity of $C_{g,p}$ we get

$$C_{g,p}(F_j) \le \sum_{k=j}^{\infty} C_{g,p}(E_k) \le \sum_{k=j}^{\infty} 2^{kp} \|f_{n_k} - f_{n_{k+1}}\|_p^p$$

for any $j \in \mathbb{N}$, whence $\lim_{j\to\infty} C_{g,p}(F_j) = 0$. Moreover, for all $\overline{m} > j$ and $l > m > \overline{m}$, we have

$$\sup_{x \in F_j^c} |Gf_{n_l}(x) - Gf_{n_m}(x)| \le \sup_{x \in F_j^c} \left(\sum_{i=m}^{l-1} |Gf_{n_i}(x) - Gf_{n_{i+1}}(x)| \right) \le \sum_{i=m}^{l-1} \frac{1}{2^i} \le \frac{1}{2^{\overline{m}-1}},$$

which implies the uniform convergence of $\{Gf_{n_k}\}$ in F_j^c.

To sum up, we have shown that for any $\delta > 0$, we can choose $j \in \mathbb{N}$ so large that $C_{g,p}(F_j) < \delta$ and $\{Gf_{n_k}\}$ converges uniformly in F_j^c. Hence $\{Gf_{n_k}\}$ converges (g,p)-quasi-uniformly. Then by Proposition 2.1.13(i) the conclusion follows. □

Let us now characterize the closure of the set $\mathcal{V}_{E,p}$ in (3.25).

Lemma 3.2.6. *Let g be a kernel, and let $p \in (1,\infty)$. Then*

$$\overline{\mathcal{V}_{E,p}} = \{f \in L^p(\mathbb{R}^N) \mid f \ge 0,\ Gf \ge 1\ (g,p)\text{-q. e. on } E\}.$$

Proof. Set $\mathcal{V}'_{E,p} := \{f \in L^p(\mathbb{R}^N) \mid f \ge 0,\ Gf \ge 1\ (g,p)\text{-q. e. in } E\}$. Clearly, $\mathcal{V}_{E,p} \subseteq \mathcal{V}'_{E,p}$. We will prove below that $\mathcal{V}'_{E,p}$ is closed, and thus $\overline{\mathcal{V}_{E,p}} \subseteq \mathcal{V}'_{E,p}$.

On the other hand, let $f \in \mathcal{V}'_{E,p}$. Then $f \in L^p(\mathbb{R}^N)$, $f \ge 0$, and there exists a (g,p)-null set $F \subseteq \mathbb{R}^N$ such that $Gf(x) \ge 1$ for all $x \in E \setminus F$. The proof of Lemma 3.2.4 can be easily modified to show that for any $\epsilon > 0$, there exists $g \in L^p(\mathbb{R}^N)$, $g \ge 0$, such that $\|g\|_p < \epsilon$ and $Gg = \infty$ on F. Then $f_1 := f + g \in \mathcal{V}_{E,p}$ and $\|f_1 - f\|_p = \|g\|_p < \epsilon$, and thus $f \in \overline{\mathcal{V}_{E,p}}$. Then $\mathcal{V}'_{E,p} \subseteq \overline{\mathcal{V}_{E,p}}$, and the result follows.

It remains to prove that $\mathcal{V}'_{E,p}$ is closed. Let $\{f_n\} \subseteq \mathcal{V}'_{E,p}$ and $f \in L^p(\mathbb{R}^N)$ be such that $\lim_{n\to\infty} \|f_n - f\|_p = 0$. Then $f \ge 0$, and by Proposition 3.2.5 there exists a subsequence $\{f_{n_k}\} \subseteq \{f_n\}$ such that $\lim_{k\to\infty} Gf_{n_k} = Gf$ (g,p)-q. e. on E. By usual arguments this implies that $Gf \ge 1$ (g,p)-q. e. on E. This proves that $f \in \mathcal{V}'_{E,p}$ and completes the proof. □

Now we can prove that the infimum in (3.25) is in fact attained.

Proposition 3.2.7. *Let g be a kernel, and let $p \in (1,\infty)$. Let $E \subseteq \mathbb{R}^N$ be such that $C_{g,p}(E) < \infty$. Then there exists a unique nonnegative $f_E \in L^p(\mathbb{R}^N)$ such that $(Gf_E)(x) \ge 1$ (g,p)-q. e. in E and $C_{g,p}(E) = \|f_E\|_p^p$.*

Proof. For any $p \in (1,\infty)$, the space $L^p(\mathbb{R}^N)$ is uniformly convex, and thus in the closed convex subset $\overline{\mathcal{V}_{E,p}} \subseteq L^p(X)$, there exists a unique element f_E of the least norm (e. g., see [2, Corollary 1.3.4]). Then by Lemma 3.2.6 the result follows. □

The minimizer f_E given by Proposition 3.2.7 and Gf_E are called the *capacitary function* and *capacitary potential* of the set E, respectively.

The following result is analogous to Proposition 1.3.1(iv) for measures.

Proposition 3.2.8. *Let g be a kernel, and let $p \in (1,\infty)$. Then for any nondecreasing sequence $\{E_k\}$ of subsets of $\mathbb{R}^N$, we have*

$$C_{g,p}\left(\bigcup_{k=1}^{\infty} E_k\right) = \lim_{k\to\infty} C_{g,p}(E_k). \tag{3.29}$$

Proof. Set $E := \bigcup_{k=1}^{\infty} E_k$. By monotonicity we have $\lim_{k\to\infty} C_{g,p}(E_k) \le C_{g,p}(E)$. To prove the reverse inequality, we may assume that $\lim_{k\to\infty} C_{g,p}(E_k) < \infty$. Let $f_k \equiv f_{E_k}$ be the capacitary function of the set E_k given by Proposition 3.2.7. If $k < l$, then $E_k \subseteq E_l$, and thus $Gf_l \ge 1$ (g,p)-q. e. on E_k. By Lemma 3.2.6 it follows that $f_l \in \overline{\mathscr{V}}_{E_k,p}$, and thus

$$\left\| \frac{f_k + f_l}{2} \right\|_p^p \ge C_{g,p}(E_k).$$

By this inequality and the uniform convexity of $L^p(\mathbb{R}^N)$ the sequence $\{f_k\}$ is a Cauchy sequence (e. g., see [2, Corollary 1.3.3]), and hence there exists $f \in L^p(\mathbb{R}^N)$ such that $\lim_{k\to\infty} \|f_k - f\|_p = 0$. Then by Proposition 3.2.7 we have

$$\|f\|_p^p = \lim_{k\to\infty} \|f_k\|_p^p = \lim_{k\to\infty} C_{g,p}(E_k). \tag{3.30}$$

On the other hand, by Proposition 3.2.5 for any fixed $k \in \mathbb{N}$, we have $\lim_{l\to\infty} Gf_l = Gf \ge 1$ (g,p)-q. e. on E_k, and thus $Gf \ge 1$ (g,p)-q. e. on E. It follows that $f \in \overline{\mathscr{V}}_{E,p}$, whence

$$\|f\|_p^p \ge C_{g,p}(E). \tag{3.31}$$

From (3.30)–(3.31) we obtain that $\lim_{k\to\infty} C_{g,p}(E_k) \ge C_{g,p}(E)$. Hence the conclusion follows. □

The following result shows that every Borel subset of $\mathbb{R}^N$ is (g,p)-capacitable.

Proposition 3.2.9. *Let g be a kernel, and let $p \in (1,\infty)$. Then for any Borel set $E \subseteq \mathbb{R}^N$,*

$$C_{g,p}(E) = \inf\{C_{g,p}(A) \mid A \supseteq E,\ A \text{ open}\} = \sup\{C_{g,p}(K) \mid K \subseteq E,\ K \text{ compact}\}.$$

Proof. By Theorem 3.2.2 $C_{g,p}$ is an outer capacity on $\mathbb{R}^N$, and thus Proposition 1.6.2 applies. Therefore, in view of Proposition 3.2.8, the assumptions of Theorem 1.6.1 are satisfied. Hence the result follows. □

3.2.3 Dependence of $C_{g,p}$ on $p \in [1,\infty)$

Fix $\mu \in \mathfrak{R}^+(\mathbb{R}^N)$. Unless explicitly stated, in this subsection we write $C_{\mu,g,p}$ instead of $C_{g,p}$; similar notations are used in analogous cases. In particular, for any $f \in L^p(\mathbb{R}^N, \mathcal{B}(\mathbb{R}^N), \mu)$, we write $\|f\|^p_{\mu,p} \equiv \int_{\mathbb{R}^N} |f|^p\, d\mu$ (instead of $\|f\|^p_p$ as we made so far).

Let us investigate the behavior of $C_{\mu,g,p}(E)$ as a function of p, $E \subseteq \mathbb{R}^N$ being fixed. To this purpose, we need the following two lemmas.

Lemma 3.2.10. *Let $\mu \in \mathfrak{R}^+(\mathbb{R}^N)$, $\nu_R \in \mathfrak{R}^+_f(\mathbb{R}^N)$, and $\nu_R := \mu \llcorner B(0,R)$ $(R > 0)$. Let g be a kernel, and let $p \in (1,\infty)$. Then for all $E \subseteq \mathbb{R}^N$,*

$$C_{\mu,g,p}(E) \leq C_{\nu_R,g,p}(E). \tag{3.32}$$

Proof. Set $\nu \equiv \nu_R$ for simplicity. Let

$$f \in \mathscr{V}_{E,\nu,g,p} := \{f \in L^p(\mathbb{R}^N, \mathcal{B}(\mathbb{R}^N), \nu) \mid f \geq 0,\ (G_\nu f)(x) \geq 1\ \forall x \in E\}$$

(see (3.25b)). Set $F := f\chi_{B(0,R)}$. Then $F \in L^p(\mathbb{R}^N, \mathcal{B}(\mathbb{R}^N), \mu)$, $F \geq 0$, and $(G_\mu F)(x) = (G_\nu f)(x) \geq 1$ for all $x \in E$. It follows that $F \in \mathscr{V}_{E,\mu,g,p}$, and thus

$$C_{\mu,g,p}(E) \leq \|F\|_{L^p(\mathbb{R}^N,\mathcal{B}(\mathbb{R}^N),\mu)} = \|f\|_{L^p(\mathbb{R}^N,\mathcal{B}(\mathbb{R}^N),\nu)}.$$

Hence the result follows. □

Lemma 3.2.11. *Let $\mu \in \mathfrak{R}^+_f(\mathbb{R}^N)$, let g be a kernel, and let $p \in [1,\infty)$. Then for any $E \subseteq \mathbb{R}^N$, the map $p \mapsto (\frac{C_{g,p}(E)}{\|\mu\|})^{\frac{1}{p}}$ is nondecreasing.*

Proof. By Hölder's inequality, for any $f \in \mathscr{V}_{E,q}$ and $1 < p < q < \infty$, we have

$$\int_{\mathbb{R}^N} f^p\, d\mu \leq \Big(\int_{\mathbb{R}^N} f^q\, d\mu\Big)^{\frac{p}{q}} \mu(\mathbb{R}^N)^{\frac{q-p}{q}},$$

whence $(\frac{C_{\mu,g,p}(E)}{\|\mu\|})^{\frac{1}{p}} \leq (\frac{C_{\mu,g,q}(E)}{\|\mu\|})^{\frac{1}{q}}$.

Let $1 = p < q < \infty$ and $f \in \mathscr{V}_{E,q}$. Then by (3.22) the measure $\sigma = f\mu$, $\sigma(F) := \int_F f\, d\mu$ $(F \in \mathcal{B}(\mathbb{R}^N))$ belongs to $\mathscr{V}_{E,1}$, and

$$\|\sigma\| = \int_{\mathbb{R}^N} f\, d\mu \leq \Big(\int_{\mathbb{R}^N} f^q\, d\mu\Big)^{\frac{1}{q}} \mu(\mathbb{R}^N)^{\frac{q-1}{q}}.$$

From this inequality we get $\frac{C_{g,1}(E)}{\|\mu\|} \leq (\frac{C_{g,q}(E)}{\|\mu\|})^{\frac{1}{q}}$, and thus the result follows. □

Let us denote by $\mathscr{S}_0(\mathbb{R}^N) \subseteq \mathscr{S}_+(\mathbb{R}^N)$ the set of nonnegative simple functions that vanish outside a subset of finite measure μ. For any $p \in (1, \infty)$, set

$$S_{\mu,g,p}(E) := \inf_{s \in \Sigma_E} \|s\|^p_{\mu,p}, \tag{3.33a}$$

where

$$\Sigma_E \equiv \Sigma_{E,\mu,g} := \{s \in \mathscr{S}_0(\mathbb{R}^N) \mid (G_\mu s)(x) \geq 1 \; \forall x \in E\}. \tag{3.33b}$$

If $\Sigma_E = \emptyset$, then $S_{\mu,g,p}(E) := \infty$.

Proposition 3.2.12. *Let $\mu \in \mathfrak{R}^+(\mathbb{R}^N)$. Let g be a kernel, and let $p \in (1, \infty)$. Then for any compact $K \subseteq \mathbb{R}^N$:*
(i) *$S_{\mu,g,p}(K) = C_{\mu,g,p}(K)$;*
(ii) *the map $p \mapsto C_{\mu,g,p}(K)$ is upper semicontinuous. Moreover, if $\mu \in \mathfrak{R}^+_f(\mathbb{R}^N)$, it is continuous from the right in $(1, \infty)$.*

Proof. (i) By definition we have $\Sigma_K \subseteq \mathscr{V}_{K,p}$ (see (3.25b)), and hence $S_{\mu,g,p}(K) \geq C_{\mu,g,p}(K)$. If $C_{\mu,g,p}(K) = \infty$, then the conclusion follows. Otherwise, let $\epsilon \in (0, 1)$, and let $f \in \mathscr{V}_{K,p}$ satisfy $\|f\|^p_{\mu,p} < C_{\mu,g,p}(K) + \epsilon$. Set $f_\epsilon := \frac{f}{1-\epsilon}$. By standard results (see Lemma 2.8.1) there exists $\{s_n\} \subseteq \mathscr{S}_0(\mathbb{R}^N)$ such that $\lim_{n\to\infty} \|s_n - f_\epsilon\|_{\mu,p} = 0$. We will prove the following:

Claim. There exists $\bar{n} \in \mathbb{N}$ such that $G_\mu s_n \geq 1$ in K for all $n > \bar{n}$.

It follows from this claim that for every $n > \bar{n}$, we have $\{s_n\} \subseteq \Sigma_K$, and thus $S_{\mu,g,p}(K) \leq \|s_n\|^p_{\mu,p}$. Letting $n \to \infty$ in the previous inequality, we get

$$S_{\mu,g,p}(K) \leq \frac{1}{(1-\epsilon)^p} \|f\|^p_{\mu,p} \leq \frac{1}{(1-\epsilon)^p} [C_{\mu,g,p}(K) + \epsilon],$$

whence $S_{\mu,g,p}(K) \leq C_{\mu,g,p}(K)$ by the arbitrariness of ϵ.

To prove the claim, suppose by contradiction that there exist $\{s_{n_k}\} \subseteq \{s_n\}$ and $\{x_k\} \subseteq K$ such that $(G_\mu s_{n_k})(x_k) < 1$ for each $k \in \mathbb{N}$. Since K is compact, there exist a subsequence of $\{x_k\}$ (not relabeled) and $\bar{x} \in K$ such that $x_k \to \bar{x}$ as $k \to \infty$. Arguing as in the proof of Lemma 3.2.1(i), this implies that

$$\begin{aligned}(G_\mu f_\epsilon)(\bar{x}) &= \int_{\mathbb{R}^N} g(\bar{x}, y) f_\epsilon(y)\, d\mu(y) \leq \liminf_{k\to\infty} \int_{\mathbb{R}^N} g(x_k, y)\, s_{n_k}(y)\, d\mu(y) \\ &= \liminf_{k\to\infty} (G_\mu s_{n_k})(x_k) \leq 1,\end{aligned}$$

which is a contradiction since $G_\mu f_\epsilon \geq \frac{1}{1-\epsilon} > 1$ on K. Hence both the claim and claim (i) follow.

(ii) If $\Sigma_K = \emptyset$, by (i) we have $C_{\mu,g,p}(K) = \infty$ for all p, and hence the claim is obvious. Otherwise, for every $s \in \Sigma_K$, we have $s = \sum_{k=1}^{j_s} c_{k,s} \chi_{E_{k,s}}$ for some $j_s \in \mathbb{N}$, $c_{k,s} > 0$, and

$E_{k,s} \in \mathcal{B}(\mathbb{R}^N)$. Hence by (i) we have

$$C_{\mu,g,p}(K) = \inf_{s\in\Sigma_K} \|s\|^p_{\mu,p} = \inf_{s\in\Sigma_K} \sum_{k=1}^{j_s} c^p_{k,s}\mu(E_{k,s}).$$

By this equality $C_{\mu,g,p}(K)$ is the infimum of continuous functions of p and thus is an upper semicontinuous function of p. Therefore

$$\limsup_{q\to p^+} C_{\mu,g,q}(K) \le C_{\mu,g,p}(K). \tag{3.34a}$$

On the other hand, by Lemma 3.2.11 we have $(\frac{C_{\mu,g,p}(K)}{\|\mu\|})^{\frac{q}{p}} \le \frac{C_{\mu,g,q}(K)}{\|\mu\|}$ for every $q > p$, whence

$$C_{\mu,g,p}(K) \le \liminf_{q\to p^+} C_{\mu,g,q}(K). \tag{3.34b}$$

From (3.34) the conclusion follows. □

Proposition 3.2.13. *Let $\mu \in \mathfrak{R}^+(\mathbb{R}^N)$. Let g be a kernel, and let $p \in (1,\infty)$. Let $K \subseteq \mathbb{R}^N$ be compact, and let the following hold:*

$$\lim_{R\to\infty} \int_{\{|y|\ge R\}} \Big(\sup_{x\in K} g(x,y)\Big)\, d\mu(y) = 0, \tag{3.35a}$$

$$\limsup_{|y|\to\infty}\Big(\sup_{x\in K} g(x,y)\Big) =: M_0 < \infty. \tag{3.35b}$$

Then the map $p \mapsto C_{\mu,g,p}(K)$ is continuous from the right in $(1,\infty)$.

Proof. We may assume that $C_{\mu,g,p}(K) < \infty$ for any $p \in (1,\infty)$, since otherwise by Proposition 3.2.12(i) $C_{\mu,g,p}(K) = \infty$ for all $p \in (1,\infty)$, and the claim is obviously satisfied. It suffices to prove the result for $p \in [p_1, q_1]$ with $p_1 \in (1,2)$ and $q_1 := \frac{p_1}{p_1-1}$. By Proposition 3.2.12(ii) the map $p \mapsto C_{\mu,g,p}(K)$ is upper semicontinuous, and thus it attains a maximum value in the closed interval $[p_1, q_1]$, that is, there exists

$$\max_{p\in[p_1,q_1]} C_{\mu,g,p}(K) =: M_1 < \infty. \tag{3.36}$$

Set

$$\epsilon_1 := \frac{1}{[(M_0+1)(M_1+1)]^{q_1/p_1}} \in (0,1),$$

and fix $\epsilon \in (0,\epsilon_1)$. For any $p \in [p_1, q_1]$, there exists $f_p \in \mathscr{V}_{K,\mu,g,p}$ such that

$$\|f_p\|^p_{\mu,p} < C_{\mu,g,p}(K) + \epsilon \le M_1 + \epsilon \tag{3.37}$$

(see (3.36)). Let $\nu_R \in \mathfrak{R}_f^+(\mathbb{R}^N)$ and $\nu_R := \mu \llcorner B(0,R)$ $(R > 0)$. We will prove the following:

Claim. For any $\epsilon \in (0, \epsilon_1)$, there exists $\bar{R} > 0$ such that $f_{\epsilon,p} := f_p[1-(\frac{\epsilon}{\epsilon_1})^{1/q_1}]^{-1} \in \mathscr{V}_{K,\nu_R,g,p}$ for all $R > \bar{R}$ and $p \in [p_1, q_1]$.

If so, then from (3.37) it follows that for any $\epsilon \in (0, \epsilon_1)$ and $R > \bar{R}$,

$$C_{\nu_R,g,p}(K) \le \|f_{\epsilon,p}\|^p_{\nu_R,p} = \int_{B_r} [f_{\epsilon,p}(y)]^p \, d\mu(y)$$
$$\le \frac{\|f_p\|^p_{\mu,p}}{[1-(\frac{\epsilon}{\epsilon_1})^{1/q_1}]^p} \le \frac{C_{\mu,g,p}(K) + \epsilon}{[1-(\frac{\epsilon}{\epsilon_1})^{1/q_1}]^{q_1}}. \tag{3.38}$$

By (3.38) and (3.32) we obtain

$$0 \le \sup_{p\in[p_1,q_1]} C_{\nu_R,g,p}(K) - C_{\mu,g,p}(K) \le \left(\left[1-\left(\frac{\epsilon}{\epsilon_1}\right)^{1/q_1}\right]^{-q_1} - 1\right) M_1 + \epsilon\left[1-\left(\frac{\epsilon}{\epsilon_1}\right)^{1/q_1}\right]^{-q_1}.$$

Now let $\epsilon \to 0^+$ and thus $R \to \infty$. It follows that $C_{\nu_R,g,p}(K) \to C_{\mu,g,p}(K)$ uniformly in $[p_1, q_1]$ as $R \to \infty$. On the other hand, by Proposition 3.2.12(ii) the map $p \mapsto C_{\nu_R,g,p}(K)$ is continuous from the right on $[p_1, q_1]$ for each $R > 0$. Hence the result follows.

To prove Claim, observe that, clearly, $f_{\epsilon,p} \in L^p(\mathbb{R}^N, \mathcal{B}(\mathbb{R}^N), \nu_R)$ and $f_{\epsilon,p} \ge 0$. Set $B_R \equiv B(0,R)$, let $q = \frac{p}{p-1}$ be the conjugate exponent of p, and fix $x \in K$. In view of (3.35) and (3.37), for any $\epsilon \in (0, \epsilon_1)$, there exists $\bar{R} > 0$ such that for all $R > \bar{R}$,

$$\int_{B_R^c} g(x,y) f_p(y)\, d\mu(y) \le \|f_p\|_{\mu,p} \left\{ \int_{B_R^c} \left(\sup_{x\in K} g(x,y)\right)^q d\mu(y)\right\}^{1/q}$$
$$\le \|f_p\|_{\mu,p} \left(\sup_{x\in K} g(x,y)\right)^{\frac{1}{p}} \left\{ \int_{B_R^c} \sup_{x\in K} g(x,y)\, d\mu(y)\right\}^{1/q}$$
$$\le [(M_0+1)(M_1+1)]^{1/p_1} \epsilon^{1/q} \le \left(\frac{\epsilon}{\epsilon_1}\right)^{1/q_1},$$

since $q \le q_1$ and $\epsilon \in (0,1)$. Therefore, for all ϵ, R as above and for any $x \in K$,

$$1 \le (G_\mu f_p)(x) = \int_{B_R} g(x,y) f_p(y)\, d\mu(y) + \int_{B_R^c} g(x,y) f_p(y)\, d\mu(y)$$
$$\le (G_{\nu_R} f_p)(x) + \left(\frac{\epsilon}{\epsilon_1}\right)^{1/q_1}.$$

From the above inequality we get $(G_{\nu_R} f_{\epsilon,p})(x) \ge 1$ for all $x \in K$, and thus $f_{\epsilon,p} \in \mathscr{V}_{K,\nu_R,g,p}$. This completes the proof. □

In the remaining part of this subsection, we always take $\mu = \lambda_N$, the Lebesgue measure on $\mathbb{R}^N$, and omit the dependence on μ in our notations. Hence we set $C_{g,p} \equiv C_{\lambda_N,g,p}$ (see definitions (3.25) and (3.26)). We also set $L^p(\mathbb{R}^N) \equiv L^p(\mathbb{R}^N, \mathcal{B}(\mathbb{R}^N), \lambda_N)$, $d\mu(y) \equiv dy$, and $\|\cdot\|_p \equiv \|\cdot\|_{\lambda_N,p}$.

Let us prove that for any compact $K \subseteq \mathbb{R}^N$, the map $p \mapsto C_{g,p}(K)$ ($p \in [1,\infty)$) is continuous from the right at $p = 1$. For this purpose, set

$$S_{g,1}(E) := \inf_{s \in \Sigma'_E} \|s\|_1, \tag{3.39a}$$

where

$$\Sigma'_E \equiv \Sigma'_{E,g} := \left\{ s \in \mathscr{S}_0(\mathbb{R}^N) \mid \int_{\mathbb{R}^N} g(x,y)s(y)\,dy \geq 1 \ \forall x \in E \right\}. \tag{3.39b}$$

If $\Sigma'_E = \emptyset$, then $S_{g,1}(E) := \infty$.

The following result is analogous to Proposition 3.2.12.

Proposition 3.2.14. *Let g be a kernel. Then for any compact $K \subseteq \mathbb{R}^N$:*
(i) $S_{g,1}(K) = C_{g,1}(K)$;
(ii) *the map $p \mapsto C_{g,p}(K)$ is upper semicontinuous in $[1,\infty)$.*

Proof. (i) Since $\Sigma'_K \subseteq \mathscr{V}_{K,1}$, we have $S_{g,1}(K) \geq C_{g,1}(K)$. If $C_{g,1}(K) = \infty$, then the conclusion follows. Otherwise, let $\epsilon \in (0,1)$, and let $\mu \in \mathscr{V}_{K,1}$ satisfy $\|\mu\| < C_{g,1}(K) + \epsilon$ (see (3.26)). Set $\mu_\epsilon := \frac{\mu}{1-\epsilon}$. Let $\rho \in C_c^\infty(\mathbb{R}^N)$, $\int_{\mathbb{R}^N} \rho\,dy = 1$, $\rho_k(y) := k^N \rho(ky)$ ($k \in \mathbb{N}, y \in \mathbb{R}^N$), and $f_k := \mu_\epsilon * \rho_k$. Then $\{f_k\} \subseteq C^\infty(\mathbb{R}^N)$, $f_k \geq 0$, $\|f_k\|_1 \leq \|\mu_\epsilon\|$, and $\int_{\mathbb{R}^N} f_k \varphi\,dy \to \int_{\mathbb{R}^N} \varphi\,d\mu_\epsilon$ for every $\varphi \in C_c(\mathbb{R}^N)$. On the other hand, since f_k is nonnegative and belongs to $L^1(\mathbb{R}^N)$, it can be approximated strongly in $L^1(\mathbb{R}^N)$ and from below by a sequence of functions in $\mathscr{S}_0(\mathbb{R}^N)$. Therefore there exists a sequence $\{s_k\} \subseteq \mathscr{S}_0(\mathbb{R}^N)$ such that $\|s_k\|_1 \leq \|\mu_\epsilon\|$ and $\int_{\mathbb{R}^N} s_k \varphi\,dy \to \int_{\mathbb{R}^N} \varphi\,d\mu_\epsilon$ for every $\varphi \in C_c(\mathbb{R}^N)$. As in the proof of Proposition 3.2.12(i), we can prove that $\{s_k\} \subseteq \Sigma'_K$ for all k large enough, whence $S_{g,1}(K) \leq C_{g,1}(K)$, and claim (i) follows.

Concerning (ii), it suffices to observe that by (i) and Proposition 3.2.12(i) we have

$$C_{g,p}(K) = \inf_{s \in \Sigma'_K} \sum_{k=1}^{j_s} c_{k,s}^p \lambda_N(E_{k,s}) \qquad \text{for all } p \in [1,\infty)$$

(where $s = \sum_{k=1}^{j_s} c_{k,s} \chi_{E_{k,s}}$ for some $j_s \in \mathbb{N}$, $c_{k,s} > 0$, and $E_{k,s} \in \mathcal{B}(\mathbb{R}^N)$). Then arguing as in the proof of Proposition 3.2.12(i), the conclusion follows. □

Remark 3.2.2. Let $\nu_R := \lambda_N \llcorner B_R$ ($B_R \equiv B(0,R), R > 0$). In view of Lemma 3.2.11 and Proposition 3.2.14(ii), the map $p \mapsto C_{\nu_R,g,p}(K)$ is continuous from the right in $[1,\infty)$ for any $R > 0$.

Now we have the following analogue of Proposition 3.2.13.

Proposition 3.2.15. *Let g be a kernel. Let $K \subseteq \mathbb{R}^N$ be compact, and let*

$$\int_{\{|y|\geq R_0\}} \left(\sup_{x\in K}[g(x,y)]^\alpha\right) dy =: M_2 < \infty \tag{3.40a}$$

for some $R_0 > 0$ (depending on K) and

$$\lim_{|y|\to\infty}\left(\sup_{x\in K} g(x,y)\right) = 0 \tag{3.40b}$$

for some $\alpha \in [1,\infty)$. Then

$$\lim_{p\to 1^+} C_{g,p}(K) = C_{g,1}(K). \tag{3.41}$$

Proof. Let $\beta \in (1,\infty]$ be the conjugate exponent of α. Fix any $p_1 \in (1,\beta)$, and let $p \in [1,p_1]$. By Proposition 3.2.14(ii) the map $p \mapsto C_{g,p}(K)$ is upper semicontinuous in $[1,p_1]$, and thus there exists

$$\max_{p\in[1,p_1]} C_{g,p}(K) =: M_3 < \infty. \tag{3.42}$$

Set

$$\epsilon_2 := \frac{1}{[(M_2+1)(M_3+1)]^{1-\alpha/q_1}} \in (0,1),$$

where $q_1 := \frac{p_1}{p_1-1}$ (observe that $\alpha < q_1$), and fix $\epsilon \in (0,\epsilon_2)$. For any $p \in [1,p_1]$, there exists $s_p \in \Sigma_K' \subseteq \mathscr{V}_{K,p}$ such that

$$\|s_p\|_p^p < C_{g,p}(K) + \epsilon \leq M_3 + \epsilon \tag{3.43}$$

(this follows from Propositions 3.2.14(i) and 3.2.12(i) if $p = 1$ and $p \in (1,\infty)$, respectively). Let $\nu_R \in \mathfrak{R}_f^+(\mathbb{R}^N)$, $\nu_R := \lambda_N \llcorner B_R$ ($B_R \equiv B(0,R), R > 0$). We will prove the following: for any $\epsilon \in (0,\epsilon_2)$, there exists $\bar{R} > 0$ such that for all $R > \bar{R}$ and $p \in [1,p_1]$, we have $s_{\epsilon,p} := \frac{s_p}{1-(\frac{\epsilon}{\epsilon_2})^{1-\alpha/q_1}} \in \Sigma_{K,\nu_R} \subseteq \mathscr{V}_{K,\nu_R,p}$.

If so, then by (3.43) for any $\epsilon \in (0,\epsilon_2)$ and $R > \bar{R}$, we have

$$\begin{aligned} C_{\nu_R,g,p}(K) &\leq \|s_{\epsilon,p}\|_{\nu_R,p}^p = \int_{B_R} [s_{\epsilon,p}(y)]^p\, dy \\ &\leq \frac{\|s_p\|_p^p}{[1-(\frac{\epsilon}{\epsilon_2})^{1-\alpha/q_1}]^p} \leq \frac{C_{g,p}(K)+\epsilon}{[1-(\frac{\epsilon}{\epsilon_2})^{1-\alpha/q_1}]^{p_1}}. \end{aligned} \tag{3.44}$$

On the other hand, arguing as in the proof of Lemma 3.2.10, we easily see that for any $p \in [1,\infty)$, $E \subseteq \mathbb{R}^N$, and $R > 0$, we have

$$C_{g,p}(E) \le C_{V_R,g,p}(E). \tag{3.45}$$

Letting $\epsilon \to 0^+$ and thus $R \to \infty$, from (3.44)–(3.45) we obtain that $C_{V_R,g,p}(K) \to C_{g,p}(K)$ uniformly in $[1,p_1]$. Since the map $p \mapsto C_{V_R,g,p}(K)$ is continuous from the right on $[1,p_1]$ for each $R > 0$ (see Remark 3.2.2), the result follows.

To complete the proof, we must prove the above claim. Let q be the conjugate exponent of p, and fix $x \in K$. By (3.40) and (3.43), for any $\epsilon \in (0,\epsilon_2)$, there exists $\bar{R} > R_0$ such that for all $R > \bar{R}$:

(a) if $p = 1$, then

$$\int_{B_R^c} g(x,y)s_p(y)\,dy \le \|s_p\|_1 \sup_{x\in K,\,|y|\ge R} g(x,y) < (M_3+1)\,\epsilon; \tag{3.46a}$$

(b) if $p \in (1,p_1]$, then

$$\begin{aligned}\int_{B_R^c} g(x,y)s_p(y)dy &\le \|s_p\|_p \left\{\int_{B_R^c}\Big(\sup_{x\in K} g(x,y)\Big)^q dy\right\}^{1/q}\\ &\le \|s_p\|_p \left\{\int_{B_R^c}\Big(\sup_{x\in K} g(x,y)\Big)^\alpha dy\right\}^{1/q}\Big(\sup_{x\in K,\,|y|\ge R} g(x,y)\Big)^{1-\frac{\alpha}{q}}\\ &\le (M_2+1)(M_3+1)\,\epsilon^{1-\frac{\alpha}{q}} \le \left(\frac{\epsilon}{\epsilon_2}\right)^{1-\alpha/q_1},\end{aligned} \tag{3.46b}$$

since $\epsilon \in (0,1)$ and $1-\frac{\alpha}{q} \ge 1-\frac{\alpha}{q_1} > 0$.

By (3.46), for all ϵ, R as above and for all $p \in [1,p_1]$ and $x \in K$, we have

$$\int_{B_R^c} g(x,y)s_p(y)\,dy \le \left(\frac{\epsilon}{\epsilon_2}\right)^{1-\alpha/q_1}.$$

Therefore, for any $x \in K$,

$$\begin{aligned}1 \le (Gs_p)(x) &= \int_{B_R} g(x,y)s_p(y)\,dy + \int_{B_R^c} g(x,y)s_p(y)\,dy\\ &\le (G_{V_R}s_p)(x) + \left(\frac{\epsilon}{\epsilon_2}\right)^{1-\alpha/q_1}.\end{aligned}$$

From the above inequality we get $(G_{V_R}s_{\epsilon,p})(x) \ge 1$ for all $x \in K$, and thus $s_{\epsilon,p} \in \Sigma_{K,V_R}$. This completes the proof. □

Remark 3.2.3. If condition (3.35a) in Proposition 3.2.13 is replaced by

$$\lim_{R\to\infty} \int_{\{|y|\geq R\}} \left(\sup_{x\in K}[g(x,y)]^{\alpha}\right) dy = 0 \tag{3.47}$$

for some $\alpha \in [1,\infty)$, arguing as in the proof of Proposition 3.2.13, we get that for any compact $K \subseteq \mathbb{R}^N$, the map $p \mapsto C_{\mu,g,p}(K)$ is continuous from the right in $(1,\alpha')$, where α' denotes the conjugate exponent of α. On the other hand, condition (3.47) implies (3.40a). Hence, if also (3.40b) is satisfied, then by Proposition 3.2.15 the conclusion holds in $[1,\alpha')$.

3.3 Bessel, Riesz, and Sobolev capacities

The results of this section rely on the general theory of capacities outlined in Section 3.2.

3.3.1 Bessel and Riesz capacities

Let $g_r = g_r(x)$ and $\gamma_r = \gamma_r(x)$ ($r > 0$, $x \in \mathbb{R}^N$) be the Bessel and Riesz kernels, respectively (see Subsection 3.1.3). As a particular case of Definition 3.2.1 and (3.26) (see Remark 3.2.1), we have the following definition.

Definition 3.3.1. Let $r > 0$ and $E \subseteq \mathbb{R}^N$.

(i) Let $p \in (1,\infty)$. The *Bessel capacity of E, of order r and degree p*, is

$$B_{r,p}(E) := \inf_{f\in V_{E,p}} \|f\|_p^p, \quad V_{E,p} := \{f \in L^p(\mathbb{R}^N) \mid (g_r * f)(x) \geq 1\ \forall x \in E\}. \tag{3.48}$$

If $V_{E,p} = \emptyset$, then $B_{r,p}(E) := \infty$.

(ii) The *Bessel capacity of E, of order r and degree* 1, is

$$B_{r,1}(E) := \inf_{\sigma\in V_{E,1}} \|\mu\|, \tag{3.49a}$$

$$V_{E,1} := \left\{\sigma \in \mathfrak{R}_f(\mathbb{R}^N) \mid \int_{\mathbb{R}^N} g_r(x-y)\, d\sigma(y) \geq 1\ \forall x \in E\right\}. \tag{3.49b}$$

If $V_{E,1} = \emptyset$, then $B_{r,1}(E) := \infty$.

Let $r \in (0,N)$ and $p \in [1,\infty)$. The *Riesz capacity* $R_{r,p}(E)$ is defined replacing g_r by γ_r.

Proposition 3.3.1. (i) *Let $p \in (1,\infty)$ and $E \subseteq \mathbb{R}^N$. Then $B_{r,p}(E)$ and $R_{r,p}(E)$ are invariant under orthogonal transformations and translations of the set E.*

(ii) *For any* $p \in [1,\infty)$, $B_{r,p}$ *and* $R_{r,p}$ *are outer capacities on* $\mathbb{R}^N$. *Moreover, for any* $p \in (1,\infty)$, *every Borel subset of* $\mathbb{R}^N$ *is* $B_{r,p}$*-capacitable:*

$$B_{r,p}(E) = \inf\{B_{r,p}(A) \mid A \supseteq E,\ A \text{ open}\} = \sup\{B_{r,p}(K) \mid K \subseteq E,\ K \text{ compact}\},$$

and similarly for $R_{r,p}$.

(iii) *For any compact* $K \subseteq \mathbb{R}^N$, *the map* $p \mapsto B_{r,p}(K)$ *is continuous from the right in* $[1,\infty)$.

(iv) *For any compact* $K \subseteq \mathbb{R}^N$ *the map* $p \mapsto R_{r,p}(K)$ $(r \in (0,N))$ *is continuous from the right in* $[1,\frac{N}{r})$.

Proof. Claim (i) is easily checked, whereas (ii) follows from Theorem 3.2.2 and Proposition 3.2.9. Since $g(x,y) = g_r(x-y)$ satisfies conditions (3.35) and (3.40), claim (iii) follows from Propositions 3.2.12 and 3.2.15. Concerning (iv), it is easily checked that the Riesz kernel γ_r $(r \in (0,N))$ satisfies condition (3.47) for any $\alpha > \frac{N}{N-r}$. Hence by Remark 3.2.3 the conclusion follows. □

3.3.2 Metric properties of the Bessel capacity

In this subsection, we always set $V_E \equiv V_{E,p}$ (see Definition 3.3.1) for notational simplicity. It is informative to point out the relationship between Bessel capacity and Lebesgue measure. Concerning this point, we have the following:

Proposition 3.3.2. *Let* $r > 0$ *and* $p \in (1,\infty)$. *Then there exists a constant* $M > 0$ *(depending on r, p, and N) such that for any* $E \subseteq \mathbb{R}^N$:

(i) *if* $rp < N$, *then* $B_{r,p}(E) \geq M[\lambda_N^*(E)]^{\frac{N-rp}{N}}$;

(ii) *if* $rp = N$, *then* $B_{r,p}(E) \geq M[\lambda_N^*(E)]^{\epsilon}$ *for all* $\epsilon \in (0,1]$;

(iii) *if* $rp > N$, *then* $B_{r,p}(E) \geq M$ *if* $E \neq \emptyset$.

Proof. Since $B_{r,p}$ is an outer capacity (see Proposition 3.3.1(ii)), it suffices to prove the claims when $E = A$ is open, bounded, and nonempty.

(i) For any $f \in V_A$, we have

$$\lambda_N(A) \leq \int_A (g_r * f)(x)\,dx \leq \|g_r * f\|_{p^*}\,[\lambda_N(A)]^{1-\frac{1}{p^*}}, \tag{3.50}$$

where $p^* := \frac{Np}{N-rp}$. By Sobolev embedding results there exists $M_0 > 0$ (depending on r, p, and N) such that $\|g_r * f\|_{p^*} \leq M_0\|f\|_p$. Then from (3.50) we get

$$[\lambda_N(A)]^{\frac{p}{p^*}} \leq M_0^p\|f\|_p^p, \tag{3.51}$$

whence by the arbitrariness of $f \in V_A$ the claim follows.

(ii) Arguing as in (*i*), instead of (3.51)m we now have

$$[\lambda_N(A)]^{\frac{p}{q}} \leq M_0^p \|f\|_p^p \quad \text{for all } q \geq p.$$

The claim follows by setting $\epsilon := \frac{p}{q}$.

(iii) Using (3.5)m we easily check that in this case, $g_r \in L^{\frac{p}{p-1}}(\mathbb{R}^N)$. Let $A \ni 0$. Then for any $f \in V_A$,

$$1 \leq (g_r * f)(0) \leq \|g_r\|_{\frac{p}{p-1}} \|f\|_p,$$

whence $B_{r,p}(A) \geq B_{r,p}(\{0\}) \geq [\|g_r\|_{\frac{p}{p-1}}]^{-p}$. Then by Proposition 3.3.1(i) the claim follows. This completes the proof. □

Remark 3.3.1. Let $r > 0$ and $p \in (1, \infty)$. By Proposition 3.3.2 $B_{r,p}$-null sets in $\mathbb{R}^N$ can only exist if $rp \leq N$, and in such case, they are λ_N-null. Therefore a property that is true $B_{r,p}$-quasi-everywhere also holds λ_N-a. e. In particular, functions that are $B_{r,p}$-q. e. equal are also λ_N-a. e. equal, and $f_n \to f$ λ_N-a. e. if $f_n \to f$ $B_{r,p}$-q. e. Hence, for any fixed f, a representative of its equivalence class with respect to (1.33) is also equal to f a. e. and thus is a representative of its equivalence class with respect to (1.3).

Set, for convenience, $\beta_\rho \equiv B(x_0, \rho)$ ($x_0 \in \mathbb{R}^N$, $\rho > 0$).

Proposition 3.3.3. *Let $r > 0$, $p \in (1, \infty)$, and $rp < N$. Then there exists a constant $C > 0$ (depending on r, p, and N) such that*

$$C^{-1}\rho^{N-rp} \leq B_{r,p}(\beta_\rho) \leq C\rho^{N-rp} \quad \textit{for all } \rho \in (0, 1]. \tag{3.52a}$$

Proposition 3.3.4. *Let $r > 0$, $p \in (1, \infty)$, and $rp = N$. Let $\rho_0 \in (0, 1)$. Then there exists a constant $\tilde{C} > 0$ (depending on N and ρ_0) such that*

$$\tilde{C}^{-1}\left(-\log \frac{\rho}{\rho_0}\right)^{1-p} \leq B_{r,p}(\beta_\rho) \leq \tilde{C}\left(-\log \frac{\rho}{\rho_0}\right)^{1-p} \quad \textit{for all } \rho \in (0, 1]. \tag{3.52b}$$

We only prove Proposition 3.3.3, referring the reader to [109, Theorem 2.6.14] for the proof of Proposition 3.3.4.

Proof of Proposition 3.3.3. The first inequality in (3.52a) follows from Proposition 3.3.2(i). As for the second, let $f \in V_{\beta_4}$, and thus $g_r * f \geq 1$ on β_4. By a change of variable this implies that

$$\rho^{-N} \int_{\mathbb{R}^N} g_r\left(\frac{x-z}{\rho}\right) f\left(\frac{z}{\rho}\right) dz \geq 1 \quad \text{on } \beta_{4\rho}. \tag{3.53}$$

On the other hand, by (3.5) and (3.6) there exists $C_0 > 0$ (depending on r and N) such that

$$C_0^{-1}|x|^{r-N}e^{-2|x|} \le g_r(x) \le C_0\,|x|^{r-N}e^{-\frac{|x|}{2}} \quad (x \in \mathbb{R}^N)$$

(recall that g_r is a decreasing function of $|x|$ and observe that $r < rp < N$). It follows that

$$\begin{aligned} g_r\left(\frac{x}{\rho}\right) &\le C_0\,\rho^{N-r}|x|^{r-N}e^{-\frac{|x|}{2\rho}} \le C_0\,\rho^{N-r}|x|^{r-N}e^{-2|x|} \\ &\le C_0^2\,\rho^{N-r}g_r(x) \quad \text{for all } \rho \in (0, \tfrac{1}{4}]. \end{aligned} \tag{3.54}$$

From (3.53)–(3.54) we obtain

$$C_0^2\rho^{-r}\int_{\mathbb{R}^N} g_r(x-z)f\left(\frac{z}{\rho}\right)dz \ge 1 \quad \text{on } \beta_{4\rho} \text{ for all } \rho \in (0, \tfrac{1}{4}]. \tag{3.55}$$

Inequality (3.55) shows that $C_0^2\rho^{-r}f(\frac{z}{\rho}) \in V_{\beta_{4\rho}}$ for all $\rho \in (0, \frac{1}{4}]$. Since

$$\left\|C_0^2\rho^{-r}f\left(\frac{\cdot}{\rho}\right)\right\|_p^p = C_0^{2p}\rho^{-rp}\int_{\mathbb{R}^N}\left[f\left(\frac{x}{\rho}\right)\right]^p dx = C_0^{2p}\rho^{N-rp}\,\|f\|_p^p,$$

we clearly get

$$B_{r,p}(\beta_{4\rho}) \le C_0^{2p}\rho^{N-rp}\,B_{r,p}(\beta_4) \quad \text{for all } \rho \in (0, \tfrac{1}{4}].$$

This inequality yields

$$B_{r,p}(\beta_\rho) \le 4^{rp-N}C_0^{2p}B_{r,p}(\beta_4)\,\rho^{N-rp} \quad \text{for all } \rho \in (0,1], \tag{3.56}$$

whence by a proper definition of C the conclusion follows. □

Proposition 3.3.5. *Let $r > 0$, $p \in (1,\infty)$, and $rp \le N$. Then for every countable $E \subseteq \mathbb{R}^N$, we have $B_{r,p}(E) = 0$.*

Proof. Letting $\rho \to 0^+$ in (3.52), we obtain that $B_{r,p}(\{x_0\}) = 0$ for all $x_0 \in \mathbb{R}^N$. Since $B_{r,p}$ is σ-subadditive, the result follows. □

Remark 3.3.2. (i) Let $r > 0$, $p \in (1,\infty)$, and $rp < N$. By the monotonicity and Proposition 1.6.2 from (3.56) we get

$$B_{r,p}(\overline{\beta}_\rho) \le 4^{rp-N}C_0^{2p}B_{r,p}(\overline{\beta}_4)\,\rho^{N-rp} \quad \text{for all } \rho \in (0,1), \tag{3.57}$$

whence letting $p \to 1^+$ (see Proposition 3.3.1(iii)), we get

$$B_{r,1}(\overline{\beta}_\rho) \le 4^{r-N}C_0^2B_{r,1}(\overline{\beta}_4)\,\rho^{N-r} \quad \text{for all } \rho \in (0,1).$$

Since $r < N$, it follows from this inequality that $B_{r,1}(\{x_0\}) = 0$ for each $x_0 \in \mathbb{R}^N$ (thus $B_{r,1}(E) = 0$ for every countable $E \subseteq \mathbb{R}^N$). If $r = N$, then from (3.52b) we similarly obtain that $B_{r,1}(\{x_0\}) > 0$ for all $x_0 \in \mathbb{R}^N$.

(ii) By (i) and Proposition 3.3.2(iii) no $B_{r,p}$-null set in $\mathbb{R}^N$ exists if either $p = 1$ and $r = N$, or $p \in (1, \infty)$ and $rp > N$. In the remaining cases the considerations of Remark 3.3.1 hold.

3.3.3 Sobolev capacity

Definition 3.3.2. Let $m \in \mathbb{N}$ and $p \in [1, \infty)$.

(i) The *Sobolev capacity, of order m and degree p*, of any compact $K \subseteq \mathbb{R}^N$ is

$$C_{m,p}(K) := \inf_{f \in W_K} \|f\|_{m,p}^p, \quad W_K := \{f \in C_c^\infty(\mathbb{R}^N) \mid f(x) \ge 1 \ \forall x \in K\}. \tag{3.58a}$$

(ii) For any open $A \subseteq \mathbb{R}^N$, we set

$$C_{m,p}(A) := \sup\{C_{m,p}(K) \mid K \subseteq A, \ K \text{ compact}\}. \tag{3.58b}$$

(iii) For any $E \subseteq \mathbb{R}^N$, we set

$$C_{m,p}(E) := \inf\{C_{m,p}(A) \mid A \supseteq E, \ A \text{ open}\}. \tag{3.58c}$$

By definition, $C_{m,p}$ is an outer capacity on $\mathbb{R}^N$. Sometimes, we will speak of (m,p)-capacity instead of $C_{m,p}$-capacity. We say that a property holds (m,p)-quasi-everywhere if it is true $C_{m,p}$-quasi-everywhere (see Subsection 1.6.1). When $m = 1$, we set $C_p \equiv C_{1,p}$, and thus we say that a set is p-null instead of $(1,p)$-null, and so on.

Remark 3.3.3. The well-posedness of Definition 3.3.2 requires that for any compact $K \subseteq \mathbb{R}^N$,

$$C_{m,p}(K) := \inf_{f \in W_K} \|f\|_{m,p}^p = \inf\{C_{m,p}(A) \mid A \supseteq K, \ A \text{ open}\}. \tag{3.59}$$

Let us prove equality (3.59). Let $\epsilon > 0$. It is easily seen that there exists $f \in W_K$ such that $f(x) > 1$ for any $x \in K$ and $\|f\|_{m,p}^p < C_{m,p}(K) + \epsilon$. Set $K_1 := \{f \ge 1\}$, ad thus $K \subseteq \mathring{K}_1 \subseteq K_1$, where $\mathring{K}_1$ denotes the interior of K_1. Since K_1 is compact, it easily follows from (3.58a) that for any compact $K \subseteq K_1$, we have $C_{m,p}(K) \le C_{m,p}(K_1)$. Therefore, since $\mathring{K}_1$ is open and $K \subseteq \mathring{K}_1 \subseteq K_1$, by (3.58b) we have that

$$C_{m,p}(K) \le C_{m,p}(\mathring{K}_1) \le C_{m,p}(K_1).$$

On the other hand, f belongs to W_{K_1}, and thus

$$C_{m,p}(K_1) \le \|f\|_{m,p}^p < C_{m,p}(K) + \epsilon.$$

By the arbitrariness of ϵ, from the above inequalities we get $C_{m,p}(K) = C_{m,p}(\mathring{K}_1)$, which implies (3.59) since $\mathring{K}_1$ is open and $\mathring{K}_1 \supseteq K$.

It is often useful to use a different version of Definition 3.3.2, where the set W_K is replaced by

$$W'_K := \{f \in \mathcal{S}(\mathbb{R}^N) \mid f(x) \geq 1 \; \forall x \in K\}. \tag{3.60}$$

Then we set $C'_{m,p}(K) := \inf_{f \in W'_K} \|f\|^p_{m,p}$ for every compact $K \subseteq \mathbb{R}^N$, subsequently extending the definition of $C'_{m,p}$ as in (3.58b)–(3.58c).

We say that two capacities C, C' on $\mathbb{R}^N$ are *equivalent* if there exists $A > 0$ such that $A^{-1}C(E) \leq C'(E) \leq A\,C(E)$ for all $E \subseteq \mathbb{R}^N$. It is easily seen that $C_{m,p}$ and $C'_{m,p}$ are equivalent capacities.

Remark 3.3.4. Other definitions of Sobolev capacity are present in the literature; in particular:

(a) Definition 3.3.2 with W_K replaced by

$$W''_K := \{f \in C^\infty_c(\mathbb{R}^N) \mid 0 \leq f \leq 1 \text{ in } \mathbb{R}^N, f = 1 \text{ in } K\}; \tag{3.61}$$

(b) alternatively, for any $E \subseteq \mathbb{R}^N$, set

$$C''_{m,p}(E) := \inf_{f \in Z_E} \|f\|^p_{m,p}, \tag{3.62a}$$

where

$$Z_E := \{f \in W^{m,p}(\mathbb{R}^N) \mid f \geq 1 \text{ a. e. in a neighborhood of } E\}, \tag{3.62b}$$

and $C''_{m,p}(E) := \infty$ if $Z_E = \emptyset$;

(c) instead of the set Z_E, consider in (3.62a) the set

$$Y_E := \{f \in Z_E \mid 0 \leq f \leq 1 \text{ a. e. in } \mathbb{R}^N, f = 1 \text{ a. e. in a neighborhood of } E\}. \tag{3.62c}$$

Addressing their (possibly, partial) equivalence is lengthy, and thus we omit it. Let us only mention that $C_{m,p}$ and $C''_{m,p}$ are equivalent if $p \in (1, \infty)$.

Let us state the following related definition (see [27]).

Definition 3.3.3. For any compact $K \subseteq \mathbb{R}^N$, the *Laplacian capacity of* K is

$$C_{\Delta,1}(K) := \inf_{f \in \Omega_K} \|\Delta f\|_1, \tag{3.63a}$$

where

$$\Omega_K := \{f \in C^\infty_c(\mathbb{R}^N) \mid f(x) \geq 1 \text{ in a neighbourhood of } K\}. \tag{3.63b}$$

The following characterization of (m,p)-null sets is an immediate consequence of Definition 3.3.2 and Remark 3.3.4.

Proposition 3.3.6. *Let $m \in \mathbb{N}$ and $p \in (1,\infty)$.*
(i) *Let $K \subseteq \mathbb{R}^N$ be compact. Then the following statements are equivalent:*
 (i_1) *$C_{m,p}(K) = 0$;*
 (i_2) *there exists a sequence $\{f_k\} \subseteq C_c^\infty(\mathbb{R}^N)$ such that for all $k \in \mathbb{N}$, $f_k \geq 1$ a.e. in a neighborhood of K (depending on k), and $\lim_{k\to\infty} \|f_k\|_{m,p} = 0$.*
(ii) *Let $E \subseteq \mathbb{R}^N$. Then the following statements are equivalent:*
 (ii_1) *$C_{m,p}(E) = 0$;*
 (ii_2) *there exists a sequence $\{f_k\} \subseteq W^{m,p}(\mathbb{R}^N)$ such that for all $k \in \mathbb{N}$, $f_k \geq 1$ a.e. in a neighborhood of E (depending on k), and $\lim_{k\to\infty} \|f_k\|_{m,p} = 0$.*

Remark 3.3.5. Let $m \in \mathbb{N}$ and $p \in (1,\infty)$, and let $K \subseteq \mathbb{R}^N$ be compact and (m,p)-null. It is easily seen that for any open $A \supseteq K$, there exists a sequence $\{f_k\} \subseteq C_c^\infty(A)$ such that $0 \leq f_k \leq 1$ in $\mathbb{R}^N$, $f_k = 1$ in a neighborhood of K, and $\lim_{k\to\infty} \|f_k\|_{m,p} = 0$ (see [10, Lemma 2.1]).

We finish this subsection with a result concerning (m,p)-quasi-continuous functions.

Theorem 3.3.7. *Let $m \in \mathbb{N}$ and $p \in (1,\infty)$. Then every $h \in W^{m,p}(\mathbb{R}^N)$ has an (m,p)-quasi-continuous representative $\hat{h}$. Moreover, $\hat{h}$ is (m,p)-essentially unique.*

Proof. By standard density results, for any $h \in W^{m,p}(\mathbb{R}^N)$, there exists $\{h_n\} \subseteq W^{m,p}(\mathbb{R}^N) \bigcap C(\mathbb{R}^N)$ such that $\|h_n - h\|_{m,p} \to 0$ as $n \to \infty$. Then by Theorem 3.1.5 there exist $f \in L^p(\mathbb{R}^N)$ and $\{f_n\} \subseteq L^p(\mathbb{R}^N)$ such that $h = g_m * f$, $h_n = g_m * f_n$, and $\|f_n - f\|_p \to 0$. It follows from Proposition 3.2.5 and Theorem 3.4.8 that there exists a subsequence $\{h_{n_k}\} \subseteq \{h_n\}$ that converges in $C_{m,p}$-capacity and (m,p)-quasi-uniformly to some $\hat{h} \in W^{m,p}(\mathbb{R}^N)$. Then $\hat{h}$ is an (m,p)-representative of h, and by Proposition 3.2.3 it is (m,p)-quasi-continuous. This proves the existence claim.

To prove the uniqueness, let $u \in W^{m,p}(\mathbb{R}^N)$, and let $\hat{u}$ be an (m,p)-quasi-continuous representative. Then by (3.28) for any $\lambda > 0$, we have

$$\lambda^p C_{m,p}(\{|\hat{h} - \hat{u}| \geq \lambda\}) \leq \int_{\mathbb{R}^N} |h - u|^p \, d\mu,$$

whence $C_{m,p}(\{|\hat{h} - \hat{u}| \geq \lambda\}) = 0$ if $h = u$. By the σ-subadditivity of $C_{m,p}$ this implies that

$$C_{m,p}(\{|f - g| > 0\}) \leq \sum_{k=1}^\infty C_{m,p}\left(\left\{|f - g| > \frac{1}{k}\right\}\right) = 0.$$

Hence the conclusion follows. □

Remark 3.3.6. It is customary to identify every function $h \in W^{m,p}(\mathbb{R}^N)$ with its (m,p)-quasi-continuous representative $\hat{h}$, as we do in the following.

3.4 Relationship between different concepts of capacity

We now address the relationships between the concepts of capacity introduced so far.

3.4.1 Bessel versus Hausdorff

Let $r > 0$ and $p \in (1,\infty)$. Consider the function $h_{r,p} : (0,\infty) \to (0,\infty]$ defined as

$$h_{r,p}(\rho) := \begin{cases} \rho^{N-rp} & \text{if } rp < N, \\ [-\log \rho]_+^{1-p} & \text{if } rp = N. \end{cases} \tag{3.64}$$

Let $\mathcal{H}^*_{h_{r,p},\delta}$ ($\delta \in (0,\infty]$) and $\mathcal{H}^*_{h_{r,p}}$ be the Hausdorff capacities, and let $\mathcal{H}_{h_{r,p}}$ be the Hausdorff measure relative to the gauge function (3.64) (see Definition 1.7.1).

Proposition 3.4.1. *Let $r > 0$, $p \in (1,\infty)$, and $rp \le N$. Then there exists $c_1 > 0$ (depending on r, p, and N) such that for any $E \subseteq \mathbb{R}^N$, we have*

$$B_{r,p}(E) \le c_1\, \mathcal{H}^*_{h_{r,p},1}(E) \le c_1\, \mathcal{H}^*_{h_{r,p}}(E). \tag{3.65}$$

Moreover,

$$\mathcal{H}^*_{h_{r,p}}(E) < \infty \quad \Rightarrow \quad B_{r,p}(E) = 0. \tag{3.66}$$

Proof. We only prove (3.65), referring the reader to [2, Theorem 5.1.9] for the proof of (3.66). If $\mathcal{H}^*_{h_{r,p},1}(E) = \infty$, then the result is obvious. Otherwise, let $E \subseteq \bigcup_{n=1}^{\infty} B(x_n,\rho_n)$ with $\rho_n \le \frac{\delta}{2}$ for all $n \in \mathbb{N}$ and $\delta \in (0,1)$. Then by the σ-subadditivity of $B_{r,p}$ we have

$$B_{r,p}(E) \le \sum_{n=1}^{\infty} B_{r,p}(B(x_n,\rho_n)).$$

If $rp = N$, then from this inequality and (3.52b) we get

$$B_{r,p}(E) \le \tilde{C}_0 \sum_{n=1}^{\infty} [-\log(2\rho_n)]^{1-p} = \tilde{C}_0 \sum_{n=1}^{\infty} h_{r,p}(2\rho_n)$$

with $\tilde{C}_0 := \tilde{C}(N,\frac{1}{2}) > 0$. If $rp < N$, then from (3.57) we similarly get

$$B_{r,p}(E) \le 4^{rp-N} C_0^{2p} B_{r,p}(\overline{\beta}_4) \sum_{n=1}^{\infty} \rho_n^{N-rp} = 8^{rp-N} C_0^{2p} B_{r,p}(\overline{\beta}_4) \sum_{n=1}^{\infty} h_{r,p}(2\rho_n).$$

Set $c_1 := \tilde{C}_0$ if $rp = N$ and $c_1 := 8^{rp-N} C_0^{2p} B_{r,p}(\overline{\beta}_4)$ if $rp < N$. Taking the infimum over all coverings of E by balls as above gives the first inequality in (3.65), whereas the second follows from (1.46). □

The following proposition provides a lower estimate of $B_{r,p}$ in terms of the Hausdorff capacity with $\delta = \infty$ (see [2, Corollary 5.1.14] for the proof).

Proposition 3.4.2. *Let $r, s > 0$, $p, q \in (1, \infty)$, and $0 < sq \le rp \le N$. Then there exists $c_2 > 0$ (depending on s, q, r, p, and N) such that for any compact set $K \subseteq \mathbb{R}^N$,*

$$[\mathcal{H}^*_{N-sq,\infty}(K)]^{N-rp} \le c_2 [B_{r,p}(K)]^{N-sq} \quad \text{if } sq < rp < N, \tag{3.67a}$$

$$(1 + [-\log \mathcal{H}^*_{N-sq,\infty}(K)]_+)^{1-p} \le c_2 B_{r,p}(K) \quad \text{if } sq < rp = N, \tag{3.67b}$$

$$[\mathcal{H}^*_{h_{s,q},\infty}(K)]^{p-1} \le c_2 [B_{r,p}(K)]^{q-1} \quad \text{if } sq = rp = N \text{ and } p < q. \tag{3.67c}$$

Corollary 3.4.3. *Let $r > 0$, $p \in (1, \infty)$, and $0 < rp < N$. Let $E \subseteq \mathbb{R}^N$ have Hausdorff dimension $N - n$ ($n \in (0, N)$), and let $\mathcal{H}^*_{N-n}(E) < \infty$. Then $B_{r,p}(E) = 0$ if and only if $rp \le n$.*

Proof. If $rp \le n$, then

$$N - rp \ge N - n = \dim_H(E) = \sup\{s > 0 \mid \mathcal{H}^*_s(E) = \infty\}$$

(see Definition 1.7.2). This inequality and the assumption $\mathcal{H}^*_{N-n}(E) < \infty$ imply that for $rp \le n$, we have $\mathcal{H}^*_{N-rp}(E) < \infty$, and thus $B_{r,p}(E) = 0$ by (3.66).

Conversely, let $B_{r,p}(E) = 0$. Then by (3.67a) and the second inequality in (1.47) $\mathcal{H}^*_{N-sq}(E) = 0$ for every $s > 0$ and $q \in (1, \infty)$ such that $sq < rp$. Since by Definition 1.7.2 $\dim_H(E) = \inf\{s > 0 \mid \mathcal{H}^*_s(E) = 0\}$, we obtain that $N - rp \ge \dim_H(E) = N - n$. Hence the result follows. □

From Propositions 3.4.1–3.4.2 we obtain a relationship between Bessel capacities with different order and/or degree:

Theorem 3.4.4. *Let $r, s > 0$, $p, q \in (1, \infty)$, and $0 < sq \le rp \le N$. Let $E \subseteq \mathbb{R}^N$ be any Borel set with $\operatorname{diam} E \le d$ ($d > 0$). Then there exists $c_3 > 0$ (depending on s, q, r, p, d, and N) such that*

$$[B_{s,q}(E)]^{N-rp} \le c_3 [B_{r,p}(E)]^{N-sq} \quad \text{if } sq < rp < N, \tag{3.68a}$$

$$\left(1 + \left[\log \frac{c_3}{B_{s,q}(E)}\right]_+\right)^{1-p} \le c_3 B_{r,p}(E) \quad \text{if } sq < rp = N, \tag{3.68b}$$

$$[B_{s,q}(E)]^{p-1} \le c_3 [B_{r,p}(E)]^{q-1} \quad \text{if } sq = rp = N \text{ and } p < q, \tag{3.68c}$$

$$B_{s,q}(E) \le c_3 B_{r,p}(E) \quad \text{if } sq = rp < N \text{ and } p < q, \tag{3.68d}$$

Moreover, in all cases of (3.68), *there exists a Borel set $E \subseteq \mathbb{R}^N$ such that $B_{s,q}(E) = 0$ and $B_{r,p}(E) > 0$.*

Proof. By Proposition 1.7.2 there exist $x_0 \in \mathbb{R}^N$ and $C_1 = C_1(s,q,N) > 0$ such that

$$\mathcal{H}^*_{N-sq,1}(E) \le C\,\mathcal{H}^*_{N-sq,\infty}(E), \tag{3.69a}$$

where

$$C \equiv C(s,q,N) := C_1 \frac{\mathcal{H}^*_{N-sq,1}(B(x_0,\frac{d}{2}))}{\mathcal{H}^*_{N-sq,\infty}(B(x_0,\frac{d}{2}))} \max\{1, (\tfrac{d}{2})^{sq}\}.$$

Moreover, by Proposition 3.4.1 there exists $c_1 = c_1(s,q,N) > 0$ such that

$$B_{s,q}(E) \le c_1\,\mathcal{H}^*_{N-sq,1}(E). \tag{3.69b}$$

Inequality (3.68a) follows from (3.67a) and (3.69) by setting $c_3 := (c_1 C)^{N-rp} c_2$. The proof of inequalities (3.68b)–(3.68c) is similar using (3.67b)–(3.67c) and (3.69). We refer the reader to [2, Theorem 5.5.1] for the proof of the remaining claims. □

Theorem 3.4.4 suggests the following definition (which, in particular, extends to capacities the notion of absolute continuity; see Definition 1.8.7(ii)).

Definition 3.4.1. Let $r, s > 0$ and $p, q \in [1,\infty)$. The Bessel capacity $B_{r,p}$ is:
(i) *stronger* than $B_{s,q}$ if $B_{r,p}(E) = 0 \Rightarrow B_{s,q}(E) = 0$ for all $E \subseteq \mathbb{R}^N$;
(ii) *strictly stronger* than $B_{s,q}$ if, moreover, there exists $E \subseteq \mathbb{R}^N$ such that $B_{s,q}(E) = 0$ and $B_{r,p}(E) > 0$.

We say that $B_{s,q}$ is *weaker* than $B_{r,p}$ if $B_{r,p}$ is stronger than $B_{s,q}$. The Bessel capacities $B_{r,p}$ and $B_{s,q}$ are *equivalent* if $B_{r,p}(E) = 0 \Leftrightarrow B_{s,q}(E) = 0$ for all $E \subseteq \mathbb{R}^N$.

The same holds for Riesz and Sobolev capacities.

Then we have the following result (see [3] for the proof).

Theorem 3.4.5 (du-Plessis, Fuglede). *Let $r, s > 0$ and $p, q \in (1,\infty)$. Let $rp \le N$ and either $sq < rp$, or $sq = rp$ and $p < q$. Then $B_{r,p}$ is strictly stronger than $B_{s,q}$.*

Corollary 3.4.6. *Let $r > 0$, $1 < q < p < \infty$, and $0 < rp \le N$. Then $B_{r,p}$ is stronger than $B_{r,q}$.*

Proof. Fix $p \in (1,\infty)$ such that $rp \le N$. By Theorem 3.4.5, for any $E \subseteq \mathbb{R}^N$ such that $B_{r,p}(E) = 0$, we have $B_{r,q}(E) = 0$ for all $q \in (1,p)$. □

Remark 3.4.1. Let $r > 0$, $p \in (1,\infty)$, and $0 < rp \le N$. By Corollary 3.4.6 and Proposition 3.3.1(iii), if $K \subseteq \mathbb{R}^N$ is compact and $B_{r,p}(K) = 0$, then $B_{r,1}(K) = 0$.

3.4.2 Bessel versus Riesz

In view of (3.10), it is apparent that

$$R_{r,p}(E) \le B_{r,p}(E) \quad \text{for all } E \subseteq \mathbb{R}^N \quad (r \in (0,N);\ p \in [1,\infty)). \tag{3.70}$$

Proposition 3.4.7. *Let $r \in (0,N)$, $p \in (1,\infty)$, and $rp < N$.*
(i) *Let $E \subseteq \mathbb{R}^N$ have* diam $E \le 2P$ *$(P > 0)$. Then there exists $M > 0$ such that*

$$B_{r,p}(E) \le M R_{r,p}(E). \tag{3.71}$$

(ii) *Let $E \subseteq \mathbb{R}^N$ be bounded. Then $B_{r,p}(E) = 0$ if and only if $R_{r,p}(E) = 0$.*

Proof. Claim (ii) follows by (3.70)–(3.71). To prove (i), it is not restrictive to assume that $E \subseteq B(0,P)$ (see Proposition 3.3.1(i)). Let $h \in L^p(\mathbb{R}^N)$ such that $h \ge 0$ and $\gamma_r * h \ge 1$ on E satisfy $\|h\|_p^p \le 2R_{r,p}(E)$. Then by (3.70), the monotonicity of $B_{r,p}$, and (3.52a) we have

$$\|h\|_p^p \le 2B_{r,p}(\beta_{2P}) \le 2^{N-rp+1} C P^{N-rp}.$$

Then there exists $M_1 > 0$ (depending on r, p, and N) such that for all $x \in E$,

$$0 \le (\gamma_r * h)(x) - (\gamma_{r,\rho} * h)(x) \le \|h\|_p \|\gamma_r - \gamma_{r,\rho}\|_{\frac{p}{p-1}} \le M_1 \left(\frac{P}{\rho}\right)^{\frac{N-rp}{p}}, \tag{3.72}$$

where $\gamma_{r,\rho}$ denotes the inhomogeneous Riesz kernel (see (3.11)). Since $\gamma_r * h \ge 1$ on E, by (3.72) we have $\gamma_{r,\rho} * h \ge \frac{1}{2}$ on E for all $\rho \ge (2M_1)^{\frac{p}{N-rp}} P$. Moreover, for any $\rho \ge (2M_1)^{\frac{p}{N-rp}} P$, there exists $M_2 > 0$ such that $\gamma_{r,\rho} \le M_2 g_r$, and thus $g_r * (2M_2 h) \ge 1$ on E. Hence for any h as above, there exists $f := 2M_2 h \in V_{E,r,p}$ (see (3.48)), and thus $B_{r,p}(E) \le (2M_2)^p R_{r,p}(E)$. Then inequality (3.71) and the result follow. □

Remark 3.4.2. Let $r \in (0,N)$ and $1 < p < \frac{N}{r}$. Then there exists $M > 0$ such that for any $E \subseteq \mathbb{R}^N$,

$$B_{r,p}(E) \le M\big[R_{r,p}(E) + R_{r,p}(E)^{\frac{N}{N-rp}}\big] \tag{3.73}$$

(see [2, Subsection 5.6.1] for the proof). By (3.70) and (3.73) the second statement of Proposition 3.4.7 in fact holds for any $E \subseteq \mathbb{R}^N$ (namely, $B_{r,p}$ and $R_{r,p}$ are equivalent; see Definition 3.4.1).

3.4.3 Bessel versus Sobolev

Let us prove that, under the assumptions of the Calderón theorem (see Theorem 3.1.5), the capacities $B_{m,p}$ and $C_{m,p}$ are equivalent.

Theorem 3.4.8. *Let $m \in \mathbb{N}$ and $p \in (1,\infty)$. Then for any Borel set $E \subseteq \mathbb{R}^N$,*

$$M^{-p}C_{m,p}(E) \leq B_{m,p}(E) \leq M^p C_{m,p}(E) \tag{3.74}$$

with $M > 0$ as in (3.8).

Proof. By Proposition 3.3.1(ii) and Definition 3.3.2 it suffices to prove (3.74) when $E = K$ is compact. Let us first prove that $B_{m,p}(K) \leq M^p C_{m,p}(K)$. Indeed, let $h \in W_K$. Then $h \in C_c^\infty(\mathbb{R}^N) \subseteq W^{m,p}(\mathbb{R}^N) = L^{m,p}(\mathbb{R}^N)$ by the Calderón theorem. Hence there exists $f \in L^p(\mathbb{R}^N)$ such that $g_m * f = h$ and $M\|h\|_{m,p} \geq \|g_m * f\|_{L^{m,p}(\mathbb{R}^N)} = \|f\|_p \geq \|f^+\|_p$. Moreover, $g_m * f^+ \geq g_m * f = h \geq 1$ on K, and thus f^+ belongs to $V_{K,m,p}$. It follows that

$$B_{m,p}(K) \leq \|f^+\|_p^p \leq M^p\|h\|_{m,p}^p \qquad \text{for all } h \in W_K.$$

Taking the infimum over $h \in W_K$ of the right-hand side of the above inequality proves that $B_{m,p}(K) \leq M^p\, C_{m,p}(K)$.

Conversely, let us prove that $M^{-p}\, C_{m,p}(K) \leq B_{m,p}(K)$. Let $f \in V_K$ satisfy $g_m * f > 1$ in K and $\|f\|_p^p < B_{m,p}(K) + \epsilon$ for some $\epsilon > 0$. Set

$$f_n(x) := \begin{cases} \min\{f(x), n\} & \text{if } |x| \leq n, \\ 0 & \text{otherwise} \end{cases} \qquad (n \in \mathbb{N}).$$

By definition, for any $n \in \mathbb{N}$, we have $f_n \in L^\infty(\mathbb{R}^N)$, $\operatorname{supp} f_n \subseteq B(0,n)$, and $0 \leq f_n \leq f_{n+1} \leq f$ in $\mathbb{R}^N$, whence by the monotone convergence theorem we get $\lim_{n\to\infty}\|f_n\|_p^p = \|f\|_p^p$ and $\lim_{n\to\infty} g_m * f_n = g_m * f$ pointwise in $\mathbb{R}^N$. Therefore, since $g_m * f$ is lower semicontinuous (see Lemma 3.2.1-(*i*)) and $g_m * f > 1$ in K, there exist $\bar{n} \in \mathbb{N}$ and $\eta > 0$ such that $g_m * f_{\bar{n}} \geq 1 + \eta$ on K and $\|f_{\bar{n}}\|_p^p < B_{m,p}(K) + \epsilon$.

Now observe that for any $q \in [1,\infty)$, there exists $h \in C_c^\infty(\mathbb{R}^N)$ with $\operatorname{supp} h \subseteq B(0,\bar{n})$ such that $\|f_{\bar{n}} - h\|_q < \epsilon$. Then $g_m * h \in \mathcal{S}(\mathbb{R}^N)$, and for $q > \max\{\frac{N}{m}, p\}$, we have that:

a) by Sobolev embedding results $g_m * h$ approximates uniformly $g_m * f_{\bar{n}}$, and hence by a proper choice of h we have $g_m * h \geq 1$ on K;
b) by Hölder's inequality we have

$$\|f_{\bar{n}} - h\|_p \leq \|f_{\bar{n}} - h\|_q[\lambda_N(B(0,\bar{n}))]^{\frac{q-p}{pq}} < \epsilon[\lambda_N(B(0,\bar{n}))]^{\frac{q-p}{pq}}. \tag{3.75}$$

By the above remarks it follows that $g_m * h \in W_K'$ (see (3.60)). Then by inequality (3.75) we get

$$\begin{aligned} C_{m,p}(K) &\leq \|g_m * h\|_{m,p}^p \leq M^p\|g_m * h\|_{L^{m,p}(\mathbb{R}^N)}^p = M^p\|h\|_p^p \\ &\leq M^p\{\|f_{\bar{n}}\|_p + \epsilon[\lambda_N(B(0,\bar{n}))]^{\frac{q-p}{pq}}\}^p \\ &\leq M^p\{[B_{m,p}(K) + \epsilon]^{\frac{1}{p}} + \epsilon[\lambda_N(B(0,\bar{n}))]^{\frac{q-p}{pq}}\}^p. \end{aligned}$$

By the arbitrariness of ϵ from the above inequality we get $C_{m,p}(K) \le M^p B_{m,p}(K)$. This completes the proof. □

3.4.4 Sobolev versus Hausdorff

In view of Theorem 3.4.8, if $m \in \mathbb{N}$ and $p \in (1, \infty)$, then the metric properties of the Bessel capacity $B_{m,p}$ proved in Subsection 3.3.2 and their consequences in Subsections 3.4.1–3.4.2 hold for the Sobolev capacity $C_{m,p}$. In particular, results analogous to Propositions 3.4.1–3.4.2 hold with $r = m$ and $B_{m,p}$ replaced by $C_{m,p}$; we leave their formulation to the reader.

When $p = 1$, we have the following (see [102, Theorem 3.5.5], where the definition of $C_{m,1}$ is the same as in Definition 3.3.2(i) with W_K replaced by the set W_K'' defined in (3.61)).

Theorem 3.4.9. *Let $m \in \mathbb{N}$, $m \in [1, N)$. Then there exist M_1, $M_2 > 0$ (depending on m and N) such that for any compact $K \subseteq \mathbb{R}^N$,*

$$M_1 \mathcal{H}^*_{N-m,1}(K) \le C_{m,1}(K) \le M_2 \mathcal{H}^*_{N-m,1}(K). \tag{3.76}$$

3.4.5 Laplacian capacity versus Sobolev

The following result was proven in [27].

Proposition 3.4.10. *For any compact $K \subseteq \mathbb{R}^N$, we have*

$$C_{\Delta,1}(K) = 2\, C_{1,2}(K). \tag{3.77}$$

Proof. We only prove that for any compact $K \subseteq \mathbb{R}^N$, $C_{\Delta,1}(K) \ge 2\, C_{1,2}(K)$, referring the reader to [27, Lemma E.1] for the inverse inequality. Let $f \in \Omega_K$. Set $h := \min\{f^+, 1\}$. Then $0 \le h \le 1$ in $\mathbb{R}^N$, and $h = 1$ in some open $A \supseteq K$. By standard regularization arguments there exists $\hat{h} \in C_c^\infty(\mathbb{R}^N)$ such that $0 \le h \le 1$ in $\mathbb{R}^N$ and $\hat{h} = 1$ in A. Then $\hat{h} \in W_K'$ (see (3.60)), whence

$$C_{1,2}(K) \le \int_{\mathbb{R}^N} |\nabla \hat{h}|^2\, dx \le \int_{\mathbb{R}^N} |\nabla h|^2\, dx = \int_{\mathbb{R}^N} \nabla h \cdot \nabla f\, dx = -\int_{\mathbb{R}^N} h \Delta f\, dx.$$

Since $f \in C_c^\infty(\mathbb{R}^N)$, we have $\int_{\mathbb{R}^N} \Delta f\, dx = 0$. Then from the above inequality we get

$$C_{1,2}(K) \le -\int_{\mathbb{R}^N} \left(h - \frac{1}{2}\right) \Delta f\, dx \le \frac{1}{2} \|\Delta f\|_1.$$

Then by the arbitrariness of $f \in \Omega_K$ the result follows. □

3.4.6 (m,p)-concentrated and (m,p)-diffuse measures

Let U be an open bounded subset of $\mathbb{R}^N$. We say that a signed measure on $\mathcal{B}(U)$ is (m,p)-concentrated if it is $C_{m,p}$-concentrated, and (m,p)-diffuse if it is $C_{m,p}$-diffuse (see Definition 1.8.10).

Proposition 3.4.11. *Let $m \in \mathbb{N}$.*

(i) *Let either $p = 1$ and $m = N$, or $p \in (1,\infty)$ and $mp > N$. Then every $\mu \in \mathfrak{R}_f(U)$ is $B_{m,p}$-diffuse.*

(ii) *Let $p \in (1,\infty)$ and $mp > N$. Then every $\mu \in \mathfrak{R}_f(U)$ is (m,p)-diffuse.*

Proof. By Remark 3.3.2(ii) no $B_{m,p}$-null set in $\mathbb{R}^N$ exists if either $p = 1$ and $m = N$, or $p \in (1,\infty)$ and $mp > N$. Therefore by Theorem 3.4.8 no (m,p)-null set in $\mathbb{R}^N$ exists if $p \in (1,\infty)$ and $mp > N$. Hence the result follows. □

We denote by $\mathfrak{R}_{c,m,p}(U)$ and $\mathfrak{R}_{d,m,p}(U)$ the sets of (m,p)-concentrated and (m,p)-diffuse finite signed Radon measures on U, respectively. Clearly, $\mathfrak{R}_{c,m,p}(U) \bigcap \mathfrak{R}_{d,m,p}(U) = \{0\}$. A common parlance is that "measures belonging to $\mathfrak{R}_{d,m,p}(U)$ do not charge (m,p)-null sets" (e. g., see [10]). We always set $\mathfrak{R}_{c,p}(U) \equiv \mathfrak{R}_{c,1,p}(U)$ and $\mathfrak{R}_{d,p}(U) \equiv \mathfrak{R}_{d,1,p}(U)$.

Remark 3.4.3. (i) If $\mu \in \mathfrak{R}_{d,m,p}(U)$, then every (m,p)-quasi-continuous function is μ-measurable. In fact, if v is (m,p)-quasi-continuous, then there exist an (m,p)-null set $E \subseteq U$ and a sequence $\{v_n\} \subseteq C(U)$ such that $v_n \to v$ pointwise in $U \setminus E$ (see [2, Section 7.1]). Since $\mu \in \mathfrak{R}_{d,m,p}(U)$, the set E is μ-null, and thus the claim follows.

(ii) If $\mu \in \mathfrak{R}_{d,m,p}(U)$, then every function $v \in W_0^{m,p}(U) \bigcap L^\infty(U)$ also belongs to $L^\infty(U,\mu)$. In fact, v can be identified with its (m,p)-quasi-continuous representative $\hat{v}$ (see Remark 3.3.6), which is μ-measurable by (i).

By Proposition 1.8.12 we have the following:

Proposition 3.4.12. *Let $m \in \mathbb{N}$ and $p \in [1,\infty)$. Then there exists a unique couple $(\mu_{c,m,p}, \mu_{d,m,p})$ such that $\mu_{c,m,p} \in \mathfrak{R}_{c,m,p}(U)$, $\mu_{d,m,p} \in \mathfrak{R}_{d,m,p}(U)$, and s*

$$\mu = \mu_{d,m,p} + \mu_{c,m,p}. \tag{3.78}$$

Definition 3.4.2. The measures $\mu_{c,m,p}$ and $\mu_{d,m,p}$ given by Proposition 3.4.12 are called the *(m,p)-concentrated* and *(m,p)-diffuse parts* of μ, respectively.

In the following, we set $\mu_{c,p} \equiv \mu_{c,1,p}$ and $\mu_{d,p} \equiv \mu_{d,1,p}$. Set

$$\mathfrak{R}_s(U) := \{\mu \in \mathfrak{R}_f(U) \mid \mu \perp \lambda_N\}, \quad \mathfrak{R}_{ac}(U) := \{\mu \in \mathfrak{R}_f(U) \mid \mu \ll \lambda_N\}.$$

By Proposition 3.3.2 and Theorem 3.4.8 every (m,p)-null set is Lebesgue measurable and λ_N-null. This plainly implies that for every $m \in \mathbb{N}$ and $p \in (1,\infty)$,

$$\mathfrak{R}_{c,m,p}(U) \subseteq \mathfrak{R}_s(U), \quad \mathfrak{R}_{ac}(U) \subseteq \mathfrak{R}_{d,m,p}(U). \tag{3.79}$$

Combining the Lebesgue decomposition of μ (see (1.53) written with $\nu = \mu$) with (3.78) and using (3.79) gives

$$\mu_{c,m,p} = [\mu_s]_{c,m,p}, \tag{3.80}$$

$$\mu_{d,m,p} = \mu_{ac} + [\mu_s]_{d,m,p} \tag{3.81}$$

for every $\mu \in \mathfrak{R}_f(U)$. From (3.78)–(3.81) we obtain the decomposition

$$\mu = \mu_{ac} + [\mu_s]_{d,m,p} + \mu_{c,m,p}. \tag{3.82}$$

Remark 3.4.4. Analogously to equalities (1.54), we have

$$[\mu_{d,m,p}]^{\pm} = [\mu^{\pm}]_{d,m,p}, \quad [\mu_{c,m,p}]^{\pm} = [\mu^{\pm}]_{c,m,p}. \tag{3.83}$$

In particular, by (3.80) and (3.83) we have

$$[\mu_s^{\pm}]_{c,m,p} = [\mu^{\pm}]_{c,m,p} = [\mu_{c,m,p}]^{\pm}. \tag{3.84}$$

The following result is a simple consequence of Theorem 3.4.5.

Proposition 3.4.13. *Let $m,n \in \mathbb{N}$ and $p,q \in (1,\infty)$. Let $mp \le d$ and either $nq < mp$, or $nq = mp$ and $p < q$. Then* (i) $\mathfrak{R}_{d,n,q}(U) \subseteq \mathfrak{R}_{d,m,p}(U)$ *and* (ii) $\mathfrak{R}_{c,m,p}(U) \subseteq \mathfrak{R}_{c,n,q}(U)$.

It has been already observed that $\mathfrak{R}_f(U) \subseteq \mathcal{D}^*(U)$ (see Remark 3.1.1), and thus, in particular, a measure $\mu \in \mathfrak{R}_f(U)$ can belong to the dual space $W^{-m,q}(U) = (W_0^{m,p}(U))^*$ (where $m \in \mathbb{N}$, $p \in [1,\infty)$, and $q \in (1,\infty]$ is the conjugate exponent of p). If $p \in (1,\infty)$, then measures of this kind are diffuse with respect to the (m,p)-capacity. In fact, let $p \in (1,\infty)$, and let $\mu \in W^{-m,q}(U)$ be a nonnegative Radon measure. Let $K \subseteq U$ be compact, and let $f \in L^p(\mathbb{R}^N)$ be such that $f \ge 0$ and $g_m * f \ge 1$ in K. Then

$$\begin{aligned}\mu(K) &\le \int_U d\mu(x) \int_{\mathbb{R}^N} g_m(x-y)f(y)\,dy = \int_{\mathbb{R}^N} dy\, f(y) \int_U g_m(x-y)d\mu(x)\\ &\le \|g_m * \mu\|_q \|f\|_p.\end{aligned}$$

By the arbitrariness of f this implies that $\mu(K) \le M\,\|g_m * \mu\|_q [C_{m,p}(K)]^{\frac{1}{p}}$ with $M > 0$ as in (3.8) (see (3.48) and Theorem 3.4.8). The latter inequality can be extended to any Borel subset of U by regularity of μ and capacitability of $C_{m,p}$, and hence the claim follows.

The following proposition extends the above remark removing the requirement of nonnegativity of μ (see [2, Subsection 7.6.1], [56]).

Proposition 3.4.14 (Grun–Rehomme). *Let $m \in \mathbb{N}$ and $p \in [1,\infty)$. Then*

$$\mathfrak{R}_f(U) \cap W^{-m,q}(U) \subseteq \mathfrak{R}_{d,m,p}(U).$$

Proof. Let $\mu \in \mathfrak{R}_f(U) \cap W^{-m,q}(U)$, and let $E \in \mathcal{B}(U)$ be such that $C_{m,p}(E) = 0$. By the Hahn decomposition (see Definition 1.8.3) there exists $E^+ \in \mathcal{B}(U)$, $E^+ \subseteq E$, such that $\mu^+(E^+) = \mu(E^+)$ and $\mu^-(E^+) = 0$. Moreover, since $\mu^\pm \in \mathfrak{R}_f^+(U)$, for any $0 < k < \mu^+(E^+) = \mu^+(E)$ and any $\epsilon > 0$, there exist a compact set $K \subseteq E^+$ and an open set $A \supseteq E^+$, $A \subseteq U$, such that $\mu^+(K) > k$ and $\mu^-(A) < \epsilon$. By the monotonicity of $C_{m,p}$ we have $C_{m,p}(K) \le C_{m,p}(E^+) \le C_{m,p}(E) = 0$. Then there exists a sequence $\{f_j\} \subseteq C_c^\infty(U)$ such that for any $j \in \mathbb{N}$, we have $0 \le f_j \le 1$ in U, $f_j = 1$ in K, and $f_j \to 0$ strongly in $W_0^{m,p}(U)$ as $j \to \infty$ (see Proposition 3.3.6). Clearly, the sequence $\{g_j\}$, $g_j := f_j\psi$ with $\psi \in C_c^\infty(A)$ such that $0 \le \psi \le 1$ and $\psi = 1$ in K, has the same properties. It follows that

$$k < \mu^+(K) \le \int_U g_j \, d\mu^+ = \int_U g_j \, d\mu + \int_U g_j \, d\mu^- \le \int_U g_j \, d\mu + \mu^-(A) < \int_U g_j \, d\mu + \epsilon.$$

Since $g_j \to 0$ strongly in $W_0^{m,p}(U)$ and $\mu \in W^{-m,q}(U)$, letting $j \to \infty$ in the above inequality, we obtain $k \le \epsilon$ for any $k < \mu^+(E)$, whence $\mu^+(E) \le \epsilon$. By the arbitrariness of ϵ it follows that $\mu^+(E) = 0$. It is similarly proven that $\mu^-(E) = 0$, and hence the claim follows. □

The above result is a part of the following characterization of $\mathfrak{R}_{d,m,p}(U)$, where, as usual, we denote $L^1(U) + W^{-m,q}(U) \equiv (W_0^{m,p}(U) \bigcap L^\infty(U))^*$.

Theorem 3.4.15. *Let $m \in \mathbb{N}$ and $p \in (1,\infty)$. Let $\mu \in \mathfrak{R}_f(U)$. Then the following statements are equivalent:*
(i) $\mu \in L^1(U) + W^{-m,q}(U)$;
(ii) $\mu \in \mathfrak{R}_{d,m,p}(U)$.

Theorem 3.4.15 is stated for $m = 2$ in [10] and proven in [52, Theorem B] (see [23, Theorem 2.1] for the case $m = 1$). Clearly, $L^1(U) \subseteq \mathfrak{R}_{ac}(U) \subseteq \mathfrak{R}_{d,m,p}(U)$ (see (3.79)), and thus the implication (i)⇒(ii) follows by Proposition 3.4.14. To prove the opposite implication, we need the following lemma (see [10, Lemma 4.2] for the proof).

Lemma 3.4.16. *Let $m \in \mathbb{N}$ and $p \in (1,\infty)$, and let $\mu \in \mathfrak{R}_{d,m,p}(U)$ be nonnegative. Then there exists a sequence $\{\mu_k\} \subseteq W^{-m,q}(U) \cap \mathfrak{R}_f(U)$ such that $0 \le \mu_k \le \mu_{k+1} \le \mu$ for all $k \in \mathbb{N}$ and $\lim_{k\to\infty} \|\mu_k - \mu\|_{\mathfrak{R}_f(U)} = 0$.*

Proof of Theorem 3.4.15. We only must prove that (ii)⇒(i). It is not restrictive to assume that μ is nonnegative. For any $k \in \mathbb{N}$, set $\nu_k := \mu_k - \mu_{k-1}$, where $\{\mu_k\} \subseteq W^{-m,q}(U)$ is given by Lemma 3.4.16, and $\mu_0 := 0$. Then

$$\sum_{k=1}^{\infty} \|\nu_k\| = \sum_{k=1}^{\infty} \nu_k(U) = \lim_{k\to\infty} \mu_k(U) = \mu(U) < \infty. \tag{3.85}$$

Let $\{\rho_j\} \subseteq C_c^\infty(U)$ be a standard mollifier. For any fixed $k \in \mathbb{N}$, we have $\lim_{j\to\infty} \|\nu_k - \nu_k * \rho_j\|_{W^{-m,q}(U)} = 0$, and thus there exists $j_k \in \mathbb{N}$ such that

$$\|\nu_k - \nu_k * \rho_{j_k}\|_{W^{-m,q}(U)} \leq \frac{1}{2^k}. \tag{3.86}$$

Set

$$\nu_k = (\nu_k - \nu_k * \rho_{j_k}) + \nu_{j_k} * \rho_{j_k} =: \omega_k + f_k \quad (k \in \mathbb{N}). \tag{3.87}$$

Observe that by Lemma 3.4.16 $\nu_k \in \mathfrak{R}_f^+(U)$, and thus $f_k \geq 0$. By (3.86) we have that $\sum_{k=1}^{\infty} \|\omega_k\|_{W^{-m,q}(U)} \leq 1$, and hence there exists $\omega := \sum_{k=1}^{\infty} \omega_k \in W^{-m,q}(U)$. On the other hand, by (3.85) we have $\sum_{k=1}^{\infty} \|f_k\|_{L^1(U)} \leq \|\mu\| + 1$, and hence there exists $f := \sum_{k=1}^{\infty} f_k \in L^1(U)$. Since $\lim_{n\to\infty} \| \sum_{k=1}^{n} \nu_k - \mu\|_{\mathfrak{R}_f(U)} = 0$, from (3.87) we obtain $\mu = f + \omega$. Then the result follows. □

Remark 3.4.5. In view of Theorem 3.4.15, the duality symbol $\langle \mu, \zeta \rangle$ makes sense for any $\mu \in \mathfrak{R}_{d,m,p}(U)$ and $\zeta \in W_0^{m,p}(U) \bigcap L^\infty(U)$.

4 Vector integration

4.1 Measurability of vector functions

4.1.1 Measurability

Let $(X, \mathcal{A})$ be a measurable space, and let Y be a Banach space with norm $\|\cdot\|_Y$. The following concept of measurability (where $\mathcal{B}(Y)$ denotes the Borel σ-algebra on Y generated by the norm topology) is a particular case of Definition 2.1.1.

Definition 4.1.1. A function $f : X \to Y$ is *measurable* (or *$(\mathcal{A}, \mathcal{B}(Y))$-measurable*) if $f^{-1}(E) \in \mathcal{A}$ for all $E \in \mathcal{B}(Y)$.

The following proposition partly generalizes Corollary 2.1.5(ii).

Proposition 4.1.1. *Let $f_n : X \to Y$ be measurable ($n \in \mathbb{N}$), and*

$$\lim_{n\to\infty} \|f_n(x) - f(x)\|_Y = 0 \quad \text{for every } x \in X. \tag{4.1}$$

Then the function f is measurable.

Proof. Let $U \subseteq Y$ be open. It is easily seen that

$$f^{-1}(U) \subseteq \bigcup_{k\geq m} f_k^{-1}(U) \quad \text{for every } m \in \mathbb{N}. \tag{4.2}$$

In fact, fix $x \in f^{-1}(U)$. Then $f(x) \in U$, and thus there exists $r > 0$ such that the open ball $B(f(x), r)$ is contained in U. On the other hand, by (4.1) there exists $\bar{k} \in \mathbb{N}$ such that $f_k(x) \in B(f(x), r)$ for all $k > \bar{k}$, whence (4.2) immediately follows.

By the arbitrariness of m in (4.2) we get

$$f^{-1}(U) \subseteq \bigcap_{m=1}^{\infty} \bigcup_{k\geq m} f_k^{-1}(U), \tag{4.3}$$

whereas for every closed set $C \subseteq Y$,

$$\bigcap_{m=1}^{\infty} \bigcup_{k\geq m} f_k^{-1}(C) \subseteq f^{-1}(C). \tag{4.4}$$

Set $U_n := \{y \in U \mid d(y, U^c) > \frac{1}{n}\}$ and $C_n := \overline{U}_n$ ($n \in \mathbb{N}$). Clearly, U_n is open, and $U = \bigcup_{n=1}^{\infty} U_n = \bigcup_{n=1}^{\infty} C_n$. From (4.3)–(4.4) we get

$$f^{-1}(U) = \bigcup_{n=1}^{\infty} f^{-1}(U_n) \subseteq \bigcup_{n=1}^{\infty} \bigcap_{m=1}^{\infty} \bigcup_{k\geq m} f_k^{-1}(U_n) \subseteq \bigcup_{n=1}^{\infty} \bigcap_{m=1}^{\infty} \bigcup_{k\geq m} f_k^{-1}(C_n) \subseteq \bigcup_{n=1}^{\infty} f^{-1}(C_n) = f^{-1}(U).$$

https://doi.org/10.1515/9783110556902-005

It follows that $f^{-1}(U) = \bigcup_{n=1}^{\infty}\bigcap_{m=1}^{\infty}\bigcup_{k\geq m} f_k^{-1}(U_n) \in \mathcal{A}$, and thus by Lemma 2.1.2 the conclusion follows. □

Remark 4.1.1. If $f : X \to Y$ is measurable, then the real-valued function

$$\|f\|_Y : X \to [0,\infty), \quad \|f\|_Y(x) := \|f(x)\|_Y \quad (x \in X) \tag{4.5}$$

is measurable. In fact, the map $\|\cdot\|_Y : Y \mapsto \mathbb{R}$, $y \mapsto \|y\|_Y$ $(y \in Y)$ is continuous and thus Borel measurable (see Remark 2.1.3). Then the claim follows from Proposition 2.1.1.

Definition 4.1.2. A function $s : X \to Y$ is *simple* if there exist $y_1, \dots, y_p \in Y$ and a partition $\{E_1, \dots, E_p\} \subseteq \mathcal{A}$ of X, with $y_i = 0$ if $\mu(E_i) = \infty$, such that $s = \sum_{i=1}^{p} y_i \chi_{E_i}$.

Simple functions are measurable. We denote the set of simple functions $s : X \to Y$ by $\mathscr{S}(X; Y)$, with $\mathscr{S}(X) \equiv \mathscr{S}(X; \mathbb{R})$.

4.1.2 μ-measurability

Let $(X, \mathcal{A}, \mu)$ be a measure space.

Definition 4.1.3. A function $f : X \to Y$ is *μ-measurable* if there exists a sequence $\{s_n\} \subseteq \mathscr{S}(X; Y)$ such that

$$\lim_{n\to\infty} \|s_n(x) - f(x)\|_Y = 0 \quad \text{for } \mu\text{-a.e. } x \in X. \tag{4.6}$$

Remark 4.1.2. (i) Let $f, g : X \to Y$. If f is μ-measurable and $g = f$ μ-a.e. in X, then g is also μ-measurable. In fact, let $N_1, N_2 \in \mathcal{A}$ be μ-null such that $\lim_{n\to\infty} \|s_n(x) - f(x)\|_Y = 0$ for all $x \in N_1^c$ and $f(x) = g(x)$ for all $x \in N_2^c$. Then $N := N_1 \cup N_2$ is μ-null, and $\lim_{n\to\infty} \|s_n(x) - g(x)\|_Y = 0$ for all $x \in N^c$. Thus by Definition 4.1.3 the claim follows.

(ii) If $f : X \to Y$ is μ-measurable, then there exist a μ-null set $N \in \mathcal{A}$ and a sequence $\{s_n\} \subseteq \mathscr{S}(X; Y)$ such that $\lim_{n\to\infty} \|s_n(x) - f(x)\|_Y = 0$ for all $x \in N^c$. Since simple functions are measurable, by Proposition 4.1.1 the function f is $(\mathcal{A} \cap N^c, \mathcal{B}(Y))$-measurable and thus measurable in the sense of functions defined μ-a.e. (see Definitions 2.1.2 and 4.1.1). Moreover, it is measurable in the usual sense if μ is complete (see Proposition 2.1.6). We tacitly summarize these remarks by saying that "μ-measurable functions are measurable".

(iii) By (ii) and Remark 4.1.1, if $f : X \to Y$ is μ-measurable, then the function $\|f\|_Y$ defined by (4.5) is measurable.

The set of μ-measurable functions is a vector space. The following result (analogous to Proposition 4.1.1) shows that if μ is σ-finite, then it is closed with respect to μ-a.e. convergence.

Proposition 4.1.2. *Let μ be σ-finite, and let $f_n : X \to Y$ be μ-measurable ($n \in \mathbb{N}$). Then every function $f : X \mapsto Y$ such that*

$$\lim_{n\to\infty} \|f_n(x) - f(x)\|_Y = 0 \quad \text{for } \mu\text{-a. e. } x \in X$$

is μ-measurable.

To prove Proposition 4.1.2, we need the following general form of the Egorov theorem (see Theorem 2.1.12). The proof is almost verbatim the same as in the case $Y = \mathbb{R}$ and thus is omitted.

Theorem 4.1.3 (Egorov). *Let $\mu(X) < \infty$. Let $f_n : X \to Y$ be μ-measurable ($n \in \mathbb{N}$), and let $\lim_{n\to\infty} \|f_n(x) - f(x)\|_Y = 0$ for μ-a. e. $x \in X$. Then for every $\delta > 0$, there exists a measurable set $E \subseteq X$ such that $\mu(E^c) < \delta$ and*

$$\lim_{n\to\infty} \sup_{x\in E} \|f_n(x) - f(x)\|_Y = 0.$$

Proof of Proposition 4.1.2. (i) Let us first prove the result when $\mu(X) < \infty$. We will prove the following:

Claim. Let $(Z, \mathcal{A}, \mu)$ be a finite measure space. Let $g_n : Z \to Y$ be μ-measurable ($n \in \mathbb{N}$), and let $g : Z \to Y$ satisfy $\lim_{n\to\infty} \|g_n(z) - g(z)\|_Y = 0$ for μ-a. e. $z \in Z$. Then for every $\delta > 0$, there exist $Z_\delta \subseteq Z$ such that $\mu(Z_\delta) < \delta$ and a sequence $\{s_{\delta,m}\} \subseteq \mathscr{S}(Z; Y)$ such that

$$\sup_{z\in Z_\delta^c} \|s_{\delta,m}(z) - g(z)\|_Y \le \frac{1}{m} \quad \text{for all } m \in \mathbb{N}. \tag{4.7}$$

Using the claim, we can argue as follows. For every $k \in \mathbb{N}$, choose $Z = Z_{k-1}$ ($Z_0 := X$), $g_n = f_n$, $g = f$, and $\delta = \frac{1}{k}$ (by abuse of notation we set $Z_k \equiv Z_{\frac{1}{k}}$). Then there exist $Z_k \subseteq Z_{k-1}$ such that $\mu(Z_k) < \frac{1}{k}$ and a sequence $\{s_{k,m}\} \subseteq \mathscr{S}(Z_{k-1}; Y)$ such that

$$\sup_{x\in X_k} \|s_{k,m}(x) - f(x)\|_Y \le \frac{1}{m} \quad \text{for all } m \in \mathbb{N}, \tag{4.8}$$

where $X_k := Z_{k-1} \setminus Z_k$ ($k \in \mathbb{N}$). Observe that the sequence $\{Z_k\}$ is nonincreasing, the sets X_k are disjoint, and $Z_k = \bigcap_{j=1}^k X_j^c = (\bigcup_{j=1}^k X_j)^c$ for all $k \in \mathbb{N}$. Without loss of generality, we can assume that $s_{k,m}(x) = 0$ for all $x \in X_k^c$, and thus $s_{k,m} = s_{k,m}\chi_{X_k}$ in X ($k, m \in \mathbb{N}$).

Set $t_m := \sum_{k=1}^m s_{k,m} = \sum_{k=1}^m s_{k,m}\chi_{X_k}$ ($m \in \mathbb{N}$). Clearly, $\{t_m\} \subseteq \mathscr{S}(X; Y)$. Let $x \in \bigcup_{k=1}^\infty X_k$, and thus $x \in X_j$ for some $j \in \mathbb{N}$. By (4.8) we have

$$\|t_m(x) - f(x)\|_Y \le \sup_{x\in X_j} \|s_{j,m}(x) - f(x)\|_Y \le \frac{1}{m} \quad \text{for all } m \in \mathbb{N},$$

whence

$$\lim_{m\to\infty}\|t_m(x)-f(x)\|_Y=0 \quad \text{for all } x\in \bigcup_{k=1}^{\infty} X_k. \tag{4.9}$$

On the other hand, since $\{Z_k\}$ is nonincreasing, and $Z_k = (\bigcup_{j=1}^{k} X_j)^c$, $\mu(Z_k) < \frac{1}{k}$ for all $k\in\mathbb{N}$, by Proposition 1.3.1 we have $\mu((\bigcup_{j=1}^{\infty} X_j)^c) = \lim_{k\to\infty}\mu(Z_k) = 0$. Therefore the convergence in (4.9) takes place μ-a. e. in X, and thus f is μ-measurable. Hence the result follows in this case.

It remains to prove the claim. By Theorem 4.1.3, for any $\delta>0$, there exists $Z_1\subseteq Z$ with $\mu(Z_1)<\frac{\delta}{2}$ such that

$$\sup_{z\in Z_1^c}\|g_n(z)-g(z)\|_Y\to 0 \quad \text{as } n\to\infty.$$

Hence for every $m\in\mathbb{N}$, there exists $n_m\in\mathbb{N}$ such that

$$\sup_{z\in Z_1^c}\|g_{n_m}(z)-g(z)\|_Y\le\frac{1}{2m}. \tag{4.10}$$

Since $g_{n_m}:Z\to Y$ is μ-measurable, there exists a sequence $\{u_{m,k}\}\subseteq\mathscr{S}(Z;Y)$ such that

$$\lim_{k\to\infty}\|u_{m,k}(z)-g_{n_m}(z)\|_Y=0 \quad \text{for } \mu\text{-a. e. } z\in Z.$$

Then by Theorem 4.1.3 there exists $Z_{2,m}\subseteq Z$ such that $\mu(Z_{2,m})<\frac{\delta}{2^{m+1}}$ and

$$\sup_{z\in Z_{2,m}^c}\|u_{m,k}-g_{n_m}(z)\|_Y\to 0 \quad \text{as } k\to\infty.$$

Hence there exists $k_m\in\mathbb{N}$ such that

$$\sup_{z\in Z_{2,m}^c}\|u_{m,k_m}(z)-g_{n_m}(z)\|_Y\le\frac{1}{2m}. \tag{4.11}$$

Set $Z_\delta := Z_1\cup(\bigcup_{m=1}^{\infty} Z_{2,m})$ and $s_{\delta,m} := u_{m,k_m}$ $(m\in\mathbb{N})$. Then $\mu(Z_\delta)<\delta$ and $\{s_{\delta,m}\}\subseteq\mathscr{S}(Z;Y)$, and by (4.10)–(4.11) we get (4.7). Hence the claim follows.

(ii) In the general case, let $\{E_k\}\subseteq\mathcal{A}$ satisfy $\mu(E_k)<\infty$ for all $k\in\mathbb{N}$, $E_k\cap E_l=\emptyset$ for $k\ne l$, and $X=\bigcup_{k=1}^{\infty}E_k$. For every $k,n\in\mathbb{N}$, the restriction $f_n|_{E_k}$ is μ-measurable, and $f_n|_{E_k}\to f|_{E_k}$ μ-a. e. in E_k as $n\to\infty$. Since $\mu(E_k)<\infty$, it follows by (i) that $f|_{E_k}$ is μ-measurable $(k\in\mathbb{N})$. Then there exists a sequence $\{v_{k,m}\}\subseteq\mathscr{S}(E_k;Y)$ of simple functions such that

$$\lim_{m\to\infty}\|v_{k,m}(x)-f(x)\|_Y=0 \quad \text{for every } k\in\mathbb{N} \text{ and } \mu\text{-a. e. } x\in E_k.$$

Set $v_{k,m}(x) := 0$ for $x \in E_k^c$ and $s_m := \sum_{k=1}^m v_{k,m} = \sum_{k=1}^m v_{k,m}\chi_{E_k}$ $(m \in \mathbb{N})$. Then $\{s_m\} \subseteq \mathscr{S}(X;Y)$, and

$$\lim_{m\to\infty} \|s_m(x) - f(x)\|_Y \to 0 \quad \text{for } \mu\text{-a. e. } x \in \bigcup_{k=1}^\infty E_k = X.$$

Hence the conclusion follows. □

The following proposition gives conditions under which the converse of Remark 4.1.2(ii) is true.

Proposition 4.1.4. *Let μ be σ-finite, and let Y be separable. Then every measurable function $f : X \to Y$ is μ-measurable.*

Proof. Let us prove the claim when $\mu(X) < \infty$, the general case following as in the proof of Proposition 4.1.2. Fix $n \in \mathbb{N}$, and let $D_n \equiv \{y_{n,k}\}$ be a countable dense set in the open ball $B(0,n) \subseteq Y$. Consider the open balls $B(y_{n,k}, \frac{1}{n}) \subseteq Y$, and set

$$W_{n,1} := B\left(y_{n,1}, \frac{1}{n}\right) \cap B(0,n),$$
$$W_{n,k} := \left[B\left(y_{n,k}, \frac{1}{n}\right) \cap B(0,n)\right] \setminus \left(\bigcup_{j=1}^{k-1} W_{n,j}\right) \quad (k \geq 2).$$

Then

$$W_{n,k} \cap W_{n,l} = \emptyset \quad \text{for } k \neq l, \quad \bigcup_{k=1}^\infty W_{n,k} = B(0,n). \tag{4.12}$$

In fact, the sets $W_{n,k}$ are disjoint, and $\bigcup_{k=1}^\infty W_{n,k} \subseteq B(0,n)$. Conversely, for every $y \in B(0,n)$, there exists a sequence $\{y_{n,k_j}\} \subseteq D_n$ such that $\|y_{n,k_j} - y\|_Y \to 0$ as $j \to \infty$. Hence for some $j_0 \in \mathbb{N}$, $\|y_{n,k_{j_0}} - y\|_Y < \frac{1}{n}$, that is, $y \in \bigcup_{k=1}^\infty W_{n,k}$. Therefore $\bigcup_{k=1}^\infty W_{n,k} \supseteq B(0,n)$. Observe that by the second equality in (4.12) we have $Y = \bigcup_{n,k=1}^\infty W_{n,k}$.

Now set $C_{n,k} := f^{-1}(W_{n,k}) \subseteq X$. Since $W_{n,k} \in \mathcal{B}(Y)$ and f is measurable, we have holds $C_{n,k} \in \mathcal{A}$. Hence for all $n, p \in \mathbb{N}$, the function

$$s_{n,p} : X \mapsto Y, \quad s_{n,p}(x) := \sum_{k=1}^p y_{n,k}\chi_{C_{n,k}}(x) \quad (x \in X)$$

is simple. Moreover, by (4.12) we have

$$C_{n,k} \cap C_{n,l} = f^{-1}(W_{n,k} \cap W_{n,l}) = f^{-1}(\emptyset) = \emptyset \quad \text{for } k \neq l,$$
$$X = f^{-1}(Y) = \bigcup_{n,k=1}^\infty C_{n,k}.$$

Set

$$f_n : X \mapsto Y, \quad f_n(x) := \sum_{k=1}^{\infty} y_{n,k}\chi_{C_{n,k}}(x) \quad (n \in \mathbb{N},\, x \in X);$$

observe that for every $x \in X$, the above sum reduces to a single term, since the sets $C_{n,k}$ are disjoint.

By the above remarks we have

$$\lim_{p\to\infty} \|s_{n,p}(x) - f_n(x)\|_Y = 0 \quad \text{for every } x \in X,$$

and thus f_n is μ-measurable. Then by Proposition 4.1.2 the conclusion follows if we prove that

$$\lim_{n\to\infty} \|f_n(x) - f(x)\|_Y = 0 \quad \text{for every } x \in X. \tag{4.13}$$

To this purpose, fix $x_0 \in X$, and let $\bar{n}$ be the least integer greater than $\|f(x_0)\|_Y$. Then $f(x_0) \in B(0,n)$ for every $n \geq \bar{n}$, and thus by (4.12) there exists a unique $k_n \in \mathbb{N}$ such that $f(x_0) \in W_{n,k_n}$, that is, $x_0 \in C_{n,k_n}$. By the definition of f_n this implies that $f_n(x_0) = y_{n,k_n}$. It follows that

$$\|f(x_0) - f_n(x_0)\|_Y = \|f(x_0) - y_{n,k_n}\|_Y < \frac{1}{n}.$$

Letting $n \to \infty$ in this inequality, by the arbitrariness of x_0 we obtain (4.13). This proves the result. □

Let us state the following definition.

Definition 4.1.4. A function $f : X \to Y$ is:

(i) *separably valued* if the range $f(X) \subseteq Y$ is separable;

(ii) *μ-a. e. separably valued* if there exists a μ-null set $N \subseteq X$ such that $f(N^c) \subseteq Y$ is separable.

Proposition 4.1.5. *Let μ be σ-finite. Let $f : X \to Y$ be separably valued, and let $D \equiv \{y_n\} \subseteq f(X)$ be countable and dense in $f(X)$. Then f is μ-measurable if and only if for every $n \in \mathbb{N}$, the map $F_n : X \to [0,\infty)$, $x \mapsto F_n(x) := \|f(x) - y_n\|_Y$, is measurable.*

Proof. We only prove the result when μ is finite. If f is μ-measurable, then there exists a sequence $\{s_k\} \subseteq \mathscr{S}(X;Y)$ such that $\lim_{k\to\infty} \|s_k(x) - f(x)\|_Y = 0$ for μ-a. e. $x \in X$. Fix $n \in \mathbb{N}$. Since

$$\big|F_n(x) - \|s_k(x) - y_n\|_Y\big| \leq \|s_k(x) - f(x)\|_Y \quad \text{for all } k \in \mathbb{N},$$

we have $F_n(x) = \lim_{k\to\infty} \|s_k(x) - y_n\|_Y$ for μ-a. e. $x \in X$. On the other hand, $\{s_k - y_n\} \subseteq \mathscr{S}(X;Y)$ for all $n \in \mathbb{N}$, and thus the map $x \mapsto \|s_k(x) - y_n\|_Y$ is measurable. Hence by Corollary 2.1.5(ii) the claim follows.

Conversely, let F_n be measurable for every $n \in \mathbb{N}$. Since $f^{-1}(B(y_n, \frac{1}{k})) = F_n^{-1}((0, \frac{1}{k}))$ $(k \in \mathbb{N})$, we have $f^{-1}(B(y_n, \frac{1}{k})) \in \mathcal{A}$ (see Proposition 2.1.4; here $B(y_n, \frac{1}{k}) := \{y \in Y \mid \|y - y_n\|_Y < \frac{1}{k}\}$). On the other hand, since D is dense in $f(X)$, the countable family $\{B(y_n, \frac{1}{k}) \mid n \in \mathbb{N}, k \in \mathbb{N}\}$ is a basis for the relative topology on $f(X)$, and thus by Lemma 2.1.2 $f^{-1}(F) \in \mathcal{A}$ for all $F \in \mathcal{B}(Y)$. Hence f is measurable and thus μ-measurable by Proposition 4.1.4. Then the result follows. □

Let us mention the following consequence of Remark 2.1.3(i) and Proposition 4.1.4.

Corollary 4.1.6. *Let $(X, \mathcal{T})$ be a topological space, let μ be a σ-finite measure on $(X, \mathcal{B}(X))$, and let Y be separable. Then every continuous function $f : X \to Y$ is μ-measurable.*

The following result can be viewed as a partial converse of Proposition 4.1.4. As its proof shows, it is a natural consequence of the definition of μ-measurability.

Proposition 4.1.7. *Let μ be σ-finite, and let $f : X \mapsto Y$ be μ-measurable. Then f is μ-a. e. separably valued.*

Proof. Suppose first that $\mu(X) < \infty$. By (4.6) and Theorem 4.1.3, for every $k \in \mathbb{N}$, there exists a μ-measurable set $E_k \subseteq X$ such that $\mu(E_k) < \frac{1}{k}$ and

$$\lim_{n\to\infty} \sup_{x\in E_k^c} \|s_n(x) - f(x)\|_Y = 0. \tag{4.14}$$

Since the range $s_n(X) \subseteq Y$ is finite for all $n \in \mathbb{N}$, the set $\bigcup_{n=1}^{\infty} s_n(X)$ is countable. Then by (4.14) the set $f(E_k^c) \subseteq Y$ is separable for every $k \in \mathbb{N}$, and thus $f(\bigcup_{k=1}^{\infty} E_k^c) = \bigcup_{k=1}^{\infty} f(E_k^c)$ is also separable. Setting $N := \bigcap_{k=1}^{\infty} E_k$, we have $f(N^c) = f(\bigcup_{k=1}^{\infty} E_k^c)$ and $\mu(N) = \lim_{k\to\infty} \mu(E_k) = 0$. Hence the claim follows in this case.

In the general case, let $\{E_k\}$ be a disjoint sequence of measurable sets such that $X = \bigcup_{k=1}^{\infty} E_k$ and $\mu(E_k) < \infty$ for all $k \in \mathbb{N}$. By the above considerations, for every $k \in \mathbb{N}$, the restriction $f|_{E_k}$ is μ-a. e. separably valued, that is, there exists a μ-null subset $N_k \subseteq E_k$ such that the set $f(E_k \setminus N_k)$ is separable. Therefore the set $\bigcup_{k=1}^{\infty} f(E_k \setminus N_k) = f(\bigcup_{k=1}^{\infty}(E_k \setminus N_k))$ is separable. Set $N := \bigcup_{k=1}^{\infty} N_k$. Then N is μ-null, and $f(N^c) = f(\bigcup_{k=1}^{\infty}(E_k \setminus N_k))$. Hence the result follows. □

4.1.3 Weak and weak* measurability

Let $(X, \mathcal{A}, \mu)$ be a measure space, let Y be a Banach space, and let Y^* be the dual space of Y. Henceforth in this chapter, we denote by $\langle \cdot, \cdot \rangle \equiv \langle \cdot, \cdot \rangle_{Y^*,Y}$ the duality map between Y and Y^*.

Definition 4.1.5. (i) A function $f : X \mapsto Y$ is *weakly measurable* if for every $y^* \in Y^*$, the function from X to $\mathbb{R}$, $x \mapsto \langle y^*, f(x)\rangle$, is measurable.
(ii) A function $f : X \mapsto Y^*$ is *weakly* measurable* if for every $y \in Y$, the function from X to $\mathbb{R}$, $x \mapsto \langle f(x), y\rangle$, is measurable.

Remark 4.1.3. (i) A μ-measurable function $f : X \to Y$ is weakly measurable. In fact, let $\{s_n\} \subseteq \mathscr{S}(X; Y)$ satisfy (4.6). Then for all $y^* \in Y^*$,

$$\lim_{n\to\infty} \langle y^*, s_n(x)\rangle = \langle y^*, f(x)\rangle \quad \text{for } \mu\text{-a. e. } x \in X. \tag{4.15}$$

For every $n \in \mathbb{N}$, the function $x \mapsto \langle y^*, s_n(x)\rangle : X \mapsto \mathbb{R}$ is simple and thus measurable. Then by (4.15) and Corollary 2.1.5(ii) for all $y^* \in Y^*$, the real-valued map $x \in X \mapsto \langle y^*, f(x)\rangle$ is measurable.
(ii) Clearly, a weakly measurable function $f : X \to Y^*$ is weakly* measurable. Then by (i) a μ-measurable function $f : X \to Y^*$ is weakly measurable and thus weakly* measurable.

Remark 4.1.4. The set of weakly measurable functions (respectively, of weakly* measurable functions) is a vector space. If $\{f_n\}$ is a sequence of weakly measurable functions and $f_n(x) \rightharpoonup f(x)$ for all $x \in X$, $f : X \to Y$ is also weakly measurable. Similarly, if $\{f_n\}$ is a sequence of weakly* measurable functions and $f_n \overset{*}{\rightharpoonup} f$, then $f : X \to Y^*$ is weakly* measurable.

Let us recall the following definition (e. g., see [44]).

Definition 4.1.6. (i) Let $Z \subseteq Y$. The *polar set of* Z is the set

$$Z_p := \{y^* \in Y^* \mid |\langle y^*, y\rangle| \le 1 \text{ for all } y \in Z\} \subseteq Y^*.$$

(ii) Let $Z^* \subseteq Y^*$. The *polar set of* Z^* is the set

$$Z_p^* := \{y \in Y \mid |\langle y^*, y\rangle| \le 1 \text{ for all } y^* \in Z^*\} \subseteq Y.$$

Remark 4.1.5. *Let*

$$B_1 := \{y \in Y \mid \|y\|_Y \le 1\} \quad \textit{and} \quad B_1^* := \{y^* \in Y^* \mid \|y^*\|_{Y^*} \le 1\}$$

be the closed unit balls in Y and Y^*, respectively. Then

$$(B_1)_p = B_1^* \quad \textit{and} \quad (B_1^*)_p = B_1. \tag{4.16}$$

Let us prove a preliminary result (see [44, Section 8.15, Lemma 1]).

Proposition 4.1.8. *Let S be a separable subspace of Y, let B_1 be the unit closed ball of Y, and let $(B_1)_p$ be the polar set of B_1. Then there exists a countable set $W \subseteq (B_1)_p$ such that*

$$\|y\|_Y = \sup_{a\in W} |\langle a, y\rangle| \quad \textit{for all } y \in S. \tag{4.17}$$

Proof. Since $\|y\|_Y = \sup_{a\in(B_1)_p} |\langle a, y\rangle|$ (see (4.16)), it suffices to prove the existence of a countable set $W \subseteq (B_1)_p$ such that

$$\|y\|_Y \le \sup_{a\in W} |\langle a, y\rangle| \quad \text{for all } y \in S. \tag{4.18}$$

Let $D \equiv \{y_n\}$ be a dense countable subset in S, and fix $n \in \mathbb{N}$. Since $\|y_n\|_Y = \sup_{a\in(B_1)_p} |\langle a, y_n\rangle|$, there exists a sequence $\{a_{n,j}\} \subseteq (B_1)_p$ such that

$$\|y_n\|_Y = \sup_{j\in\mathbb{N}} |\langle a_{n,j}, y_n\rangle|. \tag{4.19}$$

Consider the countable set $W := \{a_{k,j} \mid k, j \in \mathbb{N}\}$. Then by (4.19) we have that

$$\|y_n\|_Y \le \sup_{k,j\in\mathbb{N}} |\langle a_{k,j}, y_n\rangle| = \sup_{a\in W} |\langle a, y_n\rangle|. \tag{4.20}$$

Now fix $z \in S$. By the denseness of D, for any $\epsilon > 0$, there exists $y_m \in D$ such that $\|z - y_m\|_Y < \epsilon$. Then by (4.20)

$$\|z\|_Y \le \|z - y_m\|_Y + \|y_m\|_Y \le \sup_{k,j\in\mathbb{N}} |\langle a_{k,j}, y_m\rangle| + \epsilon. \tag{4.21}$$

On the other hand, since $W \subseteq (B_1)_p$, we have

$$\begin{aligned}\sup_{k,j\in\mathbb{N}} |\langle a_{k,j}, y_m\rangle| &\le \sup_{k,j\in\mathbb{N}} |\langle a_{k,j}, z\rangle| + \sup_{k,j\in\mathbb{N}} |\langle a_{k,j}, z - y_m\rangle| \\ &\le \sup_{k,j\in\mathbb{N}} |\langle a_{k,j}, z\rangle| + \|z - y_m\|_Y \le \sup_{k,j\in\mathbb{N}} |\langle a_{k,j}, z\rangle| + \epsilon. \end{aligned} \tag{4.22}$$

From (4.21)–(4.22) we obtain

$$\|z\|_Y \le \sup_{a\in W} |\langle a, z\rangle| + 2\epsilon,$$

whence by the arbitrariness of ϵ inequality (4.18) follows. This proves the result. □

Proposition 4.1.9. (i) *Let $f : X \to Y$ be weakly measurable and μ-a. e. separably valued. Then the function $x \mapsto \|f(x)\|_Y$ is measurable.*

(ii) *Let $f : X \to Y^*$ be weakly* measurable and μ-a. e. separably valued. Then the function $x \mapsto \|f(x)\|_{Y^*}$ is measurable.*

Proof. (i) It is not restrictive to assume that $f(X)$ itself is separable. Then by Proposition 4.1.8 and the first equality in (4.16) there exists a countable set $W \subseteq B_1^*$ such that

$$\|f(x)\|_Y = \sup_{a\in W} |\langle a, f(x)\rangle| \quad \text{for all } x \in X. \tag{4.23}$$

Since f is weakly measurable and $|\cdot|$ is continuous, for any $a \in W$, the map $x \mapsto |\langle a, f(x)\rangle|$ is measurable. Then by (4.23) and Corollary 2.1.5(ii) the claim follows.

(ii) Arguing as in (i) and using the second equality in (4.16), there exists a countable set $W \subseteq B_1$ such that

$$\|f(x)\|_{Y^*} = \sup_{a\in W} |\langle f(x), a\rangle| \quad \text{for all } x \in X. \tag{4.24}$$

Since f is weakly* measurable and $|\cdot|$ is continuous, for any $a \in W$, the map $x \mapsto |\langle a, f(x)\rangle|$ is measurable. Then by (4.24) and Corollary 2.1.5(ii) the conclusion follows. □

Remark 4.1.6. Observe that Y^* is the topological dual both of Y with the weak topology $\sigma(Y, Y^*)$ and of Y^{**} with the weak* topology $\sigma(Y^{**}, Y^*)$. Therefore, since the embedding $Y \subseteq Y^{**}$ is isometric, applying Proposition 4.1.9(ii) to $Y^{**} = (Y^*)^*$ is in agreement with Proposition 4.1.9(i).

In particular, by Proposition 4.1.9(ii) the function $x \mapsto \|f(x)\|_{Y^*}$ is measurable if Y^* is separable, and $f : X \to Y^*$ is weakly* measurable. The same conclusion holds if Y is separable, as the following proposition shows.

Proposition 4.1.10. *Let Y be separable, and let $f : X \to Y^*$ be weakly* measurable. Then the function $x \mapsto \|f(x)\|_{Y^*}$ is measurable.*

Proof. Let $D \equiv \{y_n\} \subseteq Y$ be countable and dense. For every $n \in \mathbb{N}$, the function $l_n : X \to \mathbb{R}$, $l_n(x) := |\langle f(x), y_n\rangle|$, is measurable, and thus the map $l : X \to \mathbb{R}$, $l(x) := \sup_{y_n\in D, \|y_n\|_Y\le 1} l_n(x)$, is measurable (see Corollary 2.1.5(ii)). Then the conclusion will follow if we prove that for all $x \in X$,

$$l(x) = \|f(x)\|_{Y^*}. \tag{4.25}$$

To this purpose, observe first that

$$\|f(x)\|_{Y^*} = \sup_{y\in Y, \|y\|_Y\le 1} |\langle f(x), y\rangle| \ge \sup_{y_n\in D, \|y_n\|_Y\le 1} |\langle f(x), y_n\rangle| = l(x).$$

To prove the reverse inequality, set $\alpha := \|f(x)\|_{Y^*} > 0$ (if $\|f(x)\|_{Y^*} = 0$, then equality (4.25) is obvious). For any $k \in \mathbb{N}$, choose $u_k \in Y$ such that $\|u_k\|_Y \le 1$ and

$$|\langle f(x), u_k\rangle| \ge \alpha - \frac{1}{2k}. \tag{4.26}$$

By the denseness of D there exists $y_k \in D$ such that $\|u_k - y_k\|_Y < \frac{1}{4\alpha k}$, whence

$$\|y_k\|_Y \leq \|y_k - u_k\|_Y + \|u_k\|_Y \leq \frac{1}{4\alpha k} + 1, \tag{4.27}$$

$$|\langle f(x), u_k - y_k\rangle| \leq \frac{\alpha}{4\alpha k} = \frac{1}{4k}. \tag{4.28}$$

Set

$$\bar{y}_k := \begin{cases} \frac{y_k}{\|y_k\|_Y} & \text{if } \|y_k\|_Y > 1, \\ y_k & \text{if } \|y_k\|_Y \leq 1. \end{cases}$$

If $\|y_k\|_Y > 1$, then by (4.27)–(4.28)

$$\begin{aligned} |\langle f(x), u_k - \bar{y}_k\rangle| &\leq |\langle f(x), u_k - y_k\rangle| + |\langle f(x), y_k - \bar{y}_k\rangle| \\ &\leq \frac{1}{4k} + \alpha(\|y_k\|_Y - 1) \leq \frac{1}{4k} + \frac{\alpha}{4\alpha k} = \frac{1}{2k}. \end{aligned} \tag{4.29}$$

By (4.26) and (4.28)–(4.29), for every $k \in \mathbb{N}$, there exists $\bar{y}_k \in D$ such that $\|\bar{y}_k\|_Y \leq 1$ and

$$|\langle f(x), \bar{y}_k\rangle| \geq |\langle f(x), u_k\rangle| - |\langle f(x), \bar{y}_k - u_k\rangle| \geq \alpha - \frac{1}{k}.$$

This proves (4.25), and thus the conclusion follows. □

The relationship between μ-measurability and weak measurability is clarified by the following theorem.

Theorem 4.1.11 (Pettis). *Let μ be σ-finite, and let $f : X \to Y$. Then the following statements are equivalent:*
(i) *f is μ-measurable;*
(ii) *f is weakly measurable and μ-a. e. separably valued.*

Proof. (i)⇒(ii). By Remark 4.1.3 f is weakly measurable, whereas the fact that f is μ-a. e. separably valued follows from Proposition 4.1.7. Hence the claim follows.

(ii)⇒(i). It is not restrictive to assume that $f(X)$ is separable. Let $D \equiv \{y_n\}$ be a dense countable subset in $f(X)$. By Proposition 4.1.9(i) the function $x \mapsto \|f(x) - y_n\|_Y$ is measurable for any $y_n \in D$, and thus by Proposition 4.1.5 f is μ-measurable. Hence the result follows. □

In the same way, Proposition 4.1.9(ii) gives the following result.

Proposition 4.1.12. *Let μ be σ-finite, and let $f : X \mapsto Y^*$. Then the following statements are equivalent:*
(i) *f is μ-measurable;*
(ii) *f is weakly* measurable and μ-a. e. separably valued.*

Remark 4.1.7. Let μ be σ-finite, and let Y^* be separable. By Theorem 4.1.11 and Proposition 4.1.12, for every $f : X \to Y^*$, the three notions of μ-measurability, weak measurability, and weak* measurability coincide.

4.2 Integration of vector functions

4.2.1 Bochner integrability

Definition 4.2.1. Let $s : X \mapsto Y$ be a simple function, $s = \sum_{i=1}^{p} y_i \chi_{E_i}$ with $y_1, \dots, y_p \in Y$ and a partition $\{E_1, \dots, E_p\} \subseteq \mathcal{A}$ of X, such that $y_i = 0$ if $\mu(E_i) = \infty$. The quantity

$$\int_X s\, d\mu := \sum_{i=1}^{p} y_i\, \mu(E_i)$$

is called the *integral of* s on X.

Remark 4.2.1. (i) It is easily seen that the above definition is well posed, since the integral $\int_X s\, d\mu$ does not depend on the choice of the partition $\{E_1, \dots, E_p\}$. A similar calculation shows that the map from $\mathcal{S}(X;Y)$ to Y, $s \mapsto \int_X s\, d\mu$, is linear: if $s = \sum_{i=1}^{p} y_i \chi_{E_i}$ and $t = \sum_{j=1}^{q} y'_j \chi_{F_j}$ are simple functions, then for all $\alpha, \beta \in \mathbb{R}$,

$$\int_X (\alpha s + \beta t) d\mu = \alpha \int_X s\, d\mu + \beta \int_X t\, d\mu. \tag{4.30}$$

(ii) By the triangular inequality we have

$$\left\| \int_X s\, d\mu \right\|_Y \le \int_X \|s\|_Y\, d\mu, \tag{4.31}$$

where $\|s\|_Y := \sum_{i=1}^{p} \|y_i\|_Y \chi_{E_i}$.

Definition 4.2.2. A μ-measurable function $f : X \mapsto Y$ is called *Bochner integrable* if there exists a sequence $\{s_n\} \subseteq \mathcal{S}(X;Y)$ such that $s_n \to f$ μ-a. e. in X and

$$\lim_{n\to\infty} \int_X \|s_n - f\|_Y\, d\mu = 0. \tag{4.32}$$

Every such sequence $\{s_n\} \subseteq \mathcal{S}(X;Y)$ is called an *approximating sequence* of f. The quantity

$$\int_X f\, d\mu := \lim_{n\to\infty} \int_X s_n\, d\mu, \tag{4.33}$$

the limit existing in the sense of strong convergence in Y, is called the *Bochner integral of f on X*.

If $f : X \mapsto Y$ is Bochner integrable, then we set

$$\int_E f\,d\mu := \int_X f\chi_E\,d\mu \quad \text{for } E \in \mathcal{A}. \tag{4.34}$$

The set of Bochner-integrable functions $f : X \to Y$ is denoted by $L^1(X;Y) \equiv L^1(X,\mathcal{A},\mu;Y)$; if $Y = \mathbb{R}$, then we set $L^1(X;\mathbb{R}) \equiv L^1(X) \equiv L^1(X,\mathcal{A},\mu)$. It is easily seen that if $Y = \mathbb{R}$, then the notions of Bochner integrability and Bochner integral coincide with those in Definitions 2.2.2–2.2.3.

Remark 4.2.2. (i) For every $n \in \mathbb{N}$, the integral in the right-hand side of (4.32) is well defined (see Remark 4.1.1 and Definition 2.2.2). Also, the Bochner integral is well defined, since the limit in (4.33) exists and does not depend on the choice of the sequence $\{s_n\}$. Indeed, if $\{s_n\}$ is an approximating sequence of f, then by (4.30)–(4.31) we have

$$\left\| \int_X (s_m - s_n)\,d\mu \right\|_Y \le \int_X \|s_m - s_n\|_Y d\mu \le \int_X \|s_m - f\|_Y\,d\mu + \int_X \|s_n - f\|_Y\,d\mu.$$

Hence $\{\int_X s_n\,d\mu\} \subseteq Y$ is a Cauchy sequence and thus has a strong limit in Y. Moreover, let $\{s_n\}$ and $\{t_n\}$ be two approximating sequences for f. Then, as $n \to \infty$, we get

$$\left\| \int_X (s_n - t_n)\,d\mu \right\|_Y \le \int_X \|s_n - t_n\|_Y\,d\mu \le \int_X \|s_n - f\|_Y\,d\mu + \int_X \|t_n - f\|_Y\,d\mu \to 0,$$

and thus the sequences $\{\int_X s_n\,d\mu\}$ and $\{\int_X t_n\,d\mu\}$ have the same limit in Y.

(ii) If $f : X \mapsto Y$ is Bochner integrable and $E \in \mathcal{A}$, then the function $f\chi_E : X \to Y$ is Bochner integrable. Then the definition in (4.34) is well posed.

Remark 4.2.3. (i) By (4.30) and (4.33), for all $f, g \in L^1(X;Y)$ and $\alpha, \beta \in \mathbb{R}$, we have $\alpha f + \beta g \in L^1(X;Y)$, and thus $L^1(X;Y)$ is a vector space. Moreover,

$$\int_X (\alpha f + \beta g)d\mu = \alpha \int_X f\,d\mu + \beta \int_X g\,d\mu. \tag{4.35}$$

(ii) By (4.34)–(4.35) the Bochner integral is finitely additive: if $E_1, \dots, E_p \in \mathcal{A}$ are disjoint, then

$$\int_{\bigcup_{i=1}^p E_i} f\,d\mu = \sum_{i=1}^p \int_{E_i} f\,d\mu. \tag{4.36}$$

The following result provides a simple criterion for the Bochner integrability.

Proposition 4.2.1. *A μ-measurable function $f : X \to Y$ is Bochner integrable if and only if*

$$\int_X \|f\|_Y \, d\mu < \infty. \tag{4.37}$$

Proof. Let f be Bochner integrable, and let $\{s_n\}$ be an approximating sequence. Then for every $\epsilon > 0$, there exists $\bar{n} \in \mathbb{N}$ such that

$$\int_X \|f\|_Y \, d\mu \le \int_X \|f - s_{\bar{n}}\|_Y \, d\mu + \int_X \|s_{\bar{n}}\|_Y \, d\mu < \int_X \|s_{\bar{n}}\|_Y \, d\mu + \epsilon < \infty,$$

and thus (4.37) follows. Conversely, let $f : X \to Y$ be a μ-measurable function satisfying (4.37), and let $\{t_n\} \subseteq \mathscr{S}(X;Y)$ be any sequence such that $t_n \to f$ μ-a. e. in X. For every $n \in \mathbb{N}$, set $s_n := t_n \chi_{E_n}$, where

$$E_n := \{x \in X \mid \|t_n(x)\|_Y \le 2\|f(x)\|_Y\}.$$

Let us show that $s_n \to f$ μ-a. e. in X and (4.32) holds, and thus f is Bochner integrable.

Indeed, since $t_n \to f$ μ-a. e. in X, there exists a μ-null set $N \subseteq X$ such that $\lim_{n\to\infty} t_n(x) = f(x)$ for all $x \in N^c$. Hence for every $x \in N^c$, there exists $\bar{n} \in \mathbb{N}$ such that $\|t_n(x)\|_Y \le 2\|f(x)\|_Y$ for all $n \ge \bar{n}$, that is, $x \in E_n$ for all $n \ge \bar{n}$, and thus $\lim_{n\to\infty} \chi_{E_n}(x) = 1$. Hence for all $x \in N^c$, $\lim_{n\to\infty} s_n(x) = \lim_{n\to\infty} t_n(x) = f(x)$. Since $\mu(N) = 0$, it follows that $s_n \to f$ μ-a. e. in X.

By the very definition of s_n we have $\|s_n(x)\|_Y = 0$ for all $x \in E_n^c$ and $\|s_n(x)\|_Y = \|t_n(x)\|_Y \le 2\|f(x)\|_Y$ for all $x \in E_n$. Therefore

$$\|s_n(x) - f(x)\|_Y \le 3\|f(x)\|_Y \quad \text{for all } x \in X.$$

Since $s_n \to f$ μ-a. e. in X and $\|f\|_Y \in L^1(X)$ by (4.37), from the above inequality and the dominated convergence theorem we obtain (4.32). Then the result follows. □

Proposition 4.2.2. *Let f be Bochner integrable. Then:*

(i) *for all $E \in \mathcal{A}$, we have*

$$\left\| \int_E f \, d\mu \right\|_Y \le \int_E \|f\|_Y \, d\mu; \tag{4.38}$$

(ii) $\lim_{\mu(E)\to 0} \int_E f \, d\mu = 0$;

(iii) *the Bochner integral is σ-additive: for any disjoint sequence $\{E_i\} \subseteq \mathcal{A}$,*

$$\int_{\bigcup_{i=1}^\infty E_i} f \, d\mu = \sum_{i=1}^\infty \int_{E_i} f \, d\mu, \tag{4.39}$$

the series in the right-hand side being absolutely convergent in Y.

Proof. Let $\{s_n\} \subseteq \mathscr{S}(X;Y)$ be an approximating sequence of f. By (4.32), (4.34), and the triangle inequality, for any $E \in \mathcal{A}$, we get

$$\begin{aligned}\lim_{n\to\infty}\left|\int_E \|s_n\|_Y d\mu - \int_E \|f\|_Y d\mu\right| \\ \leq \lim_{n\to\infty}\int_E \big|\|s_n\|_Y - \|f\|_Y\big| d\mu \leq \lim_{n\to\infty}\int_X \|s_n - f\|_Y \, d\mu = 0. \qquad (4.40)\end{aligned}$$

On the other hand, for all $n \in \mathbb{N}$ and $E \in \mathcal{A}$, we have

$$\left\|\int_E s_n \, d\mu\right\|_Y \leq \int_E \|s_n\|_Y \, d\mu$$

(see (4.31)). By (4.33) and (4.40), letting $n \to \infty$ in the above inequality gives (4.38), and thus claim (i) follows. Claim (ii) follows from (4.38) since $\lim_{\mu(E)\to 0}\int_E \|f\|_Y \, d\mu = 0$ (see Remark 2.8.2(i)). Concerning (iii), observe that by (i) and Proposition 4.2.1

$$\sum_{i=1}^{\infty}\left\|\int_{E_i} f \, d\mu\right\|_Y \leq \sum_{i=1}^{\infty}\int_{E_i}\|f\|_Y d\mu = \int_{\bigcup_{i=1}^{\infty} E_i}\|f\|_Y \, d\mu \leq \int_X \|f\|_Y d\mu < \infty,$$

and hence the series in the right-hand side of (4.39) is absolutely convergent. As for its sum, by (4.36) and (4.38) we have

$$\begin{aligned}\left\|\int_{\bigcup_{i=1}^{\infty} E_i} f \, d\mu - \sum_{i=1}^{p}\int_{E_i} f \, d\mu\right\|_Y &= \left\|\int_{\bigcup_{i=p+1}^{\infty} E_i} f \, d\mu\right\|_Y \leq \int_{\bigcup_{i=p+1}^{\infty} E_i}\|f\|_Y \, d\mu \\ &= \sum_{i=p+1}^{\infty}\int_{E_i}\|f\|_Y \, d\mu \to 0 \quad \text{as } p \to \infty,\end{aligned}$$

since

$$\sum_{i=1}^{\infty}\int_{E_i}\|f\|_Y \, d\mu = \int_{\bigcup_{i=1}^{\infty} E_i}\|f\|_Y \, d\mu \leq \int_X \|f\|_Y \, d\mu < \infty.$$

This proves (4.39), and hence the result follows. □

Let Y, Z be Banach spaces, and let B_1 denote the closed unit ball in Y. We denote by $\mathscr{L}(Y;Z)$ the Banach space of linear continuous operators $T : Y \to Z$ endowed with the operator norm $\|T\|_{\mathscr{L}(Y;Z)} := \sup_{y\in B_1}\|Ty\|_Z$.

Proposition 4.2.3. *Let Z be a Banach space, let $T \in \mathscr{L}(Y;Z)$, and let $f : X \to Y$ be Bochner integrable. Then the function $Tf : X \to Z$, $(Tf)(x) := T(f(x))$, is Bochner inte-*

grable, and

$$T\Big(\int_X f\,d\mu\Big) = \int_X (Tf)\,d\mu. \tag{4.41}$$

Proof. Let $\{s_n\}$ be an approximating sequence of f, $s_n = \sum_{k=1}^{p_n} y_{n,k}\chi_{E_{n,k}}$ with $y_{n,k} \in Y$ and a partition $\{E_{n,1}, \dots, E_{n,p_n}\} \subseteq \mathcal{A}$ of X. Then $\{Ts_n\} \subseteq \mathcal{S}(X;Z)$, since $Ts_n = \sum_{k=1}^{p_n}(Ty_{n,k})\chi_{E_{n,k}}$. By the continuity of T we have that $Ts_n \to Tf$ μ-a. e. in X and

$$\int_X \|Ts_n - Tf\|_Z\,d\mu \le \|T\|_{\mathcal{L}(Y;Z)} \int_X \|s_n - f\|_Y\,d\mu \to 0 \quad \text{as } n \to \infty.$$

Hence $\{Ts_n\}$ is an approximating sequence for Tf, and thus Tf is Bochner integrable. Finally, for every $n \in \mathbb{N}$, we have

$$T\Big(\int_X s_n\,d\mu\Big) = \sum_{k=1}^{p_n}(Ty_k)\,\mu(E_k) = \int_X (Ts_n)\,d\mu.$$

Since $\{s_n\}$ and $\{Ts_n\}$ are approximating sequences for f and Tf, respectively, and T is continuous, letting $n \to \infty$ in the previous equality gives (4.41). This completes the proof. □

Corollary 4.2.4. *Let $f : X \to Y$ be Bochner integrable. Then for every $y^* \in Y^*$, the function $\langle y^*, f\rangle : X \to \mathbb{R}$, $\langle y^*, f\rangle(x) := \langle y^*, f(x)\rangle$, is integrable, and for every $E \in \mathcal{A}$, we have*

$$\Big\langle y^*, \int_E f\,d\mu \Big\rangle = \int_E \langle y^*, f(x)\rangle\,d\mu. \tag{4.42}$$

Let us prove the following generalization of Corollary 2.9.5.

Proposition 4.2.5. *Let $f : U \subseteq X \to Y$ be Bochner integrable. Then for μ-a. e. $x_0 \in X$,*

$$\lim_{r\to 0^+} \frac{1}{\mu(B_r(x_0))} \int_{B_r(x_0)} \|f - f(x_0)\|_Y\,d\mu = 0, \tag{4.43a}$$

$$\lim_{r\to 0^+} \Big\| \frac{1}{\mu(B_r(x_0))} \int_{B_r(x_0)} f\,d\mu - f(x_0) \Big\|_Y = 0. \tag{4.43b}$$

Proof. Equality (4.43b) follows from (4.43a) and (4.38). To prove (4.43a), observe that by Proposition 4.1.7 f is μ-a. e. separably valued, since by assumption it is Bochner integrable and thus μ-measurable. Without loss of generality, we may assume that the range $f(U)$ is separable. Let $D \equiv \{y_n\}$ be a countable dense subset of $f(U)$. By (2.96), for every $n \in \mathbb{N}$, there exists a μ-null set $N_n \subseteq U$ such that for all $x_0 \in U \setminus N_n$,

$$\lim_{r\to 0^+} \frac{1}{\mu(B_r(x_0))} \int_{B_r(x_0)} \|f - y_n\|_Y \, d\mu = \|f(x_0) - y_n\|_Y. \tag{4.44}$$

Then the set $N := \bigcup_{n=1}^{\infty} N_n$ is μ-null, and for any $x_0 \in U \setminus N$ equality (4.44), holds for every $n \in \mathbb{N}$.

Fix $x_0 \in U \setminus N$. Then for any $n \in \mathbb{N}$,

$$\begin{aligned}
&\limsup_{r\to 0^+} \frac{1}{\mu(B_r(x_0))} \int_{B_r(x_0)} \|f - f(x_0)\|_Y \, d\mu \\
&\quad \le \limsup_{r\to 0^+} \frac{1}{\mu(B_r(x_0))} \int_{B_r(x_0)} \|f - y_n\|_Y \, d\mu + \|f(x_0) - y_n\|_Y = 2\|f(x_0) - y_n\|_Y.
\end{aligned}$$

On the other hand, by the denseness of D in $f(U)$, for any $\epsilon > 0$, there exists $\bar{n} \in \mathbb{N}$ such that $\|f(x_0) - y_n\|_Y < \frac{\epsilon}{2}$ for every $n > \bar{n}$. Then by the above inequality for any $\epsilon > 0$, we get

$$\limsup_{r\to 0^+} \frac{1}{\mu(B_r(x_0))} \int_{B_r(x_0)} \|f - f(x_0)\|_Y \, d\mu \le \epsilon,$$

whence (4.43a) follows. This proves the result. □

Proposition 4.2.6. *Let $f : X \to Y$ be Bochner integrable. Then the linear operator*

$$T : L^\infty(X) \to Y, \quad Tg := \int_X fg \, d\mu \quad (g \in L^\infty(X)),$$

is compact.

Proof. Let $\{s_n\} \subseteq \mathscr{S}(X;Y)$ be an approximating sequence of f with $s_n = \sum_{k=1}^{p_n} y_{n,k}\chi_{E_{n,k}}$, $y_{n,k} \in Y$, and a partition $\{E_{n,1}, \dots, E_{n,p_n}\} \subseteq \mathcal{A}$ of X such that $y_{n,k} = 0$ if $\mu(E_{n,k}) = \infty$. For all $n \in \mathbb{N}$, define $T_n \in \mathscr{L}(L^\infty(X); Y)$ by

$$T_n g := \int_X s_n g \, d\mu = \sum_{k=1}^{p_n} y_{n,k} \int_{E_{n,k}} g \, d\mu \quad (g \in L^\infty(X)).$$

Clearly, every T_n has finite rank. Moreover, by (4.32) and (4.38) we have

$$\|T_n - T\|_{\mathscr{L}(L^\infty(X);Y)} = \sup_{g\in L^\infty(X),\, g\neq 0} \frac{\|(T_n - T)g\|_Y}{\|g\|_\infty} \le \int_X \|s_n - f\|_Y \, d\mu \to 0 \quad \text{as } n \to \infty.$$

Hence the result follows. □

The following result establishes a link between the Bochner integral and vector measures, generalizing Remark 2.2.3.

Proposition 4.2.7. *Let f be Bochner integrable. Then the map*

$$\nu : \mathcal{A} \to Y, \quad \nu(E) := \int_E f\, d\mu \quad (E \in \mathcal{A}), \tag{4.45}$$

is a vector measure absolutely continuous with respect to μ with bounded variation

$$|\nu|(E) = \int_E \|f\|_Y\, d\mu \quad (E \in \mathcal{A}). \tag{4.46}$$

Proof. By Proposition 4.2.2(iii) and Definition 1.9.2(i) the map ν is a vector measure, and by (4.46) $\nu \ll \mu$ (see Definition 1.9.4). To prove (4.46), let $E \in \mathcal{A}$, $E = \bigcup_{i=1}^n E_i$ with disjoint $E_1, \dots, E_n \in \mathcal{A}$. By (4.38) we have

$$\sum_{i=1}^n \|\nu(E_i)\|_Y = \sum_{i=1}^n \left\| \int_{E_i} f\, d\mu \right\|_Y \le \sum_{i=1}^n \int_{E_i} \|f\|_Y\, d\mu = \int_E \|f\|_Y\, d\mu,$$

whence $|\nu|(E) \le \int_E \|f\|_Y\, d\mu$ for all $E \in \mathcal{A}$ (see (1.62)). In particular, $|\nu|(X) \le \int_X \|f\|_Y\, d\mu < \infty$ since f is Bochner integrable (see Proposition 4.2.1). Then ν is of bounded variation (see Definition 1.9.2(ii)).

Let us prove the inverse inequality $|\nu|(E) \ge \int_E \|f\|_Y\, d\mu$. Let $\{s_n\} \subseteq \mathscr{S}(X;Y)$ be an approximating sequence of f, and thus for any $\epsilon > 0$, there exists $\bar{n} \in \mathbb{N}$ such that

$$\int_X \|s_{\bar{n}} - f\|_Y\, d\mu < \epsilon. \tag{4.47}$$

Let $s_{\bar{n}} = \sum_{l=1}^p y_l \chi_{G_l}$ with $y_1, \dots, y_p \in Y$ and $\{G_1, \dots, G_p\} \subseteq \mathcal{A}$ a partition of X such that $y_i = 0$ if $\mu(G_i) = \infty$. Fix $E \in \mathcal{A}$, and set $F_l := G_l \cap E$ $(l = 1, \dots, p)$. Then the sets $F_1, \dots, F_p \in \mathcal{A}$ are disjoint, and $E = \bigcup_{l=1}^p F_l$. It is easily checked that

$$\sum_{l=1}^p \left\| \int_{F_l} s_{\bar{n}}\, d\mu \right\|_Y = \int_E \|s_{\bar{n}}\|_Y\, d\mu. \tag{4.48}$$

By (4.47)–(4.48) we have

$$\begin{aligned}
&\left| \sum_{l=1}^p \left\| \int_{F_l} f\, d\mu \right\|_Y - \int_E \|s_{\bar{n}}\|_Y\, d\mu \right| \\
&= \left| \sum_{l=1}^p \left\| \int_{F_l} f\, d\mu \right\|_Y - \sum_{l=1}^p \left\| \int_{F_l} s_{\bar{n}}\, d\mu \right\|_Y \right| \le \sum_{l=1}^p \left| \left\| \int_{F_l} f\, d\mu \right\|_Y - \left\| \int_{F_l} s_{\bar{n}}\, d\mu \right\|_Y \right|
\end{aligned}$$

$$\leq \sum_{l=1}^{p} \left\| \int_{F_l} f \, d\mu - \int_{F_l} s_{\bar{n}} \, d\mu \right\|_Y \leq \sum_{l=1}^{p} \int_{F_l} \|s_{\bar{n}} - f\|_Y \, d\mu = \int_E \|s_{\bar{n}} - f\|_Y \, d\mu < \epsilon.$$

From the above inequality, (1.62), and (4.47), for all $E \in \mathcal{A}$, we get

$$|\nu|(E) \geq \sum_{l=1}^{p} \left\| \int_{F_l} f \, d\mu \right\|_Y \geq \int_E \|s_{\bar{n}}\|_Y \, d\mu - \epsilon \geq \int_E \|f\|_Y \, d\mu - 2\epsilon,$$

whence $|\nu|(E) \geq \int_E \|f\|_Y \, d\mu$ by the arbitrariness of ϵ. Then the conclusion follows. □

Remark 4.2.4. In view of Proposition 4.2.7, it is natural to ask if the converse statement holds, namely:

Let $\nu : \mathcal{A} \to Y$ be a vector measure of bounded variation such that $\nu \ll \mu$. Does there exist a Bochner-integrable function $f : X \to Y$ such that $\nu(E) = \int_E f \, d\mu$ for all $E \in \mathcal{A}$?

If $Y = \mathbb{R}$, then by the Radon–Nikodým theorem the answer is affirmative (see Theorem 2.9.1). Remarkably, this is not the case for a general infinite-dimensional Banach space Y (see Section 4.4.1).

Let us finally state the following definition, which generalizes Definition 2.2.4.

Definition 4.2.3. Let $(X, \mathcal{A})$ be a measurable space, let Y be a Banach space, and let μ be a signed measure on $\mathcal{A}$. A function $f : X \to Y$ is *Bochner integrable with respect to* μ if $f \in L^1(X, \mathcal{A}, \mu^{\pm}; Y)$. In such case, we set

$$\int_X f \, d\mu := \int_X f \, d\mu^+ - \int_X f \, d\mu^-.$$

The results of Propositions 4.2.2–4.2.3 are easily generalized to the case of Definition 4.2.3. In particular, inequality (4.38) is replaced by

$$\left\| \int_E f \, d\mu \right\|_Y \leq \int_E \|f\|_Y \, d\,|\mu| \quad \text{for all } E \in \mathcal{A},$$

which generalizes inequality (2.11).

4.2.2 Fubini theorem

Let us prove the following generalization of Theorem 2.3.3.

Theorem 4.2.8 (Fubini). *Let $(X_1, \mathcal{A}_1, \mu_1)$ and $(X_2, \mathcal{A}_2, \mu_2)$ be σ-finite measure spaces. Let $f \in L^1(X_1 \times X_2, \mathcal{A}_1 \times \mathcal{A}_2, \mu_1 \times \mu_2; Y)$. Then:*

(i) *$f_{x_1} \in L^1(X_2, \mathcal{A}_2, \mu_2; Y)$ for μ_1-a. e. $x_1 \in X_1$, and $f_{x_2} \in L^1(X_1, \mathcal{A}_1, \mu_1; Y)$ for μ_2-a. e. $x_2 \in X_2$;*

(ii) *the function*

$$X_1 \mapsto Y, \quad x_1 \to \int_{X_2} f_{x_1} d\mu_2, \tag{4.49}$$

belongs to $L^1(X_1, \mathcal{A}_1, \mu_1; Y)$, *and*

$$X_2 \to Y, \quad x_2 \mapsto \int_{X_1} f_{x_2} d\mu_1, \tag{4.50}$$

belongs to $L^1(X_2, \mathcal{A}_2, \mu_2; Y)$;

(iii) *we have*

$$\int_{X_1 \times X_2} f \, d(\mu_1 \times \mu_2) = \int_{X_1} d\mu_1 \int_{X_2} f_{x_1} \, d\mu_2 = \int_{X_2} d\mu_2 \int_{X_1} f_{x_2} \, d\mu_1. \tag{4.51}$$

Proof. (i) By assumption, f is $\mu_1 \times \mu_2$-measurable. Hence there exists a sequence $\{s_n\} \subseteq \mathscr{S}(X_1 \times X_2; Y)$, $s_n = s_n(x_1, x_2)$, such that $s_n \to f$ $(\mu_1 \times \mu_2)$-a. e. in $X_1 \times X_2$. Therefore, for μ_1-a. e. $x_1 \in X_1$, the sequence of x_1-traces $\{(s_n)_{x_1}\}$ converges to f_{x_1} μ_2-a. e. in X_2 as $n \to \infty$ (see Remark 2.3.2). Recall that for every $E \in \mathcal{A}_1 \times \mathcal{A}_2$, we have $E_{x_1} \in \mathcal{A}_2$ and that $\chi_E(x_1, \cdot) = \chi_{E_{x_1}}$ $(x_1 \in X_1)$ (see Remark 2.1.5). Hence $\{(s_n)_{x_1}\} \subseteq \mathscr{S}(X_2; Y)$, and thus f_{x_1} is μ_2-measurable for μ_1-a. e. $x_1 \in X_1$.

Now observe that by the Tonelli theorem and Proposition 4.2.1 we have

$$\int_{X_1 \times X_2} \|f\|_Y \, d(\mu_1 \times \mu_2) = \int_{X_1} d\mu_1 \int_{X_2} \|f_{x_1}\|_Y d\mu_2 < \infty,$$

and hence $\int_{X_2} \|f_{x_1}\|_Y d\mu_2 < \infty$ for μ_1-a. e. $x_1 \in X_1$. Then by Proposition 4.2.1 $f_{x_1} \in L^1(X_2, \mathcal{A}_2, \mu_2; Y)$ for μ_1-a. e. $x_1 \in X_1$. Similar arguments hold for f_{x_2}, and thus claim (i) follows.

(ii) Since $f \in L^1(X_1 \times X_2, \mathcal{A}_1 \times \mathcal{A}_2, \mu_1 \times \mu_2; Y)$ and $s_n \to f$ $(\mu_1 \times \mu_2)$-a. e. in $X_1 \times X_2$, by the dominated convergence theorem and Proposition 4.2.1 we get

$$\lim_{n\to\infty} \int_{X_1 \times X_2} \|s_n - f\|_Y \, d(\mu_1 \times \mu_2) = 0,$$

whence by the Tonelli theorem

$$\lim_{n\to\infty} \int_{X_1} d\mu_1 \int_{X_2} \|(s_n)_{x_1} - f_{x_1}\|_Y d\mu_2 = 0.$$

Therefore (possibly up to a subsequence, not relabeled for simplicity) we have that

$$\lim_{n\to\infty} \int_{X_2} \|(s_n)_{x_1} - f_{x_1}\|_Y d\mu_2 = 0 \quad \text{for } \mu_1\text{-a. e. } x_1 \in X_1,$$

whence by Proposition 4.2.2(i)

$$\lim_{n\to\infty} \int_{X_2} (s_n)_{x_1} d\mu_2 = \int_{X_2} f_{x_1} d\mu_2 \quad \text{for } \mu_1\text{-a. e. } x_1 \in X_1. \tag{4.52}$$

It is easily seen that for every $n \in \mathbb{N}$, the map $x_1 \mapsto \int_{X_2} (s_n)_{x_1} d\mu_2$ is μ_1-measurable. Indeed, let

$$s_n = \sum_{k=1}^{p_n} y_{n,k} \chi_{E_{n,k}} \quad \Rightarrow \quad (s_n)_{x_1} = \sum_{k=1}^{p_n} y_{n,k} \chi_{(E_{n,k})_{x_1}} \quad (k = 1, \dots, p_n)$$

with $y_{n,k} \in Y$ and $E_{n,k} \in \mathcal{A}_1 \times \mathcal{A}_2$. For all $n \in \mathbb{N}$ and $k = 1, \dots, p_n$, by the Tonelli theorem the real-valued function $x_1 \mapsto \int_{X_2} \chi_{(E_{n,k})_{x_1}} d\mu_2$ is $\mathcal{A}_1$-measurable, and thus plainly the map $x_1 \mapsto y_{n,k} \int_{X_2} \chi_{(E_{n,k})_{x_1}} d\mu_2 = \int_{X_2} y_{n,k} \chi_{(E_{n,k})_{x_1}} d\mu_2$ is μ_1-measurable (see (4.41)). Since the set of μ_1-measurable functions is a vector space, the claim follows.

By equality (4.52) and the above remarks the function defined by (4.49) is μ_1-measurable. Moreover, by Proposition 4.2.2(i) and the Tonelli theorem we have

$$\int_{X_1} d\mu_1 \left\| \int_{X_2} f_{x_1} d\mu_2 \right\|_Y \leq \int_{X_1} d\mu_1 \int_{X_2} \|f_{x_1}\|_Y d\mu_2 = \int_{X_1 \times X_2} \|f\|_Y \, d(\mu_1 \times \mu_2) < \infty,$$

and hence the function (4.49) belongs to $L^1(X_1, \mathcal{A}_1, \mu_1; Y)$. It is similarly seen that the function (4.50) belongs to $L^1(X_2, \mathcal{A}_2, \mu_2; Y)$, and thus claim (ii) follows.

(iii) By Corollary 4.2.4, for every $y^* \in Y^*$, we have

$$\left\langle y^*, \int_{X_1 \times X_2} f \, d(\mu_1 \times \mu_2) \right\rangle = \int_{X_1 \times X_2} \langle y^*, f \rangle \, d(\mu_1 \times \mu_2), \tag{4.53}$$

since $f \in L^1(X_1 \times X_2, \mathcal{A}_1 \times \mathcal{A}_2, \mu_1 \times \mu_2; Y)$;

$$\left\langle y^*, \int_{X_2} f_{x_1} \, d\mu_2 \right\rangle = \int_{X_2} \langle y^*, f_{x_1} \rangle \, d\mu_2 \quad \text{for } \mu_1\text{-a. e. } x_1 \in X_1, \tag{4.54}$$

since $f_{x_1} \in L^1(X_2, \mathcal{A}_2, \mu_2; Y)$ for μ_1-a. e. $x_1 \in X_1$; and

$$\left\langle y^*, \int_{X_1} d\mu_1 \int_{X_2} f_{x_1} \, d\mu_2 \right\rangle = \int_{X_1} d\mu_1 \left\langle y^*, \int_{X_2} f_{x_1} \, d\mu_2 \right\rangle, \tag{4.55}$$

since the function in (4.49) belongs to $L^1(X_1, \mathcal{A}_1, \mu_1; Y)$. From (4.53)–(4.55) and the Fubini theorem for real-valued functions (see the first equality in (2.18)), we get

$$\left\langle y^*, \int_{X_1 \times X_2} f \, d(\mu_1 \times \mu_2) \right\rangle = \left\langle y^*, \int_{X_1} d\mu_1 \int_{X_2} f_{x_1} \, d\mu_2 \right\rangle$$

for every $y^* \in Y^*$, and thus the first equality in (4.51) follows. The second equality is similarly proven, and hence the result follows. □

4.2.3 Integration with respect to vector measures

Conceptually related to the Bochner integral is the integration of scalar functions with respect to vector measures. Let us outline some results in this direction, while referring the reader to [41] for a more general treatment.

Let $(X, \mathcal{A})$ be a measurable space, let Y be a Banach space, and let $\mu : \mathcal{A} \to Y$ be a vector measure such that $|\mu|_w(X) < \infty$ (see Definition 1.9.3). Let $s \in \mathscr{S}(X)$, $s = \sum_{i=1}^n \alpha_i \chi_{E_i}$. The quantity

$$\int_X s \, d\mu := \sum_{i=1}^n \alpha_i \mu(E_i) \in Y \tag{4.56}$$

is called the *integral* of s.

Remark 4.2.5. If $Y = \mathbb{R}$, then μ is a finite signed measure, and equality (4.56) reads

$$\int_X s \, d\mu = \int_X s \, d\mu^+ - \int_X s \, d\mu^- \tag{4.57}$$

in agreement with Definition 2.2.4.

For any $y^* \in Y^*$, from (4.56) we get

$$\left\langle y^*, \int_X s \, d\mu \right\rangle = \sum_{i=1}^n \alpha_i \langle y^*, \mu(E_i) \rangle = \sum_{i=1}^n \alpha_i \langle y^*, \mu \rangle (E_i) = \int_X s \, d\langle y^*, \mu \rangle,$$

where $\langle y^*, \mu \rangle$ is the map defined in (1.65). It follows that if $\|s\|_\infty \neq 0$, then

$$\left| \left\langle y^*, \int_X s \, d\mu \right\rangle \right| \leq \sum_{i=1}^n |\alpha_i| \, |\langle y^*, \mu \rangle|(E_i) \leq \|s\|_\infty \sum_{i=1}^n \frac{|\alpha_i|}{\|s\|_\infty} |\langle y^*, \mu \rangle|(E_i) \leq \|s\|_\infty |\langle y^*, \mu \rangle|(X),$$

whence by (1.66)

$$\left\| \int_X s\, d\mu \right\|_Y \le \|s\|_\infty \, |\mu|_w(X). \tag{4.58}$$

Now let $f : X \to \mathbb{R}$ be measurable and bounded, and let $\{s_n\} \subseteq \mathscr{S}(X)$, $\|s_n - f\|_\infty \to 0$ as $n \to \infty$. By inequality (4.58)

$$\left\| \int_X s_m\, d\mu - \int_X s_n\, d\mu \right\|_Y \le \|s_m - s_n\|_\infty \, |\mu|_w(X)$$

for all $m, n \in \mathbb{N}$, and hence $\{\int_X s_n\, d\mu\}$ is a Cauchy sequence in Y. The limit

$$\lim_{n\to\infty} \int_X s_n\, d\mu =: \int_X f\, d\mu \tag{4.59}$$

is called the *integral* of f.

From (4.58)–(4.59) we obtain

$$\left\| \int_X f\, d\mu \right\|_Y \le \int_X |f|\, d|\mu|_w \le \|f\|_\infty \, |\mu|_w(X). \tag{4.60}$$

Remark 4.2.6. If $Y = \mathbb{R}$, then the integral defined in (4.59) coincides with that given by (2.10) (see (4.57)).

4.2.4 Weaker notions of integral

For clearness, in this subsection, we use the explicit duality symbols $\langle \cdot, \cdot \rangle_{Y^*,Y}$ and $\langle \cdot, \cdot \rangle_{Y^{**},Y^*}$. If Y is thought of as embedded in Y^{**}, then we set $\langle y, y^* \rangle_{Y^{**},Y^*} \equiv \langle y^*, y \rangle_{Y^*,Y}$. Let us prove the following result.

Proposition 4.2.9. (i) *Let $f : X \to Y$ be weakly measurable, and let the map $x \mapsto \langle y^*, f(x) \rangle_{Y^*,Y}$ belong to $L^1(X)$ for every $y^* \in Y^*$. Then for every $E \in \mathcal{A}$, there exists $y_E^{**} \in Y^{**}$ such that*

$$\langle y_E^{**}, y^* \rangle_{Y^{**},Y^*} = \int_E \langle y^*, f(x) \rangle_{Y^*,Y} d\mu(x) \quad \text{for all } y^* \in Y^*. \tag{4.61}$$

(ii) *Let $f : X \to Y^*$ be weakly* measurable, and let the map $x \mapsto \langle f(x), y \rangle_{Y^*,Y}$ belong to $L^1(X)$ for every $y \in Y$. Then for every $E \in \mathcal{A}$, there exists $y_E^* \in Y^*$ such that*

$$\langle y_E^*, y \rangle_{Y^*,Y} = \int_E \langle f(x), y \rangle_{Y^*,Y} d\mu(x) \quad \text{for all } y \in Y. \tag{4.62}$$

Proof. We only prove claim (i), the proof of (ii) being similar. For every $E \in \mathcal{A}$, set

$$T_E : Y^* \to L^1(X), \quad T_E y^* := \langle y^*, f\chi_E \rangle_{Y^*,Y} \quad (y^* \in Y^*).$$

We will prove that the linear operator T_E is closed and thus bounded since it is defined on the whole space Y^*. Then

$$\left| \int_E \langle y^*, f(x) \rangle_{Y^*,Y} d\mu(x) \right| \le \int_E |\langle y^*, f(x) \rangle_{Y^*,Y}| \, d\mu(x)$$
$$= \|T_E y^*\|_{L^1(X)} \le \|T_E\|_{\mathscr{L}(Y^*;L^1(X))} \|y^*\|_{Y^*}.$$

It follows that the map from Y^* to $\mathbb{R}$, $y^* \mapsto \int_E \langle y^*, f(x) \rangle_{Y^*,Y} d\mu(x)$, is linear and continuous. Hence claim (i) follows.

It remains to prove that T_E is closed. Let $\{y_n^*\} \subseteq Y^*$, $y^* \in Y^*$, and $g \in L^1(X)$ be such that $\|y_n^* - y^*\|_{Y^*} \to 0$ and $\|T_E y_n^* - g\|_{L^1(X)} \to 0$ as $n \to \infty$. Then by classical results there exists a subsequence $\{y_{n_k}^*\} \subseteq \{y_n^*\}$ such that $T_E y_{n_k}^* = \langle y_{n_k}^*, f\chi_E \rangle_{Y^*,Y} \to g$ μ-a. e. in X. On the other hand, since $y_n^* \to y^*$ strongly in Y^*, we have

$$\lim_{k\to\infty} \langle y_{n_k}^*, f\chi_E \rangle_{Y^*,Y} = \langle y^*, f\chi_E \rangle_{Y^*,Y} \quad \mu\text{-a. e. in } X.$$

It follows that $g = \langle y^*, f\chi_E \rangle_{Y^*,Y} = T_E y^*$. This completes the proof. □

Definition 4.2.4. Let the assumptions of Proposition 4.2.9 be satisfied, and let $E \in \mathcal{A}$. The element $y_E^{**} \in Y^{**}$ in (4.61) is called the *Dunford integral* of f on E. If y_E^{**} belongs to Y, that is, if there exists $y_E \in Y$ such that

$$\langle y^*, y_E \rangle_{Y^*,Y} = \int_E \langle y^*, f(x) \rangle_{Y^*,Y} d\mu(x) \quad \text{for all } y^* \in Y^*, \tag{4.63}$$

then the element $y_E \in Y$ is called the *Pettis integral* of f on E. If y_X^{**} belongs to Y, then the function f is called *Pettis integrable* on X.

The element $y_E^* \in Y^*$ in (4.62) is called the *Gelfand integral* of f on E.

To avoid confusion between the above notions, if necessary, we will use the following notation:

$$y_E^{**} \equiv \mathscr{D}\int_E f \, d\mu(x), \quad y_E \equiv \mathscr{P}\int_E f \, d\mu(x), \quad y_E^* \equiv \mathscr{G}\int_E f \, d\mu(x).$$

Hence equalities (4.61)–(4.63) read

$$\left\langle \mathscr{D}\int_E f \, d\mu, y^* \right\rangle_{Y^{**},Y^*} = \int_E \langle y^*, f(x) \rangle_{Y^*,Y} d\mu(x) \quad \text{for all } y^* \in Y^*; \tag{4.64a}$$

$$\left\langle y^*, \mathscr{P}\int_E f\,d\mu\right\rangle_{Y^*,Y} = \int_E \langle y^*, f(x)\rangle_{Y^*,Y}\,d\mu(x) \quad \text{for all } y^* \in Y^*; \tag{4.64b}$$

$$\left\langle \mathscr{G}\int_E f\,d\mu, y\right\rangle_{Y^*,Y} = \int_E \langle f(x), y\rangle_{Y^*,Y}\,d\mu(x) \quad \text{for all } y \in Y. \tag{4.64c}$$

Remark 4.2.7. (i) If f is Bochner integrable, then the Bochner integral and the Pettis integral of f coincide (see Corollary 4.2.4).

(ii) The Dunford integral and the Pettis integral of f coincide if X is reflexive; otherwise, this need not be the case (see [41, Example II.3.3]).

4.3 The spaces $L^p(X;Y)$, $L^p_w(X;Y)$ and $L^p_{w^*}(X;Y^*)$

4.3.1 Definition and general properties

Let $(X, \mathcal{A}, \mu)$ be a measure space, let Y be a Banach space with norm $\|\cdot\|_Y$, and let Y^* be the dual space of Y with norm $\|\cdot\|_{Y^*}$. Let $f : X \to Y$, and let $\|f\|_Y : X \to [0, \infty)$, $\|f\|_Y(x) := \|f(x)\|_Y$. For any $p \in [1, \infty]$, define

$$\begin{aligned} L^p(X;Y) &:= \{f : X \to Y \ \mu\text{-measurable} \mid \|f\|_Y \in L^p(X)\}; \\ L^p_w(X;Y) &:= \{f : X \to Y \ \text{weakly measurable} \mid \|f\|_Y \in L^p(X)\}; \\ L^p_{w^*}(X;Y^*) &:= \{f : X \to Y^* \ \text{weakly}^*\text{measurable} \mid \|f\|_{Y^*} \in L^p(X)\}. \end{aligned}$$

If necessary, the extended notations $L^p(X, \mathcal{A}, \mu; Y) \equiv L^p(X;Y)$ and $L^p_w(X, \mathcal{A}, \mu; Y)$, $L^p_{w^*}(X, \mathcal{A}, \mu; Y^*)$ will be used. We also set $L^p(X) \equiv L^p(X; \mathbb{R})$.

Remark 4.3.1. (i) The sets $L^p(X;Y)$, $L^p_w(X;Y)$, and $L^p_{w^*}(X;Y^*)$ are vector spaces. By Remark 4.1.3(i) we have $L^p(X;Y) \subseteq L^p_w(X;Y)$, with equality if μ is σ-finite and Y is separable (see Theorem 4.1.11). Similarly, by Remark 4.1.3(ii) we have $L^p(X;Y^*) \subseteq L^p_w(X;Y^*) \subseteq L^p_{w^*}(X;Y^*)$, with equality if μ is σ-finite and Y^* is separable (see Proposition 4.1.12).

(ii) The definition of $L^p(X;Y)$ and $L^p_w(X;Y)$ requires the map $x \mapsto \|f(x)\|_Y$ to be measurable. This is ensured by the μ-measurability (see Remark 4.1.2(iii)) or by the weak measurability of $f : X \to Y$ if Y is separable (see Proposition 4.1.9(i)). Similarly, the definition of $L^p_{w^*}(X;Y^*)$ requires the measurability of the map $x \mapsto \|f(x)\|_{Y^*}$, which is ensured by the weak* measurability of $f : X \to Y^*$ if Y is separable (see Propositions 4.1.10 and 4.1.9(ii)).

Remark 4.3.2. Let $p \in [1, \infty]$, and fix $\bar{y} \in Y$ with $\|\bar{y}\|_Y = 1$. It is easily seen that the map $T : L^p(X) \to L^p(X;Y)$, $Tf := f\bar{y}$ ($f \in L^p(X)$), is an isometric isomorphism between $L^p(X;Y)$ and its image. Hence the Lebesgue space $L^p(X)$ of real-valued functions is embedded in $L^p(X;Y)$.

Remark 4.3.3. (i) Let $U \subseteq \mathbb{R}^N$ be open. Consider the trace $\mathcal{B}^N \cap U$ of the Borel σ-algebra $\mathcal{B}^N \equiv \mathcal{B}(\mathbb{R}^N)$ and the induced Lebesgue measure $\lambda_N|_{\mathcal{B}^N \cap U}$. Generalizing Remark 2.8.1, we set $L^p(U;Y) \equiv L^p(U, \mathcal{B}^N \cap U, \lambda_N|_{\mathcal{B}^N\cap U}; Y)$ (similarly for $L^p_w(U;Y)$ and $L^p_{w^*}(U;Y^*)$). We will say that $f \in L^p_{\mathrm{loc}}(U;Y)$ if $f \in L^p(V, \mathcal{B}^N \cap V, \mu|_{\mathcal{B}^N \cap V})$ for every open subset $V \Subset U$.

(ii) If $f \in L^1_{\mathrm{loc}}(U;Y)$, then the map $\mu : \mathcal{B}^N \cap U \to Y$, $\mu(E) := \int_E f\,dx$ for any Borel subset $E \subseteq U$, is a vector measure (see Proposition 4.2.7).

Let $f : X \mapsto Y$ with $\|f\|_Y : X \to [0,\infty)$ measurable. Set

$$\|f\|_p := \left(\int_X \|f\|_Y^p \,d\mu\right)^{\frac{1}{p}} \quad \text{if } p \in [1,\infty), \tag{4.65a}$$

$$\|f\|_\infty := \operatorname{ess\,sup} \|f\|_Y. \tag{4.65b}$$

Similarly, for any $f : X \to Y^*$ with $\|f\|_{Y^*} : X \to [0,\infty)$ measurable, set

$$\|f\|_p^* := \left(\int_X \|f\|_{Y^*}^p \,d\mu\right)^{\frac{1}{p}} \quad \text{if } p \in [1,\infty), \tag{4.66a}$$

$$\|f\|_\infty^* := \operatorname{ess\,sup} \|f\|_{Y^*}. \tag{4.66b}$$

The map $f \mapsto \|f\|_p$ is a seminorm on $L^p(X;Y)$ and $L^p_w(X;Y)$, and $f \mapsto \|f\|_p^*$ is a seminorm on $L^p_{w^*}(X;Y)$ ($p \in [1,\infty]$). As in the case $Y = \mathbb{R}$, we identify functions that are equal μ-a. e. in X. Then $L^p(X;Y)$ and $L^p_w(X;Y)$ are normed vector spaces with norm $\|\cdot\|_p$, and $L^p_{w^*}(X;Y^*)$ is a normed vector space with norm $\|\cdot\|_p^*$.

Definition 4.3.1. Let $p \in [1,\infty]$. The normed vector spaces $L^p(X;Y)$, $L^p_w(X;Y)$, and $L^p_{w^*}(X;Y^*)$ are called *vector Lebesgue spaces.*

If no confusion arises, then vector Lebesgue spaces are simply called "Lebesgue spaces" for shortness.

Remark 4.3.4. If μ is σ-finite, then $L^p(X;Y)$ is a closed subspace of $L^p_w(X;Y)$ ($p \in [1,\infty]$; see Remark 4.3.1(i)). Indeed, let $\{f_n\} \subseteq L^p(X;Y)$ and $f \in L^p_w(X;Y)$, and let $\|f_n - f\|_p \to 0$ as $n \to \infty$. Then the sequence of real-valued functions $g_n := \|f_n - f\|_Y$ converges to 0 in $L^p(X)$ as $n \to \infty$, and thus there exists a subsequence $\{g_{n_k}\} \subseteq \{g_n\}$ such that $g_{n_k}(x) = \|f_{n_k}(x) - f(x)\|_Y \to 0$ as $k \to \infty$ for μ-a. e. $x \in X$. Since every f_{n_k} is μ-measurable, by Proposition 4.1.2 f is μ-measurable as well, and thus $f \in L^p(X;Y)$.

Proposition 4.3.1. *Let μ be finite, and let $1 \le p < r \le \infty$. Then $L^r(X;Y) \subseteq L^p(X;Y)$. Similarly, $L^r_w(X;Y) \subseteq L^p_w(X;Y)$ and $L^r_{w^*}(X;Y^*) \subseteq L^p_{w^*}(X;Y^*)$.*

Proof. If $r = \infty$, then the claim follows from the inequality $\|f\|_p \le [\mu(X)]^{\frac{1}{p}} \|f\|_\infty$. If $r \in [1,\infty)$, then by the Hölder inequality we have $\|f\|_p \le [\mu(X)]^{\frac{r-p}{pr}} \|f\|_r$. Hence the result follows. □

Theorem 4.3.2. *Let $p \in [1,\infty]$. Then $L^p(X;Y)$ is a Banach space.*

Proof. We only prove the result for $p \in [1,\infty)$; the proof for $p = \infty$ is the same as in the case $Y = \mathbb{R}$, replacing the absolute value in $\mathbb{R}$ by the norm $\|\cdot\|_Y$ and using Proposition 4.1.2.

Let $\{f_k\} \subseteq L^p(X;Y)$ satisfy $\sum_{k=1}^{\infty}\|f_k\|_p =: M < \infty$, and set $S_n := \sum_{k=1}^{n} f_k$ $(n \in \mathbb{N})$. By a well-known characterization of Banach spaces (e. g., see [66, Theorem 8.1]) it suffices to prove that there exists $S \in L^p(X;Y)$ such that $\lim_{n\to\infty}\|S_n - S\|_p = 0$.

For all $n \in \mathbb{N}$ and μ-a. e. $x \in X$, define $F_n : X \to [0,\infty]$, $F_n(x) := \sum_{k=1}^{n}\|f_k(x)\|_Y$. Hence by Remark 4.1.2(iii) and Corollary 2.1.5 F_n is $\mathcal{A}$-measurable, and

$$\|F_n\|_{L^p(X)} \le \sum_{k=1}^{n}\|f_k\|_p \le M \quad (n \in \mathbb{N}).$$

Then by Corollary 2.1.5 (see also Remark 2.1.2) the function $F : X \to [0,\infty]$,

$$F(x) := \lim_{n\to\infty} F_n(x) = \sum_{k=1}^{\infty}\|f_k(x)\|_Y,$$

is $\mathcal{A}$-measurable, and by the Fatou lemma we have

$$\int_X F^p\,d\mu \le \liminf_{n\to\infty}\int_X (F_n)^p\,d\mu = \liminf_{n\to\infty}\|F_n\|^p_{L^p(X)} \le M^p < \infty. \tag{4.67}$$

By (4.67) the function F is finite μ-a. e. in X, that is, the series $\sum_{k=1}^{\infty}\|f_k(x)\|_Y$ is convergent for μ-a. e. $x \in X$. As a consequence, for μ-a. e. $x \in X$ and for any $\epsilon > 0$, there exists $\bar{n} \in \mathbb{N}$ such that for all $n > \bar{n}$ and $p \in \mathbb{N}$,

$$\|S_{n+p}(x) - S_n(x)\|_Y = \left\|\sum_{k=n+1}^{n+p} f_k(x)\right\|_Y \le \sum_{k=n+1}^{n+p}\|f_k(x)\|_Y < \epsilon.$$

Hence for μ-a. e. $x \in X$, $\{S_n(x)\} \subseteq Y$ $(n \in \mathbb{N})$ is a Cauchy sequence. Since Y is a Banach space, this implies that for μ-a. e. $x \in X$, there exists $\lim_{n\to\infty} S_n(x) \in Y$.

Define

$$S : X \to Y, \quad S(x) := \lim_{n\to\infty} S_n(x) \quad \text{for } \mu\text{-a. e. } x \in X. \tag{4.68}$$

Then by Proposition 4.1.2 S is μ-measurable. Moreover, for μ-a. e. $x \in X$,

$$\|S(x)\|_Y = \lim_{n\to\infty}\|S_n(x)\|_Y \le \lim_{n\to\infty} F_n(x) = F(x). \tag{4.69}$$

By (4.67) and (4.69) $S \in L^p(X;Y)$. Moreover, by the Fatou lemma

$$\|S_n - S\|_p^p = \int_X \|S_n(x) - S(x)\|_Y^p\,d\mu \le \liminf_{m\to\infty}\int_X \left\|\sum_{k=n+1}^{m} f_k(x)\right\|_Y^p d\mu$$

$$\leq \liminf_{m\to\infty} \int_X \left(\sum_{k=n+1}^{m} \|f_k(x)\|_Y \right)^p d\mu \leq \int_X \left(\sum_{k=n+1}^{\infty} \|f_k(x)\|_Y \right)^p d\mu$$
$$= \int_X [F(x) - F_n(x)]^p \, d\mu \quad (n \in \mathbb{N}). \tag{4.70}$$

By (4.67) and the dominated convergence theorem, letting $n \to \infty$ in the above inequality proves the result. □

Remark 4.3.5. Let $p \in [1, \infty]$. By Theorem 4.3.2 and Remark 4.3.1(i), $L^p_w(X; Y)$ is a Banach space if μ is σ-finite and Y is separable. The same holds for $L^p_{w^*}(X; Y^*)$ if μ is σ-finite and Y^* is separable.

4.3.2 Spaces of continuous functions

Let $U \subseteq \mathbb{R}^N$ be open. We denote by $C(U; Y)$ the vector space of continuous functions $f : U \to Y$ and by $C_b(U; Y) \subseteq L^\infty(U; Y)$ the closed subspace of bounded continuous functions. We set $C(U; \mathbb{R}) \equiv C(U)$ and similarly for its subspaces.

Definition 4.3.2. Let $U \subseteq \mathbb{R}^N$ be open, and let $\alpha \in (0, 1)$. The space of *Hölder functions of exponent α* is

$$C^\alpha(\overline{U}; Y) := \{f : U \to Y \mid \exists C > 0 \text{ such that } \|f(x) - f(y)\|_Y \leq C|x - y|^\alpha \;\; \forall x, y \in U\}.$$

Every Hölder function is uniformly continuous and thus bounded in U. The space $C^\alpha(\overline{U}; Y)$ endowed with the norm

$$f \mapsto \|f\|_{C^{0,\alpha}(U;Y)} := \|f\|_\infty + \sup_{\substack{x,y\in U\\ x\neq y}} \frac{\|f(x) - f(y)\|_Y}{|x - y|^\alpha}$$

is a Banach space. Formally, for $\alpha = 1$, we obtain the Banach space of *Lipschitz continuous functions* $f : U \to Y$ endowed with the norm

$$f \mapsto \|f\|_{\mathrm{Lip}(U;Y)} := \|f\|_\infty + \sup_{\substack{x,y\in U\\ x\neq y}} \frac{\|f(x) - f(y)\|_Y}{|x - y|}.$$

If U is bounded, then for any $0 < \alpha < \beta < 1$,

$$\mathrm{Lip}(\overline{U}; Y) \hookrightarrow C^\beta(\overline{U}; Y) \hookrightarrow C^\alpha(\overline{U}; Y) \hookrightarrow C_b(U; Y).$$

For any $m \in \mathbb{N}$, we denote by $C^m(U; Y)$ the vector space of maps $f : U \to Y$ that are m times Fréchet differentiable in U. The mth Fréchet derivative $f^{(m)}(x_0)$ of f at any point $x_0 \in U$ is an element of

$$\underbrace{\mathscr{L}(U;\mathscr{L}(U;\ldots\mathscr{L}}_{m-1\text{ times}}(U;Y))\ldots) \sim \mathscr{L}_m(U;Y),$$

where $\mathscr{L}_m(U;Y)$ denotes the space of linear continuous maps from U^m to Y. For any $m \in \mathbb{N}$, there is a natural isometry between $\mathscr{L}_m(U;Y)$ and Y, and thus $f^{(m)} : U \to \mathscr{L}_m(U;Y)$ can be identified with a map from U to Y. We denote by $D^\alpha f \equiv \frac{\partial^{|\alpha|} f}{\partial^{\alpha_1} x_1 \ldots \partial^{\alpha_n} x_n}$ ($\alpha \equiv (\alpha_1,\ldots,\alpha_n)$ with integer α_k, $|\alpha| := \sum_{k=1}^N \alpha_k$) any partial derivative of f of order $|\alpha|$, and thus $D^\alpha f(x_0) \in Y$ for any $x_0 \in U$. We set $D^0 f \equiv f$, $C^0(U;Y) \equiv C(U;Y)$, $C^\infty(U;Y) := \bigcap_{m=0}^\infty C^m(U;Y)$, and $C_c^\infty(U;Y) := C^\infty(U;Y) \bigcap C_c(U;Y)$. The space $C^m(U;Y)$ is a Banach space with norm $\|f\|_{C^m(U;Y)} := \sum_{|\alpha|\le m} \|D^\alpha f\|_\infty$.

4.3.3 Convergence theorems

The following concepts and results generalize some of those in Subsection 2.8.3. The proofs are an easy adaptation of those for the case $Y = \mathbb{R}$ and thus are omitted.

Definition 4.3.3. Let $p \in [1,\infty)$. A subset $\mathscr{F} \subseteq L^p(X;Y)$ is called *p-uniformly integrable* if for any $\epsilon > 0$:
(i) there exists $E \in \mathcal{A}$ with $\mu(E) < \infty$ such that $\sup_{f\in\mathscr{F}} \int_{E^c} \|f\|_Y^p \, d\mu < \epsilon$;
(ii) there exists $\delta > 0$ such that $\sup_{f\in\mathscr{F}} \int_F \|f\|_Y^p \, d\mu < \epsilon$ for all $F \in \mathcal{A}$ with $\mu(F) < \delta$.

We say that $\mathscr{F} \subseteq L^1(X;Y)$ is *uniformly integrable* if it is 1-uniformly integrable.

Remark 4.3.6. Definition 4.3.3 amounts to require that the set $\tilde{\mathscr{F}} := \{\|f\|_Y \mid f \in \mathscr{F}\} \subseteq L^p(X)$ is p-uniformly integrable. Hence the criteria of uniform integrability given by Propositions 2.8.4–2.8.6 can be used in the present case.

Theorem 4.3.3 (Vitali). *Let $p \in [1,\infty)$. Let $\{f_n\} \subseteq L^p(X;Y)$ be a sequence converging μ-a. e. in X to some function $f : X \to Y$. Then the following statements are equivalent:*
(i) *$f \in L^p(X;Y)$ and $f_n \to f$ in $L^p(X;Y)$;*
(ii) *the sequence $\{f_n\}$ is p-uniformly integrable.*

Theorem 4.3.4 (Dominated convergence theorem). *Let $p \in [1,\infty)$. Let $\{f_n\} \subseteq L^p(X;Y)$ be a sequence converging μ-a. e. in X to some function $f : X \to Y$. Let there exist $g \in L^p(X)$ such that $\|f_n\|_Y \le g$ μ-a. e. in X for all $n \in \mathbb{N}$. Then $f_n \to f$ in $L^p(X;Y)$.*

Corollary 4.3.5. *Let $\mu(X) < \infty$, and let $1 \le p < r < \infty$. Let $\{f_n\} \subseteq L^r(X;Y)$ be a bounded sequence, converging μ-a. e. in X to some function $f : X \to Y$. Then $f_n \to f$ in $L^p(X;Y)$.*

4.3.4 Approximation results and separability

Proposition 4.3.6. *Let $p \in [1,\infty)$.*
(i) *The set $\mathscr{S}(X;Y)$ of simple functions is dense in $L^p(X;Y)$.*
(ii) *Let X be a σ-compact Hausdorff space, and let $\mu \in \mathfrak{R}^+(X)$. If $D \subseteq Y$ is dense in Y, then the set*

$$\mathscr{E}_D(X;Y) := \left\{ \sum_{j=1}^{m} y_j g_j \;\middle|\; y_j \in D,\, g_j \in C_c(X),\, m \in \mathbb{N} \right\} \tag{4.71}$$

is dense in $L^p(X;Y)$.

Proof. (i) Let $f \in L^p(X;Y)$. By definition there exists $\{s_n\} \subseteq \mathscr{S}(X;Y)$ such that $\lim_{n\to\infty} \|s_n(x) - f(x)\|_Y = 0$ for μ-a. e. $x \in X$. Set $t_n := s_n\chi_{E_n}$, where $E_n := \{x \in X \mid \|s_n(x)\|_Y \leq 2\|f(x)\|_Y\}$ $(n \in \mathbb{N})$. Then $\lim_{n\to\infty} \|t_n(x) - f(x)\|_Y = 0$ for μ-a. e. $x \in X$ and $\|t_n - f\|_Y \leq 3\|f\|_Y \in L^p(X)$, whence $\lim_{n\to\infty} \int_X \|t_n - f\|_Y^p \, d\mu = 0$ by the dominated convergence theorem. Hence claim (i) follows.

(ii) In view of (i), it suffices to show that for any $s \in \mathscr{S}(X;Y)$, there exists a sequence $\{f_n\} \subseteq \mathscr{E}_D(X;Y)$ that converges to s in $L^p(X;Y)$. This follows if we show that, for all $E \in \mathcal{B}(X)$ and $y \in Y$, $y\chi_E$ can be approximated in $L^p(X;Y)$ by elements of $\mathscr{E}_D(X;Y)$. Moreover, since μ is σ-finite, it suffices to assume that $\mu(E) < \infty$.

Let $E \in \mathcal{B}(X)$, $\mu(E) < \infty$, and let $y \in Y$ be fixed. By Lemma 2.8.3 there exists $\{g_n\} \subseteq C_c(X)$ such that $g_n \to \chi_E$ in $L^p(X)$. Moreover, by the denseness of D in Y there exists $\{y_n\} \subseteq D$ such that $\|y_n - y\|_Y \to 0$. Then for any $n \in \mathbb{N}$ sufficiently large,

$$\begin{aligned}\|y_n g_n - y\chi_E\|_p &\leq \|(y_n - y)\chi_E\|_p + \|y_n(g_n - \chi_E)\|_p \\ &\leq \|y_n - y\|_Y [\mu(E)]^{\frac{1}{p}} + 2\|y\|_Y \|g_n - \chi_E\|_{L^p(X)}.\end{aligned}$$

Letting $n \to \infty$ in the above inequality, the conclusion follows. □

Theorem 4.3.7. *Let X be a σ-compact Hausdorff space, let $\mu \in \mathfrak{R}^+(X)$, and let Y be separable. Then for every $p \in [1,\infty)$, the space $L^p(X;Y)$ is separable.*

Proof. By assumption there exists a nondecreasing sequence of compact subsets K_n such that $X = \bigcup_{n=1}^{\infty} K_n$. By the dominated convergence theorem, for any $f \in L^p(X;Y)$, the sequence $\{f\chi_{K_n}\}$ converges to f in $L^p(X;Y)$. Hence it suffices to prove the result when X is compact.

Let X be compact. By assumption there exists $D \subseteq Y$ countable and dense, and by Proposition 4.3.6 the corresponding set $\mathscr{E}_D(X;Y)$ (see (4.71)) is dense in $L^p(X;Y)$. On the other hand, by the Stone–Weierstrass theorem the countable set

$$\hat{\mathscr{E}}_D(X;Y) := \left\{ \sum_{j=1}^{m} y_j p_{j,k} \mid y_j \in D,\ p_{j,k} \text{ polynomial with rational coefficients; } k, m \in \mathbb{N} \right\}$$

is dense in $\mathscr{E}_D(X;Y)$ in the $\|\cdot\|_\infty$ norm, and thus in $L^p(X;Y)$ for all $p \in [1,\infty)$ since $\mu(X) < \infty$. Then the conclusion follows. □

Remark 4.3.7. It is well known that $L^\infty(X)$ is not separable, and hence by Remark 4.3.2 the same holds for $L^\infty(X;Y)$.

Using the denseness of $\mathscr{S}(X;Y)$ in $L^1(X;Y)$ (see Proposition 4.3.6(i)), we can prove the following generalization of Lemma 2.5.1. The proof is the same, and thus we omit it.

Proposition 4.3.8. *Let $(X,\mathcal{A},\mu)$ be a measure space, and let Y be a Banach space. Let $(X',\mathcal{A}')$ be a measurable space, let $f : X \to X'$ be $(\mathcal{A},\mathcal{A}')$-measurable, and let $\mu_f : \mathcal{A}' \to \overline{\mathbb{R}}_+$ be the image of μ under f. Then for any $g \in L^1(X',\mathcal{A}',\mu_f;Y)$, we have $g \circ f \in L^1(X,\mathcal{A},\mu;Y)$ and*

$$\int_F g\,d\mu_f = \int_{f^{-1}(F)} g \circ f\,d\mu \quad \text{for all } F \in \mathcal{A}'. \tag{4.72}$$

4.4 Duality of vector Lebesgue spaces

4.4.1 The Radon–Nikodým property

Let $(X,\mathcal{A},\mu)$ be a measure space, let Y be a Banach space, and let $f : X \to Y$ be Bochner integrable. By Proposition 4.2.7 the map $\nu : \mathcal{A} \to Y$, $\nu(E) := \int_E f\,d\mu$ is a vector measure of bounded variation, and $\nu \ll \mu$. As already observed (see Remark 4.2.4), it is natural to wonder if the converse is true, namely:

(Q_1) Let $\nu : \mathcal{A} \to Y$ be a vector measure of bounded variation such that $\nu \ll \mu$. Does there exist a Bochner-integrable function $f : X \to Y$ such that $\nu(E) := \int_E f\,d\mu$ for all $E \in \mathcal{A}$?

Stated equivalently, the question is whether the Radon–Nikodým theorem (see Theorem 2.9.1) can be extended to vector-valued measures. If a function f as in (Q_1) exists, then it is called the *Bochner density* of the vector measure ν.

A related interesting question can be raised. If the measure μ is σ-finite, by Theorem 2.8.14 $(L^1(X))^* = L^\infty(X)$, that is, for every continuous linear map $T \in \mathscr{L}(L^1(X);\mathbb{R})$, there exists a unique function $g \in L^\infty(X)$ such that

$$\langle T,f\rangle = \int_X fg\,d\mu \quad \text{for all } f \in L^1(X)$$

(we always denote by $\langle\cdot,\cdot\rangle$ the duality map between any Banach space and its dual space). It is natural to ask if this result can be generalized to vector-valued maps, namely:

(Q_2) Let $T \in \mathscr{L}(L^1(X); Y)$. Does there exist $g \in L^\infty(X; Y)$ such that $Tf = \int_X fg\, d\mu$ for all $f \in L^1(X)$?

In general, the answer to both above questions is negative, as the following examples show (see [41, Examples III-1, III-1']).

Example 4.4.1. Let $(X, \mathcal{A}, \mu) = (I, \mathcal{B}\cap I, \lambda|_{\mathcal{B}\cap I})$, where $I \equiv (0,1)$. For all $E \in \mathcal{A}$ and $n \in \mathbb{N}$, set

$$\nu_n(E) := \int_E \sin(2^n \pi x) dx. \tag{4.73}$$

Then $|\nu_n(E)| \leq \lambda(E) = 1$ for all $n \in \mathbb{N}$, and $\lim_{n\to\infty} \nu_n(E) = 0$ by the Riemann–Lebesgue lemma.

Let $c_0 := \{\{a_n\} \subseteq \mathbb{R} \mid \lim_{n\to\infty} a_n = 0\}$, $\|\{a_n\}\|_{c_0} := \sup_{n\in\mathbb{N}} |a_n|$. Then the map

$$\nu : \mathcal{A} \to c_0, \quad \nu(E) := \{\nu_n(E)\} \quad (E \in \mathcal{A}) \tag{4.74}$$

is well defined, and $\|\nu_n(E)\|_{c_0} \leq \lambda(E)$ for all $E \in \mathcal{A}$ and $n \in \mathbb{N}$. Hence ν is a vector measure of bounded variation (in fact, $|\nu|(I) \leq 1$; see Definition 1.9.2), and $\nu \ll \lambda$. Let there exist Bochner-integrable $g : I \to c_0$ such that

$$\nu(E) = \int_E g(x)\, dx \quad \text{for all } E \in \mathcal{A}. \tag{4.75}$$

Then $g \equiv \{g_n\}$ with $g_n : I \to \mathbb{R}$, $g_n(x) = \sin(2^n \pi x)$ for a. e. $x \in I$ (see (4.73)–(4.75)). Set

$$E_0 := \bigcap_{k=1}^{\infty} \bigcup_{n=k}^{\infty} E_n, \quad E_n := \left\{x \in I \,\middle|\, |\sin(2^n \pi x)| \geq \frac{\sqrt{2}}{2}\right\} \quad (n \in \mathbb{N}).$$

Observe that $\lambda(E_n) = \frac{1}{2}$ for every n, and thus

$$\lambda(E_0) = \lim_{k\to\infty} \lambda\left(\bigcup_{n=k}^{\infty} E_n\right) \geq \lim_{k\to\infty} \lambda(E_k) = \frac{1}{2}.$$

Now fix $x \in E_0$. By definition, for any $k \in \mathbb{N}$, there exists $n \geq k$ such that $x \in E_n$, and thus $|\sin(2^n \pi x)| = |g_n(x)| \geq \frac{\sqrt{2}}{2}$. Hence the sequence $\{g_n(x)\}$ cannot converge to 0 as $n \to \infty$, and thus $g(x) \notin c_0$ for a. e. $x \in E_0$. Since $\lambda(E_0) > 0$, we have a contradiction with (4.75). Hence for $Y = c_0$, the answer to question (Q_1) is negative.

Example 4.4.2. Let $(X, \mathcal{A}, \mu)$ be as in Example 4.4.1. Set

$$T : L^1(I) \to c_0, \quad Tf := \int_I f(x) \sin(2^n \pi x) dx \quad (f \in L^1(I)). \tag{4.76}$$

By the Riemann–Lebesgue lemma the definition is well posed. The map T is linear and continuous:

$$\|Tf\|_{c_0} \leq \sup_{n \in \mathbb{N}} \int_I |f(x) \sin(2^n \pi x)| \, dx \leq \|f\|_1.$$

By absurd, let there exist $g \in L^\infty(I; c_0)$ such that $Tf = \int_I fg \, dx$ for all $f \in L^1(I)$. In particular, $T\chi_E = \int_E g \, dx = \nu(E)$ for all $E \in \mathcal{A}$, where ν is the vector measure defined in (4.73)–(4.74). However, by Example 4.4.1 no Bochner integrable function g exists such that this equality holds, a contradiction. Hence for the operator defined in (4.76), the answer to question (Q_2) is negative.

The above examples motivate the following definitions.

Definition 4.4.1. (i) Let $(X, \mathcal{A}, \mu)$ be a σ-finite measure space. A Banach space Y *has the Radon–Nikodým property with respect to* $(X, \mathcal{A}, \mu)$ if for any vector measure $\nu : \mathcal{A} \to Y$ of bounded variation and absolutely continuous with respect to μ, there exists $g \in L^1(X; Y)$ such that $\nu(E) = \int_E g \, d\mu$ for all $E \in \mathcal{A}$. The map g is called the *Bochner density* of the vector measure ν.

(ii) A Banach space Y *has the Radon–Nikodým property* if it has the Radon–Nikodým property with respect to every σ-finite measure space $(X, \mathcal{A}, \mu)$.

Definition 4.4.2. Let $(X, \mathcal{A}, \mu)$ be a σ-finite measure space. We say that an operator $T \in \mathscr{L}(L^1(X); Y)$ is *Riesz representable* (or simply *representable*) if there exists $g \in L^\infty(X; Y)$ such that $Tf = \int_X fg \, d\mu$ for all $f \in L^1(X)$.

Weak and weak* versions of the Radon–Nikodým property involving the Pettis integral instead of the Bochner integral are considered in [101, Definitions (7-1-3)–(7-1-7)].

A simple case where the Radon–Nikodým property holds is given by the following proposition.

Proposition 4.4.1. *Let $(X, \mathcal{A}, \mu)$ be a measure space with σ-finite and purely atomic μ. Then every Banach space Y has the Radon–Nikodým property with respect to $(X, \mathcal{A}, \mu)$.*

Proof. Let $\{F_n\} \subseteq \mathcal{A}$ be a disjoint sequence of μ-atoms such that $\bigcup_{n=1}^\infty F_n \subseteq X$ and $\mu = \mu \llcorner (\bigcup_{n=1}^\infty F_n)$ (see Definition 1.8.9(iii)). Let $\nu : \mathcal{A} \to Y$ be a vector measure of bounded variation and absolutely continuous with respect to μ. Set

$$g : X \to Y, \quad g := \sum_{n=1}^\infty \frac{\nu(F_n)}{\mu(F_n)} \chi_{F_n}.$$

Clearly, we have

$$\int_X \|g\|_Y \, d\mu = \sum_{n=1}^{\infty} \|\nu(F_n)\|_Y \leq |\nu|(X) < \infty.$$

By the definition of a μ-atom, for all $E \in \mathcal{A}$ and $n \in \mathbb{N}$, either $\mu(E \cap F_n) = 0$, and thus $\nu(E \cap F_n) = 0$, or $\mu(E \cap F_n) = \mu(F_n)$, which in turn implies $\nu(F_n) = \nu(E \cap F_n)$. Therefore, for all $E \in \mathcal{A}$,

$$\int_E g \, d\mu = \sum_{n=1}^{\infty} \frac{\nu(F_n)}{\mu(F_n)} \mu(E \cap F_n) = \sum_{n=1}^{\infty} \nu(E \cap F_n) = \nu(E).$$

Hence the result follows. □

Let us prove a necessary condition for the Radon–Nikodým property to hold.

Proposition 4.4.2. *If Y has the Radon–Nikodým property, then the same holds for each its closed subspace.*

Proof. Let Y_1 be a closed subspace of Y. We only prove that Y_1 has the Radon–Nikodým property with respect to every finite measure space $(X, \mathcal{A}, \mu)$ (the general case where μ is σ-finite follows by usual arguments).

Let $\nu : \mathcal{A} \to Y_1 \subseteq Y$ be a vector measure of bounded variation such that $\nu \ll \mu$. By assumption there exists $g \in L^1(X; Y)$ such that $\nu(E) := \int_E g \, d\mu$ for all $E \in \mathcal{A}$. For every $n \in \mathbb{N}$, let $\{F_1, \dots, F_n\}$ be a partition of X with $\mu(F_i) \neq 0$ for all $i = 1, \dots, n$. Set

$$P_n : L^1(X; Y) \mapsto L^1(X; Y), \quad P_n f := \sum_{i=1}^{n} \frac{\int_{F_i} f \, d\mu}{\mu(F_i)} \chi_{F_i}.$$

Clearly, $\|P_n f\|_1 \leq \sum_{i=1}^{n} \int_{F_i} \|f\|_Y \, d\mu = \|f\|_1$, and thus P_n is a bounded linear operator with norm $\|P_n\| \leq 1$ $(n \in \mathbb{N})$. Moreover, let $s \in \mathscr{S}(X; Y)$, $s = \sum_{j=1}^{m} y_j \chi_{E_j}$ with $y_j \in Y$ and $\{E_1, \dots, E_m\}$ a partition of X. Therefore, for all $n \in \mathbb{N}$,

$$P_n s = \sum_{i=1}^{n} \sum_{j=1}^{m} \frac{\mu(E_j \cap F_i)}{\mu(F_i)} y_j \chi_{F_i}.$$

By the above remarks the sequence $\{P_n\} \subseteq \mathscr{L}(L^1(X; Y); L^1(X; Y))$ is bounded, and it is easily checked that

$$\lim_{n \to \infty} \|P_n s - s\|_1 = 0 \quad \text{for all } s \in \mathscr{S}(X; Y).$$

By the denseness of $\mathscr{S}(X; Y)$ in $L^1(X; Y)$ it follows that

$$\lim_{n \to \infty} \|P_n f - f\|_1 = 0 \quad \text{for all } f \in L^1(X; Y).$$

In particular, this holds for $f = g$, and thus there exists a subsequence of $\{P_n g\}$ (not relabeled for simplicity) such that

$$P_n g = \sum_{i=1}^{n} \frac{\int_{F_i} g \, d\mu}{\mu(F_i)} \chi_{F_i} = \sum_{i=1}^{n} \frac{\nu(F_i)}{\mu(F_i)} \chi_{F_i} \to g \quad \mu\text{-a. e. in } X \text{ as } n \to \infty.$$

Since by assumption $\nu(F_i) \in Y_1$ for $i = 1, \dots, n$ and Y_1 is a closed subspace of Y, it follows that $g(x) \in Y_1$ for μ-a. e. $x \in X$. Hence $g \in L^1(X; Y_1)$, and thus the conclusion follows. □

A partial converse of Proposition 4.4.2 holds (see [41, Theorem III.3.2]).

Proposition 4.4.3. *If every closed separable subspace of a Banach space Y has the Radon–Nikodým property, then the same holds for Y.*

The following result plays an important role in the sequel (see Theorem 4.4.12).

Theorem 4.4.4 (Dunford–Pettis). *Every separable dual space has the Radon–Nikodým property.*

Proof. Let $\nu : \mathcal{A} \to Y^*$ be a vector measure of bounded variation and absolutely continuous with respect to μ. We will prove that there exists $f \in L^\infty(X, \mathcal{A}, |\nu|; Y^*)$ such that

$$\nu(E) := \int_E f \, d|\nu| \quad \text{for all } E \in \mathcal{A} \tag{4.77}$$

(by Proposition 4.3.1 $L^\infty(X, \mathcal{A}, |\nu|; Y^*) \subseteq L^1(X, \mathcal{A}, |\nu|; Y^*)$, since $|\nu|$ is finite). On the other hand, since $\nu \ll \mu$ implies $|\nu| \ll \mu$, it is easily seen that $f \in L^\infty(X, \mathcal{A}, \mu; Y^*)$, and by the Radon–Nikodým theorem $|\nu|(E) := \int_E \frac{d|\nu|}{d\mu} \, d\mu$ $(E \in \mathcal{A})$ with $\frac{d|\nu|}{d\mu} \in L^1(X, \mathcal{A}, \mu)$. Then $g := f \frac{d|\nu|}{d\mu}$ belongs to $L^1(X; Y^*) \equiv L^1(X, \mathcal{A}, \mu; Y^*)$, and $\nu(E) = \int_E g \, d\mu$ for all $E \in \mathcal{A}$. Hence by Remark 2.9.1(i) the result follows.

Let $y \in Y$ be fixed. Define

$$\nu_y : \mathcal{A} \to \mathbb{R}, \quad \nu_y(E) := \langle \nu(E), y \rangle \quad \text{for } E \in \mathcal{A}. \tag{4.78}$$

Clearly, ν_y is a finite signed measure such that

$$|\nu_y(E)| \le \|\nu(E)\|_{Y^*} \|y\|_Y \le |\nu|(E) \|y\|_Y \quad \text{for all } y \in Y \text{ and } E \in \mathcal{A}. \tag{4.79}$$

Hence $\nu_y \ll |\nu|$, and thus by Theorem 2.9.1 there exists $f_y \in L^1(X, \mathcal{A}, |\nu|)$ such that

$$\nu_y(E) = \int_E f_y \, d|\nu| \quad \text{for all } y \in Y \text{ and } E \in \mathcal{A}. \tag{4.80}$$

By Proposition 4.2.5 and inequality (4.79), for $|\nu|$-a. e. $x_0 \in X$, we have

$$f_y(x_0) = \lim_{r\to 0^+} \frac{1}{|\nu|(B_r(x_0))}\left|\int_{B_r(x_0)} f_y\, d|\nu|\right| \le \|y\|_Y,$$

whence $f_y \in L^\infty(X, \mathcal{A}, |\nu|)$, and $\|f_y\|_\infty \le \|y\|_Y$ for all $y \in Y$. In particular, there exists a $|\nu|$-null set $N_{y,1} \in \mathcal{A}$ such that $|f_y(x)| \le \|y\|_Y$ for all $x \in (N_{y,1})^c$.

Now observe that Y is separable, since by assumption Y^* is separable. Let $D \subseteq Y$ be countable and dense. Set

$$D_S := \left\{ y = \sum_{i=1}^n \alpha_i y_i \mid \alpha_1, \dots, \alpha_n \in \mathbb{R},\ y_1, \dots, y_n \in D \text{ for some } n \in \mathbb{N} \right\},$$

$$D_\mathbb{Q} := \left\{ y = \sum_{i=1}^n q_i y_i \mid q_1, \dots, q_n \in \mathbb{Q},\ y_1, \dots, y_n \in D, \text{ for some } n \in \mathbb{N} \right\}.$$

By (4.78) and (4.80), for any $y \in D_\mathbb{Q}$, we get

$$\begin{aligned}\int_E f_y\, d|\nu| &= \left\langle \nu(E), \sum_{i=1}^n q_i y_i \right\rangle = \sum_{i=1}^n q_i \langle \nu(E), y_i\rangle \\ &= \sum_{i=1}^n q_i \int_E f_{y_i}\, d|\nu| = \int_E \left(\sum_{i=1}^n q_i f_{y_i}\right) d|\nu|.\end{aligned}$$

Therefore, for any $y = \sum_{i=1}^n q_i y_i \in D_\mathbb{Q}$, there exists a $|\nu|$-null set $N_{y,2} \in \mathcal{A}$ such that $f_{\sum_{i=1}^n q_i y_i} = \sum_{i=1}^n q_i f_{y_i}$ for all $x \in (N_{y,2})^c$.

Set $N := \bigcup_{y \in D_\mathbb{Q}} (N_{y,1} \cup N_{y,2})$. Since $D_\mathbb{Q}$ is countable and $|\nu|(N_{y,1}) = |\nu|(N_{y,2}) = 0$ for each $y \in D_\mathbb{Q}$, we also have $|\nu|(N) = 0$. In addition, for any $x \in N^c = \bigcap_{y \in D_\mathbb{Q}} ((N_{y,1})^c \cap (N_{y,2})^c)$, we have

$$\left|\sum_{i=1}^n q_i f_{y_i}(x)\right| = |f_{\sum_{i=1}^n q_i y_i}(x)| \le \left\|\sum_{i=1}^n q_i y_i\right\|_Y$$

for all $q_1, \dots, q_n \in \mathbb{Q}$, $y_1, \dots, y_n \in D$ ($n \in \mathbb{N}$), since we may use the inequality $|f_y(x)| \le \|y\|_Y$ for any $x \in (N_{y,1})^c$ with $y = \sum_{i=1}^n q_i y_i$. By the denseness of $\mathbb{Q}$ in $\mathbb{R}$, for all $x \in N^c$, we get

$$|f_{\sum_{i=1}^n \alpha_i y_i}(x)| = \left|\sum_{i=1}^n \alpha_i f_{y_i}(x)\right| \le \left\|\sum_{i=1}^n \alpha_i y_i\right\|_Y$$

for all $\alpha_1, \dots, \alpha_n \in \mathbb{R}$, $y_1, \dots, y_n \in D$ ($n \in \mathbb{N}$), and thus for all $x \in N^c$,

$$|f_y(x)| \le \|y\|_Y \quad \text{for all } y \in D_S. \tag{4.81}$$

By (4.81), for each $x \in N^c$, there exists $f(x) \in Y^*$ such that $\|f(x)\|_{Y^*} \leq 1$ and

$$\langle f(x), y \rangle = f_y(x) \quad \text{for all } y \in D_S. \tag{4.82}$$

Set $f(x) := 0$ for $x \in N$. The function $f : X \to Y^*$ thus defined satisfies (4.82) $|\nu|$-a. e. in X. By the denseness of D_S in Y f is weakly* measurable, and thus by Proposition 4.1.12 it is $|\nu|$-measurable. Moreover, $\|f(x)\|_{Y^*} \leq 1$ for $|\nu|$-a. e. $x \in X$, thus $f \in L^\infty(X, \mathcal{A}, |\nu|; Y^*)$, and by (4.80) and (4.82)

$$\nu_y(E) = \int_E \langle f(x), y \rangle \, d|\nu|(x) \quad \text{for all } y \in D_S \text{ and } E \in \mathcal{A}. \tag{4.83}$$

Now let $y \in Y$, and let $\{y_n\} \subseteq D_S$ be such that $\|y_n - y\|_Y \to 0$ as $n \to \infty$. By the above considerations, for $|\nu|$-a. e. $x \in X$, we have

$$\langle f(x), y_n \rangle \to \langle f(x), y \rangle, \quad \sup_{n \in \mathbb{N}} |\langle f(x), y_n \rangle| \leq \sup_{n \in \mathbb{N}} \|y_n\|_Y \leq 2\|y\|_Y.$$

Since $|\nu|$ is finite, by the dominated convergence theorem we obtain that $\langle f, y \rangle \in L^1(X, \mathcal{A}, |\nu|)$ and for any $E \in \mathcal{A}$,

$$\begin{aligned} \int_E \langle f(x), y \rangle \, d|\nu|(x) &= \lim_{n\to\infty} \int_E \langle f(x), y_n \rangle \, d|\nu|(x) \\ &= \lim_{n\to\infty} \nu_{y_n}(E) = \lim_{n\to\infty} \langle \nu(E), y_n \rangle = \langle \nu(E), y \rangle \end{aligned} \tag{4.84}$$

(see (4.83) and (4.78)). On the other hand, by Proposition 4.2.3 we have

$$\int_E \langle f, y \rangle \, d|\nu| = \left\langle \int_E f \, d|\nu|, y \right\rangle. \tag{4.85}$$

By the arbitrariness of $y \in Y$, from (4.84)–(4.85) we obtain (4.77). Then the conclusion follows. □

Corollary 4.4.5 (Phillips). *Reflexive Banach spaces have the Radon–Nikodým property.*

Proof. Let Y be a reflexive Banach space. Then every closed separable subspace of Y is a separable dual space and thus by Theorem 4.4.4 has the Radon–Nikodým property. Hence by Proposition 4.4.3 the result follows. □

Let us now go back to the notion of Riesz representability (see Definition 4.4.2). The relationship between Examples 4.4.1 and 4.4.2 is elucidated by the following theorem.

Theorem 4.4.6. *Let $(X, \mathcal{A}, \mu)$ be a finite measure space. Then the following statements are equivalent:*
(i) *Y has the Radon–Nikodým property with respect to $(X, \mathcal{A}, \mu)$;*
(ii) *every operator $T \in \mathscr{L}(L^1(X); Y)$ is Riesz representable.*

Proof. (i)⇒(ii). Let $T \in \mathscr{L}(L^1(X);Y)$. Set $\nu : \mathcal{A} \to Y$, $\nu(E) := T\chi_E$ $(E \in \mathcal{A})$. For every $E \in \mathcal{A}$, we have

$$\|\nu(E)\|_Y \le \|T\|_{\mathscr{L}(L^1(X);Y)}\|\chi_E\|_1 = \|T\|_{\mathscr{L}(L^1(X);Y)}\,\mu(E), \tag{4.86}$$

and thus ν is a vector measure of bounded variation (recall that μ is finite) and absolutely continuous with respect to μ. Since by assumption Y has the Radon–Nikodým property, there exists $g \in L^1(X;Y)$ such that $\nu(E) = \int_E g\,d\mu$ for all $E \in \mathcal{A}$. Then by Proposition 4.2.7, Definition 1.9.2(ii), and (4.86) we have

$$|\nu|(E) = \int_E \|g\|_Y\,d\mu \le \|T\|_{\mathscr{L}(L^1(X);Y)}\,\mu(E) \quad (E \in \mathcal{A}).$$

Then by Proposition 4.2.5 $g \in L^\infty(X;Y)$ with $\|g\|_\infty \le \|T\|_{\mathscr{L}(L^1(X);Y)}$ (see (4.43b)).

To summarize, we proved that there exists $g \in L^\infty(X;Y)$ such that $T\chi_E = \int_E g\,d\mu$ for all $E \in \mathcal{A}$. By the denseness of $\mathscr{S}(X)$ in $L^1(X)$ it follows that $Tf = \int_X fg\,d\mu$ for all $f \in L^1(X)$. This proves the claim.

(ii)⇒(i). Let ν be a vector measure of bounded variation such that $\nu \ll \mu$. Then $|\nu| \ll \mu$, and hence by the Radon–Nikodým theorem there exists $h \in L^1(X)$, $h \ge 0$, such that $|\nu|(E) = \int_E h\,d\mu$ for all $E \in \mathcal{A}$. For any $n \in \mathbb{N}$, set $F_n := \{n-1 \le h < n\}$. Thus $F_m \cap F_n = \emptyset$ if $m \ne n$ and $\bigcup_{n=1}^\infty F_n = X$. Let $s \in \mathscr{S}(X)$, $s = \sum_{i=1}^m c_i\chi_{E_i}$ with $c_1,\dots,c_m \in \mathbb{R}$ and a partition $\{E_1,\dots,E_m\}$ of X. For every $n \in \mathbb{N}$, define

$$T_n : \mathscr{S}(X) \to Y, \quad T_n s := \int_{F_n} s\,d\nu = \sum_{i=1}^m c_i\nu(E_i \cap F_n). \tag{4.87}$$

For any $n \in \mathbb{N}$ and $s \in \mathscr{S}(X)$, we have

$$\|T_n s\|_Y \le \sum_{i=1}^m |c_i||\nu|(E_i \cap F_n) = \sum_{i=1}^m |c_i| \int_{E_i \cap F_n} h\,d\mu \le n\sum_{i=1}^m |c_i|\,\mu(E_i) = n\|s\|_1,$$

and thus T_n uniquely extends to a bounded linear operator from $L^1(X)$ to Y, denoted again by T_n. Since by assumption every $T \in \mathscr{L}(L^1(X);Y)$, every T_n is Riesz representable for every $n \in \mathbb{N}$, that is, there exists $g_n \in L^\infty(X;Y)$ such that $T_n f = \int_X fg_n\,d\mu$ for all $f \in L^1(X)$. In particular (see (4.87)),

$$\nu(E \cap F_n) = T_n\chi_E = \int_E g_n\,d\mu \quad \text{for all } E \in \mathcal{A}. \tag{4.88}$$

Define $g : X \to Y$ by $g := \sum_{n=1}^\infty g_n$. It is easily checked that $g_n = 0$ μ-a. e. in F_n^c. Since $F_n \cap F_m = \emptyset$ for all $n \ne m$ and $\mu(X) < \infty$, it follows that $g \in L^\infty(X;Y) \subseteq L^1(X;Y)$.

From (4.88) for any $m \in \mathbb{N}$, we get

$$\nu\left(E \cap \left(\bigcup_{n=1}^{m} F_n\right)\right) = \int\limits_{E\cap(\bigcup_{n=1}^{m} F_n)} g\, d\mu \quad (E \in \mathcal{A}).$$

By Lemma 1.9.2, letting $m \to \infty$ in this equality, we get $\nu(E) = \int_E g\, d\mu$ for all $E \in \mathcal{A}$. Then the conclusion follows. □

4.4.2 Duality and Radon–Nikodým property

For any $f : X \to Y$ and $g : X \to Y^*$, we set

$$\langle g, f\rangle_{Y^*,Y} : X \to \mathbb{R}, \quad \langle g, f\rangle_{Y^*,Y}(x) := \langle g(x), f(x)\rangle_{Y^*,Y} \quad (x \in X).$$

Let $p \in [1, \infty]$, and let q be its Hölder conjugate,

$$q := \begin{cases} 1 & \text{if } p = \infty, \\ \frac{p}{p-1} & \text{if } p \in (1, \infty), \\ \infty & \text{if } p = 1. \end{cases} \tag{4.89}$$

Under proper assumptions, by Theorem 2.8.14 we can identify the dual space $(L^p(X))^*$ with $L^q(X)$. It is natural to ask whether the analogous statement $(L^p(X; Y))^* = L^q(X; Y^*)$ holds for vector Lebesgue spaces. This depends on the Radon–Nikodým property for the dual space Y^* by the following:

Theorem 4.4.7. *Let $(X, \mathcal{A}, \mu)$ be a finite measure space, and let Y be a Banach space. Let $p \in [1, \infty)$, and let q be its Hölder conjugate. The following statements are equivalent:*
(i) *Y^* has the Radon–Nikodým property with respect to $(X, \mathcal{A}, \mu)$;*
(ii) *the map*

$$\Theta : L^q(X; Y^*) \to (L^p(X; Y))^*, \quad \Theta g := T_g, \tag{4.90a}$$

where

$$T_g : L^p(X; Y) \to \mathbb{R}, \quad T_g f := \int\limits_X \langle g, f\rangle_{Y^*,Y}\, d\mu \quad (f \in L^p(X; Y)), \tag{4.90b}$$

is an isometric isomorphism.

Remark 4.4.1. Let $f \in L^p(X; Y)$ and $g \in L^q(X; Y^*)$. Since g is μ-measurable, there exists a sequence $\{t_n\} \subseteq \mathscr{S}(X; Y^*)$, say $t_n = \sum_{k=1}^{k_n} y_k^* \chi_{E_k}$ ($y_k^* \in Y^*, E_k \in \mathcal{A}$), such that $\|t_n(x) - g(x)\|_{Y^*} \to 0$ for μ-a. e. $x \in X$. Also, f is μ-measurable and thus weakly measurable;

hence $\langle t_n, f\rangle_{Y^*,Y} : X \to \mathbb{R}$ is $\mathcal{A}$-measurable for every $n \in \mathbb{N}$. Moreover, $\langle t_n, f\rangle_{Y^*,Y} \to \langle g, f\rangle_{Y^*,Y}$ as $n \to \infty$ μ-a. e. in X, since

$$|\langle t_n, f\rangle_{Y^*,Y} - \langle g, f\rangle_{Y^*,Y}| \leq \|t_n(x) - g(x)\|_{Y^*} \|f\|_Y.$$

Therefore the real-valued function $\langle g, f\rangle_{Y^*,Y}$ is $\mathcal{A}$-measurable, and definition (4.90b) is well posed.

By the Hölder inequality we have

$$\int_X |\langle g, f\rangle_{Y^*,Y}|\, d\mu \leq \int_X \|g\|_{Y^*} \|f\|_Y\, d\mu \leq \|g\|_{L^q(X;Y^*)} \|f\|_{L^p(X;Y)}. \tag{4.91}$$

Then the map defined in (4.90b) is linear and continuous, and thus $T_g \in (L^p(X;Y))^*$. Hence definition (4.90a) is well posed.

A direct consequence of Theorem 4.4.7 is the following result.

Proposition 4.4.8. *Let $(X, \mathcal{A}, \mu)$ be a finite measure space, and let $p \in (1, \infty)$. Then the space $L^p(X;Y)$ is reflexive if and only if Y is reflexive.*

Proof. If Y is reflexive, then Y^* is also reflexive, and by Corollary 4.4.5 both have the Radon–Nikodým property. Hence by Theorem 4.4.7

$$(L^q(X;Y))^{**} = (L^p(X;Y^*))^* = L^q(X;Y^{**}) = L^q(X;Y) \quad \text{for any } q \in (1,\infty).$$

This proves the sufficiency. As for the necessity, observe that Y is isometrically isomorphic to the closed subspace $\{y\chi_X \mid y \in Y\} \subseteq L^p(X;Y)$ and recall that every closed subspace of a reflexive space is reflexive. Hence the result follows. □

To prove Theorem 4.4.7, we need two preliminary results.

Proposition 4.4.9. *Let $(X, \mathcal{A}, \mu)$ be a finite measure space, and let Y be a Banach space. Let $p \in [1, \infty)$, and let q be its Hölder conjugate. Then $L^q(X;Y^*)$ is isometrically isomorphic to a closed subspace of $(L^p(X;Y))^*$.*

Proof. By (4.91) we have $\|T_g\|_{(L^p(X;Y))^*} \leq \|g\|_{L^q(X;Y^*)}$. The conclusion will follow if we prove the reverse inequality.

To this purpose, let $g \in L^q(X;Y^*)$, $g = \sum_{k=1}^\infty y_k^* \chi_{E_k}$, where $\{E_k\} \subseteq \mathcal{A}$ is a disjoint sequence such that $\bigcup_{k=1}^\infty E_k = X$ and $\mu(E_k) > 0$ for all k. Set $\hat{g} : X \to \mathbb{R}$, $\hat{g}(x) := \|g(x)\|_{Y^*}$ $(x \in X)$, and

$$\hat{T}_{\hat{g}} : L^p(X) \to \mathbb{R}, \quad \hat{T}_{\hat{g}} u := \int_X \hat{g}\, u\, d\mu \quad (u \in L^p(X)).$$

Clearly, $\hat{g} \in L^q(X)$, $\hat{T}_{\hat{g}} \in (L^p(X))^*$, and by Theorem 2.8.14

$$\|\hat{T}_{\hat{g}}\|_{(L^p(X))^*} = \sup_{u \in L^p(X),\, \|u\|_p \le 1} |\hat{T}_{\hat{g}} u| = \|\hat{g}\|_q = \|g\|_{L^q(X;Y^*)}.$$

Then for any $\epsilon > 0$, there exists $h \in L^p(X)$ with $0 < \|h\|_p \le 1$ and $h \ge 0$ such that

$$\|g\|_{L^q(X;Y^*)} - \frac{\epsilon}{2} < |\hat{T}_{\hat{g}} h| = \int_X \|g(x)\|_{Y^*} h \, d\mu. \tag{4.92}$$

On the other hand, let $\{y_k^*\} \subseteq Y^*$ be the sequence of coefficients of g. Plainly, there exists a sequence $\{y_k\} \subseteq Y$ such that $\|y_k\|_Y \le 1$ for all $k \in \mathbb{N}$ and

$$\|y_k^*\|_{Y^*} - \frac{\epsilon}{2\|h\|_1} < \langle y_k^*, y_k \rangle_{Y^*,Y} \tag{4.93}$$

(observe that $h \in L^p(X) \subseteq L^1(X)$ since $\mu(X) < \infty$).

Set $f : X \to Y$, $f := (\sum_{k=1}^{\infty} y_k \chi_{E_k}) h$. Thus $f \in L^p(X;Y)$ and $\|f\|_{L^p(X;Y)} \le \|h\|_p \le 1$. By (4.90b)–(4.93) we obtain

$$\begin{aligned} T_g f &= \int_X \langle g(x), f(x) \rangle_{Y^*,Y} \, d\mu(x) = \int_X h \left(\sum_{k=1}^{\infty} \langle y_k^*, y_k \rangle_{Y^*,Y} \chi_{E_k} \right) d\mu \\ &\ge \int_X h \sum_{k=1}^{\infty} \left(\|y_k^*\|_{Y^*} - \frac{\epsilon}{2\|h\|_1} \right) \chi_{E_k} \, d\mu \\ &= \int_X \|g\|_{Y^*} h \, d\mu - \frac{\epsilon}{2\|h\|_1} \int_X h \, d\mu > \|g\|_{L^q(X;Y^*)} - \epsilon. \end{aligned}$$

By the arbitrariness of $\epsilon > 0$ it follows that $T_g f = |T_g f| \ge \|g\|_{L^q(X;Y^*)}$.

To sum up, we exhibited $f \in L^p(X;Y)$ with $\|f\|_{L^p(X;Y)} \le 1$ such that $|T_g f| \ge \|g\|_{L^q(X;Y^*)}$. Hence $\|T_g\|_{(L^p(X;Y))^*} \ge \|g\|_{L^q(X;Y^*)}$, and thus the result follows. □

Remark 4.4.2. By similar methods it can be proved that for any $p \in [1, \infty)$, the dual space $(L^p(X;Y))^*$ is isometrically isomorphic to the space $V^q(X;Y^*) := \{\nu : \mathcal{A} \to Y^* \mid \nu \text{ vector measure}\}$ (here q is the Hölder conjugate of p) endowed with the norm

$$\|\nu\|_q := \sup_\pi \left(\sum_{i=1}^{n} \frac{\|\nu(E_i)\|_{Y^*}^q}{[\mu(E_i)]^{q-1}} \right)^{\frac{1}{p}} \quad \text{if } q \in (1, \infty)$$

($\pi \equiv \{E_1, \dots, E_n\}$ being any partition of X with $\mu(E_i) > 0$ for each $i = 1, \dots, n$),

$$\|\nu\|_\infty := \inf\{C > 0 \mid \|\nu(E)\|_{Y^*} \le C\mu(E) \text{ for any } E \in \mathcal{A}\}$$

(see [41, Section IV.6], [44, Section 8.20]).

Lemma 4.4.10. *Let $(X, \mathcal{A}, \mu)$ be a finite measure space, and let Y be a Banach space. Let $\nu : \mathcal{A} \to Y^*$ be a vector measure of bounded variation and absolutely continuous with respect to μ. Suppose that for any $E \in \mathcal{A}$ with $\mu(E) > 0$, there exist $F \in \mathcal{A}$, $F \subseteq E$ with $\mu(F) > 0$, and a Bochner integrable function $h_F : F \to Y^*$ such that*

$$\nu(G) = \int_G h_F \, d\mu \quad \textit{for all } G \in \mathcal{A},\ G \subseteq F. \tag{4.94}$$

Then Y^* has the Radon–Nikodým property with respect to $(X, \mathcal{A}, \mu)$.

Proof. Let us apply the exhaustion lemma (see Lemma 1.3.5) with the following property:

$$\begin{cases} \text{there exists a Bochner-integrable function } h_E : E \mapsto Y^* \\ \text{such that (4.94) is satisfied for all } G \subseteq E. \end{cases} \tag{P}$$

Then there exist a disjoint sequence $\{E_k\} \subseteq \mathcal{A}$ such that $\bigcup_{k=1}^{\infty} E_k = X$ and a sequence $\{h_k\}$ of Bochner-integrable functions $h_k : E_k \to Y^*$ such that

$$\nu(E \cap E_k) = \int_{E \cap E_k} h_k \, d\mu \quad \text{for all } E \in \mathcal{A} \text{ and } k \in \mathbb{N}.$$

Set $g : X \to Y^*$, $g := \sum_{k=1}^{\infty} h_k \chi_{E_k}$. Then g is Bochner integrable, and for all $E \in \mathcal{A}$, we have

$$\nu\left(E \cap \left(\bigcup_{k=1}^{n} E_k \right) \right) = \sum_{k=1}^{n} \nu(E \cap E_k) = \sum_{k=1}^{n} \int_{E \cap E_k} h_k \, d\mu = \int_E g \chi_{\bigcup_{k=1}^{n} E_k} \, d\mu. \tag{4.95}$$

By Lemma 1.9.2, letting $n \to \infty$ in (4.95), we obtain

$$\nu(E) = \lim_{n \to \infty} \int_E g \chi_{\bigcup_{k=1}^{n} E_k} \, d\mu = \int_E g \, d\mu. \tag{4.96}$$

By the arbitrariness of E in (4.95) and Proposition 4.2.7 we also get

$$\int_X \|g\|_Y^* \chi_{\bigcup_{k=1}^{n} E_k} \, d\mu \le |\nu|(X) \quad \text{for all } n \in \mathbb{N},$$

whence $\int_X \|g\|_Y^* \, d\mu \le |\nu|(X) < \infty$ by the monotone convergence theorem. It follows that $g \in L^1(X; Y^*)$, and thus by (4.96) the result follows. □

Now we can prove Theorem 4.4.7.

Proof of Theorem 4.4.7. (i)⇒(ii). Let Θ be defined by (4.90a). Observe preliminarily that, in view of (4.90b), for any $g \in L^q(X;Y^*)$ and $f \in L^p(X;Y)$, we have

$$\langle \Theta g, f\rangle_{(L^p(X;Y))^*,L^p(X;Y)} = \int_X \langle g,f\rangle_{Y^*,Y}\, d\mu. \tag{4.97}$$

By Proposition 4.4.9 the map Θ is isometric and thus injective. The claim will follow if we prove that it is also surjective.

To this purpose, fix any $T \in (L^p(X;Y))^*$. Define

$$\nu_T : \mathcal{A} \to Y^*, \quad \langle \nu_T(E), y\rangle_{Y^*,Y} := \langle T, \chi_E y\rangle_{(L^p(X;Y))^*,L^p(X;Y)} \quad (E \in \mathcal{A},\ y \in Y).$$

Plainly, the map ν_T is σ-additive and thus is a vector measure. Moreover,

$$|\langle \nu_T(E), y\rangle_{Y^*,Y}| \le \|T\|_{(L^p(X;Y))^*}\|\chi_E y\|_{L^p(X;Y)} = \|T\|_{(L^p(X;Y))^*}(\mu(E))^{\frac{1}{p}}\|y\|_Y,$$

whence

$$\|\nu_T(E)\|_{Y^*} \le \|T\|_{(L^p(X;Y))^*}(\mu(E))^{\frac{1}{p}} \quad \text{for all } E \in \mathcal{A}.$$

Then ν_T is of bounded variation (since $\mu(X) < \infty$ by assumption), and $\nu_T \ll \mu$.

Since Y^* has the Radon–Nikodým property with respect to $(X, \mathcal{A}, \mu)$, there exists $g \in L^1(X;Y^*)$ such that $\nu_T(E) := \int_E g\, d\mu$ for all $E \in \mathcal{A}$. Then for any $s \in \mathscr{S}(X;Y)$, $s = \sum_{i=1}^k y_i\chi_{E_i}$, there holds

$$\begin{aligned}
&\langle T, s\rangle_{(L^p(X;Y))^*,L^p(X;Y)} \\
&= \sum_{i=1}^k \langle T, y_i\chi_{E_i}\rangle_{(L^p(X;Y))^*,L^p(X;Y)} = \sum_{i=1}^k \langle \nu_T(E_i), y_i\rangle_{Y^*,Y} \\
&= \sum_{i=1}^k \Big\langle \int_{E_i} g\, d\mu, y_i\Big\rangle_{Y^*,Y} = \sum_{i=1}^k \int_X \langle g, y_i\rangle_{Y^*,Y}\chi_{E_i}\, d\mu = \int_X \langle g, s\rangle_{Y^*,Y}\, d\mu.
\end{aligned} \tag{4.98}$$

Let $\{F_k\} \subseteq \mathcal{A}$ be a nondecreasing sequence such that $\bigcup_{k=1}^\infty F_k = X$, and let the function $\|g\|_{Y^*} : X \to [0,\infty)$ is bounded on each F_k (a possible choice is $F_k = \{\|g(\cdot)\|_{Y^*} \le k\}$). Fix $\bar{k} \in \mathbb{N}$, and consider the space $L^p(F_{\bar{k}};Y) \equiv L^p(F_{\bar{k}}, \mathcal{A} \cap F_{\bar{k}}, \mu|_{\mathcal{A}\cap F_{\bar{k}}};Y)$. Set

$$T_{\bar{k}} : L^p(F_{\bar{k}};Y) \to \mathbb{R}, \quad T_{\bar{k}}f := \int_{F_{\bar{k}}} \langle g\chi_{F_{\bar{k}}}, f\rangle_{Y^*,Y}\, d\mu \quad (f \in L^p(F_{\bar{k}};Y)). \tag{4.99}$$

Since $\|g\|_{Y^*}$ is bounded on $F_{\bar{k}}$, we have $g\chi_{F_{\bar{k}}} \in L^q(F_{\bar{k}};Y^*)$. Hence by Proposition 4.4.9 $T_{\bar{k}}$ is an element of $(L^p(F_{\bar{k}};Y))^*$, and $\|T_{\bar{k}}\|_{(L^p(F_{\bar{k}};Y))^*} = \|g\chi_{F_{\bar{k}}}\|_{L^q(F_{\bar{k}};Y^*)}$.

Observe that $L^p(F_{\bar{k}};Y)$ can be identified with the subspace

$$L^p(F_{\bar{k}};Y) \simeq \{f \in L^p(X;Y) \mid f = 0 \ \mu\text{-a.e. in } F_{\bar{k}}^c\}$$

of $L^p(X;Y)$; moreover, the simple functions $s \in \mathscr{S}(F_{\bar{k}};Y)$, $s = \sum_{i=1}^k y_i\chi_{E_i\cap F_{\bar{k}}}$, are dense in $L^p(F_{\bar{k}};Y)$. Then from (4.98)–(4.99) for every $f \in L^p(X;Y)$, we obtain

$$\langle T, f\chi_{F_{\bar{k}}}\rangle_{(L^p(X;Y))^*,L^p(X;Y)} = \int_X \langle g, f\chi_{F_{\bar{k}}}\rangle_{Y^*,Y}\,d\mu = T_{\bar{k}}(f\chi_{F_{\bar{k}}}),$$

and thus $T = T_{\bar{k}}$ on $F_{\bar{k}}$. It follows that

$$\|T_{\bar{k}}\|_{(L^p(F_{\bar{k}};Y))^*} = \|g\chi_{F_{\bar{k}}}\|_{L^q(F_{\bar{k}};Y^*)} \le \|T\|_{(L^p(X;Y))^*},$$

whence $\|g\|_{L^q(X;Y^*)} \le \|T\|_{(L^p(X;Y))^*}$ by the arbitrariness of $\bar{k}$ and the monotone convergence theorem.

Therefore we have that $g \in L^q(X;Y^*)$. Moreover, by the denseness of $\mathscr{S}(X;Y)$ in $L^p(X;Y)$ (see Proposition 4.3.6(i)) from (4.98) we obtain

$$\langle T, f\rangle_{(L^p(X;Y))^*,L^p(X;Y)} = \int_X \langle g, f\rangle_{Y^*,Y}\,d\mu \quad \text{for all } f \in L^p(X;Y). \tag{4.100}$$

By the arbitrariness of f, from (4.97) and (4.100) we obtain that $T = \Theta g$. Hence the claim follows.

(ii)⟹(i). Let $\nu : \mathcal{A} \to Y^*$ be a vector measure of bounded variation such that $\nu \ll \mu$. Let us prove that for any $E \in \mathcal{A}$ with $\mu(E) > 0$, there exist $F \in \mathcal{A}$, $F \subseteq E$ with $\mu(F) > 0$, and a Bochner-integrable function $h_F : F \to Y^*$ such that (4.94) holds. Then by Lemma 4.4.10 the conclusion will follow.

Fix $E \in \mathcal{A}$ with $\mu(E) > 0$. Let us apply the Hahn decomposition (see Theorem 1.8.2) to the signed finite measure $|\nu| - \tau\mu$ ($\tau > 0$). Then there exist $\bar{\tau} > 0$ and $F \subseteq E$ with $\mu(F) > 0$ such that $|\nu|(G) \le \bar{\tau}\mu(G)$ for all $G \in \mathcal{A}$, $G \subseteq F$. Let $s \in \mathscr{S}(X;Y)$, $s = \sum_{i=1}^k y_i\chi_{E_i}$ (where $\{E_i, \dots, E_k\}$ is a partition of X such that $\mu(E_i) > 0$ for each $i = 1, \dots, k$); thus $\|s\|^p_{L^p(X;Y)} = \sum_{i=1}^k \mu(E_i)\|y_i\|^p_Y$. Define

$$T : \mathscr{S}(X;Y) \to \mathbb{R}, \quad Ts := \sum_{i=1}^k \langle \nu(F\cap E_i), y_i\rangle_{Y^*,Y} \quad (s \in \mathscr{S}(X;Y)). \tag{4.101}$$

The map T is linear, and

$$\begin{aligned} |Ts| &\le \sum_{i=1}^k |\langle \nu(F\cap E_i), y_i\rangle_{Y^*,Y}| \\ &\le \sum_{i=1}^k \|\nu(F\cap E_i)\|_{Y^*}\|y_i\|_Y \le \sum_{i=1}^k |\nu|(F\cap E_i)\|y_i\|_Y \end{aligned}$$

$$\leq \sum_{i=1}^{k} \bar{\tau}\mu(F \cap E_i)\|y_i\|_Y \leq \bar{\tau}\|s\|_{L^1(X;Y)} \leq \bar{\tau}[\mu(X)]^{\frac{1}{q}}\|s\|_{L^p(X;Y)}. \tag{4.102}$$

By (4.102) the map T is linear and bounded on $\mathscr{S}(X;Y)$, and thus by Proposition 4.3.6(i) it can be uniquely extended to an element of $(L^p(X;Y))^*$, denoted again by T.

Since by assumption $(L^p(X;Y))^* = L^q(X;Y^*)$, by the above remarks there exists $g \in L^q(X;Y^*)$ such that

$$\langle T,f\rangle_{(L^p(X;Y))^*,L^p(X;Y)} = \int_X \langle g,f\rangle_{Y^*,Y}\,d\mu \quad \text{for all } f \in L^p(X;Y). \tag{4.103}$$

Set $h_F := g\chi_F$. Let $y \in Y$ and $f = y\chi_G$ with $G \in \mathcal{A}$, $G \subseteq F$, $\mu(G) > 0$. Then from (4.101) and (4.103) we get

$$\langle \nu(G),y\rangle_{Y^*,Y} = \langle T,y\chi_G\rangle_{(L^p(X;Y))^*,L^p(X;Y)} = \int_G \langle h_F,y\rangle_{Y^*,Y}\,d\mu = \left\langle \int_G h_F\,d\mu, y\right\rangle_{Y^*,Y}$$

(see Corollary 4.2.4), whence $\nu(G) = \int_G h_F\,d\mu$ by the arbitrariness of y. This completes the proof. □

4.4.3 Duality results

This subsection is devoted to the proof of the following duality result.

Theorem 4.4.11. *Let $(X,\mathcal{A},\mu)$ be a σ-finite measure space, and let Y be a separable Banach space. Let $p \in [1,\infty)$, and let q be its Hölder conjugate. Then the map*

$$\Theta : L^q_{w^*}(X;Y^*) \to (L^p_w(X;Y))^*, \quad \Theta g := T_g \quad (g \in L^q_{w^*}(X;Y^*)), \tag{4.104a}$$

where

$$T_g f := \int_X \langle g(x),f(x)\rangle_{Y^*,Y}\,d\mu \quad (f \in L^p_w(X;Y)), \tag{4.104b}$$

is an isometric isomorphism.

Observe that $L^p_w(X;Y) = L^p(X;Y)$ in the above statement, since Y is separable (see Remark 4.3.1(i)).

Remark 4.4.3. Let $f \in L^p_w(X;Y)$ and $g \in L^q_{w^*}(X;Y^*)$. Since Y is separable, by Theorem 4.1.11 f is μ-measurable. Hence there exists a sequence $\{s_n\} \subseteq \mathscr{S}(X;Y)$, say $s_n = \sum_{k=1}^{k_n} y_k\chi_{E_k}$ $(y_k \in Y, E_k \in \mathcal{A})$ as in Definition 4.1.2, such that $\|s_n(x) - f(x)\|_Y \to 0$ for μ-a.e. $x \in X$.

Since g is weakly* measurable, the real-valued function $\langle g, s_n\rangle_{Y^*,Y} : X \to \mathbb{R}$ is $\mathcal{A}$-measurable for every $n \in \mathbb{N}$. Moreover, $\langle g, s_n\rangle_{Y^*,Y} \to \langle g, f\rangle_{Y^*,Y}$ as $n \to \infty$ μ-a. e. in X, since

$$\left|\langle g(x), s_n(x)\rangle_{Y^*,Y} - \langle g(x), f(x)\rangle_{Y^*,Y}\right| \le \|s_n(x) - f(x)\|_Y \|g(x)\|_{Y^*} \quad (x \in X).$$

Therefore the real-valued function $x \mapsto \langle g(x), f(x)\rangle_{Y^*,Y}$ is measurable, and the definition in (4.104b) is well posed. Arguing as in Remark 4.4.1 shows that the map defined in (4.104b) is linear and continuous, and thus $T_g \in (L^p_w(X;Y))^*$, and definition (4.109a) below is also well posed.

For shortness, if μ is σ-finite and Y is separable, then the conclusion of Theorem 4.4.11 can be expressed by the equality

$$(L^p(X;Y))^* = (L^p_w(X;Y))^* = L^q_{w^*}(X;Y^*), \tag{4.105}$$

where p, q are conjugate exponents. In particular, if μ is σ-finite and Y^* is separable, then equality (4.105) reads

$$(L^p(X;Y))^* = L^q(X;Y^*) \tag{4.106}$$

(see Remark 4.3.1(i)), in agreement with Theorem 4.4.12 below.

We will prove Theorem 4.4.11 in the case $p = 1$ and $q = \infty$, which is relevant for our purposes (see Subsection 4.4.5); the proof for $p \in (1, \infty)$ will be given under the stronger assumption of separability of Y^* (see Subsection 4.4.4), referring the reader to [44, Section 8.20] for the general case.

4.4.4 Duality results: separable Y^*

As already observed, equalities (4.105) and (4.106) coincide if Y^* is separable. Therefore, if we assume the separability of Y^*, then Theorem 4.4.11 is a consequence of the following result.

Theorem 4.4.12. *Let $(X, \mathcal{A}, \mu)$ be a σ-finite measure space, and let Y^* be separable. Let $p \in [1, \infty)$, and let q be its Hölder conjugate. Then (4.106) holds.*

Proof. If μ is finite, then the result follows by Theorems 4.4.4 and 4.4.7. Let us prove that it also holds if μ is σ-finite. Let $\{E_k\} \subseteq \mathcal{A}$ be a nondecreasing sequence such that $\mu(E_k) < \infty$ for every k and $\bigcup_{k=1}^\infty E_k = X$. Set $\mathcal{A}_k := \mathcal{A} \cap E_k$, $\mu_k := \mu|_{\mathcal{A}\cap E_k}$, and consider the space $L^p(E_k;Y) \equiv L^p(E_k, \mathcal{A}_k, \mu_k; Y)$ $(k \in \mathbb{N})$. Observe that $L^p(E_k;Y)$ can be identified with the subspace

$$L^p(E_k;Y) \simeq \{f \in L^p(X;Y) \mid f = 0 \ \mu\text{-a. e. in } E_k^c\}$$

of $L^p(X;Y)$.

Let $T \in (L^p(X;Y))^*$. For any $k \in \mathbb{N}$, set $T_k := T|_{L^p(E_k;Y)}$. Clearly, $T_k \in (L^p(E_k;Y))^*$, and $\|T_k\|_{(L^p(E_k;Y))^*} \le \|T\|_{(L^p(X;Y))^*}$. Since $\mu_k(E_k) < \infty$, by Theorem 4.4.7 there exists a unique $g_k \in L^q(E_k;Y^*)$ such that

$$\langle T_k, f\rangle_{(L^p(E_k;Y))^*, L^p(E_k;Y)} = \int_{E_k} \langle g_k, f\rangle_{Y^*,Y}\, d\mu_k \quad \text{for all } f \in L^p(E_k;Y), \tag{4.107}$$

$$\|g_k\|_{L^q(E_k;Y^*)} = \|T_k\|_{(L^p(E_k;Y))^*} \le \|T\|_{(L^p(X;Y))^*} \quad (k \in \mathbb{N}). \tag{4.108}$$

Without loss of generality, we can suppose that $g_{k+1}|_{E_k} = g_k$ for every k. Then the map $g : X \to Y^*$, $g|_{E_k} := g_k$ $(k \in \mathbb{N})$, is well defined and μ-measurable. Letting $k \to \infty$ in (4.108), by the monotone convergence theorem we obtain that $\|g\|_{L^q(X;Y^*)} \le \|T\|_{(L^p(X;Y))^*}$, and thus $g \in L^q(X;Y^*)$. Rewriting (4.107) as

$$\langle T, f\chi_{E_k}\rangle_{(L^p(X;Y))^*, L^p(X;Y)} = \int_X \langle g, f\chi_{E_k}\rangle_{Y^*,Y}\, d\mu \quad \text{for all } f \in L^p(X;Y)$$

and letting $k \to \infty$ give

$$\langle T, f\rangle_{(L^p(X;Y))^*, L^p(X;Y)} = \int_X \langle g, f\rangle_{Y^*,Y}\, d\mu \quad \text{for all } f \in L^p(X;Y).$$

Hence the conclusion follows. □

4.4.5 Duality results: separable Y

The assumption of separability of Y^* is too strong in many cases, e. g., where $Y = C_0(\overline{U})$ with open and bounded $U \subseteq \mathbb{R}^N$. In this case, by Theorem 2.7.1 $Y^* = \mathfrak{R}_f(U)$, which is not separable, whereas $Y = C_0(\overline{U})$ is separable. The following result (see [44, Theorem 8.17.5], [93, Corollaire (2.3)]) often provides the mathematical framework to address situations of this kind, as we will see further.

Theorem 4.4.13. *Let $(X, \mathcal{A}, \mu)$ be a σ-finite measure space, and let Y be separable. Then the map*

$$\Theta : L^\infty_{w^*}(X;Y^*) \to (L^1_w(X;Y))^*, \quad \Theta g := T_g \quad (g \in L^\infty_{w^*}(X;Y^*)), \tag{4.109a}$$

where

$$T_g f := \int_X \langle g(x), f(x)\rangle_{Y^*,Y}\, d\mu \quad (f \in L^1_w(X;Y)), \tag{4.109b}$$

is an isometric isomorphism.

To prove Theorem 4.4.13, we need some preliminary remarks. Let E and F be Banach spaces. By $\mathscr{L}_2(E,F;\mathbb{R})$ we denote the space of bilinear continuous maps from $E\times F$ to $\mathbb{R}$, endowed with the norm

$$\mathscr{L}_2(E,F;\mathbb{R}) \ni B \equiv B(e,f) \mapsto \|B\| := \sup_{\|e\|_E\le 1,\|f\|_F\le 1} |B(e,f)|.$$

Let us recall the following result (e. g., see [31, Subsection 1.9], [44, Subsection 8.17.4]).

Proposition 4.4.14. *Let E and F be Banach spaces. Then $\mathscr{L}_2(E,F;\mathbb{R})$ is isometrically isomorphic to both $\mathscr{L}(E;F^*)$ and $\mathscr{L}(F;E^*)$.*

Now we can prove the following result.

Proposition 4.4.15. *Let $(X,\mathcal{A},\mu)$ be a σ-finite measure space, and let Y be separable. Let $T: L^1(X)\to Y^*$ be linear and continuous. Then there exists a unique $g\in L^\infty_{w^*}(X;Y^*)$ such that*

$$\langle T\hat{f},y\rangle_{Y^*,Y} = \int_X \hat{f}(x)\langle g(x),y\rangle_{Y^*,Y}\,d\mu(x) \quad \textit{for all } \hat{f}\in L^1(X) \textit{ and } y\in Y. \tag{4.110}$$

Moreover, $\|T\| = \|g\|^*_\infty = \operatorname{ess\,sup}_{x\in X}\|g(x)\|_{Y^*}$.

Proof. Applying Proposition 4.4.14 with $E = L^1(X)$, $F = Y$, shows that $\mathscr{L}(L^1(X);Y^*)$ is isometrically isomorphic to $\mathscr{L}(Y;L^\infty(X))$. Then for any $T\in\mathscr{L}(L^1(X);Y^*)$, there exists a unique $S\in\mathscr{L}(Y;L^\infty(X))$ such that $\|S\| = \|T\|$ and

$$\langle Sy,\hat{f}\rangle_{L^\infty(X),L^1(X)} = \langle T\hat{f},y\rangle_{Y^*,Y} \quad \text{for all } y\in Y,\ \hat{f}\in L^1(X). \tag{4.111}$$

For μ-a. e. $x\in X$, consider the linear map from Y to $\mathbb{R}$, $y\mapsto (Sy)(x)$ $(y\in Y)$. Since

$$|(Sy)(x)| \le \|Sy\|_{L^\infty(X)} \le \|S\|\|y\|_Y, \tag{4.112}$$

this map is continuous, thus there exists $g_x\in Y^*$ such that $\langle g_x,y\rangle_{Y^*,Y} = (Sy)(x)$. Set

$$g: X\to Y^*, \quad g(x) := g_x \quad \text{for } \mu\text{-a. e. } x\in X.$$

Since $Sy\in L^\infty(X)$, the map $x\mapsto (Sy)(x) = \langle g(x),y\rangle_{Y^*,Y}$ is measurable, and hence g is weakly* measurable. Moreover, by (4.112) for μ-a. e. $x\in X$, we have

$$|\langle g(x),y\rangle_{Y^*,Y}| = |\langle g_x,y\rangle_{Y^*,Y}| = |(Sy)(x)| \le \|S\|\|y\|_Y.$$

It follows that

$$\|g\|^*_\infty = \operatorname{ess\,sup}_{x\in X}\|g(x)\|_{Y^*} \le \|S\| = \|T\|, \tag{4.113}$$

and thus $g\in L^\infty_{w^*}(X;Y^*)$.

Now observe that by (4.111) and the very definition of g, for any $\hat{f} \in L^1(X)$ and $y \in Y$, we have

$$\begin{aligned}\langle T\hat{f}, y\rangle_{Y^*,Y} &= \langle\langle g(\cdot), y\rangle_{Y^*,Y}, \hat{f}\rangle_{L^\infty(X),L^1(X)}\\ &= \int_X \hat{f}(x)\langle g(x), y\rangle_{Y^*,Y}\, d\mu(x),\end{aligned}$$

and thus (4.110) follows. By (4.110) we plainly obtain that $\|T\| \le \|g\|_\infty^*$. Since the opposite inequality has been proven (see (4.113)), it follows that $\|T\| = \|g\|_\infty^*$.

It remains to prove the uniqueness. To this purpose, let there exist $g_1, g_2 \in L^\infty_{w^*}(X; Y^*)$ satisfying (4.110). Then

$$\int_X \hat{f}(x)[\langle g_1(x), y\rangle_{Y^*,Y} - \langle g_2(x), y\rangle_{Y^*,Y}]\, d\mu(x) = 0 \quad \text{for all } \hat{f} \in L^1(X) \text{ and } y \in Y.$$

By the arbitrariness of $\hat{f}$, for any $y \in Y$, there exists a μ-null subset $N_y \subseteq X$ such that

$$\langle g_1(x), y\rangle_{Y^*,Y} = \langle g_2(x), y\rangle_{Y^*,Y} \quad \text{for all } x \in N_y^c. \tag{4.114}$$

Since Y is separable, there exists a dense countable set $D \equiv \{y_k\} \subseteq Y$. Set $N := \bigcup_{k=1}^\infty N_{y_k}$. Then N is μ-null, and by (4.114) $\langle g_1(x) - g_2(x), y_k\rangle_{Y^*,Y} = 0$ for all $x \in N^c$ and $y_k \in D$, and thus $g_1 = g_2$ μ-a. e. in X by the denseness of D. Hence the conclusion follows. □

Remark 4.4.4. Equality (4.110) can be restated as follows:

$$T\hat{f} = \mathcal{G}\int_X \hat{f} g\, d\mu \quad \text{for all } \hat{f} \in L^1(X),$$

that is, $T\hat{f}$ is the Gelfand integral of $\hat{f}g$ on X (see (4.64c); observe that $\hat{f}g : X \to Y^*$ is weakly* measurable, and the map $x \mapsto \langle \hat{f}(x)g(x), y\rangle_{Y^*,Y}$ belongs to $L^1(X)$ for every $y \in Y$). In view of Proposition 4.4.15, this amounts to saying that the operator T is Riesz representable or that Y^* has the Radon–Nikodým property in a weaker sense (see Definitions 4.4.1–4.4.2 and [101, Chapter 7]; see also [93]).

Now we can prove Theorem 4.4.13.

Proof of Theorem 4.4.13. Choosing in (4.109b) $f = \hat{f}y$ ($y \in Y, \hat{f} \in L^1(X)$) and arguing as in the uniqueness proof of Proposition 4.4.15 show that the operator Θ defined in (4.109) is injective (here the separability of Y is used). Moreover,

$$|\langle \Theta g, f\rangle_{(L^1_w(X;Y))^*, L^1_w(X;Y)}| \le \int_X |\langle g(x), f(x)\rangle_{Y^*,Y}|\, d\mu \le \int_X \|g(x)\|_{Y^*}\|f(x)\|_Y\, d\mu \le \|g\|_\infty^* \|f\|_1,$$

and thus Θ is continuous, and $\|\Theta g\|_{(L^1_w(X;Y))^*} \le \|g\|_\infty^*$.

Let us prove that Θ is surjective. For any $F \in (L^1_w(X;Y))^*$, set

$$T_F : L^1(X) \to Y^*, \quad \langle T_F \hat{f}, y\rangle_{Y^*,Y} := \langle F, \hat{f}y\rangle_{(L^1_w(X;Y))^*, L^1_w(X;Y)} \quad (\hat{f} \in L^1(X),\, y \in Y).$$

Clearly, T_F is linear, and

$$|\langle T_F \hat{f}, y\rangle_{Y^*,Y}| \le \|F\|_{(L^1_w(X;Y))^*} \|\hat{f}\|_1 \|y\|_Y;$$

thus T_F is continuous, and $\|T_F\| \le \|F\|_{(L^1_w(X;Y))^*}$.

Since $T_F \in \mathscr{L}(L^1(X); Y^*)$, by Proposition 4.4.15 there exists a unique $g_F \in L^\infty_{w^*}(X;Y^*)$ such that

$$\langle T_F \hat{f}, y\rangle_{Y^*,Y} = \int_X \langle g_F(x), \hat{f}(x) y\rangle_{Y^*,Y}\, d\mu(x) \quad \text{for all } \hat{f} \in L^1(X) \text{ and } y \in Y, \tag{4.115}$$

and $\|T_F\| = \|g_F\|^*_\infty$. By the definition of T_F and equality (4.115) we also have that

$$\langle F, \hat{f}y\rangle_{(L^1_w(X;Y))^*, L^1_w(X;Y)} = \int_X \langle g_F(x), \hat{f}(x) y\rangle_{Y^*,Y}\, d\mu(x),$$

whence plainly

$$\langle F, s\rangle_{(L^1_w(X;Y))^*, L^1_w(X;Y)} = \int_X \langle g_F(x), s(x)\rangle_{Y^*,Y}\, d\mu(x) \quad \text{for all } s \in \mathscr{S}(X;Y).$$

By the denseness of $\mathscr{S}(X;Y)$ in $L^1(X;Y) = L^1_w(X;Y)$ (see Proposition 4.3.6(i)) it follows that for any $f \in L^1_w(X;Y)$,

$$\langle F, f\rangle_{(L^1_w(X;Y))^*, L^1_w(X;Y)} = \int_X \langle g_F(x), f(x)\rangle_{Y^*,Y}\, d\mu(x). \tag{4.116}$$

By the arbitrariness of f, comparing (4.116) with (4.109) shows that $F = \Theta g_F$, and thus T is surjective. Moreover, from (4.116) we get

$$\|F\|_{(L^1_w(X;Y))^*} \le \|g_F\|^*_\infty = \|T_F\|.$$

The opposite inequality has already been proven, and thus we obtain that

$$\|\Theta g_F\|_{(L^1_w(X;Y))^*} = \|F\|_{(L^1_w(X;Y))^*} = \|g_F\|^*_\infty.$$

Hence the result follows. □

If Z is a σ-compact metric space, then the space $C_0(Z)$ is separable (see Proposition A.2 and paragraph A.7 in Appendix A). Then by Theorems 2.7.1 and 4.4.13 we have the following result.

Proposition 4.4.16. *Let $(X, \mathcal{A}, \mu)$ be a σ-finite measure space, and let Z be a σ-compact metric space. Then*

$$(L^1(X; C_0(Z)))^* = (L^1_w(X; C_0(Z)))^* = L^\infty_{w^*}(X; \mathfrak{R}_f(Z)). \tag{4.117}$$

4.5 Vector Lebesgue spaces of real-valued functions

In this section, we will often deal with the Lebesgue measure on some Euclidean space $\mathbb{R}^d$ ($d \in \mathbb{N}$). We set for shortness $dx \equiv d\lambda_d(x)$, $dy \equiv d\lambda_{d'}(x)$, and $dxdy \equiv d\lambda_{d+d'}(x, y)$.

Let $U \subseteq \mathbb{R}^M$ be open. It is interesting to characterize the space $L^p(U; Y)$ when Y is a Lebesgue or Sobolev space of real functions. A natural question is whether the space $L^p(U; L^p(V))$, $V \subseteq \mathbb{R}^N$ open, can be identified with $L^p(U \times V)$, $U \times V \subseteq \mathbb{R}^{M+N}$. The answer is given by the following results.

Proposition 4.5.1. *Let $U \subseteq \mathbb{R}^M$ and $V \subseteq \mathbb{R}^N$ be open. Let $p \in [1, \infty]$ and $r \in [1, \infty)$. Then the following statements are equivalent:*
(i) $f \in L^p(U; L^r(V))$;
(ii) $f \in L^1_{\mathrm{loc}}(U \times V)$, *and*

$$\int_U \left(\int_V |f(x,y)|^r dy \right)^{\frac{p}{r}} dx < \infty \quad \text{if } p \in [1, \infty), \tag{4.118a}$$

$$\left\| \int_V |f(\cdot, y)|^r dy \right\|_{L^\infty(U)} < \infty \quad \text{if } p = \infty. \tag{4.118b}$$

Proof. (i)⇒(ii). Let us first prove that $f : U \times V \to \mathbb{R}$, $f(x,y) := f(x)(y)$ ($x \in U, y \in V$), is $\mathcal{B}(U) \times \mathcal{B}(V)$-measurable. For any $x \in U$, set

$$\tilde{f}(x,y) := \begin{cases} f(x,y) & \text{if } y \in V, \\ 0 & \text{if } y \in \mathbb{R}^N \setminus V. \end{cases}$$

The claim will follow if we prove that $\tilde{f} : U \times \mathbb{R}^N \to \mathbb{R}$ is $\mathcal{B}(U) \times \mathcal{B}^N$-measurable.

Since $f \in L^p(U; L^r(V))$, the map $\tilde{f} : U \to L^r(\mathbb{R}^N)$, $\tilde{f}(x)(y) := \tilde{f}(x,y)$ ($y \in \mathbb{R}^N$), is λ_N-measurable. Then there exists a sequence $\{s_j\} \subseteq \mathscr{S}(U; L^r(\mathbb{R}^N))$ such that (see Definition 4.1.3)

$$\lim_{j \to \infty} \int_{\mathbb{R}^N} |s_j(x,y) - \tilde{f}(x)(y)|^r dy = 0 \quad \text{for a. e. } x \in U. \tag{4.119}$$

Let $\rho_{1/m}$ ($m \in \mathbb{N}$) be a standard mollifier. For any $(x,y) \in U \times \mathbb{R}^N$, set

$$\begin{aligned} \tilde{f}_m(x)(y) &:= \int_{\mathbb{R}^N} \tilde{f}(x, y - z)\rho_{1/m}(z)\, dz, \\ s_{j,m}(x,y) &:= \int_{\mathbb{R}^N} s_j(x, y - z)\rho_{1/m}(z)\, dz. \end{aligned} \tag{4.120}$$

By the convergence in (4.119), for any fixed $m \in \mathbb{N}$ and for a. e. $x \in U$, we have

$$\lim_{j\to\infty} \sup_{y\in\mathbb{R}^N} |s_{j,m}(x,y) - \tilde{f}_m(x)(y)| = 0. \tag{4.121}$$

Clearly, every $s_{j,m}$ is $\mathcal{B}(U) \times \mathcal{B}^N$-measurable. On the other hand, by (4.121) the function $\tilde{f}_m(x,y) : U\times\mathbb{R}^N \to \mathbb{R}, \tilde{f}_m(x,y) := \tilde{f}_m(x)(y)$, is the limit of $s_{j,m}(x,y)$ for a. e. $(x,y) \in U\times\mathbb{R}^N$. Then $\tilde{f}_m$ is also $\mathcal{B}(U) \times \mathcal{B}^N$-measurable.

Now observe that for a. e. $x \in U$, we have $\tilde{f}_m(x,\cdot) \to \tilde{f}(x,\cdot)$ a. e. in $\mathbb{R}^N$ (possibly, extracting a subsequence, not relabeled; see Proposition 4.6.1(ii)). Then by Remark 2.3.2 $\tilde{f}$ is the limit a. e. in $U\times\mathbb{R}^N$ of a sequence of $\mathcal{B}(U)\times\mathcal{B}^N$-measurable functions, and thus is $\mathcal{B}(U) \times \mathcal{B}^N$-measurable.

Now the fact that $f \in L^1_{\mathrm{loc}}(U \times V)$ follows easily from inequalities (4.118), which in turn are an obvious consequence of the assumption. Hence the claim follows.

(ii)⇒(i) By (4.118) $f(x,\cdot) \in L^r(V)$ for a. e. $x \in U$. Then the claim will follow from (4.118) if we prove that the map $x \mapsto f(x,\cdot)$ from U to $L^r(V)$ is λ_N-measurable. In turn, by the separability of $L^r(V)$ ($q \in [1,\infty)$) this follows from Proposition 4.1.5 if we show that the map $x \mapsto \int_V |f(x,y) - d(y)|^r dy$ ($x \in U$) is measurable for every $d = d(y)$ belonging to a countable and dense subset of $L^r(V)$. To this purpose, observe that the map $(x,y) \mapsto |f(x,y) - d(y)|^r$ from $U \times V$ to $\mathbb{R}$ is $\mathcal{B}(U) \times \mathcal{B}(V)$-measurable, since f is $\mathcal{B}(U) \times \mathcal{B}(V)$-measurable by assumption and $d \in L^r(V)$, and thus is $\mathcal{B}(V)$-measurable. Then by Theorem 2.3.2(i) we obtain the result. □

Corollary 4.5.2. *Let $U \subseteq \mathbb{R}^M$ and $V \subseteq \mathbb{R}^N$ be open, and let $p \in [1,\infty)$. Then the following statements are equivalent:* (i) $f \in L^p(U; L^p(V))$ *and* (ii) $f \in L^p(U \times V)$.

Proof. The proof follows immediately from Proposition 4.5.1 with $p = r \in [1,\infty)$, observing that inequality (4.118a) by the Tonelli theorem gives

$$\int_{U\times V} |f|^p\, dxdy = \int_U \left(\int_V |f(x,y)|^p\, dy \right) dx < \infty.$$

□

The separability of the space $L^r(V)$ for $r \in [1,\infty)$ was important in the proof of Proposition 4.5.1. Hence the following result is not surprising.

Proposition 4.5.3. *There exists a function $f \in L^\infty(\mathbb{R}^2)$ that does not belong to $L^\infty(\mathbb{R}; L^\infty(\mathbb{R}))$.*

Proof. Let $f : \mathbb{R}^2 \to \mathbb{R}$, $f(r,x) := \chi_{I_r}(x)$ with $I_r \equiv (-|r|,|r|)$ $((r,x) \in \mathbb{R}^2)$. It is easily seen that $f(r,x) = \chi_E$ with $E := \{(r,x) \in \mathbb{R}^2 \mid |x| < |r|\}$. Since $E \in \mathcal{B}^2$, f is $\mathcal{B}^2$-measurable, and thus $f \in L^\infty(\mathbb{R}^2)$.

Set $F : \mathbb{R} \to L^\infty(\mathbb{R})$, $F(r) := f(r,\cdot) = \chi_{I_r}$ for a. e. $r \in \mathbb{R}$. The result will follow if we prove that F is not $(\mathcal{B}(L^\infty(\mathbb{R})), \mathcal{B}(\mathbb{R}))$-measurable (and hence not λ-measurable; see Remark 4.1.2(ii)). To this purpose, let $S \subseteq \mathbb{R}$ be symmetric with respect to the origin, and suppose $S \notin \mathcal{B}(\mathbb{R})$. Consider the open subset $A \subseteq L^\infty(\mathbb{R})$, $A := \bigcup_{s\in S} B(\chi_{I_s}, \frac{1}{2})$, where

$$B\left(\chi_{I_s}, \frac{1}{2}\right) := \left\{f \in L^\infty(\mathbb{R}) \mid \|f - \chi_{I_s}\|_\infty < \frac{1}{2}\right\} \tag{4.122}$$

denotes the open ball with center χ_{I_s} and radius $\frac{1}{2}$ in $L^\infty(\mathbb{R})$. Since $\|\chi_{I_r} - \chi_{I_s}\|_\infty = 1$ for $r \neq s$, the family of balls in (4.122) is disjoint. Hence

$$\begin{aligned} F^{-1}(A) &= \left\{r \in \mathbb{R} \mid \exists s \in S \text{ such that } \|F(r) - \chi_{I_s}\|_\infty < \frac{1}{2}\right\} \\ &= \{r \in \mathbb{R} \mid |r| = |s|\} = S, \end{aligned}$$

since the condition

$$\|F(r) - \chi_{I_s}\|_\infty = \|\chi_{I_r} - \chi_{I_s}\|_\infty < \frac{1}{2}$$

implies that $\chi_{I_r} \in B(\chi_{I_s}, \frac{1}{2})$, and thus $\chi_{I_r} = \chi_{I_s}$ since the family of balls in (4.122) is disjoint. Therefore there exists an open subset $A \subseteq L^\infty(\mathbb{R})$ such that $F^{-1}(A) \notin \mathcal{B}(\mathbb{R})$. This proves the result. □

Let us now address the case where $Y = W^{m,r}(U)$ $(m \in \mathbb{N}, r \in [1,\infty))$. Let $U \subseteq \mathbb{R}^M$ and $V \subseteq \mathbb{R}^N$ be open, and let $f \in L^1_{\mathrm{loc}}(U \times V)$, $f = f(x,y)$ $(x \in U, y \in V)$. For any $k = 1,\dots,N$, we distinguish the distributional derivative

$$\left\langle \frac{\partial f}{\partial y_k}, \zeta \right\rangle = -\int\limits_{U\times V} f \frac{\partial \zeta}{\partial y_k}\, dxdy \quad \text{for } \zeta \in C_c^\infty(U \times V) \tag{4.123a}$$

from the distributional derivative $\partial_{y_k} f(x,\cdot)$ defined for a. e. $x \in U$ as follows:

$$\langle \partial_{y_k} f(x,\cdot), \eta\rangle = -\int\limits_V f(x,y) \frac{\partial \eta}{\partial y_k}(y)\, dy \quad \text{for } \eta \in C_c^\infty(V). \tag{4.123b}$$

For simplicity, we only state the result for $Y = W^{1,r}(U)$.

Proposition 4.5.4. *Let $U \subseteq \mathbb{R}^M$ and $V \subseteq \mathbb{R}^N$ be open. Let $p \in [1,\infty]$ and $r \in [1,\infty)$. Then the following statements are equivalent:*

(i) $f \in L^p(U; W^{1,r}(V))$;

(ii) *$f \in L^1_{\mathrm{loc}}(U \times V)$, and for all $k = 1, \dots, N$,*

$$\int_U \left(\int_V |f(x,y)|^r dy \right)^{\frac{p}{r}} dx < \infty, \quad \int_U \left(\int_V \left| \frac{\partial f}{\partial y_k}(x,y) \right|^r dy \right)^{\frac{p}{r}} dx < \infty \tag{4.124a}$$

if $p \in [1, \infty)$, and

$$\left\| \int_V |f(\cdot,y)|^r dy \right\|_{L^\infty(U)} < \infty, \quad \left\| \int_V \left| \frac{\partial f}{\partial y_k}(x,y) \right|^r dy \right\|_{L^\infty(U)} < \infty \tag{4.124b}$$

if $p = \infty$. Moreover, for a. e. $x \in U$, we have

$$\partial_{y_k} f(x,\cdot) = \frac{\partial f}{\partial y_k}(x,\cdot) \quad a.e.\ in\ V. \tag{4.125}$$

Proof. (i)⇒(ii). By Proposition 4.5.1 $f \in L^1_{\mathrm{loc}}(U \times V)$, and the first inequality in either statement (4.124) is satisfied. Moreover, by Proposition 4.5.1 we also have that for all $k = 1, \dots, N$,

$$\int_U \left(\int_V |\partial_{y_k} f(x,y)|^r dy \right)^{\frac{p}{r}} dx < \infty \quad \text{if } p \in [1, \infty),$$

$$\left\| \int_V |\partial_{y_k} f(\cdot,y)|^r dy \right\|_{L^\infty(U)} < \infty \quad \text{if } p = \infty.$$

Hence the claim follows if we prove equality (4.125). In (4.123a), set $\zeta(x,y) = \xi(x)\eta(y)$ with $\xi \in C^\infty_c(U)$ and $\eta \in C^\infty_c(V)$. Then by (4.123b) and the Fubini theorem

$$\begin{aligned} \left\langle \frac{\partial f}{\partial y_k}, \zeta \right\rangle &= - \int_{U \times V} f \xi \frac{\partial \eta}{\partial y_k} \, dxdy = - \int_U dx\, \xi(x) \int_V f(x,y) \frac{\partial \eta}{\partial y_k}(y)\, dy \\ &= \int_U dx\, \xi(x) \langle \partial_{y_k} f(x,\cdot), \eta \rangle = \langle \partial_{y_k} f(x,\cdot), \zeta \rangle, \end{aligned}$$

whence by standard arguments we obtain (4.125).

(ii)⇒(i). By (4.124)–(4.125) we have that $f, \partial_{y_k} f \in L^p(U; L^r(V))$ ($k = 1, \dots, N$) if we prove that the map from U to $W^{1,r}(V)$, $x \mapsto f(x,\cdot)$, is λ_N-measurable. By the separability of $W^{1,r}(V)$ ($r \in [1, \infty)$) this follows from Proposition 4.1.5 arguing as in the proof of Proposition 4.5.1. This completes the proof. □

4.6 Vector Sobolev spaces

The main purpose of this section is to prove two useful embedding results (Theorems 4.6.7–4.6.8).

4.6.1 Vector distributions

Let $U \subseteq \mathbb{R}^N$ be open. Let $\mathscr{D}(U)$ be the space of test functions on U, and let Y be a Banach space with norm $\|\cdot\|_Y$.

Definition 4.6.1. The space of *vector distributions* from U to Y (denoted $\mathscr{D}'(U;Y)$) is the space of continuous linear maps $T : \mathscr{D}(U) \mapsto Y$. Namely, $T \in \mathscr{D}'(U;Y)$ if $T : \mathscr{D}(U) \to Y$ is linear, and for any sequence $\zeta_k \to \zeta$ in $\mathscr{D}(U)$, we have $\|\langle T,\zeta_k\rangle - \langle T,\zeta\rangle\|_Y \to 0$.

Hereafter we use the symbol $T\zeta \equiv \langle T,\zeta\rangle \in Y$ ($T \in \mathscr{D}'(U;Y)$, $\zeta \in \mathscr{D}(U)$). We also set $\mathscr{D}'(U;\mathbb{R}) \equiv \mathscr{D}^*(U)$.

Remark 4.6.1. To every vector measure $\mu : \mathcal{B}(U) \mapsto Y$, there corresponds $T_\mu \in \mathscr{D}'(U;Y)$ defined by

$$\langle T_\mu,\zeta\rangle \equiv \langle \mu,\zeta\rangle = \int_U \zeta \, d\mu \quad (\zeta \in C_c^\infty(U)),$$

and by inequality (4.60) we have

$$\|\langle T_\mu,\zeta\rangle\|_Y \leq \int_U |\zeta| \, d|\mu|_w \leq \|\zeta\|_\infty \, |\mu|_w(U).$$

In particular, by Remark 4.3.3(ii) to every $f \in L^1_{\mathrm{loc}}(U;Y)$, there corresponds $T_f \in \mathscr{D}'(U;Y)$ such that

$$\langle T_f,\zeta\rangle = \int_U f\,\zeta \, dx \quad (\zeta \in C_c^\infty(U)),$$

$$\|\langle T_f,\zeta\rangle\|_Y \leq \int_U \|f\|_Y |\zeta| \, dx \leq \|\zeta\|_\infty \int_U \|f\|_Y \, dx.$$

Definition 4.6.2. A sequence $\{T_k\} \subseteq \mathscr{D}'(U;Y)$ *converges in* $\mathscr{D}'(U;Y)$ to $T \in \mathscr{D}'(U;Y)$ (written $T_k \to T$ in $\mathscr{D}'(U;Y)$) if $\|\langle T_k,\zeta\rangle - \langle T,\zeta\rangle\|_Y \to 0$ for all $\zeta \in C_c^\infty(U)$.

Definition 4.6.3. Let $T \in \mathscr{D}'(U;Y)$. By the *αth distributional derivative* of T, denoted $D^\alpha T$ ($\alpha \in \overline{\mathbb{N}}^N$), we mean the vector distribution $D^\alpha T \in \mathscr{D}'(U;Y)$,

$$\langle D^\alpha T,\zeta\rangle := (-1)^{|\alpha|}\langle T, D^\alpha \zeta\rangle \quad \text{for all } \zeta \in C_c^\infty(U).$$

We will use the same notations for distributional derivatives as in the scalar case $Y = \mathbb{R}$, e. g., $\frac{dT}{dx} \equiv D^1 T$ if $N = 1$.

Definition 4.6.4. The *convolution* of $T \in \mathscr{D}'(\mathbb{R}; Y)$ with $\rho \in C_c^\infty(\mathbb{R})$ is

$$\langle T * \rho, \zeta \rangle := \langle T, \zeta * \hat{\rho} \rangle \quad \text{for all } \zeta \in C_c^\infty(U),$$

where $\hat{\rho}(x) := \rho(-x)$ $(x \in \mathbb{R})$.

If $T = f \in L^1_{\mathrm{loc}}(\mathbb{R}; Y)$, then it is easily seen that

$$T * \rho = \int_{\mathbb{R}} \rho(\cdot - y) f(y)\, dy.$$

If ρ_ϵ is a standard mollifier, then we have the following:

Proposition 4.6.1. (i) *Let $T \in \mathscr{D}'(\mathbb{R}; Y)$. Then $T * \rho_\epsilon \in C^\infty(\mathbb{R}; Y)$, and $(D^\alpha T) * \rho_\epsilon \to D^\alpha T$ in $\mathscr{D}'(\mathbb{R}; Y)$ as $\epsilon \to 0^+$ $(\alpha \in \overline{\mathbb{N}}^N)$.*

(ii) *Let $f \in L^p(\mathbb{R}^N; Y)$ $(p \in [1, \infty))$. Then $\{\rho_\epsilon * f\} \subseteq L^p(\mathbb{R}^N; Y) \cap C^\infty(\mathbb{R}^N; Y)$, $\|\rho_\epsilon * f\|_p \le \|f\|_p$ for all $\epsilon > 0$, and $\rho_\epsilon * f \to f$ in $L^p(\mathbb{R}^N; Y)$ as $\epsilon \to 0^+$.*

Definition 4.6.5. Let Y and Z be Banach spaces, and let $L \in \mathscr{L}(Y; Z)$. The *image* of $T \in \mathscr{D}'(U; Y)$ under L is the vector distribution $L(T) \in \mathscr{D}'(U; Z)$ defined by

$$\langle L(T), \zeta \rangle := L(\langle T, \zeta \rangle) \quad \text{for all } \zeta \in C_c^\infty(U).$$

Remark 4.6.2. The following claims are easily checked:
(i) if $T_k \to T$ in $\mathscr{D}'(U; Y)$, then $L(T_k) \to L(T)$ in $\mathscr{D}'(U; Z)$;
(ii) for any $T \in \mathscr{D}'(U; Y)$, $D^\alpha(L(T)) = L(D^\alpha T) \in \mathscr{D}'(U; Z)$.

4.6.2 Definition and general properties

Definition 4.6.6. Let $U \subseteq \mathbb{R}^N$ be open, and let $m \in \overline{\mathbb{N}}$ and $p \in [1, \infty]$. The function $f \in L^p(U; Y)$ belongs to the *Sobolev space* $W^{m,p}(U; Y)$ if $D^\alpha f \in L^p(U; Y)$ for all $\alpha \in \overline{\mathbb{N}}^N$, $|\alpha| \le m$. The space $W^{m,p}_{\mathrm{loc}}(U; Y)$ is defined by replacing $L^p(U; Y)$ by $L^p_{\mathrm{loc}}(U; Y)$.

We set $W^{m,p}(U; \mathbb{R}) \equiv W^{m,p}(U)$. We will use the same notations as in the scalar case (see Subsection 3.1.2), e. g., we set $H^m(U; Y) \equiv W^{m,2}(U; Y)$

The space $W^{m,p}(U; Y)$ is a normed vector space endowed formally with the same norm $\|\cdot\|_{m,p}$ as in the scalar case (see (3.2)), where now

$$\|D^\alpha f\|_p^p = \int_U \|D^\alpha f\|_Y^p \, dx \quad \text{if } p \in [1, \infty),$$

$$\|D^\alpha f\|_\infty = \operatorname{ess\,sup} \|D^\alpha f\|_Y \quad \text{if } p = \infty.$$

The space $W^{m,p}(U;Y)$ endowed with this norm is a Banach space, as it is easily proved.

The following result is the counterpart of Theorem 3.1.3.

Theorem 4.6.2. *Let Y be separable. Then:*
(i) *if $p \in [1,\infty)$, then $W^{m,p}(U;Y)$ is separable;*
(ii) *if $p \in (1,\infty)$ and Y is reflexive, then $W^{m,p}(U;Y)$ is also reflexive.*

Proof. We only consider the case $N = m = 1$ for simplicity. Let $I \subseteq \mathbb{R}$ be an open interval. As in the proof of Proposition 3.1.4, set $X := [L^p(I;Y)]^2$ with norm

$$(f,g) \mapsto \|(f,g)\|_X := \left(\|f\|_p^p + \|g\|_p^p\right)^{1/p} \quad (p \in [1,\infty)).$$

Clearly, the map $T : W^{1,p}(I;Y) \to [L^p(I;Y)]^2$, $T(f) := (f, \frac{df}{dx})$, is an isometric isomorphism, and $Z := T(W^{1,p}(I;Y)) \subseteq X$ is a closed subspace and thus a Banach space with norm $\|\cdot\|_X$. Also, observe that $(f,g) \in Z$ if and only if $\frac{df}{dx} = g$.

(i) Since Y is separable, by Theorem 4.3.7 so is the space $L^p(I;Y)$, and hence X and Z are separable. Let $D \subseteq Z$ be countable and dense, and set $E := T^{-1}(D)$. Let $h \in W^{1,p}(I;Y)$. Then $Th = (h, \frac{dh}{dx}) \in Z$. By the denseness of D there exists a sequence $\{(f_k,g_k)\} \subseteq D$ such that $\lim_{k\to\infty} \|(f_k,g_k) - Th\|_X = 0$. Then for any $k \in \mathbb{N}$, we have $\frac{df_k}{dx} = g_k$, $\{f_k\} \subseteq E$, and

$$\lim_{k\to\infty} \|(f_k,g_k) - Th\|_X^p = \lim_{k\to\infty}\left(\|f_k - h\|_p^p + \left\|\frac{df_k}{dx} - \frac{dh}{dx}\right\|_p^p\right) = 0,$$

hence $f_k \to h$ in $W^{1,p}(I;Y)$, and claim (i) follows.

(ii) If Y is reflexive and $p \in (1,\infty)$, then the spaces $L^p(I;Y)$ and hence X and Z are also reflexive. Since $W^{1,p}(I;Y)$ is isomorphic to Z, claim (ii) follows. □

Remark 4.6.3. Let Y and Z be Banach spaces, and let $L \in \mathscr{L}(Y;Z)$. For any $f \in W^{m,p}(U;Y)$ ($m \in \overline{\mathbb{N}}$, $p \in [1,\infty]$), set

$$l(f)(x) := L(f(x,\cdot)) \quad \text{for a. e. } x \in U.$$

Since $D_y^\alpha(l(f)(x)) = L(D_y^\alpha f(x,\cdot))$, we have $l(f) \in W^{m,p}(U;Z)$ (see Remark 4.6.2(ii)). It follows that $l \in \mathscr{L}(W^{m,p}(U;Y); W^{m,p}(U;Z))$.

For instance, let $U = I \subseteq \mathbb{R}$ be an open interval, let $V \subseteq \mathbb{R}^N$ be open, and let

$$L : L^1(V) \to \mathbb{R}, \quad L(g) = \int_V g\,dx \quad (g \in L^1(V)).$$

Then the map

$$l : W^{1,p}(I;L^1(V)) \to W^{1,p}(I), \quad l(f)(x) := \int_V f(x,y)\,dy \quad \text{for } \lambda\text{-a. e. } x \in I,$$

is linear and continuous. Moreover, for λ-a. e. $x \in I$,

$$\frac{d}{dx}\int_V f(x,y)\,dy = \int_V \frac{\partial f}{\partial x}(x,y)\,dy.$$

When Y is a Lebesgue space, we have the following characterization of $W^{m,p}(U;Y)$, analogous to Propositions 4.5.1 and 4.5.4. We only deal with the case where U is an open interval $I \subseteq \mathbb{R}$. Let $V \subseteq \mathbb{R}^N$ be open. By Corollary 4.5.2 the space $L^1_{\mathrm{loc}}(I \times V)$ can be identified with $L^1_{\mathrm{loc}}(I; L^1_{\mathrm{loc}}(V))$. If $f \in L^1_{\mathrm{loc}}(I \times V)$, then we distinguish the distributional derivative $\frac{\partial f}{\partial x} \in \mathscr{D}^*(I \times V)$,

$$\left\langle \frac{\partial f}{\partial x}, \zeta \right\rangle = -\int_{I\times V} f \frac{\partial \zeta}{\partial x}\,dxdy \quad \text{for all } \zeta \in C_c^\infty(I \times V), \tag{4.126}$$

from the distributional derivative $\frac{df}{dx} \in \mathscr{D}'(I; L^1_{\mathrm{loc}}(V))$,

$$\left\langle \frac{df}{dx}, \xi \right\rangle = -\int_I f(x,\cdot)\,\xi'(x)\,dx \quad \text{for all } \xi \in C_c^\infty(I). \tag{4.127}$$

The proof of the following result is analogous to that of Proposition 4.5.4, thus is omitted (see [43, Proposition 2.6.1] for details).

Proposition 4.6.3. *Let $I \subseteq \mathbb{R}$ be an open interval, and let $V \subseteq \mathbb{R}^N$ be open. Let $p, r \in [1,\infty)$. Then the following statements are equivalent:*
(i) $f \in W^{1,p}(I; L^r(V))$;
(ii) $f, \frac{\partial f}{\partial x} \in L^1_{\mathrm{loc}}(I \times V)$, *and*

$$\int_I \left(\int_V |f(x,y)|^r\,dy\right)^{\frac{p}{r}} dx < \infty, \quad \int_I \left(\int_V \left|\frac{\partial f}{\partial x}(x,y)\right|^r dy\right)^{\frac{p}{r}} dx < \infty. \tag{4.128}$$

Moreover, for λ-a. e. $x \in I$,

$$\frac{df}{dx}(x) = \frac{\partial f}{\partial x}(x,\cdot) \quad \text{a. e. in } V. \tag{4.129}$$

4.6.3 Continuous embedding

Let W and Y be Banach spaces. We write $W \hookrightarrow Y$ if W is continuously embedded in Y, and $W \overset{c}{\hookrightarrow} Y$ if the embedding is compact. The following result will be proven.

Theorem 4.6.4. *Let $I \subseteq \mathbb{R}$ be an open interval. Then:*
(i) $W^{1,p}(I;Y) \hookrightarrow C^{1-\frac{1}{p}}(\bar{I};Y)$ *for any* $p \in (1,\infty)$;
(ii) $W^{1,\infty}(I;Y) \hookrightarrow \mathrm{Lip}(\bar{I};Y)$.

To prove Theorem 4.6.4, we must show that:

(i) for any $p \in (1,\infty)$, every $f \in W^{1,p}(I;Y)$ has a representative $\tilde{f} \in C^{1-\frac{1}{p}}(\bar{I};Y)$, and there exists $M_0 > 0$ such that

$$\|\tilde{f}\|_{C^{0,1-\frac{1}{p}}(I;Y)} \le M_0 \|f\|_{1,p}; \tag{4.130a}$$

(ii) every $f \in W^{1,\infty}(I;Y)$ has a representative $\tilde{f} \in \mathrm{Lip}(\bar{I};Y)$, and there exists $M_0 > 0$ such that

$$\|\tilde{f}\|_{\mathrm{Lip}(I;Y)} \le M_0 \|f\|_{1,\infty}. \tag{4.130b}$$

To this purpose, we need elementary Lemma 4.6.5 below, where for any $g \in L^1_{\mathrm{loc}}(I;Y)$, we set

$$\int_{x_1}^{x_2} g(x)\,dx := \begin{cases} \int_{(x_1,x_2)} g\,dx & \text{if } x_1 < x_2, \\ -\int_{(x_2,x_1)} g\,dx & \text{if } x_1 > x_2. \end{cases}$$

It is easily checked that

$$\sum_{j=1}^{p-1} \int_{x_j}^{x_{j+1}} g(x)\,dx = \int_{x_1}^{x_p} g(x)\,dx \quad (x_1,\dots,x_p \in I). \tag{4.131}$$

Lemma 4.6.5. *Let $f \in W^{1,p}(I;Y)$ $(p \in [1,\infty])$. Then*

(i) *f has a uniformly continuous representative $\tilde{f}$;*

(ii) *for any $x_1, x_2 \in I$, we have*

$$\tilde{f}(x_2) - \tilde{f}(x_1) = \int_{x_1}^{x_2} \frac{df}{dx}(x)\,dx. \tag{4.132}$$

Proof. Set $g(\xi) := \int_{\bar{x}}^{\xi} \frac{df}{dx}(x)\,dx$ $(\bar{x},\xi \in I)$. The definition is well posed since $\frac{df}{dx} \in L^p(I;Y) \subseteq L^1_{\mathrm{loc}}(I;Y)$. Moreover, by (4.131) we have

$$\|g(x_2) - g(x_1)\|_Y \le \left| \int_{x_1}^{x_2} \left\| \frac{df}{dx} \right\|_Y (x)\,dx \right|.$$

Since $\|\frac{df}{dx}\|_Y \in L^1_{\mathrm{loc}}(I)$, by the above inequality g is uniformly continuous in I, and thus, in particular, $g \in L^1_{\mathrm{loc}}(I;Y)$. It is easily checked that its distributional derivative $\frac{dg}{dx}$ is equal to $\frac{df}{dx}$. Plainly, this implies that there exists $y \in Y$ such that $f = g + y$ is in $\mathscr{D}'(I;Y)$ (e. g., see [43, Proposition 2.1.2]) and thus in $L^1_{\mathrm{loc}}(I;Y)$. Then $\tilde{f} := g + y$ is a uniformly continuous representative of f, which clearly satisfies (4.132). □

Proof of Theorem 4.6.4. Let $p \in (1,\infty)$. From (4.132) we get

$$\|\tilde{f}(x_2) - \tilde{f}(x_1)\|_Y \le \left|\int_{x_1}^{x_2} \left\|\frac{df(x)}{dx}\right\|_Y dx\right| \le \left\|\frac{df}{dx}\right\|_p |x_2 - x_1|^{1-\frac{1}{p}}. \tag{4.133}$$

Now observe that the function $x \mapsto \|\tilde{f}(x)\|_Y$ is uniformly continuous in I, and thus $\|\tilde{f}(\cdot)\|_Y \in C_b(I)$. Let $x_0 \in I$ be a maximum point, that is, $\|\tilde{f}(x_0)\|_Y = \|\tilde{f}\|_\infty$. Then from (4.132) we get

$$\tilde{f}(x_0) = \tilde{f}(x) + \int_{x_0}^{x} \frac{df}{dx}(x)\,dx \quad \text{for all } x \in I. \tag{4.134}$$

Fix $a \in (0, \min\{1, \frac{\lambda(I)}{2}\})$. At least one of the intervals $(x_0 - a, x_0)$ and $(x_0, x_0 + a)$ is contained in I. Suppose that $(x_0, x_0 + a) \subseteq I$. Then from (4.134) we get

$$\|\tilde{f}\|_\infty \le \|\tilde{f}(x)\|_Y + \int_{x_0}^{x} \left\|\frac{df}{dx}(x)\right\|_Y dx \quad \text{for all } x \in (x_0, x_0 + a).$$

Integrating the above inequality over $I_a := (x_0, x_0 + a)$, for any $p \in (1,\infty)$, we get

$$\|\tilde{f}\|_\infty \le \frac{1}{a}\|f\|_{L^1(I_a;Y)} + \left\|\frac{df}{dx}\right\|_{L^1(I_a;Y)} \le a^{-\frac{1}{p}}\|f\|_p + a^{1-\frac{1}{p}}\left\|\frac{df}{dx}\right\|_p. \tag{4.135}$$

By a suitable choice of $M_0 > 0$ from (4.133) and (4.135) we obtain (4.130a). This proves the result for $p \in (1,\infty)$. If $p = \infty$, then from (4.132) we get

$$\|\tilde{f}(x_2) - \tilde{f}(x_1)\|_Y \le \left|\int_{x_1}^{x_2} \left\|\frac{df}{dx}\right\|_Y (x)\,dx\right| \le \left\|\frac{df}{dx}\right\|_\infty |x_2 - x_1| \quad \text{if } p = \infty. \tag{4.136}$$

From (4.136) we obtain (4.130b) with $M_0 = 1$, and thus the result follows. □

For further purposes, let us state the following prolongation result (see [43, Corollaire 2.3.1] for the proof).

Proposition 4.6.6. *Let $I \subseteq \mathbb{R}$ be a bounded open interval. Then there exists a linear bounded operator $T : L^1(I;Y) \to L^1(\mathbb{R};Y)$ such that:*
(i) *$(Pf)(x) = f(x)$ for a. e. $x \in I$;*
(ii) *there exists $M_1 > 0$ (only depending on $\lambda(I)$) such that for all $p \in [1,\infty]$ and for any Banach space Z,*

$$f \in L^1(I;Y)\bigcap L^p(I;Z) \quad \Rightarrow \quad Pf \in L^p(\mathbb{R};Z), \quad \|Pf\|_{L^p(\mathbb{R};Z)} \le M_1\|f\|_{L^p(I;Z)}, \tag{4.137a}$$

$$f \in L^1(I;Y)\bigcap W^{1,p}(I;Z) \quad \Rightarrow \quad Pf \in W^{1,p}(\mathbb{R};Z), \quad \|Pf\|_{W^{1,p}(\mathbb{R};Z)} \le M_1\|f\|_{W^{1,p}(I;Z)}. \tag{4.137b}$$

4.6.4 Compact embedding

Let us prove the following results (see [94]).

Theorem 4.6.7 (Aubin). *Let $I \subseteq \mathbb{R}$ be a bounded open interval, and let $p \in (1,\infty]$ and $r \in [1,\infty]$. Let W, Y, Z be Banach spaces such that $W \overset{c}{\hookrightarrow} Y \hookrightarrow Z$. Let $E \subseteq W^{1,p}(I;Z) \bigcap L^r(I;W)$ be bounded in both spaces, that is,*

$$\sup_{f\in E}\big(\|f\|_{W^{1,p}(I;Z)} + \|f\|_{L^r(I;W)}\big) < \infty. \tag{4.138}$$

Then E is relatively compact in both $C(\bar{I};Z)$ and $L^r(I;Y)$.

Theorem 4.6.8 (Simon). *Let $I \subseteq \mathbb{R}$ be a bounded open interval, and let $r \in [1,\infty)$. Let W, Y, Z be Banach spaces such that $W \overset{c}{\hookrightarrow} Y \hookrightarrow Z$. Let $E \subseteq W^{1,1}(I;Z) \bigcap L^r(I;W)$ be bounded in both spaces, that is,*

$$\sup_{f\in E}\big(\|f\|_{W^{1,1}(I;Z)} + \|f\|_{L^r(I;W)}\big) < \infty. \tag{4.139}$$

Then E is relatively compact in both $L^p(I;Z)$ $(p \in [1,\infty))$ and $L^r(I;Y)$.

To prove Theorems 4.6.7–4.6.8, we need two preliminary lemmas.

Lemma 4.6.9. *Let $I \subseteq \mathbb{R}$ be a bounded open interval, and let $p \in [1,\infty]$. Let W, Y, Z be Banach spaces such that $W \overset{c}{\hookrightarrow} Y \hookrightarrow Z$. Let $E \subseteq W^{1,p}(I;Z) \bigcap L^r(I;W)$ satisfy (4.138). For every $f \in E$ and $k \in \mathbb{N}$, set*

$$f_k : \bar{I} \to Z, \quad f_k := (Pf) * \rho_{1/k}|_{\bar{I}} \quad (k \in \mathbb{N}),$$

where $P : L^1(I;Z) \mapsto L^1(\mathbb{R};Z)$ is the prolongation operator given by Proposition 4.6.6, and $\rho_{1/k}$ is a standard mollifier. Then for all $k \in \mathbb{N}$, the set $E_k := \{f_k \mid f \in E\}$ is relatively compact in $C(\bar{I};Z)$.

Proof. Fix $k \in \mathbb{N}$. Observe preliminarily that for any $f \in E \subseteq W^{1,p}(I;Z) \bigcap L^r(I;W)$, we have $Pf \in W^{1,p}(\mathbb{R};Z) \bigcap L^r(\mathbb{R};W)$, and hence $f_k \in C^\infty(I;Z)$. The result will follow by the Ascoli–Arzelà theorem if we prove that (a) the set $E_k \subseteq C(\bar{I};Z)$ is equicontinuous and (b) for any $x \in I$, the set $E_{k,x} := \{f_k(x) \mid f \in E\}$ is relatively compact in Z.

(a) We must show that for any $\epsilon > 0$ there exists $\delta > 0$ such that $|f_k(x_1) - f_k(x_2)| < \epsilon$ for all $f \in E$ and any $x_1, x_2 \in I$ such that $|x_1 - x_2| < \delta$. To this aim, it suffices to prove that the set $F_k := \{f_k' \mid f_k \in E_k\}$ is bounded in $C(\bar{I};Z)$, that is, there exists $C_k > 0$ such that

$$\|f_k'\|_{C(\bar{I};Z)} \le C_k \quad \text{for all } f \in Z. \tag{4.140}$$

In fact, for any fixed $f \in E$, by definition $f_k(x) = \int_{\mathbb{R}}(Pf)(y)\rho_{1/k}(x-y)\,dy$, and hence $f_k'(x) = \int_{\mathbb{R}}(Pf)(y)\rho_{1/k}'(x-y)\,dy$. Therefore by (4.137a)

$$\|f_k'(x)\|_Z \le \|\rho_{1/k}'\|_{L^q(\mathbb{R})}\|Pf\|_{L^p(\mathbb{R};Z)} \le C\|\rho_{1/k}'\|_{L^q(\mathbb{R})}\|f\|_{L^p(I;Z)},$$

where q denotes the Hölder conjugate of p. Taking the supremum over $x \in I$ and using (4.138), from the above inequality we obtain that (a) is satisfied.

(b) By (4.137a) for any $x \in I$, we have

$$\|f_k(x)\|_W \le \|\rho_{1/k}\|_{L^s(\mathbb{R})}\|Pf\|_{L^r(\mathbb{R};Z)} \le C\|\rho_{1/k}\|_{L^s(\mathbb{R})}\|f\|_{L^r(I;W)},$$

where s denotes the Hölder conjugate of r. Then by (4.138) the set $E_{k,x} := \{f_k(x) \mid f \in E\}$ is bounded in W and hence relatively compact in Z. This proves (b), and thus the result follows. □

Lemma 4.6.10. *Let W, Y, Z be Banach spaces such that $W \overset{c}{\hookrightarrow} Y \hookrightarrow Z$. Then for any $\eta > 0$, there exists $C_\eta > 0$ such that for all $f \in W$,*

$$\|f\|_Y \le \eta\,\|f\|_W + C_\eta\,\|f\|_Z. \tag{4.141}$$

Proof. By contradiction, let there exist $\eta > 0$ such that for every $k \in \mathbb{N}$, there exists $f_k \in W$ satisfying

$$\|f_k\|_Y > \eta\,\|f_k\|_W + k\|f_k\|_Z. \tag{4.142}$$

Set $g_k := \frac{f_k}{\|f_k\|_Y}$ ($k \in \mathbb{N}$). Then $\|g_k\|_Y = 1$ for all $k \in \mathbb{N}$, and thus by (4.142)

$$\|g_k\|_W \le \frac{1}{\eta}, \quad \|g_k\|_Z \le \frac{1}{k}. \tag{4.143}$$

By the first inequality in (4.143) the sequence $\{g_k\}$ is bounded in W. Then since $W \overset{c}{\hookrightarrow} Y \hookrightarrow Z$, there exist a subsequence (not relabeled) of $\{g_k\}$ and $g \in Z$ such that $\lim_{k\to\infty}\|g_k - g\|_Y = \lim_{k\to\infty}\|g_k - g\|_Z = 0$. By the second inequality in (4.143) $g = 0$, whence $\lim_{k\to\infty}\|g_k\|_Y = 0$. This is absurd since $\|g_k\|_Y = 1$ for all $k \in \mathbb{N}$, and hence the result follows. □

Proof of Theorem 4.6.7. (i) Let us prove that E is relatively compact in $C(\bar{I};Z)$. To this purpose, we first prove that

$$\lim_{k\to\infty}\sup_{f\in E}\|f_k - f\|_{C(\bar{I};Z)} = 0, \tag{4.144}$$

where $\{f_k\}$ is the sequence associated with f by Lemma 4.6.9. For any $x \in I$, we have $(Pf)(x) = f(x)$, and thus

$$\|f_k(x) - f(x)\|_Z = \left\| \int_{\mathbb{R}} [(Pf)(x-y) - (Pf)(x)]\,\rho_{1/k}(y)\,dy \right\|_Z \leq \int_{\mathbb{R}} \|(Pf)(x-y) - (Pf)(x)\|_Z\,\rho_{1/k}(y)\,dy. \tag{4.145}$$

By (4.130a) and (4.137b), from (4.145) we get

$$\|f_k(x) - f(x)\|_Z \leq M_0 \|Pf\|_{W^{1,p}(\mathbb{R};Z)} \int_{(-\frac{1}{k},\frac{1}{k})} |y|^{1-\frac{1}{p}} \rho_{1/k}(y)\,dy \leq \frac{M_0 M_1}{k^{1-\frac{1}{p}}} \|f\|_{W^{1,p}(I;Z)}.$$

Therefore

$$\sup_{f\in E} \|f_k - f\|_{C(\overline{I};Z)} \leq \frac{M_0 M_1}{k^{1-\frac{1}{p}}} \sup_{f\in E} \|f\|_{W^{1,p}(I;Z)}.$$

In view of (4.138), letting $k \to \infty$ in the above inequality, we obtain (4.144).

Fix $\epsilon > 0$. By (4.144) there exists $m \in \mathbb{N}$ such that

$$\|f\|_{C(\overline{I};Z)} \leq \|f_m\|_{C(\overline{I};Z)} + \frac{\epsilon}{3} \quad \text{for all } f \in E. \tag{4.146}$$

It follows that $E \subseteq F_m := \{g \in C(\overline{I};Z) \mid d(g, E_m) < \frac{\epsilon}{3}\}$ (recall that $E_m := \{f_m \mid f \in E\}$; see Lemma 4.6.9). On the other hand, since by Lemma 4.6.9 E_m is relatively compact in $C(\overline{I};Z)$, there exist $n_m \in \mathbb{N}$ and $f_m^{(1)},\dots,f_m^{(n_m)} \in E_m$ (corresponding to $f^{(1)},\dots,f^{(n_m)} \in E$) such that

$$E_m \subseteq \bigcup_{l=1}^{n_m} B\left(f_m^{(l)}, \frac{\epsilon}{3}\right) \subseteq \bigcup_{l=1}^{n_m} B\left(f^{(l)}, \frac{2\epsilon}{3}\right), \tag{4.147}$$

where $B(g,r) \subseteq C(\overline{I};Z)$ is the open ball with center g and radius r, and (4.146) was used. By (4.147) we have

$$E \subseteq F_m \subseteq \bigcup_{l=1}^{n_m} B(f^{(l)}, \epsilon) \quad \text{with } f^{(1)},\dots,f^{(n_m)} \in E,$$

and thus the claim follows.

(ii) Let us now prove that E is relatively compact in $L^r(I;Y)$. By (i) E is relatively compact in $C(\overline{I};Z)$ and thus in $L^r(I;Z)$. Then for any $\{f_i\} \subseteq E$, there exists a subsequence (not relabeled for simplicity) converging in $L^r(I;Z)$ ($r \in [1,\infty]$). On the other hand, for any fixed $\epsilon > 0$, set $\eta := \frac{\epsilon}{4\sup_{f\in E}\|f\|_{L^r(I;W)}}$. Then by Lemma 4.6.10 there exists $C_\eta > 0$ such that for all $i, j \in \mathbb{N}$,

$$\|f_i - f_j\|_{L^r(I;Y)} \leq \eta\,\|f_i - f_j\|_{L^r(I;W)} + C_\eta\,\|f_i - f_j\|_{L^r(I;Z)}. \tag{4.148}$$

Since $\{f_i\}$ is converging in $L^r(I;Z)$, there exists $l \in \mathbb{N}$ such that $\|f_i - f_j\|_{L^r(I;Z)} < \frac{\epsilon}{2C_\eta}$ for all $i, j \geq l$. Then from (4.148) we obtain that $\|f_i - f_j\|_{L^r(I;Y)} < \epsilon$ for $i, j \geq l$, and thus $\{f_i\}$ converges in $L^r(I;Y)$. This completes the proof. □

Proof of Theorem 4.6.8. (i) Let us prove that E is relatively compact in $L^1(I;Z)$. To this purpose, observe that by Lemma 4.6.9 the set $E_k := \{f_k \mid f \in E\}$ $(k \in \mathbb{N})$ is relatively compact in $C(\bar{I};Z)$. Since $C(\bar{I};Z) \hookrightarrow L^1(I;Z)$, E_k s relatively compact in $L^1(I;Z)$. Let us prove that

$$\lim_{k\to\infty}\sup_{f\in E}\|f_k - f\|_{L^1(I;Z)} = 0. \tag{4.149}$$

Arguing as for (4.145), we have that

$$\begin{aligned}\|f_k - f\|_{L^1(I;Z)} &\leq \int_I dx \int_{\mathbb{R}} \|(Pf)(x-y) - (Pf)(x)\|_Z\,\rho_{1/k}(y)\,dy \\ &= \int_{(-\frac{1}{k},\frac{1}{k})} dy\,\rho_{1/k}(y) \int_I \|(Pf)(x-y) - (Pf)(x)\|_Z\,dx \\ &\leq \sup_{|y|\leq\frac{1}{k}} \int_I \|(Pf)(x-y) - (Pf)(x)\|_Z\,dx.\end{aligned} \tag{4.150}$$

On the other hand, we have

$$\begin{aligned}\int_I \|(Pf)(x-y) - (Pf)(x)\|_Z\,dx &\leq \int_I dx \int_{\mathbb{R}} \|(Pf)'(\xi)\|_Z \chi_{J_{x-y,y}}(\xi)\,d\xi \\ = \int_{\mathbb{R}} d\xi\,\|(Pf)'(\xi)\|_Z \int_I \chi_{K_{\xi,\xi+y}}(\eta)\,d\eta &\leq |y| \int_{\mathbb{R}} d\xi\,\|(Pf)'(\xi)\|_Z,\end{aligned} \tag{4.151}$$

where $J_{x-y,y}$ and $K_{\xi,y}$ denote the intervals with extremes $x-y$, y and ξ, $\xi+y$, respectively. By (4.150)–(4.151), using (4.137b), we get

$$\|f_k - f\|_{L^1(I;Z)} \leq \frac{M_1}{k}\,\|f\|_{W^{1,1}(I;Z)}.$$

In view of (4.139), letting $k \to \infty$ in this inequality, we obtain (4.149).

Fix $\epsilon > 0$. By (4.149) there exists $m \in \mathbb{N}$ such that

$$\|f\|_{L^1(I;Z)} \leq \|f_m\|_{L^1(I;Z)} + \frac{\epsilon}{3} \quad \text{for all } f \in E. \tag{4.152}$$

Hence $E \subseteq F_m := \{g \in L^1(I;Z) \mid d(g,E_m) < \frac{\epsilon}{3}\}$. On the other hand, since E_m is relatively compact in $L^1(I;Z)$, there exist $n_m \in \mathbb{N}$ and $f_m^{(1)},\dots,f_m^{(n_m)} \in E_m$ (corresponding to $f^{(1)},\dots,f^{(n_m)} \in E$) such that

$$E_m \subseteq \bigcup_{l=1}^{n_m} B\left(f_m^{(l)}, \frac{\epsilon}{3}\right) \subseteq \bigcup_{l=1}^{n_m} B\left(f^{(l)}, \frac{2\epsilon}{3}\right), \tag{4.153}$$

where $B(g,r) \subseteq L^1(I;Z)$ denotes the open ball with center g and radius r, and inequality (4.152) was used. By (4.153) we have

$$E \subseteq F_m \subseteq \bigcup_{l=1}^{n_m} B(f^{(l)},\epsilon) \quad \text{with } f^{(1)},\dots,f^{(n_m)} \in E,$$

and thus the claim follows.

(ii) Now we can prove that E is relatively compact in $L^p(I;Z)$ for all $p \in (1,\infty)$. By (i) E is relatively compact in $L^1(I;Z)$, and thus for any $\{f_i\} \subseteq E$, there exists a subsequence (not relabeled for simplicity) converging in $L^1(I;Z)$ and hence λ-a. e. in I. On the other hand, since E is bounded in $W^{1,1}(I;Z) \hookrightarrow C(\bar{I};Z) \hookrightarrow L^p(I;Z)$ ($p \in (1,\infty)$), by Corollary 4.3.5 the sequence $\{f_i\}$ converges in $L^p(I;Z)$ for all $p \in (1,\infty)$. Hence the claim follows.

(iii) To prove that E is relatively compact in $L^r(I;Y)$, observe that by (ii) E is relatively compact in $L^r(I;Z)$ ($r \in [1,\infty)$). Then for any $\{f_i\} \subseteq E$, there exists a subsequence (not relabeled for simplicity) converging in $L^r(I;Z)$. Arguing as in the proof of Theorem 4.6.7 shows that $\{f_i\}$ converges in $L^r(I;Y)$. Then the result follows. □

5 Sequences of finite Radon measures

5.1 Notions of convergence

Let X be a locally compact Hausdorff space. The vector space $\mathfrak{R}_f(X)$ of finite signed Radon measures on X, endowed with the norm $\|\mu\| := |\mu|(X)$ ($|\mu|$ being the variation of the measure μ), is a Banach space; in fact, it is a closed subspace of the Banach space $(\mathfrak{M}_f(X), \|\cdot\|)$ of finite signed measures on X. In this chapter, we first discuss different types of convergence in this space, and then we establish their relationship with analogous concepts of convergence for Young measures. Once this is made, we draw conclusions about the convergence of bounded sequences in $L^1(X)$.

5.1.1 Strong convergence

Proposition 1.8.4 suggests the following definition, which in particular applies to $\mathfrak{R}_f(X)$ if X is a locally compact Hausdorff space endowed with the Borel σ-algebra.

Definition 5.1.1. Let $(X, \mathcal{A})$ be a measurable space. A sequence $\{\mu_k\} \subseteq \mathfrak{M}_f(X)$ *strongly converges* to $\mu \in \mathfrak{M}_f(X)$ (written $\mu_k \to \mu$) if

$$\lim_{k\to\infty} \|\mu_k - \mu\| = \lim_{k\to\infty} |\mu_k - \mu|(X) = 0.$$

The *strong topology* on $\mathfrak{M}_f(X)$ is the metric topology associated with the norm $\|\cdot\|$.

For clearness, we denote by $\|\cdot\|_{\mathfrak{R}_f(X)}$ the restriction of the above norm to the subspace $\mathfrak{R}_f(X) \subseteq \mathfrak{M}_f(X)$.

5.1.2 Weak* convergence

Theorem 2.7.1 suggests the following definition.

Definition 5.1.2. Let X be a locally compact Hausdorff space. A sequence $\{\mu_k\} \subseteq \mathfrak{R}_f(X)$ converges *weakly** to $\mu \in \mathfrak{R}_f(X)$ (written $\mu_k \overset{*}{\rightharpoonup} \mu$) if

$$\lim_{k\to\infty} \int_X g\, d\mu_k = \int_X g\, d\mu \quad \text{for all } g \in C_0(X). \tag{5.1}$$

The *weak* topology* on $\mathfrak{R}_f(X)$, denoted $\mathcal{T}_{w^*}$, is the weakest topology that makes all maps $\mu \mapsto \int_X g\, d\mu$, $g \in C_0(X)$, continuous.

Remark 5.1.1. Since $C_0(X) \subseteq C_b(X) \subseteq L^1(X, \mathcal{B}(X), |\mu|)$ and $C_0(X) \subseteq C_b(X) \subseteq L^1(X, \mathcal{B}(X), |\mu_k|)$ for all $k \in \mathbb{N}$, the integrals in (5.1) and (5.3) are well defined (see Definition 2.2.4).

https://doi.org/10.1515/9783110556902-006

A basis for $\mathcal{T}_{w^*}$ is the following family of subsets of $\mathfrak{R}_f(X)$:

$$B_{g_1,\dots,g_n;\epsilon}(\mu_0) := \left\{\mu \in \mathfrak{R}_f(X) \,\middle|\, \left|\int_X g_j\,d\mu - \int_X g_j\,d\mu_0\right| < \epsilon \quad \forall j = 1,\dots,n\right\}, \tag{5.2}$$

where $g_1,\dots,g_n \in C_0(X)$, $\mu_0 \in \mathfrak{R}_f(X)$, $n \in \mathbb{N}$ and $\epsilon > 0$.

Under suitable assumptions on X, the weak* topology $\mathcal{T}_{w^*}$ restricted to bounded subsets of $\mathfrak{R}_f(X)$ is metrizable:

Proposition 5.1.1. *Let X be a locally compact Hausdorff space with countable basis, and let $\mathcal{M} \subseteq \mathfrak{R}_f(X)$ be bounded. Then the relative weak* topology $\mathcal{T}_{w^*} \cap \mathcal{M}$ is metrizable.*

Proof. The space $C_0(X)$ endowed with the $\|\cdot\|_\infty$ norm is separable (see Appendix A, Subsection A.7). Then by Theorem 2.7.1 and Proposition A.11 the result follows. □

5.1.3 Narrow convergence

Choosing in Definition 5.1.2 a different class of test functions leads to another notion of convergence on $\mathfrak{R}_f(X)$.

Definition 5.1.3. Let X be a locally compact Hausdorff space. A sequence $\{\mu_k\} \subseteq \mathfrak{R}_f(X)$ converges *narrowly* to $\mu \in \mathfrak{R}_f(X)$ (written $\mu_k \overset{n}{\rightharpoonup} \mu$) if

$$\lim_{k\to\infty} \int_X g\,d\mu_k = \int_X g\,d\mu \quad \text{for all } g \in C_b(X). \tag{5.3}$$

The *narrow topology* on $\mathfrak{R}_f(X)$, denoted $\mathcal{T}_n$, is the weakest topology that makes all maps $\mu \mapsto \int_X g\,d\mu$, $g \in C_b(X)$, continuous.

The narrow topology is called "weak topology" by some authors (e. g., see [45]). A basis for $\mathcal{T}_n$ is the family of subsets $B_{g_1,\dots,g_n;\epsilon}(\mu_0) \subseteq \mathfrak{R}_f(X)$ defined by (5.2), yet with $g_1,\dots,g_n \in C_b(X)$.

Remark 5.1.2. The *vague convergence* of a sequence $\{\mu_k\} \subseteq \mathfrak{R}_f(X)$ to $\mu \in \mathfrak{R}_f(X)$ is defined by

$$\lim_{k\to\infty} \int_X g\,d\mu_k = \int_X g\,d\mu \quad \text{for all } g \in C_c(X)$$

(e. g., see [42]). Clearly, in $\mathfrak{R}_f(X)$, strong convergence ⇒ narrow convergence ⇒ weak* convergence ⇒ vague convergence.

Remark 5.1.3. From the lower semicontinuity of the norm it follows that if $\mu_k \overset{*}{\rightharpoonup} \mu$ (in particular, if $\mu_k \overset{n}{\rightharpoonup} \mu$), then

$$|\mu|(X) = \|\mu\|_{\mathfrak{R}_f(X)} \leq \liminf_{k\to\infty} \|\mu_k\|_{\mathfrak{R}_f(X)} = \liminf_{k\to\infty} |\mu_k|(X). \tag{5.4}$$

In particular, $\mu(X) \leq \liminf_{k\to\infty} \mu_k(X)$ if $\{\mu_k\} \subseteq \mathfrak{R}_f^+(X)$. Specifically, if $\mu_k \overset{n}{\rightharpoonup} \mu$, then choosing $g = 1$ in (5.3) gives $\lim_{k\to\infty} \mu_k(X) = \mu(X)$, and hence "no mass is lost". This need not be the case if $\mu_k \overset{*}{\rightharpoonup} \mu$: if $X = \mathbb{R}$ and $\mu_k = \delta_k$ ($k \in \mathbb{N}$), then $\mu_k \overset{*}{\rightharpoonup} 0$ and $\mu_k(\mathbb{R}) = 1$ for all k. Hence in this case, inequality (5.4) is strict. Similar phenomena were encountered for sequences of functions (see Example 2.8.1(i)).

Remark 5.1.4. Let X be a locally compact Hausdorff space, and let $\mu \in \mathfrak{R}_f^+(X)$. Let $\{h_k\} \subseteq L^1(X, \mathcal{B}(X), \mu)$, and for any $k \in \mathbb{N}$, set

$$\nu_k(E) := \int_E h_k \, d\mu \quad (E \in \mathcal{B}(X)). \tag{5.5}$$

Clearly, $\{\nu_k\} \subseteq \mathfrak{R}_f(X)$, and $\nu_k \ll \mu$ for all $k \in \mathbb{N}$. There are obvious connections between the convergence of $\{h_k\}$ in $L^1(X, \mathcal{B}(X), \mu)$ and that of $\{\nu_k\}$ in $\mathfrak{R}_f(X)$:

- if $h_k \to h$ strongly in $L^1(X)$ and $\nu(E) := \int_E h \, d\mu$ ($E \in \mathcal{B}(X)$), then $\|\nu_k - \nu\|_{\mathfrak{R}_f(X)} = |\nu_k - \nu|(X) = \int_X |h_k - h| \, d\mu \to 0$, and hence $\nu_k \to \nu$ strongly in $\mathfrak{R}_f(X)$;
- if $h_k \to h$ weakly in $L^1(X)$, then, in particular,

  $$\lim_{k\to\infty} \int_X g h_k \, d\mu = \int_X g h \, d\mu \quad \text{for all } g \in C_b(X),$$

  and thus

  $$\lim_{k\to\infty} \int_X g \, d\nu_k = \int_X g \, d\nu \quad \text{for all } g \in C_b(X)$$

  with ν as above. Hence $\nu_k \overset{n}{\rightharpoonup} \nu$.

More generally, even if the sequence $\{h_k\}$ does not converge in $L^1(X)$, the sequence $\{\nu_k\}$ defined by (5.5) can converge (weakly* or narrowly) to some $\nu \in \mathfrak{R}_f(X)$ (see Example 5.1.1(i)). By abuse of notation, in such cases, we occasionally write $h_k \overset{*}{\rightharpoonup} \nu$ or $h_k \overset{n}{\rightharpoonup} \nu$.

Having Remark 5.1.4 in mind, let us point out some phenomena related to convergence in $\mathfrak{R}_f(X)$.

Example 5.1.1. (i) Every $\nu \in \mathfrak{R}_f^+(\mathbb{R})$ is the weak* limit of a sequence $\{h_k\} \subseteq C_c^\infty(\mathbb{R})$ bounded in $L^1(\mathbb{R}) \equiv L^1(\mathbb{R}, \mathcal{B}, \lambda)$ (more exactly, of the sequence of measures ν_k with density h_k defined in (5.5)). In fact, let $\rho_{1/k}$ be a mollifier, $\rho_{1/k}(x) := k\rho(kx)$ with $\rho \in C_c^\infty(\mathbb{R})$, $\rho \geq 0$, $\int_{\mathbb{R}} \rho \, d\lambda = 1$ $(k \in \mathbb{N})$. For any $k \in \mathbb{N}$, consider the convolution $\nu * \rho_{1/k} \in C_c^\infty(\mathbb{R})$,

$$\int_{\mathbb{R}} g(x)(\nu * \rho_{1/k})(x)\, d\lambda(x) := \int_{\mathbb{R}} d\nu(x) \int_{\mathbb{R}} \rho_{1/k}(y-x)(x) g(y)\, d\lambda(y) \quad (g \in C_c^\infty(\mathbb{R}))$$

(see Definition 3.1.4(ii) and Remark 3.1.1). Set $h_k := \nu * \rho_{1/k}$ and $\nu_k(E) := \int_E h_k \, d\lambda$ for $E \in \mathcal{B}$. By Proposition 3.1.1(i)

$$\lim_{k\to\infty} \int_{\mathbb{R}} g\, d\nu_k = \lim_{k\to\infty} \int_{\mathbb{R}} g h_k \, d\lambda = \int_{\mathbb{R}} g \, d\nu \quad \text{for all } g \in C_c^\infty(\mathbb{R}).$$

It is easily seen that $\|h_k\|_{L^1(\mathbb{R})} \leq \|\nu\| = \nu(X) < \infty$, and hence by standard arguments the above equality holds for any $g \in C_0(\mathbb{R})$.

(ii) In particular, the above considerations apply if ν is an atomic measure, e. g., if $\nu = \delta_0$. Hence a sequence $\{\nu_k\} \subseteq \mathfrak{R}_f^+(\mathbb{R})$, $\nu_k \ll \lambda$ for all $k \in \mathbb{N}$, can converge weakly* to a measure $\nu \perp \lambda$. Also, in this case, we say that *mass concentration* occurs (see Example 2.8.1). Conversely, consider the measure space $(I, \mathcal{B} \cap I, \lambda|_{\mathcal{B}\cap I})$, where $I \equiv (0,1)$. For any fixed $h \in C_b(I)$, set $\nu(E) := \int_E h \, d\lambda$ $(E \in \mathcal{B} \cap I)$ and consider the sequence of atomic measures $\nu_k := \frac{1}{k}\sum_{j=1}^k h(\frac{j}{k})\delta_{\frac{j}{k}}$ $(k \in \mathbb{N})$. By elementary results we have

$$\lim_{k\to\infty} \int_I g\, d\nu_k = \lim_{k\to\infty} \frac{1}{k} \sum_{j=1}^k g\left(\frac{j}{k}\right) h\left(\frac{j}{k}\right) = \int_I g h \, d\lambda = \int_I g\, d\nu \quad \text{for all } g \in C_b(I).$$

Hence in this case, $\nu_k \perp \lambda$ for all $k \in \mathbb{N}$, $\nu \ll \lambda$, and $\nu_k \overset{n}{\rightharpoonup} \nu$.

Henceforth we will mostly deal with sequences of finite (positive) Radon measures. Results for sequences of finite signed Radon measures can be easily obtained from those proven below (see Remark 2.7.2).

The narrow convergence can be characterized as follows.

Theorem 5.1.2 (Portmanteau theorem). *Let X be a locally compact Hausdorff space with countable basis, and let $\{\mu_k\} \subseteq \mathfrak{R}_f^+(X)$, $\mu \in \mathfrak{R}_f^+(X)$. Then the following statements are equivalent:*

(i) $\mu_k \overset{n}{\rightharpoonup} \mu$;

(ii) *for every uniformly continuous $g \in C_b(X)$,*

$$\lim_{k\to\infty} \int_X g \, d\mu_k = \int_X g \, d\mu; \tag{5.6}$$

(iii) $\mu_k(X) \to \mu(X)$, *and for any closed subset* $C \subseteq X$,

$$\limsup_{k\to\infty} \mu_k(C) \le \mu(C); \tag{5.7}$$

(iv) $\mu_k(X) \to \mu(X)$, *and for any open subset* $A \subseteq X$,

$$\liminf_{k\to\infty} \mu_k(A) \ge \mu(A); \tag{5.8}$$

(v) *for all* $E \in \mathcal{B}(X)$ *such that* $\mu(\partial E) = 0$,

$$\lim_{k\to\infty} \mu_k(E) = \mu(E). \tag{5.9}$$

Proof. (i)⇒(ii). Obvious from Definition 5.1.3.

(ii)⇒(iii). Choosing $g = 1$ shows that $\mu_k(X) \to \mu(X)$. Let X be endowed with metric d compatible with the given topology. Inequality (5.7) is obvious if $C = \emptyset$. Otherwise, for any $n \in \mathbb{N}$, set $C_{1/n} := \{x \in X \mid d(x, C) < 1/n\}$. Clearly, $C_{1/n}$ is open, and $C = \bigcap_{n=1}^{\infty} C_{1/n}$, and thus $\mu(C) = \lim_{n\to\infty} \mu(C_{1/n})$. Set $f_n : X \mapsto [0, \infty)$, $f_n(x) := \max\{1 - nd(x, C), 0\}$ ($n \in \mathbb{N}$). Plainly, for any $n \in \mathbb{N}$, the function f_n is uniformly continuous in X, and $\chi_C \le f_n \le \chi_{C_{1/n}}$. Then by (5.6) for any fixed $n \in \mathbb{N}$, we get

$$\limsup_{k\to\infty} \mu_k(C) \le \lim_{k\to\infty} \int_X f_n \, d\mu_k = \int_X f_n \, d\mu \le \mu(C_{1/n}).$$

Letting $n \to \infty$ in this inequality, we get (5.7), and thus the claim follows.

(iii)⇔(iv). By (5.7), for any open subset $A \subseteq X$,

$$\liminf_{k\to\infty} \mu_k(A) = \lim_{k\to\infty} \mu_k(X) - \limsup_{k\to\infty} \mu_k(A^c) \ge \mu(X) - \mu(A^c) = \mu(A),$$

and thus (5.8) follows. Inverting the argument proves the claim.

(iv)⇒(v). By assumption we have $\mu(\partial E) = \mu(\overline{E}) - \mu(E^\circ) = 0$, and thus $\mu(\overline{E}) = \mu(E^\circ) = \mu(E)$. Then by (5.7)–(5.8)

$$\begin{aligned} \mu(E) = \mu(E^\circ) &\le \liminf_{k\to\infty} \mu_k(E^\circ) \le \liminf_{k\to\infty} \mu_k(E) \\ &\le \limsup_{k\to\infty} \mu_k(E) \le \limsup_{k\to\infty} \mu_k(\overline{E}) \le \mu(\overline{E}) = \mu(E). \end{aligned}$$

Hence (5.9) follows.

(v)⇒(i). For any $g \in C_b(X)$ and $t > 0$, set $E_t := \{g > t\}$. Since g is continuous, for λ-a. e. $t \in (0, \infty)$, we have $\mu(\partial E_t) \le \mu(\{g = t\}) = 0$, and thus by (5.9)

$$\lim_{k\to\infty} \mu_k(\{g > t\}) = \mu(\{g > t\}) \quad \text{for } \lambda\text{-a. e. } t \in (0, \infty). \tag{5.10}$$

Then by equalities (2.19) (with $\alpha = 1$), (5.10), and the dominated convergence theorem we get that for any $g \in C_b(X)$

$$\lim_{k\to\infty} \int_X g \, d\mu_k = \lim_{k\to\infty} \int_{[0,\infty)} \mu_k(\{g > t\}) \, d\lambda(t) = \int_{[0,\infty)} \mu(\{g > t\}) \, d\lambda(t) = \int_X g \, d\mu.$$

This completes the proof. □

Remark 5.1.5. As already mentioned in Remark 5.1.3, Theorem 5.1.2 shows that if $\{\mu_k\} \subseteq \mathfrak{R}_f^+(X)$, $\mu \in \mathfrak{R}_f^+(X)$, and $\mu_k \xrightarrow{n} \mu$, then $\|\mu_k\|_{\mathfrak{R}_f(X)} \to \|\mu\|_{\mathfrak{R}_f(X)}$.

5.1.4 Prokhorov distance

The results of this subsection are formulated for a metric space (X, d). However, by Proposition A.2 they also hold if X is a locally compact Hausdorff space with countable basis and d is a metric compatible with its topology.

Let (X, d) be a metric space. For any $E \subseteq X$ and $r > 0$, set $E_r := \{x \in X \mid d(x, E) < r\}$. For any $\mu_1, \mu_2 \in \mathfrak{R}_f^+(X)$, define

$$d_P(\mu_1, \mu_2) := \inf\{r > 0 \mid \mu_1(E) \leq \mu_2(E_r) + r, \ \mu_2(E) \leq \mu_1(E_r) + r \ \ \forall E \in \mathcal{B}(X)\}. \tag{5.11}$$

Observe that every $r \geq \max\{\|\mu_1\|_{\mathfrak{R}_f(X)}, \|\mu_2\|_{\mathfrak{R}_f(X)}\}$ is contained in the set in the right-hand side, and thus $d_P(\mu_1, \mu_2) < \infty$.

Proposition 5.1.3. *$(\mathfrak{R}_f^+(X), d_P)$ is a metric space.*

Proof. Let us prove that the map $(\mu_1, \mu_2) \mapsto d_P(\mu_1, \mu_2)$ is a metric on $\mathfrak{R}_f^+(X)$. Clearly, $d_P(\mu_1, \mu_2) = d_P(\mu_2, \mu_1)$. Moreover:

(i) for all $E \in \mathcal{B}(X)$, we have $E \subseteq E_r$, and thus $\mu(E) \leq \mu(E_r)$ for all $\mu \in \mathfrak{R}_f^+(X)$ and $r > 0$. Therefore

$$d_P(\mu, \mu) := \inf\{r > 0 \mid \mu(E) \leq \mu(E_r) + r \ \ \forall E \in \mathcal{B}(X)\} = 0;$$

(ii) if $d_P(\mu_1, \mu_2) = 0$, then there is a sequence $r_n \to 0^+$ such that $\mu_1(E) \leq \mu_2(E_{r_n}) + r_n$, $\mu_2(E) \leq \mu_1(E_{r_n}) + r_n$ for all n, and $E \in \mathcal{B}(X)$. Since $\bigcap_{n=1}^{\infty} E_{r_n} = \overline{E}$, it follows that $\mu_1(E) \leq \mu_2(\overline{E})$ and $\mu_2(E) \leq \mu_1(\overline{E})$ for all $E \in \mathcal{B}(X)$, and hence $\mu_1(C) = \mu_2(C)$ for all closed sets $C \in \mathcal{B}(X)$. In particular, $\mu_1(X) = \mu_2(X)$, whence $\mu_1(A) = \mu_2(A)$ for any open set $A \in \mathcal{B}(X)$. Then by the regularity of μ_1, μ_2 (see Lemma 1.3.2(i)) we get $\mu_1 = \mu_2$;

(iii) let $\mu_1, \mu_2, \mu_3 \in \mathfrak{R}_f^+(X)$ and $r, s > 0$ be such that $d_P(\mu_1, \mu_2) < r$ and $d_P(\mu_2, \mu_3) < s$. Then for all $E \in \mathcal{B}(X)$, we have

$$\mu_1(E) \leq \mu_2(E_r) + r \leq \mu_3(E_{r+s}) + r + s, \quad \mu_3(E) \leq \mu_2(E_s) + s \leq \mu_1(E_{r+s}) + r + s,$$

whence $d_P(\mu_1,\mu_3) \le r+s$. Taking the infimum over r, s, we obtain that

$$d_P(\mu_1,\mu_3) \le d_P(\mu_1,\mu_2) + d_P(\mu_2,\mu_3),$$

that is, the triangle inequality. Hence the result follows. □

Definition 5.1.4. The metric d_P on $\mathfrak{R}_f^+(X)$ defined in (5.11) is called the *Prokhorov metric.*

Let us now prove the following result.

Lemma 5.1.4. *Let (X,d) be a metric space, and let $\mu_1,\mu_2 \in \mathfrak{R}_f^+(X)$ satisfy*

$$\mu_1(E) \le \mu_2(E_r) + r \quad \textit{for all } E \in \mathcal{B}(X)\ (r>0). \tag{5.12}$$

Then for all $F \in \mathcal{B}(X)$,

$$\mu_2(F) \le \mu_1(F_r) + r + \|\mu_2\|_{\mathfrak{R}_f(X)} - \|\mu_1\|_{\mathfrak{R}_f(X)}. \tag{5.13}$$

Remark 5.1.6. In view of Lemma 5.1.4, if (5.12) is satisfied and $\|\mu_1\|_{\mathfrak{R}_f(X)} = \|\mu_2\|_{\mathfrak{R}_f(X)}$, then

$$\begin{aligned} d_P(\mu_1,\mu_2) &= \inf\{r>0 \mid \mu_1(E) \le \mu_2(E_r) + r \ \ \forall E \in \mathcal{B}(X)\} \\ &= \inf\{r>0 \mid \mu_2(E) \le \mu_1(E_r) + r \ \ \forall E \in \mathcal{B}(X)\}. \end{aligned} \tag{5.14}$$

Proof of Lemma 5.1.4. Choosing $E = (F_r)^c$ in (5.12), we plainly get

$$\begin{aligned} \mu_1(F_r) &= \|\mu_1\|_{\mathfrak{R}_f(X)} - \mu_1((F_r)^c) = \|\mu_1\|_{\mathfrak{R}_f(X)} - \mu_1(E) \ge \|\mu_1\|_{\mathfrak{R}_f(X)} - \mu_2(E_r) - r \\ &= \|\mu_1\|_{\mathfrak{R}_f(X)} - \|\mu_2\|_{\mathfrak{R}_f(X)} + \mu_2((E_r)^c) - r \ge \|\mu_1\|_{\mathfrak{R}_f(X)} - \|\mu_2\|_{\mathfrak{R}_f(X)} + \mu_2(F) - r. \end{aligned}$$

(observe that $E \subseteq (F_r)^c \Leftrightarrow F \subseteq (E_r)^c$). □

Example 5.1.2. Let (X,d) be a metric space. Then for all $x_1,x_2 \in X$,

$$d_P(\delta_{x_1},\delta_{x_2}) = \min\{1, d(x_1,x_2)\}. \tag{5.15}$$

To prove (5.15), observe that by (5.14)

$$d_P(\delta_{x_1},\delta_{x_2}) = \inf\{r>0 \mid \chi_E(x_1) \le \chi_{E_r}(x_2) + r \ \ \forall E \in \mathcal{B}(X)\}.$$

Since $\chi_E(x_1) \le 1$, we have $d_P(\delta_{x_1},\delta_{x_2}) \le 1$. Moreover, if $r > d(x_1,x_2)$, then $\chi_E(x_1) \le \chi_{E_r}(x_2) + r$ for all $E \in \mathcal{B}(X)$: indeed, if $x_1 \in E$ (otherwise, the inequality is trivial) and $r > d(x_1,x_2)$, then $x_2 \in E_r$, and thus $\chi_{E_r}(x_2) = 1$, and the inequality is satisfied. To sum up, we have $d_P(\delta_{x_1},\delta_{x_2}) \le \min\{1, d(x_1,x_2)\}$.

To prove the reverse inequality, choose $E = \{x_1\}$, and thus $\chi_E(x_1) = 1$. If $d(x_1, x_2) \geq 1$ and $r \in (0, 1)$, then $x_2 \in (E_r)^c$, and thus $\chi_{E_r}(x_2) = 0$. It follows that

$$\chi_E(x_1) = 1 > r = \chi_{E_r}(x_2) + r, \tag{5.16}$$

and hence $d_P(\delta_{x_1}, \delta_{x_2}) \geq 1 = \min\{1, d(x_1, x_2)\}$. Similarly, if $d(x_1, x_2) < 1$ and $r \in (0, d(x_1, x_2)]$, then $x_2 \in (E_r)^c$, and thus inequality (5.16) again holds. Hence $d_P(\delta_{x_1}, \delta_{x_2}) \geq d(x_1, x_2) = \min\{1, d(x_1, x_2)\}$. Hence equality (5.15) follows.

Remark 5.1.7. By Example 5.1.2 the map from X to $\mathfrak{R}_f^+(X)$, $x \mapsto \delta_x$, is an isometric injection of $(X, \min\{1, d\})$ into $(\mathfrak{R}_f^+(X), d_P)$. Observe that $\min\{1, d\}$ is a metric equivalent to d on X.

Now we can prove that the narrow topology on $\mathfrak{R}_f^+(X)$ is metrizable.

Proposition 5.1.5. *Let (X, d) be a separable metric space. Then the narrow topology on $\mathfrak{R}_f^+(X)$ and the metric topology associated with the Prokhorov metric coincide.*

Remark 5.1.8. In view of Proposition 5.1.5, if (X, d) is a separable metric space, then:
(i) there holds $\mu_k \overset{n}{\rightharpoonup} \mu$ if and only if $d_P(\mu_k, \mu) \to 0$;
(ii) a subset $\mathscr{M} \subseteq \mathfrak{R}_f^+(X)$ is relatively sequentially compact in the narrow topology if and only if it is relatively compact in $(\mathfrak{R}_f^+(X), d_P)$.

To prove Proposition 5.1.5, we first establish the following lemma.

Lemma 5.1.6. *Let (X, d) be a metric space, and let $\{\mu_k\} \subseteq \mathfrak{R}_f^+(X)$, $\mu \in \mathfrak{R}_f^+(X)$ satisfy $d_P(\mu_k, \mu) \to 0$. Then $\mu_k \overset{n}{\rightharpoonup} \mu$.*

Proof. Let $\{\epsilon_k\} \subseteq (0, \infty)$ satisfy $\epsilon_k \to 0^+$ as $k \to \infty$ and $d_P(\mu_k, \mu) < \epsilon_k$ for all $k \in \mathbb{N}$. For any closed subset $C \subseteq X$, set $C_k \equiv C_{\epsilon_k} := \{x \in X \mid d(x, C) < \epsilon_k\}$, and thus $C = \bigcap_{k=1}^{\infty} C_k$ and $\mu(C) = \lim_{k\to\infty} \mu(C_k)$. Since $d_P(\mu_k, \mu) < \epsilon_k$, by definition (5.11) for any $k \in \mathbb{N}$, we have

$$\mu_k(C) \leq \mu(C_k) + \epsilon_k, \quad \mu(C) \leq \mu_k(C_k) + \epsilon_k. \tag{5.17}$$

Since $\lim_{k\to\infty} \mu(C_k) = \mu(C)$ and $\lim_{k\to\infty} \epsilon_k = 0$, letting $k \to \infty$ in the first inequality in (5.17), we obtain (5.7). On the other hand, choosing $C = X$ both in the second inequality in (5.17) and in (5.7) gives

$$\mu(X) \leq \liminf_{k\to\infty} \mu_k(X) \leq \limsup_{k\to\infty} \mu_k(X) \leq \mu(X),$$

and thus $\lim_{k\to\infty} \mu_k(X) = \mu(X)$. Hence by Theorem 5.1.2 (see claims (i) and (iii)) the result follows. □

Proof of Proposition 5.1.5. In view of Lemma 5.1.6, we only must prove that $\mu_k \overset{n}{\rightharpoonup} \mu$ implies $d_P(\mu_k, \mu) \to 0$. To this purpose, let $D \equiv \{x_k\}$ be a countable dense subset of X.

For any fixed $\epsilon > 0$, set $E_1 := B_{\epsilon/2}(x_1)$ and $E_{m+1} := B_{\epsilon/2}(x_{m+1}) \setminus (\bigcup_{k=1}^{m} E_k)$ ($m \in \mathbb{N}$). Clearly, $\operatorname{diam}(E_m) \leq \epsilon$ for all $m \in \mathbb{N}$, $E_l \cap E_m = \emptyset$ for all $l, m \in \mathbb{N}$, $l \neq m$, and $\bigcup_{m=1}^{\infty} E_m = X$.

Since $\mu(X) = \sum_{m=1}^{\infty} \mu(E_m) < \infty$, there exists $\bar{m} \in \mathbb{N}$ such that

$$\mu\left(\bigcup_{m=\bar{m}+1}^{\infty} E_m\right) = \sum_{m=\bar{m}+1}^{\infty} \mu(E_m) < \epsilon. \tag{5.18}$$

Consider the finite family

$$\mathbf{S} := \{U = E_{m_1} \cup E_{m_2} \cup \cdots \cup E_{m_p} \mid 1 \leq m_1 < m_2 < \cdots m_p \leq \bar{m}\} \subseteq \mathcal{P}(X),$$

and for any $U \in \mathbf{S}$, set $U_\epsilon := \{x \in X \mid d(x, U) < \epsilon\}$. Since U_ϵ is open and by assumption $\mu_k \overset{n}{\rightharpoonup} \mu$, by the portmanteau theorem $\liminf_{k\to\infty} \mu_k(U_\epsilon) \geq \mu(U_\epsilon)$. Therefore, since $\mathbf{S}$ is finite, there exists $\bar{k} \in \mathbb{N}$ such that

$$\mu_k(U_\epsilon) \geq \mu(U_\epsilon) - \epsilon \quad \text{for all } k \geq \bar{k} \text{ and } U \in \mathbf{S}. \tag{5.19}$$

Now observe that for all $F \in \mathcal{B}(X)$,

$$F = \left[F \cap \left(\bigcup_{m=1}^{\bar{m}} E_m\right)\right] \cup \left[F \cap \left(\bigcup_{m=\bar{m}+1}^{\infty} E_m\right)\right] \subseteq [F \cap U_\epsilon] \cup \left[F \cap \left(\bigcup_{m=\bar{m}+1}^{\infty} E_m\right)\right] \tag{5.20}$$

for some $U = E_{m_1} \cup E_{m_2} \cup \cdots \cup E_{m_p} \in \mathbf{S}$ such that $E_{m_j} \cap F \neq \emptyset$ for all $j = m_1, \ldots, m_p$ (observe that this implies $U_\epsilon \subseteq F_{2\epsilon}$, since $\operatorname{diam}(E_m) \leq \epsilon$ for any $m \in \mathbb{N}$). From (5.18)–(5.20) we get

$$\mu(F) \leq \mu(U_\epsilon) + \epsilon \leq \mu_k(U_\epsilon) + 2\epsilon \leq \mu_k(F_{2\epsilon}) + 2\epsilon \quad \text{for all } k \geq \bar{k}. \tag{5.21}$$

By the arbitrariness of $F \in \mathcal{B}(X)$ and Lemma 5.1.4, from (5.21) we get

$$\mu_k(G) \leq \mu(G_{2\epsilon}) + 2\epsilon + \|\mu_k\|_{\mathfrak{R}_f(X)} - \|\mu\|_{\mathfrak{R}_f(X)} \leq \mu(G_{3\epsilon}) + 3\epsilon \tag{5.22}$$

for all $G \in \mathcal{B}(X)$ and all $k \in \mathbb{N}$ large enough, since $\lim_{k\to\infty} \|\mu_k\|_{\mathfrak{R}_f(X)} = \|\mu\|_{\mathfrak{R}_f(X)}$ (see Remark 5.1.5).

On the other hand, by (5.21) we also have that for all $k \in \mathbb{N}$ large enough and any $G \in \mathcal{B}(X)$,

$$\mu(G) \leq \mu_k(G_{3\epsilon}) + 3\epsilon. \tag{5.23}$$

In view of definition (5.11), from (5.22)–(5.23) we obtain that for all $\epsilon > 0$ and $k \in \mathbb{N}$ large enough, $d_p(\mu_k, \mu) < \epsilon$. Hence the conclusion follows. □

5.1.5 Narrow convergence and tightness

It is interesting to characterize relatively sequentially compact subsets of $\mathfrak{R}_f(X)$, in both weak* and narrow topologies. Concerning the weak* convergence, we have the following:

Theorem 5.1.7. *Let X be a locally compact Hausdorff space with countable basis, and let $\mathcal{M} \subseteq \mathfrak{R}_f(X)$. Then the following statements are equivalent:*
(i) *$\mathcal{M}$ is relatively sequentially compact in the weak* topology;*
(ii) *$\mathcal{M}$ is bounded.*

Proof. (i)⇒(ii). By Theorem 2.7.1 the Banach space $\mathfrak{R}_f(X)$ is the dual space of $C_0(X)$. Were $\mathcal{M}$ unbounded, there would exist $\{\mu_k\} \subseteq \mathcal{M}$ such that $\|\mu_k\|_{\mathfrak{R}_f(X)} \to \infty$. On the other hand, by assumption every $\{\mu_k\} \subseteq \mathcal{M}$ contains a subsequence that converges weakly* in $\mathfrak{R}_f(X)$ and thus is bounded by the uniform boundedness principle. The contradiction proves the claim.

(ii)⇒(i) follows from the Banach–Alaoglu theorem. □

Concerning the narrow convergence, we begin by the following:

Definition 5.1.5. Let X be a locally compact Hausdorff space. A subset $\mathcal{M} \subseteq \mathfrak{R}_f(X)$ is called *tight* if for any $\epsilon > 0$, there exists a compact subset $K \subseteq X$ such that $|\mu|(K^c) < \epsilon$ for all $\mu \in \mathcal{M}$.

Example 5.1.3. Let $X = \mathbb{R}$ and $\mu = \delta_{\bar{x}}$ ($\bar{x} \in \mathbb{R}$). Thus $\mu(E) = \chi_E(\bar{x})$ for all $E \in \mathcal{B}(\mathbb{R})$. Clearly, for any $F \subseteq \mathbb{R}$, the set $\mathcal{M} = \{\delta_{\bar{x}} \mid \bar{x} \in F\}$ is tight if and only if F is bounded. Observe that if F is bounded, then the set $\{\alpha\delta_{\bar{x}} \mid \bar{x} \in F,\ \alpha > 0\}$ is tight but unbounded.

Let us prove for future reference the following result.

Proposition 5.1.8. *Let X be a locally compact Hausdorff space with countable basis, and let $\mathcal{M} \subseteq \mathfrak{R}^+_f(X)$ be tight. Then the narrow topology on $\mathcal{M}$ and the weak* topology on $\mathcal{M}$ coincide, that is, $\mathcal{T}_n \cap \mathcal{M} = \mathcal{T}_{w^*} \cap \mathcal{M}$.*

Proof. It suffices to prove that for any sequence $\{\mu_k\} \subseteq \mathcal{M}$ such that $\mu_k \overset{*}{\rightharpoonup} \mu$, we have $\mu_k \overset{n}{\rightharpoonup} \mu$. For fixed $\epsilon > 0$, by Definition 5.1.5 there exists a compact subset $K \subseteq X$ such that $\mu_k(K^c) < \epsilon$ for all $k \in \mathbb{N}$. On the other hand, by Proposition A.9 for any open set $A \subseteq X$, $A \supseteq K$, there exists $f \in C_c(X)$ such that $f(X) \subseteq [0,1]$, $f|_K = 1$, and $\operatorname{supp} f \subseteq A$.

Fix $g \in C_b(X)$. Then $fg \in C_0(X)$, and

$$\begin{aligned}
&\left|\int_X g\,d\mu_k - \int_X g\,d\mu\right| \\
&\quad \le \left|\int_K g\,d\mu_k - \int_K g\,d\mu\right| + \left|\int_{K^c} g\,d\mu_k - \int_{K^c} g\,d\mu\right|
\end{aligned}$$

$$\le \left| \int_X fg\, d\mu_k - \int_X fg\, d\mu \right| + \left| \int_X [\chi_K - f] g\, d\mu_k \right| + \left| \int_X [\chi_K - f] g\, d\mu \right| + \left| \int_{K^c} g\, d\mu_k \right| + \left| \int_{K^c} g\, d\mu \right|$$
$$=: I_1 + I_2 + I_3 + I_4 + I_5.$$

Since by assumption $\mu_k \overset{*}{\rightharpoonup} \mu$, there exists $\bar{k} \in \mathbb{N}$ such that for all $k > \bar{k}$,

$$I_1 = \left| \int_X fg\, d\mu_k - \int_X fg\, d\mu \right| < \epsilon.$$

We also have that for all $k \in \mathbb{N}$,

$$I_2 + I_4 = \left| \int_X [\chi_K - f] g\, d\mu_k \right| + \left| \int_{K^c} g\, d\mu_k \right| \le 2\|g\|_\infty \mu_k(K^c) < 2\|g\|_\infty \epsilon.$$

Moreover,

$$I_3 + I_5 = \left| \int_X [\chi_K - f] g\, d\mu \right| + \left| \int_{K^c} g\, d\mu \right| \le 2\|g\|_\infty \mu(K^c) \le 2\|g\|_\infty \epsilon,$$

since $\mu(K^c) \le \liminf_{k\to\infty} \mu_k(K^c)$ by the lower semicontinuity of the norm (see Remark 5.1.3).

To summarize, we have proved that for any fixed $g \in C_b(X)$ and for any $\epsilon > 0$, there exists $\bar{k} \in \mathbb{N}$ such that for all $k > \bar{k}$,

$$\left| \int_X g\, d\mu_k - \int_X g\, d\mu \right| < (1 + 4\|g\|_\infty)\epsilon.$$

Hence the result follows. □

The following result, which is the counterpart of Theorem 5.1.7 for weak* convergence, shows that narrow convergence and tightness are deeply intertwined.

Theorem 5.1.9 (Prokhorov). *Let X be a locally compact Hausdorff space with countable basis, and let $\mathcal{M} \subseteq \mathfrak{R}_f^+(X)$. Then the following statements are equivalent:*
(i) *$\mathcal{M}$ is relatively sequentially compact in the narrow topology;*
(ii) *$\mathcal{M}$ is bounded and tight.*

Remark 5.1.9. More generally, Theorem 5.1.9 holds if X is a Polish space (see Subsection A6, Appendix A). The same holds for Proposition 5.1.10.

The proof of Theorem 5.1.9 will be given in two steps, which correspond to Propositions 5.1.10–5.1.11.

Proposition 5.1.10 (Ulam). *Let X be a locally compact Hausdorff space with countable basis, and let $\mathcal{M} \subseteq \mathfrak{R}_f^+(X)$ be relatively sequentially compact in the narrow topology. Then $\mathcal{M}$ is bounded and tight.*

Proof. By assumption, for any sequence $\{\mu_k\} \subseteq \mathcal{M}$, there exist $\{\mu_{k_l}\} \subseteq \{\mu_k\}$ and $\mu \in \mathfrak{R}_f(X)$ such that $\mu_{k_l} \overset{n}{\rightharpoonup} \mu$, and thus $\|\mu_{k_l}\|_{\mathfrak{R}_f(X)} \to \|\mu\|_{\mathfrak{R}_f(X)}$ (see Remark 5.1.5). Plainly, this implies that $\mathcal{M}$ is bounded.

To prove that $\mathcal{M}$ is tight, observe preliminarily that X is a Polish space, and hence there exists a compatible metric d such that (X, d) is complete. Fix $\epsilon > 0$, and let $D \equiv \{x_k\}$ be a countable dense subset of X. For any fixed $p \in \mathbb{N}$, set $E_{n,p} := \bigcup_{k=1}^n B_{1/p}(x_k)$ (where $B_{1/p}(x_k) := \{x \in X \mid d(x_k, x) < 1/p\}$; $n \in \mathbb{N}$). Clearly, every set $E_{n,p}$ is open, and $E_{n,p} \subseteq E_{n+1,p}$ for all $n \in \mathbb{N}$, $\bigcup_{n=1}^\infty E_{n,p} = X$. We will prove the following:

Claim. Let $\{A_n\} \subseteq \mathcal{P}(X)$ be an increasing sequence of open subsets such that $\bigcup_{n=1}^\infty A_n = X$. Then for every $\epsilon > 0$, there exists $\bar{n} \in \mathbb{N}$ such that $\mu(A_{\bar{n}}^c) < \epsilon$ for all $\mu \in \mathcal{M}$.

By the claim, for any $p \in \mathbb{N}$, there exists $n_p \in \mathbb{N}$ such that

$$\mu((E_{n_p,p})^c) < \frac{\epsilon}{2^p} \quad \text{for all } \mu \in \mathcal{M},$$

and thus, in particular,

$$\mu((\overline{E}_{n_p,p})^c) < \frac{\epsilon}{2^p} \quad \text{for all } \mu \in \mathcal{M}. \tag{5.24}$$

Set

$$K := \bigcap_{p=1}^\infty \overline{E}_{n_p,p} \subseteq \bigcap_{p=1}^\infty \bigcup_{k=1}^{n_p} \overline{B}_{1/p}(x_k).$$

By (5.24) $\mu(K^c) < \epsilon$ for all $\mu \in \mathcal{M}$, and hence the result will follow if we prove that K is compact. To this purpose, observe that for every $p \in \mathbb{N}$, the set K is contained in the finite union $\bigcup_{k=1}^{n_p} \overline{B}_{1/p}(x_k)$ with $\operatorname{diam} \overline{B}_{1/p}(x_k) = 2/p$ ($p \in \mathbb{N}$). By a diagonal argument it follows that every sequence $\{x'_j\} \subseteq K$ contains a Cauchy subsequence $\{x'_{j_l}\} \subseteq \{x'_j\}$. Then since K is closed and X is complete, there exists $x' \in K$ such that $d(x'_{j_l}, x') \to 0$ as $l \to \infty$. Therefore K is sequentially compact and thus compact. Hence the result follows.

It remains to prove the claim. By contradiction let there exist an increasing sequence $\{A_n\}$ of open subsets with $\bigcup_{n=1}^\infty A_n = X$ and $\epsilon > 0$ with the following property: for all $n \in \mathbb{N}$, there exists $\mu_n \in \mathcal{M}$ such that $\mu_n(A_n^c) \geq \epsilon$. Since $\mathcal{M}$ is relatively sequentially compact in the narrow topology, there exist $\{\mu_{n_k}\} \subseteq \{\mu_n\}$ and $\mu \in \mathfrak{R}_f(X)$ such that $\mu_{n_k} \overset{n}{\rightharpoonup} \mu$. Then by Theorem 5.1.2 and the monotonicity of $\{A_n\}$, for every fixed $n \in \mathbb{N}$, we have

$$\mu(A_n^c) \geq \limsup_{k\to\infty} \mu_{n_k}(A_n^c) \geq \limsup_{k\to\infty} \mu_{n_k}(A_{n_k}^c) \geq \epsilon.$$

However, since μ is finite, we get $\lim_{n\to\infty} \mu(A_n^c) = 0$, a contradiction. Hence the claim follows, which completes the proof. □

Proposition 5.1.11. *Let X be a locally compact Hausdorff space with countable basis space, and let $\mathscr{M} \subseteq \mathfrak{R}_f^+(X)$ be bounded and tight. Then $\mathscr{M}$ is relatively sequentially compact in the narrow topology.*

Proof. By assumption we have

$$\sup_{\mu\in\mathscr{M}} \|\mu\|_{\mathfrak{R}_f(X)} = \sup_{\mu\in\mathscr{M}} \mu(X) =: C < \infty. \tag{5.25}$$

Recall that by Proposition A.2 the space X is σ-compact. Let us first prove the result assuming that it is compact. Let $D \equiv \{f_k\}$ be a countable dense subset of $C_b(X)$ (which is separable since X is compact; see Proposition A.3). Then

$$\sup_{\mu\in\mathscr{M}} \left| \int_X f_1 \, d\mu \right| \leq \|f_1\|_\infty \sup_{\mu\in\mathscr{M}} \|\mu\|_{\mathfrak{R}_f(X)} = \|f_1\|_\infty C < \infty,$$

and thus the set $\{\int_X f_1 \, d\mu \mid \mu \in \mathscr{M}\} \subseteq \mathbb{R}$ is bounded. Then there exist a sequence $\{\mu_j^1\} \subseteq \mathscr{M}$ and $a_1 \in \mathbb{R}$ such that

$$\lim_{j\to\infty} \int_X f_1 \, d\mu_j^1 = a_1. \tag{5.26}$$

Since

$$\sup_{j\in\mathbb{N}} \left| \int_X f_2 \, d\mu_j^1 \right| \leq \|f_2\|_\infty \sup_{j\in\mathbb{N}} \|\mu_j^1\| \leq \|f_2\|_\infty C < \infty,$$

the set $\{\int_X f_2 \, d\mu_j^1 \mid j \in \mathbb{N}\} \subseteq \mathbb{R}$ is also bounded. Then there exist a subsequence $\{\mu_j^2\} \subseteq \{\mu_j^1\}$ and $a_2 \in \mathbb{R}$ such that

$$\lim_{j\to\infty} \int_X f_2 \, d\mu_j^2 = a_2.$$

Clearly, by (5.26) we also have $\lim_{j\to\infty} \int_X f_1 \, d\mu_j^2 = a_1$. Iterating the argument, for every $k \in \mathbb{N}$, $k \geq 3$, there exist a subsequence $\{\mu_j^k\} \subseteq \{\mu_j^{k-1}\}$ and $a_k \in \mathbb{R}$ such that

$$\lim_{j\to\infty} \int_X f_l \, d\mu_j^k = a_l \quad \text{for all } l = 1, \dots, k.$$

By a diagonal argument it follows that the sequence $\{\mu_j\} \subseteq \mathscr{M}$, where $\mu_j := \mu_j^j$ ($j \in \mathbb{N}$), satisfies

$$\lim_{j\to\infty} \int_X f_k \, d\mu_j = a_k \quad \text{for all } k \in \mathbb{N}. \tag{5.27}$$

Now consider the map $F_0 : D \to \mathbb{R}$, $F_0(f_k) := a_k$ ($k \in \mathbb{N}$). By (5.27) F_0 is linear, positive, and continuous, since

$$|F_0(f_k)| = |a_k| \le \lim_{j\to\infty} \left| \int_X f_k \, d\mu_j \right| \le \|f_k\|_\infty C \quad \text{for all } k \in \mathbb{N}.$$

Since D is dense in $C_b(X)$, F_0 can be uniquely extended to a functional $F \in (C_b(X))^*$, which is positive and satisfies $\|F\| \le C$. We will prove the following:

Claim. For any $\epsilon > 0$, there exists a compact subset $K \subseteq X$ such that $\langle F, f \rangle < \epsilon$ for all $f \in C_b(X)$ such that $f(X) \subseteq [0,1]$ and $f|_K = 0$.

Relying on the claim, we can complete the proof when X is compact. In fact, by Proposition 2.6.6 there exists a unique $\mu \in \mathfrak{R}_f^+(X)$ such that

$$\langle F, f \rangle = \int_X f \, d\mu \quad \text{for all } f \in C_b(X) \tag{5.28}$$

(see (2.49)); moreover, by Theorem 2.6.1(ii) $\|\mu\|_{\mathfrak{R}_f(X)} = \|F\| \le C$. Let $g \in C_b(X)$, and let $\{f_l\} \subseteq D$ be a sequence such that $\lim_{l\to\infty} \|f_l - g\|_\infty = 0$. Fix $\epsilon > 0$, and let $l_0 \in \mathbb{N}$ be so large that $\|f_{l_0} - g\|_\infty < \frac{\epsilon}{4C}$. By (5.27)–(5.28) we have

$$\langle F, f_{l_0} \rangle = F_0(f_{l_0}) = a_{l_0} = \int_X f_{l_0} \, d\mu = \lim_{j\to\infty} \int_X f_{l_0} \, d\mu_j,$$

and hence there exists $j_0 \in \mathbb{N}$ such that

$$\left| \int_X f_{l_0} \, d\mu_j - \int_X f_{l_0} \, d\mu \right| < \frac{\epsilon}{2} \quad \text{for all } j > j_0.$$

Then for all $j > j_0$,

$$\begin{aligned} &\left| \int_X g \, d\mu_j - \int_X g \, d\mu \right| \\ &\le \left| \int_X (g - f_{l_0}) \, d\mu_j \right| + \left| \int_X (g - f_{l_0}) \, d\mu \right| + \left| \int_X f_{l_0} \, d\mu_j - \int_X f_{l_0} \, d\mu \right| \\ &\le \|f_{l_0} - g\|_\infty \Big(\sup_{j\in\mathbb{N}} \|\mu_j\| + \|\mu\|_{\mathfrak{R}_f(X)} \Big) + \frac{\epsilon}{2} < 2C\Big(\frac{\epsilon}{4C}\Big) + \frac{\epsilon}{2} = \epsilon. \end{aligned}$$

To sum up, we proved that there exist a sequence $\{\mu_j\} \subseteq \mathscr{M}$ and $\mu \in \mathfrak{R}_f^+(X)$ such that $\mu_j \overset{n}{\rightharpoonup} \mu$, and hence the result follows.

Let us now prove the above claim. Since by assumption $\mathscr{M}$ is tight, for any $\epsilon > 0$, there exists a compact subset $K \subseteq X$ such that $\sup_{\mu \in \mathscr{M}} \mu(K^c) < \frac{\epsilon}{6}$. Let $f \in C_b(X)$ satisfy $f(X) \subseteq [0,1]$ and $f|_K = 0$. Since D is dense in $C_b(X)$, there exists a sequence $\{f_l\} \subseteq D$, $f_l \geq 0$ for all $l \in \mathbb{N}$, such that $\lim_{l\to\infty} \|f_l - f\|_\infty = 0$; thus, in particular,

$$\|f_l\|_\infty < 1 + \frac{\epsilon}{6}, \quad \|f_l|_K\|_\infty < \frac{\epsilon}{6C} \quad \text{for all } l \in \mathbb{N} \text{ sufficiently large.}$$

Then for any $l \in \mathbb{N}$ sufficiently large,

$$\begin{aligned}\langle F, f_l\rangle = a_l = \lim_{j\to\infty} \int_X f_l \, d\mu_j &\leq \limsup_{j\to\infty} \int_K f_l \, d\mu_j + \limsup_{j\to\infty} \int_{K^c} f_l \, d\mu_j \\ &\leq \frac{\epsilon}{6C} \sup_{j\in\mathbb{N}} \mu_j(X) + \left(1 + \frac{\epsilon}{6}\right) \sup_{j\in\mathbb{N}} \mu_j(K^c) < \frac{\epsilon}{6} + \left(1 + \frac{\epsilon}{6}\right)\frac{\epsilon}{6} < \frac{\epsilon}{2}.\end{aligned}$$

Since $\|f_l - f\|_\infty \to 0$ and F is continuous, from this inequality we get

$$\langle F, f\rangle = \lim_{l\to\infty} \langle F, f_l\rangle \leq \frac{\epsilon}{2} < \epsilon,$$

and thus the claim follows. This completes the proof when X is compact.

In the general case, let $\{K_l\}$ be an increasing sequence of compact subsets of X such that $\bigcup_{l=1}^\infty K_l = X$. By (5.25) we have

$$\sup_{\mu\in\mathscr{M}} \|\mu\|_{\mathfrak{R}_f(K_l)} \leq \sup_{\mu\in\mathscr{M}} \|\mu\|_{\mathfrak{R}_f(X)} = C \quad \text{for every } l \in \mathbb{N}.$$

By the above considerations, for every $l \in \mathbb{N}$, there exist a sequence $\{\mu_j^l\} \subseteq \mathscr{M}$ and $\mu^l \in \mathfrak{R}_f^+(X)$ such that $\mu_j^l \overset{n}{\rightharpoonup} \mu^l$ as $j \to \infty$. Then by a diagonal argument the conclusion follows. □

In view of Theorem 5.1.9, it is useful to have criteria for tightness. Let us state the following definition.

Definition 5.1.6. Let X be a locally compact Hausdorff space. A function $\varphi : X \to (-\infty, \infty]$ is called *inf-compact* if the set $\{\varphi \leq r\} := \{x \in X \mid \varphi(x) \leq r\}$ is compact for every $r \in \mathbb{R}$.

Example 5.1.4. If $X = \mathbb{R}$, then every continuous function $\varphi : \mathbb{R} \to [0, \infty)$ such that $\varphi(x) \to \infty$ as $|x| \to \infty$ is inf-compact. Indeed, for every $r \in \mathbb{R}$, the set $\{\varphi \leq r\}$ is closed and bounded, and thus compact.

Definition 5.1.6 leads to the following characterization of tight subsets of $\mathfrak{R}_f^+(X)$.

Proposition 5.1.12. *Let X be a locally compact Hausdorff space with countable basis. Then the following statements are equivalent:*

(i) $\mathcal{M} \subseteq \mathfrak{R}_f^+(X)$ *is tight;*

(ii) *there exists an inf-compact* $\varphi : X \mapsto [0, \infty)$ *such that*

$$\sup_{\mu \in \mathcal{M}} \int_X \varphi \, d\mu < \infty. \tag{5.29}$$

Proof. (i)⇒(ii). By Proposition A.2 there exists an increasing sequence $\{K_n\}$ of compact subsets of X such that $\mu(K_n^c) < \frac{1}{2^n}$ for all $\mu \in \mathcal{M}$. Set $\varphi := \sum_{n=1}^{\infty} \chi_{K_n^c}$. Then for every $\mu \in \mathcal{M}$,

$$\sup_{\mu \in \mathcal{M}} \int_X \varphi \, d\mu \le \sum_{n=1}^{\infty} \frac{1}{2^n} < \infty,$$

and thus inequality (5.29) is satisfied.

To show that φ is inf-compact, let $x \in X$ be fixed. For every $n \in \mathbb{N}$, we have either (a) $x \in K_n$ or (b) $x \in K_n^c$. Since $K_n^c \supseteq K_{n+1}^c$ for each n,

- in case (a), $\varphi(x) \le n - 1$;
- in case (b), $\varphi(x) = n$ if $x \in K_n^c \setminus K_{n+1}^c = K_n^c \cap K_{n+1}$ and $\varphi(x) \ge n + 1$ if $x \in K_{n+1}^c$.

It follows that $\{\varphi \le n\} = K_{n+1}$, whence plainly $\{\varphi \le r\} = K_{[r]+1}$ for all $r > 0$, $[r]$ denoting the largest integer $n < r$. Since $K_{[r]+1}$ is compact, the claim follows.

(ii)⇒(i). By assumption there exists $\varphi : X \to [0, \infty)$ such that the set $K_r \equiv \{\varphi \le r\}$ is compact for every $r > 0$ and (5.29) is satisfied. Then by the Chebyshev inequality, for every $\mu \in \mathcal{M}$,

$$r\mu(K_r^c) < \int_{K_r^c} \varphi \, d\mu \le \int_X \varphi \, d\mu \le \sup_{\mu \in \mathcal{M}} \int_X \varphi \, d\mu,$$

whence by the arbitrariness of r the conclusion follows. □

Remark 5.1.10. In view of the above proof, if $X = \mathbb{R}$, then we can choose $K_n = \overline{B}_n(0)$, the closed ball with center 0 and radius $n \in \mathbb{N}$. In this case, φ can be rewritten as $\varphi(x) := \sum_{n=0}^{\infty} n\chi_{(n,n+1]}(|x|)$ $(x \in X)$; observe that it is radial, nondecreasing, and diverging at infinity.

5.2 Parameterized measures and disintegration

Let X and Y be locally compact Hausdorff spaces, and let $\mathcal{B}(X)$ and $\mathcal{B}(Y)$ be the associated Borel σ-algebras. Recall that by Theorem 2.7.1 the Banach space $\mathfrak{R}_f(X)$ is the dual space of $C_0(X)$. Then we can state the following definition.

Definition 5.2.1. Let X and Y be locally compact Hausdorff spaces. A weakly* measurable map from X to $\mathfrak{R}_f(Y)$ is called a *parameterized measure* on Y.

For parameterized measures on Y, we will use the notation $\{\nu_x\}_{x\in X}$ with $\nu_x \in \mathfrak{R}_f(Y)$ for every $x \in X$.

Remark 5.2.1. Let us point out two alternative formulations of Definition 5.2.1 (e. g., see [5]). Firstly, by Definition 4.1.5(ii) and Theorem 2.7.1 a family $\{\nu_x\}_{x\in X} \subseteq \mathfrak{R}_f(Y)$ is a parameterized measure on Y if and only if

$$x \mapsto \langle \nu_x, g\rangle = \int_Y g(y)\, d\nu_x(y) \quad \text{is } \mathcal{B}(X)\text{-measurable for every } g \in C_0(Y). \tag{5.30}$$

Secondly, condition (5.30) can be rephrased as follows:

$$x \mapsto \nu_x(F) \quad \text{is } \mathcal{B}(X)\text{-measurable for every } F \in \mathcal{B}(Y). \tag{5.31}$$

In fact, by (5.31) the map $x \mapsto \int_Y \chi_F(y)\, d\nu_x(y)$ is $\mathcal{B}(X)$-measurable for every $F \in \mathcal{B}(Y)$, and thus the same holds with χ_F replaced by any simple function $s \in \mathcal{S}(Y)$. On the other hand, for every $g \in C_0(Y)$, there exists a sequence $\{s_n\} \subseteq \mathcal{S}(Y)$ such that $\|s_n - g\|_\infty \to 0$ as $n \to \infty$ (see Theorem 2.1.7(ii)). Hence

$$\lim_{n\to\infty} \int_Y [s_n(y) - g(y)]d\nu_x(y) = 0,$$

for all $x \in X$, whence by Corollary 2.1.5(ii) we obtain (5.30).

Conversely, let (5.30) hold, and let $F \in \mathcal{B}(Y)$. By Proposition 2.1.16 (see also Remark 2.1.7) there exists a sequence $\{\zeta_m\} \subseteq C_c(Y)$ such that $\zeta_m \to \chi_F$ μ-a. e. in Y and $\|\zeta_m\|_\infty \leq 1$ for all $m \in \mathbb{N}$. Then writing (5.30) with $g = \zeta_m$, letting $m \to \infty$, and using the dominated convergence theorem, we obtain (5.31).

Remark 5.2.2. If $\{\nu_x\}_{x\in X}$ is a parameterized measure on Y, then the real-valued function $x \mapsto |\nu_x|(F)$ is also $\mathcal{B}(X)$-measurable for every $F \in \mathcal{B}(Y)$. Indeed, by Corollary 2.1.5(i) the function $x \mapsto \sum_{i=1}^n |\nu_x(F_i)|$ is $\mathcal{B}(X)$-measurable for any partition $\{F_1, \ldots, F_n\} \subseteq \mathcal{B}(Y)$ of F, and thus by Proposition 1.8.3 (see (1.52)) and Corollary 2.1.5(ii) the claim follows.

Let X and Y be locally compact Hausdorff spaces with countable bases. Then the topological product $X \times Y$ is a locally compact Hausdorff space with countable basis, and by Theorem 1.2.3

$$\mathcal{B}(X \times Y) = \mathcal{B}(X) \times \mathcal{B}(Y).$$

By Lemma 2.1.8 the projection

$$p_X : X \times Y \to X, \quad p_X(x, y) := x,$$

is $(\mathcal{B}(X \times Y), \mathcal{B}(X))$-measurable, and the projection

$$p_Y : X \times Y \to Y, \quad p_Y(x,y) := y,$$

is $(\mathcal{B}(X \times Y), \mathcal{B}(Y))$-measurable. Hence for any $\mu \in \mathfrak{M}_f(X \times Y)$, the image measures (see (2.28))

$$\mu_{p_X} : \mathcal{B}(X) \mapsto \mathbb{R}, \quad \mu_{p_X}(E) := \mu(p_X^{-1}(E)) = \mu(E \times Y) \quad \text{for all } E \in \mathcal{B}(X), \tag{5.32}$$

$$\mu_{p_Y} : \mathcal{B}(Y) \mapsto \mathbb{R}, \quad \mu_{p_Y}(F) := \mu(p_Y^{-1}(F)) = \mu(X \times F) \quad \text{for all } F \in \mathcal{B}(Y), \tag{5.33}$$

are well defined. If $\mu \in \mathfrak{R}_f(X \times Y)$, then by Proposition 1.3.3 $\mu_{p_X} \in \mathfrak{R}_f(X)$ and $\mu_{p_Y} \in \mathfrak{R}_f(Y)$.

Theorem 5.2.1 (Disintegration). *Let X and Y be locally compact Hausdorff spaces with countable bases. Let $\nu \in \mathfrak{R}_f(X \times Y)$, and let $|\nu|_{p_X} \in \mathfrak{R}_f^+(X)$ be the projection of its variation $|\nu|$ onto the space X. Then there exists a parameterized measure $\nu \equiv \{\nu_x\}_{x \in X}$ on Y with the following properties:*
(i) *$|\nu_x|(Y) = 1$ for $|\nu|_{p_X}$-a. e. $x \in X$;*
(ii) *for every $h \in L^1(X \times Y, \mathcal{B}(X \times Y), |\nu|)$,*

$$h(x,\cdot) \in L^1(Y, \mathcal{B}(Y), |\nu_x|) \quad \textit{for } |\nu|_{p_X}\textit{-a. e. } x \in X, \tag{5.34}$$

$$\textit{the map } x \mapsto \int_Y h(x,y)\, d\nu_x(y) \textit{ belongs to } L^1(X, \mathcal{B}(X), |\nu|_{p_X}), \tag{5.35}$$

and

$$\int_{X \times Y} h(x,y)\, d\nu(x,y) = \int_X d\,|\nu|_{p_X}(x) \int_Y h(x,y)\, d\nu_x(y). \tag{5.36}$$

Moreover, the parameterized measure $\{\nu_x\}_{x \in X}$ is unique in the following sense: let $\{\nu'_x\}_{x \in X}$ be a parameterized measure on Y such that

$$\textit{the map } x \mapsto \int_Y h(x,y)\, d\nu'_x(y) \textit{ belongs to } L^1(X, \mathcal{B}(X), |\nu|_{p_X}) \tag{5.37}$$

and

$$\int_{X \times Y} h(x,y)\, d\nu(x,y) = \int_X d\,|\nu|_{p_X}(x) \int_Y h(x,y)\, d\nu'_x(y) \tag{5.38}$$

for every bounded $\mathcal{B}(X \times Y)$-measurable $h : X \times Y \to \mathbb{R}$ with compact support. Then $\nu'_x = \nu_x$ for $|\nu|_{p_X}$-a. e. $x \in X$.

Definition 5.2.2. The parameterized measure $\{\nu_x\}_{x\in X}$ is called the *disintegration* of the measure ν.

Remark 5.2.3. (i) Let $G \in \mathcal{B}(X \times Y)$, and let $G_x := \{y \in Y \mid (x,y) \in G\} \in \mathcal{B}(Y)$ be its x-section (see Proposition 1.2.2). Setting $h = \chi_G$ in (5.36) and recalling that $\chi_G(x,\cdot) = \chi_{G_x}$ (see Remark 2.1.5), we get

$$\nu(G) = \int_X \nu_x(G_x)\, d\,|\nu|_{p_X}(x) \quad \text{for all } G \in \mathcal{B}(X \times Y). \tag{5.39}$$

If ν is positive, then by (5.39) and the arbitrariness of G it follows that ν_x is positive for $|\nu|_{p_X}$-a. e. $x \in X$.

(ii) Let $\nu_{p_Y} \in \mathfrak{R}_f(Y)$ be the projection of ν onto Y (see (5.33)). For any $F \in \mathcal{B}(Y)$ from (5.39) with $G = X \times F$ (thus $G_x = F$ for every $x \in X$), we get

$$\nu_{p_Y}(F) = \nu(X \times F) = \int_X \nu_x(F)\, d\,|\nu|_{p_X}(x). \tag{5.40}$$

Proof of Theorem 5.2.1. For any $g \in C_0(Y)$, $g = g(y)$, set

$$\mu_g : \mathcal{B}(X) \to [0,\infty), \quad \mu_g(E) := \int_{E\times Y} g\, d\nu \quad \text{for all } E \in \mathcal{B}(X). \tag{5.41}$$

Clearly, $\mu_g \in \mathfrak{R}_f(X)$, and

$$|\mu_g(E)| \le \|g\|_\infty\, |\nu|(E \times Y) = \|g\|_\infty\, |\nu|_{p_X}(E) \quad \text{for all } E \in \mathcal{B}(X). \tag{5.42}$$

By inequality (5.42) we have $\mu_g \ll |\nu|_{p_X}$, and thus by the Radon–Nikodým theorem there exists $(\mu_g)_r \in L^1(X, \mathcal{B}(X), |\nu|_{p_X})$ such that

$$\mu_g(E) = \int_E (\mu_g)_r d\,|\nu|_{p_X} \quad \text{for all } E \in \mathcal{B}(X). \tag{5.43}$$

Moreover, by (5.42) and Proposition 4.2.5 (see (4.43b)) $(\mu_g)_r \in L^\infty(X, \mathcal{B}(X), |\nu|_{p_X})$, and $\|(\mu_g)_r\|_\infty \le \|g\|_\infty$.

Clearly, the map $g \mapsto \mu_g$ with $g \in C_0(Y)$ and μ_g defined in (5.41) is linear. Let $D \subseteq C_0(Y)$ be countable and dense (observe that the space $C_0(Y)$ is separable; see paragraph A.7 in Appendix A). Then by (5.43) there exists a $|\nu|_{p_X}$-null set $N \in \mathcal{B}(X)$ such that

$$(\mu_{\alpha g+\beta g'})_r(x) = \alpha(\mu_g)_r(x) + \beta(\mu_{g'})_r(x) \quad \text{for all } x \in N^c,\ \alpha,\beta \in \mathbb{R}, \text{ and } g, g' \in D. \tag{5.44}$$

Set

$$T_x : D \to \mathbb{R}, \quad T_x g := (\mu_g)_r(x) \quad \text{for all } x \in N^c \text{ and } g \in D.$$

By (5.44) T_x is linear. Moreover,

$$|T_x g| = |(\mu_g)_r(x)| \leq \|(\mu_g)_r\|_\infty \leq \|g\|_\infty \quad (x \in N^c), \tag{5.45}$$

and thus T_x is bounded. Since D is dense in $C_0(Y)$, T_x admits a unique continuation, denoted again T_x, to the whole space $C_0(Y)$. Hence $T_x \in (C_0(Y))^*$ for all $x \in N^c$, and thus by Theorem 2.7.1 there exists $\nu_x \in \mathfrak{R}_f(Y)$ such that

$$T_x g = (\mu_g)_r(x) = \int_Y g(y)\, d\nu_x(y) \quad \text{for all } g \in C_0(Y) \quad (x \in N^c) \tag{5.46}$$

and

$$\|\nu_x\| = |\nu_x|(Y) = \|T_x\| \leq 1 \tag{5.47}$$

(see (5.45)). We can complete the definition of ν_x in X by setting $\nu_x := \delta_{\bar{y}}$ for any $x \in N$ with arbitrary fixed $\bar{y} \in Y$.

Let us show that the function $x \mapsto \nu_x(F)$ is $\mathcal{B}(X)$-measurable for every $F \in \mathcal{B}(Y)$ and thus the family $\{\nu_x\}_{x \in X}$ is a parameterized measure on Y (see Remark 5.2.1). Indeed, by definition we have $T_x g = (\mu_g)_r(x)$ for all $x \in N^c$; since $(\mu_g)_r$ belongs to $L^\infty(X, \mathcal{B}(X), |\nu|_{p_X})$ and thus is $\mathcal{B}(X)$-measurable, the map $x \mapsto T_x g$ is $(\mathcal{B}(X) \cap N^c)$-measurable for any $g \in C_0(Y)$. Now observe that for every $F \in \mathcal{B}(Y)$, there is a sequence $\{g_n\} \subseteq C_0(Y)$ such that $g_n \to \chi_F$ in $L^1(Y, \mathcal{B}(Y), \nu_x)$ as $n \to \infty$ (see Proposition 2.1.16 and Remark 2.1.7). Then

$$T_x g_n = \int_Y g_n(y)\, d\nu_x(y) \to \nu_x(F) \quad \text{as } n \to \infty,$$

and thus by Corollary 2.1.5(ii) the map $x \mapsto \nu_x(F)$ is $\mathcal{B}(X) \cap N^c$-measurable. On the other hand, clearly, the map $x \mapsto \nu_x(F) = \delta_{\bar{y}}(F)$ $(x \in N)$ is $(\mathcal{B}(X) \cap N)$-measurable for all $F \in \mathcal{B}(Y)$. Then by Proposition 2.1.3 the claim follows.

In view of the above remarks, it is clear that every map $h(x,y) := s(x)g(y)$ with $s \in \mathcal{S}(X)$ and $g \in C_0(Y)$ satisfies (5.34)–(5.35) (see Remark 5.2.1). Moreover, equality (5.36) also holds with this choice of $h(x,y)$. In fact, by (5.41), (5.43), and (5.46) for any $g \in C_0(Y)$ and $E \in \mathcal{B}(X)$, we have

$$\begin{aligned}\mu_g(E) &= \int_{X \times Y} \chi_E(x) g(y)\, d\nu = \int_E (\mu_g)_r(x)\, d|\nu|_{p_X}(x) \\ &= \int_X d|\nu|_{p_X}(x) \int_Y \chi_E(x) g(y)\, d\nu_x(y),\end{aligned}$$

whence

$$\int_{X\times Y} s(x)g(y)\,d\nu = \int_X d\,|\nu|_{p_X}(x) \int_Y s(x)g(y)\,d\nu_x(y)$$

for all $s \in \mathcal{S}(X)$ and $g \in C_0(Y)$. By Lemma 2.8.3 and the dominated convergence theorem the above equality also holds for all $s \in \mathcal{S}(X)$ and $g \in \mathcal{S}(Y)$. Since the set $\{s(x)g(y) \mid s \in \mathcal{S}(X), g \in \mathcal{S}(Y)\}$ is dense in $L^1(X\times Y, \mathcal{B}(X\times Y), \nu)$, claim (ii) follows.

By Remark 5.2.2, Proposition 1.8.3, and Corollary 2.1.5(ii), for every $F \in \mathcal{B}(Y)$, the function $x \mapsto |\nu_x|(F)$ is $\mathcal{B}(X)$-measurable. Then from (5.39) we get

$$|\nu(G)| \le \int_X |\nu_x|(G_x)\,d\,|\nu|_{p_X}(x) \quad \text{for all } G \in \mathcal{B}(X\times Y),$$

whence by the arbitrariness of G and the definition of $|\nu|$ we have

$$|\nu|(G) \le \int_X |\nu_x|(G_x)\,d\,|\nu|_{p_X}(x).$$

Choosing $G = X\times Y$ in this inequality and using (5.47), we obtain

$$|\nu|(X\times Y) \le \int_X |\nu_x|(Y)\,d\,|\nu|_{p_X}(x) \le |\nu|(X\times Y)$$

(see (5.32)), and thus claim (i) follows.

Finally, let $\{\nu_x'\}_{x\in X}$ be a parameterized measure on Y satisfying (5.37)–(5.38) for all bounded $\mathcal{B}(X\times Y)$-measurable h with compact support. From (5.36) and (5.38) we obtain

$$\int_X d\,|\nu|_{p_X}(x) \int_Y h(x,y)\,d\nu_x(y) = \int_X d\,|\nu|_{p_X}(x) \int_Y h(x,y)\,d\nu_x'(y)$$

for all h as above, whence plainly $\nu_x' = \nu_x$ for $|\nu|_{p_X}$-a. e. $x \in X$. This completes the proof. □

Remark 5.2.4. Let X and Y be as in Theorem 5.2.1, let $\nu \in \mathfrak{R}_f^+(X\times Y)$, and let $\{\nu_x\}_{x\in X}$ be its disintegration. By equality (5.40), for every $F \in \mathcal{B}(Y)$ such that $\nu_{p_Y}(F) = 0$, there exists a ν_{p_X}-null set $N \subseteq X$ such that $\nu_x(F) = 0$ for all $x \in N^c$. Since Y has a countable basis, the measurable space $(Y, \mathcal{B}(Y))$ is separable (see Definition 1.2.5). Then by a standard argument the choice of the ν_{p_X}-null set N can be made independent of $F \in \mathcal{B}(Y)$. It follows that $\nu_x \ll \nu_{p_Y}$ for ν_{p_X}-a. e. $x \in X$.

5.3 Young measures revisited

Let X be a locally compact Hausdorff space, and let $\{u_k\} \subseteq L^1(X, \mathcal{B}(X), \mu)$ be weakly convergent. It has already been observed (see Example 2.8.1(i)) that the same need not hold for $\{h \circ u_k\}$ with nonlinear h such that $\{h \circ u_k\} \subseteq L^1(X, \mathcal{B}(X), \mu)$. To get information about the behavior of $\{h \circ u_k\}$, Remark 5.1.4 suggests to address the sequence $\{\nu_k\} \subseteq \mathfrak{R}_f(X)$ associated with $\{h \circ u_k\}$ as in (5.5). In this framework the sequence $\{\nu_{u_k}\}$ of Young measures associated with $\{u_k\}$ plays an important role.

Let $\mu \in \mathfrak{R}_f^+(X)$, and let $\nu \in \mathfrak{Y}^+(X, \mathcal{B}(X), \mu; Y) \equiv \mathfrak{Y}^+(X; Y)$ be a Young measure. Then by definition $\nu_{p_X} = \mu$ (see Definition 2.5.3), and since $\mathfrak{Y}^+(X; Y) \subseteq \mathfrak{R}_f^+(X \times Y)$, from Theorem 5.2.1 we obtain the following result.

Proposition 5.3.1 (Disintegration of Young measures). *Let X and Y be locally compact Hausdorff spaces with countable bases. Let $\mu \in \mathfrak{R}_f^+(X)$, and let $\nu \in \mathfrak{Y}^+(X; Y)$. Then there exists a parameterized measure $\nu \equiv \{\nu_x\}_{x \in X}$ on Y with the following properties:*

(i) *$\nu_x \in \mathfrak{P}(Y) \subseteq \mathfrak{R}_f^+(Y)$ for μ-a. e. $x \in X$;*

(ii) *for every $h \in L^1(X \times Y, \mathcal{B}(X \times Y), \nu)$, we have*

$$h(x, \cdot) \in L^1(Y, \mathcal{B}(Y), \nu_x) \text{ for } \mu\text{-a. e. } x \in X, \tag{5.48}$$

$$\text{the map } x \mapsto \int_Y h(x,y)\, d\nu_x(y) \text{ belongs to } L^1(X, \mathcal{B}(X), \mu), \tag{5.49}$$

and

$$\int_{X \times Y} h(x,y)\, d\nu(x,y) = \int_X d\mu(x) \int_Y h(x,y)\, d\nu_x(y). \tag{5.50}$$

If $\{\nu'_x\}_{x \in X}$ is a parameterized measure on Y such that

$$\text{the map } x \mapsto \int_Y h(x,y)\, d\nu'_x(y) \text{ belongs to } L^1(X, \mathcal{B}(X), \mu) \tag{5.51}$$

and

$$\int_{X \times Y} h(x,y)\, d\nu(x,y) = \int_X d\mu(x) \int_Y h(x,y)\, d\nu'_x(y) \tag{5.52}$$

for every bounded $\mathcal{B}(X \times Y)$-measurable $h : X \times Y \to \mathbb{R}$ with compact support, then $\nu'_x = \nu_x$ for μ-a. e. $x \in X$.

It is interesting to characterize the disintegration of Young measures associated with functions (see Definition 2.5.3). Recall that the Young measure ν_u associated with

u is defined as

$$\nu_u(E \times F) := \mu(E \cap u^{-1}(F)) \quad \text{for all } E \in \mathcal{B}(X) \text{ and } F \in \mathcal{B}(Y).$$

Proposition 5.3.2. *Let X and Y be locally compact Hausdorff spaces with countable bases. Let $\mu \in \mathfrak{R}_f^+(X)$, let $u : X \to Y$ be $(\mathcal{B}(X), \mathcal{B}(Y))$-measurable, and let $\nu_u \in \mathfrak{Y}^+(X; Y)$ be the Young measure associated with u. Then*

$$(\nu_u)_x = \delta_{u(x)} \quad \text{for } \mu\text{-a. e. } x \in X. \tag{5.53}$$

Proof. For any $F \in \mathcal{B}(Y)$ from (5.36), we get

$$\nu_u(X \times F) = \int_{X\times Y} \chi_F(y)\, d\nu_u(x,y) = \int_X (\nu_u)_x(F)\, d\mu(x).$$

On the other hand, by definition

$$\nu_u(X \times F) = \mu(u^{-1}(F)) = \int_X \chi_{u^{-1}(F)}(x)\, d\mu(x).$$

From (2.29), for every $F \in \mathcal{B}(Y)$, there exists a μ-null set $N \subseteq X$ such that for all $x \in N^c$,

$$(\nu_u)_x(F) = \chi_{u^{-1}(F)}(x) = \begin{cases} 1 & \text{if } u(x) \in F, \\ 0 & \text{otherwise.} \end{cases}$$

Since X has a countable basis, the measurable space $(X, \mathcal{B}(X))$ is separable (see Definition 1.2.5), and thus by a standard argument the choice of the μ-null set N can be made independent of $F \in \mathcal{B}(Y)$. Then the result follows. □

From equalities (5.36) and (5.53) we immediately obtain the following result.

Corollary 5.3.3. *Let X and Y be locally compact Hausdorff spaces with countable bases. Let $\mu \in \mathfrak{R}_f^+(X)$, let $u : X \to Y$ be $(\mathcal{B}(X), \mathcal{B}(Y))$-measurable, and let $\nu_u \in \mathfrak{Y}^+(X, \mathcal{B}(X), \mu; Y)$ be the Young measure associated with u. Then for every $h \in L^1(X \times Y, \mathcal{B}(X \times Y), \nu_u)$, we have*

$$\int_{X\times Y} h(x,y)\, d\nu_u(x,y) = \int_X h(x, u(x))\, d\mu(x). \tag{5.54}$$

In view of Proposition 5.3.1, it is interesting to consider the set of weakly* measurable functions from X to $\mathfrak{R}_f(Y)$, which for μ-a. e. $x \in X$ take values in the set $\mathfrak{P}(Y) \subseteq \mathfrak{R}_f^+(Y)$ of probability measures on Y. We denote by $\mathfrak{P}(X; Y) \equiv \mathfrak{P}(X, \mathcal{B}(X), \mu; Y)$ this set, called the set of *parameterized probabilities* on Y. It is easily seen that $\mathfrak{P}(X; Y)$ is a closed convex subset of the unit ball of the space $L^\infty_{w^*}(X; \mathfrak{R}_f(Y))$,

$$\mathscr{U} := \{\nu \in L^\infty_{w^*}(X; \mathfrak{R}_f(Y)) \mid \operatorname{ess\,sup}_{x\in X} \|\nu_x\|_{\mathfrak{R}_f(Y)} \leq 1\}. \tag{5.55}$$

Corollary 5.3.4. *Let X and Y be locally compact Hausdorff spaces with countable bases, and let $\mu \in \mathfrak{R}^+_f(X)$. Then there is a one-to-one correspondence between $\mathfrak{Y}^+(X;Y)$ and $\mathfrak{P}(X;Y)$.*

Proof. By Proposition 5.3.1 to each $\nu \in \mathfrak{Y}^+(X;Y)$ corresponds a parameterized probability on Y. Conversely, let $\{\nu_x\}_{x\in X} \in \mathfrak{P}(X;Y)$, and set

$$\nu : \mathcal{B}(X\times Y) \to [0,\infty), \quad \nu(G) := \int_X \nu_x(G_x)\, d\mu(x) \quad (G \in \mathcal{B}(X\times Y))$$

(observe that by Remark 5.2.1 the definition is well posed). By Proposition 1.3.3 we have $\nu \in \mathfrak{R}^+_f(X\times Y)$, and choosing $G = E\times Y$ with $E \in \mathcal{B}(X)$, we get

$$\nu(E\times Y) = \int_E \nu_x(Y)\, d\mu(x) = \mu(E).$$

Hence $\nu \in \mathfrak{Y}^+(X;Y)$ (see Definition 2.5.3), and thus the result follows. □

By Corollary 5.3.4 and previous remarks it is natural to endow $\mathfrak{Y}^+(X;Y)$ with the topology of weak* convergence on $L^\infty_{w^*}(X;\mathfrak{R}_f(Y))$ (see Subsection 5.3.1). Another topology on $\mathfrak{Y}^+(X;Y)$, related to narrow convergence, will be introduced in Subsection 5.3.2. To this purpose, we need the following definition.

Definition 5.3.1. Let $(X, \mathcal{A}, \mu)$ be a finite measure space, and let Y be a topological space.

(i) By a *Carathéodory function* we mean any $h : X\times Y \to \mathbb{R}$ such that

$$\begin{cases} \text{(a)} \quad h(x,\cdot) \in C(Y) \ \text{ for } \mu\text{-a.e. } x \in X; \\ \text{(b)} \quad h(\cdot,y) \text{ is } \mathcal{A}\text{-measurable } \text{ for all } y \in Y. \end{cases}$$

(ii) By a *bounded Carathéodory integrand* we mean any Carathéodory function such that

$$\begin{cases} \text{(a)} \quad h(x,\cdot) \in C_b(Y) \ \text{ for } \mu\text{-a.e. } x \in X; \\ \text{(b)} \quad \text{the map } x \mapsto \|h(x,\cdot)\|_\infty \text{ is } \mathcal{A}\text{-measurable, and } \int_X \|h(x,\cdot)\|_\infty\, d\mu(x) < \infty. \end{cases}$$

C_0-*Carathéodory integrands* are defined by replacing $C_b(Y)$ by $C_0(Y)$ in (a).
Two Carathéodory functions h and h' belong to the same *equivalence class* if $h(x,\cdot) = h'(x,\cdot)$ for μ-a.e. $x \in X$.

The vector space of (equivalence classes of) Carathéodory functions will be denoted by $\mathscr{C}(X \times Y)$, and those of bounded Carathéodory integrands and C_0-Carathéodory integrands will be denoted by $\mathscr{C}_b(X \times Y)$ and $\mathscr{C}_0(X \times Y)$, respectively.

It is immediately seen that the map from $\mathscr{C}_b(X \times Y)$ to $[0, \infty)$,

$$h \mapsto \|h\|_{\mathscr{C}_b(X \times Y)} := \int_X \|h(x, \cdot)\|_\infty \, d\mu(x), \tag{5.56}$$

is a norm. Then we have the following result.

Proposition 5.3.5. *Let $(X, \mathcal{A}, \mu)$ be a finite measure space, and let Y be a locally compact Hausdorff space with countable basis. Then the map*

$$T : \mathscr{C}_0(X \times Y) \to L^1(X; C_0(Y)), \quad (Th)(x) := h(x, \cdot) \quad \textit{for } \mu\textit{-a. e. } x \in X, \tag{5.57}$$

is an isometric isomorphism of $(\mathscr{C}_0(X \times Y), \|\cdot\|_{\mathscr{C}_0(X \times Y)})$ onto $L^1(X; C_0(Y))$. In particular, $\mathscr{C}_0(X \times Y)$ endowed with the norm $\|\cdot\|_{\mathscr{C}_b(X \times Y)}$ is a Banach space.

Proof. Let us first prove that the operator T is well defined. For any $h \in \mathscr{C}_0(X \times Y)$ and $g \in C_0(Y)$, the function $(x, y) \mapsto h(x, y) - g(y)$ belongs to $\mathscr{C}_0(X \times Y)$, and thus for every $\alpha \in \mathbb{R}$, the set $\{x \in X \mid \|h(x, \cdot) - g\|_\infty < \alpha\}$ belongs to $\mathcal{A}$ (see Definition 5.3.1(ii)). On the other hand, $C_0(Y)$ is separable (see Appendix A, Subsection A.7), and thus the Borel σ-algebra $\mathcal{B} \equiv \mathcal{B}(C_0(Y))$ is generated by a countable basis of open balls $B(g_k, \alpha) := \{g \in C_0(Y) \mid \|g - g_k\|_\infty < \alpha\}$ $(k \in \mathbb{N})$. It follows that the map $x \mapsto h(x, \cdot) = (Th)(x)$ is $(\mathcal{A}, \mathcal{B})$-measurable and thus μ-measurable by Proposition 4.1.4, since $C_0(Y)$ is separable. Moreover,

$$\begin{aligned} \|h\|_{\mathscr{C}_b(X \times Y)} &= \int_X \|h(x, \cdot)\|_\infty \, d\mu(x) \\ &= \int_X \|(Th)(x)\|_\infty \, d\mu(x) = \|Th\|_{L^1(X; C_0(Y))} < \infty. \end{aligned} \tag{5.58}$$

Hence $Th \in L^1(X; C_0(Y))$, and thus the claim follows.

Since T is linear and isometric (see (5.58)), it is injective. To prove the surjectivity, fix any $\hat{h} \in L^1(X; C_0(Y))$. Then $\hat{h}(x) \in C_0(Y)$ for μ-a. e. $x \in X$, and the map $x \mapsto \|\hat{h}(x)\|_\infty$ belongs to $L^1(X)$. Moreover, for any $y \in Y$, the map $x \mapsto \hat{h}(x)(y)$ is $\mathcal{A}$-measurable, since the map $x \mapsto \|\hat{h}(x)\|_\infty$ is $\mathcal{A}$-measurable and the map from $C_0(Y)$ to $\mathbb{R}$, $g \mapsto g(y)$ $(g \in C_0(Y), y \in Y)$, is continuous. Therefore the function $(x, y) \mapsto \hat{h}(x)(y)$ belongs to $\mathscr{C}_0(X \times Y)$, and thus T is surjective. Hence the result follows. □

By Proposition 5.3.5, Theorem 4.3.7, and Proposition 4.4.16 we have the following:

Corollary 5.3.6. *Let $(X, \mathcal{A}, \mu)$ be a finite measure space, and let Y be a locally compact Hausdorff space with countable basis. Then:*

(i) *the space* $\mathscr{C}_0(X \times Y)$ *is separable;*
(ii) *we have*

$$(\mathscr{C}_0(X \times Y))^* = L^\infty_{w^*}(X; \mathfrak{R}_f(Y)).$$

Remark 5.3.1. Let $(X, \mathcal{A}, \mu)$ be a finite measure space, and let Y be a compact space. Then $C_b(Y)$ is separable, and the proof of Proposition 5.3.5 shows that the map

$$T : \mathscr{C}_b(X \times Y) \to L^1(X; C_b(Y)), \quad (Th)(x) := h(x, \cdot) \quad \text{for } \mu\text{-a.e. } x \in X, \tag{5.59}$$

is an isometric isomorphism of $(\mathscr{C}_b(X \times Y), \| \cdot \|_{\mathscr{C}_b(X\times Y)})$ onto $L^1(X; C_b(Y))$.

5.3.1 Weak* convergence

As already observed, by Corollary 5.3.4 and previous remarks it is natural to endow $\mathfrak{Y}^+(X; Y)$ with the topology of weak* convergence on $L^\infty_{w^*}(X; \mathfrak{R}_f(Y))$. Then by Corollary 5.3.6(ii) we have the following definition (recall that by definition $\mathfrak{Y}^+(X; Y) \subseteq \mathfrak{R}^+_f(X \times Y)$).

Definition 5.3.2. Let X and Y be locally compact Hausdorff spaces with countable bases. A sequence $\{\nu_k\} \subseteq \mathfrak{Y}^+(X; Y)$ converges *weakly** to $\nu \in L^\infty_{w^*}(X; \mathfrak{R}_f(Y))$ (written $\nu_k \overset{*}{\rightharpoonup} \nu$) if

$$\lim_{k\to\infty} \int_{X\times Y} h \, d\nu_k = \int_{X\times Y} h \, d\nu \quad \text{for all } h \in \mathscr{C}_0(X \times Y).$$

The *weak* topology* on $\mathfrak{Y}^+(X; Y)$, denoted $\mathcal{T}_{w^*}$, is the weakest topology that makes continuous all maps $\nu \mapsto \int_{X\times Y} h \, d\nu$ with $h \in \mathscr{C}_0(X \times Y)$.

A basis for $\mathcal{T}_{w^*}$ is the family of subsets

$$B_{h_1,\dots,h_n;\epsilon}(\nu_0) := \left\{ \nu \in \mathfrak{Y}^+(X; Y) \,\middle|\, \left| \int_{X\times Y} h_j \, d\nu - \int_{X\times Y} h_j \, d\nu_0 \right| < \epsilon \quad \forall j = 1, \dots, n \right\}, \tag{5.60}$$

where $h_1, \dots, h_n \in \mathscr{C}_0(X \times Y)$, $\nu_0 \in \mathfrak{Y}^+(X; Y)$, $n \in \mathbb{N}$, and $\epsilon > 0$.

Remark 5.3.2. By Definition 5.3.2 and the lower semicontinuity of the norm in $L^\infty_{w^*}(X; \mathfrak{R}_f(Y))$ the weak* limit ν belongs to the unit ball $\mathscr{U} \subseteq L^\infty_{w^*}(X; \mathfrak{R}_f(Y))$ (see (5.55)). However, due to possible "loss of mass", ν need not be a parameterized probability on Y (see Remark 5.1.3). Also, observe that $\nu \in L^\infty_{w^*}(X; \mathfrak{R}^+_f(Y))$ since $\{\nu_k\} \subseteq \mathfrak{Y}^+(X; Y)$.

Theorem 5.3.7. *Let X and Y be locally compact Hausdorff spaces with countable bases. Then for every sequence* $\{\nu_k\} \subseteq \mathfrak{Y}^+(X; Y)$, *there exist a subsequence* $\{\nu_{k_l}\} \subseteq \{\nu_k\}$ *and* $\nu \in \mathscr{U} \subseteq L^\infty_{w^*}(X; \mathfrak{R}_f(Y))$ *such that* $\nu_{k_l} \overset{*}{\rightharpoonup} \nu$.

Proof. By Theorem 4.3.7 the space $L^1(X; C_0(Y))$ is separable, and hence by Proposition 4.4.16 and the Banach theorem the unit ball $\mathscr{U} \subseteq L^\infty_{w^*}(X; \mathfrak{R}_f(Y))$ is weakly* compact. Then by Corollary 5.3.6 the result follows. □

Remark 5.3.3. Since $L^1(X; C_0(Y)) = \mathscr{C}_0(X\times Y)$ is separable, the relative weak* topology $\mathcal{T}_{w^*} \bigcap \mathscr{U}$ is metrizable. In fact, by Proposition A.11 it coincides with the metric topology induced on $L^\infty_{w^*}(X; \mathfrak{R}_f(Y))$ by the norm

$$\nu \mapsto |||\nu||| := \sum_{k=1}^{\infty} \frac{1}{2^k} \left(\frac{|\int_{X\times Y} h_k \, d\nu|}{1 + \int_X \|h_k(x,\cdot)\|_\infty \, d\mu(x)} \right), \tag{5.61}$$

where $D \equiv \{h_k\} \subseteq \mathscr{C}_0(X \times Y)$ is a dense countable subset.

5.3.2 Narrow convergence and tightness

Definition 5.3.3. Let X and Y be locally compact Hausdorff spaces with countable bases. A sequence $\{\nu_k\} \subseteq \mathfrak{Y}^+(X; Y)$ converges *narrowly* to $\nu \in \mathfrak{Y}^+(X; Y)$ (written $\nu_k \overset{n}{\rightharpoonup} \nu$) if

$$\lim_{k\to\infty} \int_{X\times Y} h \, d\nu_k = \int_{X\times Y} h \, d\nu \quad \text{for all } h \in \mathscr{C}_b(X \times Y).$$

The *narrow topology* on $\mathfrak{Y}^+(X; Y)$, denoted $\mathcal{T}_n$, is the weakest topology that makes continuous all maps $\nu \mapsto \int_{X\times Y} h \, d\nu$ with $h \in \mathscr{C}_b(X \times Y)$.

A basis for $\mathcal{T}_n$ is the family (5.60), yet with $h_1, \dots, h_n \in \mathscr{C}_b(X \times Y)$.

The following lower semicontinuity result will be used.

Proposition 5.3.8. *Let X and Y be locally compact Hausdorff spaces with countable bases, and let $h : X \times Y \to [0, \infty)$ be a Carathéodory function. Then the map from $\mathfrak{Y}^+(X; Y)$ to $[0, \infty]$, $\nu \mapsto \int_{X\times Y} h \, d\nu$, is lower semicontinuous in the narrow topology.*

Proof. We must prove that if $\nu_k \overset{n}{\rightharpoonup} \nu$ in $\mathfrak{Y}^+(X; Y)$, then

$$\int_{X\times Y} h(x,y) \, d\nu(x,y) \le \liminf_{k\to\infty} \int_X d\mu(x) \int_Y h(x,y) \, d(\nu_k)_x(y). \tag{5.62}$$

Let d denote a compatible metric on the space Y, which by Proposition A.2 is metrizable. For every $n \in \mathbb{N}$, define $h_n : X \times Y \to [0, \infty)$ by

$$h_n(x,y) := \min\left\{ \inf_{\zeta\in Y} [h(x,\zeta) + n\, d(\zeta, y)],\ n \right\} \quad (x \in X, y \in Y).$$

It is easily seen that for every $n \in \mathbb{N}$, h_n is $\mathcal{B}(X \times Y)$-measurable, $h_n \in \mathscr{C}_b(X \times Y)$, $h_n \le h_{n+1}$, and $h_n \to h$ pointwise in $X \times Y$. Then by the monotone convergence theorem

$$\int_{X\times Y} h\, d\nu = \lim_{n\to\infty} \int_{X\times Y} h_n\, d\nu = \sup_{n\in\mathbb{N}} \int_{X\times Y} h_n\, d\nu.$$

For every $n \in \mathbb{N}$, the map $\nu \mapsto \int_{X\times Y} h_n\, d\nu$ is continuous by Definition 5.3.3, since $h_n \in \mathscr{C}_b(X \times Y)$. Hence the map $\nu \mapsto \int_{X\times Y} h\, d\nu$ is lower semicontinuous as the supremum of a sequence of continuous functions. This proves the result. □

For Young measures, the definition of tightness is as follows.

Definition 5.3.4. Let X and Y be locally compact Hausdorff spaces with countable basis. A subset $\mathcal{N} \subseteq \mathfrak{Y}^+(X;Y)$ is called *tight* if for any $\epsilon > 0$, there exists a compact subset $K \subseteq Y$ such that $\nu(X \times K^c) < \epsilon$ for all $\nu \in \mathcal{N}$.

Remark 5.3.4. By Definition 5.3.4 a subset $\mathcal{N} \subseteq \mathfrak{Y}^+(X;Y)$ is tight if and only if the set of projections onto Y, $\{\nu_{p_Y} \mid \nu \in \mathcal{N}\} \subseteq \mathfrak{R}_f^+(Y)$, is tight in the sense of Definition 5.1.5.

The proof of the following result is strictly analogous to that of Proposition 5.1.8, and thus we omit it.

Proposition 5.3.9. *Let X and Y be locally compact Hausdorff spaces with countable bases, and let $\mathcal{N} \subseteq \mathfrak{Y}^+(X;Y)$ be tight. Then the narrow topology on $\mathcal{N}$ and the weak* topology on $\mathcal{N}$ coincide, that is, $\mathcal{T}_n \cap \mathcal{N} = \mathcal{T}_{w^*} \cap \mathcal{N}$.*

As a criterion for tightness, from Remark 5.3.4 and Proposition 5.1.12 we get the following result.

Proposition 5.3.10. *Let X and Y be locally compact Hausdorff spaces with countable bases. Then the following statements are equivalent:*
(i) *$\mathcal{N} \subseteq \mathfrak{Y}^+(X;Y)$ is tight;*
(ii) *there exists an inf-compact $\varphi : Y \to [0,\infty)$ such that*

$$\sup_{\nu\in\mathcal{N}} \int_Y \varphi\, d\nu_{p_Y} < \infty. \tag{5.63}$$

For Young measures, we have the following analogue of Theorem 5.1.9.

Theorem 5.3.11. *Let X and Y be locally compact Hausdorff spaces with countable bases. Let $\mathcal{N} \subseteq \mathfrak{Y}^+(X;Y)$. Then the following statements are equivalent:*
(i) *$\mathcal{N}$ is relatively sequentially compact in the narrow topology;*
(ii) *$\mathcal{N}$ is tight.*

Proof. (i)⇒(ii). By assumption, for any sequence $\{\nu_k\} \subseteq \mathscr{N}$, there exist a subsequence $\{\nu_{k_j}\} \subseteq \{\nu_k\}$ and $\nu \in \mathfrak{Y}^+(X;Y)$ such that

$$\lim_{j\to\infty} \int_{X\times Y} h\, d\nu_{k_j} = \int_{X\times Y} h\, d\nu \quad \text{for all } h \in \mathscr{C}_b(X\times Y). \tag{5.64}$$

Let us choose $h(x,y) = g(y)$ with $g \in C_b(Y)$. Since

$$\int_{X\times Y} g(y)\, d\nu(x,y) = \int_Y g(y)\, d\nu_{p_Y}(y),$$
$$\int_{X\times Y} g(y)\, d\nu_{k_j}(x,y) = \int_Y g(y)\, d(\nu_{k_j})_{p_Y}(y)$$

(see Remark 2.5.1(i)), from (5.64) we get

$$\lim_{j\to\infty} \int_Y g(y)\, d(\nu_{k_j})_{p_Y}(y) = \int_Y g(y)\, d\nu_{p_Y}(y) \quad \text{for all } g \in C_b(Y).$$

It follows that the set $\{\nu_{p_Y} \mid \nu \in \mathscr{N}\}$ is relatively sequentially compact in the narrow topology on $\mathfrak{R}_f^+(Y)$ and thus by Theorem 5.1.9 is tight. Then by Remark 5.3.4 the claim follows.

(ii)⇒(i). By Remark 2.5.1 $\mathscr{N}$ is a bounded subset of $\mathfrak{R}_f^+(X\times Y)$. Let us show that it is also a tight subset of $\mathfrak{R}_f^+(X\times Y)$, that is, for any $\epsilon > 0$, there exists a compact subset $\tilde{K} \subseteq X\times Y$ such that $\nu(\tilde{K}^c) < \epsilon$ for all $\nu \in \mathscr{N}$ (see Definition 5.1.5). By assumption, for any $\epsilon > 0$, there exists a compact subset $K \subseteq Y$ such that $\nu_{p_Y}(K^c) = \nu(X\times K^c) < \frac{\epsilon}{2}$ for all $\nu \in \mathscr{N}$. Moreover, by Proposition A.2 X is σ-compact, and hence there exists a compact subset $K_0 \subseteq X$ such that $\mu(K_0^c) < \frac{\epsilon}{2}$. Observe that, by the definition of Young measure, $\nu_{p_X}(K_0^c) = \nu(K_0^c\times Y) = \mu(K_0^c)$ for every $\nu \in \mathscr{N}$. Set $\tilde{K} := K_0 \times K \subseteq X\times Y$. Clearly, $\tilde{K}$ is compact, and $\tilde{K}^c \subseteq (X\times K^c)\cup(K_0^c\times Y)$. Then for all $\nu \in \mathscr{N}$,

$$\nu(\tilde{K}^c) \le \nu(X\times K^c) + \nu(K_0^c\times Y) = \nu_{p_Y}(K^c) + \mu(K_0^c) < \epsilon,$$

whence the claim follows.

By the above remarks and Proposition 5.1.11 (see also Remark 5.1.9), for any sequence $\{\nu_k\} \subseteq \mathscr{N}$, there exist a subsequence $\{\nu_{k_j}\} \subseteq \{\nu_k\}$ and $\nu \in \mathfrak{R}_f^+(X\times Y)$ such that

$$\lim_{j\to\infty} \int_{X\times Y} g\, d\nu_{k_j} = \int_{X\times Y} g\, d\nu \quad \text{for all } g \in C_b(X\times Y). \tag{5.65}$$

It is easily seen that $\nu \in \mathfrak{Y}^+(X;Y)$. Indeed, since $\{\nu_{k_j}\} \subseteq \mathcal{N} \subseteq \mathfrak{Y}^+(X;Y)$, by the disintegration theorem equality (5.65) reads

$$\lim_{j\to\infty} \int_X d\mu(x) \int_Y g(x,y)\, d(\nu_{k_j})_x(y) = \int_X d\nu_{p_X}(x) \int_Y g(x,y)\, d\nu_x(y)$$

for all $g \in C_b(X \times Y)$. Choosing $g = g(x)$ in this equality, we obtain

$$\int_X g(x) d\mu(x) = \int_X g(x)\, d\nu_{p_X}(x) \quad \text{for all } g \in C_b(X).$$

By Proposition 4.3.6(ii) (see also Proposition A.2), from this equality we plainly get $\nu_{p_X}(E) = \mu(E)$ for all $E \in \mathcal{B}(X)$. Hence the claim follows.

In particular, the equality in (5.65) holds for all $g \in C_0(X \times Y)$. Since the space $C_0(X \times Y)$ is dense in $L^1(X; C_0(Y)) = \mathscr{C}_0(X \times Y)$ (see Proposition 4.3.6(ii) and Proposition 5.3.5), for any $h \in L^1(X; C_0(Y))$, there exists a sequence $\{g_m\} \subseteq C_0(X \times Y)$ such that $\lim_{m\to\infty} \|g_m - h\|_{L^1(X;C_0(Y))} = 0$. Fix $\epsilon > 0$, and let $m_0 \in \mathbb{N}$ be so large that $\|g_{m_0} - h\|_{L^1(X;C_0(Y))} < \frac{\epsilon}{3}$. On the other hand, by (5.65) there exists $j_0 \in \mathbb{N}$ such that

$$\left| \int_{X\times Y} g_{m_0}\, d\nu_{k_j} - \int_{X\times Y} g_{m_0}\, d\nu \right| < \frac{\epsilon}{3} \quad \text{for all } j > j_0.$$

Then for all $j > j_0$,

$$\begin{aligned}
&\left| \int_{X\times Y} h\, d\nu_{k_j} - \int_{X\times Y} h\, d\nu \right| \\
&\quad \le \left| \int_{X\times Y} (h - g_{m_0})\, d\nu_{k_j} \right| + \left| \int_{X\times Y} (h - g_{m_0})\, d\nu \right| + \left| \int_{X\times Y} g_{m_0}\, d\nu_{k_j} - \int_{X\times Y} g_{m_0}\, d\nu \right| \\
&\quad < \left| \int_{X\times Y} (h - g_{m_0})\, d\nu_{k_j} \right| + \left| \int_{X\times Y} (h - g_{m_0})\, d\nu \right| + \frac{\epsilon}{3}.
\end{aligned}$$

Since $\{\nu_{k_j}\} \subseteq \mathcal{N} \subseteq \mathfrak{Y}^+(X;Y)$, by the disintegration theorem we get

$$\left| \int_{X\times Y} (h - g_{m_0})\, d\nu_{k_j} \right| = \left| \int_X d\mu(x) \int_Y (h - g_{m_0})(x,y)\, d(\nu_{k_j})_x(y) \right| \le \|g_{m_0} - h\|_{L^1(X;C_0(Y))} < \frac{\epsilon}{3}.$$

Since $\nu \in \mathfrak{Y}^+(X;Y)$, we similarly obtain

$$\left| \int_{X\times Y} (h - g_{m_0})\, d\nu \right| = \left| \int_X d\mu(x) \int_Y (h - g_{m_0})(x,y)\, d\nu_x(y) \right| \le \|g_{m_0} - h\|_{L^1(X;C_0(Y))} < \frac{\epsilon}{3}.$$

In view of the above inequalities, for any $\epsilon > 0$, there exists $j_0 \in \mathbb{N}$ such that for all $j > j_0$,

$$\left| \int_{X\times Y} h\, d\nu_{k_j} - \int_{X\times Y} h\, d\nu \right| < \epsilon \quad \text{for all } h \in \mathscr{C}_0(X \times Y).$$

To sum up, we proved that for any sequence $\{\nu_k\} \subseteq \mathscr{N}$, there exist a subsequence $\{\nu_{k_j}\} \subseteq \{\nu_k\}$ and $\nu \in \mathfrak{Y}^+(X;Y)$ such that $\nu_{k_j} \overset{*}{\rightharpoonup} \nu$. Since $\mathscr{N}$ is tight, by Proposition 5.3.9 the conclusion follows. □

The following result is similarly proven.

Proposition 5.3.12. *Let X and Y be locally compact Hausdorff spaces with countable bases. Let $\{\nu_k\} \subseteq \mathfrak{Y}^+(X;Y)$ and $\nu \in \mathfrak{Y}^+(X;Y)$ satisfy $(\nu_k)_x \overset{n}{\rightharpoonup} \nu_x$ in $\mathfrak{R}_f^+(Y)$ for μ-a. e. $x \in X$. Then $\nu_k \overset{n}{\rightharpoonup} \nu$ in $\mathfrak{Y}^+(X;Y)$.*

Proof. Let $h \in \mathscr{C}_b(X \times Y)$. By Definition 5.3.1(ii) we have $h(x,\cdot) \in C_b(Y)$ for μ-a. e. $x \in X$. Then for μ-a. e. $x \in X$,

$$\lim_{k\to\infty} \int_Y h(x,y)\, d(\nu_k)_x(y) = \int_Y h(x,y)\, d\nu_x(y)$$

(see Definition 5.1.3). On the other hand, for all $k \in \mathbb{N}$,

$$\left| \int_Y h(x,y)\, d(\nu_k)_x(y) \right| \le \|h(x,\cdot)\|_\infty (\nu_k)_x(Y) = \|h(x,\cdot)\|_\infty,$$

and by Definition 5.3.1(ii) the map $x \mapsto \|h(x,\cdot)\|_\infty$ belongs to $L^1(X)$. Then by the dominated convergence theorem we have

$$\lim_{k\to\infty} \int_X d\mu(x) \int_Y h(x,y)\, d(\nu_k)_x(y) = \int_X d\mu(x) \int_Y h(x,y)\, d\nu_x(y),$$

whence the result follows. □

5.4 Sequences of Young measures associated with functions

Let X and Y be locally compact Hausdorff spaces with countable bases, and let $u_j : X \to Y$ be $(\mathcal{B}(X), \mathcal{B}(Y))$-measurable ($j \in \mathbb{N}$). Applying the above results to the sequence $\{\nu_{u_j}\} \subseteq \mathfrak{Y}^+(X;Y)$ of Young measures associated with $\{u_j\}$, we can obtain information about the convergence of the sequence $\{u_j\}$ itself. A first result in this direction is the following:

Proposition 5.4.1. *Let X and Y be a locally compact Hausdorff spaces with countable bases. Let $u_j, u : X \to Y$ be $(\mathcal{B}(X), \mathcal{B}(Y))$-measurable, and let $\nu_{u_j}, \nu_u \in \mathfrak{Y}^+(X;Y)$ be the Young measures associated with u_j, respectively, u ($j \in \mathbb{N}$). Then:*

(i) *if $u_j \to u$ μ-a. e. in X, then $\nu_{u_j} \overset{n}{\rightharpoonup} \nu_u$ in $\mathfrak{Y}^+(X;Y)$;*

(ii) *if $\nu_{u_j} \overset{n}{\rightharpoonup} \nu_u$ in $\mathfrak{Y}^+(X;Y)$, then there exists a subsequence $\{u_{j_k}\} \subseteq \{u_j\}$ such that $u_{j_k} \to u$ μ-a. e. in X.*

Proof. Observe preliminarily that by Proposition A.2 the space Y is metrizable, and hence there exists a compatible metric d.

(i) By equality (5.53) we have $(\nu_{u_j})_x = \delta_{u_j(x)}$ and $(\nu_u)_x = \delta_{u(x)}$ for μ-a. e. $x \in X$. Moreover, by equality (5.15)

$$d_P((\nu_{u_j})_x, (\nu_u)_x) = d_P(\delta_{u_j(x)}, \delta_{u(x)}) = \min\{1, d(u_j(x), u(x))\},$$

where d_P denotes the Prokhorov metric on $\mathfrak{R}_f^+(Y)$. Since $u_j \to u$ μ-a. e. in X, it follows that

$$d_P((\nu_{u_j})_x, (\nu_u)_x) \to 0 \quad \text{for } \mu\text{-a. e. } x \in X.$$

Then by Lemma 5.1.6 and Proposition 5.3.12 the claim follows.

(ii) Plainly, the map $h : X \times Y \to \mathbb{R}$, $h(x,y) := \min\{1, d(y, u(x))\}$ belongs to $\mathscr{C}_b(X \times Y)$. Since by assumption $\nu_{u_j} \overset{n}{\rightharpoonup} \nu_u$, we have

$$\lim_{j\to\infty} \int_X d\mu(x) \int_Y \min\{1, d(y, u(x))\}\, d(\nu_{u_j})_x(y) = \int_X d\mu(x) \int_Y \min\{1, d(y, u(x))\}\, d(\nu_u)_x(y),$$

which by (5.54) reads

$$\lim_{j\to\infty} \int_X \min\{1, d(u_j(x), u(x))\}\, d\mu(x) = \int_X \min\{1, d(u(x), u(x))\}\, d\mu(x) = 0.$$

Hence (possibly extracting a subsequence, not relabeled) $d(u_j(x), u(x)) \to 0$ for μ-a. e. $x \in X$. Then the result follows. □

5.4.1 Weak* convergence of $\{\nu_{u_j}\}$

Henceforth we are dealing with the case $Y = \mathbb{R}$ (similar results hold for $Y = \mathbb{R}^d$ with $d \in \mathbb{N}$, $d \geq 2$). Let $\mathscr{F}$ be a family of $\mathcal{B}(X)$-measurable functions from X to $\mathbb{R}$, and let $\mathscr{N}_{\mathscr{F}} := \{\nu_u \mid u \in \mathscr{F}\} \subseteq \mathfrak{Y}^+(X;\mathbb{R})$ denote the set of the associated Young measures. By Corollary 5.3.4 $\mathscr{N}_{\mathscr{F}}$ is contained in the unit ball $\mathscr{U}$ of the space $L^\infty_{w^*}(X; \mathfrak{R}_f(\mathbb{R}))$,

$$\mathscr{U} := \{\nu \in L^\infty_{w^*}(X; \mathfrak{R}_f(\mathbb{R})) \mid \operatorname{ess\,sup}_{x\in X} \|\nu_x\|_{\mathfrak{R}_f(\mathbb{R})} \leq 1\}.$$

Moreover, $L^{\infty}_{w^*}(X;\mathfrak{R}_f(\mathbb{R})) = (L^1(X;C_0(\mathbb{R})))^* = (\mathscr{C}_0(X\times\mathbb{R}))^*$ (see Propositions 4.4.16 and 5.3.5), and by Theorem 4.3.7 the space $L^1(X;C_0(\mathbb{R}))$ is separable. Then we have the following result (see [8]).

Theorem 5.4.2 (Ball). *Let X be a locally compact Hausdorff space with countable basis, and let $\mu \in \mathfrak{R}^+_f(X)$. Let $\mathscr{F}$ be a family of $\mathcal{B}(X)$-measurable functions from X to $\mathbb{R}$. Then for any sequence $\{u_j\} \subseteq \mathscr{F}$:*

(i) *there exist a subsequence $\{u_{j_k}\} \subseteq \{u_j\}$ and $\nu \equiv \{\nu_x\}_{x\in X} \in \mathscr{U}$ such that*

$$g\circ u_{j_k} \overset{*}{\rightharpoonup} g^* \quad \text{in } L^\infty(X) \text{ for all } g\in C_0(\mathbb{R}), \tag{5.66}$$

where

$$g^*(x) := \int_{\mathbb{R}} g(y)\,d\nu_x(y) \quad \text{for } \mu\text{-a. e. } x\in X; \tag{5.67}$$

(ii) *if for some closed subset $K\subseteq\mathbb{R}$,*

$$\lim_{j\to\infty}\mu(u_j^{-1}(A^c)) = 0 \quad \text{for any open neighborhood } A\supset K, \tag{5.68}$$

then $\operatorname{supp}\nu_x \subseteq K$ *for μ-a. e. $x\in X$.*

Proof. (i) By the Banach–Alaoglu theorem, for any $\{\nu_{u_j}\}\subseteq\mathscr{N}_{\mathscr{F}}\subseteq\mathscr{U}$, there exist $\{\nu_{u_{j_k}}\}\subseteq\{\nu_{u_j}\}$ and $\nu\in\mathscr{U}$ such that $\nu_{u_{j_k}} \overset{*}{\rightharpoonup} \nu$ in $L^\infty_{w^*}(X;\mathfrak{R}_f(\mathbb{R}))$, that is,

$$\lim_{k\to\infty}\int_X d\mu(x)\int_{\mathbb{R}} h(x,y)\,d(\nu_{u_{j_k}})_x(y) = \int_X d\mu(x)\int_{\mathbb{R}} h(x,y)\,d\nu_x(y) \tag{5.69}$$

for all $h\in L^1(X;C_0(\mathbb{R}))$. Choosing in (5.69) $h(x,y) = f(x)g(y)$ with $f\in L^1(X)$ and $g\in C_0(\mathbb{R})$ gives

$$\lim_{k\to\infty}\int_X d\mu(x)f(x)\int_{\mathbb{R}} g(y)\,d(\nu_{u_{j_k}})_x(y) = \int_X d\mu(x)f(x)\int_{\mathbb{R}} g(y)\,d\nu_x(y),$$

whence by equality (5.54)

$$\lim_{k\to\infty}\int_X f(x)g(u_{j_k}(x))\,d\mu(x) = \int_X d\mu(x)f(x)\int_{\mathbb{R}} g(y)\,d\nu_x(y)$$

for all $f\in L^1(X)$. By (5.49) the map $x\mapsto\int_{\mathbb{R}} g(y)\,d\nu_x(y)$ is $\mathcal{B}(X)$-measurable; moreover, $\operatorname{ess\,sup}_{x\in X}|\int_{\mathbb{R}} g(y)\,d\nu_x(y)| \le \|g\|_\infty$, and thus the function g^* defined in (5.67) belongs to $L^\infty(X)$. Hence the claim follows.

(ii) If $K = \mathbb{R}$, then the claim is obvious. Let $K \subset \mathbb{R}$ satisfy (5.68), and let $g \in C_0(\mathbb{R})$, $g|_K = 0$. Clearly, for any $\epsilon > 0$, the set $A_\epsilon := \{y \in \mathbb{R} \mid |g(y)| < \epsilon\}$ is open, and $A_\epsilon \supset K$. Then by assumption we have

$$\lim_{j\to\infty} \mu(u_j^{-1}((A_\epsilon)^c)) = \lim_{j\to\infty} \{x \in X \mid |g(u_j(x))| \geq \epsilon\} = 0,$$

and thus $g(u_j) \to 0$ in measure. Since the sequence $\{g(u_j)\}$ is bounded in $L^\infty(X)$ and μ is finite, by the dominated convergence theorem (see Proposition 2.8.8(i)) we have

$$\lim_{j\to\infty} \int_X f(x)\, g(u_j(x))\, d\mu(x) = 0 \quad \text{for all } f \in L^1(X).$$

From this equality and (5.66)–(5.67) we obtain that $\int_{\mathbb{R}} g(y)\, d\nu_x(y) = 0$ for μ-a. e. $x \in X$. Since by assumption $g|_K = 0$, the conclusion follows. □

5.4.2 Narrow convergence of $\{\nu_{u_j}\}$

Let us now apply the results of Subsection 5.3.2 to sequences of Young measures associated with functions. A preliminary step in this direction is the following:

Proposition 5.4.3. *Let X be a locally compact Hausdorff space with countable basis, let $\mu \in \mathfrak{R}_f^+(X)$, and let $\mathscr{F} \subseteq L^1(X)$ be bounded. Then the set $\mathscr{N}_{\mathscr{F}} := \{\nu_u \in \mathfrak{Y}^+(X;\mathbb{R}) \mid u \in \mathscr{F}\}$ of the associated Young measures is tight.*

Proposition 5.4.3 is an easy consequence of the following lemma.

Lemma 5.4.4. *Let X be a locally compact Hausdorff space with countable basis, let $\mu \in \mathfrak{R}_f^+(X)$, and let $\mathscr{F} \subseteq L^1(X)$. Then the following statements are equivalent:*
(i) *the set $\mathscr{N}_{\mathscr{F}} := \{\nu_u \in \mathfrak{Y}^+(X;\mathbb{R}) \mid u \in \mathscr{F}\}$ is tight;*
(ii) *there exists an inf-compact $\psi : [0,\infty) \to [0,\infty)$ such that*

$$\sup_{u\in\mathscr{F}} \int_X \psi(|u|)(x)\, d\mu(x) < \infty; \tag{5.70}$$

(iii) *we have*

$$\lim_{k\to\infty} \Big(\sup_{u\in\mathscr{F}} \mu(\{|u| \geq k\})\Big) = 0.$$

Proof. (i)⇒(ii). By Proposition 5.3.10 there exists an inf-compact $\varphi : \mathbb{R} \to [0,\infty)$ such that

$$\sup_{\nu_u\in\mathscr{N}_{\mathscr{F}}} \int_{\mathbb{R}} \varphi(y)\, d(\nu_u)_{p_{\mathbb{R}}}(y) < \infty, \tag{5.71}$$

where $(\nu_u)_{p_\mathbb{R}}$ denotes the projection of ν_u onto $\mathbb{R}$ (see (5.33)). By Remark 5.1.10 we can assume that $\varphi(y) = \psi(|y|)$ with $\psi(t) := \sum_{n=0}^{\infty} n\chi_{(n,n+1]}(t)$ $(t \in [0,\infty))$. Clearly, ψ is inf-compact, and by (5.71)

$$\sup_{\nu_u \in \mathcal{N}_\mathcal{F}} \int_\mathbb{R} \psi(|y|)\, d(\nu_u)_{p_\mathbb{R}}(y) < \infty. \tag{5.72}$$

On the other hand, by (5.53) and (5.36), for any $u \in \mathcal{F}$, we have

$$\begin{aligned}\int_X \psi(|u|)(x)\, d\mu(x) &= \int_X d\mu(x) \int_\mathbb{R} \psi(|y|)\, d(\nu_u)_x(y) \\ &= \int_{X\times\mathbb{R}} \psi(|y|)\, d\nu_u(x,y) = \int_\mathbb{R} \psi(|y|)\, d(\nu_u)_{p_\mathbb{R}}(y).\end{aligned} \tag{5.73}$$

From (5.72)–(5.73) we obtain (5.70), and hence the claim follows.

(ii)⇒(i). If $\psi : [0,\infty) \to [0,\infty)$ is inf-compact, then the map $\varphi : \mathbb{R} \mapsto [0,\infty)$, $\varphi(y) := \psi(|y|)$ $(y \in \mathbb{R})$ is also inf-compact. Now from (5.70) and (5.73) we get (5.72) and thus (5.71). Then by Proposition 5.3.10 the claim follows.

(ii)⇒(iii). By Remark 5.1.10 it is not restrictive to assume that ψ nondecreasing and diverging at infinity. Then by the Chebyshev inequality

$$\sup_{u\in\mathcal{F}} \mu(\{|u| \geq k\}) \leq \frac{1}{\psi(k)} \sup_{u\in\mathcal{F}} \int_X \psi(|u|)(x)\, d\mu(x) \quad \text{for all } k > 0,$$

whence by (5.70), as $k \to \infty$, the claim follows.

(iii)⇒(ii). By assumption there exists a nondecreasing diverging sequence $\{k_l\} \subseteq [0,\infty)$ such that

$$\sup_{u\in\mathcal{F}} \mu(\{|u| \geq k_l\}) \leq \frac{1}{l^3} \quad \text{for all } l \in \mathbb{N}.$$

Define $\psi(t) := \sum_{l=0}^{\infty} l\chi_{(k_l,k_{l+1}]}(t)$ $(k_0 := 0; t \in [0,\infty))$. Clearly, $\psi : [0,\infty) \to [0,\infty)$ is inf-compact, and for all $u \in \mathcal{F}$,

$$\int_X \psi(|u|)(x)\, d\mu(x) = \sum_{l=0}^{\infty} l\mu(\{k_l \leq |u| < k_{l+1}\}) \leq \sum_{l=1}^{\infty} \frac{1}{l^2} < \infty.$$

This completes the proof. □

Proof of Proposition 5.4.3. By assumption we have

$$\sup_{u\in\mathcal{F}} \|u\|_{L^1(X)} = \sup_{u\in\mathcal{F}} \int_X |u|\, d\mu < \infty,$$

and thus in (5.70) we can choose $\psi : [0,\infty) \to [0,\infty)$, $\psi(y) = y$ (which is obviously inf-compact). Then by Lemma 5.4.4 the result follows. □

Remark 5.4.1. Observe that condition (ii) of Lemma 5.4.4 is weaker than the de la Vallée-Poussin criterion (see Proposition 2.8.6(iii)). This is in agreement with Propositions 2.8.6 and 5.4.3; indeed, the boundedness of $\mathscr{F}$ implies condition (ii) of Lemma 5.4.4, whereas the de la Vallée-Poussin criterion is equivalent to both the boundedness and the uniform integrability of $\mathscr{F}$.

By Proposition 5.4.3 and Theorem 5.3.11, if $\mathscr{F} \subseteq L^1(X)$ is bounded, then the set $\mathscr{N}_{\mathscr{F}}$ of the associated Young measures is relatively sequentially compact in the narrow topology. Hence we have the following:

Theorem 5.4.5. *Let X be a locally compact Hausdorff space with countable basis, let $\mu \in \mathfrak{R}_f^+(X)$, and let $\mathscr{F} \subseteq L^1(X)$ be bounded. Then for any $\{u_j\} \subseteq \mathscr{F}$, there exist a subsequence $\{u_{j_k}\} \subseteq \{u_j\}$ and a Young measure $\nu \in \mathfrak{Y}^+(X;\mathbb{R})$ such that for all $h \in \mathscr{C}_b(X \times \mathbb{R})$,*

$$\lim_{k\to\infty} \int_X h(x, u_{j_k}(x))\, d\mu(x) = \int_X d\mu(x) \int_{\mathbb{R}} h(x,y)\, d\nu_x(y). \tag{5.74}$$

Proof. Fix arbitrary $\{u_j\} \subseteq \mathscr{F}$. By Proposition 5.4.3 the set $\mathscr{N}_{\mathscr{F}}$ of the associated Young measures is tight, and hence there exist a subsequence $\{\nu_{u_{j_k}}\} \subseteq \{\nu_{u_j}\}$ and a Young measure $\nu \in \mathfrak{Y}^+(X;\mathbb{R})$ such that $\nu_{u_{j_k}} \overset{n}{\rightharpoonup} \nu$, that is, for all $h \in \mathscr{C}_b(X \times \mathbb{R})$, we have

$$\lim_{k\to\infty} \int_{X\times\mathbb{R}} h(x,y)\, d\nu_{u_{j_k}}(x,y) = \int_{X\times\mathbb{R}} h(x,y)\, d\nu(x,y).$$

By (5.54) we have that

$$\int_{X\times\mathbb{R}} h(x,y)\, d\nu_{u_{j_k}}(x,y) = \int_X h(x, u_{j_k}(x))\, d\mu(x),$$

whereas by (5.36)

$$\int_{X\times\mathbb{R}} h(x,y)\, d\nu(x,y) = \int_X d\mu(x) \int_{\mathbb{R}} h(x,y)\, d\nu_x(y).$$

Hence equality (5.74) follows. □

By Proposition 5.4.3, Corollary 5.3.3, and Proposition 5.3.8 we have the following result.

Corollary 5.4.6. *Let X be a locally compact Hausdorff space with countable basis, let $\mu \in \mathfrak{R}_f^+(X)$, and let $\mathscr{F} \subseteq L^1(X)$ be bounded. Then for any $\{u_j\} \subseteq \mathscr{F}$, there exist a subsequence $\{u_{j_k}\} \subseteq \{u_j\}$ and a Young measure $\nu \in \mathfrak{Y}^+(X;\mathbb{R})$ such that for any Carathéodory* *function* $h : X \times \mathbb{R} \to [0, \infty)$,

$$\int_{X\times\mathbb{R}} h(x,y)\,d\nu(x,y) \le \liminf_{k\to\infty} \int_X h(x, u_{j_k}(x))\,d\mu(x). \tag{5.75}$$

Proof. For any $\{u_j\} \subseteq \mathscr{F}$, there exist a subsequence $\{\nu_{u_{j_k}}\} \subseteq \{\nu_{u_j}\}$ and a Young measure $\nu \in \mathfrak{Y}^+(X;\mathbb{R})$ such that $\nu_{u_{j_k}} \overset{n}{\rightharpoonup} \nu$. Then by inequality (5.62) and equality (5.54) the result follows. □

It is natural to regard the limiting measure $\nu \in \mathfrak{Y}^+(X;\mathbb{R})$ given by Theorem 5.4.5 (or by Theorem 5.4.2, since narrow convergence implies weak* convergence) as associated with the subsequence $\{u_{j_k}\}$ mentioned in the same theorem. It is sometimes possible to calculate explicitly this measure, as the following result shows.

Proposition 5.4.7. *Let $I \subseteq \mathbb{R}$ be an interval of unit length, and let $u : \mathbb{R} \to K$ with compact $K \subseteq \mathbb{R}$ be $\mathcal{B}$-measurable and 1-periodic. Set $u_k(x) := u(kx)$ ($k \in \mathbb{N}, x \in I$), and let $\{\nu_{u_k}\} \subseteq \mathfrak{Y}^+(I;K)$ be the sequence of the associated Young measures. Then ν_{u_k} converges narrowly in $\mathfrak{Y}^+(I;K)$ to the product measure $\nu = \lambda|_I \times \lambda_u$, where λ_u denotes the image of the Lebesgue measure λ under u:*

$$\int_I g \circ u\,d\lambda = \int_{\mathbb{R}} g\,d\lambda_u \quad \textit{for all } g \in L^1(\mathbb{R}, \mathcal{B}, \lambda_u).$$

Proof. For simplicity, set $I = (0,1)$. Since $K \subseteq \mathbb{R}$ is compact, $C_b(K)$ is separable, and by Remark 5.3.1 $\mathscr{C}_b(X \times K) = L^1(X; C_b(K))$. Hence by Proposition 4.3.6(ii) it suffices to prove that for all $f \in C_c(I)$ and $g \in C_b(K)$, we have

$$\lim_{k\to\infty} \int_I f(x)\,g(u_k(x))\,d\lambda(x) = \int_I f(x)\,d\lambda(x) \int_I g(u(x))\,d\lambda(x). \tag{5.76}$$

For any fixed $k \in \mathbb{N}$, let $\{I_1, \dots, I_k\}$ be a partition of I, I_m being the interval with extremes $\frac{m-1}{k}, \frac{m}{k}$ ($m = 1, \dots, k$). Then

$$\begin{aligned}\int_I f(x)\,g(u_k(x))\,d\lambda(x) &= \int_I f(x)\,g(u(kx))\,d\lambda(x)\\ &= \frac{1}{k}\sum_{m=1}^{k} \int_{I_k} f\left(\frac{\xi}{k}\right) g(u(\xi)) d\lambda(\xi) = \frac{1}{k}\sum_{m=1}^{k} \int_I f\left(\frac{\xi+m}{k}\right) g(u(\xi)) d\lambda(\xi).\end{aligned} \tag{5.77}$$

Since

$$\lim_{k\to\infty} \frac{1}{k}\sum_{m=1}^{k} f\left(\frac{\xi+m}{k}\right) = \int_I f(x)\,d\lambda(x)$$

for all $\xi \in I$, by the dominated convergence theorem, letting $k \to \infty$ in (5.77), we obtain equality (5.76). Hence the result follows. □

Example 5.4.5. (i) Let $I = (-\frac{1}{2}, \frac{1}{2})$, $K = [-1, 1]$, and $u(x) = \text{sen}(\pi x)$ $(x \in I)$. By elementary results on the Riemann integral we have

$$\int_I (g \circ u)(x)\, d\lambda(x) = \int_{-\frac{1}{2}}^{\frac{1}{2}} g(u(x))\, dx = \pi \int_{-1}^{1} \frac{g(y)}{\sqrt{1-y^2}}\, dy.$$

Hence for λ-a. e. $x \in I$ and all $F \in \mathcal{B} \cap [-1, 1]$,

$$\nu_x(F) = \pi \int_F \frac{d\lambda(y)}{\sqrt{1-y^2}} = \text{constant}.$$

(ii) Let $I = (0, 1)$, $K = \{-1, 1\}$ and $u(x) = \text{sgn}[\text{sen}(2\pi x)]$ $(x \in I)$. Then

$$\int_I (g \circ u)(x)\, d\lambda(x) = \int_0^{\frac{1}{2}} g(u(x))\, dx + \int_{\frac{1}{2}}^{1} g(u(x))\, dx,$$

and hence $\lambda_u = \frac{\delta_1 + \delta_{-1}}{2}$. Then for λ-a. e. $x \in I$, we have

$$\nu_x = \frac{1}{2}(\delta_1 + \delta_{-1}) = \text{constant}.$$

Observe carefully that the limiting measure $\nu \in \mathfrak{Y}^+(X; \mathbb{R})$ given by Theorem 5.4.5 or 5.4.2 need not be associated with any $\mathcal{B}(X)$-measurable function (for instance, this was the case in Example 5.4.5(ii)). This point is made clear by the following proposition, whose proof is deferred until next subsection.

Proposition 5.4.8. *Let X be a locally compact Hausdorff space with countable basis, and let $\mu \in \mathfrak{R}_f^+(X)$. Let $\{u_j\} \subseteq L^1(X)$ and $u \in L^1(X)$ satisfy $u_j \rightharpoonup u$ but $u_j \not\to u$ strongly in $L^1(X)$. Let $\{\nu_{u_j}\}$ be the sequence of the associated Young measures. Then there exist a subsequence $\{\nu_{u_{j_k}}\} \subseteq \{\nu_{u_j}\}$ and a Young measure $\nu \in \mathfrak{Y}^+(X; \mathbb{R})$ such that:*

(i) $\nu_{u_{j_k}} \overset{n}{\rightharpoonup} \nu$;
(ii) *ν is not associated with any $\mathcal{B}(X)$-measurable function.*

5.4.3 Uniform integrability of $\{u_j\}$

If $\mathscr{F} \subseteq L^1(X)$ is both bounded and uniformly integrable, further information is gathered combining the above results with the Dunford–Pettis theorem (Theorem 2.8.18).

In this direction, let us first show that uniform integrability allows us to drop the sign condition on the function h in Corollary 5.4.6.

Lemma 5.4.9. *Let X be a locally compact Hausdorff space with countable basis, let $\mu \in \mathfrak{R}_f^+(X)$, and let $\mathscr{F} \subseteq L^1(X)$ be bounded. Suppose that for some $h \in \mathscr{C}(X \times \mathbb{R})$, the set $\mathscr{G}_h := \{h \circ u \mid u \in \mathscr{F}\}$ is bounded in $L^1(X)$ and uniformly integrable. Then for any $\{u_j\} \subseteq \mathscr{F}$, there exist a subsequence $\{u_{j_k}\} \subseteq \{u_j\}$ and a Young measure $\nu \in \mathfrak{Y}^+(X; \mathbb{R})$, both independent of h, such that inequality* (5.75) *holds.*

Proof. Let $\{u_{j_k}\} \subseteq \{u_j\}$ be the subsequence and $\nu \in \mathfrak{Y}^+(X; \mathbb{R})$ the Young measure given by Corollary 5.4.6. Since $\mathscr{G}_h$ is bounded and uniformly integrable, its modulus $\eta(\mathscr{G}_h)$ of uniform integrability is equal to zero (see Definition 2.8.2 and Remark 2.8.4). Thus, in particular,

$$\limsup_{t\to\infty} \sup_{k\in\mathbb{N}} \int_{\{|h\circ u_{j_k}|\geq t\}} |h \circ u_{j_k}|\, d\mu = 0.$$

Let h_- denote the negative part of h, $h_- := \max\{-h, 0\}$. Since $\eta(\mathscr{G}_h) = 0$, using Proposition 2.8.6, it is easily seen that $\eta(\mathscr{G}_{h_-}) = 0$ as well. Hence for any $\epsilon > 0$, there exists $\bar{t} > 0$ such that for any $t \geq \bar{t}$,

$$\int_{\{h_-\circ u_{j_k}\geq t\}} (h_- \circ u_{j_k})d\mu = -\int_{A_{kt}} (h \circ u_{j_k})d\mu < \epsilon \quad \text{for all } k \in \mathbb{N}, \tag{5.78}$$

where $A_{kt} := \{h \circ u_{j_k} \leq -t\}$ $(k \in \mathbb{N}, t > 0)$.

For any $t > 0$, set $h_t := \max\{h, -t\}$. On the complementary set $(A_{kt})^c$, we have $h \circ u_{j_k} > -t$, and hence $h \circ u_{j_k} = h_t \circ u_{j_k}$, whereas on A_{kt}, we have that $h_t \circ u_{j_k} = -t$. Therefore, using inequality (5.78), we get

$$\begin{aligned}
\int_X (h \circ u_{j_k})d\mu &= \int_{A_{kt}} (h \circ u_{j_k})d\mu + \int_{(A_{kt})^c} (h \circ u_{j_k})d\mu \\
&= \int_{A_{kt}} (h \circ u_{j_k})d\mu + \int_{(A_{kt})^c} (h_t \circ u_{j_k})d\mu \\
&= \int_X (h_t \circ u_{j_k})d\mu + \int_{A_{kt}} (h \circ u_{j_k})d\mu - \int_{A_{kt}} (h_t \circ u_{j_k})d\mu \\
&= \int_X (h_t \circ u_{j_k})d\mu + \int_{A_{kt}} (h \circ u_{j_k})d\mu + t\mu(A_{kt}) \\
&\geq \int_X (h_t \circ u_{j_k})d\mu + \int_{A_{kt}} (h \circ u_{j_k})d\mu > \int_X (h_t \circ u_{j_k})d\mu - \epsilon.
\end{aligned} \tag{5.79}$$

Now observe that $h_t + t \geq 0$ in $X \times \mathbb{R}$ for all $t > 0$, and thus by (5.75)

$$\int_{X\times\mathbb{R}} [h_t(x,y) + t]dv(x,y) \leq \liminf_{k\to\infty} \int_X [h_t(x, u_{j_k}(x)) + t]d\mu(x).$$

Since $\mu(X) < \infty$ and $h \leq h_t$ $(t > 0)$, we get

$$\int_{X\times\mathbb{R}} h(x,y)\, dv(x,y) \leq \int_{X\times\mathbb{R}} h_t(x,y)\, dv(x,y) \leq \liminf_{k\to\infty} \int_X h_t(x, u_{j_k}(x))\, d\mu(x). \tag{5.80}$$

From (5.80)–(5.79), for any $\epsilon > 0$, we obtain

$$\int_{X\times\mathbb{R}} h(x,y)\, dv(x,y) \leq \liminf_{k\to\infty} \int_X h(x, u_{j_k}(x))\, d\mu(x) + \epsilon,$$

whence by the arbitrariness of ϵ inequality (5.75) follows. This completes the proof. □

Then we have the following result.

Proposition 5.4.10. *Let X be a locally compact Hausdorff space with countable basis, let $\mu \in \mathfrak{R}_f^+(X)$, and let $\mathscr{F} \subseteq L^1(X)$ be bounded. Suppose that for some $h \in \mathscr{C}(X \times \mathbb{R})$, the set $\mathscr{G}_h := \{h \circ u \mid u \in \mathscr{F}\}$ is bounded in $L^1(X)$ and uniformly integrable. Then for any $\{u_j\} \subseteq \mathscr{F}$, there exist a subsequence $\{u_{j_k}\} \subseteq \{u_j\}$ and a Young measure $v \in \mathfrak{Y}^+(X;\mathbb{R})$, both independent of h, such that $h \circ u_{j_k} \rightharpoonup h^*$ in $L^1(X)$ with*

$$h^*(x) = \int_{\mathbb{R}} h(x,y)\, dv_x(y) \quad \text{for } \mu\text{-a.e. } x \in X. \tag{5.81}$$

Proof. Let $\{u_{j_k}\} \subseteq \{u_j\}$ be the subsequence and $v \in \mathfrak{Y}^+(X;\mathbb{R})$ the Young measure given by Lemma 5.4.9. Since by assumptions $\mathscr{G}_h$ is bounded in $L^1(X)$ and uniformly integrable, by the Dunford–Pettis theorem it is relatively sequentially compact in the weak topology of $L^1(X)$. Then we can extract a subsequence of $\{h \circ u_{j_k}\} \subseteq \mathscr{G}_h$, denoted again $h \circ u_{j_k}$ for simplicity, which converges weakly to some $h^* \in L^1(X)$, that is, for all $f \in L^\infty(X)$,

$$\lim_{k\to\infty} \int_X f(x)\, h(x, u_{j_k}(x))\, d\mu(x) = \int_X f(x)\, h^*(x)\, d\mu(x). \tag{5.82}$$

It remains to prove equality (5.81). To this purpose, recall that by Lemma 5.4.9 we have both (5.75) and

$$-\int_{X\times\mathbb{R}} h(x,y)\, dv(x,y) \leq \liminf_{k\to\infty}\left(-\int_X h(x, u_{j_k}(x))\, d\mu(x)\right),$$

that is,

$$\int_{X\times\mathbb{R}} h(x,y)\,d\nu(x,y) \geq \limsup_{k\to\infty}\left(\int_X h(x,u_{j_k}(x))\,d\mu(x)\right).$$

It follows that

$$\limsup_{k\to\infty}\int_X h(x,u_{j_k}(x))\,d\mu(x) \leq \int_{X\times\mathbb{R}} h(x,y)\,d\nu(x,y) \leq \liminf_{k\to\infty}\int_X h(x,u_{j_k}(x))\,d\mu(x),$$

and thus

$$\lim_{k\to\infty}\int_X h(x,u_{j_k}(x))\,d\mu(x) = \int_{X\times\mathbb{R}} h(x,y)\,d\nu(x,y). \tag{5.83}$$

Now observe that for any $f \in L^\infty(X)$ and $h \in \mathscr{C}(X\times\mathbb{R})$, the function $\hat{h}(x,y) := f(x)h(x,y)$ belongs to $\mathscr{C}(X\times\mathbb{R})$. Since by assumption the set $\mathscr{G}_h$ is bounded in $L^1(X)$, the same holds for the set $\mathscr{G}_{\hat{h}} := \{\hat{h}\circ u \mid u \in \mathscr{F}\}$. Moreover, since by assumption $\mathscr{G}_h$ is uniformly integrable, using Remark 2.8.3 and Proposition 2.8.6, it is easily checked that $\mathscr{G}_{\hat{h}}$ is uniformly integrable as well. Therefore we can replace h by $\hat{h}$ in (5.83), thus obtaining, for any $f \in L^\infty(X)$,

$$\lim_{k\to\infty}\int_X f(x)\,h(x,u_{j_k}(x))\,d\mu(x) = \int_X d\mu(x)f(x)\int_{\mathbb{R}} h(x,y)\,d\nu_x(y). \tag{5.84}$$

By the arbitrariness of f, from (5.82) and (5.84) we obtain (5.81). This completes the proof. □

Remark 5.4.2. (i) If $\mathscr{F} \subseteq L^1(X)$ is bounded and uniformly integrable, the same holds for the set $\mathscr{F}_{k,f} := \{k|u| + f \mid u \in \mathscr{F}\}$ with $k > 0$ and $f \in L^1(X)$ (see Remark 2.8.3). It plainly follows that the conclusions of Proposition 5.4.10 hold if: (a) $\mathscr{F} \subseteq L^1(X)$ is bounded and uniformly integrable; (b) h is a Carathéodory function such that $y \mapsto h(x,y)$ has at most linear growth, that is, there exist $k > 0$ and $f \in L^1(X)$ such that

$$|h(x,y)| \leq k\,|y| + f(x) \quad \text{for } \mu\text{-a. e. } x \in X \text{ and all } y \in \mathbb{R}.$$

(ii) Let $\{u_j\} \subseteq L^1(X)$ be bounded and uniformly integrable. Applying (i) with $h(y) = y$ shows that for some subsequence $\{u_{j_k}\} \subseteq \{u_j\}$, we have $u_{j_k} \rightharpoonup u^*$ in $L^1(X)$, where

$$u^*(x) := \int_{\mathbb{R}} y\,d\nu_x(y) \quad \text{for } \mu\text{-a. e. } x \in X. \tag{5.85}$$

The element u^* is called the *barycenter* of the disintegration $\{\nu_x\}_{x\in X}$.

Similarly, applying (i) with $h(y) = y^{\pm} := \max\{\pm y, 0\}$ shows that for some subsequence of $\{u_{j_k}\}$ (not relabeled for simplicity), we have $u^{\pm}_{j_k} \rightharpoonup u_{\pm}$ in $L^1(X)$, where

$$u_{\pm}(x) := \int_{\mathbb{R}} y^{\pm}\, d\nu_x(y) \quad \text{for } \mu\text{-a. e. } x \in X. \tag{5.86}$$

Clearly,

$$u^* = u_+ - u_- \quad \text{in } L^1(X). \tag{5.87}$$

We finish this subsection by proving Proposition 5.4.8.

Proof of Proposition 5.4.8. Since $u_j \nrightarrow u$ strongly in $L^1(X)$, there exist $\epsilon > 0$ and a subsequence $\{u_{j_k}\} \subseteq \{u_j\}$ such that

$$\|u_{j_k} - u\|_{L^1(X)} \geq \epsilon \quad \text{for all } k \in \mathbb{N}. \tag{5.88}$$

Since $u_{j_k} \rightharpoonup u$, by the Dunford–Pettis theorem the subsequence $\{u_{j_k}\}$ is bounded and uniformly integrable. Therefore by Proposition 5.4.10 and Remark 5.4.2 there exist $\{u_{j_l}\} \equiv \{u_{j_{k_l}}\} \subseteq \{u_{j_k}\}$, a subsequence $\{\nu_{u_{j_l}}\}$ of associated Young measures, and a Young measure ν such that $\nu_{u_{j_l}} \overset{n}{\rightharpoonup} \nu$ and $u_{j_l} \rightharpoonup u^*$, $u^* \in L^1(X)$ denoting the barycenter of the disintegration $\{\nu_x\}_{x\in X}$ (see (5.85)).

By contradiction suppose that for any subsequence $\{\nu_{u_{j_k}}\}$ and for any Young measure $\tau \in \mathfrak{Y}^+(X;\mathbb{R})$ such that $\nu_{u_{j_k}} \overset{n}{\rightharpoonup} \tau$, there exists a $\mathcal{B}(X)$-measurable function with which τ is associated. In particular, this would hold for $\nu_{u_{j_l}}$ and ν above, i. e., $\nu = \nu_f$ for some f. Then by (5.53) $\nu_x = (\nu_f)_x = \delta_{f(x)}$ for μ-a. e. $x \in X$, and thus from (5.85) we get $u^* = f$. Since by assumption $u_{j_k} \rightharpoonup u$ in $L^1(X)$, by the uniqueness of the limit we get that $u = f$. Moreover, since $\nu_{u_{j_l}} \overset{n}{\rightharpoonup} \nu_f$ in $\mathfrak{Y}^+(X;Y)$, by Proposition 5.4.1 (possibly up to a subsequence, not relabeled) $u_{j_l} \to f = u$ μ-a. e. in X.

To sum up, by the absurd assumption there would exist a uniformly integrable subsequence $\{u_{j_l}\} \subseteq \{u_{j_k}\}$ such that $u_{j_l} \to u$ μ-a. e. in X. However, by the Vitali theorem this would imply $\lim_{l\to\infty} \|u_{j_l} - u\|_{L^1(X)} = 0$, which contradicts (5.88), from which the result follows. □

5.4.4 Biting lemma

If a sequence $\{u_j\} \subseteq L^1(X)$ is bounded but not uniformly integrable, it is possible to associate with it a uniformly integrable subsequence by removing sets of small measure. This is the content of the following theorem (see [78, Theorem 6.6] and [103, Theorem 23]).

Theorem 5.4.11 (Biting lemma). *Let X be a locally compact Hausdorff space with countable basis, let $\mu \in \mathfrak{R}_f^+(X)$, and let $\mathscr{F} \subseteq L^1(X)$ be bounded. Then for any $\{u_j\} \subseteq \mathscr{F}$, there exist $\{u_{j_k}\} \subseteq \{u_j\}$ and a nonincreasing sequence $\{E_m\} \subseteq \mathcal{B}(X)$ with $\mu(\bigcap_{m=1}^{\infty} E_m) = 0$ such that the sequence $\{u_{j_k}\chi_{E_k^c}\}$ is uniformly integrable.*

Proof. Let $\{u_j\} \subseteq \mathscr{F}$, and let

$$\eta \equiv \eta(\{u_j\}) := \lim_{t\to\infty} \sup_{j\in\mathbb{N}} \int_{\{|u_j|\geq t\}} |u_j|\, d\mu$$

be its modulus of uniform integrability (see Definition 2.8.2). If $\eta(\{u_j\}) = 0$, then by Remark 2.8.4 the sequence $\{u_j\}$ is uniformly integrable, and hence choosing $E_m = \emptyset$ for every $m \in \mathbb{N}$ the result follows. If $\eta(\{u_j\}) > 0$, then set

$$g_m(t) := \sup_{j\geq m} \int_{\{|u_j|\geq t\}} |u_j|\, d\mu \quad (m \in \mathbb{N},\, t \geq 0). \tag{5.89}$$

For every $m \in \mathbb{N}$, the function g_m is nonincreasing, and thus there exists $\lim_{t\to\infty} g_m(t) =: L_m \geq 0$, and $g_m \geq g_{m+1}$ in $[0, \infty)$. Moreover,

$$\lim_{t\to\infty} g_m(t) = \eta \quad \text{for all } m \in \mathbb{N}. \tag{5.90}$$

In fact, equality (5.90) with $m = 1$ holds by the very definition of η. For any $m \geq 2$, set

$$\bar{g}_m(t) := \max_{j=1,\dots,m-1} \int_{\{|u_j|\geq t\}} |u_j|\, d\mu,$$

and thus $g_1(t) = \max\{\bar{g}_m(t), g_m(t)\}$ $(m \geq 2, t \geq 0)$. By Remark 2.8.2 and Proposition 2.8.6 $\lim_{t\to\infty} \bar{g}_m(t) = 0$ for any fixed $m \geq 2$. Were also $L_m = 0$, we would have $\lim_{t\to\infty} g_1(t) = \eta = 0$, a contradiction. It follows that $L_m > 0$, and thus $g_m(t) > \bar{g}_m(t)$ for t sufficiently large, whence $g_1(t) = g_m(t)$. Therefore, for every $m \in \mathbb{N}$,

$$\lim_{t\to\infty} g_m(t) = \lim_{t\to\infty} g_1(t) = \eta.$$

By the above remarks there exists a nondecreasing diverging sequence $\{t_k\} \subseteq (0, \infty)$ such that

$$\eta \leq g_m(t_k) \leq g_1(t_k) < \eta + \frac{1}{k} \quad \text{for all } k, m \in \mathbb{N}. \tag{5.91}$$

Fix $k \in \mathbb{N}$. Since by definition $g_1(t_k) = \sup_{j\in\mathbb{N}} \int_{\{|u_j|\geq t_k\}} |u_j|\,d\mu$, there exists $j_k \in \mathbb{N}$ such that

$$\int_{\{|u_{j_k}|\geq t_k\}} |u_{j_k}|\,d\mu > g_1(t_k) - \frac{1}{k}$$

(observe that the sequence $\{j_k\}$ is nondecreasing and diverging). From this inequality and the first inequality in (5.91) we get

$$\int_{\{|u_{j_k}|\geq t_k\}} |u_{j_k}|\,d\mu > \eta - \frac{1}{k} \quad \text{for all } k \in \mathbb{N}. \tag{5.92}$$

Now set $F_k := \{x \in X \mid |u_{j_k}(x)| \geq t_k\}$. We shall prove the following:

Claim. The sequence $\{u_{j_k}\chi_{F_k^c}\}$ is uniformly integrable.

Using the claim, we can complete the proof. By the Chebyshev inequality, for all $k \in \mathbb{N}$, we have

$$t_k\,\mu(F_k) \leq \int_X |u_{j_k}(x)|\,d\mu \leq \sup_{j\in\mathbb{N}} \|u_j\|_{L^1(X,\mathbb{R})} < \infty,$$

and thus $\lim_{k\to\infty}\mu(F_k) = 0$. Hence there exists a subsequence of $\{F_k\}$ (not relabeled for simplicity) such that $\mu(F_k) < \frac{1}{2^k}$ for all $k \in \mathbb{N}$. Define $E_m := \bigcup_{k=m}^{\infty} F_k$ ($m \in \mathbb{N}$). Then for any m, we have $E_m \in \mathcal{B}(X)$, $E_m \supseteq E_{m+1}$, and $\mu(E_m) \leq \sum_{k=m}^{\infty}\frac{1}{2^k}$, thus $\mu(\bigcap_{m=1}^{\infty} E_m) = \lim_{m\to\infty}\mu(E_m) = 0$. Moreover, since $E_m^c = \bigcap_{k=m}^{\infty} F_k^c \subseteq F_m^c$, for all $k \in \mathbb{N}$, we have

$$|u_{j_k}|\chi_{F_k^c} \geq |u_{j_k}|\chi_{E_k^c} \quad \Rightarrow \quad \{|u_{j_k}|\chi_{F_{k^c}} \geq t\} \supseteq \{|u_{j_k}|\chi_{E_{k^c}} \geq t\} \quad (t \geq 0). \tag{5.93}$$

Since by the claim the sequence $\{u_{j_k}\chi_{F_k^c}\}$ is uniformly integrable, by Proposition 2.8.6 and (5.93) we have

$$0 = \limsup_{t\to\infty}\sup_{k\in\mathbb{N}} \int_{\{|u_{j_k}|\chi_{F_k^c}\geq t\}} |u_{j_k}|\chi_{F_k^c}\,d\mu \geq \limsup_{t\to\infty}\sup_{k\in\mathbb{N}} \int_{\{|u_{j_k}|\chi_{E_k^c}\geq t\}} |u_{j_k}|\chi_{E_k^c}\,d\mu.$$

Then by Proposition 2.8.6 the sequence $\{u_{j_k}\chi_{E_k^c}\}$ is uniformly integrable, and thus the conclusion follows.

It remains to prove the claim. To this purpose, observe that the function

$$t \mapsto \int_{\{|u_{j_k}|\geq t\}} |u_{j_k}|\chi_{F_k^c}\,d\mu = \int_{\{t\leq|u_{j_k}|<t_k\}} |u_{j_k}|\,d\mu \quad (k \in \mathbb{N},\ t \geq 0)$$

is nonincreasing. The same holds for the map

$$h_m(t) := \sup_{k\geq m} \int_{\{|u_{j_k}|\geq t\}} |u_{j_k}|\chi_{F_k^c}\, d\mu \quad (m \in \mathbb{N},\, t \geq 0),$$

and thus there exists $\lim_{t\to\infty} h_m(t) =: \tilde{L}_m$. Since $\{|u_{j_k}|\chi_{F_k^c} \geq t\} \subseteq \{|u_{j_k}| \geq t\}$ $(t \geq 0)$, it follows that

$$\sup_{k\in\mathbb{N}} \int_{\{|u_{j_k}|\chi_{F_k^c}\geq t\}} |u_{j_k}|\chi_{F_k^c}\, d\mu \leq h_1(t).$$

Hence the claim follows by Proposition 2.8.6 if we show that $\tilde{L}_1 = 0$.

To this purpose, for any $m \geq 2$, set

$$\bar{h}_m(t) := \max_{k=1,\dots,m-1} \int_{\{|u_{j_k}|\geq t\}} |u_{j_k}|\chi_{F_k^c}\, d\mu,$$

and thus $h_1(t) = \max\{\bar{h}_m(t), h_m(t)\}$ $(m \geq 2, t \geq 0)$. Since $\bar{h}_m \leq \bar{g}_m$ in $(0,\infty)$ (see (5.89)), we have $\lim_{t\to\infty} \bar{h}_m(t) = 0$. Plainly, this implies $\tilde{L}_1 = \tilde{L}_m$ for any $m \in \mathbb{N}$. In particular, for any $m \in \mathbb{N}$,

$$\lim_{l\to\infty} h_1(t_l) = \lim_{l\to\infty} h_m(t_l). \tag{5.94}$$

By (5.91)–(5.92), for any fixed $l \in \mathbb{N}$, we have

$$\begin{aligned} h_m(t_l) &= \sup_{k\geq m}\Bigg(\int_{\{|u_{j_k}|\geq t_l\}} |u_{j_k}|\, d\mu - \int_{\{|u_{j_k}|\geq t_k\}} |u_{j_k}|\, d\mu \Bigg) \\ &\leq \sup_{k\geq m}\Bigg(g_1(t_l) - \int_{\{|u_{j_k}|\geq t_k\}} |u_{j_k}|\, d\mu \Bigg) \\ &\leq \sup_{k\geq m}\left(\eta + \frac{1}{l} - \eta + \frac{1}{k} \right) = \frac{1}{l} + \frac{1}{m}. \end{aligned} \tag{5.95}$$

From (5.94)–(5.95) we get

$$\tilde{L}_1 = \lim_{l\to\infty} h_m(t_l) \leq \frac{1}{m} \quad \text{for all } m \in \mathbb{N},$$

and thus $\tilde{L}_1 = 0$ by the arbitrariness of $m \in \mathbb{N}$. This proves the claim, whence the result follows. □

Remark 5.4.3. Let X be a locally compact Hausdorff space, and let $\mu \in \mathfrak{R}_f^+(X)$. In view of the biting lemma, some authors (e. g., see [53]) say that a sequence $\{u_j\} \subseteq L^1(X)$

converges *in the biting sense* in $L^1(X)$ (written $u_j \overset{b}{\rightharpoonup} u$) if there exist $u \in L^1(X)$ and a nonincreasing sequence $\{E_m\} \subseteq \mathcal{B}(X)$ such that $\mu(\bigcap_{m=1}^{\infty} E_m) = 0$ and $u_{j_k}\chi_{E_m^c} \rightharpoonup u\chi_{E_m^c}$ in $L^1(X)$ for every $m \in \mathbb{N}$.

From the biting lemma we get the following result (see [92, Theorem 4.5] and [4, Section 2]).

Theorem 5.4.12. *Let X be a locally compact Hausdorff space with countable basis, let $\mu \in \mathfrak{R}_f^+(X)$, and let $\{u_j\} \subseteq L^1(X)$ be bounded. Then there exist $\{u_{j_k}\} \subseteq \{u_j\}$, $\nu \in \mathfrak{Y}^+(X;\mathbb{R})$, $\sigma \in \mathfrak{R}_f(X)$, a nonincreasing sequence $\{E_m\} \subseteq \mathcal{B}(X)$ with $\mu(\bigcap_{m=1}^{\infty} E_m) = 0$, and $F \in \mathcal{B}(X)$ (E_m, F possibly empty) with the following properties:*

(i) $u_{j_k}\chi_{E_m^c} \rightharpoonup u^*\chi_{E_m^c}$ *in $L^1(X)$ for every fixed $m \in \mathbb{N}$, with u^* given by (5.85) belonging to $L^1(X)$;*

(ii) $u_{j_k}\chi_{E_k^c} \rightharpoonup u^*$ *in $L^1(X)$, and* $u_{j_k}\chi_{E_k} \overset{*}{\rightharpoonup} \sigma$ *in $\mathfrak{R}_f(X)$;*

(iii) $u_{j_k}\chi_{F^c} \to u^*\chi_{F^c}$ *μ-a. e. in X, and* $u_{j_k}\chi_{F^c\setminus E_m} \to u^*\chi_{F^c\setminus E_m}$ *strongly in $L^1(X)$ for every fixed $m \in \mathbb{N}$;*

(iv) $\eta(\{u_{j_k}\}) = \eta$, *where $\eta(\{u_{j_k}\})$ and $\eta \equiv \eta(\{u_j\})$ are the moduli of uniform integrability of $\{u_{j_k}\}$ and $\{u_j\}$, respectively;*

(v) *for all $u \in L^1(X)$,*

$$\lim_{k\to\infty} \|u_{j_k} - u\|_{L^1(X)} = \eta + \int_X d\mu(x) \int_{\mathbb{R}} |y - u(x)|\, d\nu_x(y)$$
$$= \eta + \|(u - u^*)\chi_{F^c}\|_{L^1(X)} + \int_F d\mu(x) \int_{\mathbb{R}} |y - u(x)|\, d\nu_x(y). \tag{5.96}$$

Here $\eta \equiv \eta(\{u_j\})$ is the modulus of uniform integrability of $\{u_j\}$, and $\{\nu_x\}_{x\in X}$ is the disintegration of the Young measure $\nu \in \mathfrak{Y}^+(X;\mathbb{R})$ associated with the subsequence $\{u_{j_k}\} \subseteq \{u_j\}$ by Theorem 5.4.5.

Proof. (i)–(ii) Without loss of generality, we may assume that the sequence $\{u_j\}$ has the properties stated in Theorem 5.4.5, Corollary 5.4.6, and Proposition 5.4.10 for some Young measure $\nu \in \mathfrak{Y}^+(X;\mathbb{R})$. Hence the function u^* defined in (5.85) belongs to $L^1(X)$, since $\{u_j\}$ is bounded in $L^1(X)$ (see Corollary 5.4.6 with $h(x,y) = h(y) = y$).

Let $\{u_{j_k}\} \subseteq \{u_j\}$ and $\{E_m\} \subseteq \mathcal{B}(X)$ be as in the proof of Theorem 5.4.11. Then there exists $\tilde{u} \in L^1(X)$ such that (possibly up to a subsequence, not relabeled) $u_{j_k}\chi_{E_k^c} \rightharpoonup \tilde{u}$ in $L^1(X)$. Clearly, it is not restrictive to assume that the sequence $\{u_{j_k}\chi_{E_k^c}\}$ has the properties stated in Theorem 5.4.5, Corollary 5.4.6, and Proposition 5.4.10 for some $\tilde{\nu} \in \mathfrak{Y}^+(X;\mathbb{R})$. Then from (5.85) we get

$$\tilde{u}(x) = \int_{\mathbb{R}} y\, d\tilde{\nu}_x(y) \quad \text{for } \mu\text{-a. e. } x \in X.$$

Moreover, by (5.74) for every $h \in \mathscr{C}_b(X \times \mathbb{R}) \cap L^\infty(X \times \mathbb{R})$, we have

$$
\begin{aligned}
\int_{X\times\mathbb{R}} h(x,y)\, d\tilde{\nu}(x,y) &= \int_X d\mu(x) \int_{\mathbb{R}} h(x,y)\, d\tilde{\nu}_x(y) \\
&= \lim_{k\to\infty} \int_X h(x, u_{j_k}\chi_{E_k^c}(x))\, d\mu(x) \\
&= \lim_{k\to\infty} \left\{ \int_{E_k^c} h(x, u_{j_k}(x))\, d\mu(x) + \int_{E_k} h(x,0)\, d\mu(x) \right\} \\
&= \lim_{k\to\infty} \int_X h(x, u_{j_k}(x))\, d\mu(x) + \lim_{k\to\infty} \int_{E_k} [h(x,0) - h(x, u_{j_k}(x))]\, d\mu(x) \\
&\le \int_X d\mu(x) \int_{\mathbb{R}} h(x,y)\, d\nu_x(y) + 2\|h\|_\infty \lim_{k\to\infty} \mu(E_k) = \int_{X\times Y} h(x,y)\, d\nu(x,y),
\end{aligned}
$$

since $\lim_{k\to\infty} \mu(E_k) = 0$. By the arbitrariness of h in the above equality, we obtain that $\nu = \tilde{\nu}$, and thus $u^* = \tilde{u}$ in $L^1(X)$. Then the first convergence in (ii) follows.

Fix $m \in \mathbb{N}$. Since the sequence $\{E_m^c\}$ is nondecreasing, $E_m^c \subseteq E_k^c$ for all $k > m$, and thus $u_{j_k}\chi_{E_m^c} = u_{j_k}\chi_{E_k^c}$ μ-a.e. in E_m^c. Hence claim (i) follows from the first convergence in (ii).

To complete the proof of claim (ii), consider the sequence $\{\sigma_k\} \subseteq \mathfrak{R}_f(X)$, $\sigma_k(E) := \int_E u_{j_k}\chi_{E_k}\, d\mu$ $(E \in \mathcal{B}(X))$. Since

$$
\|\sigma_k\|_{\mathfrak{R}_f(X)} = |\sigma_k|(X) \le \mu(X) \sup_{j\in\mathbb{N}} \|u_j\|_{L^1(X)} < \infty,
$$

by the Banach–Alaoglu theorem there exists $\sigma \in \mathfrak{R}_f(X)$ such that (possibly up to a subsequence, not relabeled) $\sigma_k \overset{*}{\rightharpoonup} \sigma$. Hence the claim follows.

(iii) Set $F := \{x \in X \mid \nu_x \ne \delta_{u^*(x)}\}$. Then (possibly extracting a subsequence, not relabeled) $u_{j_k} \to u^*$ μ-a. e. in F^c by Proposition 5.4.1. Moreover, since for every fixed $m \in \mathbb{N}$, the sequence $\{u_{j_k}\chi_{E_m^c}\}$ is uniformly integrable, the assumptions of the Vitali theorem are satisfied in $F^c \cap E_m^c = F^c \setminus E_m$. Hence $u_{j_k}\chi_{F^c\setminus E_m} \to u^*\chi_{F^c\setminus E_m}$ strongly in $L^1(X)$, and thus claim (iii) follows.

(iv) Clearly, $\eta(\{u_{j_k}\}) \le \eta$ (see (2.62)). Hence the claim will follow if we prove that for any subsequence $\{u_{j_l}\} \equiv \{u_{j_{k_l}}\} \subseteq \{u_{j_k}\}$, there holds $\eta(\{u_{j_l}\}) \ge \eta$. To this purpose, observe that since $\{u_{j_l}\} \subseteq \{u_{j_k}\}$, by (5.92) we have

$$
\int_{\{|u_{j_l}|\ge t_l\}} |u_{j_l}|\, d\mu > \eta - \frac{1}{l} \quad \text{for all } l \in \mathbb{N}. \tag{5.97}
$$

On the other hand, the sequence $\{t_l\}$ is nondecreasing and diverging (see the proof of Theorem 5.4.11), and thus for any $t > 0$, there exists $t_{\bar{l}} > t$. Then from (5.97) we obtain

$$\int_{\{|u_{j_{\bar{l}}}|\geq t\}} |u_{j_{\bar{l}}}| \, d\mu > \eta - \frac{1}{\bar{l}} \quad \text{for all } l \in \mathbb{N},$$

whence

$$\sup_{l\in\mathbb{N}} \int_{\{|u_{j_l}|\geq t\}} |u_{j_l}| \, d\mu \geq \eta.$$

Letting $t \to \infty$ in this inequality, we get $\eta(\{u_{j_l}\}) \geq \eta$, and thus the claim follows.

(v) It suffices to prove the first equality in (5.96), whence the second follows by the definition of the set F. For any fixed $u \in L^1(X)$, set

$$\|u_{j_k} - u\|_{L^1(X)} - \left(\eta + \int_X d\mu(x) \int_{\mathbb{R}} |y - u(x)| \, d\nu_x(y)\right) = \sum_{k=1}^{4} I_k,$$

where

$$I_1 := -\int_{E_m} d\mu(x) \int_{\mathbb{R}} |y - u(x)| \, d\nu_x(y),$$

$$I_2 := \|u_{j_k} - u\|_{L^1(E_m;\mathbb{R})} - \|u_{j_k}\|_{L^1(E_m;\mathbb{R})},$$

$$I_3 := \|u_{j_k} - u\|_{L^1(E_m^c;\mathbb{R})} - \int_{E_m^c} d\mu(x) \int_{\mathbb{R}} |y - u(x)| \, d\nu_x(y),$$

$$I_4 := \|u_{j_k}\|_{L^1(E_m;\mathbb{R})} - \eta.$$

Fix $\epsilon > 0$. Concerning I_1, by inequality (5.75) we have that

$$\begin{aligned}\int_{X\times\mathbb{R}} |y - u(x)| \, d\nu(x,y) &\leq \liminf_{k\to\infty} \int_X |u_{j_k}(x) - u(x)| \, d\mu(x) \\ &\leq \sup_{j\in\mathbb{N}} \|u_j\|_{L^1(X)} + \|u\|_{L^1(X)} < \infty.\end{aligned}$$

Hence the function $h(x,y) = |y - u(x)|$ belongs to $L^1(X \times \mathbb{R}, \mathcal{B}(X) \times \mathcal{B}(\mathbb{R}), \nu)$. Since $I_1 = -\int_{X\times\mathbb{R}} \chi_{E_m}(x)|y - u(x)|d\nu(x,y)$ and $\chi_{E_m} \to 0$ as $m \to \infty$, for j_k sufficiently large, by the dominated convergence theorem we get $|I_1| < \frac{\epsilon}{4}$ for all $m \in \mathbb{N}$ sufficiently large. Similarly, for any $m \in \mathbb{N}$ large enough, we have $|I_2| \leq \|u\|_{L^1(E_m;\mathbb{R})} < \frac{\epsilon}{4}$.

Fix $m \in \mathbb{N}$ so large that $|I_1| + |I_2| < \frac{\epsilon}{2}$. Concerning I_3, observe that

$$\|u_{j_k} - u\|_{L^1(E_m^c;\mathbb{R})} = \int_{E_m^c} |u_{j_k}(x) - u(x)| \, d\mu(x)$$

$$= \int_{E_m^c} d\mu(x) \int_{\mathbb{R}} |y - u(x)|\, d(\nu_{j_k})_x(x).$$

As shown above, the sequence $\{u_{j_k}\chi_{E_m^c}\}$ is uniformly integrable, and hence the same holds for the sequence $\{|u_{j_k}(x) - u(x)|\chi_{E_m^c}\}$. Then applying Proposition 5.4.10 with $h(x,y) = |y - u(x)|$, we obtain that $|I_3| < \frac{\epsilon}{4}$ for all j_k sufficiently large.

Let us now address I_4. Recall from the proof of Theorem 5.4.11 that by definition $E_m := \bigcup_{k=m}^{\infty} F_k$, where $F_k := \{|u_{j_k}(x)| \geq t_k\}$, and thus $F_k \subseteq E_m$ for all $k \geq m$. Then for any fixed $m \in \mathbb{N}$, by (5.92) we have

$$\int_{E_m} |u_{j_k}|\, d\mu \geq \int_{\{|u_{j_k}| \geq t_k\}} |u_{j_k}|\, d\mu > \eta - \frac{1}{k} \quad \text{for all } k \geq m.$$

If $m \geq [\frac{4}{\epsilon}] + 1$, where $[\frac{4}{\epsilon}]$ denotes the largest integer not exceeding $\frac{4}{\epsilon}$, then from the above inequality we get

$$\|u_{j_k}\|_{L^1(E_m;\mathbb{R})} \geq \eta - \frac{\epsilon}{4} \quad \text{for all } k \geq \left[\frac{4}{\epsilon}\right] + 1. \tag{5.98}$$

On the other hand, by Lemma 2.8.7 $\eta = \lim_{\delta \to 0^+} H_\delta$, where

$$H_\delta := \sup\left\{ \int_E |u_j|\, d\mu \mid j \in \mathbb{N},\ E \in \mathcal{B} \text{ such that } \mu(E) < \delta \right\}.$$

Let $\delta > 0$ be so small that $H_\delta < \eta + \frac{\epsilon}{4}$. Since $\lim_{m\to\infty} \mu(E_m) = 0$, we can choose m so large that $\mu(E_m) < \delta$, and thus for each j_k, we obtain

$$\|u_{j_k}\|_{L^1(E_m;\mathbb{R})} \leq H_\delta < \eta + \frac{\epsilon}{4}. \tag{5.99}$$

By such a choice of m and j_k from (5.98)–(5.99) we get $|I_4| < \frac{\epsilon}{4}$.

To summarize, for any $\epsilon > 0$, there exists $\bar{k} \in \mathbb{N}$ such that for all $k > \bar{k}$,

$$\left| \|u_{j_k} - u\|_{L^1(X)} - \eta - \int_X d\mu(x) \int_{\mathbb{R}} |y - u(x)|\, d\nu_x(y) \right| < \epsilon.$$

This completes the proof. □

Remark 5.4.4. Let $\{u_j\} \subseteq L^1(X)$ be bounded. Then by Theorem 5.4.12(ii) there exists $\{u_{j_k}\} \subseteq \{u_j\}$ such that

$$u_{j_k} = u_{j_k}\chi_{E_k^c} + u_{j_k}\chi_{E_k} \overset{*}{\rightharpoonup} u \quad \text{in } \mathfrak{R}_f(X), \tag{5.100a}$$

where

$$u(E) := \int_E u^* \, d\mu + \sigma(E) \quad \text{for all } E \in \mathcal{B}(X) \tag{5.100b}$$

with $u^* \in L^1(X)$ and $\sigma \in \mathfrak{R}_f(X)$ given by the same theorem. By abuse of notation, instead of (5.100), we often write $u_{j_k} \overset{*}{\rightharpoonup} u := u^* + \sigma$, and similarly in analogous cases.

Since the measure $E \mapsto \int_E u^* \, d\mu$ $(E \in \mathcal{B}(X))$ is absolutely continuous with respect to μ, a natural question is whether the measure σ is singular with respect to μ or not. In general, the answer is negative, as the following example shows (see [4, 9]).

Example 5.4.6. Let $X = (0,1)$, $\mu = \lambda \llcorner X$, and

$$u_k = \sum_{i=1}^{k} k \chi_{(\frac{i}{k} - \frac{i}{k^2}, \frac{i}{k})}.$$

It is easily seen that

$$u_k \overset{*}{\rightharpoonup} \mu \quad \text{in } \mathfrak{R}_f(X). \tag{5.101}$$

On the other hand, since $u_k \to 0$ in measure, its associated sequence of Young measures $\{\nu_k\}$ converges narrowly to a Young measure $\nu \in \mathfrak{Y}^+(X;\mathbb{R})$ with disintegration $\nu_x = \delta_0$ for λ-a. e. $x \in (0,1)$. By (5.100) (possibly up to a subsequence, not relabeled) we have

$$u_k \overset{*}{\rightharpoonup} u^* + \sigma \quad \text{in } \mathfrak{R}_f(X) \tag{5.102}$$

with $u^* = 0$. By (5.101)–(5.102) $\sigma = \mu$, and thus σ is not singular with respect to $\mu = \lambda \llcorner X$.

Remark 5.4.5. Let $\{u_j\} \subseteq L^1(X)$ be bounded, and thus the same holds for the sequences $\{u_j^\pm\}$. Hence by Theorem 5.4.12(ii) there exist $\{u_{j_k}\} \subseteq \{u_j\}$, $\nu \in \mathfrak{Y}^+(X;\mathbb{R})$, and $\sigma_\pm \in \mathfrak{R}_f^+(X)$ such that

$$u_{j_k}^\pm \overset{*}{\rightharpoonup} u_\pm + \sigma_\pm \quad \text{in } \mathfrak{R}_f(X) \tag{5.103}$$

with $u_\pm \in L^1(X)$, $u_\pm \geq 0$, given by (5.86) (without loss of generality, we may assume that $\{u_j\}$ and ν have the properties stated in Theorem 5.4.5, Corollary 5.4.6, and Proposition 5.4.10). By (5.100)–(5.103) we get $u^* = u_+ - u_-$ (in agreement with (5.87)) and $\sigma = \sigma_+ - \sigma_-$; however, the latter equality is not the Jordan decomposition of σ, since $\sigma_\pm$ need not be mutually singular.

Remark 5.4.6. Let $\{u_j\} \subseteq L^1(X)$ be bounded.

(i) If $\{u_j\}$ is uniformly integrable, hence $\eta = 0$, then from each of its subsequences we can extract a subsequence $\{u_{j_k}\}$ for which equality (5.96) reads as

$$\lim_{k\to\infty} \|u_{j_k} - u\|_{L^1(X)} = \|(u - u^*)\chi_{F^c}\|_{L^1(X)} + \int_F d\mu(x) \int_{\mathbb{R}} |y - u(x)|\, d\nu_x(y). \tag{5.104}$$

If, moreover, $u_j \to u$ μ-a. e. in X, by Proposition 5.4.1 $u = u^*$ and $F = \emptyset$. Then we obtain that $u_{j_k} \to u$ strongly in $L^1(X)$, whence plainly $u_j \to u$ strongly in $L^1(X)$. The argument can be inverted, in agreement with the Vitali theorem (see Theorem 2.8.10).

(ii) In particular, equality (5.104) holds if $u_j \rightharpoonup u$ in $L^1(X)$. If $u_j \nrightarrow u$ strongly in $L^1(X)$, then the argument used in (i) shows that some of its subsequences does not converge μ-a. e. in X. Then using Proposition 5.4.1 gives Proposition 5.4.8 again.

(iii) Clearly, concentration phenomena depend on whether $\eta > 0$ or not. For instance, if $u_j \to u$ μ-a. e. in X, then $u = u^*$ and $F = \emptyset$, and thus from (5.96) we get

$$\lim_{k\to\infty} \|u_{j_k} - u^*\|_{L^1(X)} = \eta. \tag{5.105}$$

In agreement with the Vitali theorem, this equality shows that such a sequence strongly converges to u^* if and only if it is uniformly integrable. Observe that (5.105) is satisfied by the sequence $u_k = k\chi_{(0,\frac{1}{k})}$ considered in Example 2.8.1(ii) with $u^* = 0$ and $\|u_k\|_{L^1(0,1)} = \eta = 1$ $(k \in \mathbb{N})$.

Proposition 5.4.13. *Let X be a locally compact Hausdorff space with countable basis, let $\mu \in \mathfrak{R}_f^+(X)$, and let $\{u_j\} \subseteq L^1(X)$ be bounded. Let $\{u_{j_k}\} \subseteq \{u_j\}$, $\nu \in \mathfrak{Y}^+(X;\mathbb{R})$, and $\sigma_\pm \in \mathfrak{R}_f^+(X)$ be given by Remark 5.4.5. Let $f \in C(\mathbb{R})$, and let there exist*

$$\lim_{y\to\pm\infty} \frac{f(y)}{y} =: M_f^\pm \in \mathbb{R}. \tag{5.106}$$

Then (possibly extracting a subsequence, not relabeled for simplicity) we have

$$f \circ u_{j_k} \overset{*}{\rightharpoonup} f^* + M_f^+ \sigma_+ - M_f^- \sigma_- \quad \text{in } \mathfrak{R}_f(X). \tag{5.107}$$

Here

$$f^* \in L^1(X), \quad f^*(x) := \int_{\mathbb{R}} f(y)\, d\nu_x(y) \quad \text{for a. e. } x \in X, \tag{5.108}$$

$\{\nu_x\}_{x\in X}$ being the disintegration of the Young measure $\nu \in \mathfrak{Y}^+(X;\mathbb{R})$ associated with the subsequence $\{u_{j_k}\}$ by Theorem 5.4.5.

Proof. By (5.106) there exists $L > 0$ such that

$$|f(y)| \le L\,(1 + |y|) \quad \text{for all } y \in \mathbb{R}, \tag{5.109}$$

and for every $\epsilon > 0$, there exists $y_\epsilon > 0$ such that

$$(M_f^\pm \mp \epsilon) y \le f(y) \le (M_f^\pm \pm \epsilon) y \quad \text{if } \pm y \ge y_\epsilon. \tag{5.110}$$

Let $l > y_\epsilon$ be fixed, and let $g_{0,l} \in C(\mathbb{R})$, $g_{\pm,l} \in C(\mathbb{R})$ satisfy:

a) $0 \le g_{0,l}(y) \le 1$, $0 \le g_{\pm,l}(y) \le 1$, $g_{-,l}(y) + g_{0,l}(y) + g_{+,l}(y) = 1$ for all $y \in \mathbb{R}$;
b) $\operatorname{supp} g_{-,l} \subseteq (-\infty, -l]$, $\operatorname{supp} g_{0,l} \subseteq [-l-1, l+1]$, $\operatorname{supp} g_{+,l} \subseteq [l, \infty)$.

Then plainly

$$\begin{aligned}(M_f^+ - \epsilon) y g_{+,l}(y) + f(y) g_{0,l}(y) + (M_f^- + \epsilon) y g_{-,l}(y) \le f(y) \\ \le (M_f^+ + \epsilon) y g_{+,l}(y) + f(y) g_{0,l}(y) + (M_f^- - \epsilon) y g_{-,l}(y) \quad \text{for all } y \in \mathbb{R}.\end{aligned} \tag{5.111}$$

For μ-a. e. $x \in X$, define

$$G_{0,l}^*(x) := \int_{\mathbb{R}} f(y) g_{0,l}(y)\, d\nu_x(y), \quad G_{\pm,l}^*(x) := \int_{\mathbb{R}} y g_{\pm,l}(y)\, d\nu_x(y), \tag{5.112}$$

and thus $G_{0,l}^* \in L^1(X)$, $G_{\pm,l}^* \in L^1(X)$. Since $\lim_{l\to+\infty} g_{\pm,l}(y) = 0$ for every $y \in \mathbb{R}$ and $|y g_{\pm,l}(y)| \le |y| \in L^1(\mathbb{R}, \mathcal{B}, \nu_x)$ for μ-a. e. $x \in X$ (this follows from (5.86) since $u_\pm \in L^1(X)$), by the dominated convergence theorem we get that $\lim_{l\to+\infty} G_{\pm,l}^*(x) = 0$ for μ-a. e. $x \in X$. Moreover, $|G_{\pm,l}^*(x)| \le \int_{\mathbb{R}} y^\pm d\nu_x(y)$ for μ-a. e. $x \in X$. Since the map $x \mapsto \int_{\mathbb{R}} y^\pm d\nu_x(y)$ belongs to $L^1(X)$, by the dominated convergence theorem we obtain

$$\lim_{l\to+\infty} G_{\pm,l}^* = 0 \quad \text{in } L^1(X). \tag{5.113}$$

Similarly, since $\lim_{l\to+\infty} g_{0,l}(y) = 1$ for every $y \in \mathbb{R}$ and $|f(y) g_{0,l}(y)| \le |f(y)|$ for μ-a. e. $x \in X$ with $f \in L^1(\mathbb{R}, \mathcal{B}, \nu_x)$ (this follows from (5.109) and (5.86) since $u_\pm \in L^1(X)$), we have that $\lim_{l\to+\infty} G_{0,l}^*(x) = f^*(x)$ for μ-a. e. $x \in X$, with f^* as in (5.108). On the other hand, by inequality (5.109) $|G_{0,l}^*(x)| \le L \int_{\mathbb{R}} (1 + |y|)\, d\nu_x(y)$ for μ-a. e. $x \in X$. Since the map $x \mapsto \int_{\mathbb{R}} (1 + |y|) d\nu_x(y)$ belongs to $L^1(X)$, by the dominated convergence theorem it follows that

$$\lim_{l\to+\infty} G_{0,l}^* = f^* \quad \text{in } L^1(X). \tag{5.114}$$

We will prove that as $k \to \infty$ (possibly extracting a subsequence, not relabeled), we have

$$u_{j_k} g_{\pm,l}(u_{j_k}) \overset{*}{\rightharpoonup} G_{\pm,l}^* \pm \sigma_\pm, \quad f(u_{j_k}) g_{0,l}(u_{j_k}) \overset{n}{\rightharpoonup} G_{0,l}^* \quad \text{in } \mathfrak{R}_f(X), \tag{5.115}$$

$\{u_{j_k}\} \subseteq \{u_j\}$ and $\sigma_\pm \in \mathfrak{R}_f^+(X)$ being given by Remark 5.4.5.

Using the convergence statements in (5.115), we can conclude the proof. Indeed, by inequality (5.111) μ-a. e. in X we have

$$\begin{aligned}(M_f^- + \epsilon)\, u_{j_k}\, g_{-,l}(u_{j_k}) + f(u_{j_k}) g_{0,l}(u_{j_k}) + (M_f^+ - \epsilon)\, u_{j_k}\, g_{+,l}(u_{j_k}) \le f(u_{j_k})\\ \le (M_f^- - \epsilon)\, u_{j_k}\, g_{-,l}(u_{j_k}) + f(u_{j_k}) g_{0,l}(u_{j_k}) + (M_f^+ + \epsilon)\, u_{j_k}\, g_{+,l}(u_{j_k}).\end{aligned}$$

By (5.115), letting $k \to \infty$ in the above inequality gives for every $\zeta \in C_c(X)$, $\zeta \ge 0$:

$$\begin{aligned}&(M_f^- + \epsilon)\Big(\iint_X G_{-,l}^* \zeta \, d\mu - \langle \sigma_{-,l}, \zeta\rangle\Big) + \iint_X G_{0,l}^* \zeta \, d\mu\\ &\quad + (M_f^+ - \epsilon)\Big(\iint_X G_{+,l}^* \zeta \, d\mu + \langle \sigma_{+}, \zeta\rangle\Big) \le \liminf_{k\to\infty} \iint_X f(u_{j_k}) \zeta \, d\mu\\ &\le \limsup_{k\to\infty} \iint_X f(u_{j_k}) \zeta \, d\mu \le (M_f^- - \epsilon)\Big(\iint_X G_{-,l}^* \zeta \, d\mu - \langle \sigma_{-,l}, \zeta\rangle\Big)\\ &\quad + \iint_X G_{0,l}^* \zeta \, d\mu + (M_f^+ + \epsilon)\Big(\iint_X G_{+,l}^* \zeta \, d\mu + \langle \sigma_{-}, \zeta\rangle\Big).\end{aligned}$$

By (5.113)–(5.114), letting $l \to \infty$ in this inequality, we obtain

$$\begin{aligned}&(M_f^+ - \epsilon)\langle \sigma_{+,l}, \zeta\rangle + \iint_X f^* \zeta \, d\mu - (M_f^- + \epsilon)\langle \sigma_{-}, \zeta\rangle\\ &\le \liminf_{k\to\infty} \iint_X f(u_{j_k}) \zeta \, d\mu \le \limsup_{k\to\infty} \iint_X f(u_{j_k}) \zeta \, d\mu\\ &\le (M_f^+ + \epsilon)\langle \sigma_{+}, \zeta\rangle + \iint_X f^* \zeta \, d\mu - (M_f^- - \epsilon)\langle \sigma_{-}, \zeta\rangle \end{aligned} \tag{5.116}$$

for all nonnegative $\zeta \in C_c(X)$ and $\epsilon > 0$. By the arbitrariness of ϵ and ζ, from (5.116) we obtain (5.107), and hence the result follows.

It remains to prove (5.115). We only prove the first statement with "+", since the other case is similar. To this purpose, observe that

$$y g_+(y) - y^+ \le y - y^+ \le 0 \quad \text{for all } y \in \mathbb{R}, \quad y g_+(y) - y^+ = 0 \quad \text{if } y \in (-\infty, 0] \cup [l+1, \infty),$$

whence

$$|y g_+(y) - y^+| = y^+ - y g_+(y) \le |y| \quad \text{for all } y \in \mathbb{R}.$$

Then by Proposition 5.4.10 and Remark 5.4.2, as $k \to \infty$, we have

$$u_{j_k} g_+(u_{j_k}) - u_{j_k}^+ \rightharpoonup \int_{\mathbb{R}} [y g_+(y) - y^+] d\nu_x(y) \quad \text{in } L^1(X)$$

(possibly extracting a subsequence, not relabeled), and thus, in particular,

$$u_{j_k} g_+(u_{j_k}) - u_{j_k}^+ \overset{n}{\rightharpoonup} \int_{\mathbb{R}} [y g_+(y) - y^+] d\nu_x(y) \quad \text{in } \mathfrak{R}_f(X). \tag{5.117}$$

On the other hand, by Remark 5.4.4, as $k \to \infty$,

$$(u_{j_k})^+ \overset{*}{\rightharpoonup} \int_{\mathbb{R}} y^+ d\nu_x(y) + \sigma_+ \quad \text{in } \mathfrak{R}_f(X). \tag{5.118}$$

From (5.117)–(5.118) we obtain that $u_{j_k} g_{+,l}(u_{j_k}) \overset{*}{\rightharpoonup} G^*_{+,l} + \sigma_+$.

Concerning the second statement in (5.115), observe that

$$\sup_{x \in X} |f(u_{j_k}(x))| \, g_{0,l}(u_{j_k}(x)) \le \max_{y \in [-l-1, l+1]} |f(y)| < \infty.$$

Then the sequence $\{f(u_{j_k}) g_{0,l}(u_{j_k})\}$ is bounded in $L^\infty(X)$ and thus is bounded in $L^1(X)$ and uniformly integrable (see Lemma 2.8.12). Then by Proposition 5.4.10 (possibly extracting a subsequence, not relabeled) we have

$$\begin{aligned} \lim_{k\to\infty} \int_X f(u_{j_k}(x))\, g_{0,l}(u_{j_k}(x))\, \zeta(x)\, d\mu(x) &= \int_X d\mu(x)\, \zeta(x) \int_{\mathbb{R}} f(y)\, g_{0,l}(y)\, d\nu_x(y) \\ &= \int_X G^*_{0,l}(x)\, \zeta(x)\, d\mu(x) \end{aligned}$$

for any $\zeta \in C_b(X)$. This completes the proof. □

Let us finally prove the following consequence of Theorem 5.4.12, which improves Proposition 2.8.8(iv).

Proposition 5.4.14. *Let $\{u_j\} \subseteq L^1(X)$ and $u \in L^1(X)$. Let $u_j \to u$ μ-a. e. in X, and let*

$$\limsup_{j\to\infty} \|u_j\|_{L^1(X)} \le \|u\|_{L^1(X)}. \tag{5.119}$$

Then $u_j \to u$ strongly in $L^1(X)$.

Proof. By contradiction let there exist $\epsilon > 0$ and a subsequence $\{u_{j_k}\} \subseteq \{u_j\}$ such that

$$\|u_{j_k} - u\|_{L^1(X)} \ge \epsilon \quad \text{for all } k \in \mathbb{N}. \tag{5.120}$$

By (5.119) there exists a subsequence of $\{u_{j_k}\}$ (not relabeled for simplicity) bounded in $L^1(X)$. Moreover, since $u_{j_k} \to u$ μ-a. e. in X, by Proposition 5.4.1 $\nu_{u_{j_k}} \overset{n}{\rightharpoonup} \nu_u$ with $(\nu_u)_x = \delta_{u(x)}$ for μ-a. e. $x \in X$. Then by Theorem 5.4.12 there exists a further subsequence, again not relabeled, such that

$$\limsup_{k\to\infty} \|u_{j_k}\|_{L^1(X)} = \eta + \|u\|_{L^1(X)} \tag{5.121}$$

(see (5.96)). From (5.119) and (5.121) we get $\eta = 0$, and hence $\{u_{j_k}\}$ is uniformly integrable. Then by the Vitali theorem $u_{j_k} \to u$ strongly in $L^1(X)$, a contradiction with (5.120), whence the result follows. □

Part II: **Applications**

Outline of Part II

In Chapter 6 we study the initial-boundary value problem

$$\begin{cases} \partial_t u = \Delta\phi(u) & \text{in } \Omega \times (0,T) \\ u = 0 & \text{on } \partial\Omega \times (0,T) \\ u = u_0 & \text{in } \Omega \times \{0\}, \end{cases}$$

where u_0 is a finite Radon measure on $\Omega \subseteq \mathbb{R}^N$ $(N \geq 3)$, Ω open and bounded, and the function ϕ is nondecreasing, with at most polynomial growth at infinity. We discuss well-posedness of the problem and regularity results of Radon measure-valued solutions, depending on the behaviour of ϕ at infinity and on properties of u_0 related to suitable Sobolev capacities.

Chapter 7 deals with the Cauchy problem for the scalar conservation law

$$\begin{cases} \partial_t u + \partial_x [\phi(u)] = 0 & \text{in } \mathbb{R} \times (0,T) \\ u = u_0 & \text{in } \mathbb{R} \times \{0\}, \end{cases}$$

where u_0 is a positive finite Radon measure on $\mathbb{R}$, whose singular part typically is a finite superposition of Dirac masses, and ϕ is sublinear at infinity. We discuss well-posedness of the problem and qualitative properties of entropy Radon measure-valued solutions, depending on u_0 and on the behaviour of ϕ at infinity.

Finally, Chapter 8 is devoted to study the Sobolev regularization of the ill-posed initial-boundary value problem

$$\begin{cases} \partial_t u = \nabla \cdot [\phi(\nabla u)] & \text{in } \Omega \times (0,T) \\ u = 0 & \text{in } \partial\Omega \times (0,T) \\ u = u_0 & \text{in } \Omega \times \{0\}. \end{cases}$$

Here $\Omega \subseteq \mathbb{R}^N$ is open and bounded, u_0 belongs to some Sobolev space and ϕ satisfies suitable growth conditions at infinity. Studying the vanishing viscosity limit of the regularization, the existence of a Young measure-valued solution is proven. Moreover, the limiting Young measure is characterized in terms of the properties of ϕ, and the asymptotic behaviour of solutions for large times is investigated.

https://doi.org/10.1515/9783110556902-007

6 Case study 1: quasilinear parabolic equations

6.1 Statement of the problem and preliminary results

In this chapter, we study the problem

$$\begin{cases} \partial_t u = \Delta\phi(u) & \text{in } \Omega \times (0,T) =: Q, \\ u = 0 & \text{on } \partial\Omega \times (0,T) =: \Gamma, \\ u = u_0 & \text{in } \Omega \times \{0\}, \end{cases} \tag{P}$$

where $\Omega \subseteq \mathbb{R}^N$ ($N \geq 3$) is a bounded open set with smooth boundary $\partial\Omega$, $T > 0$, u_0 is a finite Radon measure, and

$$u_0 \in \mathfrak{R}_f(\Omega). \tag{A_0}$$

As for the diffusion function $\phi : \mathbb{R} \to \mathbb{R}$, we will use the following assumptions:

$$\phi \in C_b(\mathbb{R}) \cap C^\infty(\mathbb{R}) \text{ is nondecreasing and nonconstant in } \mathbb{R}, \text{ with } \phi(0) = 0; \tag{A_1}$$

$$\phi \in W^{1,\infty}(\mathbb{R}). \tag{A_2}$$

More general hypotheses will be used in Section 6.7.

By assumption (A_1) there exist

$$\lim_{s \to \pm\infty} \phi(s) =: \phi_{\pm\infty} \in \mathbb{R}. \tag{6.1}$$

If ϕ' has a limit at $\pm\infty$, then this limit is zero, and thus problem (P) is *degenerate parabolic at* $\pm\infty$.

Henceforth for any $u \in \mathfrak{R}_f(Q)$, we denote by $u_r \in L^1(Q)$ and $u_s \in \mathfrak{R}_f(Q)$ the densities of the absolutely continuous part u_{ac} and, respectively, the singular part of u with respect to the Lebesgue measure (see Definition 1.8.7). Similar notations are used for the space $\mathfrak{R}_f(\Omega)$, e. g., for u_0. We will denote by $u(\cdot,t)$ ($t \in (0,T)$) the disintegration $\{u_t\}_{t\in(0,T)}$ of any measure $u \in \mathfrak{R}_f(Q)$ (see Definition 5.2.2). "Almost everywhere" is always meant with respect to the Lebesgue measure. We set for shortness $dt \equiv d\lambda(t)$, $dx \equiv d\lambda_N(x)$, and $dx\,dt \equiv d\lambda_{N+1}(x,t)$. As usual, by $\langle\cdot,\cdot\rangle$ we denote the duality map between $\mathfrak{R}_f(\Omega)$ and $C_0(\Omega)$. By abuse of notation, we also set

$$\langle\mu,\rho\rangle \equiv \int_\Omega \rho\,d\mu \quad \text{for } \rho \in C_b(\overline{\Omega}).$$

6.1.1 Weak solutions

Let us introduce our main concept of solution.

https://doi.org/10.1515/9783110556902-008

Definition 6.1.1. Let (A_0)–(A_1) hold. By a *weak solution* of problem (P) we mean any $u \in L^{\infty}_{w*}(0,T;\mathfrak{R}_f(\Omega))$ such that:

(i) $\phi(u_r) \in L^2(0,T;H^1_0(\Omega))$;

(ii) for every $\zeta \in C^1([0,T];C^1_c(\Omega))$ such that $\zeta(\cdot,T) = 0$ in Ω, we have

$$\int_0^T \langle u(\cdot,t), \partial_t\zeta(\cdot,t)\rangle\, dt = \iint_Q \nabla\phi(u_r)\cdot\nabla\zeta\, dxdt - \langle u_0, \zeta(\cdot,0)\rangle. \tag{6.2}$$

Set $C^{1,2}(\overline{Q}) := \{\zeta \in C(\overline{Q}) \mid \exists\, \partial_t\zeta, \partial_{x_i}\zeta, \partial^2_{x_ix_j}\zeta \in C(\overline{Q})\ \ \forall\, i,j = 1,\dots,N\}$. A weaker notion of solution is the following:

Definition 6.1.2. Let (A_0)–(A_1) hold. A *very weak solution* of (P) is any $u \in L^{\infty}_{w*}(0,T;\mathfrak{R}_f(\Omega))$ such that

$$\int_0^T \langle u(\cdot,t), \partial_t\zeta(\cdot,t)\rangle dt = -\iint_Q \phi(u_r)\,\Delta\zeta\, dxdt - \langle u_0, \zeta(\cdot,0)\rangle \tag{6.3}$$

with $\zeta \in C^{1,2}(\overline{Q})$ vanishing on $\overline{\Gamma} \cup (\Omega \times \{T\})$.

Remark 6.1.1. By standard approximation arguments we can choose in (6.2) test functions $\zeta \in C^1(\overline{Q})$ that vanish on $\overline{\Gamma} \cup (\Omega \times \{T\})$. Therefore every weak solution of (P) is a very weak solution.

Claim (ii) of the following proposition explains how the initial condition in (P) is satisfied.

Proposition 6.1.1. *Let (A_0)–(A_1) hold, and let u be a very weak solution of problem (P). Then:*

(i) *there exists a null set $F^* \subseteq (0,T)$ such that for every $t \in (0,T)\setminus F^*$ and all $\rho \in C^2_c(\Omega)$,*

$$\langle u(\cdot,t),\rho\rangle - \langle u_0,\rho\rangle = \int_0^t ds \int_\Omega \phi(u_r(x,t))\,\Delta\rho(x)dx, \tag{6.4}$$

(ii) *for every $\rho \in C_0(\overline{\Omega})$, we have*

$$\operatorname*{ess\,lim}_{t\to 0^+} \langle u(\cdot,t),\rho\rangle = \langle u_0,\rho\rangle, \tag{6.5a}$$

$$\operatorname*{ess\,lim}_{t\to t_0} \langle u(\cdot,t),\rho\rangle = \langle u(\cdot,t_0),\rho\rangle \quad \textit{for a. e. } t_0 \in (0,T). \tag{6.5b}$$

Proof. (i) We will show that there exists a null set $F^* \subseteq (0,T)$ such that for all $\rho \in C_b(\overline{\Omega})$ and $t \in (0,T) \setminus F^*$,

$$\lim_{q\to\infty}\left(2q\int_{t-\frac{1}{q}}^{t+\frac{1}{q}} |\langle u(\cdot,s),\rho\rangle - \langle u(\cdot,t),\rho\rangle|\,ds\right) = 0. \tag{6.6}$$

Using (6.3) and (6.6), we can prove equality (6.4). Let $\rho \in C_c^2(\Omega)$ and $t_1 \in (0,T)\setminus F^*$. By standard regularization arguments we can choose $\zeta(x,t) = \rho(x)k_q(t)$ in (6.3) with $q \in \mathbb{N}$, $q \geq \frac{1}{T-t_1} + 1$, and

$$k_q(t) := \chi_{(0,t_1]}(t) + [q(t_1 - t) + 1]\chi_{(t_1,t_1+\frac{1}{q}]}(t) \quad (t \in [0,T]). \tag{6.7}$$

Then from (6.3) we get

$$q\int_{t_1}^{t_1+\frac{1}{q}} \langle u(\cdot,t),\rho\rangle dt - \langle u_0,\rho\rangle = \iint_Q \phi(u_r)\Delta\rho(x)\,k_q(t)dxdt.$$

Since $\lim_{q\to\infty} k_q(t) = \chi_{(0,t_1]}(t)$ for all $t \in (0,T)$, letting $q \to \infty$ in this equality and using (6.6), we obtain

$$\langle u(\cdot,t_1),\rho\rangle - \langle u_0,\rho\rangle = \int_0^{t_1} dt \int_\Omega \phi(u_r(x,t))\,\Delta\rho(x)dx,$$

that is, (6.4) by changing notations.

To prove equality (6.6), observe that since $u \in L^\infty_{w^*}(0,T;\mathfrak{R}_f(\Omega))$, by Theorem 5.2.1 there exists a null set $F_0 \subseteq (0,T)$ such that the disintegration $u(\cdot,t) \in \mathfrak{R}_f(\Omega)$ is defined for every $t \in (0,T) \setminus F_0$; moreover, for any $\rho \in C_b(\overline{\Omega})$, the map $t \mapsto \psi_\rho(t) := \langle u(\cdot,t),\rho\rangle$ belongs to the space $L^\infty(0,T)$. Let $D \equiv \{\rho_k\}$ be a countable dense subset of $C_b(\overline{\Omega})$. Since $\psi_{\rho_k} \in L^\infty(0,T)$, there exists a null set $F_k \subseteq (0,T)$ such that (6.6) holds for every $t \in (0,T) \setminus F_k$ (see Corollary 2.9.5(ii)). It is not restrictive to assume that $F_0 \subseteq F_k$ for every $k \in \mathbb{N}$. Set $F^* := \bigcup_{k=1}^\infty F_k$. Then for all $t \in (0,T) \setminus F^*$ and $k \in \mathbb{N}$, we have

$$\lim_{q\to\infty}\left(2q\int_{t-\frac{1}{q}}^{t+\frac{1}{q}} |\langle u(\cdot,s),\rho_k\rangle - \langle u(\cdot,t),\rho_k\rangle|\,ds\right) = 0.$$

Then by the denseness of D in $C_b(\overline{\Omega})$ equality (6.6) plainly follows. This proves the claim.

(ii) Let $\{t_n\} \subseteq (0,T) \setminus F^*$ and $t_n \to 0^+$ as $n \to \infty$. Since $\phi(u_r) \in L^\infty(Q)$, writing (6.4) with $t = t_n$ and letting $n \to \infty$ give

$$\langle u(\cdot,t_n),\rho\rangle \to \langle u_0,\rho\rangle \quad \text{for all } \rho \in C_c^2(\Omega).$$

Moreover, since $u \in L^\infty_{w*}(0,T;\mathfrak{R}_f(\Omega))$, we have $\sup_{n\in\mathbb{N}} \|u(\cdot,t_n)\|_{\mathfrak{R}_f(\Omega)} < \infty$. Hence by the Banach–Alaoglu theorem there exist $\mu_0 \in \mathfrak{R}_f(\Omega)$ and a subsequence $\{t_{n_k}\} \subseteq \{t_n\}$ such that $u(\cdot,t_{n_k}) \overset{*}{\rightharpoonup} \mu_0$ in $\mathfrak{R}_f(\Omega)$. By standard density arguments this implies that $\mu_0 = u_0$ in $\mathfrak{R}_f(\Omega)$. Hence $u(\cdot,t_n) \overset{*}{\rightharpoonup} u_0$ in $\mathfrak{R}_f(\Omega)$ along the whole sequence $\{t_n\}$, and (6.5a) follows from the arbitrariness of $\{t_n\}$. Applying the same argument to the equality

$$\langle u(\cdot,t),\rho\rangle - \langle u(\cdot,t_0),\rho\rangle = \int_{t_0}^{t} ds \int_\Omega \phi(u_r(x,t))\,\Delta\rho(x)dx,$$

which is an elementary consequence of (6.4) for a. e. $t_0, t \in (0,T)$, we prove (6.5b). Hence the conclusion follows. □

Remark 6.1.2. Relying on Proposition 6.1.1, it is easily seen that the map $t \mapsto u(\cdot,t)$ admits a representative defined for all $t \in [0,T]$, such that $u(\cdot,0) = u_0$ and for every $\rho \in C_0(\overline{\Omega})$,

$$\lim_{t\to 0^+} \langle u(\cdot,t),\rho\rangle = \langle u_0,\rho\rangle, \tag{6.8a}$$

$$\lim_{t\to t_0} \langle u(\cdot,t),\rho\rangle = \langle u(\cdot,t_0),\rho\rangle \quad \text{for all } t_0 \in [0,T]. \tag{6.8b}$$

Hereafter we refer to this continuous representative whenever properties of the map $t \mapsto u(\cdot,t)$ are stated for *every* t in some subinterval of $[0,T]$. In doing so, equality (6.4) holds for every $t \in (0,T)$, and equalities (6.5) are replaced by (6.8).

6.1.2 Weak entropy solutions

Let us introduce for future purposes another type of solutions to problem (P). Let (A_1) be satisfied. For every nondecreasing $f \in C(\mathbb{R})$, set

$$F_\phi(y) := \int_0^y f(\phi(z))dz \quad (y \in \mathbb{R}), \tag{6.9}$$

$$M^\pm_{F_\phi} := \lim_{y\to\pm\infty} \frac{F_\phi(y)}{y} = \lim_{y\to\pm\infty} f(\phi(y)) \in \mathbb{R}. \tag{6.10}$$

Definition 6.1.3. Let (A_0)–(A_1) hold. An *entropy solution* of problem (P) is any weak solution u of (P) that satisfies the *entropy inequalities*:

$$\iint_Q \{F_\phi(u_r)\partial_t\zeta - f(\phi(u_r))\nabla\phi(u_r)\cdot\nabla\zeta\}\,dxdt$$

$$
\begin{aligned}
&+ M^+_{F_\phi} \int_0^T \langle u^+_s(\cdot,t), \partial_t\zeta(\cdot,t)\rangle dt - M^-_{F_\phi} \int_0^T \langle u^-_s(\cdot,t), \partial_t\zeta(\cdot,t)\rangle dt \\
&\geq - \int_\Omega F_\phi(u_{0r})\zeta(x,0)\,dx - M^+_{F_\phi}\langle u^+_{0s}, \zeta(\cdot,0)\rangle + M^-_{F_\phi}\langle u^-_{0s}, \zeta(\cdot,0)\rangle
\end{aligned} \tag{6.11}
$$

for any nondecreasing $f \in C(\mathbb{R})$ and $\zeta \in C^1([0,T]; C^1_c(\Omega))$ such that $\zeta \geq 0$ and $\zeta(\cdot,T) = 0$ in Ω.

It is easily seen that all terms in (6.11) are well defined, and thus Definition 6.1.3 is well posed.

6.2 Persistence

A remarkable property of very weak solutions of (P) is that their 2-concentrated part (see Definition 1.8.10 and Subsection 3.3.3) is persistent in time:

Theorem 6.2.1. *Let (A_0)–(A_1) hold, and let u be a very weak solution of problem (P). Then for every $t \in (0,T)$, we have*

$$
[u(\cdot,t)]_{c,2} = [u_0]_{c,2} \quad \text{in } \mathfrak{R}_f(\Omega). \tag{6.12}
$$

Remark 6.2.1. Let u be a very weak solution of (P). In view of Theorem 6.2.1, for all $t \in (0,T)$, $u(\cdot,t)$ is 2-diffuse if and only if the same holds for u_0.

Proof of Theorem 6.2.1. Let $K \subseteq \Omega$ be any compact set such that $C_2(K) = 0$. By Proposition 3.4.10 there exists a sequence $\{f_n\} \subseteq C^\infty_c(\Omega)$ such that: (i) $f_n(\Omega) \subseteq [0,1]$, and $f_n|_K = 1$; and (ii) $\|\Delta f_n\|_{L^1(\Omega)} \to 0$ as $n \to \infty$. In particular, since

$$
\int_\Omega |\nabla f_n|^2\,dx = -\int_\Omega f_n \Delta f_n\,dx \leq \int_\Omega |\Delta f_n|\,dx,
$$

we have $\lim_{n\to\infty} \|f_n\|_{H^1_0(\Omega)} = 0$.

Let $U \subseteq \Omega$ be open and such that $U \supseteq K$. By Lemma A.9 and standard regularization arguments there exists $\rho_U \in C^\infty_c(U)$ such that $\rho_U(U) \subseteq [0,1]$ and $\rho_U|_K = 1$. For all $t \in (0,T]$ and $n \in \mathbb{N}$, from equality (6.4) we get (see (3.80)–(3.82))

$$
\begin{aligned}
&\langle [u(\cdot,t)]_{c,2}, f_n\rho_U\rangle - \langle [u_0]_{c,2}, f_n\rho_U\rangle \\
&= \int_0^t ds \int_\Omega \phi(u_r(x,t))\Delta(f_n\rho_U)(x)dx + \langle [u_0]_{d,2}, f_n\rho_U\rangle - \langle [u(\cdot,t)]_{d,2}, f_n\rho_U\rangle
\end{aligned} \tag{6.13}
$$

Since $\|f_n\|_{H^1_0(\Omega)} \to 0$, we have $f_n\rho_U \overset{*}{\rightharpoonup} 0$ in $L^\infty(\Omega)$ and $f_n\rho_U \to 0$ in $H^1_0(\Omega)$. Moreover, by Theorem 3.4.15 both $[u_0]_{d,2}$ and $[u(\cdot,t)]_{d,2}$ $(t \in (0,T))$ belong to $L^1(\Omega) + H^{-1}(\Omega)$.

It follows that

$$\lim_{n\to\infty}\langle [u(\cdot,t)]_{d,2}, f_n\rho_U\rangle = \lim_{n\to\infty}\langle [u_0]_{d,2}, f_n\rho_U\rangle = 0 \quad (t\in(0,T)). \tag{6.14}$$

On the other hand, since $\phi(u_r)\in L^\infty(Q)$, $\|\Delta f_n\|_{L^1(\Omega)}\to 0$ and $\|f_n\|_{H^1_0(\Omega)}\to 0$, we get

$$\lim_{n\to\infty}\int_0^t\int_\Omega \phi(u_r)\Delta(f_n\rho_U)\,dxds = 0. \tag{6.15}$$

Now observe that since $\chi_K\le f_n\rho_U\le\chi_U$ for all $n\in\mathbb{N}$, for all $n\in\mathbb{N}$, we have

$$[u_0]^\pm_{c,2}(K)\le\langle [u_0]^\pm_{c,2}, f_n\rho_U\rangle\le [u_0]^\pm_{c,2}(U), \tag{6.16a}$$

and

$$[u(\cdot,t)]^\pm_{c,2}(K)\le\langle [u(\cdot,t)]^\pm_{c,2}, f_n\rho_U\rangle\le [u(\cdot,t)]^\pm_{c,2}(U)\quad (t\in(0,T)). \tag{6.16b}$$

Moreover, for any $t\in(0,T)$,

$$\limsup_{n\to\infty}\langle [u(\cdot,t)]_{c,2}, f_n\rho_U\rangle\le\limsup_{n\to\infty}\langle [u_0]_{c,2}, f_n\rho_U\rangle, \tag{6.17}$$

since by (6.14)–(6.15) the right-hand side of (6.13) vanishes as $n\to\infty$. From (6.16)–(6.17), for any $t\in(0,T)$, we obtain

$$\begin{aligned}
&[u(\cdot,t)]^+_{c,2}(K)-[u(\cdot,t)]^-_{c,2}(U)\le\langle [u(\cdot,t)]_{c,2}, f_n\rho_U\rangle\\
&\quad\le\limsup_{n\to\infty}\langle [u(\cdot,t)]_{c,2}, f_n\rho_U\rangle\le\limsup_{n\to\infty}\langle [u_0]_{c,2}, f_n\rho_U\rangle\\
&\quad\le\limsup_{n\to\infty}\langle [u_0]^+_{c,2}, f_n\rho_U\rangle-\liminf_{n\to\infty}\langle [u_0]^-_{c,2}, f_n\rho_U\rangle\le [u_0]^+_{c,2}(U)-[u_0]^-_{c,2}(K).
\end{aligned} \tag{6.18}$$

It is similarly seen that

$$[u_0]^+_{c,2}(K)-[u_0]^-_{c,2}(U)\le [u(\cdot,t)]^+_{c,2}(U)-[u(\cdot,t)]^-_{c,2}(K)\quad (t\in(0,T)). \tag{6.19}$$

By inequalities (6.18)–(6.19) and the regularity of the measures $[u_0]^\pm_{c,2}$ and $[u(\cdot,t)]^\pm_{c,2}$ (see Lemma 1.3.2) we have that

$$[u(\cdot,t)]^+_{c,2}(K)+[u_0]^-_{c,2}(K)\le [u_0]^+_{c,2}(K)+[u(\cdot,t)]^-_{c,2}(K)$$

and

$$[u_0]^+_{c,2}(K)+[u(\cdot,t)]^-_{c,2}(K)\le [u(\cdot,t)]^+_{c,2}(K)+[u_0]^-_{c,2}(K).$$

By these inequalities, for all $t \in (0,T)$, we have

$$[u(\cdot,t)]_{c,2}(K) = [u_0]_{c,2}(K) \quad \text{for every 2-null compact } K \subseteq \Omega,$$

whence, by the arbitrariness of K and the regularity of $[u_0]_{c,2}$, $[u(\cdot,t)]_{c,2}$,

$$[u(\cdot,t)]_{c,2}(E) = [u_0]_{c,2}(E) \quad \text{for every 2-null } E \in \mathcal{B}(\Omega). \tag{6.20}$$

On the other hand, by definition of 2-concentrated measure, for every $t \in (0,T)$, there exists a 2-null set $E_1(t) \in \mathcal{B}(\Omega)$ such that $[u(\cdot,t)]_{c,2} = [u(\cdot,t)]_{c,2} \llcorner E_1(t)$. Similarly, there exists a 2-null set $E_2 \in \mathcal{B}(\Omega)$ such that $[u_0]_{c,2} = [u_0]_{c,2} \llcorner E_2$. Then by equality (6.20) we have

$$\begin{aligned}
&[u(\cdot,t)]_{c,2}(E_1(t) \setminus E_2) = [u_0]_{c,2}(E_1(t) \setminus E_2) = 0,\\
&[u_0]_{c,2}(E_2 \setminus E_1(t)) = [u(\cdot,t)]_{c,2}(E_2 \setminus E_1(t)) = 0.
\end{aligned}$$

Therefore both $[u(\cdot,t)]_{c,2}$ and $[u_0]_{c,2}$ are concentrated on the set $F(t) := E_1(t) \cap E_2$. Clearly, for any $E \in \mathcal{B}(\Omega)$, we have $C_2(E \cap F(t)) = C_2(F(t)) = 0$. Then by (6.20) we have

$$[u(\cdot,t)]_{c,2}(E) = [u(\cdot,t)]_{c,2}(E \cap F(t)) = [u_0]_{c,2}(E \cap F(t)) = [u_0]_{c,2}(E)$$

for every $E \in \mathcal{B}(\Omega)$. Hence the conclusion follows. □

6.3 Uniqueness

The persistence property in Theorem 6.2.1 plays an important role when proving the uniqueness of weak solutions to problem (P), since it completely characterizes their 2-concentrated part. On the other hand, since Definition 6.1.1 does not provide any prescription about the 2-diffuse part $[u_s(\cdot,t)]_{d,2}$ (see equalities (3.80)–(3.82)), unsurprisingly, weak solutions of (P) are not a uniqueness class of the problem (see Remark 6.3.2). This makes the following definition important.

Definition 6.3.1. Let (A_0)–(A_1) hold. A weak solution u of (P) *satisfies the compatibility conditions*, if for a. e. $t \in (0,T)$,

$$[u_s^\pm(\cdot,t)]_{d,2} = [u_s^\pm(\cdot,t)]_{d,2} \llcorner \mathcal{S}_\pm^t, \tag{6.21a}$$

where

$$\mathcal{S}_\pm^t := \{x \in \Omega \mid \phi^\pm(u_r(x,t)) = \pm\phi_{\pm\infty}\}. \tag{6.21b}$$

In this definition and henceforth, every statement where the symbols $\pm$ and $\mp$ appear must be read as a couple of independent statements, which correspond to the upper and lower choice of signs, respectively.

Remark 6.3.1. By Definition 6.1.1(i) $\phi(u_r(\cdot,t)) \in H_0^1(\Omega)$ for a. e. $t \in (0,T)$, and hence $\phi(u_r(\cdot,t))$ has a 2-quasi continuous representative $v(\cdot,t) \in H_0^1(\Omega)$ (see Definition 2.1.7) In (6.21b), we identify the functions $\phi^\pm(u_r(\cdot,t))$ with their 2-quasi-continuous representatives $v^\pm(\cdot,t) \in H_0^1(\Omega)$, which satisfy $0 \le v^\pm \le \pm\phi_{\pm\infty}$ 2-quasi everywhere in $\mathbb{R}$. In (6.21a), we always identify the sets $S_\pm^t$ with $\tilde{S}_\pm^t := \{x \in \Omega \mid v^\pm(x,t) = \pm\phi_{\pm\infty}\}$. Since $v^\pm(\cdot,t)$ is defined up to 2-null sets, the same holds for the sets $\tilde{S}_\pm^t$.

The uniqueness for problem (P) is the content of the following theorem, which makes use of the compatibility conditions in Definition 6.3.1.

Theorem 6.3.1. *Let (A_0)–(A_1) hold. Then for every $u_0 \in \mathfrak{R}_f(\Omega)$, there exists at most one weak solution of problem (P) that satisfies the compatibility conditions* (6.21a).

Remark 6.3.2. The existence of weak solutions of (P) satisfying the compatibility conditions will be proven below (see Theorem 6.4.2). On the other hand, it is easy to exhibit weak solutions of (P) that do not satisfy the compatibility conditions and thus cannot be those given by Theorem 6.4.2. Therefore weak solutions are not a uniqueness class for problem (P).

Let (A_0) hold with $u_{0s} \neq 0$. Define $u \in L^\infty_{w*}(0,T;\mathfrak{R}_f(\Omega))$ by

$$u(\cdot,t) := w(\cdot,t) + u_{0s} \quad \text{for a. e. } t \in (0,T), \tag{6.22}$$

where $w \in L^\infty(0,T;L^1(\Omega))$ is the unique weak solution of (P) with initial data u_{0r} (the existence and uniqueness of w is ensured by Theorem 6.4.3). Plainly, u is a weak solution of (P) with $u_r = w$ a. e. in Q and singular part $u_s(\cdot,t) = u_{0s}$ constant in time, and thus u satisfies the compatibility conditions if and only if

$$[u_{0s}^\pm]_{d,2} = [u_{0s}^\pm]_{d,2} \llcorner \{x \in \Omega \mid \phi^\pm(w(x,t)) = \pm\phi_{\pm\infty}\} \quad \text{for a. e. } t \in (0,T).$$

Examples where these equalities are not satisfied are easily given, and thus the claim follows.

Proof of Theorem 6.3.1. Let u_1 and u_2 be two weak solutions of problem (P). For shortness, set $u_{1r} \equiv (u_1)_r$ and $u_{1s} \equiv (u_1)_s$ (and similarly for u_2). Let u_1 and u_2 satisfy the compatibility conditions

$$[u_{is}^\pm(\cdot,t)]_{d,2} = [u_{is}^\pm(\cdot,t)]_{d,2} \llcorner S_{i,\pm}^t, \tag{6.23}$$

where

$$S_{i,\pm}^t := \{x \in \Omega \mid \phi(u_{ir})(x,t) = \phi_{\pm\infty}\} \quad (i = 1,2).$$

From (6.2) we immediately get

$$\int_0^T \langle u_1(\cdot,t) - u_2(\cdot,t), \partial_t \zeta(\cdot,t)\rangle dt = \iint_Q \nabla[\phi(u_{1r}) - \phi(u_{2r})] \cdot \nabla\zeta \, dxdt \tag{6.24}$$

for every $\zeta \in C^1([0,T]; C_c^1(\Omega))$ such that $\zeta(\cdot,T) = 0$ in Ω. Moreover, by equality (6.12), for all $t \in (0,T)$, we have

$$[u_1(\cdot,t)]_{c,2} = [u_0]_{c,2} = [u_2(\cdot,t)]_{c,2} \quad \text{in } \mathfrak{R}_f(\Omega). \tag{6.25}$$

By (6.24)–(6.25) we get

$$\int_0^T \langle [u_1(\cdot,t)]_{d,2} - [u_2(\cdot,t)]_{d,2}, \partial_t\zeta(\cdot,t)\rangle dt = \iint_Q \nabla[\phi(u_{1r}) - \phi(u_{2r})] \cdot \nabla\zeta \, dxdt \tag{6.26}$$

for every $\zeta \in C^1([0,T]; C_c^1(\Omega))$ such that $\zeta(\cdot,T) = 0$ in Ω.

By standard regularization arguments this equality also holds for any $\zeta \in C([0,T]; H_0^1(\Omega))$ with $\partial_t\zeta \in L^\infty(Q) \cap L^2(0,T; H_0^1(\Omega))$ and $\zeta(\cdot,T) = 0$ a. e. in Ω. This allows us to choose in (6.26) the test function

$$\zeta(x,t) := -\int_t^T [\phi(u_{1r}) - \phi(u_{2r})](x,s)\, ds \quad ((x,t) \in Q),$$

thus obtaining

$$\begin{aligned}&\int_0^T \langle [u_1(\cdot,t)]_{d,2} - [u_2(\cdot,t)]_{d,2}, [\phi(u_{1r}) - \phi(u_{2r})](\cdot,t)\rangle dt \\ &\quad = -\iint_Q \nabla[\phi(u_{1r}) - \phi(u_{2r})] \cdot \left(\int_t^T \nabla[\phi(u_{1r}) - \phi(u_{2r})](x,s)\, ds\right) dxdt.\end{aligned} \tag{6.27}$$

By equality (3.81) we have

$$\begin{aligned}&\int_0^T \langle [u_1(\cdot,t)]_{d,2} - [u_2(\cdot,t)]_{d,2}, [\phi(u_{1r}) - \phi(u_{2r})](\cdot,t)\rangle dt \\ &\quad = \int_0^T \langle [u_{1s}(\cdot,t)]_{d,2} - [u_{2s}(\cdot,t)]_{d,2}, [\phi(u_{1r}) - \phi(u_{2r})](\cdot,t)\rangle dt \\ &\qquad + \iint_Q (u_{1r} - u_{2r})\, [\phi(u_{1r}) - \phi(u_{2r})]\, dxdt,\end{aligned} \tag{6.28}$$

whereas, by the compatibility conditions (6.23),

$$\begin{aligned}
&\int_0^T \langle [u_{1s}(\cdot,t)]_{d,2} - [u_{2s}(\cdot,t)]_{d,2}, [\phi(u_{1r}) - \phi(u_{2r})](\cdot,t)\rangle dt \\
&\quad = \int_0^T \langle [u_{1s}^+(\cdot,t)]_{d,2}, \phi_{+\infty} - \phi(u_{2r})(\cdot,t)\rangle dt \\
&\quad\quad - \int_0^T \langle [u_{1s}^-(\cdot,t)]_{d,2}, \phi_{-\infty} - \phi(u_{2r})(\cdot,t)\rangle dt \\
&\quad\quad - \int_0^T \langle [u_{2s}^+(\cdot,t)]_{d,2}, \phi(u_{1r})(\cdot,t) - \phi_{+\infty}\rangle dt \\
&\quad\quad + \int_0^T \langle [u_{2s}^-(\cdot,t)]_{d,2}, \phi(u_{1r})(\cdot,t) - \phi_{-\infty}\rangle dt \geq 0.
\end{aligned} \tag{6.29}$$

By (6.28)–(6.29) we obtain

$$\begin{aligned}
&\int_0^T \langle [u_1(\cdot,t)]_{d,2} - [u_2(\cdot,t)]_{d,2}, [\phi(u_{1r}) - \phi(u_{2r})](\cdot,t)\rangle dt \\
&\quad \geq \iint_Q (u_{1r} - u_{2r}) [\phi(u_{1r}) - \phi(u_{2r})]\, dxdt.
\end{aligned} \tag{6.30}$$

Concerning the right-hand side of (6.27), we have that

$$\begin{aligned}
&-\iint_Q \nabla[\phi(u_{1r}) - \phi(u_{2r})] \cdot \left(\int_t^T \nabla[\phi(u_{1r}) - \phi(u_{2r})](x,s)\, ds \right) dxdt \\
&= \frac{1}{2} \iint_Q \frac{d}{dt} \left| \int_t^T \nabla[\phi(u_{1r}) - \phi(u_{2r})](x,s)\, ds \right|^2 dxdt \\
&= -\frac{1}{2} \int_\Omega \left| \int_0^T \nabla[\phi(u_{1r}) - \phi(u_{2r})](x,s)\, ds \right|^2 dx.
\end{aligned} \tag{6.31}$$

From (6.27) and (6.30)–(6.31) we get

$$\iint_Q (u_{1r} - u_{2r}) [\phi(u_{1r}) - \phi(u_{2r})]\, dxdt \leq -\frac{1}{2} \int_\Omega \left| \int_0^T \nabla[\phi(u_{1r}) - \phi(u_{2r})](x,t)\, dt \right|^2 dx \leq 0,$$

whence

$$\phi(u_{1r}(x,t)) = \phi(u_{2r}(x,t)) \quad \text{for a. e. } (x,t) \in Q. \tag{6.32}$$

Now let us choose in equality (6.24) $\zeta(x,t) = \rho(x)k_q(t)$ with $\rho \in C_c^2(\Omega)$ and k_q as in (6.7). Arguing as in the proof of (6.4) and using equality (6.32), for a. e. $t_1 \in (0,T)$, we obtain

$$\langle u_1(\cdot,t_1) - u_2(\cdot,t_1), \rho \rangle = 0.$$

By the arbitrariness of ρ this implies $u_1(\cdot,t_1) = u_2(\cdot,t_1)$ in $\mathfrak{R}_f(\Omega)$ for a. e. $t_1 \in (0,T)$. Hence the result follows. □

6.4 Existence and regularity results

6.4.1 Existence

The proof of our first existence result is constructive. In fact, we exhibit a weak solution of (P) by studying the sequence $\{u_n\}$ of solutions of the *approximating problems*

$$\begin{cases} \partial_t u_n = \Delta\phi_n(u_n) & \text{in } Q, \\ u_n = 0 & \text{on } \Gamma, \\ u_n(\cdot,0) = u_{0n} & \text{in } \Omega \end{cases} \tag{P_n}$$

(see Section 6.5). Here $\phi_n(z) := \phi(z) + \frac{z}{n}$ ($z \in \mathbb{R}$), and u_{0n} is a suitable regularization of the initial measure u_0 (see (6.48)–(6.49)).

Theorem 6.4.1. *Let* (A_0)–(A_2) *hold. Then there exists a weak solution of problem* (P), *which is obtained as a limiting point in the weak* topology of* $L^\infty_{w*}(0,T;\mathfrak{R}_f(\Omega))$ *of the sequence of solutions to problems* (P_n). *Moreover:*
(i) *for a. e.* $0 < t_1 < t_2 < T$,

$$u_s^\pm(\cdot,t_2) \le u_s^\pm(\cdot,t_1) \le u_{0s}^\pm \quad \textit{in } \mathfrak{R}_f(\Omega). \tag{6.33}$$

(ii) $u \in L^\infty(0,T;L^1(\Omega))$ *if* $u_0 \in L^1(\Omega)$.

In view of Theorem 6.3.1, for the well-posedness of (P), it is important to prove the existence of weak solutions *satisfying the compatibility conditions*. This is ensured by the following result.

Theorem 6.4.2. *Let* (A_0)–(A_2) *hold. Then weak solutions of problem* (P) *given by Theorem 6.4.1 satisfy the compatibility conditions.*

By Theorems 6.3.1, 6.4.1, and 6.4.2 we have the following result.

Theorem 6.4.3. *Let (A_0)–(A_2) hold. Then for every $u_0 \in \mathfrak{R}_f(\Omega)$, there exists a unique weak solution u of problem (P) satisfying the compatibility conditions. Moreover, claims (i)–(ii) of Theorem 6.4.1 hold.*

Weak solutions of (P) constructed in the proof of Theorem 6.4.1 have several important properties (in particular, by Theorem 6.4.2 they satisfy the compatibility conditions). This motivates the following definition.

Definition 6.4.1. Weak solutions of problem (P) given by Theorem 6.4.1 are called *constructed solutions.*

Another interesting property of constructed solutions is as follows:

Proposition 6.4.4. *Let (A_0)–(A_2) hold. Then every constructed solution of problem (P) is an entropy solution.*

6.4.2 Regularity

An interesting feature of entropy solutions is that their singular parts can neither appear spontaneously nor increase in time:

Theorem 6.4.5. *Let (A_0)–(A_1) hold. Assume that*

$$\text{there exists } L_0 \geq 0 \text{ such that } \phi'(s) > 0 \text{ if } |s| > L_0. \tag{6.34}$$

Let u be an entropy solution of problem (P). Then for a. e. $t_1, t_2 \in (0, T)$ such that $t_1 < t_2$, inequality (6.33) holds.

Proof. Let u be an entropy solution u of problem (P). Let us first prove that for all $\rho \in C^1_c(\Omega)$, $\rho \geq 0$, and $h \in C^1([0, T])$ such that $h(T) = 0$, we have

$$\int_0^T \langle u_s^{\pm}(\cdot, t), \rho\rangle h'(t)\, dt \geq -h(0)\,\langle u_{0s}^{\pm}, \rho\rangle. \tag{6.35}$$

We only prove (6.35) with "+". Since ϕ is strictly increasing in $[L_0, +\infty)$ (see (6.34)), by standard approximation results we can choose $F_\phi(y) = [y - m]^+$ with $m > L_0$ in (6.11) (this corresponds to choosing $f = \chi_{[\phi(m), +\infty)}$ in (6.9)). Then $M^+_{F_\phi} = 1$ and $M^-_{F_\phi} = 0$, and thus we get

$$\iint_Q \{[u_r - m]^+ \zeta_t - \chi_{[m,+\infty)}(u_r)\nabla\phi(u_r) \cdot \nabla\zeta\}\, dxdt + \int_0^T \langle u_s^+(\cdot, t), \zeta_t(\cdot, t)\rangle\, dt$$

$$\geq -\int_{\Omega} [u_{0r} - m]^{+} \zeta(x, 0)\, dx - \langle u_{0s}^{+}, \zeta(\cdot, 0) \rangle \tag{6.36}$$

for every $\zeta \in C^1([0, T]; C_c^1(\Omega))$ such that $\zeta \geq 0$ and $\zeta(\cdot, T) = 0$ in Ω. Choosing in this inequality $\zeta(x, t) = \rho(x)h(t)$ with ρ, h as above and letting $m \to \infty$ give (6.35). Now fix $t_1 \in (0, T)$. For any $p \in \mathbb{N}$ large enough, set

$$h_p(t) := \chi_{[0,t_1]}(t) + p\left(t_1 + \frac{1}{p} - t\right)\chi_{(t_1, t_1+\frac{1}{p}]}(t) \quad (t \in [0, T]).$$

Choosing $h = h_p$ in (6.35) gives, for all nonnegative $\rho \in C_c^1(\Omega)$,

$$p \int_{t_1}^{t_1+\frac{1}{p}} \langle u_s^{\pm}(\cdot, t), \rho \rangle\, dt \leq \langle u_{0s}^{\pm}, \rho \rangle. \tag{6.37}$$

Since by Definition 6.1.3 the mapping $t \mapsto \langle u_s^{\pm}(\cdot, t), \rho \rangle$ belongs to $L^{\infty}(0, T)$, by Corollary 2.9.5(i) we have

$$\lim_{p \to \infty} p \int_{t_1}^{t_1+\frac{1}{p}} \langle u_s^{\pm}(\cdot, t), \rho \rangle\, dt = \langle u_s^{\pm}(\cdot, t_1), \rho \rangle.$$

Then letting $p \to \infty$ in (6.37), for a. e. $t_1 \in (0, T)$, we get $\langle u_s^{\pm}(\cdot, t_1), \rho \rangle \leq \langle u_{0s}^{\pm}, \rho \rangle$ for all $\rho \in C_c^1(\Omega)$, $\rho \geq 0$, whence the second inequality in (6.33) plainly follows. The first is similarly proven with the following choice of $h = h_p$ in (6.35):

$$h_p(t) := p(t - t_1)\chi_{[t_1, t_1+\frac{1}{p}]}(t) + \chi_{(t_1+\frac{1}{p}, t_2)}(t) + p\left(t_2 + \frac{1}{p} - t\right)\chi_{[t_2, t_2+\frac{1}{p}]}(t) \quad (t \in [0, T]).$$

Hence the result follows. □

Theorem 6.4.5 can be regarded as a regularity result for entropy solutions. In particular, in view of Proposition 6.4.4, it applies to constructed solutions. Remarkably, constructed solutions of (*P*) display a number of further regularizing effects. To discuss this point, we state the following definition.

Definition 6.4.2. A weak solution u of problem (*P*) undergoes an $\mathfrak{R}_f$-L^1 *regularizing effect*, if there exists a *waiting time* $\tau \in [0, T)$ such that $u(\cdot, t) \in L^1(\Omega)$ for all $t \in (\tau, T)$. If $\tau = 0$, *then instantaneous* $\mathfrak{R}_f$-L^1 *regularization* occurs.

Sufficient conditions for the occurrence of $\mathfrak{R}_f$-L^1 regularizing effects are provided by the following proposition, which relies on a suitable convergence rate of ϕ to the saturation values $\phi_{\pm\infty}$ (see [87, Proposition 3.8] for the proof).

Proposition 6.4.6. *Let (A_0)–(A_2) hold. Let there exist $s_\pm \in (0,1)$ and $M_\pm \in (0, \phi_{\pm\infty}]$ such that*

$$\pm \phi_{\pm\infty} \mp \phi(y) \geq \frac{M_\pm}{(1+|y|)^{s_\pm}} \quad \text{for } \pm y \geq 0. \tag{6.38}$$

Moreover, let $u_0^\pm \in \mathfrak{R}^+_{d,p}(\Omega)$ with $p \in (1, \frac{2}{1+s_\pm})$, and let u be the unique weak solution of (P) satisfying the compatibility conditions. Then $u^\pm \in L^\infty(0,T;L^1(\Omega))$.

On the contrary, the following two propositions point out cases where instantaneous $\mathfrak{R}_f$-L^1 regularization does not occur. This happens either for suitable $u_0 \in \mathfrak{R}_{d,2}(\Omega)$ if ϕ is strictly increasing or for any constructed solution in the most singular case where ϕ is constant near $\pm\infty$. We refer the reader to [87, Propositions 3.9, 3.10] for the proofs.

Proposition 6.4.7. *Let (A_0)–(A_2) hold with strictly increasing ϕ. Then there exists $u_0 \in \mathfrak{R}^+_{d,2}(\Omega)$, with $u_{0s} \neq 0$, such that for the unique weak solution of (P) satisfying the compatibility conditions, we have*

$$\operatorname*{ess\,lim}_{t\to 0^+} \|u_s(\cdot,t) - u_{0s}\|_{\mathfrak{R}_f(\Omega)} = 0. \tag{6.39}$$

Proposition 6.4.8. *Let (A_0)–(A_2) hold, and let u be the unique weak solution of (P) satisfying the compatibility conditions. Let there exist $\pm c_\pm \geq 0$ and $\pm d_\pm \geq 0$ such that $\phi(y) = c_\pm$ if $\pm y \geq \pm d_\pm$. Then for every open subset $\tilde{\Omega} \Subset \Omega$, we have*

$$\operatorname*{ess\,lim}_{t\to 0^+} \|u_s^\pm(\cdot,t) - u_{0s}^\pm\|_{\mathfrak{R}_f(\tilde{\Omega})} = 0. \tag{6.40}$$

6.5 Proof of existence results: the approximating problems (P_n)

6.5.1 Approximation of the initial data

Let (A_0) hold. To address problem (P), we need a proper approximation of the initial data u_0. To this purpose, we will use the following technical lemma.

Lemma 6.5.1. *For any $\mu \in \mathfrak{R}_f^+(\Omega)$, there exists a sequence $\{(\alpha_n, \beta_n, \gamma_n)\} \subseteq [C_c^\infty(\Omega)]^3$, $\alpha_n \geq 0$, $\beta_n \geq 0$, $\gamma_n \geq 0$ in Ω, with the following properties:*

$$\alpha_n \to \mu_r \quad \text{in } L^1(\Omega), \tag{6.41a}$$

$$\alpha_n \stackrel{*}{\rightharpoonup} \mu_{ac}, \quad \beta_n \stackrel{*}{\rightharpoonup} [\mu_s]_{d,2}, \quad \gamma_n \stackrel{*}{\rightharpoonup} \mu_{c,2} \quad \text{in } \mathfrak{R}_f(\Omega), \tag{6.41b}$$

$$\int_\Omega \beta_n f_n \rho \, dx \to \langle [\mu_s]_{d,2}, f\rho \rangle \tag{6.41c}$$

for all $\rho \in C_c(\Omega)$ *and all* $\{f_n\} \subseteq H^1(\Omega) \cap L^\infty(\Omega)$ *such that* $\sup_{n\in\mathbb{N}} \|f_n\|_\infty < \infty$ *and* $f_n \rightharpoonup f$ *in* $H^1(\Omega)$.

Set $\mu_n := \alpha_n + \beta_n + \gamma_n \in C_c^\infty(\Omega)$. *Then* α_n, β_n, *and* γ_n *can be chosen so that*

$$\|\mu_n\|_{L^1(\Omega)} \le \|\mu\|_{\mathfrak{R}_f(\Omega)}, \quad \|\mu_n\|_{L^\infty(\Omega)} \le M\, \|\mu\|_{\mathfrak{R}_f(\Omega)} \sqrt[4]{n} \quad \textit{for some } M > 0, \tag{6.41d}$$

$$\mu_n \to \mu_r \quad \textit{a.e. in } \Omega, \quad \mu_n \overset{*}{\rightharpoonup} \mu \quad \textit{in } \mathfrak{R}_f(\Omega). \tag{6.41e}$$

Proof. Set $\tilde{\mu} := \tilde{\mu}_r + \tilde{\mu}_s \in \mathfrak{R}_f^+(\mathbb{R}^N)$,

$$\tilde{\mu}_r(x) := \mu_r(x)\chi_\Omega(x) \quad \text{for } x \in \mathbb{R}^N, \quad \tilde{\mu}_s(E) := \mu_s(\Omega \cap E) \quad \text{for } E \in \mathcal{B}^d.$$

By definition $\tilde{\mu} = \tilde{\mu} \llcorner \Omega$, and thus $\tilde{\mu}(E) = \mu(E)$ for every $E \in \mathcal{B}^d \cap \Omega$. Consider the sequence $\{\zeta_n\} \subseteq C_c^\infty(\mathbb{R}^N)$ defined by

$$\zeta_n(x) := \frac{n^{d\theta}}{\int_{\mathbb{R}^N} \zeta(x)\, dx} \zeta(n^\theta x) \quad (x \in \mathbb{R}^N), \tag{6.42}$$

where $\zeta \in C_c^\infty(\mathbb{R}^N)$ is a standard mollifier, and $\theta > 0$ is to be chosen. For any $n \in \mathbb{N}$, set

$$\tilde{\alpha}_n := \tilde{\mu}_{ac} * \zeta_n, \quad \tilde{\beta}_n := [\tilde{\mu}_s]_{d,2} * \zeta_n, \quad \tilde{\gamma}_n := \tilde{\mu}_{c,2} * \zeta_n, \quad \tilde{\mu}_n := \tilde{\alpha}_n + \tilde{\beta}_n + \tilde{\gamma}_n = \tilde{\mu} * \zeta_n$$

(see (3.82) with $p = 2$ and $m = 1$). Let Ω_n be open, $\Omega_n \Subset \Omega_{n+1} \Subset \Omega$ for every $n \in \mathbb{N}$, $\bigcup_{n=1}^\infty \Omega_n = \Omega$, and let $\eta_n \in C_c^\infty(\Omega_{n+1})$ be such that $0 \le \eta_n \le 1$ and $\eta_n = 1$ in $\overline{\Omega}_n$ $(n \in \mathbb{N})$. Let us prove that the functions

$$\alpha_n := \tilde{\alpha}_n\eta_n, \quad \beta_n := \tilde{\beta}_n\eta_n, \quad \gamma_n := \tilde{\gamma}_n\eta_n, \quad \mu_n := \tilde{\mu}_n\eta_n = \alpha_n + \beta_n + \gamma_n \quad (n \in \mathbb{N})$$

(which clearly are nonnegative in Ω) have the properties stated in (6.41).

The convergence statements in (6.41a)–(6.41b) follow by standard convolution results. To prove (6.41c), fix $\rho \in C_c^1(\Omega)$. Set $F_n := (f_n\rho) * \zeta_n$ and choose n so large that $\operatorname{supp}\rho \subseteq \overline{\Omega}_n$ and $\operatorname{supp} F_n \subseteq \Omega$. Recalling that $\eta_n = 1$ in $\overline{\Omega}_n$, we get

$$\langle [\mu_s]_{d,2}, F_n \rangle = \langle [\tilde{\mu}_s]_{d,2} * \zeta_n, \eta_n f_n \rho \rangle = \int_\Omega \beta_n f_n \rho \, dx. \tag{6.43}$$

Now observe that by the Banach–Alaoglu theorem $f_n \overset{*}{\rightharpoonup} f$ in $L^\infty(\Omega)$, since by assumption $\{f_n\}$ is bounded in $L^\infty(\Omega)$. Plainly, it follows that $F_n \overset{*}{\rightharpoonup} f\rho$ in $L^\infty(\Omega)$. Indeed, for every $\zeta \in L^1(\Omega)$, we have

$$\lim_{n\to\infty} \int_\Omega F_n \zeta \, dx = \lim_{n\to\infty} \int_\Omega (f_n\rho) * \zeta_n \zeta \, dx$$

$$= \lim_{n\to\infty} \int_\Omega \zeta(x) \left(\int_\Omega f_n(y)\rho(y)\, \zeta_n(x-y)\, dy \right) dx$$

$$= \lim_{n\to\infty} \int_\Omega f_n(y)\rho(y)\, dy \int_\Omega \zeta(x)\, \zeta_n(x-y) dx = \int_\Omega f\,\rho\,\zeta\, dx. \tag{6.44}$$

By (6.44) we also have that $F_n \rightharpoonup f\rho$ in $L^2(\Omega)$. Since $f_n\rho \in H^1_0(\Omega)$, we have $F_n \in H^1_0(\Omega)$ for every n sufficiently large, with weak gradient $\nabla F_n = [\nabla(f_n\rho)] * \zeta_n$. Then we plainly get that $F_n \rightharpoonup f\rho$ in $H^1_0(\Omega)$, since by assumption $f_n \rightharpoonup f$ in $H^1(\Omega)$.

Since $[\mu_s]_{d,2}$ belongs to $L^1(\Omega) + H^{-1}(\Omega)$ (see Theorem 3.4.15), by the above convergence results we obtain

$$\lim_{n\to\infty} \langle [\mu_s]_{d,2}, F_n \rangle = \langle [\mu_s]_{d,2}, f\rho \rangle.$$

Then letting $n \to \infty$ in (6.43) gives (6.41c) for all $\rho \in C^1_c(\Omega)$, whence by standard approximation results the claim follows.

The first inequality in (6.41d) follows from the very definition of μ_n, whereas the second follows by choosing $\theta \in (0, \frac{1}{4N}]$ in (6.42):

$$\|\mu_n\|_{L^\infty(\Omega)} \le C n^{N\theta} \|\mu\|_{\mathfrak{R}_f(\Omega)} \le C \sqrt[4]{n}\, \|\mu\|_{\mathfrak{R}_f(\Omega)}.$$

Let us address (6.41e). Set

$$[\mu_s]_n := \beta_n + \gamma_n \equiv (\tilde{\mu}_s * \zeta_n)\eta_n, \tag{6.45}$$

and observe that by the Radon–Nikodým theorem for a. e. $x \in \Omega$, we have

$$\lim_{n\to\infty} \alpha_n(x) = \mu_r(x), \tag{6.46}$$

$$\limsup_{n\to\infty} [\mu_s]_n(x) \le \frac{1}{\int_{\mathbb{R}^N} \zeta(x) dx} \lim_{n\to\infty} n^{N\theta} \tilde{\mu}_s\left(B\left(x, \frac{1}{n^\theta}\right)\right) = 0 \tag{6.47}$$

(see the second inequality in (2.98), Corollary 2.9.6). Since

$$\mu_n = \alpha_n + [\mu_s]_n,$$

combining (6.46) and (6.47) gives

$$\mu_n(x) \to \mu_r(x) \quad \text{for a. e. } x \in \Omega,$$

and the proof of the first convergence in (6.41e) is completed (see also (6.41b) for the second convergence in (6.41e)). □

For any $u_0 \in \mathfrak{R}_f(\Omega)$, consider the Jordan decomposition $u_0 = u_0^+ - u_0^-$, where $u_0^\pm \in \mathfrak{R}_f^+(\Omega)$ and $u_0^+ \perp u_0^-$. Let $u_{0r}^\pm \in L^1(\Omega)$ be the density of $[u_0^\pm]_{ac}$, and let $\delta^\pm := [u_0^\pm]_{d,2}$,

$\gamma^\pm := [u_0^\pm]_{c,2}$ be, respectively, the diffuse and concentrated parts of $u_0^\pm$ with respect to the 2-capacity. Then by (3.80)–(3.82) we have $u_0^\pm = \gamma^\pm + \delta^\pm$. Let

$$[u_0^\pm]_n := \gamma_n^\pm + \delta_n^\pm \geq 0, \quad u_{0n} := [u_0^+]_n - [u_0^-]_n, \tag{6.48}$$

with $\{\gamma_n^\pm\} \subseteq C_c^\infty(\Omega)$, $\{\delta_n^\pm\} \subseteq C_c^\infty(\Omega)$, $\gamma_n^\pm \geq 0$, $\delta_n^\pm \geq 0$ such that

$$\|[u_0^\pm]_n\|_{L^1(\Omega)} \leq \|u_0^\pm\|_{\mathcal{M}(\Omega)}, \quad \|[u_0^\pm]_n\|_{L^\infty(\Omega)} \leq M\|u_0^\pm\|_{\mathcal{M}(\Omega)} \sqrt[4]{n}, \tag{6.49a}$$

$$[u_0^\pm]_n \to u_{0r}^\pm \quad \text{a. e. in } \Omega, \tag{6.49b}$$

$$[u_0^\pm]_n \overset{*}{\rightharpoonup} u_0^\pm \quad \text{in } \mathcal{M}(\Omega), \tag{6.49c}$$

$$\delta_n^\pm \overset{*}{\rightharpoonup} [u_0^\pm]_{d,2}, \quad \gamma_n^\pm \overset{*}{\rightharpoonup} [u_0^\pm]_{c,2} \quad \text{in } \mathcal{M}(\Omega), \tag{6.49d}$$

$$\int_\Omega \delta_n^\pm f_n \rho \, dx \to \langle [u_0^\pm]_{d,2}, f\rho \rangle \tag{6.49e}$$

for all $\rho \in C_c(\Omega)$ and $\{f_n\} \subseteq H^1(\Omega) \cap L^\infty(\Omega)$ such that $\sup_{n\in\mathbb{N}} \|f_n\|_{L^\infty(\Omega)} < \infty$ and $f_n \rightharpoonup f$ in $H^1(\Omega)$. Hereafter we always assume that the initial data u_{0n} of (P_n) have properties (6.49).

Remark 6.5.1. For any $u_0 \in \mathfrak{R}_f(\Omega)$, consider the decomposition (see (3.82) with $p = 2$ and $m = 1$)

$$u_0^\pm = [u_0^\pm]_{ac} + [u_{0s}^\pm]_{d,2} + [u_0^\pm]_{c,2}. \tag{6.50}$$

Applying Lemma 6.5.1 to $\mu = u_0^\pm$, we get a sequence $\{(\alpha_n^\pm, \beta_n^\pm, \gamma_n^\pm)\} \subseteq [C_c^\infty(\Omega)]^3$ satisfying (6.41). Then, setting $\delta_n^\pm := \alpha_n^\pm + \beta_n^\pm$, properties (6.49) follow at once.

Lemma 6.5.2. *Let $\{u_{0n}\}$ satisfy* (6.48) *and* (6.49a)–(6.49c). *Let $f \in C(\mathbb{R})$ satisfy* (5.106). *Then there exists a subsequence of $\{u_{0n}\}$ (not relabeled) such that*

$$f \circ u_{0n} \overset{*}{\rightharpoonup} f \circ u_{0r} + M_f^+ u_{0s}^+ - M_f^- u_{0s}^- \quad \text{in } \mathfrak{R}_f(\Omega). \tag{6.51}$$

In particular, $u_{0n} \overset{*}{\rightharpoonup} u_0$ in $\mathfrak{R}_f(\Omega)$.

Proof. Let $\{\nu_n\} \subseteq \mathfrak{Y}^+(\Omega; \mathbb{R})$ be the sequence of Young measures associated with $\{u_{0n}\}$. By the first inequality in (6.49a) the sequence $\{u_{0n}\}$ is bounded in $L^1(\Omega)$, and thus by Theorem 5.4.5 there exists a Young measure $\nu \in \mathfrak{Y}^+(\Omega; \mathbb{R})$ such that (possibly extracting a subsequence, not relabeled) $\nu_n \to \nu$ narrowly. By the convergence in (6.49b), Proposition 5.4.1, and equality (5.53) the disintegration $\{\nu_x\}_{x\in\Omega}$ of ν satisfies $\nu_x = \delta_{u_{0r}(x)}$ for a. e. $x \in \Omega$. It follows that

$$\int_{\mathbb{R}} y^\pm d\nu_x(y) = u_{0r}^\pm(x) \quad \text{for a. e. } x \in \Omega.$$

Therefore, arguing as in Remark 5.4.4, there exist a subsequence of $\{u_{0n}\}$ (not relabeled) and $\sigma_\pm \in \mathfrak{R}_f^+(\Omega)$ such that

$$[u_{0n}]^\pm \overset{*}{\rightharpoonup} u_{0r}^\pm + \sigma_\pm \quad \text{in } \mathfrak{R}_f(\Omega). \tag{6.52}$$

By the definition of u_{0n} (see (6.48)) and (6.49c) we have $u_{0n} \overset{*}{\rightharpoonup} u_0^+ - u_0^- = u_0$, whereas by (6.52) $u_{0n} \overset{*}{\rightharpoonup} u_{0r} + \sigma$ with $\sigma := \sigma_+ - \sigma_-$.

Now observe that by the second equality in (6.48) $[u_{0n}]^\pm \le [u_0^\pm]_n$ for all $n \in \mathbb{N}$, whence by (6.49b)–(6.49c) and (6.52) we get $\sigma_\pm \le u_{0s}^\pm$ in $\mathfrak{R}_f(\Omega)$. This implies that $\sigma_\pm$ are singular with respect to the Lebesgue measure and mutually singular. Hence the equality $u_0 = u_{0r}+\sigma$ and the uniqueness of Lebesgue decomposition imply that $\sigma = u_{0s}$ and $\sigma_\pm = u_{0s}^\pm$. Then as in the proof of Proposition 5.4.13 the result follows. □

6.5.2 A priori estimates

By classical results, for any $n \in \mathbb{N}$, there exists a unique classical solution $u_n \in C^\infty(\overline{Q})$ of problem (P_n) (e. g., see [63]); moreover, for any $n \in \mathbb{N}$, we have

$$\|u_n\|_{L^\infty(0,T;L^1(\Omega))} \le \|u_{0n}\|_{L^1(\Omega)}, \qquad \|u_n\|_{L^\infty(Q)} \le \|u_{0n}\|_{L^\infty(\Omega)}. \tag{6.53}$$

We need a priori estimates for both sequences $\{u_n\}$ and $\{\phi(u_n)\}$. This is the content of the following statements.

Lemma 6.5.3. *Let (A_0)–(A_1) hold, and let $\{u_n\}$ be the sequence of solutions of problems (P_n). Then for all $n \in \mathbb{N}$, we have*

$$\|u_n\|_{L^\infty(0,T;L^1(\Omega))} \le \|u_0\|_{\mathfrak{R}_f(\Omega)}, \tag{6.54}$$

$$\|u_n\|_{L^\infty(Q)} \le M\|u_0\|_{\mathfrak{R}_f(\Omega)} \sqrt[4]{n}, \tag{6.55}$$

whence

$$\|\phi_n(u_n) - \phi(u_n)\|_{L^\infty(Q)} \le \frac{\|u_n\|_{L^\infty(Q)}}{n} \le M\|u_0\|_{\mathfrak{R}_f(\Omega)}\, n^{-3/4}, \tag{6.56}$$

$$\iint_Q |\nabla[\phi(u_n)]|^2 dxdt + \frac{1}{n}\iint_Q \phi'(u_n)|\nabla u_n|^2\, dxdt \le 2\,\|\phi\|_\infty \|u_0\|_{\mathfrak{R}_f(\Omega)}. \tag{6.57}$$

Proof. Inequalities (6.49a) and (6.53) immediately give (6.54) and (6.55), whence (6.56) follows by the very definition of ϕ_n (recall that $\phi_n(z) := \phi(z)+\frac{z}{n}, z \in \mathbb{R}$). To prove (6.57), we multiply the first equation of (P_n) by $\phi(u_n)$ and integrate over Q. Using (6.54), we obtain

$$\iint_Q |\nabla[\phi(u_n)]|^2\, dxdt + \frac{1}{n}\iint_Q \phi'(u_n)|\nabla u_n|^2\, dxdt = -\iint_Q \phi(u_n)\, \partial_t u_n\, dxdt$$

$$= \int_\Omega dx \int_0^{u_{0n}(x)} \phi(y)\,dy - \int_\Omega dx \int_0^{u_n(x,T)} \phi(y)\,dy$$
$$\leq \|\phi\|_\infty [\|u_{0n}\|_{L^1(\Omega)} + \|u_n(\cdot,T)\|_{L^1(\Omega)}] \leq 2\|\phi\|_\infty \|u_0\|_{\mathfrak{R}_f(\Omega)}.$$

Hence the result follows. □

For every $h \in C(\mathbb{R})$, set

$$H(y) := \int_0^y h(z)\,dz \quad (y \in \mathbb{R}). \tag{6.58}$$

Lemma 6.5.4. *Let (A_1) be satisfied, and let $\{u_n\}$ be the sequence of solutions of problems (P_n).*

(i) *For all $h \in \mathrm{Lip}(\mathbb{R})$, $\zeta \in C^1([0,T];C_c^1(\Omega))$, and $t \in (0,T]$, we have*

$$\begin{aligned}\int_\Omega & H(u_n(x,t))\,\zeta(x,t)\,dx \\ &+ \iint_{Q_t} h'(u_n)\nabla u_n \cdot \nabla\phi_n(u_n)\,\zeta\,dxds - \iint_{Q_t} H(u_n)\,\partial_s\zeta\,dxds \\ &= \int_\Omega H(u_{0n}(x))\,\zeta(x,0)\,dx - \iint_{Q_t} h(u_n)\nabla\phi_n(u_n)\cdot\nabla\zeta\,dxds,\end{aligned} \tag{6.59}$$

where $Q_t := \Omega \times (0,t)$. If $h(0) = 0$, then (6.59) holds for all $\zeta \in C^1(\overline{Q})$.

(ii) *For every nondecreasing $h \in C(\mathbb{R})$, $\zeta \in C^1([0,T];C_c^1(\Omega))$, $\zeta \geq 0$, and $t \in (0,T]$,*

$$\begin{aligned}\iint_{Q_t} & H(u_n)\,\partial_s\zeta\,dxds - \iint_{Q_t} h(u_n)\nabla\phi_n(u_n)\cdot\nabla\zeta\,dxds \\ &\geq \int_\Omega H(u_n(x,t))\,\zeta(x,t)\,dx - \int_\Omega H(u_{0n}(x))\,\zeta(x,0)\,dx.\end{aligned} \tag{6.60}$$

If $h(0) = 0$, then (6.60) holds for all $\zeta \in C^1(\overline{Q})$, $\zeta \geq 0$.

Proof. Since u_n is smooth, equality (6.59) plainly follows by multiplying the first equation in (P_n) by $h(u_n)\,\zeta$ with $h \in \mathrm{Lip}(\mathbb{R})$, $h(0) = 0$, and $\zeta \in C^1(\overline{Q})$, and integrating over Q. Moreover, the condition $h(0) = 0$ can be omitted if $\zeta \in C^1([0,T];C_c^1(\Omega))$. Hence claim (i) follows.

Since $\nabla u_n \cdot \nabla\phi_n(u_n) \geq 0$ for all $n \in \mathbb{N}$, from (6.59) we get (6.60) for every nondecreasing $h \in \mathrm{Lip}(\mathbb{R})$. Since every nondecreasing $h \in C(\mathbb{R})$ can be locally approximated by a sequence of nondecreasing functions in $\mathrm{Lip}(\mathbb{R})$, the conclusion follows. □

Remark 6.5.2. Let (A_0)–(A_1) hold, let $f \in C(\mathbb{R})$ be nondecreasing, and let F_ϕ be defined by (6.9). Setting $g = f \circ \phi$ in (6.60), we obtain the inequality

$$\int_\Omega F_\phi(u_n(x,t_2))\,\zeta(x,t_2)\,dx - \int_\Omega F_\phi(u_n(x,t_1))\,\zeta(x,t_1)\,dx \leq \int_{t_1}^{t_2}\int_\Omega \{F_\phi(u_n)\,\partial_t\zeta - f(\phi(u_n)\nabla[\phi_n(u_n)]\cdot\nabla\zeta\}\,dxdt \tag{6.61}$$

for all $t_1, t_2 \in [0,T]$, $t_1 < t_2$, and $\zeta \in C^1([0,T]; C^1_c(\Omega))$, $\zeta \geq 0$.

Lemma 6.5.5. *Let (A_0)–(A_1) hold, and let $\{u_n\}$ be the sequence of solutions of problems (P_n). Then:*

(i) *there exists $C_1 = C_1(\lambda_N(\Omega), \|u_0\|_{\mathfrak{R}_f(\Omega)}) > 0$ such that for all $n \in \mathbb{N}$,*

$$\iint_Q |\nabla u_n|^2\,dxdt \leq C_1\,n^{\frac{3}{2}}, \tag{6.62}$$

(ii) *there exists $C_2 = C_2(\lambda_N(\Omega), \|\phi\|_\infty, \|u_0\|_{\mathfrak{R}_f(\Omega)}) > 0$ such that for all $n \in \mathbb{N}$,*

$$\iint_Q \big|\nabla[\phi_n(u_n)]\big|^2 dxdt \leq C_2. \tag{6.63}$$

Proof. Multiplying $\partial_t u_n = \Delta\phi_n(u_n)$ by $\frac{u_n}{n}$ and integrating over Q plainly give

$$\frac{1}{2n}\int_\Omega u_n^2(x,T)\,dx - \frac{1}{2n}\int_\Omega u_{0n}^2(x)dx \leq -\iint_Q \frac{|\nabla u_n|^2}{n^2}\,dxdt.$$

From this inequality and the second inequality in (6.49a) we get

$$\iint_Q \frac{|\nabla u_n|^2}{n^2}\,dxdt \leq \frac{1}{2n}\int_\Omega u_{0n}^2 dx \leq \frac{M^2\lambda_N(\Omega)}{2\sqrt{n}}\,\|u_0\|^2_{\mathfrak{R}_f(\Omega)} =: \frac{C_1}{\sqrt{n}}.$$

This proves (6.62). Combining (6.57) and (6.62) plainly gives (6.63), and thus the result follows. □

Lemma 6.5.6. *Let (A_1) be satisfied, and let $\{u_n\}$ be the sequence of solutions of problems (P_n). Let $h \in \mathrm{Lip}(\mathbb{R})$ be bounded and nondecreasing with $h(0) = 0$, and let H be the function in (6.58). Then for any $\rho \in C^1_c(\Omega)$, there exists $M_{h,\rho} > 0$ such that for all $n \in \mathbb{N}$,*

$$\int_0^T \left|\frac{d}{dt}\int_\Omega H(u_n(x,t))\,\rho\,dx\right| dt \leq M_{h,\rho}. \tag{6.64}$$

Proof. For any $\rho \in C_c^1(\Omega)$, set

$$S_{n,\rho}(t) := \int_\Omega H(u_n(x,t))\,\rho(x)\,dx \quad (t \in (0,T)).$$

Choosing in (6.59) $\zeta(x,t) = \beta(t)\rho(x)$ with $\beta \in C_c^1(0,T)$ shows that $S_{n,\rho}$ has the weak derivative

$$\begin{aligned} S'_{n,\rho}(t) = &-\int_\Omega h(u_n(x,t))\nabla\phi_n(u_n(x,t)) \cdot \nabla\rho(x)\,dx \\ &-\int_\Omega h'(u_n(x,t))\,\nabla u_n(x,t) \cdot \nabla\phi_n(u_n(x,t))\,\rho(x)\,dx \end{aligned} \tag{6.65}$$

for a. e. $t \in (0,T)$. By (6.63) we have that

$$\begin{aligned} &\iint_Q |h(u_n)\nabla\phi_n(u_n) \cdot \nabla\rho|\,dxdt \\ &\le \|h\|_{L^\infty(\mathbb{R})}\|\rho\|_{C^1(\overline{\Omega})} \iint_Q |\nabla\phi_n(u_n)|\,dxdt \le C_2\|h\|_{L^\infty(\mathbb{R})}\|\rho\|_{C^1(\overline{\Omega})}. \end{aligned} \tag{6.66}$$

Let us estimate the second term in the right-hand side of (6.65). From (6.59) with $\zeta = 1$ and $t = T$, using the first inequality in (6.49a), we plainly get

$$\begin{aligned} 0 &\le \iint_Q h'(u_n)\nabla u_n \cdot \nabla\phi_n(u_n)\,dxdt \\ &\le \int_\Omega H(u_{0n})\,dx \le \|h\|_{L^\infty(\mathbb{R})}\|u_{0n}\|_{L^1(\Omega)} \le \|h\|_{L^\infty(\mathbb{R})}\|u_0\|_{\mathfrak{R}_f(\Omega)} \end{aligned} \tag{6.67}$$

(observe that $H(u_n(\cdot,T)) \ge 0$ a. e. in Ω, since h is nondecreasing and $h(0) = 0$). By (6.65)–(6.67) we have $S'_{n,\rho} \in L^1(0,T)$ and

$$\int_0^T |S'_{n,\rho}(t)|\,dt \le (C_2\|\rho\|_{C^1(\overline{\Omega})} + \|\rho\|_{C(\overline{\Omega})}\|u_0\|_{\mathfrak{R}_f(\Omega)})\|h\|_{L^\infty(\mathbb{R})}\,T. \tag{6.68}$$

Hence the result follows. □

Lemma 6.5.7. *Let (A_0)–(A_1) hold, and let $\{u_n\}$ be the sequence of solutions of problems (P_n). Then:*

(i) *for every $t \in (0,T)$,*

$$\int_\Omega |\nabla[\phi_n(u_n)]|^2(x,s)\,dx \le \frac{C_2}{t} \quad \text{for all } s \in (t,T] \text{ and } n \in \mathbb{N} \tag{6.69}$$

with $C_2 > 0$ given by Lemma 6.5.5(ii);

(ii) *for all* $t \in (0, T)$ *and* $n \in \mathbb{N}$,

$$\int_t^T \int_\Omega [\phi_n'(u_n)](\partial_\tau u_n)^2 dx d\tau \le \frac{C_2}{2t}. \tag{6.70a}$$

Moreover, if (A_2) *is satisfied, then for every* $t \in (0, T)$,

$$\int_t^T \int_\Omega (\partial_\tau [\phi_n(u_n)])^2 dx d\tau \le \frac{C_2(1 + \|\phi'\|_\infty)}{2t}. \tag{6.70b}$$

Proof. (i) For any fixed $t \in (0, T)$, set $h(s) := \frac{s}{t}\chi_{[0,t)}(s) + \chi_{[t,T)}(s)$ $(s \in (t, T])$. Multiplying $\partial_t u_n = \Delta\phi_n(u_n)$ by $h(t)\partial_t[\phi_n(u_n)]$ and integrating over $Q_s = \Omega \times (0, s)$, for every $s \in (t, T)$, we obtain

$$\begin{aligned} 0 &\le \iint_{Q_s} h(\tau)\phi_n'(u_n)(\partial_\tau u_n)^2 \, dx d\tau \\ &= -\int_0^s d\tau \, h(\tau) \int_\Omega \partial_\tau\{\nabla[\phi_n(u_n)]\} \cdot \nabla[\phi_n(u_n)] \, dx \\ &= -\frac{1}{2}\int_0^s h(\tau)\frac{d}{d\tau}\left(\int_\Omega |\nabla[\phi_n(u_n)]|^2 \, dx\right) d\tau \\ &= -\frac{1}{2}\int_\Omega |\nabla[\phi_n(u_n)]|^2(x, s) \, dx + \frac{1}{2t}\int_0^t d\tau \int_\Omega |\nabla[\phi_n(u_n)]|^2 \, dx. \end{aligned} \tag{6.71}$$

Using (6.63), from the above inequality we get (6.69). Hence claim (i) follows.

(ii) From (6.71) with $s = T$ we get

$$\begin{aligned} \int_t^T \int_\Omega [\phi_n'(u_n)](\partial_\tau u_n)^2 dx d\tau &\le \iint_Q h(\tau)\phi_n'(u_n)(\partial_\tau u_n)^2 \, dx d\tau \\ &\le \frac{1}{2t}\int_0^t d\tau \int_\Omega |\nabla[\phi_n(u_n)]|^2 \, dx \le \frac{1}{2t}\iint_Q |\nabla[\phi_n(u_n)]|^2 dx dt \le \frac{C_2}{2t}. \end{aligned}$$

This proves (6.70a), whence (6.70b) plainly follows. The proof is complete. □

Remark 6.5.3. Since

$$|\nabla[\phi_n(u_n)]|^2 = \left(\phi'(u_n) + \frac{1}{n}\right)^2 |\nabla u_n|^2 \ge (\phi'(u_n))^2 |\nabla u_n|^2 = |\nabla\phi(u_n)|^2,$$

from (6.69), for all $0 < t < s \leq T$, we obtain

$$\int_{\Omega} |\nabla \phi(u_n)|^2(x,s)\, dx \leq \frac{C_2}{t}. \tag{6.72}$$

Similarly, since $|\partial_\tau \phi(u_n)|^2 \leq |\partial_\tau [\phi_n(u_n)]|^2$, from (6.70b) we get

$$\int_t^T \int_{\Omega} (\partial_\tau [\phi(u_n)])^2 dx d\tau \leq \frac{C_2(1 + \|\phi'\|_\infty)}{2t}. \tag{6.73}$$

Lemma 6.5.8. *Let (A_0)–(A_1) hold, and let $\{u_n\}$ be the sequence of solutions of problems (P_n). Then:*

(i) *for all $k > 0$ and $n \in \mathbb{N}$, we have*

$$\iint_{\{\pm u_n > k\}} |\nabla \phi(u_n)|^2\, dxdt \leq \|u_0^\pm\|_{\mathfrak{R}_f(\Omega)} [\pm \phi_{\pm\infty} \mp \phi(\pm k)], \tag{6.74}$$

(ii) *there exists $M_0 > 0$ such that for all $k > 0$ and $n \in \mathbb{N}$,*

$$\iint_{\{\pm u_n > k\}} |\nabla \phi_n(u_n)|^2\, dxdt \leq M_0[\pm \phi_{\pm\infty} \mp \phi(\pm k) + n^{-1/2}]. \tag{6.75}$$

Proof. We only prove (6.74) with the choice of the upper signs, since the other case is similar. For any $k > 0$, set

$$F(y) := \chi_{(k,+\infty)}(y) \int_k^y [\phi(z) - \phi(k)]\, dz \quad (y \in \mathbb{R}).$$

Thus $F' \in W^{1,\infty}_{loc}(\mathbb{R})$, and

$$0 \leq F(y) \leq [\phi_{+\infty} - \phi(k)](y - k)\chi_{(k,+\infty)}(y). \tag{6.76}$$

Multiplying $\partial_t u_n = \Delta \phi_n(u_n)$ by $F'(u_n)$ and integrating in Q, by (6.76) and (6.49a) we obtain

$$\begin{aligned}
&\iint_{\{u_n > k\}} |\nabla \phi(u_n)|^2 + \frac{1}{n} \iint_{\{u_n > k\}} \phi'(u_n)|\nabla u_n|^2\, dxdt \\
&= \int_{\Omega} F(u_{0n}(x))\, dx - \int_{\Omega} F(u_n(x,T))\, dx \\
&\leq [\phi_{+\infty} - \phi(k)] \int_{\{u_{0n} \geq k\}} [u_{0n}(x) - k]\, dx \leq \|u_0^+\|_{\mathfrak{R}_f(\Omega)} [\phi_{+\infty} - \phi(k)].
\end{aligned}$$

Then (6.74) follows. Concerning (6.75), by (6.74) and (6.62) we have that

$$\begin{aligned}\iint\limits_{\{\pm u_n>k\}} |\nabla\phi_n(u_n)|^2\,dxdt &\le 2\iint\limits_{\{\pm u_n>k\}} \left(|\nabla\phi(u_n)|^2 + \frac{|\nabla u_n|^2}{n^2}\right)dxdt \\ &\le 2\|u_0\|_{\mathfrak{R}_f(\Omega)}\,[\pm\phi_{\pm\infty} \mp \phi(\pm k)] + 2C_1 n^{-1/2}\end{aligned}$$

with $C_1 > 0$ as in Lemma 6.5.5(i). Setting $M_0 := 2\max\{\|u_0\|_{\mathfrak{R}_f(\Omega)}, C_1\}$, we obtain (6.75), and thus the result follows. □

6.6 Proof of existence results: letting $n \to \infty$

The proof of Theorem 6.4.1 relies on the following result, whose proof requires several steps.

Theorem 6.6.1. *Let (A_0)–(A_2) hold, and let $\{u_n\}$ be the sequence of solutions of problems (P_n). Then there exists a subsequence $\{u_{n_k}\} \subseteq \{u_n\}$ with the following properties:*

(i) *there exists $u \in L^\infty_{w*}(0,T;\mathfrak{R}_f(\Omega))$ such that $\phi(u_r) \in L^2(0,T;H^1_0(\Omega))$ and*

$$u_{n_k} \overset{*}{\rightharpoonup} u \quad \text{in } L^\infty_{w*}(0,T;\mathfrak{R}_f(\Omega)), \tag{6.77}$$

(ii) *the sequence $\{\nu_{n_k}\} \subseteq \mathfrak{Y}^+(Q;\mathbb{R})$ of the Young measures associated with $\{u_{n_k}\}$ converges narrowly to a Young measure $\nu \in \mathfrak{Y}^+(Q;\mathbb{R})$, and for a. e. $(x,t) \in Q$, we have*

$$u_r(x,t) = \int\limits_{\mathbb{R}} y\,d\nu_{(x,t)}(y), \tag{6.78}$$

$$\operatorname{supp}\nu_{(x,t)} \subseteq \phi^{-1}(\{\phi(u_r(x,t))\}), \tag{6.79}$$

where $\{\nu_{(x,t)}\}_{(x,t)\in Q}$ is the disintegration of ν;

(iii) *for every $f \in C(\mathbb{R})$ satisfying (5.106) and for all $\zeta \in C([0,T];C_c(\Omega))$, we have*

$$\begin{aligned}&\lim_{k\to\infty} \iint\limits_Q f(u_{n_k})\,\zeta\,dxdt \\ &= \iint\limits_Q f^*\zeta\,dxdt + M_f^+ \int\limits_0^T \langle u_s^+(\cdot,t),\zeta(\cdot,t)\rangle dt - M_f^- \int\limits_0^T \langle u_s^-(\cdot,t),\zeta(\cdot,t)\rangle dt,\end{aligned} \tag{6.80}$$

where $f^* \in L^\infty(0,T;L^1(\Omega))$ is defined by

$$f^*(x,t) := \int\limits_{\mathbb{R}} f(y)\,d\nu_{(x,t)}(y) \quad \text{for a. e. } (x,t) \in Q. \tag{6.81}$$

To prove Theorem 6.6.1, we need the following result.

Proposition 6.6.2. *Let* (A_0)–(A_2) *hold, and let* $\{u_n\}$ *be the sequence of solutions of problems* (P_n). *Then there exist a subsequence* $\{u_{n_k}\} \subseteq \{u_n\}$ *and a Young measure* $\nu \in \mathfrak{Y}^+(Q;\mathbb{R})$ *such that:*

$$\phi(u_{n_k}) \overset{*}{\rightharpoonup} \phi^* \quad \text{in } L^\infty(Q), \quad \phi(u_{n_k}) \rightharpoonup \phi^* \quad \text{in } L^2(Q), \tag{6.82}$$

where

$$\phi^*(x,t) := \int_{\mathbb{R}} \phi(y)\, d\nu_{(x,t)}(y) \in L^2(0,T;H_0^1(\Omega)), \tag{6.83}$$

and $\{\nu_{(x,t)}\}_{(x,t)\in Q}$ *is the disintegration of* ν;

$$\nabla\phi(u_{n_k}),\ \nabla\phi_{n_k}(u_{n_k}) \rightharpoonup \nabla\phi^* \quad \text{in } [L^2(Q)]^N; \tag{6.84}$$

for any $\tau \in (0,T)$,

$$\partial_t[\phi(u_{n_k})],\ \partial_t[\phi_{n_k}(u_{n_k})] \rightharpoonup \partial_t\phi^* \quad \text{in } L^2(\Omega\times(\tau,T)), \tag{6.85}$$

$$\phi(u_{n_k}),\ \phi_{n_k}(u_{n_k}) \rightharpoonup \phi^* \quad \text{in } H^1(\Omega\times(\tau,T)), \tag{6.86}$$

$$\phi(u_{n_k}),\ \phi_{n_k}(u_{n_k}) \to \phi^* \quad \text{a.e. in } Q; \tag{6.87}$$

and for a. e. $t \in (0,T)$,

$$\phi(u_{n_k}(\cdot,t)),\ \phi_{n_k}(u_{n_k}(\cdot,t)) \to \phi^*(\cdot,t) \quad \text{a.e. in } \Omega, \tag{6.88}$$

$$\phi(u_{n_k}(\cdot,t)) \rightharpoonup \phi^*(\cdot,t) \quad \text{in } H_0^1(\Omega). \tag{6.89}$$

Proof. Since the sequence $\{\phi(u_n)\}$ is bounded in $L^\infty(Q)$, by the Banach–Alaoglu theorem there exist a subsequence $\{u_{n_k}\} \subseteq \{u_n\}$ and $v \in L^\infty(Q)$ such that $\phi(u_{n_k}) \overset{*}{\rightharpoonup} v$ in $L^\infty(Q)$. At the same time the sequence $\{u_n\}$ is bounded in $L^1(Q)$, and hence its associated sequence of Young measures $\{\nu_n\}$ is relatively compact in the narrow topology of Young measures over $Q\times\mathbb{R}$, and the existence of the limiting Young measure ν follows at once. Moreover, since Q is bounded, the subsequence $\{\phi(u_{n_k})\}$ is bounded in $L^1(Q)$ and uniformly integrable (see Lemma 2.8.12). Hence by Proposition 5.4.10 there exist a subsequence of $\{\phi(u_{n_k})\}$ (not relabeled) such that $\phi(u_{n_k}) \rightharpoonup \phi^*$ in $L^1(Q)$ with ϕ^* as in (6.83). Then $v = \phi^*$, and (6.82) follows.

Concerning (6.84), observe that by (6.57) and (6.63) there exist two subsequences $\{\nabla\phi(u_{n_k})\}$ and $\{\nabla\phi_{n_k}(u_{n_k})\}$ weakly convergent in $[L^2(Q)]^N$. By (6.82) we obtain that $\nabla\phi(u_{n_k}) \rightharpoonup \nabla\phi^*$ in $[L^2(Q)]^N$. Since

$$|\phi_{n_k}(u_{n_k}) - \phi(u_{n_k})| = \frac{|u_{n_k}|}{n_k} \to 0 \quad \text{in } L^1(Q) \tag{6.90}$$

(see (6.56)), we also get the second convergence in (6.84). By (6.84) the second convergence in (6.82) and the semicontinuity of the norm we have $\phi^* \in L^2(0,T;H_0^1(\Omega))$.

Let us prove (6.85). By the second convergence in (6.82), for all $\zeta \in C_c^1(Q)$, we have

$$\lim_{k\to\infty} \iint_Q \partial_t[\phi(u_{n_k})]\,\zeta\, dxdt = -\lim_{k\to\infty} \iint_Q \phi(u_{n_k})\,\partial_t\zeta\, dxdt = -\iint_Q \phi^*\,\partial_t\zeta\, dxdt. \tag{6.91}$$

On the other hand, by (6.70b) for any $\tau \in (0,T)$, the sequence $\{\partial_t[\phi(u_{n_k})]\}$ is bounded in $L^2(\Omega\times(\tau,T))$, and thus there exist a subsequence of $\{\partial_t[\phi(u_{n_k})]\}$ (not relabeled) and $g_\tau \in L^2(\Omega\times(\tau,T))$ such that for all $\zeta \in L^2(\Omega\times(\tau,T))$,

$$\lim_{k\to\infty} \iint_{\Omega\times(\tau,T)} \partial_t[\phi(u_{n_k})]\,\zeta\, dxdt = \iint_{\Omega\times(\tau,T)} g_\tau\,\zeta\, dxdt. \tag{6.92}$$

In view of (6.91)–(6.92), for every $\zeta \in C_c^1(\Omega\times(\tau,T))$, we have

$$\iint_{\Omega\times(\tau,T)} g_\tau\,\zeta\, dxdt = -\iint_Q \phi^*\,\partial_t\zeta\, dxdt. \tag{6.93}$$

Then the distributional derivative $\partial_t\phi^*$ can be identified with an element of $L^2(\Omega\times(\tau,T))$, and thus the first convergence in (6.85) follows. The second is similarly proven.

The convergence in (6.86) is obvious by (6.84)–(6.85). Hence both sequences $\{\phi(u_{n_k})\}$ and $\{\phi_{n_k}(u_{n_k})\}$ are bounded in $H^1(\Omega\times(\tau,T))$. Then by embedding results (possibly extracting a subsequence, not relabeled for simplicity), for every $\tau \in (0,T)$, we have $\phi_{n_k}(u_{n_k})$, $\phi(u_{n_k}) \to \phi^*$ a. e. in $\Omega\times(\tau,T)$. By the arbitrariness of τ and a standard diagonal argument we can construct a subsequence (not relabeled for simplicity) along which both convergences in (6.87) take place.

Finally, the convergences in (6.88) follow from (6.87), whereas (6.89) is a plain consequence of (6.88) and (6.69). This completes the proof. □

Proposition 6.6.3. *Let (A_0)–(A_1) hold, let $\{u_n\}$ be the sequence of solutions of problems (P_n), and let $\nu \in \mathfrak{Y}^+(Q;\mathbb{R})$ be the Young measure given in Proposition 6.6.2. Then there exist a subsequence $\{u_{n_k}\} \subseteq \{u_n\}$ and $\sigma_\pm \in L^\infty_{w^*}(0,T;\mathfrak{R}_f^+(\Omega))$ such that*

$$u_{n_k}^\pm \overset{*}{\rightharpoonup} u_{b,\pm} + \sigma_\pm \quad \text{in } L^\infty_{w*}(0,T;\mathfrak{R}_f(\Omega)), \tag{6.94}$$

where $u_{b,\pm} \in L^\infty(0,T;L^1(\Omega))$,

$$u_{b,\pm}(x,t) := \int_{\mathbb{R}} y^\pm\, d\nu_{(x,t)}(y) \ge 0 \quad \text{for a. e. } (x,t) \in Q. \tag{6.95}$$

Proof. By inequality (6.54) the sequence $\{u_n\}$ is bounded in $L^\infty(0,T;L^1(\Omega)) \subseteq L^\infty_{w^*}(0,T;\mathfrak{R}_f(\Omega))$. Then by Proposition 4.4.16 and the Banach–Alaoglu theorem there exist $\mu_\pm \in L^\infty_{w^*}(0,T;\mathfrak{R}^+_f(\Omega))$ such that

$$u^\pm_{n_k} \overset{*}{\rightharpoonup} \mu_\pm \quad \text{in } L^\infty_{w*}(0,T;\mathfrak{R}_f(\Omega)). \tag{6.96}$$

On the other hand, since the sequence $\{u_{n_k}\}$ is bounded in $L^1(Q)$, by Remark 5.4.4 there exist a subsequence of $\{u_{n_k}\}$ (not relabeled) and $\sigma_\pm \in \mathfrak{R}^+_f(Q)$ such that

$$u^\pm_{n_k} \overset{*}{\rightharpoonup} u_{b,\pm} + \sigma_\pm \quad \text{in } \mathfrak{R}_f(Q). \tag{6.97}$$

By (6.96)–(6.97) $\mu_\pm = u_{b,\pm} + \sigma_\pm$, and hence the result follows. □

Proposition 6.6.4. *Let (A_0)–(A_1) hold, and let $\{u_n\}$ be the sequence of solutions of problems (P_n). Let $\{u_{n_k}\} \subseteq \{u_n\}$, $\nu \in \mathfrak{Y}^+(Q;\mathbb{R})$, and $\sigma_\pm \in L^\infty_{w^*}(0,T;\mathfrak{R}^+_f(\Omega))$ be given by Proposition 6.6.3. Then for every $f \in C(\mathbb{R})$ satisfying (5.106) and for all $\zeta \in C([0,T];C_c(\Omega))$, we have*

$$\lim_{k\to\infty} \iint_Q f(u_{n_k})\zeta\,dxdt$$
$$= \iint_Q f^*\zeta\,dxdt + M^+_f \int_0^T \langle\sigma_+(\cdot,t),\zeta(\cdot,t)\rangle dt - M^-_f \int_0^T \langle\sigma_-(\cdot,t),\zeta(\cdot,t)\rangle dt \tag{6.98}$$

with $f^* \in L^\infty(0,T;L^1(\Omega))$ defined by

$$f^*(x,t) := \int_{\mathbb{R}} f(y)\,d\nu_{(x,t)}(y) \quad \text{for a.e. } (x,t) \in Q. \tag{6.99}$$

Proof. By (5.106) there exists $L > 0$ such that $|f(y)| \le L(1+|y|)$ for all $y \in \mathbb{R}$, whence by inequality (6.54)

$$\sup_{n\in\mathbb{N}} \|f(u_n)\|_{L^\infty(0,T;L^1(\Omega))} = \sup_{n\in\mathbb{N}} \|f(u_n)\|_{L^\infty_{w*}(0,T;\mathfrak{R}_f(\Omega))} \le L[\lambda_N(\Omega) + \|u_0\|_{\mathfrak{R}_f(\Omega)}].$$

Then by the Banach–Alaoglu theorem there exist a subsequence $\{f(u_{n_k})\} \subseteq \{f(u_n)\}$ and $\mu \in L^\infty_{w^*}(0,T;\mathfrak{R}_f(\Omega))$ such that

$$f(u_{n_k}) \overset{*}{\rightharpoonup} \mu \quad \text{in } L^\infty_{w*}(0,T;\mathfrak{R}_f(\Omega)). \tag{6.100}$$

Moreover, arguing as in the proof of Proposition 5.4.13, we obtain that

$$f(u_{n_k}) \overset{*}{\rightharpoonup} f^* + M^+_f\sigma_+ - M^-_f\sigma_- \quad \text{in } \mathfrak{R}_f(Q). \tag{6.101}$$

By (6.100)–(6.101) the conclusion follows. □

Set

$$u_b(x,t) := u_{b,+}(x,t) - u_{b,-}(x,t) = \int_{\mathbb{R}} y\, d\nu_{(x,t)}(y) \quad \text{for a. e. } (x,t) \in Q, \tag{6.102a}$$

$$u := u_b + \sigma_+ - \sigma_- . \tag{6.102b}$$

By Propositions 6.6.3 and 6.6.4 we have the following result.

Proposition 6.6.5. *Let (A_0)–(A_1) hold, and let $\{u_n\}$ be the sequence of solutions of problems (P_n). Let $\{u_{n_k}\} \subseteq \{u_n\}$ and $\sigma_\pm \in L^\infty_{w^*}(0,T;\mathfrak{R}_f^+(\Omega))$ be given by Proposition 6.6.3, and let $u \in L^\infty_{w^*}(0,T;\mathfrak{R}_f(\Omega))$ be defined by (6.102b). Then the convergence in (6.77) holds. Moreover,*

$$u_r = u_b \quad a.e.\ in\ Q, \tag{6.103a}$$

$$u_s^\pm = \sigma_\pm \quad in\ L^\infty_{w^*}(0,T;\mathfrak{R}_f(\Omega)). \tag{6.103b}$$

Proof. The convergence in (6.77) is a direct consequence of (6.94)–(6.95) and (6.102). Moreover, the remaining claims (6.103a)–(6.103b) will follow if we prove that $\sigma_\pm$ are both mutually singular and singular with respect to the Lebesgue measure.

To this aim, we shall prove the following:

Claim. For any $k \in \mathbb{N}$, $p > 0$, and $\zeta \in C^1([0,T];C_c^2(\Omega))$ with $\zeta \geq 0$ in Q and $\zeta(\cdot,T) = 0$, we have

$$-\iint_Q [u_{n_k} \mp p]^\pm \partial_t \zeta \, dxdt \leq \int_\Omega [u_{0n_k} \mp p]^\pm \zeta(x,0)\, dx + \|\zeta\|_{C^1(\overline{\Omega})} \left(\frac{C_2\, T\, \|u_0\|_{\mathfrak{R}_f(\Omega)}}{p} \right)^{\frac{1}{2}}, \tag{6.104}$$

where $C_2 = C_2(\lambda_N(\Omega), \|\phi\|_\infty, \|u_0\|_{\mathfrak{R}_f(\Omega)}) > 0$ is the constant given in Lemma 6.5.5(ii).

Relying on the above claim, we can prove that

$$\sigma_\pm(\cdot,t) \leq u_{0s}^\pm \quad \text{in } \mathfrak{R}_f^+(\Omega) \text{ for a. e. } t \in (0,T). \tag{6.105}$$

Since the measures $\sigma_\pm$ are nonnegative, this implies that $\sigma_\pm(\cdot,t)$ are absolutely continuous with respect to $u_{0s}^\pm$ for a. e. $t \in (0,T)$. Hence the conclusion follows.

We only prove (6.105) with "+", the proof with "−" being analogous. Using (6.51) with $f(y) = [y-p]^+$ $(y \in \mathbb{R},\ p > 0)$ gives

$$[u_{0n_k} - p]^+ \overset{*}{\rightharpoonup} [u_{0r} - p]^+ + u_{0s}^+ \quad \text{in } \mathfrak{R}_f(\Omega), \tag{6.106}$$

whereas with the same choice of f from (6.98)–(6.99), we get

$$\lim_{k\to\infty} \iint_Q [u_{n_k} - p]^+ \partial_t \zeta\, dxdt = \iint_Q \left(\int_{\mathbb{R}} [y-p]^+ d\nu_{(x,t)}(y) \right) \partial_t \zeta\, dxdt + \int_0^T \langle \sigma_+(\cdot,t), \partial_t \zeta(\cdot,t) \rangle dt$$

for all $\zeta \in C^1([0,T]; C_c^2(\Omega))$. By this equality, choosing the upper signs in (6.104), letting $k \to \infty$, and using (6.106), we get

$$-\iint_Q \left(\int_{\mathbb{R}} [y-p]^+ d\nu_{(x,t)}(y) \right) \partial_t \zeta \, dxdt - \int_0^T \langle \sigma_+(\cdot,t), \partial_t \zeta(\cdot,t) \rangle dt$$

$$\leq \int_\Omega [u_{0r} - p]^+ \zeta(x,0)\, dx + \langle u_{0s}^+, \zeta(\cdot,0) \rangle + \|\zeta\|_{C^1(\overline{\Omega})} \left(\frac{C_2 T \|u_0\|_{\mathfrak{R}_f(\Omega)}}{p} \right)^{\frac{1}{2}} \tag{6.107}$$

for every $\zeta \in C^1([0,T]; C_c^2(\Omega))$ with $\zeta \geq 0$ in Q, $\zeta(\cdot, T) = 0$ in Ω.

Since $\lim_{p\to+\infty} [y-p]^+ = 0$ for every $y \in \mathbb{R}$ and $[y-p]^+ \leq y^+ \in L^1(\mathbb{R}, \mathcal{B}, \nu_{(x,t)})$ for a. e. $(x,t) \in Q$ (see Proposition 5.3.1-(*i*)), by the dominated convergence theorem we have that

$$\lim_{p\to\infty} \int_{\mathbb{R}} [y-p]^+ d\nu_{(x,t)}(y) = 0 \quad \text{for a. e. } (x,t) \in Q.$$

Moreover, for a. e. $(x,t) \in Q$, we have $\int_{\mathbb{R}} [y-p]^+ d\nu_{(x,t)}(y) \leq \int_{\mathbb{R}} y^+ d\nu_{(x,t)}(y) = u_{b,+}(x,t)$ (see (6.95)). Since $u_{b,+} \in L^\infty(0,T; L^1(\Omega)) \subseteq L^1(Q)$, by the dominated convergence theorem we get

$$\lim_{p\to\infty} \iint_Q \left(\int_{\mathbb{R}} [y-p]^+ d\nu_{(x,t)}(y) \right) \partial_t \zeta \, dxdt = 0. \tag{6.108a}$$

It is similarly seen that

$$\lim_{p\to\infty} \int_\Omega [u_{0r} - p]^+ \zeta(x,0)\, dx = 0. \tag{6.108b}$$

By (6.108), as above, letting $p \to \infty$ in (6.107), for every ζ, we get

$$-\int_0^T \langle \sigma_+(\cdot,t), \partial_t \zeta(\cdot,t) \rangle dt \leq \langle u_{0s}^+, \zeta(\cdot,0) \rangle. \tag{6.109}$$

Let $t_0 \in (0,T)$, and for any $q > 0$ sufficiently large, set

$$h_q(t) := \chi_{[0,t_0-\frac{1}{q}]}(t) + q(t_0 - t)\chi_{(t_0-\frac{1}{q},t_0]}(t) \quad (t \in (0,T)).$$

By standard approximation arguments we can choose in (6.109) $\zeta(x,t) = \rho(x)\,h_q(t)$ with $\rho \in C_c^2(\Omega)$, $\rho \geq 0$. Then we get

$$q \int_{t_0-\frac{1}{q}}^{t_0} \langle \sigma_+(\cdot,t),\rho\rangle dt \leq \langle u_{0s}^+,\rho\rangle,$$

whence, as $q \to \infty$,

$$\langle \sigma_+(\cdot,t_0),\rho\rangle \leq \langle u_{0s}^+,\rho\rangle \quad \text{for a. e. } t_0 \in (0,T),.$$

Then by the arbitrariness of ρ inequality (6.105) with "+" follows.

Lets us finally prove the claim. By standard approximation arguments, for every $p > 0$, we can choose $h = \chi_{(p,\infty)}$ in (6.58), and thus $H(y) = [y-p]^+$ in (6.59). Then for every nonnegative $\zeta \in C^1([0,T]; C_c^1(\Omega))$ with $\zeta(\cdot,T) = 0$ in Ω, we get (see (6.60))

$$-\iint_Q [u_{n_k} - p]^+ \partial_t\zeta\, dxdt \leq \int_\Omega [u_{0n_k} - p]^+ \zeta(x,0)\, dx - \iint_{\{u_{n_k} > p\}} \nabla\phi_{n_k}(u_{n_k}) \cdot \nabla\zeta\, dxdt. \quad (6.110)$$

By Lemma 6.5.5(ii) there exists $C_2 = C_2(\lambda_N(\Omega), \|\phi\|_\infty, \|u_0\|_{\mathfrak{R}_f(\Omega)}) > 0$ such that

$$\iint_{\{u_{n_k} > p\}} |\nabla\phi_{n_k}(u_{n_k})|\,|\nabla\zeta|\, dxdt \leq C_2^{\frac{1}{2}} \|\zeta\|_{C^1(\overline{Q})} [\lambda_{N+1}(\{u_{n_k} > p\})]^{\frac{1}{2}},$$

whence by the Chebyshev inequality

$$\iint_{\{u_{n_k} > p\}} |\nabla\phi_{n_k}(u_{n_k})|\,|\nabla\zeta|\, dxdt \leq \|\zeta\|_{C^1(\overline{\Omega})} \left(\frac{C_2\,\|u_{n_k}\|_{L^1(Q)}}{p}\right)^{\frac{1}{2}}. \quad (6.111)$$

From (6.54) and (6.110)–(6.111) we obtain (6.104) with the upper signs. Hence the result follows. □

Proposition 6.6.6. *Let (A_0)–(A_2) hold, and let $\{u_n\}$ be the sequence of solutions of problems (P_n). Let $\{u_{n_k}\} \subseteq \{u_n\}$ and $\nu \in \mathfrak{Y}^+(Q;\mathbb{R})$ be given by Proposition 6.6.3, and let $u \in L^\infty_{w^*}(0,T;\mathfrak{R}_f(\Omega))$ be defined by (6.102b). Then for a. e. $(x,t) \in Q$:*

(i) *we have*

$$\operatorname{supp} \nu_{(x,t)} \subseteq \phi^{-1}(\{\phi^*(x,t)\}) \quad (6.112)$$

with ϕ^* given by (6.83);

(ii) *we have*

$$\phi^*(x,t) = \phi(u_r(x,t)), \quad \phi(u_r) \in L^2(0,T; H_0^1(\Omega)), \quad (6.113)$$

$$\operatorname{supp} \nu_{(x,t)} \subseteq \phi^{-1}(\{\phi(u_r(x,t))\}). \tag{6.114}$$

Proof. For every $f \in C(\mathbb{R})$, the sequence $\{(f \circ \phi)(u_{n_k})\}$ is bounded in $L^1(Q)$ and uniformly integrable, and thus (possibly extracting a subsequence, not relabeled) by Proposition 5.4.10 we get

$$f(\phi(u_{n_k})) \rightharpoonup (f \circ \phi)^* \quad \text{in } L^1(Q), \quad (f \circ \phi)^*(x,t) := \int_{\mathbb{R}} f(\phi(y))\, d\nu_{(x,t)}(y). \tag{6.115}$$

Now observe that by the convergence in (6.87) we have

$$f(\phi(u_{n_k})) \to f(\phi^*) \quad \text{a. e. in } Q,$$

whence (recall that ϕ is bounded)

$$f(\phi(u_{u_n})) \to f(\phi^*) \quad \text{in } L^1(Q).$$

From the above convergence and (6.115) we get

$$f(\phi^*(x,t)) = \int_{\mathbb{R}} f(\phi(y))\, d\nu_{(x,t)}(y) \quad \text{for all } (x,t) \in Q \setminus N, \tag{6.116}$$

where $N \subseteq Q$ is a null set.

Fix $(x,t) \in Q \setminus N$, and let $B = \mathbb{R} \setminus \phi^{-1}(\{\phi^*(x,t)\})$. Choosing $f \in C(\mathbb{R})$ such that $f(z) > 0$ for all $z \neq \phi^*(x,t)$ and $f(z) = 0$ if $z = \phi^*(x,t)$, from (6.116) we obtain

$$0 = \int_B f(\phi(y))\, d\nu_{(x,t)}(y).$$

Since $f(\phi(y)) > 0$ for all $y \in B$, the above equality ensures that $\nu_{(x,t)}(B) = 0$. Since $B = \mathbb{R} \setminus \phi^{-1}(\{\phi^*(x,t)\})$ is open, claim (i) follows.

Concerning (ii), observe that by (6.102a), (6.103a), and (6.112)

$$u_r(x,t) = \int_{\mathbb{R}} y\, d\nu_{(x,t)}(y) = \int_{\phi^{-1}(\{\phi^*(x,t)\})} y\, d\nu_{(x,t)}(y) \in \phi^{-1}(\{\phi^*(x,t)\}),$$

since $\phi^{-1}(\{\phi^*(x,t)\})$ is a closed interval and $\nu_{(x,t)}$ is a probability measure. Then for a. e. $(x,t) \in Q$, we have $\phi(u_r(x,t)) = \phi^*(x,t)$, whence the second statement in (6.113) follows, since by (6.82) and (6.84) $\phi^* \in L^2(0,T; H_0^1(\Omega))$. Since $\phi(u_r) = \phi^*$, from (6.112) we obtain (6.114). This completes the proof. □

Proof of Theorem 6.6.1. Let $u \in L^\infty_{w^*}(0,T; \mathfrak{R}_f(\Omega))$ be defined by (6.102b). Then by Propositions 6.6.3–6.6.6 the result follows. □

Now we can prove Theorem 6.4.1.

Proof of Theorem 6.4.1. Let us show that the measure $u \in L^{\infty}_{w*}(0,T;\mathfrak{R}_f(\Omega))$ given by Theorem 6.6.1 is a weak solution of problem (P). Since $\phi(u_r) \in L^2(0,T;H^1_0(\Omega))$, condition (i) of Definition 6.1.1 is satisfied. Let $\{u_{n_k}\}$ be a subsequence of solutions of approximating problems such that $u_{n_k} \overset{*}{\rightharpoonup} u$ in $L^{\infty}_{w*}(0,T;\mathfrak{R}_f(\Omega))$ (see (6.77)). Consider the weak formulation of (P_{n_k}),

$$\iint_Q \{u_{n_k}\partial_t\zeta - \nabla\phi_{n_k}(u_{n_k})\cdot\nabla\zeta\}\,dxdt = -\int_\Omega u_{0n_k}(x)\,\zeta(x,0)\,dx$$

with $\zeta \in C^1([0,T];C^1_c(\Omega))$ such that $\zeta(\cdot,T) = 0$ in Ω. Letting $k \to \infty$ and using (6.77), (6.84) with (6.113), and (6.49c) we get (6.2). Then the first part of the result follows.

Let us prove claims (i) and (ii) with the upper sign "+". To this purpose, observe that by (6.103b) and inequality (6.109)

$$-\int_0^T \langle u^+_s(\cdot,t),\zeta(\cdot,t)\rangle\,dt \le \langle u^+_{0s},\zeta(\cdot,0)\rangle$$

for all $\rho \in C([0,T];C_c(\Omega)$ such that $\zeta \ge 0$ in Q and $\zeta(\cdot,T) = 0$ in Ω. Set $\zeta(x,t) = \rho(x)h(t)$ with $\rho \in C_c(\Omega)$, $\rho \ge 0$, in the previous inequality. As in the proof of Theorem 6.4.5, by proper choices of the function h both claims (i) and (ii) easily follow. □

Assume that $\pm\phi_{\pm\infty} > 0$ and let $f_{c,\pm} : [0,\pm\phi_{\pm\infty}] \to [0,1]$ and $f_c \in C^1([0,\pm\phi_{\pm\infty}])$ satisfy

$$f_{c,\pm}(y) = \begin{cases} 0 & \text{if } y \in [0,c], \\ 1 & \text{if } y \in [\frac{c\pm\phi_{\pm\infty}}{2},\pm\phi_{\pm\infty}] \end{cases} \tag{6.117}$$

for some $c \in (0,\pm\phi_{+\infty})$. To prove Theorem 6.4.2, we need the following lemma.

Lemma 6.6.7. *Let* (A_0)–(A_2) *hold with* $\pm\phi_{\pm\infty} > 0$, *and let u be the weak solution of problem (P) given by Theorem 6.4.1. Then for a. e.* $t \in (0,T)$, *for any* $c \in (0,\pm\phi_{\pm\infty})$, *and for any* $\rho \in C^1_c(\Omega)$, $\rho \ge 0$, *we have*

$$\langle [u^\pm_s(\cdot,t)]_{d,2},\rho\rangle \le \langle [u^\pm_{0s}]_{d,2}, f_c(\phi^\pm(u_r(\cdot,t)))\rho\rangle. \tag{6.118}$$

Proof of Theorem 6.4.2. We only prove equality (6.21a) with "+", the proof with "−" being analogous. Observe that $\phi_{+\infty} = 0$ implies $\phi(y) = 0$ for all $y \in [0,\infty)$, and thus in this case, $\phi^+ \equiv 0$, $S^t_+ = \Omega$, and (6.21) is obviously satisfied.

Let $\phi_{+\infty} > 0$, and set $f_c \equiv f_{c,+}$ for simplicity (see (6.117)). Fix any $t \in (0,T)$ such that

$$\langle [u^+_s(\cdot,t)]_{d,2},\rho\rangle \le \langle [u^+_{0s}]_{d,2}, f_c(\phi^+(u_r(\cdot,t)))\rho\rangle$$

for all nonnegative $\rho \in C_c^1(\Omega)$, and let $g_c \in C^1(\mathbb{R})$ be any function such that $g_c \geq 0$ and $\operatorname{supp} g_c \subseteq [0,c)$. Then by the definition of f_c (see (6.117)) we have $f_c g_c = 0$ in $[0,\infty)$. Now recall that for a. e. $t \in (0,T)$, the function $\phi^+(u_r(\cdot,t))$ is identified with its 2-quasi-continuous representative, which is defined up to 2-null sets (see Remark 6.3.1). Hence for a. e. $t \in (0,T)$ and for any $c \in (0,\phi_{+\infty})$, we have

$$f_c(\phi^+(u_r(x,t)))\,g_c(\phi^+(u_r(x,t))) = 0 \quad \text{for all } x \in \Omega_*, \tag{6.119}$$

for some subset $\Omega_* \subseteq \Omega$ such that $C_2(\Omega \setminus \Omega_*) = 0$.

Since $g_c((\phi^+(u_r)(\cdot,t)) \in H^1(\Omega) \cap L^\infty(\Omega)$, the function $g_c((\phi^+(u_r)(\cdot,t))\tilde{\rho}$ with $\tilde{\rho} \in C_c^1(\Omega)$, $\tilde{\rho} \geq 0$, belongs to $H_0^1(\Omega) \cap L^\infty(\Omega)$ and thus can be used as a test function in inequality (6.118). Then by (6.119) we obtain

$$\begin{aligned}&\langle [u_s(x,t)]_{d,2}, g_c(\phi^+(u_r(x,t)))\tilde{\rho}\rangle\\&\quad \leq \langle [u_{0s}^{\pm}]_{d,2}, f_c(\phi^{\pm}(u_r(x,t)))\,g_c(\phi^+(u_r(x,t)))\tilde{\rho}\rangle = 0\end{aligned} \tag{6.120}$$

for every g_c and $\tilde{\rho}$ as above and $x \in \Omega_*$.

Choose $c = c_q := \phi_{+\infty} - \frac{1}{q}$, so that $\operatorname{supp} g_{c_q} \subseteq [0, \phi_{+\infty} - \frac{1}{q})$ $(q \in \mathbb{N})$. Plainly, this implies that

$$[u_s(\cdot,t)]_{d,2}(E_q^t) = 0, \tag{6.121}$$

where

$$E_q^t := \left\{x \in \Omega_* \mid \phi^+(u_r(x,t)) \leq \phi_{+\infty} - \frac{2}{q}\right\}.$$

Since $E_q^t \subseteq E_{q+1}^t$ for all $q \in \mathbb{N}$ and

$$F^t := \{x \in \Omega_* \mid \phi^+(u_r(x,t)) < \phi_{+\infty}\} = \bigcup_{q=1}^{\infty} E_q^t,$$

from equality (6.121) we obtain

$$[u_s(\cdot,t)]_{d,2}(F^t) = \lim_{q\to\infty} [u_s(\cdot,t)]_{d,2}(E_q^t) = 0.$$

It follows that

$$[u_s(\cdot,t)]_{d,2} \llcorner \Omega_* = [u_s(\cdot,t)]_{d,2} \llcorner \{x \in \Omega_* \mid \phi^+(u_r(x,t)) = \phi_{+\infty}\}. \tag{6.122}$$

On the other hand, $[u_s(\cdot,t)]_{d,2}(\Omega \setminus \Omega_*) = 0$ since $C_2(\Omega \setminus \Omega_*) = 0$. Therefore by (6.122) the result follows. □

Proof of Lemma 6.6.7. We only prove the result with "+" in (6.21a), the proof in the other case being analogous. Let $\phi_{+\infty} \in (0,\infty)$. For every $m \in \mathbb{N}$, set

$$h_m(y) := (y-m)\chi_{[m,m+1]}(y) + \chi_{(m+1,+\infty)}(y), \quad H_m(y) := \int_0^y h_m(z)\,dz \quad (y \in \mathbb{R}).$$

Let $\{u_{n_k}\}$ and u be given by Theorem 6.6.1. Since $\|h_m\|_\infty = 1$, by (6.54) and (6.64) for all $\rho \in C_c^1(\Omega)$ and $m \in \mathbb{N}$, the sequence $\{\int_\Omega H_m(u_{n_k}(x,\cdot))\rho(x)\,dx\}$ is bounded in $BV(0,T)$. Then there exist a null set $N_1 \subseteq (0,T)$ and a subsequence of $\{u_{n_k}\}$ (not relabeled for simplicity) such that the sequence $\{\int_\Omega H_m(u_{n_k}(x,t))\rho(x)\,dx\}$ converges in $L^1((0,T))$ for all $t \in (0,T) \setminus N_1$. By the separability of $C_0(\overline{\Omega})$ and a standard diagonal argument the null set N_1 can be chosen independently both of $\rho \in C_c^1(\Omega)$ and of $m \in \mathbb{N}$. Since $\{\int_\Omega H_m(u_{n_k}(x,t))\rho(x)\,dx\}$ converges for $t \in (0,T)\setminus N_1$, since $\lim_{y\to+\infty} \frac{H_m(y)}{y} = 1$, and $\lim_{y\to-\infty} \frac{H_m(y)}{y} = 0$, by (6.98) with $\zeta(x,t) = \rho(x)h(t)$ and (6.103b) we obtain, for all $t \in (0,T)\setminus N_1$, $\rho \in C_c^1(\Omega)$, and $m \in \mathbb{N}$,

$$\lim_{k\to\infty} \int_\Omega H_m(u_{n_k}(x,t))\,\rho\,dx = \int_\Omega \Big(\int_{\mathbb{R}} H_m(y)\,d\nu_{(x,t)}(y) \Big) \rho(x)\,dx + \langle u_s^+(\cdot,t), \rho \rangle. \tag{6.123}$$

In view of (6.88)–(6.89) and (6.113), there exists a null set $N_2 \subseteq (0,T)$ such that for all $t \in (0,T)\setminus N_2$,

$$\phi(u_{n_k}(\cdot,t)) \to \phi^+(u_r(\cdot,t)) \quad \text{a. e. in } \Omega, \tag{6.124a}$$

$$\phi(u_{n_k}(\cdot,t)) \rightharpoonup \phi^+(u_r(\cdot,t)) \quad \text{in } H_0^1(\Omega). \tag{6.124b}$$

Set $N := N_1 \cup N_2$, and fix $t_0 \in (0,T)\setminus N$. Observe that for every $m \in \mathbb{N}$ large enough, we have

$$H_m(s)[1 - f_c(\phi(s))] = 0 \quad \text{for every } s \in \mathbb{R}$$

with $f_c \equiv f_{c,+}$ given by (6.117). Then for all $m \in \mathbb{N}$ large enough and $\rho \in C_c^1(\Omega)$, $\rho \geq 0$, by (6.60), (6.62), (6.74), and (6.124) we have

$$\begin{aligned}
&\int_\Omega H_m(u_{n_k}(x,t_0))\,\rho(x)\,dx \\
&= \int_\Omega H_m(u_{n_k}(x,t_0)) f_c(\phi(u_{n_k}(x,t_0)))\,\rho(x)\,dx \\
&\leq \Big| \iint_{\{u_{n_k} > m\}} h_m(u_{n_k}) \nabla\phi_{n_k}(u_{n_k}) \cdot \nabla(f_c(\phi(u_{n_k}(x,t_0)))\rho)\,dxdt \Big|
\end{aligned}$$

$$
\begin{aligned}
&+ \int_\Omega H_m(u_{0n_k}) f_c(\phi(u_{n_k}(x,t_0)))\,\rho(x)\,dx \\
&\le \|f_c(\phi(u_{n_k}(\cdot,t_0)))\rho\|_{H_0^1(\Omega)} \Bigg(\iint_{\{u_{n_k}>m\}} |\nabla\phi_{n_k}(u_{n_k})|^2\,dxdt \Bigg)^{\frac{1}{2}} \\
&\quad + \int_\Omega H_m(u_{0n_k}) f_c(\phi(u_{n_k}(x,t_0)))\,\rho(x)\,dx \\
&\le M_c[\phi_{+\infty} - \phi(m) + n_k^{-1/2}]^{\frac{1}{2}} + \int_\Omega [u_0^+]_{n_k} f_c(\phi(u_{n_k}(x,t_0)))\rho(x)\,dx \\
&\le M_c[\phi_{+\infty} - \phi(m) + n_k^{-1/2}]^{\frac{1}{2}} + \int_\Omega [\delta^+]_{n_k} f_c(\phi(u_{n_k}(x,t_0)))\rho(x)\,dx + \int_\Omega [\gamma^+]_{n_k}\,\rho(x)\,dx.
\end{aligned}
$$

The above constant M_c is chosen so that $M_0\|f_c(\phi(u_{n_k}(\cdot,t_0)))\rho\|_{H_0^1(\Omega)} \le M_c$ for all $k \in \mathbb{N}$, where M_0 is the constant in (6.75). By the second convergence in (6.49d), (6.49e), and (6.123)–(6.124), letting first $k \to +\infty$ and then $m \to +\infty$ in the previous inequality, we get, for all $c \in (0, \phi_{+\infty})$ and ρ as above,

$$
\langle u_s^+(\cdot,t_0), \rho\rangle \le \langle [u_0^+]_{d,2}, f_c(\phi^+(u_r(\cdot,t_0)))\,\rho\rangle + \langle [u_0^+]_{c,2}, \rho\rangle. \tag{6.125}
$$

By Theorem 6.2.1 (see (6.12)) and equality (3.80), for a. e. $t \in (0,T)$, we have $[u_s^\pm(\cdot,t)]_{c,2} = [u^\pm(\cdot,t)]_{c,2} = [u_0^\pm]_{c,2}$. Therefore inequality (6.125) gives (6.118). This proves the result. □

Proof of Proposition 6.4.4. By Definition 6.4.1 every constructed solution u of (P) is the weak* limit in $L^\infty_{w*}(0,T;\mathcal{M}(\Omega))$ of a subsequence $\{u_{n_k}\}$ of solutions of the approximating problems (see Theorem 6.4.1). From inequality (6.61) with $t_1 = 0$ and $t_2 = T$ we get

$$
\begin{aligned}
&\iint_Q \{F_\phi(u_{n_k})\partial_t\zeta - f(\phi(u_{n_k}))\nabla\phi_{n_k}(u_{n_k})\cdot\nabla\zeta\}\,dxdt \\
&\qquad \ge -\int_\Omega F_\phi(u_{0n_k}(x))\zeta(x,0)\,dx
\end{aligned} \tag{6.126}
$$

for all $\zeta \in C^1([0,T]; C_c^1(\Omega))$ such that $\zeta \ge 0$ in Q and $\zeta(\cdot,T) = 0$ in Ω and for all nondecreasing $f \in C(\mathbb{R})$, the function F_ϕ being defined by (6.9).

By the convergence in (6.84), (6.87), and the equality in (6.113), we plainly get

$$
\lim_{k\to\infty} \iint_Q f(\phi(u_{n_k}))\nabla\phi_{n_k}(u_{n_k})\cdot\nabla\zeta\,dxdt = \iint_Q f(\phi(u_r))\nabla\phi(u_r)\cdot\nabla\zeta\,dxdt, \tag{6.127}
$$

whereas by (6.51) we have

$$\lim_{k\to\infty}\int_\Omega F_\phi(u_{0n_k}(x))\zeta(x,0)\,dx = \int_\Omega F_\phi(u_{0r})\zeta(x,0)\,dx + M^+_{F_\phi}\langle u^+_{0s},\zeta(\cdot,0)\rangle - M^-_{F_\phi}\langle u^-_{0s},\zeta(\cdot,0)\rangle \tag{6.128}$$

with $M^\pm_{F_\phi}\in\mathbb{R}$ as in (6.10). On the other hand, by Propositions 6.6.4 and 6.6.5 we have that

$$\lim_{k\to\infty}\iint_Q F_\phi(u_{n_k})\partial_t\zeta\,dxdt = \iint_Q F^*_\phi\partial_t\zeta\,dxdt + M^+_{F_\phi}\int_0^T\langle u^+_s(\cdot,t),\partial_t\zeta(\cdot,t)\rangle_\Omega\,dt - M^-_{F_\phi}\int_0^T\langle u^-_s(\cdot,t),\partial_t\zeta(\cdot,t)\rangle_\Omega\,dt \tag{6.129}$$

(see (6.102)), where for a. e. $(x,t)\in Q$,

$$F^*_\phi(x,t) = \int_{\mathbb{R}} F_\phi(y)\,d\nu_{(x,t)}(y),$$

and the Young measure $\nu\in\mathfrak{Y}^+(Q;\mathbb{R})$ is given by Proposition 6.6.4. By (6.114) we have

$$\int_{\mathbb{R}} F_\phi(y)\,d\nu_{(x,t)}(y) = \int_{\operatorname{supp}\nu_{(x,t)}} F_\phi(y)\,d\nu_{(x,t)}(y) = \int_{\operatorname{supp}\nu_{(x,t)}} d\nu_{(x,t)}(y)\int_0^y f(\phi(z))\,dz = F_\phi(u_r(x,t)) \tag{6.130}$$

for a. e. $(x,t)\in Q$.

By the above remarks, letting $k\to\infty$ in (6.126) and using (6.127)–(6.130), we prove that u satisfies the entropy inequality (6.11). Hence the result follows. □

6.7 The case of unbounded ϕ

This section is devoted to the study of problem (P) under more general assumptions, in particular, allowing the function ϕ to be unbounded. Apart from the short proof of Theorem 6.7.1, we only describe the main results and give an outline of the proofs (for complete proofs, we refer the interested reader to [87]).

Henceforth assumptions (A_1)–(A_2) are replaced by the following two hypotheses:

$$\phi \in C(\mathbb{R}) \text{ is nondecreasing and nonconstant in } \mathbb{R}, \text{ with } \phi(0) = 0; \tag{A_1'}$$

$$\text{there exist } M_0 > 0 \text{ and } \alpha \in [1, \infty) \text{ such that } |\phi(y)| \leq M_0(1 + |y|^\alpha) \quad (y \in \mathbb{R}). \tag{A_2'}$$

By assumption $(A_1{}')$ there exist

$$\lim_{y \to \pm\infty} \phi(y) =: \phi_{\pm\infty} \in \overline{\mathbb{R}}. \tag{6.131}$$

Both cases where $\phi_{\pm\infty}$ are finite or infinite are allowed, and either $\phi_{-\infty} \leq 0 < \phi_{+\infty}$, or $\phi_{-\infty} < 0 \leq \phi_{+\infty}$. The only restriction on ϕ is that it grows at most like a power; no estimates from below are used (except for the regularization result in Proposition 6.7.9), and no assumptions about existence of the diffusivity ϕ' are made.

6.7.1 Definition of solution

Let us denote by $T_K(y) := \max\{-K, \min\{y, K\}\}$ $(y \in \mathbb{R}, K > 0)$ the standard truncation function.

Definition 6.7.1. Let (A_0) hold. By a *weak solution* of problem (P) we mean any $u \in L^\infty_{w*}(0, T; \mathfrak{R}_f(\Omega))$ such that:

(i) $\phi(u_r) \in L^q(0, T; W_0^{1,q}(\Omega))$ for any $q \in [1, 1 + \frac{1}{\alpha N+1})$, and $T_K(\phi(u_r)) \in L^2(0, T; H_0^1(\Omega))$ for any $K > 0$;

(ii) for every $\zeta \in C^1([0, T]; C_c^1(\Omega))$ such that $\zeta(\cdot, T) = 0$ in Ω, we have

$$\int_0^T \langle u(\cdot, t), \partial_t \zeta(\cdot, t) \rangle dt = \iint_Q \nabla\phi(u_r) \cdot \nabla\zeta \, dxdt - \langle u_0, \zeta(\cdot, 0) \rangle. \tag{6.132}$$

It is easily seen that also in the present case equality (6.5a) holds, that is, the initial condition is satisfied in the weak* sense of $\mathfrak{R}_f(\Omega)$.

Remark 6.7.1. (i) Since $[T_K(\phi)]^\pm = T_K(\phi^\pm)$, for every weak solution of (P), we have $\phi^\pm(u_r) \in L^2(0, T; H_0^1(\Omega))$ if $\pm\phi_{\pm\infty} < +\infty$. Hence in such a case, Definition 6.7.1 reduces to Definition 6.1.1.

(ii) If ϕ is bounded, then every weak solution u of (P) has the following properties:

$$\phi(u_r) \in L^2(0, T; H_0^1(\Omega)), \tag{6.133a}$$

$$u(\cdot, t) - u_0 \in H^{-1}(\Omega) \quad \text{for a. e. } t \in (0, T), \tag{6.133b}$$

$$\operatorname*{ess\,lim}_{t \to 0^+} \|u(\cdot, t) - u_0\|_{H^{-1}(\Omega)} = 0. \tag{6.133c}$$

In fact, (6.133a) has already been pointed out in (i), whereas from equality (6.132) we get

$$u(\cdot,t) - u_0 = \Delta\left(\int_0^t \phi(u_r(\cdot,s))\,ds\right) \quad \text{in } \mathscr{D}^*(\Omega) \text{ for a. e. } t \in (0,T). \tag{6.134}$$

In view of (6.133a), equality (6.134) gives

$$\|u(\cdot,t) - u_0\|^2_{H^{-1}(\Omega)} = \int_\Omega \left|\nabla \int_0^t \phi(u_r(x,s))\,ds\right|^2 dx \le \int_0^t \int_\Omega |\nabla\phi(u_r)|^2\,dxds,$$

whence (6.133b) and (6.133c) follow. This proves the claim.

6.7.2 Persistence and uniqueness

A persistence property for ϕ sublinear at infinity (analogous to Theorem 6.2.1) is given by the following result.

Theorem 6.7.1. *Let (A_0) and $(A_1{}')$ hold, and let there exist $M > 0$ and $\alpha \in (0, \frac{N-2}{N}]$ such that $|\phi(y)| \le M(1+|y|)^\alpha$ $(y \in \mathbb{R})$. Let u be a weak solution of problem (P). Then there exists a null set $F \subseteq (0,T)$ such that for every $t \in (0,T) \setminus F$, the $C_{2,\frac{1}{1-\alpha}}$-concentrated part of $u(\cdot,t)$ is constant in time:*

$$[u(\cdot,t)]_{c,2,\frac{1}{1-\alpha}} = [u_0]_{c,2,\frac{1}{1-\alpha}}. \tag{6.135}$$

Remark 6.7.2. Observe that Theorem 6.7.1 is in agreement with the results in [26, 80] concerning the Cauchy problem for the fast diffusion porous medium equation. In particular, let $u_0 \in \mathfrak{R}^+_{c,2}(\mathbb{R}^N) \cap \mathfrak{R}^+_{d,2,\frac{1}{1-\alpha}}(\mathbb{R}^N)$ (recall that $\mathfrak{R}_{c,2,\frac{1}{1-\alpha}}(\mathbb{R}^N) \subseteq \mathfrak{R}_{c,2}(\mathbb{R}^N)$ by Proposition 3.4.13). Then a solution in the sense of [26, 80] satisfies $u(\cdot,t) \in L^1(\mathbb{R}^N)$ for all $t \in (0,T)$, and thus in this case, equality (6.12) is not satisfied. This suggests the following interpretation of Theorem 6.7.1: when diffusivity near infinity is stronger, only measures that are concentrated with respect to stronger capacities remain constant in time.

In this connection, observe that for any $\alpha, \beta \in (0, \frac{N-2}{N}]$, $\alpha < \beta$,

(a) the $C_{2,\frac{1}{1-\beta}}$-capacity is stronger than the $C_{2,\frac{1}{1-\alpha}}$-capacity;
(b) measures concentrated with respect to the $C_{2,\frac{1}{1-\beta}}$-capacity are also concentrated with respect to the $C_{2,\frac{1}{1-\alpha}}$-capacity (see Theorem 3.4.5 and Proposition 3.4.13).

Moreover, a measure concentrated with respect to the $C_{2,\frac{1}{1-\alpha}}$-capacity, $\alpha \in (0, \frac{N-2}{N}]$, is also concentrated with respect to the Newtonian capacity (see Proposition 3.4.13).

Observe that no analogue of Theorem 6.7.1 exists for $\alpha \in (\frac{N-2}{N}, 1)$, since in this case, every measure is diffuse with respect to the $C_{2,\frac{1}{1-\alpha}}$-capacity (see Proposition 3.4.11).

Proof of Theorem 6.7.1. Let u be a weak solution of problem (P). By (6.132), for every $\rho \in C_c^2(\Omega)$ and $h \in C^1([0,T])$ such that $h(0) = 1$ and $h(T) = 0$, we get

$$\int_0^T \langle u(\cdot,t), \rho\rangle h'(t)\,dt = -\iint_Q \phi(u_r)\Delta\rho\, h(t)\,dxdt - \langle u_0, \rho\rangle.$$

By standard approximation arguments, from the above equality for a. e. $t \in (0,T)$ we get

$$\begin{aligned} |\langle u(\cdot,t) - u_0, \rho\rangle| &= \left|\int_0^t \int_\Omega \phi(u_r)\Delta\rho\, dxdt\right| \\ &\leq M\,T^{1-\alpha}\left(\iint_Q (1 + |u_r|)\,dxdt\right)^\alpha \left(\int_\Omega |\Delta\rho|^{\frac{1}{1-\alpha}} dx\right)^{1-\alpha}. \end{aligned} \tag{6.136}$$

By the above inequality there exists a null set $F \subseteq (0,T)$ such that $u(\cdot,t) - u_0 \in W^{-2,\frac{1}{\alpha}}(\Omega) = (W_0^{2,\frac{1}{1-\alpha}}(\Omega))^*$ for all $t \in (0,T) \setminus F$. By [10, Lemma 4.1] this implies that for any such t, the measure $u(\cdot,t) - u_0$ is $C_{2,\frac{1}{1-\alpha}}$-diffuse, and thus the $C_{2,\frac{1}{1-\alpha}}$-concentrated part of $u(\cdot,t) - u_0$ is zero. Hence equality (6.135) follows. □

As already observed, persistence properties of solutions are connected with uniqueness, and additional conditions are needed to detect a uniqueness class for problem (P). In the present case, these conditions are given by the following definition, which generalizes Definition 6.3.1.

Definition 6.7.2. Let (A_0) and (A_1') hold. A weak solution u of problem (P) *satisfies the compatibility conditions* if for a. e. $t \in (0,T)$:
(i) if $\phi_{\pm\infty} \in \mathbb{R}$, then condition (6.21) is satisfied;
(ii) if $\phi_{\pm\infty} = \pm\infty$, then

$$[u_s^\pm(\cdot,t)]_{d,2} = 0. \tag{6.137}$$

Remark 6.3.1 also applies in the present case to $\phi^\pm(u_r)$ if $\phi_{\pm\infty} \in \mathbb{R}$.

The importance of the compatibility conditions is highlighted by the following uniqueness result, which by Remark 6.7.1(ii) includes Theorem 6.3.1 as a particular case.

Theorem 6.7.2. *Let (A_0) and (A_1') hold. Let u_1 and u_2 be two weak solutions of problem (P) satisfying the compatibility conditions, such that*

$$\phi(u_{1r}) - \phi(u_{2r}) \in L^2(\tau, T; H_0^1(\Omega)) \quad \textit{for all } \tau \in (0,T), \tag{6.138a}$$

$$u_1(\cdot,t) - u_2(\cdot,t) \in H^{-1}(\Omega) \quad \text{for a.e. } t \in (0,T), \tag{6.138b}$$

$$\operatorname{ess}\lim_{t\to 0^+} \|u_1(\cdot,t) - u_2(\cdot,t)\|_{H^{-1}(\Omega)} = 0. \tag{6.138c}$$

Then $u_1 = u_2$ in $L^\infty_{w*}(0,T;\mathfrak{R}_f(\Omega))$.

6.7.3 Existence

Again, we follow the constructive approach to the existence outlined in Section 6.4, and thus our first existence is the counterpart of Theorem 6.4.1.

Theorem 6.7.3. *Let (A_0) and $(A_1{}')$–$(A_2{}')$ hold. Then there exists a weak solution of problem (P), which is obtained as a limiting point in the weak* topology of $\mathfrak{R}_f(Q)$ of the sequence of solutions to problems (P_n).*

To prove the existence of weak solutions satisfying the compatibility conditions, now we use a two-step procedure, which uses the following additional assumptions on u_0:

$$[u_0^\pm]_{d,2} \in H^{-1}(\Omega), \tag{A_3}$$

$$[u_0^\pm]_{c,2} = [u_0^\pm]_{c,2} \llcorner K_0^\pm \quad \text{with } K_0^\pm \subseteq \Omega \text{ compact such that } C_2(K_0^\pm) = 0 \tag{A_4}$$

First, we prove the following result (see [87, Proposition 3.4]).

Proposition 6.7.4. *Let (A_0) and $(A_1{}')$–$(A_2{}')$ hold. If $\phi_{\pm\infty} = \pm\infty$, then let $u_0^\pm$ also satisfy (A_3)–(A_4). Then there exists a weak solution of problem (P) that satisfies the compatibility conditions.*

By Proposition 6.7.4, if $(A_1{}')$–$(A_2{}')$ hold with ϕ bounded, then there exists a weak solution of (P) that satisfies (6.21), a result already known by Proposition 6.4.2. If ϕ is unbounded, then a further approximation procedure is needed to get rid of assumptions (A_3)–(A_4). For every $i,j \in \mathbb{N}$, consider the problem

$$\begin{cases} \partial_t u_{i,j} = \Delta\phi(u_{i,j}) & \text{in } Q, \\ u_{i,j} = 0 & \text{on } \Gamma, \\ u_{i,j} = u_{0,i,j} & \text{in } \Omega \times \{0\}, \end{cases} \tag{$P_{i,j}$}$$

where the initial data $u_{0,i,j} \in \mathfrak{R}_f(\Omega)$ satisfy conditions (A_3)–(A_4) if $\phi_{\pm\infty} = \pm\infty$, have suitable monotonicity properties with respect to i, j, and converge to u_0 in a suitable sense as $i,j \to \infty$ (see [87, Section 7]; the existence of such a sequence is ensured by [10, Lemma 4.2]). By Proposition 6.7.4, for all $i,j \in \mathbb{N}$, there exists a weak solution $u_{i,j}$ of $(P_{i,j})$ that satisfies the compatibility conditions. By monotonicity methods it is proven that the limit in $L^\infty_{w*}(0,T;\mathfrak{R}_f(\Omega))$ of $u_{i,j}$ as $i,j \to \infty$ exists and is a weak solution of (P) satisfying the compatibility conditions. Then we have the following result.

Theorem 6.7.5. *Let* (A_0) *and* (A_1')–(A_2') *hold. Then there exists a weak solution of problem (P) that satisfies the compatibility conditions.*

Now we call *constructed solutions* weak solutions of problem (P) given by Theorem 6.7.5. If ϕ is bounded, then the constructed solutions are given by Proposition 6.7.4, and this definition agrees with Definition 6.4.1. We conclude that also in the present case the constructed solutions satisfy the compatibility conditions.

By Theorems 6.7.2 and 6.7.5 we have the following result, which generalizes Theorem 6.4.3 (recall that by Remark 3.1.1 a measure $\mu \in \mathfrak{R}_f(\Omega)$ can belong to the dual space $H^{-1}(\Omega) = (H_0^1)^*$).

Theorem 6.7.6. *Let* (A_0) *and* (A_1')–(A_2') *hold.*

(i) *Let ϕ be bounded. Then for every $u_0 \in \mathfrak{R}_f(\Omega)$, there exists a unique weak solution u of problem (P) satisfying* (6.21).

(ii) *Let ϕ be unbounded. Then for every $u_0 \in \mathfrak{R}_f(\Omega) \cap H^{-1}(\Omega)$, there exists a unique weak solution u of problem (P) satisfying the compatibility conditions. Moreover,*

$$\phi(u_r) \in L^2(\tau, T; H_0^1(\Omega)) \quad \textit{for all } \tau \in (0,T), \tag{6.139a}$$

$$u(\cdot,t) \in H^{-1}(\Omega) \quad \textit{for a. e. } t \in (0,T), \tag{6.139b}$$

$$\operatorname{ess}\lim_{t\to 0^+} \|u(\cdot,t) - u_0\|_{H^{-1}(\Omega)} = 0. \tag{6.139c}$$

6.7.4 Regularization

Also, in the present case the singular parts of constructed solutions can neither appear spontaneously nor increase in time.

Theorem 6.7.7. *Let* (A_0) *and* (A_1')–(A_2') *hold, and let u be a constructed solution of problem (P). Then for a. e. $t_1, t_2 \in (0,T)$, $t_1 < t_2$, inequality* (6.33) *holds.*

Let us finally mention two $\mathfrak{R}_f$-L^1 regularizing effects (see [87, Remark 3.7 and Proposition 3.11]).

Proposition 6.7.8. *Let* (A_0) *and* (A_1')–(A_2') *hold with $\phi_{\pm\infty} = \pm\infty$. Then for every constructed solution of problem (P) with 2-diffuse initial data, instantaneous regularization occurs.*

Proposition 6.7.9. *Let* (A_0) *and* (A_1')–(A_2') *hold, and let u be a constructed solution of problem (P). Let there exist $\alpha_\pm > \frac{N-2}{N}$ and $a_\pm, b_\pm > 0$ such that*

$$\pm\,\phi(z) \ge a_\pm |z|^{\alpha_\pm} - b_\pm \quad \textit{for } \pm z \ge 0. \tag{6.140}$$

Then $u^\pm \in L^\infty(0,T; L^1(\Omega))$.

7 Case study 2: hyperbolic conservation laws

7.1 Statement of the problem

In this chapter, we consider the Cauchy problem

$$\begin{cases} u_t + [\phi(u)]_x = 0 & \text{in } \mathbb{R} \times (0,T) =: S, \\ u = u_0 & \text{in } \mathbb{R} \times \{0\}, \end{cases} \tag{P}$$

where $T > 0$, and u_0 is a positive finite Radon measure on $\mathbb{R}$. We refer the reader to [19] for the case where u_0 is a finite signed Radon measure on $\mathbb{R}$.

7.1.1 Assumptions and preliminary remarks

We suppose that

$$u_0 \in \mathfrak{R}_f^+(\mathbb{R}), \tag{A_0}$$

$$\phi \in \operatorname{Lip}([0,\infty)), \quad \text{and} \quad \lim_{y\to\infty} \frac{\phi(y)}{y} = 0 \tag{A_1}$$

(without loss of generality we can assume that $\phi(0) = 0$). Let us mention that the general case where $\lim_{y\to\infty} \frac{\phi(y)}{y} = C_\phi \neq 0$ can be treated replacing x by $x - C_\phi t$ (see [18] for details). By (A_1) (and the condition $\phi(0) = 0$) there exists $M > 0$ such that

$$|\phi'(y)| \le M, \quad |\phi(y)| \le My \quad \text{for every } y > 0. \tag{7.1}$$

We will also use the following stronger assumption (see Section 7.2):

$$\text{the singular part } u_{0s} \text{ of } u_0 \text{ is a finite superposition of Dirac masses,}$$
$$u_{0s} = \sum_{j=1}^{p} c_j \delta_{x_j} \quad (x_1 < x_2 < \cdots < x_p;\ c_j > 0 \text{ for } 1 \le j \le p). \tag{A_0'}$$

For specific purposes, we make the following assumptions (see Section 7.3):

$$\begin{cases} \phi \in C^1([0,\infty)), \text{ and for every } y_1 > 0, \text{ there exist } a, b \ge 0,\ a + b > 0, \\ \text{such that } \phi' \text{ is strictly monotone in } [y_1 - a, y_1 + b], \end{cases} \tag{A_2}$$

$$\begin{cases} \phi \in C^\infty([0,\infty));\ \lim_{y\to\infty} \frac{\phi(y)}{y} = 0;\ \text{there exist } \alpha \ge -1 \text{ and } \beta \in \mathbb{R} \text{ such that} \\ \phi''(y)\,[\alpha\,\phi(y) + \beta] \le -[\phi'(y)]^2 < 0 \text{ for all } y \in [0,\infty). \end{cases} \tag{A_2'}$$

Remark 7.1.1. By (A_2') the map $y \mapsto \phi''(y)\,[\alpha\,\phi(y) + \beta]$ is strictly negative and continuous in $[0,\infty)$, and hence two cases are possible: either (a) $\alpha\,\phi + \beta > 0$ and $\phi'' < 0$,

https://doi.org/10.1515/9783110556902-009

or (b) $\alpha\phi + \beta < 0$ and $\phi'' > 0$ in $[0,\infty)$. In case (a), $0 < \phi' \leq \phi'(0)$ in $[0,\infty)$, since $\phi'' < 0$ and $\lim_{y\to\infty}\phi'(y) = 0$. Similarly, in case (b), we plainly have $\phi'(0) \leq \phi' < 0$ in $[0,\infty)$. In particular, in both cases, (A_2') implies (A_2) and (A_1). Moreover, if $\phi(0) = 0$, then $\alpha\phi + \beta > 0$ in $[0,\infty)$ if and only if $\beta > 0$.

Problem (P) with a superlinear ϕ of the type $\phi(y) = y^p$, $p > 1$, was studied in [67], where the existence and uniqueness of nonnegative entropy solutions was proved (see also [33]). By definition, in that paper, for positive times, the solution takes values in $L^1(\mathbb{R})$, although u_0 is a finite Radon measure. Interesting results concerning (P) with ϕ *at most linear at infinity* can be found in the pioneering paper [39].

When $\phi(y) = Cy$ $(C \in \mathbb{R})$, problem (P) is the Cauchy problem for the linear transport equation

$$\begin{cases} u_t + Cu_x = 0 & \text{in } S, \\ u = u_0 & \text{in } \mathbb{R}\times\{0\}, \end{cases}$$

whose the solution is trivially u_0 translated along the lines $x = Ct + x_0$ $(x_0 \in \mathbb{R})$. In particular, the singular part $u_s(\cdot,t)$ of the solution is nonzero for $t > 0$ if and only if it is nonzero for $t = 0$.

It is natural to ask what happens if ϕ is sublinear, as suggested by several mathematical models (see [50, 88, 89]). In this case the natural space to seek solutions of (P) is $L^\infty_{w^*}(0,T;\mathfrak{R}_f^+(\mathbb{R}))$, as it is assumed in the following definitions.

7.1.2 Definition of solution

Throughout this subsection, we assume that assumptions (A_0)–(A_1) are satisfied.

Definition 7.1.1. A *Young measure solution* of problem (P) is a pair (u,ν) such that:

(i) $u \in L^\infty_{w^*}(0,T;\mathfrak{R}_f^+(\mathbb{R}))$, $\nu \in \mathfrak{Y}(S;\mathbb{R})$;

(ii) $\operatorname{supp}\nu_{(x,t)} \subseteq [0,\infty)$ for a. e. $(x,t) \in S$, and

$$u_r(x,t) = \int_{[0,\infty)} y\, d\nu_{(x,t)}(y), \tag{7.2}$$

where $\{\nu_{(x,t)}\}_{(x,t)\in S}$ is the disintegration of ν;

(iii) for every $\zeta \in C^1([0,T];C_c^1(\mathbb{R}))$ with $\zeta(\cdot,T) = 0$ in $\mathbb{R}$, we have

$$\iint_S [u_r\partial_t\zeta + \phi^*\partial_x\zeta]\,dxdt + \int_0^T \langle u_s(\cdot,t),\partial_t\zeta(\cdot,t)\rangle\,dt = -\langle u_0,\zeta(\cdot,0)\rangle, \tag{7.3}$$

where

$$\phi^*(x,t) := \int_{[0,\infty)} \phi(y)\, d\nu_{(x,t)}(y) \quad \text{for a. e. } (x,t) \in S. \tag{7.4}$$

Set $H_+(y) := \chi_{(0,\infty)}(y)$, $H_-(y) := -\chi_{(-\infty,0)}(y)$, and $\operatorname{sgn}(y) := H_+(y) + H_-(y)$ $(y \in \mathbb{R})$.

Definition 7.1.2. A *Young measure entropy solution* of problem (P) is a Young measure solution such that

$$\iint_S [E_l^*\, \partial_t\zeta + F_l^*\, \partial_x\zeta]\, dxdt + \int_0^T \langle u_s(\cdot,t), \partial_t\zeta(\cdot,t)\rangle\, dt \\ \geq -\int_{\mathbb{R}} E_l(u_{0r}(x))\, \zeta(x,0)\, dx - \langle u_{0s}, \zeta(\cdot,0)\rangle \tag{7.5}$$

for any ζ as above, $\zeta \geq 0$, and $l \in [0,\infty)$, where $E_l(y) = |y-l|$, $F_l(y) = \operatorname{sgn}(y-l)[\phi(y) - \phi(l)]$ $(y \in [0,\infty))$, and

$$E_l^* := \int_{[0,\infty)} E_l(y)\, d\nu(y), \quad F_l^* := \int_{[0,\infty)} F_l(y)\, d\nu(y).$$

Young measure entropy subsolutions of (P) are defined by requiring (7.5) to hold for any ζ and l as above, with $E_l(y) = [y-l]_+$ and $F_l(y) = H_+(y-l)[\phi(y) - \phi(l)]$. A similar definition holds for *Young measure entropy supersolution.*

Remark 7.1.2. (i) By (7.1), (7.2), and (7.4) we have

$$|\phi^*(x,t)| \leq M \int_{[0,\infty)} y\, d\nu_{(x,t)}(y) = M u_r(x,t) \quad \text{for a. e. } (x,t) \in S. \tag{7.6}$$

Since $u_r \in L^\infty(0,T;L^1(\mathbb{R}))$, by (7.6) we have that $\phi^* \in L^\infty(0,T;L^1(\mathbb{R}))$.

For any open $\Omega \subseteq \mathbb{R}$ and $\tau \in (0,T)$, set $Q_\tau := \Omega \times (0,\tau)$.

Definition 7.1.3. A measure $u \in L^\infty_{w^*}(0,T;\mathfrak{R}_f^+(\Omega))$ is called a *solution of* (P) *in* Q_τ if for any $\zeta \in C^1([0,\tau];C_c^1(\Omega))$, $\zeta(\cdot,\tau) = 0$ in Ω, we have

$$\iint_{Q_\tau} [u_r\partial_t\zeta + \phi(u_r)\,\partial_x\zeta]\, dxdt + \int_0^\tau \langle u_s(\cdot,t), \partial_t\zeta(\cdot,t)\rangle\, dt = -\langle u_0, \zeta(\cdot,0)\rangle. \tag{7.7}$$

Solutions of (P) in S are referred to as *solutions of* (P).

Definition 7.1.4. A solution of (P) in Q_τ is called an *entropy solution in* Q_τ if it satisfies the *entropy inequality*

$$\begin{aligned}\iint_{Q_\tau} \{E_l(u_r)\,\partial_t\zeta + F_l(u_r)\,\partial_x\zeta\}\,dxdt + \int_0^\tau \langle u_s(\cdot,t), \partial_t\zeta(\cdot,t)\rangle\,dt \\ \geq -\int_\Omega E_l(u_{0r}(x))\,\zeta(x,0)\,dx - \langle u_{0s}, \zeta(\cdot,0)\rangle\end{aligned} \tag{7.8}$$

for all $\zeta \in C^1([0,\tau]; C_c^1(\Omega))$ such that $\zeta \geq 0$ and $\zeta(\cdot,\tau) = 0$ in Ω and for all $l \in [0,\infty)$.

Definition 7.1.5. *Entropy subsolutions and supersolutions* of (P) in Q_τ are defined by requiring the following inequalities to hold:

$$\begin{aligned}\iint_{Q_\tau} \{[u_r - l]_+\,\partial_t\zeta + H_+(u_r - l)[\phi(u_r) - \phi(l)]\partial_x\zeta\}\,dxdt + \int_0^\tau \langle u_s(\cdot,t), \partial_t\zeta(\cdot,t)\rangle_\Omega\,dt \\ \geq -\int_\Omega [u_{0r} - l]_+\,\zeta(x,0)\,dx - \langle u_{0s}, \zeta(\cdot,0)\rangle_\Omega,\end{aligned} \tag{7.9}$$

respectively,

$$\iint_{Q_\tau} \{[u_r - l]_-\,\partial_t\zeta + H_-(u_r - l)[\phi(u_r) - \phi(l)]\partial_x\zeta\}\,dxdt \geq -\int_\Omega [u_{0r} - l]_-\,\zeta(x,0)\,dx \tag{7.10}$$

for all ζ and l as above.

It is easily seen that every entropy solution is both an entropy subsolution and an entropy supersolution of (P).

Remark 7.1.3. (i) Equality (7.7) also reads

$$\langle u, \partial_t\zeta\rangle + \iint_S \phi(u_r)\,\partial_x\zeta\,dxdt = -\langle u_0, \zeta(\cdot,0)\rangle,$$

showing that $u_t = -[\phi(u_r)]_x$ in $\mathcal{D}^*(S)$.

(ii) A solution of problem (P) is also a Young measure solution. Moreover, it follows from (7.1) that $\phi(u_r) \in L^\infty(0,T; L^1(\mathbb{R}))$. Similar remarks hold for entropy solutions, subsolutions, and supersolutions.

The following result is analogous to Proposition 6.1.1. The proof is similar, and thus we omit it (see [17, Proposition 3.5]).

Proposition 7.1.1. *Let* (A_1) *hold, let* (u, ν) *be a Young measure solution of* (P). *Then:*

(i) *there exists a null set $F^* \subseteq (0,T)$ such that for all $t \in (0,T) \setminus F^*$ and $\rho \in C_c^1(\mathbb{R})$,*

$$\langle u(\cdot,t),\rho\rangle - \langle u_0,\rho\rangle = \int_0^t \int_{\mathbb{R}} \phi^*(x,s)\rho'(x)\,dxds; \tag{7.11}$$

(ii) *for all $\rho \in C_c(\mathbb{R})$,*

$$\operatorname{ess}\lim_{t\to 0^+} \langle u(\cdot,t),\rho\rangle = \langle u_0,\rho\rangle, \tag{7.12}$$

$$\operatorname{ess}\lim_{t\to t_0} \langle u(\cdot,t),\rho\rangle = \langle u(\cdot,t_0),\rho\rangle \quad \textit{for a.e. } t_0 \in (0,T). \tag{7.13}$$

Remark 7.1.4. As in Remark 6.1.2, it is easily seen that the map $t \mapsto u(\cdot,t)$ has a representative defined for any $t \in [0,T]$ and such that

$$\lim_{t\to t_0} \langle u(\cdot,t),\rho\rangle = \langle u(\cdot,t_0),\rho\rangle \quad \text{for all } t_0 \in [0,T] \text{ and } \rho \in C_c(\mathbb{R}). \tag{7.14}$$

Now we refer to this continuous representative whenever properties of the map $t \mapsto u(\cdot,t)$ are stated for every t in some subinterval of $[0,T]$.

As in the parabolic problem dealt with in Chapter 6 (see Section 6.3), information about regularity will be important to prove the uniqueness. A first result in this direction, partly reminiscent of inequality (6.33), is as follows.

Proposition 7.1.2. *Let (A_0)–(A_1) hold.*

(i) *Let u be a Young measure entropy solution of (P). Then*

$$u_s(\cdot,t_2) \le u_s(\cdot,t_1) \quad \textit{in } \mathfrak{R}_f^+(\mathbb{R}) \quad \textit{for a.e. } 0 \le t_1 \le t_2 \le T. \tag{7.15}$$

In particular,

$$u_s(\cdot,t) \le u_{0s} \quad \textit{in } \mathfrak{R}_f^+(\mathbb{R}) \quad \textit{for a.e. } t \in (0,T). \tag{7.16}$$

(ii) *Let u be a solution of problem (P). Then there is* conservation of mass*:*

$$\|u(\cdot,t)\|_{\mathfrak{R}_f(\mathbb{R})} = \|u_0\|_{\mathfrak{R}_f(\mathbb{R})} \quad \textit{for a.e. } t \in (0,T). \tag{7.17}$$

7.2 Uniqueness

In this section, we consider the case where ϕ is bounded (see Remark 7.3.1 for a uniqueness result when ϕ is unbounded).

Let us define the *waiting time* $t_0 \in [0,T]$ for solutions u of (CL):

$$t_0 := \inf\{\tau \in (0,T] \mid u_s(\cdot,t) = 0,\ u_r(\cdot,t) \in L^\infty(\mathbb{R}) \text{ for a.e. } t \in (\tau,T)\}. \tag{7.18}$$

Proposition 7.2.1. *Let ϕ be bounded, and let assumptions* (A_0)–(A_1) *hold. Let* $u_{0s}(\{x_0\}) > 0$ *for some* $x_0 \in \mathbb{R}$, *and let u be a solution of problem (P). Then the waiting time* t_0 *defined by* (7.18) *satisfies*

$$t_0 \geq \min\left\{T, \frac{u_{0s}(\{x_0\})}{2\|\phi\|_\infty}\right\} > 0. \tag{7.19}$$

Remark 7.2.1. In connection with equality (7.12), observe that, if $u_{0s} \neq 0$ and the waiting time t_0 is equal to 0, then the map $t \mapsto u(\cdot, t)$ cannot be continuous at $t = 0$ in the strong topology of $\mathfrak{R}_f(\mathbb{R})$. Instead, the strong continuity may occur if the waiting time is positive (see [18, Theorem 4.1]).

An upper bound of the waiting time in terms of ϕ and u_0 is the content of the following proposition. We refer the reader to [18, Theorem 3.11(ii)] for its proof, which relies on estimates of the density u_r of the Aronson–Bénilan type. Refined estimates of the waiting time can also be found in [20].

Proposition 7.2.2. *Let* $(A_0{}')$ *and* $(A_2{}')$ *hold with bounded* ϕ *bounded,* $\alpha > -1$, *and* $|\beta| < \lim_{y\to\infty} |\phi(y)| =: \gamma$. *Let u be the entropy solution of problem (P) given further by Theorem 7.3.4. Then the waiting time* t_0 *defined by* (7.18) *satisfies*

$$t_0 \leq \min\left\{T, \frac{(\alpha+1)\ \|u_0\|_{\mathfrak{R}_f(\mathbb{R})}}{\gamma - |\beta|}\right\}. \tag{7.20}$$

Remark 7.2.2. If $\phi(y) = 1 - (1 + y)^p$ $(p < 0)$, then the proof of Proposition 0.2.4 shows that the waiting time defined in (7.18) is $t_0 = 1$. Hence, in this case, estimate (7.20) is sharp, since assumption $(A_2{}')$ is satisfied with $\alpha = p/(1-p)$ and $\beta = -p/(1-p)$, whence

$$\frac{(\alpha+1)\ \|\delta_0\|_{\mathfrak{R}_f(\mathbb{R})}}{\gamma - |\beta|} = \frac{(p/(1-p)+1)\ \|\delta_0\|_{\mathfrak{R}_f(\mathbb{R})}}{1 + p/(1-p)} = 1.$$

Relying on Propositions 7.1.2 and 7.2.1, whose proofs will be given in Section 7.6, we can now address the uniqueness. Let $(A_0{}')$ and (A_1) be satisfied (the latter with ϕ bounded). Then $u_{0s} = \sum_{j=1}^p c_j \delta_{x_j}$, and we set $I_1 := (-\infty, x_1)$, $I_j := (x_{j-1}, x_j)$ for $j = 2, \ldots, p$, $I_{p+1} := (x_p, \infty)$, and $S_j := I_j \times (0, T)$ for $j = 1, \ldots, p+1$. In view of Proposition 7.2.1, if u is a solution of problem (P), then

$$\forall x_j\ \exists t_j \in (0, T] \quad \text{such that} \quad \begin{cases} u_s(\cdot, t)(\{x_j\}) > 0 & \text{for a. e. } t \in [0, t_j), \\ u_s(\cdot, t)(\{x_j\}) = 0 & \text{for a. e. } t \in (t_j, T). \end{cases}$$

Then we can state the following definition.

Definition 7.2.1. Let ϕ be bounded, let $(A_0{}')$–(A_1) hold, let $j = 1, \ldots, p$, and let $\tau \in (0, t_j]$. An entropy solution of (P) satisfies the *compatibility condition at* x_j *in* $[0, \tau]$ if for all $\beta \in C_c^1(0,\tau)$, $\beta \geq 0$, and $l \in [0,\infty)$,

$$\pm \operatorname{ess}\lim_{x \to x_j^\pm} \int_0^T H_-(u_r(x,t)-l)[\phi(u_r(x,t))-\phi(l)]\beta(t)\,dt \leq 0. \tag{$C_\pm$}$$

The following lemma shows that Definition 7.2.1 is well posed.

Lemma 7.2.3. *Let u be an entropy supersolution of (P), and let $\beta \in C_c^1(0,T)$, $\beta \geq 0$. Then:*

(i) *for every $l \in [0,\infty)$, the distributional derivative of the function*

$$x \mapsto -\int_0^T H_-(u_r(x,t)-l)[\phi(u_r(x,t))-\phi(l)]\beta(t)\,dt + l\,T\,\|\beta'\|_\infty\, x \tag{7.21}$$

is nonnegative;

(ii) *for all $x_0 \in \mathbb{R}$ and $l \in [0,\infty)$, there exist finite limits*

$$\operatorname{ess}\lim_{x \to x_0^\pm} \int_0^T H_-(u_r(x,t)-l)[\phi(u_r(x,t))-\phi(l)]\beta(t)\,dt. \tag{7.22}$$

Proof. Let $\alpha \in C_c^1(\mathbb{R})$, $\alpha \geq 0$. Choosing $\zeta(x,t) = \alpha(x)\beta(t)$ in (7.10) with $Q_\tau = S$ gives

$$\iint_S \{[u_r(x,t)-l]_-\,\alpha(x)\beta'(t) + H_-(u_r(x,t)-l)[\phi(u_r(x,t))-\phi(l)]\alpha'(x)\beta(t)\}\,dxdt \geq 0.$$

Since $0 \leq [u_r - l]_- \leq l$, from the above inequality we get

$$-\int_{\mathbb{R}}\left(\int_0^T H_-(u_r(x,t)-l)[\phi(u_r(x,t))-\phi(l)]\beta(t)\,dt\right)\alpha'(x)\,dx \leq l\,T\,\|\beta'\|_\infty \int_{\mathbb{R}} \alpha(x)\,dx,$$

whence claim (i) follows.

Therefore the distributional derivative of function (7.21) is a Radon measure. Clearly, the same holds for the distributional derivative, say μ, of the function $\tilde{H} \in L^1_{\mathrm{loc}}(\mathbb{R})$,

$$\tilde{H}(x) := -\int_0^T H_-(u_r(x,t)-l)[\phi(u_r(x,t))-\phi(l)]\beta(t)\,dt.$$

Fix any $\bar{x} \in \mathbb{R}$ and set $f_\mu(x) := \mu((\bar{x},x])$ if $x \geq \bar{x}$ and $f_\mu(x) := -\mu((x,\bar{x}])$ if $x < \bar{x}$. Then f_μ is continuous from the right and coincides a. e. with $\tilde{H}$ on every compact $K \subset \mathbb{R}$ up to a constant, possibly depending on K (e. g., see [5, Theorem 3.28]). Hence the claim follows. □

Now we can state a uniqueness result.

Theorem 7.2.4. *Let ϕ be bounded, and let $(A_0{}')$–(A_1) hold. Then there exists at most one entropy solution of problem (P), which belongs to $C([0,T];\mathfrak{R}_f^+(\mathbb{R}))$ and satisfies the compatibility condition at x_j in $[0,t_j]$ for all $j=1,\dots,p$.*

To prove Theorem 7.2.4, we need a few lemmas.

Lemma 7.2.5. *Let ϕ be bounded, let $(A_0{}')$–(A_1) hold, and let u be an entropy solution of (P). Then for all $\beta \in C_c^1(0,T)$, $\beta \geq 0$, $l \in [0,\infty)$, and $j=1,\dots,p$, there exist the finite limits*

$$\operatorname*{ess\,lim}_{x\to x_j^\pm} \int_0^T \operatorname{sgn}(u_r(x,t)-l)[\phi(u_r(x,t))-\phi(l)]\beta(t)\,dt. \tag{7.23}$$

Proof. We only prove the claim for the limit from the right, the proof being similar for that from the left. Let $j=1,\dots,p$ be fixed. In view of Proposition 7.1.2, the singular part of every entropy solution of (P) is nonincreasing in time, and hence by $(A_0{}')$ we have $u_s(\cdot,t)(I_{j+1})=0$ for all $t\in[0,T]$. Let $\alpha\in C_c^1(I_{j+1})$, $\alpha\geq 0$. Choosing $\zeta(x,t)=\alpha(x)\beta(t)$ in (7.8) with $Q_T=S$ gives

$$\iint_{S_{j+1}} \{|u_r(x,t)-l|\alpha(x)\beta'(t)+\operatorname{sgn}(u_r(x,t)-l)[\phi(u_r(x,t))-\phi(l)]\alpha'(x)\beta(t)\}\,dxdt \geq 0.$$

Since $0\leq |u_r-l|\leq u_r+l$, we have

$$\begin{aligned}
&-\int_{I_{j+1}} \alpha'(x)\left(\int_0^T \operatorname{sgn}(u_r(x,t)-l)[\phi(u_r(x,t))-\phi(l)]\beta(t)\,dt\right)dx\\
&\leq \|\beta'\|_\infty \int_{I_{j+1}} \alpha(x)\left(\int_0^T u_r(x,t)\,dt + lT\right)dx\\
&= -\|\beta'\|_\infty \int_{I_{j+1}} \alpha'(x)\left(\int_0^T\int_{x_j}^x u_r(y,t)\,dydt + lTx\right)dx.
\end{aligned}$$

This inequality implies that the distributional derivative of the map

$$\begin{aligned}
x \mapsto &-\int_0^T \operatorname{sgn}(u_r(x,t)-l)[\phi(u_r(x,t))-\phi(l)]\beta(t)\,dt\\
&+\|\beta'\|_\infty\left(\int_0^T\int_{x_j}^x u_r(y,t)\,dydt + lTx\right)
\end{aligned}$$

is nonnegative in I_{j+1}. Arguing as in the proof of Lemma 7.2.3, the claim follows. □

Lemma 7.2.6. *Let (A_0')–(A_1') hold, and let u be an entropy solution of (P). Then for every $j = 1, \dots, p$:*

(i) *there exist $h_j^-, h_j^+ \in L^\infty(0,T)$ such that for all $\beta \in C_c^1(0,T)$,*

$$\operatorname{ess}\lim_{x\to x_j^\pm} \int_0^T \phi(u_r(x,t))\beta(t)\,dt = \int_0^T h_j^\pm(t)\beta(t)\,dt; \tag{7.24}$$

(ii) *if u satisfies the compatibility condition $(C_\pm)$ at x_j in $[0,\tau]$, then*

$$h_j^- \le \liminf_{l\to\infty} \phi(l) \le \limsup_{l\to\infty} \phi(l) \le h_j^+ \quad a.e.\ in\ (0,\tau). \tag{7.25}$$

Remark 7.2.3. By standard density arguments, from (7.24) we get

$$\operatorname{ess}\lim_{x\to x_j^\pm} \int_0^T \phi(u_r(x,t))\zeta(x,t)\,dt = \int_0^T h_j^\pm(t)\zeta(x_j,t)\,dt \tag{7.26}$$

for every $\zeta \in L^1(0,T;C_c(U_j))$ with $x_j \in U_j \subseteq \mathbb{R}$, U_j open.

Proof of Lemma 7.2.6(i). We only prove the limit from the right. Since $\operatorname{sgn} y = 1+2H_-(y)$ for $y \in \mathbb{R}$, by (7.22)–(7.23) the limit in the left-hand side of (7.24) exists and is finite. On the other hand, for every sequence $\{x_n\}$ converging to x_j^+, the sequence $\{\phi(u_r(x_n,\cdot))\}$ is bounded in $L^\infty(0,T)$, and hence there exist a subsequence $\{x_{n_k}\} \subseteq \{x_n\}$ and a function $h_j^+ \in L^\infty(0,T)$ such that $\{\phi(u_r(x_{n_k},\cdot))\} \overset{*}{\rightharpoonup} h_j^+$ in $L^\infty(0,T)$. Since the limit in the left-hand side of (7.24) exists for all $\beta \in C_c^1(0,T)$, by separability arguments it is easily seen that the function h_j^+ is independent of the choice of the sequence $\{x_n\}$. Hence claim (i) follows.

(ii) Since u is a solution of (P) in $I_{j+1}\times(0,\tau)$ and $u_s \llcorner (I_{j+1}\times(0,\tau)) = 0$, by (7.7) we have

$$\int_0^\tau \int_{I_{j+1}} \{(u_r - l)y_t + [\phi(u_r) - \phi(l)]\,y_x\}\,dxdt = -\int_{I_{j+1}} [u_{0r}(x) - l]\,y(x,0)\,dx$$

for all $l \in [0,\infty)$ and $y \in C^1([0,\tau];C_c^1(I_{j+1}))$ such that $y(\cdot,\tau) = 0$ in I_{j+1}. Let

$$\eta_\epsilon(x) := \frac{2(x-x_j)-\epsilon}{\epsilon}\chi_{[x_j+\epsilon/2,x_j+\epsilon]}(x) + \chi_{(x_j+\epsilon,x_{j+1}]}(x) \quad (x \in I_{j+1}), \tag{7.27}$$

and let $\zeta \in C^1([0,\tau]; C_c^1([x_j, x_{j+1}))$ be such that $\zeta(\cdot,\tau) = 0$ in I_{j+1} (here $x_{j+1} = \infty$ if $j = p$). By standard arguments we can choose $y = \zeta\eta_\epsilon$ in the above equality and obtain

$$\int_0^\tau \int_{I_{j+1}} \{(u_r - l)\zeta_t\eta_\epsilon + [\phi(u_r) - \phi(l)]\,\zeta_x\eta_\epsilon\}\,dxdt + \int_{I_{j+1}} [u_{0r}(x) - l]\zeta(x,0)\eta_\epsilon(x)\,dx$$

$$= -\frac{2}{\epsilon}\int_0^\tau \int_{x_j+\epsilon/2}^{x_j+\epsilon} [\phi(u_r) - \phi(l)]\zeta\,dxdt.$$

Letting $\epsilon \to 0^+$ in the above equality plainly gives (see (7.26))

$$\begin{aligned}\int_0^\tau \int_{I_{j+1}} \{(u_r - l)\zeta_t + [\phi(u_r) - \phi(l)]\,\zeta_x\}\,dxdt + \int_{I_{j+1}} [u_{0r}(x) - l]\,\zeta(x,0)\,dx \\ = -\operatorname{ess}\lim_{x\to x_j^+} \int_0^\tau [\phi(u_r(x,t)) - \phi(l)]\,\zeta(x,t)\,dt \\ = -\int_0^\tau [h_j^+(t) - \phi(l)]\,\zeta(x_j,t)\,dt. \qquad (7.28)\end{aligned}$$

Since u is an entropy solution of (P) in $I_{j+1} \times (0,\tau)$, arguing as before, we obtain

$$\begin{aligned}\int_0^\tau \int_{I_{j+1}} \{|u_r - l|\,\zeta_t + \operatorname{sgn}(u_r - l)[\phi(u_r) - \phi(l)]\zeta_x\}\,dxdt + \int_{I_{j+1}} |u_{0r}(x) - l|\zeta(x,0)\,dx \\ \geq -\operatorname{ess}\lim_{x\to x_j^+} \int_0^\tau \operatorname{sgn}(u_r(x,t) - l)[\phi(u_r(x,t)) - \phi(l)]\zeta(x,t)\,dt\end{aligned}$$

for all ζ as above, $\zeta \geq 0$. Choosing $\zeta(x,t) = \alpha(x)\beta(t)$ with $\alpha \in C_c^1([x_j, x_{j+1}))$, $\alpha \geq 0$, and $\beta \in C^1([0,\tau])$, $\beta \geq 0$, $\beta(\tau) = 0$, by the compatibility condition $(C_\pm)$ we have

$$\begin{aligned}\int_0^\tau \int_{I_{j+1}} \{|u_r - l|\,\zeta_t + \operatorname{sgn}(u_r - l)[\phi(u_r) - \phi(l)]\zeta_x\}\,dxdt \\ + \int_{I_{j+1}} |u_{0r}(x) - l|\zeta(x,0)\,dx + \operatorname{ess}\lim_{x\to x_j^+} \int_0^\tau [\phi(u_r(x,t)) - \phi(l)]\zeta(x,t)\,dt \\ \geq -2\operatorname{ess}\lim_{x\to x_j^+} \int_0^\tau H_-(u_r(x,t) - l)[\phi(u_r(x,t)) - \phi(l)]\zeta(x,t)\,dt\end{aligned}$$

$$= -2\alpha(x_j)\,\mathrm{ess}\lim_{x\to x_j^+}\int_0^\tau H_-(u_r(x,t)-l)[\phi(u_r(x,t))-\phi(l)]\beta(t)\,dt \geq 0, \tag{7.29}$$

since $\mathrm{sgn}(y) = 1 + 2H_-(y)$. From inequalities (7.28) and (7.29) we obtain

$$\int_0^\tau\int_{I_{j+1}}\{[u_r-l]_+\zeta_t + H_+(u_r-l)[\phi(u_r)-\phi(l)]\,\zeta_x\}\,dxdt + \int_{I_{j+1}}[u_{0r}(x)-l]_+\,\zeta(x,0)\,dx$$
$$\geq -\int_0^\tau[h_j^+(t)-\phi(l)]\,\zeta(x_j,t)\,dt.$$

Letting $l\to\infty$ in the above inequality gives

$$\liminf_{l\to\infty}\int_0^\tau[h_j^+(t)-\phi(l)]\,\zeta(x_j,t)\,dt = \int_0^\tau\Big[h_j^+(t)-\limsup_{l\to\infty}\phi(l)\Big]\,\zeta(x_j,t)\,dt \geq 0,$$

whence the last inequality in (7.25) follows by the arbitrariness of ζ.

Replacing $I_{j+1}\times(0,\tau)$ by $I_j\times(0,\tau)$, we obtain, similarly to (7.28) and (7.29),

$$\int_0^\tau\int_{I_j}\{(u_r-l)\zeta_t + [\phi(u_r)-\phi(l)]\,\zeta_x\}\,dxdt + \int_{I_j}[u_{0r}(x)-l]\,\zeta(x,0)\,dx$$
$$= \mathrm{ess}\lim_{x\to x_j^-}\int_0^\tau[\phi(u_r(x,t))-\phi(l)]\,\zeta(x,t)\,dt = \int_0^\tau[h_j^-(t)-\phi(l)]\,\zeta(x_j,t)\,dt, \tag{7.30}$$

$$\int_0^\tau\int_{I_j}\{|u_r-l|\,\zeta_t + \mathrm{sgn}(u_r-l)[\phi(u_r)-\phi(l)]\zeta_x\}\,dxdt$$
$$+\int_{I_j}|u_{0r}(x)-l|\,\zeta(x,0)\,dx - \mathrm{ess}\lim_{x\to x_j^-}\int_0^\tau[\phi(u_r(x,t))-\phi(l)]\zeta(x,t)\,dt \geq 0, \tag{7.31}$$

whence

$$\int_0^\tau\int_{I_j}\{[u_r-l]_+\zeta_t + H_+(u_r-l)[\phi(u_r)-\phi(l)]\,\zeta_x\}\,dxdt + \int_{I_j}[u_{0r}(x)-l]_+\,\zeta(x,0)\,dx$$
$$\geq \int_0^\tau[h_j^-(t)-\phi(l)]\,\zeta(x_j,t)\,dt$$

and

$$\limsup_{l\to\infty}\int_0^T [h_j^-(t)-\phi(l)]\,\zeta(x_j,t)\,dt=\int_0^T\left[h_j^-(t)-\liminf_{l\to\infty}\phi(l)\right]\zeta(x_j,t)\,dt\le 0.$$

Since ζ is arbitrary, we obtain the first inequality in (7.25). □

Remark 7.2.4. By standard density arguments and (7.26) it follows from (7.31) that

$$\begin{aligned}\int_0^T\int_{I_1}\{|u_r-l|\,\zeta_t+\operatorname{sgn}(u_r-l)[\phi(u_r)-\phi(l)]\zeta_x\}\,dxdt+\int_{I_1}|u_{0r}(x)-l|\,\zeta(x,0)\,dx\\ \ge\int_0^T[h_1^-(t)-\phi(l)]\zeta(x_1,t)\,dt\end{aligned}\tag{7.32}$$

if $\zeta\in C^1([0,T];C_c^1((-\infty,x_1]))$, $\zeta\ge 0$, and $\zeta(\cdot,T)=0$ in $(-\infty,x_1]$, and from (7.29) that

$$\begin{aligned}\int_0^T\int_{I_{p+1}}\{|u_r-l|\,\zeta_t+\operatorname{sgn}(u_r-l)[\phi(u_r)-\phi(l)]\zeta_x\}\,dxdt+\int_{I_{p+1}}|u_{0r}(x)-l|\,\zeta(x,0)\,dx\\ \ge-\int_0^T[h_p^+(t)-\phi(l)]\zeta(x_p,t)\,dt\end{aligned}$$

for all $\zeta\in C^1([0,T];C_c^1([x_p,\infty)))$ such that $\zeta\ge 0$ and $\zeta(\cdot,T)=0$ in $[x_p,\infty)$. Moreover, arguing as in the proof of Lemma 7.2.6 with η_ϵ in (7.27) replaced by

$$\frac{2(x-x_j)-\epsilon}{\epsilon}\chi_{[x_j+\epsilon/2,x_j+\epsilon]}+\chi_{[x_j+\epsilon,x_{j+1}-\epsilon]}+\frac{2(x_{j+1}-x)-\epsilon}{\epsilon}\chi_{[x_{j+1}-\epsilon,x_{j+1}-\epsilon/2]},$$

we obtain that, for any $j=1,\dots,p-1$,

$$\begin{aligned}\int_0^T\int_{I_{j+1}}\{|u_r-l|\,\zeta_t+\operatorname{sgn}(u_r-l)[\phi(u_r)-\phi(l)]\zeta_x\}\,dxdt+\int_{I_{j+1}}|u_{0r}(x)-l|\,\zeta(x,0)\,dx\\ \ge-\int_0^T[h_j^+(t)-\phi(l)]\zeta(x_j,t)\,dt+\int_0^T[h_{j+1}^-(t)-\phi(l)]\zeta(x_{j+1},t)\,dt\end{aligned}$$

for all $\zeta\in C^1([0,T];C_c^1([x_j,x_{j+1}]))$ such that $\zeta\ge 0$ and $\zeta(\cdot,T)=0$ in $[x_j,x_{j+1}]$.

Remark 7.2.5. Let us mention the following inequalities, which hold for all $\zeta \in C^1([0,\tau]; C_c^1((-\infty, x_1]))$ such that $\zeta \geq 0$ and $\zeta(\cdot,\tau) = 0$ in $(-\infty, x_1]$:

$$\int_0^\tau \int_{I_1} \{[u_r - l]_+ \zeta_t + H_+(u_r - l)[\phi(u_r) - \phi(l)]\zeta_x\}\, dxdt + \int_{I_1} [u_{0r}(x) - l]_+ \zeta(x,0)\, dx \geq \int_0^\tau [h_1^-(t) - \phi(l)]\zeta(x_1, t)\, dt, \tag{7.33}$$

$$\int_0^\tau \int_{I_1} \{[u_r - l]_- \zeta_t + H_-(u_r - l)[\phi(u_r) - \phi(l)]\zeta_x\}\, dxdt + \int_{I_1} [u_{0r}(x) - l]_- \zeta(x,0)\, dx \geq 0. \tag{7.34}$$

The proof is analogous to that of (7.32), starting from (7.9) and (7.10) instead of (7.8). Similar inequalities hold in S_j for $j = 2, \ldots, p+1$.

Now we can prove Theorem 7.2.4.

Proof of Theorem 7.2.4. Let $u, v \in C([0,T]; \mathfrak{R}_f^+(\Omega))$ be entropy solutions of (P) satisfying the compatibility condition at every x_j in $[0, t_j]$, and let

$$\tau := \min\{t_u, t_v\}, \quad \text{where} \quad \begin{cases} t_u := \sup\{t \in [0,T) \mid \operatorname{supp} u_s(\cdot,t) = \operatorname{supp} u_{0s}\} \\ t_v := \sup\{t \in [0,T) \mid \operatorname{supp} v_s(\cdot,t) = \operatorname{supp} u_{0s}\}. \end{cases} \tag{7.35}$$

It suffices to show that

$$u = v \quad \text{in } \mathfrak{R}_f(S_\tau), \tag{7.36}$$

since if $\tau < T$, then a standard iteration procedure proves the result. Plainly, (7.36) follows if we prove that

$$u_r = v_r \quad \text{a.e. in } S_\tau. \tag{7.37}$$

In fact, equalities (7.7) and (7.37) imply that

$$\int_0^\tau \langle u_s(\cdot,t) - v_s(\cdot,t), \partial_t\zeta(\cdot,t)\rangle\, dt = -\iint_{S_\tau} \{(u_r - v_r)\partial_t\zeta + [\phi(u_r) - \phi(u_r)]\partial_x\zeta\}\, dxdt = 0$$

for all $\zeta \in C^1([0,\tau]; C_c^1(\mathbb{R}))$ such that $\zeta(\cdot,\tau) = 0$ in $\mathbb{R}$. Hence $\langle u_s(\cdot,t) - v_s(\cdot,t), \alpha\rangle = 0$ for a. e. $t \in (0,\tau)$ and all $\alpha \in C_c^1(\mathbb{R})$. Therefore $u_s = v_s$ in $L^\infty(0,\tau; \mathfrak{R}_f(\mathbb{R}))$, and (7.36) follows from (7.37).

It remains to prove (7.37), which is equivalent to showing that

$$u_r = v_r \quad \text{a.e. in } I_j \times (0,\tau) \text{ for any } j = 1,\dots,p+1. \tag{7.38}$$

We only prove (7.38) for $j = 1$, since in the other cases the proof is similar. Set $Q_1 := (-\infty, x_1]\times(0,\tau)$. We apply the Kružkov method of doubling variables adapted to boundary value problems (see [68, 75, 91]). Let $\xi = y(x,t,y,s) \geq 0$ defined in $Q_1 \times Q_1$ be such that $\xi(\cdot,\cdot,y,s) \in C_c^1(Q_1)$ for every $(y,s) \in Q_1$ and $\xi(x,t,\cdot,\cdot) \in C_c^1(Q_1)$ for every $(x,t) \in Q_1$. It follows from (7.32) that

$$\iint_{Q_1} \{\operatorname{sgn}(u_r(x,t) - v(y,s))[\phi(u_r(x,t)) - \phi(v_r(y,s))]\xi_x(x,t,y,s)$$
$$+ |u_r(x,t) - v_r(y,s)|\xi_t(x,t,y,s)\}\, dxdt \geq \int_0^\tau [h_1^-(t) - \phi(v_r(y,s))]\xi(x_1,t,y,s)\, dt,$$
$$\iint_{Q_1} \{\operatorname{sgn}(u_r(x,t) - v(y,s))[\phi(u_r(x,t)) - \phi(v_r(y,s))]\xi_y(x,t,y,s)$$
$$+ |u_r(x,t) - v_r(y,s)|\xi_s(x,t,y,s)\}\, dyds \geq \int_0^\tau [g_1^-(s) - \phi(u_r(x,t))]\xi(x,t,x_1,s)\, ds,$$

where, by Lemma 7.2.6(i), $g_j^\pm \in L^\infty(0,T)$ satisfies, for each $j = 1,\dots,p$,

$$\operatorname*{ess\,lim}_{x\to x_j^\pm} \int_0^T \phi(v_r(x,t))\beta(t)\, dt = \int_0^T g_j^\pm(t)\beta(t)\, dt \quad \text{if } \beta \in C_c^1(0,T). \tag{7.39}$$

Let ρ_ϵ $(\epsilon > 0)$ be a symmetric mollifier in $\mathbb{R}$, and in the previous inequalities, set

$$\xi(x,t,y,s) = \eta\left(\frac{x+y}{2}, \frac{t+s}{2}\right)\rho_\epsilon(x-y)\rho_\epsilon(t-s) \tag{7.40}$$

with $\eta \in C_c^1((-\infty,x_1]\times(0,\tau))$, $\eta \geq 0$. Then we obtain

$$\iint_{Q_1\times Q_1} \rho_\epsilon(x-y)\rho_\epsilon(t-s)\left\{|u_r(x,t) - v_r(y,s)|\,\eta_t\left(\frac{x+y}{2}, \frac{t+s}{2}\right)\right.$$
$$\left. + \operatorname{sgn}(u_r(x,t) - v_r(y,s))[\phi(u_r(x,t)) - \phi(v_r(y,s))]\eta_x\left(\frac{x+y}{2}, \frac{t+s}{2}\right)\right\} dxdtdyds$$
$$\geq \int_0^\tau \iint_{Q_1} [g_1^-(s) - \phi(u_r(x,t))]\,\eta\left(\frac{x+x_1}{2}, \frac{t+s}{2}\right)\rho_\epsilon(x_1-x)\rho_\epsilon(t-s)\, dxdtds$$

$$+\int_0^\tau \iint_{Q_1} [h_1^-(t) - \phi(v_r(y,s))]\eta\Big(\frac{x_1+y}{2}, \frac{t+s}{2}\Big)\rho_\epsilon(y-x_1)\rho_\epsilon(t-s)\,dydsdt. \tag{7.41}$$

Concerning the right-hand side of (7.41), by the well-known properties of mollifiers we have

$$\int_0^\tau \iint_{Q_1} g_1^-(s)\,\eta\Big(\frac{x+x_1}{2}, \frac{t+s}{2}\Big)\rho_\epsilon(x_1-x)\rho_\epsilon(t-s)\,dxdtds \to \frac{1}{2}\int_0^\tau g_1^-(s)\,\eta(x_1,s)\,ds$$

and

$$\int_0^\tau \iint_{Q_1} h_1^-(t)\,\eta\Big(\frac{x_1+y}{2}, \frac{t+s}{2}\Big)\rho_\epsilon(y-x_1)\rho_\epsilon(t-s)\,dydsdt \to \frac{1}{2}\int_0^\tau h_1^-(t)\,\eta(x_1,t)\,dt$$

as $\epsilon \to 0^+$. Moreover, since $\iint_{Q_1}\rho_\epsilon(x_1-x)\rho_\epsilon(t-s)\,dxds = \frac{1}{2}$ for $\epsilon < \min\{t, \tau-t\}$,

$$\begin{aligned}
&\left|\int_0^\tau \iint_{Q_1} \phi(u_r(x,t))\eta\Big(\frac{x+x_1}{2}, \frac{t+s}{2}\Big)\rho_\epsilon(x_1-x)\rho_\epsilon(t-s)\,dxdtds - \frac{1}{2}\int_0^\tau h_1^-(t)\,\eta(x_1,t)\,dt\right| \\
&\quad \le \iint_{Q_1}\rho_\epsilon(x_1-x)\,dxds \int_0^\tau \rho_\epsilon(t-s)\,\phi(u_r(x,t))\left|\eta\Big(\frac{x+x_1}{2}, \frac{t+s}{2}\Big) - \eta(x_1,t)\right| dt \\
&\qquad + \left|\int_{I_1}\rho_\epsilon(x_1-x)\,dx \int_0^\tau [\phi(u_r(x,t)) - h_1^-(t)]\,\eta\,(x_1,t)\,dt \int_0^\tau \rho_\epsilon(t-s)\,ds\right| \\
&\quad \le \frac{\|\phi\|_{L^\infty}}{2} \sup_{0\le x_1-x\le\epsilon} \int_0^\tau \sup_{|s-t|\le\epsilon}\left|\eta\Big(\frac{x_1+x}{2}, \frac{t+s}{2}\Big) - \eta(x_1,t)\right| dt \\
&\qquad + \|\rho_1\|_\infty \frac{1}{\epsilon}\int_{x_1-\epsilon}^{x_1}\left|\int_0^\tau [\phi(u_r(x,t)) - h_1^-(t)]\eta\,(x_1,t)\,dt\right| dx.
\end{aligned}$$

By the smoothness of η and equality (7.26) the right-hand side of the above inequality vanishes as $\epsilon \to 0^+$. Therefore

$$\int_0^\tau \iint_{Q_1} \phi(u_r(x,t))\eta\Big(\frac{x+x_1}{2}, \frac{t+s}{2}\Big)\rho_\epsilon(x_1-x)\rho_\epsilon(t-s)\,dxdtds \to \frac{1}{2}\int_0^\tau h_1^-(t)\,\eta(x_1,t)\,dt.$$

It is similarly seen that

$$\int_0^\tau \iint_{Q_1} \phi(v_r(y,s))\eta\Big(\frac{x_1+y}{2}, \frac{t+s}{2}\Big)\rho_\epsilon(y-x_1)\rho_\epsilon(t-s)\,dydsdt \to \frac{1}{2}\int_0^\tau g_1^-(s)\eta\,(x_1,s)\,ds.$$

Letting $\epsilon \to 0^+$ in (7.41), we obtain that, for every $\eta \in C_c^1((-\infty, x_1] \times (0,\tau))$, $\eta \geq 0$, we have

$$\iint_{Q_1} \{|u_r(x,t) - v_r(x,t)|\, \eta_t(x,t) + \operatorname{sgn}(u_r(x,t) - v_r(x,t))[\phi(u_r(x,t)) - \phi(v_r(x,t))]\eta_x(x,t)\}\, dxdt \geq 0. \tag{7.42}$$

Now fix t' and t'' such that $0 < t' < t'' < \tau$, and let $x_0 < x_1$. Let $\alpha_\delta = \alpha_\delta(x)$ and $\beta_\vartheta = \beta_\vartheta(t)$ be two families of mollifiers such that $0 < \delta < 1$ and $0 < \vartheta < \min\{t', \tau - t''\}$. In (7.42), set

$$\eta(x,t) = \eta_{\delta,\vartheta}(x,t) := \int_{t-t''}^{t-t'} \beta_\vartheta(z)\, dz \int_{\|\phi'\|_\infty(t-t'')+x_0}^{x_1+\delta} \alpha_\delta(x-y)\, dy \quad (\delta > 0)$$

with $x \in (\|\phi'\|_\infty(t-t'') + x_0, x_1]$ and $t \in (t' - \vartheta, t'' + \vartheta)$ (clearly, $\eta_{\delta,\vartheta}$ is nonnegative and belongs to $C_c^\infty((-\infty, x_1] \times (0,\tau))$). Since $\alpha_\delta(x - x_1 - \delta) = 0$ if $x \in (-\infty, x_1]$,

$$\begin{aligned}&\iint_{Q_1} |u_r - v_r|\, [\beta_\vartheta(t-t') - \beta_\vartheta(t-t'')]\left(\int_{\|\phi'\|_\infty(t-t'')+x_0}^{x_1+\delta} \alpha_\delta(x-y)\, dy\right) dxdt \\ &\quad - \iint_{Q_1} \{\|\phi'\|_\infty |u_r - v_r| + \operatorname{sgn}(u_r - v_r)[\phi(u_r) - \phi(v_r)]\} \\ &\quad \times \alpha_\delta(x - \|\phi'\|_\infty (t-t'') - x_0)\left(\int_{t-t''}^{t-t'} \beta_\vartheta(z)\, dz\right) dxdt \geq 0.\end{aligned}$$

Since $\|\phi'\|_\infty |y_1 - y_2| + \operatorname{sgn}(y_1 - y_2)[\phi(y_1) - \phi(y_2)] \geq 0$ for all $y_1, y_2 \geq 0$, it follows that

$$\iint_{Q_1} |u_r - v_r|\, [\beta_\vartheta(t-t') - \beta_\vartheta(t-t'')]\left(\int_{\|\phi'\|_\infty(t-t'')+x_0}^{x_1+\delta} \alpha_\delta(x-y)\, dy\right) dxdt \geq 0.$$

Let $\delta \to 0^+$ in this inequality. Then by the dominated convergence theorem we get

$$\int_0^\tau \int_{\|\phi'\|_\infty(t-t'')+x_0}^{x_1} |u_r(x,t) - v_r(x,t)|\, [\beta_\vartheta(t-t') - \beta_\vartheta(t-t'')]\, dxdt \geq 0,$$

whence as $\vartheta \to 0^+$ (recall that $u_r, v_r \in C([0,T]; L^1(\mathbb{R}))$ as $u, v \in C([0,T]; \mathfrak{R}_f^+(\Omega))$),

$$\int_{x_0}^{x_1} |u_r(x,t'') - v_r(x,t'')|\, dx \leq \int_{x_0 - \|\phi'\|_\infty(t''-t')}^{x_1} |u_r(x,t') - v_r(x,t')|\, dx. \tag{7.43}$$

Since $u, v \in C([0,T]; \mathfrak{R}_f^+(\Omega))$, letting $t' \to 0^+$ in (7.43), for all $(x_0, t'') \in Q_1$, we obtain

$$\int_{x_0}^{x_1} |u_r(x,t'') - v_r(x,t'')|\, dx \le \int_{x_0-\|\phi'\|_\infty t''}^{x_1} |u_r(x,0) - v_r(x,0)|\, dx = 0$$

for $u_r(\cdot,0) = v_r(\cdot,0) = u_{0r}$. By the arbitrariness of $t'' \in (0,\tau)$ it follows that $u_r = v_r$ in Q_1. □

Remark 7.2.6. Arguing as in Remark 6.3.2, it is easy to exhibit entropy solutions of problem (P) that do not satisfy $(C_\pm)$ thus are not *constructed solutions* (see Definition 7.3.1 and Theorem 7.3.3). Hence without compatibility conditions uniqueness fails.

7.3 Existence and regularity results

The existence of solutions is proven by an approximation procedure. Consider the approximating problem

$$\begin{cases} \partial_t u_n + \partial_x[\phi(u_n)] = 0 & \text{in } S, \\ u_n = u_{0n} & \text{in } \mathbb{R} \times \{0\} \quad (n \in \mathbb{N}). \end{cases} \tag{P_n}$$

Studying the limiting points of the sequence $\{u_n\}$, we will prove the following result.

Theorem 7.3.1. *Let (A_0)–(A_1) hold. Then problem (P) has a solution u, which is obtained as a limiting point of the sequence $\{u_n\}$ of entropy solutions to problems (P_n). In addition, u is a Young measure entropy solution of (P).*

Theorem 7.3.2. *Let (A_0)–(A_2) hold, and let u be the solution of (P) given by Theorem 7.3.1. Then u is an entropy solution of (P).*

The following definition is the counterpart of Definition 6.4.1.

Definition 7.3.1. Solutions of (P) given by Theorems 7.3.1 and 7.3.2 are called *constructed solutions* and *constructed entropy solutions*, respectively.

As in Section 6.4, constructed entropy solutions of (P) have several important properties. In particular, under the additional requirement $(A_2{}')$ (which in turn implies (A_1) and (A_2)), they satisfy the compatibility conditions. This is the content of the following proposition.

Theorem 7.3.3. *Let (A_0) and $(A_2{}')$ hold. Let u be a constructed entropy solution u of problem (P). Then for every $\tau \in (0,T)$:*

(i) *if ϕ is bounded, then u satisfies the compatibility conditions $(C_\pm)$ in $[0,\tau]$ at every point $x_j \in \operatorname{supp} u_s(\cdot,\tau)$;*
(ii) *if ϕ is unbounded, then $u = u_r \in L^\infty(\mathbb{R} \times (\tau,T))$.*

Remark 7.3.1. If ϕ is unbounded and satisfies assumption $(A_2{}')$, then by [67, Theorem 1.1] and Theorem 7.3.3(ii) for every $u_0 \in \mathfrak{R}_f^+(\mathbb{R}))$ there exists a unique entropy solution of problem (P) with waiting time $t_0 = 0$. In fact, by Theorem 7.3.3(ii) every constructed entropy solution u is a solution in the sense of [67], since $u = u_r \in L^\infty(\mathbb{R} \times (\tau, T))$ for every $\tau \in (0, T)$ and $\operatorname{ess\,lim}_{t\to 0^+} u(\cdot, t) = u_0$ narrowly in $\mathfrak{R}_f(\mathbb{R})$ (see Definition 5.1.3). This follows from (7.12) and Proposition 7.1.2(ii) (see also [53, Proposition 2, p. 38]).

When ϕ is bounded and u_0 is as in $(A_0{}')$, by [17, Proposition 3.20] every constructed entropy solution of (P) satisfies $u \in C([0, T]; \mathfrak{R}_f^+(\Omega))$ with $u(\cdot, 0) = u_0$. Thus by Theorems 7.3.3 and 7.2.4 we have the following existence and uniqueness result.

Theorem 7.3.4. *Let ϕ be bounded, and let $(A_0{}')$–$(A_2{}')$ hold. Then there exists a unique entropy solution $u \in C([0, T]; \mathfrak{R}_f^+(\Omega))$ of problem (P) that satisfies the compatibility conditions.*

Let us finally mention the following regularization result.

Proposition 7.3.5. *Let (A_0) and $(A_2{}')$ hold. Then for a.e. $t \in (0, T)$, the support of the singular part $u_s(\cdot, t)$ is a null set.*

7.4 Proof of existence results: the approximating problems

We assume that the initial data $u_{0n} \in L^1(\mathbb{R}) \cap L^\infty(\mathbb{R})$ of (P_n) have the following properties (see Lemma 6.5.1 and (6.49b)–(6.49c)):

$$u_{0n} \ge 0 \quad \text{in } \mathbb{R}, \quad \|u_{0n}\|_{L^1(\mathbb{R})} \le \|u_0\|_{\mathfrak{R}_f(\mathbb{R})}, \tag{7.44}$$

$$u_{0n} \overset{*}{\rightharpoonup} u_0, \quad u_{0n} \to u_{0r} \quad \text{a. e. in } \mathbb{R}, \quad \|u_{0n} - u_{0r}\|_{L^1_{\mathrm{loc}}(\mathbb{R}\setminus \operatorname{supp} u_{0s})} \to 0. \tag{7.45}$$

Let $\{u_{0n}^\epsilon\} \subseteq C_c^\infty(\mathbb{R})$, $u_{0n}^\epsilon \ge 0$, be any family such that

$$\|u_{0n}^\epsilon\|_{L^1(\mathbb{R})} \le \|u_{0n}\|_{L^1(\mathbb{R})} \le \|u_0\|_{\mathfrak{R}_f(\mathbb{R})}, \quad \|u_{0n}^\epsilon\|_{L^\infty(\mathbb{R})} \le \|u_{0n}\|_{L^\infty(\mathbb{R})}, \tag{7.46}$$

$$u_{0n}^\epsilon \to u_{0n} \quad \text{in } L^1(\mathbb{R}), \quad u_{0n}^\epsilon \overset{*}{\rightharpoonup} u_{0n} \quad \text{in } L^\infty(\mathbb{R}) \text{ as } \epsilon \to 0^+. \tag{7.47}$$

Let $\eta \in C_c^\infty(\mathbb{R})$ be a standard mollifier. Set $\eta_\epsilon(y) := \frac{1}{\epsilon}\eta(\frac{y}{\epsilon})$ $(\epsilon > 0)$ and

$$\phi_\epsilon(y) := (\eta_\epsilon * \overline{\phi})(y) - (\eta_\epsilon * \overline{\phi})(0) \quad (y \in \mathbb{R}), \tag{7.48}$$

where $\overline{\phi} := \phi\chi_{[0,\infty)}$. The regularized problem associated with (P_n),

$$\begin{cases} \partial_t u_n^\epsilon + \partial_x[\phi_\epsilon(u_n^\epsilon)] = \epsilon\, \partial_{xx} u_n^\epsilon & \text{in } S, \\ u_n^\epsilon = u_{0n}^\epsilon & \text{in } \mathbb{R} \times \{0\}, \end{cases} \tag{P_n^ϵ}$$

has a unique strong solution $u_n^\epsilon \in C([0,T];H^2(\mathbb{R})) \cap L^\infty(S)$ such that $\partial_t u_n^\epsilon \in L^2(S)$ and $u_n^\epsilon \geq 0$ (e. g., see [68]).

Some properties of the family $\{u_n^\epsilon\}$ are collected in the following lemmas. The proof of the first is almost standard (e. g., see [34]), and thus we omit it.

Lemma 7.4.1. *Let u_n^ϵ be the solution of problem (P_n^ϵ) ($\epsilon > 0$, $n \in \mathbb{N}$). Then:*

$$u_n^\epsilon \geq 0 \quad in\, S, \quad \|u_n^\epsilon\|_{L^\infty(S)} \leq \|u_{0n}\|_{L^\infty(\mathbb{R})}, \tag{7.49}$$

$$\|u_n^\epsilon(\cdot,t)\|_{L^1(\mathbb{R})} = \|u_{0n}^\epsilon\|_{L^1(\mathbb{R})} \quad (t \in (0,T)), \tag{7.50}$$

$$\|u_n^\epsilon\|_{L^\infty(0,T;L^1(\mathbb{R}))} \leq \|u_{0n}\|_{L^1(\mathbb{R})} \leq \|u_0\|_{\mathfrak{R}_f(\mathbb{R})}, \tag{7.51}$$

$$\|u_n^\epsilon(\cdot+h,\cdot) - u_n^\epsilon\|_{L^\infty(0,T;L^1(\mathbb{R}))} \leq \|u_{0n}^\epsilon(\cdot+h) - u_{0n}^\epsilon\|_{L^1(\mathbb{R})} \quad \textit{for all } h \in \mathbb{R}. \tag{7.52}$$

Lemma 7.4.2. *Let (7.1) hold. Then there exists $C > 0$ (only depending on $\|u_0\|_{\mathfrak{R}_f(\mathbb{R})}$) such that for any $n \in \mathbb{N}$, $\epsilon \in (0,1)$, and $p \in (0,1)$,*

$$\epsilon \iint_S (1+u_n^\epsilon)^{p-2} (\partial_x u_n^\epsilon)^2\, dxdt \leq \frac{C}{p(1-p)}. \tag{7.53}$$

Proof. Let $U \in C^2([0,\infty))$ with $U' \geq 0$ in $(0,\infty)$. Set

$$\Lambda_{U,\epsilon}(y) := \int_0^y U'(z)\,\phi_\epsilon'(z)\,dz + \vartheta_U \quad (\vartheta_U \in \mathbb{R}). \tag{7.54}$$

By (7.1) and (7.48), for all $y \geq 0$,

$$|\Lambda_{U,\epsilon}(y)| \leq \int_0^y U'(z)\,|\phi_\epsilon'(z)|\,ds + |\vartheta_U| \leq M\,[U(y) - U(0)] + |\vartheta_U|. \tag{7.55}$$

Multiplying the first equation in (P_n^ϵ) by $U'(u_n^\epsilon)$ gives

$$\partial_t[U(u_n^\epsilon)] + \partial_x[\Lambda_{U,\epsilon}(u_n^\epsilon)] = \epsilon\,\partial_{xx}[U(u_n^\epsilon)] - \epsilon\,U''(u_n^\epsilon)(\partial_x u_n^\epsilon)^2 \quad \text{in } S. \tag{7.56}$$

Hence for all $\zeta \in C^1([0,T];C_c^2(\mathbb{R}))$,

$$\begin{aligned} &\epsilon \iint_S U''(u_n^\epsilon)(\partial_x u_n^\epsilon)^2 \zeta\,dxdt + \int_{\mathbb{R}} U(u_n^\epsilon(x,T))\zeta(x,T)\,dx \\ &= \int_{\mathbb{R}} U(u_{0n}^\epsilon)\zeta(x,0)\,dx + \iint_S [U(u_n^\epsilon)\,\partial_t\zeta + \Lambda_{U,\epsilon}(u_n^\epsilon)\,\partial_x\zeta + \epsilon\,U(u_n^\epsilon)\,\partial_{xx}\zeta]\,dxdt. \end{aligned} \tag{7.57}$$

Choose $\vartheta_U = 0$, $U(y) = (1+u)^p - 1$ with $p \in (0,1)$, and

$$\zeta = \rho_k := \chi_{[-k,k]} + \rho(\cdot - k)\chi_{[k,k+1)} + \rho(\cdot + k)\chi_{\{(-k-1,-k]} \quad (k \in \mathbb{N})$$

with any $\rho \in C_c^2((-1,1))$ such that $\rho(0) = 1$, $0 \le \rho \le 1$, and $\rho'(0) = \rho''(0) = 0$. Then $0 \le U(y) \le y$ for $y \ge 0$, and, by (7.51), (7.55), and (7.57) we have

$$\begin{aligned}\epsilon\, p(1-p) &\iint_S (1+u_n^\epsilon)^{p-2} (\partial_x u_n^\epsilon)^2 \rho_k \, dxdt \\ &\le \int_{\mathbb{R}} u_{0n}^\epsilon(x)\, dx + \iint_S [M\, u_n^\epsilon\, |\rho_k'| + \epsilon u_n^\epsilon\, |\rho_k''|]\, dxdt \\ &\le [1 + (M+1) T \|\rho\|_{C^2([-1,1])}]\, \|u_0\|_{\mathfrak{R}_f(\mathbb{R})} =: C\end{aligned}$$

for all $\epsilon \in (0,1)$ and $k \in \mathbb{N}$. Letting $k \to \infty$ in this inequality gives (7.53), and hence the result follows. □

Lemma 7.4.3. *Let (7.1) hold, and let $U \in C^2([0,\infty))$. Let there exist $K \ge 0$ and $p \in (0,1)$ such that*

$$|U''(y)| \le K\,(1+y)^{p-2} \quad \text{for all } y \in [0,\infty). \tag{7.58}$$

Then there exists $C_p > 0$ such that for all $n \in \mathbb{N}$ and $\epsilon > 0$,

$$\epsilon \iint_S |U''(u_n^\epsilon)|\, (\partial_x u_n^\epsilon)^2\, dxdt \le C_p. \tag{7.59}$$

Proof. The result immediately follows from (7.53) and (7.58). □

Lemma 7.4.4. *Let (7.1) hold, and let $U \in C^2([0,\infty))$ with $U' \in L^\infty(0,\infty)$ satisfy (7.58). For any $\rho \in C_c^2(\mathbb{R})$ and $t \in (0,T)$, set*

$$U_{n,\rho}^\epsilon(t) := \int_{\mathbb{R}} U(u_n^\epsilon(x,t))\rho(x)\, dx, \quad U_{n,\rho}(t) := \int_{\mathbb{R}} U(u_n(x,t))\rho(x)\, dx. \tag{7.60}$$

Then the family $\{U_{n,\rho}^\epsilon\}$ is bounded in $BV(0,T)$.

Proof. By equality (7.56) we have

$$(U_{n,\rho}^\epsilon)'(t) = \int_{\mathbb{R}} [\Lambda_{U,\epsilon}(u_n^\epsilon)\rho' + \epsilon\, U(u_n^\epsilon)\rho'' - \epsilon\, U''(u_n^\epsilon)\, (\partial_x u_n^\epsilon)^2 \rho](x,t)\, dx. \tag{7.61}$$

Since $U' \in L^\infty(0,\infty)$, there exists $N > 0$ such that

$$|U(y)| \le N\,(1+y) \quad \text{for all } y \ge 0. \tag{7.62}$$

Hence by (7.1), (7.48), and (7.54) we have

$$|\Lambda_{U,\epsilon}(u_n^\epsilon)| \le M\, \|U'\|_\infty\, |u_n^\epsilon| + |\vartheta_U| =: \tilde{M} u_n^\epsilon + |\vartheta_U|,$$

whence by (7.61)

$$|(U^{\epsilon}_{n,\rho})'(t)| \le \|\rho\|_{C^2(\mathbb{R})} \int\limits_{\operatorname{supp}\rho} \{(\tilde{M}+\epsilon N)u^{\epsilon}_n(x,t) + \epsilon N + |\vartheta_U|\}\,dx$$

$$+\,\epsilon\,\|\rho\|_{L^\infty(\mathbb{R})} \int\limits_{\mathbb{R}} [|U''(u^{\epsilon}_n)|\;(\partial_x u^{\epsilon}_n)^2](x,t)\,dx. \tag{7.63}$$

By (7.63), (7.51), and (7.59) there exists $C_{p,\rho} > 0$ such that

$$\left\|(U^{\epsilon}_{n,\rho})'\right\|_{L^1(0,T)} \le \|\rho\|_{C^2(\mathbb{R})}[(\tilde{M}+N)\,T\,\|u_0\|_{\mathfrak{R}_f(\mathbb{R})} + C_{p,\rho}]. \tag{7.64}$$

On the other hand, by (7.51) and (7.62) we have

$$\left\|U^{\epsilon}_{n,\rho}\right\|_{L^1(0,T)} \le NT\|\rho\|_{L^\infty(\mathbb{R})}\,[\|u_0\|_{\mathfrak{R}_f(\mathbb{R})} + \lambda(\operatorname{supp}\rho)] \tag{7.65}$$

(as usual, λ denotes the Lebesgue measure in $\mathbb{R}$). From (7.64)–(7.65) the claim follows. □

Set $R_L := I_L \times (0,T)$, $I_L := (-L,L)$. From the above lemmas we get the following convergence results.

Lemma 7.4.5. *Let $\phi \in C([0,\infty))$. Then there exist a subsequence $\{u^{\epsilon_j}_n\} \subseteq \{u^{\epsilon}_n\}$, with $\epsilon_j \to 0^+$ as $j \to \infty$, and $u_n \in L^\infty(S) \cap L^\infty(0,T;L^1(\mathbb{R}))$, $u_n \ge 0$, such that, as $j \to \infty$,*

$$u^{\epsilon_j}_n \overset{*}{\rightharpoonup} u_n \quad \text{in } L^\infty(S), \tag{7.66}$$

$$u^{\epsilon_j}_n(\cdot,t) \overset{*}{\rightharpoonup} u_n(\cdot,t) \quad \text{in } L^\infty(\mathbb{R}) \quad \text{for a.e. } t \in (0,T), \tag{7.67}$$

$$u^{\epsilon_j}_n(\cdot,t) \to u_n(\cdot,t) \quad \text{in } L^1(I_L) \text{ for a.e. } t \in (0,T) \text{ and all } L > 0, \tag{7.68}$$

$$u^{\epsilon_j}_n \to u_n \quad \text{in } L^1(R_L) \text{ for all } L > 0, \tag{7.69}$$

$$u^{\epsilon_j}_n \to u_n, \quad \phi_{\epsilon_j}(u^{\epsilon_j}_n) \to \phi(u_n) \quad \text{a.e. in } S. \tag{7.70}$$

Moreover,

$$\|u_n\|_{L^\infty(S)} \le \|u_{0n}\|_{L^\infty(\mathbb{R})}, \quad \|u_n\|_{L^\infty(0,T;L^1(\mathbb{R}))} \le \|u_{0n}\|_{L^1(\mathbb{R})} \le \|u_0\|_{\mathfrak{R}_f(\mathbb{R})}. \tag{7.71}$$

Proof. The convergence in (7.66), the nonnegativity of u_n, and the first inequality in (7.71) immediately follow from (7.49). To prove (7.67), for any $\rho \in C^2_c(\mathbb{R})$, set

$$I^{\epsilon_j}_{n,\rho}(t) := \int\limits_{\mathbb{R}} u^{\epsilon_j}_n(x,t)\,\rho(x)\,dx \quad (t \in (0,T)).$$

By Lemma 7.4.4 with $U(y) = y$ the sequence $\{I^{\epsilon_j}_{n,\rho}\}$ is bounded in $BV(0,T)$. Hence there exist a subsequence $\{I^{\epsilon_j}_{n,\rho}\}$ (not relabeled for simplicity) and $I_{n,\rho} \in BV(0,T)$ such that

$$I^{\epsilon_j}_{n,\rho} \to I_{n,\rho} \quad \text{in } L^1(0,T) \text{ and } a.e. \text{ in } (0,T). \tag{7.72}$$

On the other hand, by (7.66)

$$\int_0^T I^{\epsilon_j}_{n,\rho}(t)\,dt = \iint_S u^{\epsilon_j}_n(x,t)\rho(x)\,dxdt \to \iint_S u_n(x,t)\rho(x)\,dxdt = \int_0^T dt \int_{\mathbb{R}} u_n(x,t)\rho(x)\,dx,$$

whence $I_{n,\rho} = \int_{\mathbb{R}} u_n(x,t)\rho(x)\,dx$ for a. e. $t \in (0,T)$), and (7.72) holds for the whole sequence $\{I^{\epsilon_j}_{n,\rho}\}$. To sum up, for any $\rho \in C^2_c(\mathbb{R})$, there exists a null set $N \subseteq (0,T)$ such that

$$\int_{\mathbb{R}} u^{\epsilon_j}_n(x,t)\rho(x)\,dx \to \int_{\mathbb{R}} u_n(x,t)\rho(x)\,dx$$

for all $t \in (0,T) \setminus N$. Since $C^2_c(\mathbb{R})$ is separable, the choice of the set N can be made independent of ρ, and thus (7.67) follows.

By (7.51), (7.52), and the Fréchet–Kolmogorov theorem, for all $t \in (0,T)$ and $L > 0$, the sequence $\{u^{\epsilon_j}_n(\cdot,t)\}$ is relatively compact in $L^1(I_L)$. Arguing as before, by (7.67) we plainly obtain (7.68). From (7.51) and (7.68) we get

$$\|u_n\|_{L^\infty(0,T;L^1(I_L))} \le \|u_{0n}\|_{L^1(\mathbb{R})} \le \|u_0\|_{\mathfrak{R}_f(\mathbb{R})} \quad \text{for all } L > 0,$$

whence by the arbitrariness of L the second inequality in (7.71) follows. From (7.51) and (7.68) by the dominated convergence theorem we also get the convergence in (7.69).

Finally, the first convergence in (7.70) follows from (7.69) (possibly extracting a subsequence, not relabeled). Since for any compact subset $K \subseteq \mathbb{R}$, $\|\phi_\epsilon - \phi\|_{C(K)} \to 0$ as $\epsilon \to 0^+$, the second convergence in (7.70) also follows. This completes the proof. □

Lemma 7.4.6. *Let (7.1) hold, and let $U \in C^2([0,\infty))$ with $U' \in L^\infty(0,\infty)$ satisfy (7.58). For any $\rho \in C^2_c(\mathbb{R})$ and $t \in (0,T)$, set*

$$U_{n,\rho}(t) := \int_{\mathbb{R}} U(u_n(x,t))\rho(x)\,dx \tag{7.73}$$

with $u_n \in L^\infty(S) \cap L^\infty(0,T;L^1(\mathbb{R}))$ given by Lemma 7.4.5. Then:

(i) *there exists a subsequence $\{U^{\epsilon_j}_{n,\rho}\} \subseteq \{U^{\epsilon}_{n,\rho}\}$ with $\epsilon_j \to 0^+$ as $j \to \infty$ ($U^{\epsilon}_{n,\rho}$ being defined by (7.60)) such that*

$$U^{\epsilon_j}_{n,\rho} \to U_{n,\rho} \quad \text{in } L^1(0,T); \tag{7.74}$$

(ii) *the family $\{U_{n,\rho}\}$ is bounded in $BV(0,T)$.*

Proof. Concerning claim (i), let $\{u_n^{\epsilon_j}\}$ be the subsequence in Lemma 7.4.5. Then for a. e. $t \in (0,T)$, by the first convergence in (7.70) we have $U(u_n^{\epsilon_j}(\cdot,t)) \to U(u_n(\cdot,t))$ a. e. in $\mathbb{R}$, whence (7.74) follows by the dominated convergence theorem. As for (ii), by (7.74) and (7.65) we get

$$\|U_{n,\rho}\|_{L^1(0,T)} = \lim_{j\to\infty} \|U_{n,\rho}^{\epsilon_j}\|_{L^1(0,T)} \leq NT\|\rho\|_{L^\infty(\mathbb{R})}\left[\|u_0\|_{\mathfrak{R}_f(\mathbb{R})} + \lambda(\operatorname{supp}\rho)\right]. \tag{7.75}$$

On the other hand, by (7.64) and the lower semicontinuity of the total variation in the $L^1(0,T)$-topology (see Remark 3.1.4(ii)) we have

$$\|U'_{n,\rho}\|_{\mathfrak{R}_f(0,T)} \leq \|\rho\|_{C^2(\mathbb{R})}[(\tilde{M}+N)\,T\,\|u_0\|_{\mathfrak{R}_f(\mathbb{R})} + C_{p,\rho}]. \tag{7.76}$$

From (7.75)–(7.76) claim (iii) follows. This completes the proof. □

Let us recall the following definition (e. g., see [34]).

Definition 7.4.1. A function $u_n \in L^\infty(0,T;L^1(\mathbb{R})) \cap L^\infty(S)$ is called an *entropy solution* of problem (P_n) if for every $\zeta \in C^1([0,T];C_c^1(\mathbb{R}))$, $\zeta(\cdot,T) = 0$ in $\mathbb{R}$, $\zeta \geq 0$ and for any $l \in [0,\infty)$ there holds

$$\iint_S [E_l(u_n)\,\partial_t\zeta + F_l(u_n)\,\partial_x\zeta]\,dxdt + \int_{\mathbb{R}} E_l(u_{0n})\,\zeta(x,0)\,dx \geq 0, \tag{7.77}$$

where $E_l(y) = |y-l|$, $F_l(y) = \operatorname{sgn}(y-l)[\phi(y)-\phi(l)]$ $(y \in [0,\infty))$.

Remark 7.4.1. Entropy solutions are weak solutions: if $\zeta \in C^1([0,T];C_c^1(\mathbb{R}))$ with $\zeta(\cdot,T) = 0$ in $\mathbb{R}$, then

$$\iint_S [u_n\partial_t\zeta + \phi(u_n)\partial_x\zeta]\,dxdt + \int_{\mathbb{R}} u_{0n}\,\zeta(x,0)\,dx = 0. \tag{7.78}$$

Proposition 7.4.7. *Let $\phi \in C([0,\infty))$. Then the function u_n given by Lemma 7.4.5 is an entropy solution of problem (P_n) $(n \in \mathbb{N})$. Moreover,*

$$\|u_n(\cdot,t)\|_{L^1(\mathbb{R})} = \|u_{0n}\|_{L^1(\mathbb{R})} \quad \textit{for a. e. } t \in (0,T) \tag{7.79}$$

and

$$\|u_n(\cdot+h,\cdot) - u_n\|_{L^\infty(0,T;L^1(\mathbb{R}))} \leq \|u_{0n}(\cdot+h) - u_{0n}\|_{L^1(\mathbb{R})} \quad \textit{for all } h \in \mathbb{R}. \tag{7.80}$$

If ϕ is locally Lipschitz continuous, then u_n is the unique entropy solution of (P_n).

Proof. Let ζ be as in Definition 7.4.1, and let $E_l(y) = |y - l|$, $F_{l,\epsilon}(y) = \operatorname{sgn}(y-l)[\phi_\epsilon(y) - \phi_\epsilon(l)]$ $(y, l \in [0,\infty))$. Then by a standard calculation from (P_n^ϵ) we get

$$\iint_S \{E_l(u_n^\epsilon)(\partial_t\zeta + \epsilon\,\zeta_{xx}) + F_{l,\epsilon}(u_n^\epsilon)\,\partial_x\zeta\}\,dxdt + \int_{\mathbb{R}} E_l(u_{0n}^\epsilon)\,\zeta(x,0)\,dx \geq 0. \tag{7.81}$$

Let $\{u_n^{\epsilon_j}\} \subseteq \{u_n^\epsilon\}$ be given by Lemma 7.4.5. By (7.48)–(7.49) we have

$$\|\phi_{\epsilon_j}(u_n^{\epsilon_j})\|_{L^\infty(S)} \leq \sup_{|y|\leq\|u_{0n}\|_{L^\infty(\mathbb{R})}} |\phi_{\epsilon_j}(y)| \leq \sup_{|y|\leq\|u_{0n}\|_{L^\infty(\mathbb{R})}} |\phi(y)|.$$

Then by (7.70), the above inequality, and the dominated convergence theorem

$$\iint_S F_{l,\epsilon_j}(u_n^{\epsilon_j})\,\partial_x\zeta\,dxdt \to \iint_S F_l(u_n)\,\partial_x\zeta\,dxdt.$$

By (7.69) and (7.47) we also have that

$$\iint_S E_l(u_n^{\epsilon_j})(\partial_t\zeta + \epsilon_j\,\zeta_{xx})\,dxdt \to \iint_S E_l(u_n)\,\partial_t\zeta\,dxdt$$

and

$$\int_{\mathbb{R}} E_l(u_{0n}^{\epsilon_j})\,\zeta(x,0)\,dx \to \int_{\mathbb{R}} E_l(u_{0n})\,\zeta(x,0)\,dx,$$

respectively. Then letting $j \to \infty$ in (7.81) (written with $\epsilon = \epsilon_j$), we obtain (7.77), and thus u_n is an entropy solution of problem (P_n). The uniqueness claim follows from Kružkov's uniqueness theorem (e. g., see [91]).

Inequality (7.80) follows from (7.52) and (7.68). Concerning (7.79), from (7.78) by a standard argument it follows that

$$\int_{\mathbb{R}} u_n(x,t)\,\rho(x)\,dx - \int_{\mathbb{R}} u_{0n}(x)\,\rho(x)\,dx = \int_0^t\int_{\mathbb{R}} \phi(u_n)(x,s)\,\rho'(x)\,dxds \tag{7.82}$$

for a. e. $t \in (0,T)$ and all $\rho \in C_c^1(\mathbb{R})$. Let $\rho_k \in C_c^1(\mathbb{R})$ $(k \in \mathbb{N})$ satisfy

$$\rho_k(x) = \begin{cases} 1 & \text{if } |x| \leq k, \\ 0 & \text{if } |x| \geq k+1, \end{cases}$$

and $\|\rho_k'\|_{L^\infty(\mathbb{R})} \le 2$. Since $u_n \in L^1(S)$, letting $k \to \infty$ in (7.82) with $\rho = \rho_k$ and using (7.1), we get

$$\left|\int_0^t \int_{\mathbb{R}} \phi(u_n)(x,s)\,\rho_k'(x)\,dxds\right| \le 2M \int_0^t \int_{\{k\le|x|\le k+1\}} |u_n(x,s)|\,dxds \to 0.$$

On the other hand, by the monotone convergence theorem we have

$$\int_{\mathbb{R}} u_n(x,t)\,\rho_k(x)\,dx \to \int_{\mathbb{R}} u_n(x,t)\,dx, \quad \int_{\mathbb{R}} u_{0n}(x)\,\rho_k(x)\,dx \to \int_{\mathbb{R}} u_{0n}(x)\,dx.$$

Then equality (7.79) follows from (7.82). This completes the proof. □

Let us prove the following result for future reference.

Lemma 7.4.8. *Let (7.1) hold, and let $U \in C^2([0,\infty)) \cap L^\infty(0,\infty)$ satisfy (7.58). Set*

$$\Lambda_U(y) := \int_0^y U'(z)\,\phi'(z)\,dz + \vartheta_U \quad (y, \vartheta_U \in \mathbb{R}), \tag{7.83}$$

and let $\Lambda_U \in L^\infty(0,\infty)$. Let u_n be the entropy solution of problem (P_n) $(n \in \mathbb{N})$ given by Proposition 7.4.7. Then for any $n \in \mathbb{N}$, there exists $\sigma_n \in \mathfrak{R}_f(S)$ such that for all $\zeta \in C_c^1(S)$,

$$\iint_S \{U(u_n)\,\partial_t\zeta + \Lambda_U(u_n)\,\partial_x\zeta\}\,dxdt = \langle \sigma_n, \zeta\rangle. \tag{7.84}$$

In addition, we have

$$\|\sigma_n\|_{\mathfrak{R}_f(S)} \le C_p \tag{7.85}$$

with $C_p > 0$ as in (7.59).

Proof. By inequality (7.59), for any fixed $n \in \mathbb{N}$, the family $\{\epsilon\, U''(u_n^\epsilon)(\partial_x u_n^\epsilon)^2\}$ is bounded in $L^1(S) \subseteq \mathfrak{R}_f(S)$. Hence by the Banach–Alaoglu theorem there exist a sequence $\{\epsilon_k\}$ such that $\epsilon_k \to 0^+$ as $k \to \infty$ and $\sigma_n \in \mathfrak{R}_f(S)$ such that

$$\epsilon_k\, U''(u_n^{\epsilon_k})(\partial_x u_n^{\epsilon_k})^2 \overset{*}{\rightharpoonup} \sigma_n \quad \text{as } k \to \infty. \tag{7.86}$$

Inequality (7.85) follows from (7.86) by the lower semicontinuity of the norm (see Remark 5.1.3).

Let $\zeta \in C_c^2(S)$. Then by (7.56) we have

$$\epsilon \iint_S U''(u_n^\epsilon)(\partial_x u_n^\epsilon)^2\zeta\,dxdt = \iint_S [U(u_n^\epsilon)\,\partial_t\zeta + \Lambda_{U,\epsilon}(u_n^\epsilon)\,\partial_x\zeta + \epsilon\, U(u_n^\epsilon)\,\partial_{xx}\zeta]\,dxdt \tag{7.87}$$

with $\Lambda_{U,\epsilon}$ defined by (7.54). By (7.1) and the second inequality in (7.49), for any fixed $n \in \mathbb{N}$, we have

$$|\Lambda_{U,\epsilon}(u_n^\epsilon)| \leq M \int_0^{\|u_{0n}\|_{L^\infty(\mathbb{R})}} |U'(z)|\, ds + |\vartheta_U| < \infty,$$

and thus for fixed $n \in \mathbb{N}$, the family $\{\Lambda_{U,\epsilon}(u_n^\epsilon)\}$ is bounded in $L^\infty(S)$. Then letting $k \to \infty$ in (7.87) written with $\epsilon = \epsilon_k$ and arguing as in the proof of Proposition 7.4.7, equality (7.84) follows. □

7.5 Proof of existence results

The proof of Theorem 7.3.1 relies on the following result, which is the counterpart of Theorem 6.6.1. As in the case of Theorem 6.6.1, the proof requires several preliminary results.

Theorem 7.5.1. *Let (A_0)–(A_1) hold, and let $\{u_n\}$ be the sequence of entropy solutions of problems (P_n). Then there exists a subsequence $\{u_{n_k}\} \subseteq \{u_n\}$ with the following properties:*

(i) *there exists $u \in L^\infty_{w*}(0, T; \mathfrak{R}_f^+(\mathbb{R}))$ such that*

$$u_{n_k} \overset{*}{\rightharpoonup} u \quad \text{in } L^\infty_{w*}(0, T; \mathfrak{R}_f(\mathbb{R})); \tag{7.88}$$

(ii) *the sequence $\{\nu_{n_k}\} \subseteq \mathfrak{Y}^+(S; \mathbb{R})$ of the Young measures associated with $\{u_{n_k}\}$ converges narrowly to a Young measure $\nu \in \mathfrak{Y}^+(S; \mathbb{R})$, and for a. e. $(x,t) \in S$, we have*

$$u_r(x,t) = \int_{[0,\infty)} y\, d\nu_{(x,t)}(y) \tag{7.89}$$

and

$$\phi(u_r(x,t)) = \int_{[0,\infty)} \phi(y)\, d\nu_{(x,t)}(y), \tag{7.90}$$

where $\{\nu_{(x,t)}\}_{(x,t)\in S}$ is the disintegration of ν;

(iii) *for every $U \in C([0,\infty))$ satisfying*

$$\lim_{y\to\infty} \frac{U(y)}{y} =: M_U \in \mathbb{R} \tag{7.91}$$

and every $\zeta \in C([0,T];C_c(\mathbb{R}))$*, we have*

$$\lim_{k\to\infty}\iint\limits_S U(u_{n_k})\zeta\,dxdt = \iint\limits_S U^*\zeta\,dxdt + M_U\int_0^T\langle u_s(\cdot,t),\zeta(\cdot,t)\rangle\,dt, \tag{7.92}$$

where $U^* \in L^\infty(0,T;L^1_{\mathrm{loc}}(\mathbb{R}))$ *is defined as*

$$U^*(x,t) := \int\limits_{[0,\infty)} U(y)\,d\nu_{(x,t)}(y) \quad \textit{for a.e. } (x,t)\in S. \tag{7.93}$$

To prove Theorem 7.5.1, we need the following.

Proposition 7.5.2. *Let* (A_0)*–*(A_1) *hold, and let* $\{u_n\}$ *be the sequence of entropy solutions of problems* (P_n)*. Then there exist a subsequence* $\{u_{n_k}\}\subseteq\{u_n\}$*, a Young measure* $\nu\in\mathfrak{Y}^+(S;\mathbb{R})$*, and* $\sigma\in L^\infty_{w^*}(0,T;\mathfrak{R}^+_f(\mathbb{R}))$ *such that*

$$u_{n_k}\overset{*}{\rightharpoonup} u_b+\sigma \quad \textit{in } L^\infty_{w*}(0,T;\mathfrak{R}_f(\mathbb{R})), \tag{7.94}$$

where $u_b\in L^\infty(0,T;L^1(\mathbb{R}))$ *is defined as*

$$u_b(x,t) := \int\limits_{[0,\infty)} y\,d\nu_{(x,t)}(y) \quad \textit{for a.e. } (x,t)\in S. \tag{7.95}$$

Proof. By the second inequality in (7.71) the sequence $\{u_n\}$ is bounded in $L^\infty(0,T;L^1(\mathbb{R}))\subseteq L^\infty_{w^*}(0,T;\mathfrak{R}_f(\mathbb{R}))$. Then by Proposition 4.4.16 and the Banach–Alaoglu theorem there exist $\mu\in L^\infty_{w^*}(0,T;\mathfrak{R}^+_f(\mathbb{R}))$ such that

$$u_{n_k}\overset{*}{\rightharpoonup}\mu \quad \text{in } L^\infty_{w*}(0,T;\mathfrak{R}_f(\mathbb{R})). \tag{7.96}$$

On the other hand, since the sequence $\{u_{n_k}\}$ is bounded in $L^1(S)$, by Remark 5.4.4(ii) there exist a subsequence of $\{u_{n_k}\}$ (not relabeled) and $\sigma\in\mathfrak{R}^+_f(S)$ such that

$$u_{n_k}\overset{*}{\rightharpoonup} u_b+\sigma \quad \text{in } \mathfrak{R}_f(S). \tag{7.97}$$

By (7.96)–(7.97) we have $\mu = u_b+\sigma$, and hence the result follows. □

Proposition 7.5.3. *Let* (A_0)*–*(A_1) *hold, and let* $\{u_n\}$ *be the sequence of entropy solutions of problems* (P_n)*. Let* $\{u_{n_k}\}\subseteq\{u_n\}$*,* $\nu\in\mathfrak{Y}^+(S;\mathbb{R})$*, and* $\sigma\in L^\infty_{w^*}(0,T;\mathfrak{R}^+_f(\mathbb{R}))$ *be given by Proposition 7.5.2. Let* $U\in C([0,\infty))$ *satisfy (7.91). Then for any* $\zeta\in C([0,T];C_c(\mathbb{R}))$*, we have*

$$\lim_{k\to\infty}\iint\limits_S U(u_{n_k})\zeta\,dxdt = \iint\limits_S U^*\zeta\,dxdt + M_U\int_0^T\langle\sigma(\cdot,t),\zeta(\cdot,t)\rangle\,dt \tag{7.98}$$

with $U^* \in L^\infty(0,T;L^1_{\mathrm{loc}}(\mathbb{R}))$ *given by* (7.93).

Proof. Fix arbitrary $L>0$. By (7.91), for any $\epsilon>0$, there exists $y_\epsilon>0$ such that

$$-\epsilon y < U(y) - M_U y < \epsilon y \quad \text{for all } y > y_\epsilon. \tag{7.99}$$

For any $j \in \mathbb{N}$, $j \geq [y_\epsilon]+1$, let $f_{j1}, f_{j2} \in C([0,\infty))$ satisfy

$$\begin{cases} 0 \leq f_{j1} \leq 1,\ 0 \leq f_{j2} \leq 1,\ f_{j1}+f_{j2} = 1 \text{ in } [0,\infty), \\ \operatorname{supp} f_{j1} \subseteq [0,j+1],\ \operatorname{supp} f_{j2} \subseteq [j,\infty). \end{cases} \tag{7.100}$$

Then, using (7.99) and (7.71), for any fixed $k \in \mathbb{N}$ and $t \in (0,T)$, we get

$$\begin{aligned} &\int_{I_L} |U(u_{n_k}(x,t))|\,dx \\ &\quad = \int_{I_L} f_{j1}(u_{n_k}(x,t))\,|U(u_{n_k}(x,t))|\,dx + \int_{I_L} f_{j2}(u_{n_k}(x,t))\,|U(u_{n_k}(x,t))|\,dx \\ &\quad \leq 2L\ \|U\|_{L^\infty([0,j+1])} + \int_{\{u_{n_k}(\cdot,t)>y_\epsilon\}} |U(u_{n_k}(x,t))|\,dx \\ &\quad \leq 2L\ \|U\|_{L^\infty([0,j+1])} + (|M_U|+\epsilon)\int_{I_L} |u_{n_k}(x,t)|\,dx \\ &\quad \leq 2L\ \|U\|_{L^\infty([0,j+1])} + (|M_U|+\epsilon)\|u_0\|_{\mathfrak{R}_f(\mathbb{R})} \end{aligned}$$

(recall that $I_L = (-L,L)$), whence

$$\|U(u_{n_k})\|_{L^\infty(0,T;L^1(I_L))} \leq 2L\ \|U\|_{L^\infty([0,j+1])} + (|M_U|+\epsilon)\|u_0\|_{\mathfrak{R}_f(\mathbb{R})}. \tag{7.101}$$

In view of this inequality, by the Banach–Alaoglu theorem there exist a subsequence of $\{U(u_{n_k})\}$ (not relabeled) and $\mu \in L^\infty_{w^*}(0,T;\mathfrak{R}_f(I_L))$ such that

$$U(u_{n_k}) \overset{*}{\rightharpoonup} \mu \quad \text{in } L^\infty_{w*}(0,T;\mathfrak{R}_f(I_L)). \tag{7.102}$$

Moreover, arguing as in the proof of Proposition 5.4.13, we obtain that

$$U(u_{n_k}) \overset{*}{\rightharpoonup} U^* + M_U\sigma \quad \text{in } \mathfrak{R}_f(R_L). \tag{7.103}$$

By (7.102)–(7.103) we obtain that for any $\zeta \in C([0,T];C_c(I_L))$,

$$\lim_{k\to\infty} \iint_{R_L} U(u_{n_k})\,\zeta\,dxdt = \iint_{R_L} U^*\zeta\,dxdt + M_U\int_0^T \langle\sigma(\cdot,t),\zeta(\cdot,t)\rangle\,dt \tag{7.104}$$

with U^* defined by (7.93) a. e. in R_L.

Equality (7.98) follows from (7.104), since for any $\zeta \in C([0,T];C_c(\mathbb{R}))$, we can choose $L>0$ so large that $\operatorname{supp}\zeta \subseteq ((-L,L)\times[0,T])$. In addition, by (7.91) there exists $\overline{M}>0$ such that $|U(y)| \le \overline{M}(y+1)$ for all $y \ge 0$, and hence $|U^*| \le \overline{M}(u_b+1)$ a. e. in S (see (7.93) and (7.95)). Since $u_b \in L^\infty(0,T;L^1(\mathbb{R}))$, it follows that $U^* \in L^\infty(0,T;L^1_{\mathrm{loc}}(\mathbb{R}))$. Then the result follows. □

To proceed, we need the following lemma, whose proof is given at the end of the section.

Lemma 7.5.4. *Let (A_0)–(A_1) hold, and let $U \in C([0,\infty))$ satisfy (7.91). Let $\{u_{n_k}\}$, σ, and $U^* \in L^\infty(0,T;L^1_{\mathrm{loc}}(\mathbb{R}))$ be as in Proposition 7.5.3. Then for every $\zeta \in C([0,T];C_c(\mathbb{R}))$,*

$$\lim_{k\to\infty}\int_0^T\left|\int_{\mathbb{R}} U(u_{n_k}(x,t))\,\zeta(x,t)\,dx-\int_{\mathbb{R}} U^*(x,t)\,\zeta(x,t)\,dx-M_U\langle\sigma(\cdot,t),\zeta(\cdot,t)\rangle\right|dt=0. \tag{7.105}$$

Remark 7.5.1. Choosing $\zeta(x,t)=\rho(x)$ with $\rho \in C_c(\mathbb{R})$, from (7.105) we obtain

$$\lim_{k\to\infty}\int_0^T\left|\int_{\mathbb{R}} U(u_{n_k}(x,t))\,\rho(x)\,dx-U^*_\rho(t)-M_U\langle\sigma(\cdot,t),\rho\rangle\right|dt=0, \tag{7.106a}$$

where

$$U^*_\rho(t) := \int_{\mathbb{R}} U^*(x,t)\rho(x)\,dx. \tag{7.106b}$$

By (7.106a) and the separability of $C_c(\mathbb{R})$, for every $L>0$ and U as in Lemma 7.5.4, there exist a null set $N \subseteq (0,T)$ and a subsequence of $\{u_{n_k}\}$ (not relabeled) such that for all $t \in (0,T)\setminus N$,

$$U(u_{n_k})(\cdot,t) \overset{*}{\rightharpoonup} U^*(\cdot,t)+M_U\,\sigma(\cdot,t) \quad \text{in } \mathfrak{R}_f(I_L). \tag{7.107}$$

Proposition 7.5.5. *Let (A_0)–(A_1) hold, and let $\{u_n\}$ be the sequence of entropy solutions of problems (P_n). Let $\{u_{n_k}\} \subseteq \{u_n\}$ and $\sigma \in L^\infty_{w^*}(0,T;\mathfrak{R}^+_f(\mathbb{R}))$ be given by Proposition 7.5.2, and let $u \in L^\infty_{w^*}(0,T;\mathfrak{R}^+_f(\mathbb{R}))$ be given as*

$$u := u_b+\sigma. \tag{7.108}$$

Then

$$u_r = u_b \quad a.e.\ in\ S, \tag{7.109a}$$

$$u_s=\sigma \quad in\ L^\infty_{w^*}(0,T;\mathfrak{R}_f(\mathbb{R})). \tag{7.109b}$$

Proof. For any nonnegative $\zeta \in C^1([0,T]; C_c^1(\mathbb{R}))$ such that $\zeta(\cdot,T) = 0$, adding (7.77) and (7.78) gives

$$\iint_S \{U_j(u_{n_k})\,\partial_t\zeta + \Lambda_{U_j}(u_{n_k})\,\partial_x\zeta\}\,dxdt \geq -\int_{\mathbb{R}} U_j(u_{0n_k})(x)\,\zeta(x,0)\,dx,$$

where for all $j \in \mathbb{N}$,

$$U_j(y) := (y-j)\chi_{[j,\infty)}(y), \quad \Lambda_{U_j}(y) := \int_0^y U_j'(z)\,\phi'(z)\,dz = [\phi(y)-\phi(j)]\chi_{[j,\infty)}(y). \quad (7.110)$$

By standard approximation arguments, for every ζ as above and for *a.e.* $\tau \in (0,T)$, we have

$$\int_{\mathbb{R}} U_j(u_{n_k})(x,\tau)\,\zeta(x,\tau)\,dx - \int_{\mathbb{R}} U_j(u_{0n_k})(x)\,\zeta(x,0)\,dx \leq \iint_{S_\tau} \{U_j(u_{n_k})\,\partial_t\zeta + \Lambda_{U_j}(u_{n_k})\,\partial_x\zeta\}\,dxdt, \quad (7.111)$$

where $S_\tau := \mathbb{R}\times(0,\tau)$. Since $M_{U_j} = 1$ and $M_{\Lambda_{U_j}} = 0$, by (7.105) we have

$$\lim_{k\to\infty}\int_0^\tau \left|\int_{\mathbb{R}} [U_j(u_{n_k})\partial_t\zeta](x,t)\,dx - \int_{\mathbb{R}} [U_j^*\partial_t\zeta](x,t)\,dx - \langle\sigma(\cdot,t),\partial_t\zeta(\cdot,t)\rangle\right| dt = 0, \quad (7.112a)$$

$$\lim_{k\to\infty}\int_0^\tau \left|\int_{\mathbb{R}} [\Lambda_{U_j}(u_{n_k})\partial_x\zeta](x,t)\,dx - \int_{\mathbb{R}} [\Lambda_{U_j}^*\partial_x\zeta](x,t)\,dx\right| dt = 0, \quad (7.112b)$$

where

$$U_j^* := \int_{[0,\infty)} U_j(y)\,d\nu(y), \quad \Lambda_{U_j}^* := \int_{[0,\infty)} \Lambda_{U_j}(y)\,d\nu(y).$$

From (7.112) it follows that

$$\lim_{k\to\infty}\iint_{S_\tau} \{U_j(u_{n_k})\,\partial_t\zeta + \Lambda_{U_j}(u_{n_k})\,\partial_x\zeta\}\,dxdt$$
$$= \iint_{S_\tau} \{U_j^*\,\partial_t\zeta + \Lambda_{U_j}^*\partial_x\zeta\}\,dxdt + \int_0^\tau \langle\sigma(\cdot,t),\partial_t\zeta(\cdot,t)\rangle\,dt. \quad (7.113)$$

On the other hand, by (7.107) and a diagonal argument there exist a null set $N \subseteq (0,T)$ and a subsequence of $\{u_{n_k}\}$ (not relabeled) such that for all $\tau \in (0,T) \setminus N$ and $j \in \mathbb{N}$,

$$\lim_{k\to\infty} \int_{\mathbb{R}} U_j(u_{n_k})(x,\tau)\,\zeta(x,\tau)\,dx = \int_{\mathbb{R}} U_j^*(x,\tau)\,\zeta(x,\tau)\,dx + \langle \sigma(\cdot,\tau), \zeta(x,\tau)\rangle. \tag{7.114}$$

Moreover, since the sequence $\{U_j(u_{0n_k}) - u_{0n_k}\}$ is bounded in $L^\infty(\mathbb{R})$ and

$$\lim_{k\to\infty} (U_j(u_{0n_k}) - u_{0n_k}) = U_j(u_{0r}) - u_{0r} \quad \text{a. e. in } \mathbb{R}$$

(see (7.45)), it follows from (7.45) that

$$\lim_{k\to\infty} \int_{\mathbb{R}} U_j(u_{0n_k})(x)\zeta(x,0)\,dx = \int_{\mathbb{R}} U_j(u_{0r})(x)\zeta(x,0)\,dx + \langle u_{0s}, \zeta(\cdot,0)\rangle. \tag{7.115}$$

Letting $k \to \infty$ in (7.111), by (7.113)–(7.115), for all $\tau \in (0,T) \setminus N$ and $j \in \mathbb{N}$, we obtain

$$\begin{aligned}\int_{\mathbb{R}} U_j^*(x,\tau)\,\zeta(x,\tau)\,dx + \langle \sigma(\cdot,\tau), \zeta(\cdot,\tau)\rangle &\le \iint_{S_\tau} \{U_j^*\,\partial_t\zeta + \Lambda^*_{U_j}\partial_x\zeta\}\,dxdt \\ &+ \int_0^\tau \langle \sigma(\cdot,t), \partial_t\zeta(\cdot,t)\rangle\,dt + \int_{\mathbb{R}} U_j(u_{0r})(x)\zeta(x,0)\,dx + \langle u_{0s}, \zeta(\cdot,0)\rangle.\end{aligned} \tag{7.116}$$

Since $0 \le U_j(y) \le y\chi_{[j,\infty)}(y)$ and $|\Lambda_{U_j}(y)| \le My\chi_{[j,\infty)}(y)$ for all $y \ge 0$ (see (7.1)), we have $|U_j^*| \le u_b$ and $|\Lambda^*_{U_j}| \le Mu_b$ ($j \in \mathbb{N}$). Since $u_b \in L^\infty(0,T;L^1(\mathbb{R}))$, it plainly follows that

$$\lim_{j\to\infty} U_j^* = \lim_{j\to\infty} \Lambda^*_{U_j} = 0 \quad \text{a. e. in } S \text{ and in } L^1(S), \tag{7.117}$$

and thus letting $j \to \infty$ in (7.116) gives, for all $\tau \in (0,T) \setminus N$ and $\zeta \in C^1([0,T];C^1_c(\mathbb{R}))$,

$$\langle \sigma(\cdot,\tau), \zeta(\cdot,\tau)\rangle \le \int_0^\tau \langle \sigma(\cdot,t), \partial_t\zeta(\cdot,t)\rangle\,dt + \langle u_{0s}, \zeta(\cdot,0)\rangle. \tag{7.118}$$

Choosing $\zeta(x,t) = \rho(x)$ in (7.118) gives $\langle \sigma(\cdot,\tau), \rho\rangle \le \langle u_{0s}, \rho\rangle$ for any nonnegative $\rho \in C^1_c(\mathbb{R})$, whence $\sigma(\cdot,\tau) \ll u_{0s}$ (recall that σ is nonnegative). Hence for any $\tau \in (0,T) \setminus N$, $\sigma(\cdot,\tau)$ is singular with respect to the Lebesgue measure. On the other hand, since $u_b \in L^\infty(0,T;L^1(\mathbb{R}))$, the measure $E \mapsto \iint_E u_b\,dxdt$ ($E \in \mathcal{B}(\mathbb{R}^2) \cap S$) is absolutely continuous with respect to the Lebesgue measure. Then from the uniqueness of the Lebesgue decomposition (see Theorem 1.8.9) the conclusion follows. □

Proposition 7.5.6. *Let* (A_0)–(A_1) *hold. Let* $\nu \in \mathfrak{Y}^+(S;\mathbb{R})$ *be given by Proposition 7.5.2, and let* $u \in L^\infty_{w^*}(0,T;\mathfrak{R}_f(\mathbb{R}))$ *be defined by (7.108). Then*

$$\phi^*(x,t) = \phi(u_r(x,t)) \quad \text{for a.e. } (x,t) \in S \tag{7.119}$$

with ϕ^* *defined by (7.4).*

Proof. Let $U, V \in C^2([0,\infty)) \cap L^\infty(0,\infty)$ satisfy (7.58), and assume that $\Lambda_U, \Lambda_V \in L^\infty(0,\infty)$ (see (7.83)). In view of Lemma 7.4.8, for any $n \in \mathbb{N}$, there exist $\sigma_n, \tau_n \in \mathfrak{R}_f(S)$ such that for all $\zeta \in C^1_c(S)$,

$$\iint_S \{U(u_n)\,\partial_t\zeta + \Lambda_U(u_n)\,\partial_x\zeta\}\,dxdt = \langle \sigma_n, \zeta\rangle, \tag{7.120a}$$

$$\iint_S \{V(u_n)\,\partial_t\zeta + \Lambda_V(u_n)\,\partial_x\zeta\}\,dxdt = \langle \tau_n, \zeta\rangle, \tag{7.120b}$$

where u_n is the entropy solution of (P_n) given by Proposition 7.4.7.

Let $A \Subset S$ be a bounded open set, and let $Y_n, Z_n : A \mapsto \mathbb{R}^2$ be defined by

$$Y_n := (\Lambda_U(u_n), U(u_n)), \quad Z_n := (V(u_n), -\Lambda_V(u_n)).$$

By (7.120) we have

$$\operatorname{div} Y_n = -\sigma_n, \quad \operatorname{curl} Z_n = -\tau_n \quad \text{in } \mathscr{D}^*(A). \tag{7.121}$$

Since U, Λ_U, V, Λ_V are bounded in $(0,\infty)$, the sequences $\{U(u_n)\}$, $\{\Lambda_U(u_n)\}$, $\{V(u_n)\}$, and $\{\Lambda_V(u_n)\}$ are bounded $L^\infty(A)$ and thus are bounded in $L^1(A)$ and uniformly integrable. Then by Proposition 5.4.10 in $L^1(A)$ we have

$$U(u_n) \rightharpoonup U^* := \int_{[0,\infty)} U(y)\,d\nu(y), \quad \Lambda_U(u_n) \rightharpoonup \Lambda_U^* := \int_{[0,\infty)} \Lambda_U(y)\,d\nu(y),$$

$$V(u_n) \rightharpoonup V^* := \int_{[0,\infty)} V(y)\,d\nu(y), \quad \Lambda_V(u_n) \rightharpoonup \Lambda_V^* := \int_{[0,\infty)} \Lambda_V(y)\,d\nu(y).$$

By the same token, $\{U(u_n)\}$, $\{\Lambda_U(u_n)\}$, $\{V(u_n)\}$, and $\{\Lambda_V(u_n)\}$ are bounded and thus weakly converging in $L^2(A)$. It follows that, as $n \to \infty$,

$$Y_n \rightharpoonup Y^* := (\Lambda_U^*, U^*), \quad Z_n \rightharpoonup Z^* := (V^*, -\Lambda_V^*) \quad \text{in } L^2(A;\mathbb{R}^2). \tag{7.122}$$

By a similar argument we have that

$$Y_n \cdot Z_n := \Lambda_U(u_n)V(u_n) - \Lambda_V(u_n)U(u_n) \rightharpoonup \int_{[0,\infty)} [\Lambda_U(y)V(y) - \Lambda_V(y)U(y)]d\nu(y) \quad \text{in } L^2(A). \tag{7.123}$$

By (7.85) (applied to both σ_n and τ_n) and (7.121), $\{\operatorname{div} Y_n\}$ and $\{\operatorname{curl} Z_n\}$ are precompact in $H^{-1}(A)$. Then by (7.122) and the div-curl lemma we get

$$Y_n \cdot Z_n \to Y^* \cdot Z^* = \Lambda_U^* V^* - \Lambda_V^* U^* \quad \text{in } \mathscr{D}^*(A) \tag{7.124}$$

(e. g., see [46, Corollary 1.3.1 and Theorem 5.2.1]). By (7.123)–(7.124) it follows that

$$\int_{[0,\infty)} [\Lambda_U(y) - \Lambda_U^*]\, V(y)\, dv(y) = \int_{[0,\infty)} [U(y) - U^*]\, \Lambda_V(y)\, dv(y) \quad \text{a. e. in } A. \tag{7.125}$$

Let U be as above, with $U' > 0$ in $(0,\infty)$. By a standard approximation argument we may choose $V(y) = |U(y) - U^*|$, and thus by a proper choice of ϑ_V we obtain

$$\Lambda_V(y) = \int_0^y V'(z)\,\phi'(z)\, dz + \vartheta_V = \operatorname{sgn}(U(y) - U^*)[\Lambda_U(y) - \Lambda_U(U^{-1}(U^*))],$$

whence by (7.125)

$$[\Lambda_U^* - \Lambda_U(U^{-1}(U^*))] \int_{[0,\infty)} |U(y) - U^*|\, dv(y) = 0 \quad \text{a. e. in } A. \tag{7.126}$$

Let $U_p \in C^2([0,\infty)) \cap L^\infty(0,\infty)$, $U_p(0) = 0$ ($p \in \mathbb{N}$), satisfy (7.58) and

$$0 < U_p' \le U_{p+1}' \le 1, \quad \lim_{p\to\infty} U_p'(y) = 1 \quad \text{in } [0,\infty), \tag{7.127}$$

and thus $\lim_{p\to\infty} U_p(y) = y$ for all $y \in [0,\infty)$. Then by the monotone convergence theorem, (7.95), and (7.109a) we get

$$\lim_{p\to\infty} U_p^* = \lim_{p\to\infty} \int_{[0,\infty)} U_p(y)\, dv(y) = u_r \quad \text{a. e. in } A. \tag{7.128}$$

We will prove that

$$\lim_{p\to\infty} [\Lambda_{U_p}^* - \Lambda_{U_p}(U_p^{-1}(U_p^*))] = \phi^* - \phi(u_r) \quad \text{a. e. in } A. \tag{7.129}$$

Using (7.128)–(7.129), we can complete the proof. Indeed, by (7.128) and the dominated convergence theorem we plainly get, for a. e. $(x,t) \in A$,

$$\lim_{p\to\infty} \int_{[0,\infty)} |U_p(y) - U_p^*(x,t)|\, dv_{(x,t)}(y) = \int_{[0,\infty)} |y - u_r(x,t)|\, dv_{(x,t)}(y). \tag{7.130}$$

Letting $p \to \infty$ in (7.126) written with $U = U_p$, by (7.129)–(7.130) we obtain that for a. e. $(x,t) \in A$,

$$[\phi^*(x,t) - \phi(u_r)(x,t)] \int_{[0,\infty)} |y - u_r(x,t)|\, d\nu_{(x,t)}(y) = 0.$$

By the arbitrariness of A the above equality implies (7.119).

It remains to prove (7.129). Observe that

$$\Lambda^*_{U_p} - \Lambda_{U_p}(U_p{}^{-1}(U_p^*)) = \int_{[0,\infty)} \left(\int_0^y U_p'(z)\, \phi'(z)\, dz \right) d\nu(y) - \int_0^{U_p{}^{-1}(U_p{}^*)} U_p'(z)\, \phi'(z)\, dz. \tag{7.131}$$

By (7.1) and (7.127) we have $U_p'(y) \to 1$ and $|U_p'(y)\phi'(y)| \le M$ for all $y \ge 0$, and hence by the dominated convergence theorem

$$\lim_{p\to\infty} \Lambda^*_{U_p}(x,t) = \phi^*(x,t) \quad \text{for a. e. } (x,t) \in A. \tag{7.132}$$

On the other hand,

$$\begin{aligned}
&\int_0^{U_p{}^{-1}(U_p^*(x,t))} U_p'(z)\, \phi'(z)\, dz - \phi(u_r)(x,t) \\
&= \int_0^{u_r(x,t)} [U_p'(z) - 1]\phi'(z)\, dz + \int_{u_r(x,t)}^{U_p{}^{-1}(U_p^*(x,t))} U_p'(z)\, \phi'(z)\, dz.
\end{aligned} \tag{7.133}$$

Arguing as for (7.132) shows that

$$\lim_{p\to\infty} \int_0^{u_r(x,t)} [U_p'(z) - 1]\phi'(z)\, dz = 0 \quad \text{for a. e. } (x,t) \in A. \tag{7.134}$$

Concerning the second term in the right-hand side of (7.133), observe that for some $\delta > 0$,

$$\begin{aligned}
&\left| \int_{u_r(x,t)}^{U_p{}^{-1}(U_p^*(x,t))} U_p'(z)\, \phi'(z)\, dz \right| \le M |u_r(x,t) - U_p{}^{-1}(U_p^*(x,t))| \\
&\le M \Big\{ |u_r(x,t) - U_p^{-1}(u_r(x,t))| \\
&\quad + \Big(\sup_{z \in (u_r(x,t)) - \delta, u_r(x,t)) + \delta)} [U_1'(z)]^{-1} \Big) |u_r(x,t) - U_p^*(x,t)| \Big\}
\end{aligned}$$

for all $p \in \mathbb{N}$ sufficiently large (see (7.127)). By (7.128), letting $p \to \infty$ in the above inequality, we get

$$\lim_{p\to\infty} \int_{u_r(x,t)}^{U_p^{-1}(U_p^*(x,t))} U_p'(z)\,\phi'(z)\,dz = 0 \quad \text{for a. e. } (x,t) \in A. \tag{7.135}$$

By (7.133)–(7.135) we get

$$\int_0^{U_p^{-1}(U_p^*(x,t))} U_p'(z)\,\phi'(z)\,dz \to \phi(u_r)(x,t) \quad \text{for a. e. } (x,t) \in A. \tag{7.136}$$

Then (7.129) follows from (7.131), (7.132), and (7.136). This completes the proof. □

Proof of Theorem 7.5.1. Let $u \in L^\infty_{w^*}(0,T;\mathfrak{R}_f(\Omega))$ be defined by (7.108). Then by Propositions 7.5.2–7.5.6 the result follows. □

Now we can prove Theorem 7.3.1.

Proof of Theorem 7.3.1. Let us prove that $u \in L^\infty_{w^*}(0,T;\mathfrak{R}_f(\Omega))$ defined by (7.108) satisfies the equality

$$\iint_S [u_r\partial_t\zeta + \phi(u_r)\,\partial_x\zeta]\,dxdt + \int_0^T \langle u_s(\cdot,t), \partial_t\zeta(\cdot,t)\rangle\,dt = -\langle u_0, \zeta(\cdot,0)\rangle \tag{7.137}$$

for all $\zeta \in C^1([0,T];C^1_c(\mathbb{R}))$, $\zeta(\cdot,T) = 0$ in $\mathbb{R}$ and thus is a solution of (P) (see Definition 7.1.3). Let $\{u_{n_k}\} \subseteq \{u_n\}$ be the subsequence given by Theorem 7.5.1. Applying Theorem 7.5.1(iii) with $U(y) = y$ and $U(y) = \phi(y)$ gives

$$\lim_{k\to\infty} \iint_S u_{n_k}\partial_t\zeta\,dxdt = \iint_S u_r\partial_t\zeta\,dxdt + \int_0^T \langle u_s(\cdot,t), \partial_t\zeta(\cdot,t)\rangle\,dt, \tag{7.138a}$$

$$\lim_{k\to\infty} \iint_S \phi(u_{n_k})\partial_x\zeta\,dxdt = \iint_S \phi(u_r)\partial_x\zeta\,dxdt \tag{7.138b}$$

(see (7.89), (7.4), and (7.119)). By the first convergence in (7.45) and (7.138), letting $k \to \infty$ in (7.78) written with $n = n_k$ gives (7.137). Inequality (7.5) is proven similarly letting $k \to \infty$ in (7.77) written with $n = n_k$, applying Theorem 7.5.1(iii) with $U(y) = E_l(y)$ and $U(y) = F_l(y)$, and arguing as in Proposition 7.5.3 to prove that $E_l(u_{0n_k}) \overset{*}{\rightharpoonup} E_l(u_{0r}) + u_{0s}$ in $\mathfrak{R}_f(\mathbb{R})$. Hence the result follows. □

To prove Theorem 7.3.2, we need the following result, which characterizes the disintegration of the Young measure ν.

Proposition 7.5.7. *Let (A_0)–(A_2) hold. Let $u \in L^\infty_{w^*}(0,T;\mathfrak{R}_f(\mathbb{R}))$ and $v \in \mathfrak{Y}^+(S;\mathbb{R})$ be given by Theorem 7.5.1. Then for a. e. $(x,t) \in S$,*

$$v_{(x,t)} = \delta_{u_r(x,t)}. \tag{7.139}$$

Proof. Fix $(x,t) \in S$. If $u_r(x,t) = 0$, then by (7.89) equality (7.139) immediately follows. Let $y_1 \equiv u_r(x,t) > 0$. For all $y_2 > y_1$ and $p \in \mathbb{N}$ sufficiently large, set

$$V_p(y) := p(y-y_1)\chi_{(y_1,y_1+\frac{1}{p})}(y) + \chi_{[y_1+\frac{1}{p},y_2)}(y) + p\Big(y_2+\frac{1}{p}-y\Big)\chi_{[y_2,y_2+\frac{1}{p})}(y),$$

$$\Lambda_{V_p}(y) := \int_0^y V_p'(z)\,\phi'(z)dz \quad (y \in [0,\infty)).$$

Then for any $y \in [0,\infty)$,

$$\lim_{p\to\infty} V_p(y) = \chi_{(y_1,y_2]}(y), \tag{7.140a}$$

$$\lim_{p\to\infty} \Lambda_{V_p}(y) = \phi'(y_1)\chi_{(y_1,y_2]}(y) + [\phi'(y_1)-\phi'(y_2)]\chi_{(y_2,\infty)}(y). \tag{7.140b}$$

Let $\{U_p\}$ be the sequence used in the proof of Proposition 7.5.6. By standard approximation arguments we can choose $U = U_p$, $V = V_p$ in equality (7.125), thus obtaining

$$\int_{[0,\infty)} [\Lambda_{U_p}(y) - \Lambda^*_{U_p}(x,t)]V_p(y)dv_{(x,t)}(y) = \int_{[0,\infty)} [U_p(y) - U_p^*(x,t)]\Lambda_{V_p}(y)dv_{(x,t)}(y).$$

By (7.128), (7.132), and (7.119) for all $y \geq 0$, we have

$$\lim_{p\to\infty} U_p(y) - U_p^*(x,t) = y - y_1, \tag{7.141a}$$

$$\lim_{p\to\infty} \Lambda^*_{U_p}(x,t) - \Lambda_{U_p}(y) = \phi(y_1) - \phi(y). \tag{7.141b}$$

From (7.140)–(7.141) we get a. e. in S

$$\lim_{p\to\infty} \int_{[0,\infty)} [\Lambda_{U_p}(y) - \Lambda^*_{U_p}]V_p(y)\,dv(y) = \int_{(y_1,y_2]} [\phi(y)-\phi(y_1)]\,dv(y),$$

$$\lim_{p\to\infty} \int_{[0,\infty)} [U_p(y) - U_p^*]\Lambda_{V_p}(y)\,dv(y)$$

$$= \int_{(y_1,y_2]} \phi'(y_1)(y-y_1)\,dv(y) + [\phi'(y_1)-\phi'(y_2)] \int_{(y_2,\infty)} (y-y_1)\,dv(y),$$

whence

$$\int_{(y_1,y_2]} [\phi(y) - \phi(y_1) - \phi'(y_1)(y - y_1)]d\nu(y) = [\phi'(y_1) - \phi'(y_2)] \int_{(y_2,\infty)} (y - y_1)d\nu(y). \quad (7.142a)$$

It is similarly seen that for any $y_0 \in (0, y_1)$,

$$\int_{(y_0,y_1)} [\phi(y) - \phi(y_1) - \phi'(y_1)(y - y_1)]d\nu_{(x,t)}(y) = [\phi'(y_1) - \phi'(y_0)] \int_{[0,y_0]} (y - y_1)d\nu_{(x,t)}(y). \quad (7.142b)$$

By assumption (A_2), ϕ' is strictly monotone in at least one of the intervals $[y_0, y_1]$ and $[y_1, y_2]$. Suppose that this occurs in $[y_1, y_2]$, the other case being similar. Then by (7.142a), for a. e. $(x, t) \in S$, we have that

$$\int_{(y_1,y_2]} |\phi(y) - \phi(y_1) - \phi'(y_1)(y - y_1)| \, d\nu_{(x,t)}(y) + |\phi'(y_1) - \phi'(y_2)| \int_{(y_2,\infty)} |y_1 - y| \, d\nu_{(x,t)}(y) = 0,$$

whence $\operatorname{supp} \nu_{(x,t)} \subseteq [0, y_1]$. Since $\nu_{(x,t)}$ is a probability measure, it follows that

$$u_r(x,t) = \int_{[0,u_r(x,t)]} y \, d\nu_{(x,t)}(y) = \int_{[0,u_r(x,t)]} [y - u_r(x,t)] \, d\nu_{(x,t)}(y) + u_r(x,t),$$

and thus

$$\int_{[0,u_r(x,t)]} |y - u_r(x,t)| \, d\nu_{(x,t)}(y) = 0.$$

By the above equality we have $\operatorname{supp} \nu_{(x,t)} = \{u_r(x,t)\}$, whence (7.139) follows since $\nu_{(x,t)}$ is a probability measure. This completes the proof. □

Remark 7.5.2. By (7.139) and Proposition 5.4.1, possibly extracting a subsequence (not relabeled), we have

$$u_{n_k} \to u_r \quad \text{a. e. in } S. \quad (7.143)$$

In particular, if ϕ is bounded, then by the dominated convergence theorem $\phi(u_{n_k}) \to \phi(u_r)$ in $L^1(R_L)$ for any $L > 0$.

Remark 7.5.3. Let (A_0)–(A_1) hold, and let ϕ' be constant in the interval $I_{a,b} = [u_r(x,t) - a, u_r(x,t) + b]$ for some $a > 0$, $b > 0$, and $(x,t) \in S$. In this case, it can be proved that $\operatorname{supp} \nu_{(x,t)} \subseteq I_{(x,t)}$, $I_{(x,t)}$ being the maximal interval where $\phi' = \phi'(u_r)$ (see [18, Proposition 5.9]).

Proof of Theorem 7.3.2. By Theorem 7.3.1 u is a Young measure entropy solution of (P), and thus inequality (7.5) is satisfied. From (7.5) and (7.139) we get (7.8), and thus the result follows. □

Let us finally prove Lemma 7.5.4.

Proof of Lemma 7.5.4. By standard approximation arguments it suffices to prove equality (7.106a).

(i) Let us first prove (7.106a) assuming that $U \in C^2([0,\infty))$ with $U' \in L^\infty(0,\infty)$ satisfies (7.58) and (7.91). Let $\rho \in C_c^2(\mathbb{R})$, $h \in C_c(0,T)$, and $\operatorname{supp}\rho \subseteq I_L$ for some $L > 0$. Let $U_{n_k,\rho}$ be defined by (7.60). Then by (7.104) we have

$$\lim_{k\to\infty} \int_0^T U_{n_k,\rho}(t)h(t)\,dt = \int_0^T U_\rho^*(t)h(t)\,dt + M_U \int_0^T \langle \sigma(\cdot,t),\rho\rangle h(t)\,dt. \tag{7.144}$$

On the other hand, by Lemma 7.4.6 the sequence $\{U_{n_k,\rho}\}$ is bounded in $BV(0,T)$, and hence there exists a subsequence (not relabeled) that clearly converges in $L^1(0,T)$ to $U_\rho^* + M_U\langle\sigma(\cdot,\cdot),\rho\rangle$ (see (7.144)). Since $\{U(u_{n_k})\}$ is bounded in $L^\infty(0,T;L^1(I_L))$ and $U^* \in L^\infty(0,T;L^1(I_L))$ (see Proposition 7.5.3 and inequality (7.101)), the condition $\rho \in C_c^2(\mathbb{R})$ may be relaxed to $\rho \in C_c(\mathbb{R})$. Hence (7.106a) follows in this case.

(ii) Let us now prove (7.106a) for all $U \in C_b([0,\infty))$ (observe that $M_U = 0$ in this case). Set

$$U_p : [0,\infty) \to \mathbb{R}, \quad U_p(y) := ((U\chi_{[0,p]}) * \theta_p)(y) \quad \text{for } y \geq 0,$$

where $\theta_p(y) := p\theta(py)$, and $\theta \in C_c^\infty(\mathbb{R})$ is a standard mollifier ($p \in \mathbb{N}$). Then $\{U_p\} \subseteq C_c^2([0,\infty))$, $U_p \to U$ uniformly on compact subsets of $[0,\infty)$, and $\|U_p\|_{L^\infty(\mathbb{R})} \leq \|U\|_{L^\infty(\mathbb{R})}$. Moreover, $U_p' \in L^\infty(0,\infty)$, and U_p satisfies (7.58) and (7.91) with $M_{U_p} = 0$, and hence by (i) we get

$$\lim_{k\to\infty} \int_0^T \left| \int_{\mathbb{R}} U_p(u_{n_k}(x,t))\,\rho(x)\,dx - U_{p,\rho}^*(t) \right| dt = 0, \tag{7.145}$$

where $U_p^* := \int_{[0,\infty)} U_p(y)\,d\nu(y)$, and $U_{p,\rho}^* := \int_{\mathbb{R}} U_p^*(x,t)\rho(x)\,dx$ for all $\rho \in C_c(\mathbb{R})$ ($p \in \mathbb{N}$). Now observe that

$$\begin{aligned}
&\int_0^T \left| \int_{\mathbb{R}} U(u_{n_k}(x,t))\,\rho(x)\,dx - U_\rho^*(t) \right| dt \\
&\quad \leq \iint_S |U(u_{n_k}(x,t)) - U_p(u_{n_k}(x,t))|\,|\rho(x)|\,dxdt \\
&\qquad + \int_0^T \left| \int_{\mathbb{R}} U_p(u_{n_k}(x,t))\,\rho(x)\,dx - U_{p,\rho}^*(t) \right| dt + \int_0^T |U_\rho^*(t) - U_{p,\rho}^*(t)|\,dt.
\end{aligned} \tag{7.146}$$

Concerning the first term in the right-hand side of (7.146), for all $p \in \mathbb{N}$ and $M > 0$, by the Chebyshev inequality and the second inequality in (7.71) we have

$$\begin{aligned}
&\iint_S |U(u_{n_k}) - U_p(u_{n_k})|\,|\rho|\,dxdt \\
&\quad \le \iint_{\{u_{n_k} \le M\}} |U(u_{n_k}) - U_p(u_{n_k})|\,|\rho|\,dxdt + \iint_{\{u_{n_k} > M\}} |U(u_{n_k}) - U_p(u_{n_k})|\,|\rho|\,dxdt \\
&\quad \le \|\rho\|_{L^\infty(\mathbb{R})}\, T\left\{\lambda(\operatorname{supp}\rho)\,\|U - U_p\|_{L^\infty(0,M)} + \frac{2}{M}\,\|u_0\|_{\mathfrak{R}_f(\mathbb{R})}\,\|U\|_{L^\infty(\mathbb{R})}\right\}. \qquad (7.147)
\end{aligned}$$

As for the third term in the right-hand side of (7.146), for all $p \in \mathbb{N}$ and $M > 0$, we have

$$\begin{aligned}
&\int_0^T |U_\rho^*(t) - U_{p,\rho}^*(t)|\,dt \\
&\quad \le \iint_S |U^* - U_p^*|\,|\rho|\,dxdt \\
&\quad \le \iint_S |\rho(x)|\,dxdt \int_{[0,\infty)} |U_p(y) - U(y)|\,d\nu_{(x,t)}(y) \\
&\quad = \iint_S |\rho(x)|\,dxdt \int_{\{0\le y\le M\}} |U_p(y) - U(y)|\,d\nu_{(x,t)}(y) \\
&\qquad + \iint_S |\rho(x)|\,dxdt \int_{\{y>M\}} |U_p(y) - U(y)|\,d\nu_{(x,t)}(y) \\
&\quad \le \|\rho\|_{L^\infty(\mathbb{R})}\Big\{\lambda(\operatorname{supp}\rho)\,T\,\|U - U_p\|_{L^\infty(0,M)} \\
&\qquad + 2\,\|U\|_{L^\infty(\mathbb{R})} \iint_{\operatorname{supp}\rho\times(0,T)} \nu_{(x,t)}(\{y > M\})\,dxdt\Big\}. \qquad (7.148)
\end{aligned}$$

By (7.145)–(7.148), for all $\rho \in C_c(\mathbb{R})$, $p \in \mathbb{N}$, and $M > 0$, we have that

$$\begin{aligned}
&\limsup_{k\to\infty} \int_0^T dt \left|\int_{\mathbb{R}} U(u_{n_k}(x,t))\,\rho(x)\,dx - U_\rho^*(t)\right| \\
&\le 2\,\|\rho\|_{L^\infty(\mathbb{R})}\Big\{\lambda(\operatorname{supp}\rho)\,T\,\|U - U_p\|_{L^\infty(0,M)} \\
&\quad + \|U\|_{L^\infty(\mathbb{R})}\left[\frac{T}{M} + \iint_{\operatorname{supp}\rho\times(0,T)} \nu_{(x,t)}(\{y > M\})\,dxdt\right]\Big\}.
\end{aligned}$$

Since $U_p \to U$ uniformly on compact sets in $[0,\infty)$, letting $p \to \infty$ in the above inequality gives

$$\limsup_{k\to\infty} \int_0^T \left| \int_{\mathbb{R}} U(u_{n_k})\rho(x)\,dx - U_\rho^*(t) \right| dt$$
$$\leq 2\,\|\rho\|_{L^\infty(\mathbb{R})}\,\|U\|_{L^\infty(\mathbb{R})} \left\{ \frac{T}{M} + \iint_{\operatorname{supp}\rho\times(0,T)} \nu_{(x,t)}(\{y > M\})\,dxdt \right\}. \tag{7.149}$$

Since $\nu_{(x,t)}$ is a probability measure, by the dominated convergence theorem we plainly have

$$\lim_{M\to\infty} \iint_{\operatorname{supp}\rho\times(0,T)} \nu_{(x,t)}(\{y > M\})\,dxdt = 0.$$

Then letting $M \to \infty$ in (7.149), we obtain (7.106a).

(iii) Now let $U \in C([0,\infty))$ be any function satisfying (7.91). As in the proof of Proposition 7.5.3, let $f_{j1}, f_{j2} \in C^2([0,\infty))$ satisfy (7.100) ($j \in \mathbb{N}$, $j \geq [y_\epsilon] + 1$). Then

$$U = U_{j1} + U_{j2}, \quad U^* = U_{j1}^* + U_{j2}^*, \quad U_\rho^* = U_{j1,\rho}^* + U_{j2,\rho}^*, \tag{7.150}$$

where $U_{jl} := f_{jl}U$, $U_{jl}^* := \int_{[0,\infty)} U_{jl}(y)\,d\nu(y)$, and $U_{jl,\rho}^* := \int_{\mathbb{R}} U_{jl}^*(x,t)\rho(x)\,dx$ for $\rho \in C_c(\mathbb{R})$ $(l = 1,2)$. Therefore

$$\int_0^T \left| \int_{\mathbb{R}} U(u_{n_k}(x,t))\rho(x)\,dx - U_\rho^*(t) - M_U\langle\sigma(\cdot,t),\rho\rangle \right| dt$$
$$\leq \int_0^T \left| \int_{\mathbb{R}} U_{j1}(u_{n_k}(x,t))\rho(x)\,dx - U_{j1,\rho}^*(t) \right| dt$$
$$+ \int_0^T \left| \int_{\mathbb{R}} U_{j2}(u_{n_k}(x,t))\rho(x)\,dx - U_{j2,\rho}^*(t) - M_U\langle\sigma(\cdot,t),\rho\rangle \right| dt =: I_{jk1} + I_{jk2}. \tag{7.151}$$

Since $U_{j1} \in C_b([0,\infty))$ and thus $M_{U_{j1}} = 0$, by (ii) from (7.106a) we get

$$\lim_{k\to\infty} I_{jk1} = 0 \quad (j \in \mathbb{N},\ j \geq [y_\epsilon] + 1). \tag{7.152}$$

On the other hand,

$$I_{jk2} \leq \int_0^T \left| \int_{\mathbb{R}} [U_{j2}(u_{n_k}(x,t)) - M_U f_{j2}(u_{n_k}(x,t))\,u_{n_k}(x,t)]\rho(x)\,dx \right| dt$$

$$+ M_U \int_0^T dt \left| \int_{\mathbb{R}} f_{j2}(u_{n_k}(x,t))\, u_{n_k}(x,t)\, \rho(x)\, dx - \int_{\mathbb{R}} F_{j2}^*(x,t)\, \rho(x)\, dx - \langle \sigma(\cdot,t), \rho \rangle \right|$$
$$+ \iint_S |U_{j2}^* - M_U F_{j2}^*|(x,t)\, |\rho(x)|\, dxdt, \tag{7.153}$$

where $F_{j2}^* := \int_{[0,\infty)} f_{j2}(y) y\, d\nu(y)$.

We address separately the three terms in the right-hand side of (7.153). By (7.99) and the second inequality in (7.71) we have

$$\int_0^T \left| \int_{\mathbb{R}} [U_{j2}(u_{n_k}(x,t)) - M_U f_{j2}(u_{n_k}(x,t))\, u_{n_k}(x,t)] \rho(x)\, dx \right| dt$$
$$\leq \epsilon \iint_S |u_{n_k}|\, |\rho|\, dxdt \leq \epsilon\, \|\rho\|_{L^\infty(\mathbb{R})}\, T\, \|u_0\|_{\mathfrak{R}_f(\mathbb{R})}\,. \tag{7.154}$$

Moreover, the map $y \mapsto f_{j2}(y)y$ belongs to $C^2([0,\infty))$, has bounded derivative, and satisfies (7.58) and (7.91) with $M_U = 1$. Then by (i)

$$\lim_{k\to\infty} \int_0^T dt \left| \int_{\mathbb{R}} f_{j2}(u_{n_k}(x,t))\, u_{n_k}(x,t)\, \rho(x)\, dx - \int_{\mathbb{R}} F_{j2}^*(x,t)\, \rho(x)\, dx - \langle \sigma(\cdot,t), \rho \rangle \right| = 0. \tag{7.155}$$

Let us show that

$$\lim_{j\to\infty} \iint_S |U_{j2}^* - M_U F_{j2}^*|(x,t)\, |\rho(x)|\, dxdt = 0. \tag{7.156}$$

For a. e. $(x,t) \in S$, we have

$$|U_{j2}^*(x,t) - M_U F_{j2}^*(x,t)| \leq \int_{[0,\infty)} |U_{j2}(y) - M_U f_{j2}(y) y|\, d\nu_{(x,t)}(y)$$
$$= \int_{[0,\infty)} f_{j2}(y)\, |U(y) - M_U y|\, d\nu_{(x,t)}(y) \leq \epsilon \int_{[j,\infty)} y\, d\nu_{(x,t)}(y) \leq \epsilon u_b(x,t)$$

(see (7.95)). Since $u_b \in L^\infty(0,T; L^1(\mathbb{R}))$ and $\lim_{j\to\infty} \int_{[j,\infty)} y\, d\nu_{(x,t)}(y) = 0$ for a. e. $(x,t) \in S$, equality (7.156) follows by the dominated convergence theorem.

By (7.153)–(7.156) we have

$$\limsup_{j\to\infty} \left(\limsup_{k\to\infty} I_{jk2} \right) \leq \epsilon\, \|\rho\|_{L^\infty(\mathbb{R})}\, T\, \|u_0\|_{\mathfrak{R}_f(\mathbb{R})}\,,$$

whence by (7.151)–(7.152)

$$\limsup_{k\to\infty} \int_0^T \left| \int_{\mathbb{R}} [U(u_{n_k}) - U^*](x,t)\,\rho(x)\,dx - M_U\langle\sigma(\cdot,t),\rho\rangle \right| dt$$
$$\leq \epsilon T \, \|u_0\|_{\mathfrak{R}_f(\mathbb{R})} \, \|\rho\|_{L^\infty(\mathbb{R})} \,.$$

By the arbitrariness of ϵ from this inequality we get (7.106a). Hence the result follows. □

7.6 Proof of regularity results

Proof of Proposition 7.1.2. (i) We only prove (7.15), the proof of (7.16) being similar. Let $\tilde{\zeta} \in C^1([0,T]; C_c^1(\mathbb{R}))$ be such that $\tilde{\zeta}(\cdot,T) = 0$ and $\tilde{\zeta} \geq 0$. Then it is easily seen that as $l \to \infty$ in (7.9),

$$\int_0^T \langle u_s(\cdot,t), \partial_t \tilde{\zeta}(\cdot,t)\rangle \, dt \geq -\langle u_{0s}, \tilde{\zeta}(\cdot,0)\rangle. \tag{7.157}$$

By the separability of $C([0,T]; C_c(\mathbb{R}))$ there exists a null set $N \subseteq (0,T)$ such that for all $\zeta \in C([0,T]; C_c(\mathbb{R}))$ and $\tau \in (0,T) \setminus N$, we have

$$\lim_{h\to 0} \frac{1}{h} \int_\tau^{\tau+h} \langle u_s(\cdot,t), \zeta(\cdot,t)\rangle \, dt = \langle u_s(\cdot,\tau), \zeta(\cdot,\tau)\rangle. \tag{7.158}$$

Let $t_1, t_2 \in (0,T) \setminus N$, $0 < t_1 < t_2 < T$. By standard approximation arguments we can choose $\tilde{\zeta}(x,t) = g_h(t)\zeta(x,t)$ in (7.157), where

$$g_h(t) := \frac{1}{h}(t - t_1)\chi_{[t_1,t_1+h]}(t) + \chi_{(t_1+h,t_2)}(t) + \frac{1}{h}(t_2 + h - t)\chi_{[t_2,t_2+h]}(t)$$

with $h \in (0, \min\{t_2 - t_1, T - t_2\})$. Letting $h \to 0$ in (7.157) and using (7.158) gives

$$\langle u_s(\cdot,t_2), \zeta(\cdot,t_2)\rangle \leq \int_{t_1}^{t_2} \langle u_s(\cdot,t), \partial_t \zeta(\cdot,t)\rangle \, dt + \langle u_s(\cdot,t_1), \zeta(\cdot,t_1)\rangle. \tag{7.159}$$

Choosing $\zeta(x,t) = \rho(x)$ in (7.159) gives (7.15), and hence the claim follows.

(ii) Arguing as for (7.158), there exists a null set $N \subseteq (0,T)$ such that for all $\rho \in C_c(\mathbb{R})$ and $\tau \in (0,T) \setminus N$,

$$\lim_{h\to 0} \frac{1}{h} \int_{\tau}^{\tau+h} \langle u(\cdot,t), \rho \rangle \, dt = \langle u(\cdot,\tau), \rho \rangle. \tag{7.160}$$

Set $\Omega_j := [-j-1, -j] \cup [j, j+1]$ and $Q_{j,\tau} := \Omega_j \times (0,\tau)$ ($\tau \in (0,T), j \in \mathbb{N}$). Let $\{\rho_j\} \subseteq C_c^1(\mathbb{R})$ be such that $\rho_j = 1$ in $[-j,j]$, $\operatorname{supp} \rho_j \subseteq [-j-1, j+1]$, $0 \le \rho_j \le 1$, and $|\rho_j'| \le 2$ in $\mathbb{R}$. Let us choose $\zeta(x,t) = f_h(t)\rho_j(x)$ in (7.7) with

$$f_h(t) := \chi_{[0,\tau)}(t) + \frac{1}{h}(\tau + h - t)\chi_{[\tau,\tau+h]}(t)$$

and h sufficiently small. Then we get

$$\iint_{Q_{j,\tau}} f_h(t)\, \phi(u_r)(x,t)\, \rho_j'(x) \, dxdt = \frac{1}{h} \int_{\tau}^{\tau+h} \langle u(\cdot,t), \rho_j \rangle \, dt - \langle u_0, \rho_j \rangle.$$

Letting $h \to 0$ in this equality and using (7.160), we obtain

$$\iint_{Q_{j,\tau}} \phi(u_r)(x,t)\, \rho_j' \, dxdt = \langle u(\cdot,\tau), \rho_j \rangle - \langle u_0, \rho_j \rangle. \tag{7.161}$$

Since $|\phi(u_r)| \le M u_r$ (see (7.1)) and $u_r \in L^\infty(0,T; L^1(\mathbb{R}))$, letting $j \to \infty$ in (7.161), we plainly obtain (7.17). Hence the result follows. □

Proof of Proposition 7.2.1. We will prove the following:

Claim 1. For a. e. $t \in (0,T)$, the map $x \mapsto Y(x,t) := \int_0^t \phi(u_r(x,s))\, ds$ belongs to $BV(\mathbb{R})$, and for any $a, b \in \mathbb{R}$, $a \le b$, we have

$$u(\cdot,t)([a,b]) - u_0([a,b]) = Y(a^-,0,t) - Y(b^+,0,t). \tag{7.162}$$

Let $a = b = x_0$. By (7.162), for a. e. $0 \le t \le T$,

$$u_s(t)(\{x_0\}) = u_{0s}(\{x_0\}) + Y(x_0^-,0,t) - Y(x_0^+,0,t) \ge u_{0s}(\{x_0\}) - 2\|\phi\|_\infty t,$$

whence $u_s(t)(\{x_0\}) > 0$ if $t \in (0, \frac{u_{0s}(\{x_0\})}{2\|\phi\|_\infty})$. Hence (7.19) follows.

It remains to prove the claim. Arguing as in the proof of (7.161), we get

$$\int_{\mathbb{R}} Y(x,\tau)\rho'(x)\, dx = \langle u(\cdot,\tau), \rho \rangle - \langle u_0, \rho \rangle. \tag{7.163}$$

Hence the distributional derivative $\partial_x Y(x,t_1,t_2)$ belongs to $\mathfrak{R}_f(\mathbb{R})$. On the other hand, by (7.1) we have $|Y(\cdot,t)| \le M \int_0^t u_r(\cdot,s)\,ds \in L^1(\mathbb{R})$, and thus for a. e. $t \in (0,T)$, $Y(\cdot,t)$ belongs to $BV(\mathbb{R})$.

To prove (7.162), observe that by standard regularization arguments in (7.163) we can choose $\rho = \rho_j$,

$$\rho_j(x) := j\left(x - a + \frac{1}{j}\right)\chi_{[a-\frac{1}{j},a]}(x) + \chi_{(a,b)}(x) + j\left(b + \frac{1}{j} - x\right)\chi_{[b,b+\frac{1}{j}]}(x)$$

with $a, b \in \mathbb{R}$, $a < b$, and $j \in \mathbb{N}$. Then we get

$$\langle u(\cdot,\tau),\rho_j\rangle - \langle u_0,\rho_j\rangle = j\int_{a-\frac{1}{j}}^{a} Y(x,\tau)\,dx - j\int_{b}^{b+\frac{1}{j}} Y(x,\tau)\,dx. \tag{7.164}$$

Letting $j \to \infty$ in (7.164) plainly gives (7.162). This completes the proof. □

Remark 7.6.1. The proof of Proposition 7.2.1 in fact relies on the correspondence (in one space dimension) between entropy solutions of hyperbolic conservation laws and viscosity solutions of Hamilton–Jacobi equations (on this subject, see [21, 20]).

It suffices to prove Theorem 7.3.3 and Proposition 7.3.5 assuming that $(A_2{}')$ holds with $\phi'' < 0$, $\phi' > 0$ in $(0,\infty)$, and $\phi(0) = 0$ (hence $\alpha\phi + \beta > 0$, $\beta > 0$, and assumption (A_1) is satisfied as well; see Remark 7.1.1). Consider the following regularization of problem (P_n) (different from the regularization (P_n^ϵ) used in Section 7.4):

$$\begin{cases} \partial_t y_n^\epsilon + \partial_x[\phi(y_n^\epsilon)] = \epsilon\,\partial_{xx}[\phi(y_n^\epsilon)] & \text{in } S, \\ y_n^\epsilon = u_{0n}^\epsilon & \text{in } \mathbb{R}\times\{0\}, \end{cases} \tag{V_n^ϵ}$$

where $\{u_{0n}^\epsilon\}$ satisfies (7.46)–(7.47). The existence, uniqueness, and regularity results recalled in Section 7.4 for problem (P_n^ϵ), as well as the a priori estimates in Lemma 7.4.1 and the convergence results in Lemma 7.4.5, continue to hold for solutions of (V_n^ϵ) (see [63]). Some properties of the family $\{y_n^\epsilon\}$ are the content of the following two lemmas.

Lemma 7.6.1. *Let (A_2'') hold, and let $\phi(0) = 0$, $\phi'' < 0$, $\phi' > 0$ in $(0,\infty)$. Let u_n be the unique entropy solution of problem (P_n) given by Proposition 7.4.7. Then there exists a subsequence $\{y_n^{\epsilon_j}\}$ of solutions of (V_n^ϵ) such that*

$$y_n^{\epsilon_j} \overset{*}{\rightharpoonup} u_n \quad \text{in } L^\infty(S), \quad y_n^{\epsilon_j} \to y_n \quad \text{in } L^1(R_L) \text{ for all } L > 0. \tag{7.165}$$

Proof. In view of (7.66) and (7.69), there exist a sequence $\{y_n^{\epsilon_j}\}$ and $y_n \in L^\infty(S) \cap L^\infty(0,T;L^1(\mathbb{R}))$ such that (7.165) holds. Moreover, for any convex function E, a function F such that $F' = E'\phi'$, and ζ as in Definition 7.4.1, we have

$$\iint_S \{E(y_n^{\epsilon_j})\,\partial_t\zeta + F(y_n^{\epsilon_j})\,\partial_x\zeta\}\,dxdt + \int_{\mathbb{R}} E(u_{0n}^{\epsilon_j})\,\zeta(x,0)\,dx \geq \epsilon_j \iint_S F'(y_n^{\epsilon_j})\partial_x y_n^{\epsilon_j}\partial_x\zeta\,dxdt.$$

Letting $j \to \infty$ in this inequality and arguing as in the proof of Proposition 7.4.7, we get

$$\iint_S [E(y_n)\,\partial_t\zeta + F(y_n)\,\partial_x\zeta]\,dxdt \geq -\int_{\mathbb{R}} E(u_{0n})\,\zeta(x,0)\,dx.$$

Then y_n is an entropy solution of (P_n), and thus by Kružkov's uniqueness theorem we get $y_n = u_n$. Hence the result follows. □

Lemma 7.6.2. *Let (A_0) and (A_2'') hold, and let $\phi(0) = 0$, $\phi'' < 0$, $\phi' > 0$ in $(0,\infty)$. Then for all $t \in (0,T)$, $\epsilon > 0$, and $n \in \mathbb{N}$, we have*

$$\pm\frac{\partial}{\partial t}\left[\frac{\alpha\,\phi(y_n^\epsilon(\cdot,t)) + \beta}{t^\alpha}\right] \leq 0 \quad \text{in } \mathbb{R} \text{ if } \pm\alpha > 0, \tag{7.166a}$$

$$\frac{\partial}{\partial t}[\phi(y_n^\epsilon(\cdot,t)) - \beta\log t] \leq 0 \quad \text{in } \mathbb{R} \text{ if } \alpha = 0. \tag{7.166b}$$

Proof. For convenience, we set $A \equiv \epsilon\frac{\partial^2}{\partial x^2} - \frac{\partial}{\partial x}$, and thus $\partial_t y_n^\epsilon = A[\phi(y_n^\epsilon)]$ in S. Let

$$z_n^\epsilon := t\,\partial_t y_n^\epsilon - g(y_n^\epsilon), \quad \text{where } g(y_n^\epsilon) := \frac{\alpha\,\phi(y_n^\epsilon) + \beta}{\phi'(y_n^\epsilon)} \quad (n \in \mathbb{N}).$$

It follows from (A_2'') and a straightforward calculation that

$$\begin{aligned}\partial_t z_n^\epsilon &= A[\phi'(y_n^\epsilon)z_n^\epsilon] + [\alpha + 1 - g'(y_n^\epsilon)]\,\frac{[z_n^\epsilon + g(y_n^\epsilon)]}{t}\\ &\leq A[\phi'(y_n^\epsilon)z_n^\epsilon] + [\alpha + 1 - g'(y_n^\epsilon)]\,\frac{z_n^\epsilon}{t} \quad \text{in } S.\end{aligned}$$

Since $z_n^\epsilon(\cdot,0) = -g(u_{0n}^\epsilon) \leq 0$ in $\mathbb{R}$, by comparison results we have $z_n^\epsilon \leq 0$ in S $(n \in \mathbb{N})$. It follows that $t\,\partial_t y_n^\epsilon(\cdot,t) \leq g(y_n^\epsilon)(\cdot,t)$ in $\mathbb{R}$ for all $t \in (0,T)$, which implies (7.166). □

To prove Theorem 7.3.3(i), we need another lemma.

Lemma 7.6.3. *Let (A_0)–(A_1) hold, and let u be a constructed solution. Let $\{u_{n_k}\}$ be as in the proof of Theorem 7.5.1. Then for a. e. $t \in (0,T)$ and every $x_0 \in \operatorname{supp} u_s(\cdot,t)$, there exist a sequence $\{x_p\} \subseteq \mathbb{R}$ and a subsequence $\{u_{n_p}\} \subseteq \{u_{n_k}\}$ such that $x_p \to x_0$ and $u_{n_p}(x_p,t) \to \infty$ as $p \to \infty$.*

Proof. By Remark 7.5.1 there exist a null set $N \subseteq (0,T)$ and a subsequence of $\{u_{n_k}\}$ (not relabeled) such that

$$u_{n_k}(\cdot,t) \overset{*}{\rightharpoonup} u_r(\cdot,t) + u_s(\cdot,t) \quad \text{in } \mathfrak{R}_f(I_L) \tag{7.167}$$

for all $t \in (0,T) \setminus N$ and $L > 0$ (see (7.107) with $U(y) = y$ and (7.109a)–(7.109b)).

Let $x_0 \in \operatorname{supp} u_s(\cdot,t)$. Then no neighborhood $I_\delta(x_0)$ exists such that the sequence $\{u_{n_k}(\cdot,t)\}$ lies in a bounded subset of $L^\infty(I_\delta(x_0))$; otherwise, there would exist a subsequence of $\{u_{n_k}(\cdot,t)\}$ (not relabeled) and $f_t \in L^\infty(I_\delta(x_0))$, $f_t \geq 0$, such that $u_{n_k}(\cdot,t) \overset{*}{\rightharpoonup} f_t$ in $L^\infty(I_\delta(x_0))$. However, in view of (7.167), this would imply that $u_s(\cdot,t) = 0$ in $I_\delta(x_0)$, a contradiction.

It follows that $\sup_{n_k \in \mathbb{N}} \|u_{n_k}(\cdot,t)\|_{L^\infty(I_{1/p}(x_0))} = \infty$ for all $p \in \mathbb{N}$. Then there exist $\{u_{n_{k_p}}\} \subseteq \{u_{n_k}\}$ and $\{x_p\}$ such that $x_p \in I_{1/p}(x_0)$ and $u_{n_p}(x_p,t) \geq p$ for every $p \in \mathbb{N}$. Hence the result follows. □

Proof of Theorem 7.3.3. As already pointed out, we only prove the result when $\phi'' < 0$ and $\phi' > 0$ in $(0,\infty)$.

(i) Let ϕ be bounded, and let $\{u_{n_k}\}$ be as in the proof of Lemma 7.6.3. By Lemma 7.6.1, for every $n_k \in \mathbb{N}$, there exists a subsequence $\{y_{n_k}^{\epsilon_j}\}$ of solutions of $(V_{n_k}^\epsilon)$ such that for a. e. $t \in (0,T)$,

$$y_{n_k}^{\epsilon_j}(\cdot,t) \to u_{n_k}(\cdot,t) \quad \text{in } L^1_{\mathrm{loc}}(\mathbb{R}) \text{ as } j \to \infty. \tag{7.168}$$

On the other hand, by the proof of Lemma 7.6.2, for any $t \in (0,T)$, we have

$$\epsilon_j\, \partial_{xx}\, \phi(y_{n_k}^{\epsilon_j}(\cdot,t)) - \partial_x\, \phi(y_{n_k}^{\epsilon_j}(\cdot,t)) = \partial_t\, y_{n_k}^{\epsilon_j}(\cdot,t) \leq \frac{g(y_{n_k}^{\epsilon_j}(\cdot,t))}{t} \quad \text{in } \mathbb{R}, \tag{7.169}$$

where $g(y) := \frac{\alpha\phi(y)+\beta}{\phi'(y)} > 0$.

Let $a, b \in \mathbb{R}$, $a < b$, be fixed, and let $\rho \in C^1_c(a,b)$, $\rho \geq 0$. Set

$$Z(y) := \int_0^y \frac{[\phi'(z)]^2}{\alpha\,\phi(z) + \beta}\, dz.$$

Then Z is increasing, and $0 \leq Z(y) \leq \phi'(0)$ for all $y \geq 0$. Multiplying inequality (7.169) by $\frac{\rho}{g(y_{n_k}^{\epsilon_j}(\cdot,t))}$ and integrating, we get

$$\begin{aligned}
&\int_a^b Z(y_{n_k}^{\epsilon_j})(x,t)[\epsilon_j\rho''(x) + \rho'(x)]\, dx \\
&\quad \leq \frac{1}{t}\int_a^b \rho(x)\, dx - \epsilon_j \int_a^b \frac{\phi'(y_{n_k}^{\epsilon_j})\, g'(y_{n_k}^{\epsilon_j})\, [\partial_x y_{n_k}^{\epsilon_j}]^2}{[g(y_{n_k}^{\epsilon_j})]^2}(x,t)\rho(x)\, dx \leq \frac{1}{t}\int_a^b \rho(x)\, dx.
\end{aligned}$$

By (7.168), letting $j \to \infty$ in the above inequality, we obtain

$$\int_a^b Z(u_{n_k}(x,t))\,\rho'(x)\,dx \le \frac{1}{t}\int_a^b \rho(x)\,dx. \tag{7.170}$$

For a. e. $t \in (0,T)$, let $x_0 \in \operatorname{supp} u_s(\cdot,t)$. Let us prove that

$$\operatorname*{ess\,lim}_{x\to x_0^+} u_r(x,t) = \infty. \tag{7.171}$$

Let $\{x_p\}$ and $\{u_{n_p}\} \subseteq \{u_{n_k}\}$ be given by Lemma 7.6.3, and let $b > x_0$ be fixed. Since $x_p \to x_0$ as $p \to \infty$, there exists $\bar{p} \in \mathbb{N}$ such that $b > x_p$ for all $p > \bar{p}$. For any such p and $m \in \mathbb{N}$ sufficiently large, set

$$\rho_{p,m}(x) := m(x-x_p)\chi_{[x_p,x_p+\frac{1}{m}]}(x) + \chi_{(x_p+\frac{1}{m},b-\frac{1}{m})}(x) + m(b-x)\chi_{[b-\frac{1}{m},b]}(x).$$

Choosing $a = x_p$ and $\rho = \rho_{p,m}$ in (7.170), we get

$$m\int_{x_p}^{x_p+\frac{1}{m}} Z(u_{n_p}(x,t))\,dx - m\int_{b-\frac{1}{m}}^{b} Z(u_{n_p}(x,t))\,dx \le \frac{1}{t}\int_{x_p}^{b}\rho_{p,m}(x)\,dx,$$

whence, as $m \to \infty$,

$$Z(u_{n_p}(x_p,t)) \le Z(u_{n_p}(b,t)) + \frac{b-x_p}{t} \quad \text{for all } p > \bar{p}.$$

By the proof of Lemma 7.6.3 we have $u_{n_p}(x_p,t) \ge p$. Then letting $p \to \infty$ and using (7.143), we get, for a. e. $b > x_0$,

$$\lim_{y\to\infty} Z(y) \le Z(u_r(b,t)) + \frac{b-x_0}{t}, \tag{7.172}$$

whence

$$\lim_{y\to\infty} Z(y) \le \operatorname*{ess\,lim}_{b\to x_0^+} Z(u_r(b,t)).$$

Since Z is continuous and increasing, from the previous inequality we obtain (7.171).

Now we can prove that u satisfies the compatibility conditions in $[0,\tau]$ at $x_0 \in \operatorname{supp} u_s(\cdot,\tau)$. Concerning (C_-), this follows from the increasing character of ϕ: in fact, $H_-(u_r - l) = 0$ if $u_r > l$, whereas if $u_r \le l$, then $H_-(u_r(x,t) - l)[\phi(u_r(x,t)) - \phi(l)] = \phi(l) - \phi(u_r(x,t)) \ge 0$, and thus (C_-) is satisfied. As for (C_+), by equality (7.171), for all $l \in [0,\infty)$, we have $\operatorname{ess\,lim}_{x\to x_0^+} H_-(u_r(\cdot,t) - l) = 0$ for *a.e.* $t \in (0,\tau)$ (recall that by monotonicity $x_0 \in \operatorname{supp} u_s(\cdot,t)$ for such t). Hence by the dominated convergence theorem (C_+) follows.

(ii) Let ϕ be unbounded. By (7.166), for all $0 < t_1 \le t \le T$ and $x \in \mathbb{R}$,

$$\int_{t_1}^{t} \phi(y_n^{\epsilon_j})(x,s)\,ds = \frac{1}{\alpha}\int_{t_1}^{t} \frac{\alpha\phi(y_n^{\epsilon_m})(x,s)+\beta}{s^\alpha} s^\alpha ds - \frac{\beta}{\alpha}(t-t_1)$$

$$\ge \frac{\alpha\phi(y_n^{\epsilon_j})(x,t)+\beta}{\alpha\, t^\alpha}\frac{t^{\alpha+1}-t_1^{\alpha+1}}{\alpha+1} - \frac{\beta}{\alpha}(t-t_1) \quad \text{if } \alpha \ne 0,$$

$$\int_{t_1}^{t} \phi(y_n^{\epsilon_j})(x,s)\,ds = \int_{t_1}^{t} [\phi(y_n^{\epsilon_j})(x,s) - \beta \log s]\,ds + \beta \int_{t_1}^{t} \log s\,ds$$

$$\ge [\phi(y_n^{\epsilon_j})(x,t) - \beta \log t](t-t_1) + \beta[t\log t - t] - \beta[t_1 \log t_1 - t_1] \quad \text{if } \alpha = 0,$$

where $\{y_n^{\epsilon_j}\}$ is the subsequence used in the proof of Lemma 7.6.1. Letting $\epsilon_j \to 0$, by (7.165) we obtain that for a. e. $t \in (t_1, T)$ and a. e. $x \in \mathbb{R}$,

$$\int_{t_1}^{t} \phi(u_n(x,s))\,ds \ge \begin{cases} \dfrac{\alpha\phi(u_n)(x,t)+\beta}{\alpha\, t^\alpha}\dfrac{t^{\alpha+1}-t_1^{\alpha+1}}{\alpha+1} - \dfrac{\beta}{\alpha}(t-t_1) & \text{if } \alpha \ne 0, \\ [\phi(u_n)(x,t) - \beta]\,(t-t_1) + \beta\, t_1 \log \frac{t}{t_1} & \text{if } \alpha = 0. \end{cases} \tag{7.173}$$

On the other hand, arguing as in the proof of Proposition 7.2.1, by the weak formulation (7.78) it is easily seen that for every $t \in (0,T]$, the map $x \mapsto Y_n(x) := \int_0^t \phi(u_n(x,s))\,ds$ belongs to $W^{1,1}(\mathbb{R})$, with weak derivative $Y_n'(\cdot) = u_{0n} - u_n(\cdot,t)$ *a.e.* in $\mathbb{R}$. Therefore

$$0 \le Y_n \le \|u_{0n}\|_{L^1(\mathbb{R})} \le \|u_0\|_{\mathfrak{R}_f(\mathbb{R})}.$$

Combining this inequality with (7.173) and letting $t_1 \to 0^+$, we get

$$\phi(u_n)(x,t) \le \frac{(\alpha+1)\|u_0\|_{\mathfrak{R}_f(\mathbb{R})}}{t} + \beta \quad \text{for a. e. } t \in (0,T) \text{ and a. e. } x \in \mathbb{R} \tag{7.174}$$

(recall that for unbounded ϕ, we have $\alpha \ge 0$, since we have assumed that $\phi' > 0$ and $\alpha\phi + \beta > 0$ in $[0,\infty)$). By (7.174) the sequence $\{u_n\}$ lies in a bounded subset of $L^\infty(\mathbb{R}\times(\tau,T))$ for every $\tau \in (0,T)$, and thus by Theorem 7.3.1, as $n \to \infty$, the conclusion follows. □

Proof of Proposition 7.3.5. By the proof of Theorem 7.3.3, for a. e. $t \in (0,T)$, inequality (7.172) is satisfied for all $x_j \in \operatorname{supp} u_s(\cdot,t)$. Let $x_1 \in \operatorname{supp} u_s(\cdot,t)$, and set $I_1 := (x_1 - \epsilon, x_1 + \epsilon)$ $(\epsilon > 0)$. Since Z is invertible, by (7.172) we have

$$\int_{I_1} u_r(x,t)\,dx \ge \int_{x_1}^{x_1+\epsilon} Z^{-1}\Big(\lim_{y\to\infty} Z(y) - \frac{1}{t}(x-x_1)\Big)\,dx$$

$$= \int_0^{\epsilon} Z^{-1}\left(\lim_{y\to\infty} Z(y) - \frac{y}{t}\right) dy =: B_\epsilon,$$

where

$$\lim_{\epsilon\to 0^+} \frac{B_\epsilon}{\epsilon} = Z^{-1}\left(\lim_{y\to\infty} Z(y)\right) = \infty. \tag{7.175}$$

If $\operatorname{supp} u_s(\cdot, t) \not\subseteq I_1$, then let $x_2 \in \operatorname{supp} u_s(\cdot, t) \setminus I_1$ and set $I_2 := (x_2 - \epsilon, x_2 + \epsilon)$. Since $(x_1, x_1 + \epsilon) \cap (x_2, x_2 + \epsilon) = \emptyset$, it follows that

$$\int_{I_1 \cup I_2} u_r(x,t)\,dx \geq \int_{x_1}^{x_1+\epsilon} u_r(x,t)\,dx + \int_{x_2}^{x_2+\epsilon} u_r(x,t)\,dx \geq 2B_\epsilon.$$

Arguing recursively, as long as $\operatorname{supp} u_s(\cdot,t) \not\subseteq \bigcup_{i=1}^{n-1} I_i$ with $I_i := (x_i - \epsilon, x_i + \epsilon)$, there exists $x_n \in \operatorname{supp} u_s(\cdot,t) \setminus \{\bigcup_{i=1}^{n-1} I_i\}$ such that

$$nB_\epsilon \leq \int_{\bigcup_{i=1}^{n} I_i} u_r(x,t)\,dx \leq \|u_0\|_{\mathfrak{R}_f(\mathbb{R})},$$

where $I_n := (x_n - \epsilon, x_n + \epsilon)$. Hence the procedure ends at some $n = n_\epsilon$ such that $n_\epsilon B_\epsilon \leq \|u_0\|_{\mathfrak{R}_f(\mathbb{R})}$. It follows that $\operatorname{supp} u_s(\cdot,t) \subseteq \bigcup_{i=1}^{n_\epsilon} I_i$, and thus

$$\lambda(\operatorname{supp}_s(\cdot,t)) \leq \lambda\left(\bigcup_{i=1}^{n_\epsilon} I_i\right) \leq 2n_\epsilon\,\epsilon \leq 2\|u_0\|_{\mathfrak{R}_f(\mathbb{R})} \frac{\epsilon}{B_\epsilon}.$$

Letting $\epsilon \to 0^+$ in this inequality and using (7.175), we get the result. □

8 Case study 3: forward–backward parabolic equations

8.1 Statement of the problem and preliminary results

In this chapter, we consider the initial-boundary value problem

$$\begin{cases} \partial_t u = \nabla \cdot [\phi(\nabla u)] & \text{in } \Omega \times (0,T) =: Q_T, \\ u = 0 & \text{in } \partial\Omega \times (0,T) =: \Gamma_T, \\ u = u_0 & \text{in } \Omega \times \{0\}. \end{cases} \tag{P}$$

Here $\Omega \subseteq \mathbb{R}^N$ is a bounded domain with smooth boundary $\partial\Omega$ if $N \geq 2$, $T \in (0,\infty]$, and the dot "$\cdot$" denotes the scalar product in $\mathbb{R}^N$.

If $N = 1$, then on $\phi : \mathbb{R} \to \mathbb{R}$ and u_0, we assume the following:

$$u_0 \in W_0^{1,\infty}(\Omega); \tag{8.1a}$$

$$\begin{cases} \text{for every } R > 0, \text{ there exists } L_R > 0 \text{ such that} \\ |\phi(y_1) - \phi(y_2)| \leq L_R\, |y_1 - y_2| \text{ for all } y_1, y_2 \in (-R,R); \end{cases} \tag{8.1b}$$

$$\begin{cases} \text{there exist } y_0 > 0,\ p \in (1,\infty), \text{ and } C_1 > 0 \text{ such that} \\ C_1 |y|^{p-1} \leq |\phi(y)| \text{ for all } |y| > y_0; \end{cases} \tag{8.1c}$$

$$\phi(y) y \geq 0 \text{ for all } y \in \mathbb{R}. \tag{8.1d}$$

By abuse of notation, in (8.1a) and hereafter, we set $W_0^{1,\infty}(\Omega) := W^{1,\infty}(\Omega) \cap C_0(\overline{\Omega})$.

If $N \geq 2$, then concerning u_0 and $\phi : \mathbb{R}^N \to \mathbb{R}^N$, $\phi \equiv (\phi_1, \ldots, \phi_N)$, we will use the following assumptions:

$$u_0 \in W_0^{1,p}(\Omega) \text{ with } p \in (1,2]; \tag{8.2a}$$

$$\begin{cases} \text{there exists } L > 0 \text{ such that} \\ |\phi(y_1) - \phi(y_2)| \leq L\, |y_1 - y_2| \text{ for all } y_1, y_2 \in \mathbb{R}^N; \end{cases} \tag{8.2b}$$

$$\begin{cases} \text{there exist } y_0 > 0,\ p \in (1,2], \text{ and } C_0 > 0 \text{ such that} \\ |\phi(y)| \leq C_0 (1 + |y|^{p-1}) \text{ for all } |y| > y_0; \end{cases} \tag{8.2c}$$

$$\text{there exists } \Phi \in C^1(\mathbb{R}^N) \text{ such that } \phi = \nabla\Phi; \tag{8.2d}$$

$$\begin{cases} \text{there exist } y_0 > 0,\ q \in (1,2], \text{ and } C_1 > 0 \text{ such that} \\ C_1 |y|^q \leq \Phi(y) \text{ for all } |y| > y_0; \end{cases} \tag{8.2e}$$

$$\phi(y) \cdot y \geq 0 \text{ for all } y \in \mathbb{R}^N. \tag{8.2f}$$

https://doi.org/10.1515/9783110556902-010

Remark 8.1.1. For $N \geq 2$, we assume the global Lipschitz continuity of ϕ, instead of local Lipschitz continuity as in the case $N = 1$; this implies the stronger restriction $p \in (1, 2]$ (instead of $p \in (1, \infty)$ as for $N = 1$) on the allowed values of p. Observe also that by (8.2c)–(8.2e) we have

$$C_1|y|^q \leq \Phi(y) \leq C_3|y|^p \quad \text{for all } |y| > y_0 \tag{8.3}$$

with some $C_3 > 0$, which implies the compatibility condition $q \leq p$. In the following, we always assume that (8.2c)–(8.2e) hold with some fixed $p = q \in (1, 2]$ and (8.2a) holds with the same p.

As explained in the Introduction, if the function ϕ is not increasing, then problem (P) is ill posed. Let us study problem (P) using the Sobolev regularization. First, we address, for $\epsilon > 0$, the initial-boundary value problem

$$\begin{cases} \partial_t u = \nabla \cdot [\phi(\nabla u)] + \epsilon \Delta \partial_t u & \text{in } Q_T, \\ u = 0 & \text{in } \Gamma_T, \\ u = u_0 & \text{in } \Omega \times \{0\}. \end{cases} \tag{P_ϵ}$$

Then we study the limit as $\epsilon \to 0^+$ of the family $\{u_\epsilon\}$ of solutions of the approximating problems (P_ϵ), proving the existence of a Young measure-valued solution of the original problem (P) (see Definition 8.1.1). We also study the asymptotic behavior of such a solution as $t \to \infty$ using compactness and ω-limit set techniques; in doing so, a major point is the proof of tightness of sequences $\{v^m\}$ of time translates of the limiting Young measure v. Finally, for $N = 1$, we calculate explicitly the limiting Young measure v, extending the characterization given in [83] for analogous cases.

8.1.1 Notions of solution

Solutions of (P) are meant in the following sense.

Definition 8.1.1. Let either $N = 1$ and (8.1) or $N \geq 2$ and (8.2a)–(8.2e) hold. By a *Young measure-valued solution* of problem (P) in Q_T we mean a pair (u, v) such that:

(i) $u \in L^\infty(0, T; W_0^{1,p}(\Omega)) \cap C([0, T); L^2(\Omega))$ for all $p \in [1, \infty)$ with $u_x \in L^\infty(\Omega \times (0, T))$ if $N = 1$, or $u \in L^\infty(0, T; W_0^{1,p}(\Omega)) \cap C([0, T); L^p(\Omega))$ $(p \in (1, 2])$ if $N \geq 2$;

(ii) $\partial_t u \in L^2(Q_T)$, $v \in \mathfrak{Y}^+(Q_T; \mathbb{R}^N)$;

(iii) for a. e. $(x, t) \in Q_T$, we have

$$\nabla u(x, t) = \int_{\mathbb{R}^N} y \, dv_{(x,t)}(y), \tag{8.4}$$

where $\{v_{(x,t)}\}_{(x,t) \in Q_T}$ denotes the disintegration of v;

(iv) for all $\zeta \in C^1([0,T]; C_c^1(\Omega))$ and $t \in (0,T)$, we have

$$\int_0^t \int_\Omega \{u\,\partial_s\zeta - \phi^* \cdot \nabla\zeta](x,s)\,dxds = \int_\Omega u(x,t)\zeta(x,t)\,dx - \int_\Omega u_0(x)\,\zeta(x,0)\,dx, \tag{8.5}$$

where $\phi^* \equiv (\phi_1^*, \dots, \phi_N^*)$,

$$\phi_i^*(x,t) := \int_{\mathbb{R}^N} \phi_i(y)\,d\nu_{(x,t)}(y) \quad (i = 1, \dots, N) \tag{8.6}$$

for a. e. $(x,t) \in Q_T$.

A Young measure-valued solution of problem (P) in Q_∞ is called *global*.

As a particular case of Definition 8.1.1, we have the following definition.

Definition 8.1.2. By a *Young measure equilibrium solution* of problem (P) we mean a pair $(\bar{u}, \bar{\nu}) \in W_0^{1,p}(\Omega) \times \mathfrak{Y}^+(Q_T; \mathbb{R}^N)$ for every $p \in (1,\infty)$ if $N = 1$ or $p \in (1,2]$ if $N \geq 2$ such that for a. e. $(x,t) \in Q_T$,

$$\int_{\mathbb{R}^N} \phi_i(y)\,d\bar{\nu}_{(x,t)}(y) = 0 \quad (i = 1, \dots, N), \tag{8.7}$$

$$\nabla\bar{u}(x) = \int_{\mathbb{R}^N} y\,d\bar{\nu}_{(x,t)}(y). \tag{8.8}$$

Concerning solutions of the regularized problem (P_ϵ), we have the following definitions. If $N = 1$, then setting

$$v := \partial_x u, \quad w := \phi(v) + \epsilon\,\partial_t v, \tag{8.9}$$

problem (P_ϵ) reads

$$\begin{cases} \partial_t u = \partial_x w & \text{in } Q_T, \\ u = 0 & \text{in } \Gamma_T, \\ u = u_0 & \text{in } \Omega \times \{0\}. \end{cases} \tag{8.10}$$

Definition 8.1.3. Let $N = 1$ and (8.1a) hold. By a *solution* of problem (P_ϵ) in Q_T we mean any function $u_\epsilon \in C^1([0,T]; W_0^{1,\infty}(\Omega))$ with $w_\epsilon \in C([0,T]; W^{2,\infty}(\Omega) \cap C(\overline{\Omega}))$ and $\partial_x w_\epsilon \in C([0,T]; W_0^{1,\infty}(\Omega))$ that satisfies (8.10) in the classical sense.

Definition 8.1.4. Let $N \geq 2$, and let $u_0 \in H_0^1(\Omega)$. By a *solution* of problem (P_ϵ) in Q_T we mean any $u_\epsilon \in C^1([0,T]; H_0^1(\Omega))$ such that $u_\epsilon(\cdot, 0) = u_{0\epsilon}$ and

$$\int_\Omega \partial_t u_\epsilon(x,t)\rho(x)\,dx + \int_\Omega \phi(\nabla u_\epsilon)(x,t) \cdot \nabla\rho(x)\,dx + \epsilon \int_\Omega \nabla\partial_t u_\epsilon(x,t) \cdot \nabla\rho(x)\,dx = 0 \quad (8.11)$$

for all $t \in (0,T)$ and $\rho \in H_0^1(\Omega)$.

Definition 8.1.5. Let $N \geq 1$. A solution of problem (P_ϵ) in Q_∞ is said to be *global* if it is a solution in Q_T for all $T \in (0,\infty)$.

8.2 The regularized problem

8.2.1 Existence

Let us prove the following well-posedness result.

Theorem 8.2.1. *Let either $N = 1$ and* (8.1) *or $N \geq 2$, $u_0 \in H_0^1(\Omega)$, and* (8.2b). *Then for each $\epsilon > 0$, there exists a unique global solution u_ϵ of (P_ϵ).*

8.2.1.1 The case $N = 1$

Let us first prove Theorem 8.2.1 when $N = 1$. A first step is the following local well-posedness result.

Lemma 8.2.2. *Let $N = 1$ and* (8.1a)–(8.1b) *hold. Then for any $\epsilon > 0$, there exists $T_\epsilon > 0$ such that problem* (8.10) *has a unique solution u_ϵ in Q_{T_ϵ}.*

Proof. Consider the problem

$$\begin{cases} \partial_t v = \partial_{xx} w & \text{in } Q_T, \\ \partial_x w = 0 & \text{in } \Gamma_T, \\ v = v_0 := u_0' & \text{in } \Omega \times \{0\}, \end{cases} \quad (8.12)$$

with w as in (8.9). Then for each $\epsilon > 0$, there exist $T_\epsilon > 0$ and a unique function $v_\epsilon \in C^1([0,T]; L^\infty(\Omega))$ with $w_\epsilon \in C([0,T]; W^{2,\infty}(\Omega) \cap C(\bar{\Omega}))$ and $\partial_{xx} w_\epsilon \in C([0,T]; L^\infty(\Omega))$ that satisfies (8.12) in Q_{T_ϵ} in the classical sense (see [74, Theorem 2.1]). Suppose $\Omega \equiv (a,b)$ for simplicity. Defining

$$u_\epsilon(x,t) := \int_a^x v_\epsilon(y,t)\,dy \quad ((x,t) \in \Omega \times [0,T_\epsilon)), \quad (8.13)$$

the conclusion follows. □

A priori estimates of the local solution u_ϵ given by Lemma 8.2.2 are now to be proved. Following [74], for any $g \in C^1(\mathbb{R})$ with $g' \geq 0$, set

$$G_\phi(y) := \int_0^y g(\phi(z))\,dz + k \quad (y, k \in \mathbb{R}). \tag{8.14}$$

By (8.9)–(8.10), in Q_{T_ϵ}, we have

$$\begin{aligned}\partial_t G_\phi(\partial_x u_\epsilon) &= g(\phi(\partial_x u_\epsilon))\,\partial_{xt} u_\epsilon = g(w_\epsilon)\,\partial_{xx} w_\epsilon + [g(\phi(\partial_x u_\epsilon)) - g(w_\epsilon)]\partial_{xx} w_\epsilon \\ &= \partial_x[g(w_\epsilon)\partial_x w_\epsilon] - g'(w_\epsilon)|\partial_x w_\epsilon|^2 + [g(\phi(\partial_x u_\epsilon)) - g(w_\epsilon)]\,\frac{w_\epsilon - \phi(\partial_x u_\epsilon)}{\epsilon}.\end{aligned}$$

Then integrating in Ω, we obtain

$$\frac{d}{dt}\int_\Omega G_\phi(\partial_x u_\epsilon(x,t))\,dx \leq 0 \quad \text{in } (0, T_\epsilon). \tag{8.15}$$

Using the above inequality (conceptually analogous to inequality (6.61); see Remark 6.5.2), we can prove the following result (see [74, Proposition 2.7]).

Proposition 8.2.3. *Let*

$$\phi(y_1) \leq \phi(y) \leq \phi(y_2) \quad \textit{for all } y \in [y_1, y_2], \tag{8.16}$$

and let $u_0'(x) \in [y_1, y_2]$ for a. e. $x \in \Omega$. Then $\partial_x u_\epsilon(x,t) \in [y_1, y_2]$ for a. e. $(x,t) \in Q_{T_\epsilon}$.

Proof. Possibly modifying the shape of ϕ, we can suppose that $\phi(y) < \phi(y_1)$ if $y < y_1$ and $\phi(y) > \phi(y_2)$ if $y > y_2$. Fix $g \in C^1(\mathbb{R})$ with $g' \geq 0$ such that $g = 0$ in $[\phi(y_1), \phi(y_2)]$, $g(s) < 0$ if $s < \phi(y_1)$, and $g(s) > 0$ if $s > \phi(y_2)$; moreover, choose the constant k in (8.14) such that $G_\phi(y) = \int_{y_1}^y g(\phi(z))\,dz$. Plainly, this implies $G \equiv 0$ in $[y_1, y_2]$ and $G > 0$ in $(-\infty, y_1) \cup (y_2, \infty)$. Then by inequality (8.15) the conclusion follows. □

Let us now prove Theorem 8.2.1 in the case $N = 1$.

Proposition 8.2.4. *Let $N = 1$ and* (8.1) *hold. Then for each $\epsilon > 0$, there exists a unique global solution u_ϵ of problem (P_ϵ).*

Proof. By assumptions (8.1c)–(8.1d) we have $\phi(y) \to \pm\infty$ as $y \to \pm\infty$, and hence there exists $C_0 > \|u_0'\|_\infty$ such that inequality (8.16) holds with $[y_1, y_2] = [-C_0, C_0]$. Then by Proposition 8.2.3 we have

$$\|u_\epsilon(\cdot, t)\|_{W_0^{1,\infty}(\Omega)} \leq C_0[\lambda(\Omega) + 1] \quad \text{for all } t \in (0, T_\epsilon), \tag{8.17}$$

where, as usual, λ denotes the Lebesgue measure in $\mathbb{R}$. Hence from Lemma 8.2.2 by standard prolongation arguments the result follows. □

8.2.1.2 The case $N \geq 2$

To prove Theorem 8.2.1 in this case, set

$$(I - \epsilon\Delta)^{-1} : H^{-1}(\Omega) \to H_0^1(\Omega), \quad (I - \epsilon\Delta)^{-1}z := v \quad \text{for } z \in H^{-1}(\Omega),$$

where $v \in H_0^1(\Omega)$ is the unique solution of the elliptic problem

$$\begin{cases} -\epsilon\Delta v + v = z & \text{in } \Omega, \\ v = 0 & \text{in } \partial\Omega, \end{cases} \tag{8.18}$$

with $z \in H^{-1}(\Omega)$. The following result is easily proven (see [30, Proposition 4.4]).

Proposition 8.2.5. *Let* (8.2b) *hold. Then the operator*

$$L : H_0^1(\Omega) \to H_0^1(\Omega), \quad L(u) := (I - \epsilon\Delta)^{-1}\nabla \cdot [\phi(\nabla u)] \quad (u \in H_0^1(\Omega)), \tag{8.19}$$

is Lipschitz continuous.

Proof of Theorem 8.2.1. We must only consider the case $N \geq 2$. By Proposition 8.2.5, for all $u_0 \in H_0^1(\Omega)$ and $\epsilon > 0$, there exists a unique solution $u_\epsilon \in C^1([0,\infty); H_0^1(\Omega))$ of the abstract Cauchy problem

$$\begin{cases} \partial_t u = L(u) & \text{in } (0,\infty), \\ u(0) = u_0 \end{cases} \tag{8.20}$$

(e. g., see [31, Théorème II.16.1 and II.1.7.1]). Therefore u_ϵ satisfies problem (P_ϵ) in Q_T (for any $T \in (0,\infty)$) in the sense of $C^1([0,T); H^{-1}(\Omega))$, and thus equality (8.11) follows for all $t \in (0,T)$ and all $\rho \in H_0^1(\Omega)$. Since $T \in (0,\infty)$ is arbitrary, u_ϵ is a global solution. Hence the result follows. □

8.2.2 A priori estimates

A priori estimates of the solution u_ϵ (uniform with respect to ϵ) are needed to study its limit as $\epsilon \to 0^+$.

8.2.2.1 The case $N = 1$

Proposition 8.2.6. *Let $N = 1$ and* (8.1) *hold. Let u_ϵ be the solution of problem (P_ϵ) given by Theorem* 8.2.1. *Then there exists $C > 0$ (possibly, depending on $\|u_0\|_{W_0^{1,\infty}(\Omega)}$) such that for all $\epsilon > 0$,*

$$\sup_{t\in(0,\infty)} \|u_\epsilon(t)\|_{W_0^{1,\infty}(\Omega)} + \|\partial_t u_\epsilon\|_{L^2(Q_\infty)} + \sqrt{\epsilon}\,\|\partial_{xt} u_\epsilon\|_{L^2(Q_\infty)} \leq C. \tag{8.21}$$

Proof. Set

$$\Phi(y) := \int_0^y \phi(z)\,dz + k \quad (y, k \in \mathbb{R}) \tag{8.22}$$

(this amounts to choose $g(s) = s$ in (8.14)). By (8.9)–(8.10) we have

$$\begin{aligned}\partial_t \Phi(\partial_x u_\epsilon) &= \phi(\partial_x u_\epsilon)\,\partial_{xt} u_\epsilon = w_\epsilon\,\partial_{xx} w_\epsilon + [\phi(\partial_x u_\epsilon) - w_\epsilon]\partial_{xx} w_\epsilon \\ &= w_\epsilon \partial_{xx} w_\epsilon - \frac{|w_\epsilon - \phi(\partial_x u_\epsilon)|^2}{\epsilon} = w_\epsilon \partial_{xx} w_\epsilon - \epsilon\,|\partial_{xt} u_\epsilon|^2,\end{aligned}$$

whence plainly

$$\iint_{Q_\infty} \{|\partial_x w_\epsilon|^2 + \epsilon\,|\partial_{xt} u_\epsilon|^2\} dxdt \le \int_\Omega \Phi(u_0)\,dx. \tag{8.23}$$

On the other hand, since the solution is global, from (8.17) we obtain the a priori estimate

$$\sup_{t\in(0,T)} \|u_\epsilon(t)\|_{W_0^{1,\infty}(\Omega)} \le C_0[\lambda(\Omega) + 1]. \tag{8.24}$$

Then from (8.23)–(8.24) the result follows. □

Remark 8.2.1. Let u_ϵ be the solution of problem (P_ϵ) given by Theorem 8.2.1, and let v_ϵ, w_ϵ be given by (8.9). Since $\partial_t u_\epsilon = \partial_x w_\epsilon$, estimate (8.21) also reads

$$\|v_\epsilon\|_{L^\infty(Q_\infty)} + \|\partial_x w_\epsilon\|_{L^2(Q_\infty)} + \sqrt{\epsilon}\,\|\partial_t v_\epsilon\|_{L^2(Q_\infty)} \le C. \tag{8.25}$$

Plainly, this implies that there exists $\tilde{C} > 0$ (only depending on $\|u_0\|_{W_0^{1,\infty}(\Omega)}$) such that for all $\epsilon > 0$ small enough,

$$\|\phi(v_\epsilon)\|_{L^\infty(Q_\infty)} \le \tilde{C}, \tag{8.26}$$

$$\|w_\epsilon\|_{L^2(0,\infty;H^1(\Omega))} \le \tilde{C}. \tag{8.27}$$

In fact, by assumption (8.1b) and inequality (8.25) there exists $C_1 > 0$ (only depending on $\|u_0\|_{W_0^{1,\infty}(\Omega)}$) such that $\|\phi(v_\epsilon)\|_{L^\infty(Q_\infty)} \le C_1$ for all $\epsilon > 0$. Then by the definition of w_ϵ and inequality (8.25) we get

$$\|w_\epsilon\|_{L^2(Q_\infty)} \le C_1 + \sqrt{\epsilon}\,C.$$

By the above inequality and (8.25) there exists a constant $\tilde{C} \ge C_1$ such that (8.26)–(8.27) hold.

8.2.2.2 The case $N \geq 2$

Let us prove analogous estimates for $N \geq 2$ and any fixed $u_0 \in W_0^{1,p}(\Omega)$ ($p \in (1,2]$). Consider a family $\{u_{0\epsilon}\} \subseteq H_0^1(\Omega)$ ($\epsilon > 0$) such that

$$\|u_{0\epsilon}\|_{W_0^{1,p}(\Omega)} \leq \|u_0\|_{W_0^{1,p}(\Omega)}, \quad u_{0\epsilon} \to u_0 \text{ in } W_0^{1,p}(\Omega). \tag{8.28}$$

Proposition 8.2.7. *Let $N \geq 2$ and (8.2a)–(8.2e) hold. Let u_ϵ be the solution of problem (P_ϵ), with Cauchy data $u_{0\epsilon} \in H_0^1(\Omega)$ as in (8.28), given by Theorem 8.2.1. Then there exists $C > 0$ (depending on $\|u_0\|_{W_0^{1,p}(\Omega)}$) such that for all $\epsilon > 0$,*

$$\|u_\epsilon\|_{L^\infty(0,\infty;W_0^{1,p}(\Omega))} + \|\partial_t u_\epsilon\|_{L^2(Q_\infty)} + \sqrt{\epsilon}\,\|\nabla \partial_t u_\epsilon\|_{L^2(Q_\infty)} \leq C. \tag{8.29}$$

Proof. Since $\partial_t u_\epsilon(\cdot,t) \in H_0^1(\Omega)$ for all $t \in (0,\infty)$, we can choose $\rho = \partial_t u_\epsilon(\cdot,t)$ in (8.11). Then by assumption (8.2d) we obtain

$$\int_0^t \int_\Omega \partial_t u_\epsilon^2(x,s)\,dxds = -\int_0^t \left(\frac{d}{ds}\int_\Omega [\Phi(\nabla u_\epsilon)](x,s)\,dx\right)ds - \epsilon \iint_{Q_t} |\nabla \partial_t u_\epsilon|^2(x,s)\,dxds,$$

whence for all $t > 0$,

$$\int_\Omega [\Phi(\nabla u_\epsilon)](x,t)\,dx + \int_0^t \int_\Omega \{\partial_t u_\epsilon^2 + \epsilon|\nabla \partial_t u_\epsilon|^2\}(x,s)\,dxds = \int_\Omega [\Phi(\nabla u_{0\epsilon})](x)\,dx. \tag{8.30}$$

On the other hand, by the first inequality in (8.3) (with $q = p$) we have

$$\begin{aligned} C_1 \int_\Omega |\nabla u_\epsilon|^p(x,t)\,dx &= C_1 \int_{\{|\nabla u_\epsilon|>y_0\}} |\nabla u_\epsilon|^p(x,t)\,dx + C_1 \int_{\{|\nabla u_\epsilon|\leq y_0\}} |\nabla u_\epsilon|^p(x,t)\,dx \\ &\leq \int_{\{|\nabla u_\epsilon|>y_0\}} [\Phi(\nabla u_\epsilon)](x,t)\,dx + C_1 y_0^p \lambda_N(\Omega) \\ &= \int_\Omega [\Phi(\nabla u_\epsilon)](x,t)\,dx - \int_{\{|\nabla u_\epsilon|\leq y_0\}} [\Phi(\nabla u_\epsilon)](x,t)\,dx + C_1 y_0^p \lambda_N(\Omega) \end{aligned} \tag{8.31}$$

with y_0 as in (8.2c). Then by equality (8.30) we have

$$\begin{aligned} C_1 \int_\Omega |\nabla u_\epsilon|^p(x,t)\,dx &+ \int_0^t \int_\Omega \{\partial_t u_\epsilon^2 + \epsilon|\nabla \partial_t u_\epsilon|^2\}(x,s)\,dxds \\ &\leq \int_\Omega [\Phi(\nabla u_{0\epsilon})](x)\,dx - \int_{\{|\nabla u_\epsilon|\leq y_0\}} [\Phi(\nabla u_\epsilon)](x,t)\,dx + C_1 y_0^p \lambda_N(\Omega) \\ &\leq \int_\Omega [\Phi(\nabla u_{0\epsilon})](x)\,dx + K_1 \end{aligned} \tag{8.32}$$

with some constant $K_1 > 0$, only depending on y_0 and $\lambda_N(\Omega)$ (here we used the continuity of Φ).

Arguing as for (8.31), by the second inequality in (8.3) and the inequality in (8.28) we get

$$\begin{aligned}
&\int_\Omega [\Phi(\nabla u_{0\epsilon})](x)\,dx \\
&\quad= \int_{\{|\nabla u_{0\epsilon}|>y_0\}} [\Phi(\nabla u_{0\epsilon})](x)\,dx + \int_{\{|\nabla u_{0\epsilon}|\le y_0\}} [\Phi(\nabla u_{0\epsilon})](x)\,dx \\
&\quad\le C_3 \int_{\{|\nabla u_{0\epsilon}|>y_0\}} |\nabla u_{0\epsilon}|^p(x)\,dx + \int_{\{|\nabla u_{0\epsilon}|\le y_0\}} [\Phi(\nabla u_{0\epsilon})](x)\,dx \\
&\quad\le C_3 \int_\Omega |\nabla u_0|^p(x)\,dx + K_2
\end{aligned} \tag{8.33}$$

with some constant $K_2 > 0$ depending on y_0 only. Since $u \to \||\nabla u|\|_{L^p(\Omega)}$ is an equivalent norm on $W_0^{1,p}(\Omega)$, from (8.32)–(8.33) we obtain, for some constant $K_3 > 0$,

$$\begin{aligned}
&\|u_\epsilon\|^p_{L^\infty(0,\infty;W_0^{1,p}(\Omega))} + \|\partial_t u_\epsilon\|^2_{L^2(Q_\infty)} + \epsilon\,\|\nabla\partial_t u_\epsilon\|^2_{L^2(Q_\infty)} \\
&\quad\le \frac{1}{\min\{1, C_1K_3\}}\big(C_3\|u_0\|^p_{W_0^{1,p}(\Omega)} + K_1 + K_2\big) =: C_4,
\end{aligned}$$

whence

$$\|u_\epsilon\|_{L^\infty(0,\infty;W_0^{1,p}(\Omega))} + \|\partial_t u_\epsilon\|_{L^2(Q_\infty)} + \sqrt{\epsilon}\,\|\nabla\partial_t u_\epsilon\|_{L^2(Q_\infty)} \le C_4 + 3.$$

Then defining

$$C := C_4 + 3,$$

we obtain inequality (8.29). This completes the proof. □

Remark 8.2.2. In view of estimates (8.21) and (8.29), the family $\{u_\epsilon\}$ is contained in a bounded subset of $L^\infty(0,\infty; W_0^{1,q}(\Omega))$ for all $1 \le q < \infty$ if $N = 1$ or of $L^\infty(0,\infty; W_0^{1,p}(\Omega))$ if $N \ge 2$. In addition, if $N \ge 2$ and $p \in (1,2]$, then by estimate (8.29) for every $T \in (0,\infty)$, there exists a constant $\bar{C}_T > 0$ (depending on $\|u_0\|_{W_0^{1,p}(\Omega)}$) such that for all $\epsilon > 0$,

$$\|u_\epsilon\|_{W^{1,p}(Q_T)} \le \bar{C}_T. \tag{8.34}$$

Similarly, if $N = 1$, then by estimate (8.21) for all $\epsilon > 0$, we have

$$\|u_\epsilon\|_{H^1(Q_T)} \le \bar{C}_T. \tag{8.35}$$

8.2.3 Letting $\epsilon \to 0^+$

If $N \geq 2$, then by (8.29) we have the following:

Proposition 8.2.8. *Let $N \geq 2$ and (8.2a)–(8.2e) hold. Let u_ϵ be the solution of problem (P_ϵ), with Cauchy data $u_{0\epsilon} \in H_0^1(\Omega)$ as in (8.28), given by Theorem 8.2.1. Then there exist a sequence $\{u_{\epsilon_k}\} \subseteq \{u_\epsilon\}$ and $u \in L^\infty(0,\infty; W_0^{1,p}(\Omega)) \cap C([0,\infty); L^p(\Omega)) \cap W^{1,p}(Q_T)$ with $\partial_t u \in L^2(Q_\infty)$ such that*

$$u_{\epsilon_k} \to u \quad \text{in } C([0,T]; L^p(\Omega)) \text{ for all } T \in (0,\infty); \tag{8.36a}$$

$$u_{\epsilon_k} \rightharpoonup u \quad \text{in } L^r(0,T; W_0^{1,p}(\Omega)) \text{ for all } T \in (0,\infty) \quad (r \in [1,\infty)); \tag{8.36b}$$

$$u_{\epsilon_k} \rightharpoonup u \quad \text{in } W^{1,p}(Q_T) \text{ for all } T \in (0,\infty); \tag{8.36c}$$

$$\partial_t u_{\epsilon_k} \rightharpoonup \partial_t u, \quad \epsilon_k \nabla \partial_t u_{\epsilon_k} \to 0 \quad \text{in } L^2(Q_\infty); \tag{8.36d}$$

Proof. Concerning (8.36a), observe that by inequality (8.34) the family $\{u_\epsilon\}$ is bounded in $W^{1,p}(Q_T)$, and hence there exists a sequence $\{u_{\epsilon_k}\} \subseteq \{u_\epsilon\}$ (possibly, depending on $T \in (0,\infty)$) that strongly converges in $L^p(Q_T)$. Let us show that $\{u_{\epsilon_k}\}$ is a Cauchy sequence in $C([0,T]; L^p(\Omega))$ for all $T \in (0,\infty)$. Indeed, for all $t \in [0,T]$, we have

$$\begin{aligned}
&\left| \int_\Omega |u_{\epsilon_k} - u_{\epsilon_m}|^p(x,t)\,dx - \int_\Omega |u_{0\epsilon_k} - u_{0\epsilon_m}|^p(x)\,dx \right| \\
&\quad \leq p \iint_{Q_t} \left[|u_{\epsilon_k} - u_{\epsilon_m}|^{p-1} \, |\partial_t u_{\epsilon_k} - \partial_t u_{\epsilon_m}| \right](x,s)\,dxds \\
&\quad \leq p \left(\iint_{Q_t} |u_{\epsilon_k} - u_{\epsilon_m}|^p dxds \right)^{1-\frac{1}{p}} \left(\iint_{Q_t} |\partial_t u_{\epsilon_k} - \partial_t u_{\epsilon_m}|^p dxds \right)^{\frac{1}{p}} \\
&\quad \leq p\,|Q_T|^{\frac{2-p}{2p}} \, \|u_{\epsilon_k} - u_{\epsilon_m}\|_{L^p(Q_T)}^{p-1} \|\partial_t u_{\epsilon_k} - \partial_t u_{\epsilon_m}\|_{L^2(Q_T)},
\end{aligned}$$

since $p \in (1,2]$. Then by inequality (8.29) we have

$$\|u_{\epsilon_k} - u_{\epsilon_m}\|_{C([0,T];L^p(\Omega))} \leq K \|u_{\epsilon_k} - u_{\epsilon_m}\|_{L^p(Q_T)}^{1-\frac{1}{p}} + \|u_{0\epsilon_k} - u_{0\epsilon_m}\|_{L^p(\Omega)}, \tag{8.37}$$

where

$$K := 2Cp\,|Q_T|^{\frac{2-p}{2p}}.$$

Then by (8.28) and (8.37) the claim follows. Hence by a diagonal argument there exist a sequence $\{u_{\epsilon_k}\} \subseteq \{u_\epsilon\}$ and $u \in C([0,\infty); L^p(\Omega))$ such that (8.36a) holds for all $T \in (0,\infty)$.

Concerning the convergence in (8.36b), observe that by (8.29) we have

$$\|u_{\epsilon_k}\|_{L^r(0,T;W_0^{1,p}(\Omega))} \leq C\,T^{\frac{1}{r}} \tag{8.38}$$

for all $T \in (0, \infty)$ and $r \in [1, \infty)$. Hence by a diagonal argument there exist a subsequence of $\{u_{\epsilon_k}\}$ (not relabeled for simplicity) and a function $u : (0, \infty) \to W_0^{1,p}(\Omega)$ such that (8.36b) holds for all $T \in (0, \infty)$ and $r \in [1, \infty)$. To prove that $u \in L^\infty(0, \infty; W_0^{1,p}(\Omega))$, observe that by (8.29)

$$\|u_{\epsilon_k}(\cdot, t)\|_{W_0^{1,p}(\Omega)} \leq C \quad \text{for a. e. } t \in (0, \infty). \tag{8.39}$$

Then there exist a subsequence of $\{u_{\epsilon_k}(\cdot, t)\}$ (possibly, depending on t), denoted again by $\{u_{\epsilon_k}(\cdot, t)\}$ for simplicity, and a function $f_t \in W_0^{1,p}(\Omega)$ such that

$$u_{\epsilon_k}(\cdot, t) \rightharpoonup f_t \quad \text{in } W_0^{1,p}(\Omega). \tag{8.40}$$

Hence by inequality (8.39) and the lower semicontinuity of the norm we have

$$\|f_t\|_{W_0^{1,p}(\Omega)} \leq C \quad \text{for all } t \in (0, \infty). \tag{8.41}$$

On the other hand, by (8.36a) and (8.40) we have $f_t = u(\cdot, t)$. Therefore by (8.41) we obtain that $u \in L^\infty(0, \infty; W_0^{1,p}(\Omega))$, as claimed.

The convergences in (8.36d) follow immediately from (8.29), since both sequences $\{\partial_t u_{\epsilon_k}\}$ and $\{\sqrt{\epsilon_k}\,\nabla \partial_t u_{\epsilon_k}\}$ belong to a bounded subset of $L^2(Q_\infty)$. This completes the proof. □

Similarly, for $N = 1$, by estimate (8.21) we have the following:

Proposition 8.2.9. *Let $N = 1$ and (8.1) hold. Let u_ϵ be the solution of problem (P_ϵ) given by Theorem 8.2.1. Then there exist a sequence $\{u_{\epsilon_k}\} \subseteq \{u_\epsilon\}$ and $u \in L^\infty(0, \infty; W_0^{1,p}(\Omega)) \cap C([0, \infty); L^2(\Omega)) \cap H^1(Q_T)$ for all $p \in [1, \infty)$ and $T \in (0, \infty)$ with $u_x \in L^\infty(Q_\infty)$ and $\partial_t u \in L^2(Q_\infty)$ such that, as $k \to \infty$,*

$$u_{\epsilon_k} \to u \quad \text{in } C([0, T]; L^2(\Omega)) \text{ for all } T \in (0, \infty); \tag{8.42a}$$

$$u_{\epsilon_k} \rightharpoonup u \quad \text{in } L^r(0, T; W_0^{1,p}(\Omega)) \text{ for all } T \in (0, \infty) \quad (r \in [1, \infty),\ p \in (1, \infty)); \tag{8.42b}$$

$$u_{\epsilon_k} \rightharpoonup u \quad \text{in } H^1(Q_T) \text{ for all } T \in (0, \infty); \tag{8.42c}$$

$$\partial_x u_{\epsilon_k} \overset{*}{\rightharpoonup} \partial_x u \quad \text{in } L^\infty(Q_\infty); \tag{8.42d}$$

$$\partial_t u_{\epsilon_k} \rightharpoonup \partial_t u, \quad \epsilon_k\, \partial_{xt} u_{\epsilon_k} \to 0 \quad \text{in } L^2(Q_\infty). \tag{8.42e}$$

Proof. The proof of (8.42a) is the same as that of (8.36a) using inequality (8.35) instead of (8.34). Concerning (8.42b), by (8.21) we have

$$\|u_{\epsilon_k}\|_{L^r(0,T;W_0^{1,p}(\Omega))} \leq C\, T^{\frac{1}{r}} [\lambda_N(\Omega)]^{\frac{1}{p}} \tag{8.43}$$

for all $T \in (0, \infty)$, $r \in [1, \infty)$, and $p \in (1, \infty)$. Hence by a diagonal argument there exist a subsequence of $\{u_{\epsilon_k}\}$, denoted again by $\{u_{\epsilon_k}\}$, and a function $u : (0, \infty) \to W_0^{1,p}(\Omega)$

such that (8.42b) holds. To prove that $u \in L^\infty(0,\infty; W_0^{1,p}(\Omega)) \cap C([0,\infty); L^2(\Omega))$ for all $p \in [1,\infty)$ with $u_x \in L^\infty(Q_\infty)$, we argue as in the proof of Proposition 8.2.8. As for (8.42d), it suffices to observe that inequality (8.21) implies

$$\|\partial_x u_\epsilon\|_{L^\infty(Q_\infty)} \le C. \tag{8.44}$$

The proof of (8.42e) is the same as that of (8.36d), and thus the result follows. □

Remark 8.2.3. In view of inequality (8.27), there exist a sequence $\{w_{\epsilon_k}\} \subseteq \{w_\epsilon\}$ and $w \in L^2(0,\infty; H^1(\Omega))$ such that

$$w_{\epsilon_k} \rightharpoonup w \quad \text{in } L^2(0,\infty; H^1(\Omega)). \tag{8.45}$$

Without loss of generality, we can assume that the convergence results of Proposition 8.2.9 hold with the same sequence of indices $\{\epsilon_k\}$. In particular (see (8.42d)),

$$v_{\epsilon_k} \overset{*}{\rightharpoonup} v \quad \text{in } L^\infty(Q_\infty) \tag{8.46}$$

with $v := \partial_x u$, where u is the limiting function given by Proposition 8.2.9.

Remark 8.2.4. Let the assumptions of Proposition 8.2.9 hold if $N = 1$ or $u_0 \in H_0^1(\Omega)$, (8.2c)–(8.2e) with $p = q = 2$, and (8.2f) hold if $N \ge 2$ (thus, in particular, let Proposition 8.2.8 hold with $p = 2$). Let u be the limiting function given by Propositions 8.2.8–8.2.9. Under these assumptions, the map $t \mapsto \|u(\cdot,t)\|^2_{L^2(\Omega)}$ is nonincreasing on $(0,\infty)$. Indeed, since the global solution u_ϵ of problem (P_ϵ) satisfies $u_\epsilon(\cdot,t) \in H_0^1(\Omega)$ for every $t \in (0,\infty)$, from (P_ϵ) by assumption (8.2f) we easily get

$$\frac{d}{dt}\left(\|u_\epsilon(\cdot,t)\|^2_{L^2(\Omega)} + \epsilon\,\|\nabla u_\epsilon(\cdot,t)\|^2_{L^2(\Omega)}\right) = -2\int_\Omega [\phi(\nabla u_\epsilon)\cdot\nabla u_\epsilon](x,t)\,dx \le 0.$$

It follows that

$$\|u_\epsilon(\cdot,t+\tau)\|^2_{L^2(\Omega)} + \epsilon\,\|\nabla u_\epsilon(\cdot,t+\tau)\|^2_{L^2(\Omega)} \le \|u_\epsilon(\cdot,t)\|^2_{L^2(\Omega)} + \epsilon\,\|\nabla u_\epsilon(\cdot,t)\|^2_{L^2(\Omega)} \tag{8.47}$$

for every $t \in (0,\infty)$, $\tau > 0$, and $\epsilon > 0$. Setting $\epsilon = \epsilon_k$ in the last inequality, with $\{u_{\epsilon_k}\} \subseteq \{u_\epsilon\}$ given by Propositions 8.2.8–8.2.9, and letting $k \to \infty$, we obtain $\|u(\cdot,t+\tau)\|^2_{L^2(\Omega)} \le \|u(\cdot,t)\|^2_{L^2(\Omega)}$ for all $t \in (0,\infty)$ and $\tau > 0$ (here we use the convergence in (8.36a) and estimate (8.29) with $p = 2$ if $N \ge 2$, respectively, the convergence in (8.42a) and estimate (8.21) if $N = 1$). Hence the claim follows.

8.3 Existence

Relying on the results of the previous section, we can prove the following existence theorem.

Theorem 8.3.1. *Let either $N = 1$ and (8.1) or $N \geq 2$ and (8.2a)–(8.2e) hold. Then there exists a global Young measure-valued solution (u, ν) of problem (P), obtained as the limit of a subsequence $\{u_{\epsilon_k}\}$ of solutions of the regularized problems (P_{ϵ_k}). Moreover,*
(i) *if $N = 1$, then $u \in L^\infty(0, \infty; W_0^{1,p}(\Omega)) \cap C([0, \infty); L^2(\Omega)) \cap H^1(Q_T)$ for all $p \in [1, \infty)$, and $u_x \in L^\infty(Q_\infty)$;*
(ii) *if $N \geq 2$, then $u \in L^\infty(0, \infty; W_0^{1,p}(\Omega)) \cap C([0, \infty); L^p(\Omega)) \cap W^{1,p}(Q_T)$ $(p \in (1, 2])$;*
(iii) *$\nu \in \mathfrak{Y}^+(Q_\infty; \mathbb{R}^N)$.*

Proof. Suppose first that $N \geq 2$. Consider the sequence $\{u_{\epsilon_k}\}$ and the function u given by Proposition 8.2.8. The weak formulation of problem (P_{ϵ_k}) reads

$$\iint_{Q_t} [u_{\epsilon_k} \partial_s \zeta - \phi(\nabla u_{\epsilon_k}) \cdot \nabla \zeta - \epsilon_k \nabla u_{\epsilon_k s} \cdot \nabla \zeta](x, s)\, dxds$$
$$= \int_\Omega u_{\epsilon_k}(x, t) \zeta(x, t)\, dx - \int_\Omega u_{0\epsilon_k}(x)\, \zeta(x, 0)\, dx \tag{8.48}$$

for all $t > 0$ and $\zeta \in C^1([0, \infty); C_c^1(\Omega))$. By (8.36a), as $k \to \infty$, for all $t > 0$, we get

$$\lim_{k\to\infty} \iint_{Q_t} u_{\epsilon_k}(x, s) \partial_s \zeta(x, s)\, dxds = \iint_{Q_t} u(x, s) \partial_s \zeta(x, s)\, dxds, \tag{8.49}$$

$$\lim_{k\to\infty} \int_\Omega u_{\epsilon_k}(x, t) \zeta(x, t)\, dx = \int_\Omega u(x, t) \zeta(x, t)\, dx, \tag{8.50}$$

whereas by (8.36d) and (8.28) we have that

$$\lim_{k\to\infty} \left(\epsilon_k \iint_{Q_t} \nabla \partial_s u_{\epsilon_k} \cdot \nabla \zeta\, dxds \right) = 0, \tag{8.51}$$

respectively,

$$\lim_{k\to\infty} \int_\Omega u_{0\epsilon_k}(x) \zeta(x, 0)\, dx = \int_\Omega u_0(x) \zeta(x, 0)\, dx. \tag{8.52}$$

Since by inequality (8.29) the sequence $\{|\nabla u_{\epsilon_k}|\}$ is bounded in $L^p(Q_T)$ with $p \in (1, 2]$, by Lemma 2.8.12 it is uniformly integrable; hence the same holds for the sequence $\{\phi(\nabla u_{\epsilon_k})\}$, since by assumption (8.2b) $|\phi(\nabla u_{\epsilon_k})| \leq L|\nabla u_{\epsilon_k}| + |\phi(0)|$. Then by Proposition 5.4.10 and a standard diagonal argument there exist a subsequence of $\{\nabla u_{\epsilon_k}\}$ (not relabeled for simplicity) and a measure $\nu \in \mathfrak{R}_f(Q_\infty \times \mathbb{R}^N)$ with $\nu \in \mathfrak{Y}^+(Q_T; \mathbb{R}^N)$ for all $T \in (0, \infty)$ such that

$$\phi(\nabla u_{\epsilon_k}) \rightharpoonup \phi^* \quad \text{in } L^1(Q_T; \mathbb{R}^N) \tag{8.53}$$

with ϕ^* defined by (8.6). Then letting $k \to \infty$ in (8.48) and using (8.49)–(8.53), we obtain equality (8.5). Similarly, applying Proposition 5.4.10 with $h = \rho_i$, where $\rho_i(y) := y_i$, we have that

$$(\nabla u_{\epsilon_k})_i \rightharpoonup \int_{\mathbb{R}^N} y_i \, d\nu_{(x,t)}(y) \quad \text{in } L^1(Q_T) \quad (i = 1, \ldots, N),$$

which, together with the convergence in (8.42b), implies equality (8.4).

Therefore the pair (u, ν) is a Young measure-valued solution of problem (P) in Q_T. It follows from Proposition 8.2.8 that $u \in L^\infty(0, \infty; W_0^{1,p}(\Omega)) \cap C([0, \infty); L^p(\Omega))$ and $\partial_t u \in L^2(Q_\infty)$. Moreover, since $\nu \in \mathfrak{Y}^+(Q_T; \mathbb{R}^N)$ for every $T \in (0, \infty)$, by elementary properties of measures we have

$$\lambda_{N+1}(E) = \lim_{k\to\infty} \lambda_{N+1}(E \cap Q_k) = \lim_{k\to\infty} \nu((E \cap Q_k) \times \mathbb{R}^N) = \nu(E \times \mathbb{R}^N)$$

for all Borel sets $E \subseteq Q_\infty$, and thus $\nu \in \mathfrak{Y}^+(Q_\infty; \mathbb{R}^N)$.

Since (u, ν) is a Young measure-valued solution of problem (P) in Q_T for all $T \in (0, \infty)$, it is a global solution; hence the result follows in the case $N \geq 2$. The proof when $N = 1$ is the same using inequality (8.21) and Proposition 8.2.9 instead of (8.29) and Proposition 8.2.8, respectively. This completes the proof. □

Remark 8.3.1. Let $N = 1$, and let $\{u_{\epsilon_k}\}$ be the sequence used in the proof of Theorem 8.3.1. By inequality (8.25) for all $T \in (0, \infty)$, we have

$$\|\phi(\partial_x u_{\epsilon_k}) - w_{\epsilon_k}\|_{L^2(Q_T)} = \epsilon_k \|\partial_{xt} u_{\epsilon_k}\|_{L^2(Q_T)} \to 0,$$

and thus by (8.45) we have that $\phi(\partial_x u_{\epsilon_k}) \rightharpoonup w$ in $L^2(Q_T)$. On the other hand, by (8.53) we have $\phi(\partial_x u_{\epsilon_k}) \rightharpoonup \phi^* := \int_{\mathbb{R}} \phi(y) d\nu(y)$ in $L^1(Q_T)$. It follows that $w = \phi^*$, and thus, in particular, $\phi^* \in L^2(0, \infty; H^1(\Omega)) \cap L^\infty(Q_T)$.

Remark 8.3.2. Arguing as in the proof of Theorem 8.3.1 shows that if $N \geq 2$ and assumption (8.2c) is satisfied, then every function ϕ_i^* defined by (8.6) belongs to $L^r(Q_T)$ for all $T \in (0, \infty)$ and $r = \frac{p}{p-1}$ with $p \in (1, 2]$ (clearly, if $N = 1$, then by (8.21) $\phi^* \in L^\infty(Q_\infty)$). Indeed, fix $T \in (0, \infty)$. By (8.2c) there exists $M > 0$ such that

$$|\phi_i^*|(x, t) \leq M \int_{\mathbb{R}^N} (1 + |y|^{p-1}) \, d\nu_{(x,t)}(y) \tag{8.54}$$

for a. e. $(x, t) \in Q_T$. By (8.54) and the Jensen inequality we have

$$|\phi_i^*|^r \leq \bar{M} \int_{\mathbb{R}^N} (1 + |y|^p) \, d\nu_{(x,t)}(y) \tag{8.55}$$

with some $\bar{M} > 0$. Let $f_j \in C_c([0,\infty))$ ($j \in \mathbb{N}$) satisfy $0 \le f_j \le 1$ and

$$f_j(z) = \begin{cases} 1 & \text{if } z \in [0,j], \\ 0 & \text{if } z \in [j+1,\infty). \end{cases} \tag{8.56}$$

Let $\{u_{\epsilon_k}\}$ be the sequence used in the proof of Theorem 8.3.1. By (8.29), for every $j \in \mathbb{N}$, the sequence $\{f_j(|\nabla u_{\epsilon_k}|)\,|\nabla u_{\epsilon_k}|^p\}$ is bounded in Q_T and thus uniformly integrable. Then by Proposition 5.4.10 there exists a subsequence (not relabeled) such that

$$\begin{aligned} &\iint\limits_{Q_T} dxdt \int\limits_{\mathbb{R}^N} (1 + f_j(|y|)\,|y|^p)\, d\nu_{(x,t)}(y) \\ &\quad = \lim_{k\to\infty} \iint\limits_{Q_T} (1 + f_j(|\nabla u_{\epsilon_k}|)\,|\nabla u_{\epsilon_k}|^p)\, dxdt \le \lim_{k\to\infty} \iint\limits_{Q_T} (1 + |\nabla u_{\epsilon_k}|^p)\, dxdt \le \tilde{C} \end{aligned} \tag{8.57}$$

for some $\tilde{C} > 0$ independent of j (recall that $\nu_{(x,t)}(\mathbb{R}^N) = 1$ for a. e. $(x,t) \in Q_T$). Since $\lim_{j\to\infty} f_j(|y|)\,|y|^p = |y|^p$ for all $y \in \mathbb{R}^N$, from (8.57) and the Fatou lemma we get

$$\iint\limits_{Q_T} dxdt \int\limits_{\mathbb{R}^N} (1 + |y|^p)\, d\nu_{(x,t)}(y) \le \tilde{C},$$

whence by (8.55) the claim follows.

For future reference, let us prove the following result (see [100, Proposition 6]).

Proposition 8.3.2. *Let $N = 1$, let $\Omega \equiv (a,b)$, and let (8.1) hold. Let (u,ν) be a Young measure-valued solution of (P) given by Theorem 8.3.1. Let G_ϕ be as in (8.14) with $g \in C_c^1(\mathbb{R})$, and let $f \in C^1(\mathbb{R}) \cap W^{1,\infty}(\mathbb{R})$. Then for a. e. $(x,t) \in Q_T$, we have*

$$\int\limits_{\mathbb{R}} f(\phi(y))\, G_\phi(y)\, d\nu_{(x,t)}(y) = \int\limits_{\mathbb{R}} f(\phi(y))\, d\nu_{(x,t)}(y) \int\limits_{\mathbb{R}} G_\phi(y)\, d\nu_{(x,t)}(y). \tag{8.58}$$

To prove Proposition 8.3.2, we need a technical lemma (we refer the reader to [83, Lemma 2.1] for the proof). Let $\{\eta_i\} \subseteq H^1(\Omega)$ be a sequence of eigenfunctions of the operator $-\Delta$ with homogeneous Neumann conditions on $\partial\Omega$, and let $\{\mu_i\}$ be the corresponding sequence of eigenvalues. For any $\epsilon > 0$, set $\mathbb{N}_\epsilon := \{i \in \mathbb{N} \mid \mu_i \le \frac{1}{\epsilon}\}$, and let $P_\epsilon, Q_\epsilon : L^2(\Omega) \to H^1(\Omega)$, $P_\epsilon + Q_\epsilon = I$, be the projection operators defined as follows:

$$P_\epsilon f := \sum_{i\in\mathbb{N}_\epsilon} f_i \eta_i, \quad Q_\epsilon f := \sum_{i\in\mathbb{N}_\epsilon^c} f_i \eta_i, \quad f_i := \int\limits_\Omega f\,\eta_i\, dx \quad (f \in L^2(\Omega)). \tag{8.59}$$

Lemma 8.3.3. *Let $N = 1$, and let* (8.1) *hold. Let u_ϵ be the solution of problem (P_ϵ) given by Theorem* 8.2.1, *and let $v_\epsilon := \partial_x u_\epsilon$. Then there exists $C > 0$ such that for all $\epsilon > 0$,*

$$\|P_\epsilon\, \phi(v_\epsilon)\|_{L^2(0,T;H_0^1(\Omega))} + \frac{1}{\sqrt{\epsilon}}\, \|Q_\epsilon \phi(v_\epsilon)\|_{L^2(Q_T)} \le C. \tag{8.60}$$

Proof of Proposition 8.3.2. Let $\{v_{\epsilon_k}\}$ be the converging sequence in (8.46). Under the present assumptions, we have $f \circ \phi \in C_b(\mathbb{R})$, whereas $|G_\phi(y)| \le \|g\|_{L^\infty(\mathbb{R})}|y|$ for all $y \in \mathbb{R}$. By (8.25) the sequence $\{v_{\epsilon_k}\}$ is bounded in $L^\infty(Q_T)$ and hence uniformly integrable in Q_T. Hence by Proposition 5.4.10 in $L^1(Q_T)$ we have

$$f(\phi(v_{\epsilon_k})) \rightharpoonup (f \circ \phi)_*, \quad G_\phi(v_{\epsilon_k}) \rightharpoonup (G_\phi)_*, \quad f(\phi(v_{\epsilon_k}))G_\phi(v_{\epsilon_k}) \rightharpoonup ((f \circ \phi)\, G_\phi)_*,$$

where

$$(f \circ \phi)_* := \int_{\mathbb{R}} f(\phi(y))\, d\nu(y), \quad (G_\phi)_* := \int_{\mathbb{R}} G_\phi(y)\, d\nu(y),$$
$$((f \circ \phi)\, G_\phi)_* := \int_{\mathbb{R}} f(\phi(y))\, G_\phi(y)\, d\nu(y).$$

Then the conclusion follows if we prove that

$$f(\phi(v_{\epsilon_k}))\, G_\phi(v_{\epsilon_k}) \rightharpoonup (f \circ \phi)_*(G_\phi)_* \quad \text{in } L^1(Q_T). \tag{8.61}$$

Set $F := f \circ \phi$, $F^{\epsilon_k} := f(P_{\epsilon_k}\, \phi(v_{\epsilon_k}))$, where P_{ϵ_k} is the projection operator defined in (8.59) with $\epsilon = \epsilon_k$. Since by assumption $\|f'\|_{L^\infty(\mathbb{R})} < \infty$, we have that

$$\begin{aligned} \|F^{\epsilon_k} - F(v_{\epsilon_k})\|_{L^2(Q_T)} &= \|f(P_{\epsilon_k}\, \phi(v_{\epsilon_k})) - f(\phi(v_{\epsilon_k}))\|_{L^2(Q_T)} \\ &\le \|f'\|_{L^\infty(\mathbb{R})}\, \|Q_{\epsilon_k} \phi(v_{\epsilon_k})\|_{L^2(Q_T)}. \end{aligned} \tag{8.62}$$

On the other hand, by (8.25) there exists $C > 0$ such that $\|G_\phi(v_{\epsilon_k})\|_{L^2(Q_T)} \le C$. Then from (8.60) and (8.62) we obtain

$$\lim_{k\to\infty} \|[F^{\epsilon_k} - F(v_{\epsilon_k})]G_\phi(v_{\epsilon_k})\|_{L^2(Q_T)} = 0.$$

Therefore, to prove (8.61), it suffices to show that for every $\zeta \in C_c^1(Q_T)$,

$$\lim_{k\to\infty} \iint_{Q_T} F^{\epsilon_k} G_\phi(v_{\epsilon_k})\, \zeta\, dxdt = \iint_{Q_T} (f \circ \phi)_*(G_\phi)_*\, \zeta\, dxdt. \tag{8.63}$$

Set $\Gamma^{\epsilon_k}(x,t) := \int_a^x G_\phi(v_{\epsilon_k}(z,t))\, dz$ $((x,t) \in Q_T)$. Then we have

$$\iint_{Q_T} F^{\epsilon_k} G_\phi(v_{\epsilon_k})\, \zeta\, dxdt = -\iint_{Q_T} \partial_x(F^{\epsilon_k}\zeta)\, \Gamma^{\epsilon_k} dxdt. \tag{8.64}$$

Since $\|f'\|_{L^\infty(\mathbb{R})} < \infty$, by inequality (8.60) there exists $\bar{C} > 0$ such that

$$\|\partial_x F^{\epsilon_k}\|_{L^2(Q_T)} \le \|f'\|_{L^\infty(\mathbb{R})} \|\partial_x [P_{\epsilon_k}[\phi(v_{\epsilon_k})]]\|_{L^2(Q_T)} \le \bar{C}\|f'\|_{L^\infty(\mathbb{R})}.$$

Since $f \in L^\infty(\mathbb{R})$, the sequence $\{F^{\epsilon_k}\}$ is bounded in $L^\infty(Q_T)$ and thus in $L^2(Q_T)$; hence by the above inequality it is also bounded in $L^2(0,T;H^1(\Omega))$. Then there exists $\bar{F} \in L^2(0,T;H^1(\Omega))$ such that $F^{\epsilon_k} \rightharpoonup \bar{F}$ in $L^2(0,T;H^1(\Omega))$. By (8.62) this implies that $F(v_{\epsilon_k}) \rightharpoonup \bar{F}$ in $L^2(Q_T)$, whence $(f \circ \phi)_* = \bar{F}$ in Q_T. Therefore $(f \circ \phi)_* \in L^2(0,T;H^1(\Omega))$, and

$$\partial_x(F^{\epsilon_k}\zeta) \rightharpoonup \partial_x[(f \circ \phi)_* \zeta] \quad \text{in } L^2(Q_T). \tag{8.65}$$

On the other hand, since the sequence $\{G_\phi(v_{\epsilon_k})\}$ is bounded in $L^\infty(Q_T)$, there exists $G \in L^\infty(Q_T)$ such that (possibly, extracting a subsequence, not relabeled) $G_\phi(v_{\epsilon_k}) \overset{*}{\rightharpoonup} G$ in $L^\infty(Q_T)$. Since $G_\phi(v_{\epsilon_k}) \rightharpoonup (G_\phi)_*$ in $L^1(Q_T)$, we have $G = (G_\phi)_*$. Then a routine proof shows that for a. e. $(x,t) \in Q_T$,

$$\Gamma^{\epsilon_k}(x,t) \to \Gamma_*(x,t) := \int_a^x (G_\phi)_*(z,t)\,dz \tag{8.66}$$

(see [100, Propositions 5 and 6] for details). Since the family $\{\Gamma^{\epsilon_k}\}$ is bounded in $L^\infty(Q_T)$, from (8.66) by the dominated convergence theorem we get

$$\Gamma^{\epsilon_k} \to \Gamma_* \quad \text{in } L^2(Q_T). \tag{8.67}$$

By (8.65) and (8.67), as $k \to \infty$, the right-hand side of equality (8.64) converges to

$$-\iint_{Q_T} \partial_x[(f \circ \phi)_* \zeta]\,\Gamma_*\,dxdt = \iint_{Q_T} (f \circ \phi)_*(G_\phi)_*\zeta\,dxdt.$$

This proves (8.63), and thus the conclusion follows. □

8.4 Asymptotic behavior

Let us now address the asymptotic behavior as $t \to \infty$ of the global Young measure-valued solution (u,v) of (P) given by Theorem 8.3.1. Set

$$u_m(x,t) := u(x,t+t_m), \quad v^m_{(x,t)} := v_{(x,t+t_m)} \tag{8.68}$$

for a. e. $(x,t) \in Q_T$, where $\{t_m\}$ is any diverging sequence, and $X_p \equiv W_0^{1,p}(\Omega)$ with $p \in (1,\infty)$ if $N = 1$ or $p \in (1,2]$ if $N \ge 2$.

Definition 8.4.1. Let $\tilde{u} \in L^\infty(0,\infty;X_p) \cap C([0,\infty);L^p(\Omega))$ (with $p = 2$ if $N = 1$ and $p \in (1,2]$ if $N \geq 2$), and let $\tilde{\nu} \in \mathfrak{Y}^+(Q_\infty;\mathbb{R}^N)$. The pair $(\tilde{u},\tilde{\nu})$ is called an ω-limit point of a global Young measure-valued solution (u,ν) of problem (P) if:
(i) there exists a diverging sequence $\{t_m\} \subseteq (0,\infty)$ such that
  a) $u_m \rightharpoonup \tilde{u}$ in $L^r(0,T;X_p)$ for all $T \in (0,\infty)$ and $r \in [1,\infty)$,
  b) $\nu^m \overset{n}{\rightharpoonup} \tilde{\nu}$ in $\mathfrak{Y}^+(Q_T;\mathbb{R}^N)$ for all $T \in (0,\infty)$;
(ii) $(\tilde{u},\tilde{\nu})$ is a global Young measure-valued solution of problem (P) with initial data function $\tilde{u}(\cdot,0)$.

The set of the ω-limit points of (u,ν) is called the ω-limit set of (u,ν).

We will prove the following results.

Theorem 8.4.1. *Let either $N = 1$ and (8.1) or $N \geq 2$ and (8.2) hold. Let $(u,\nu) \in L^\infty(0,\infty;X_p) \times \mathfrak{Y}^+(Q_\infty;\mathbb{R}^N)$ be a global Young measure-valued solution of problem (P), whose existence is ensured by Theorem 8.3.1. Then:*
(i) the ω-limit set of (u,ν) is nonempty;
(ii) if $u_0 \in H_0^1(\Omega)$ and (8.2) holds with $p = 2$ if $N \geq 2$, then for every ω-limit point $(\tilde{u},\tilde{\nu})$,

$$\operatorname{supp} \tilde{\nu}_{(x,t)} \subseteq S := \{y \in \mathbb{R}^N \mid \phi(y) \cdot y = 0\} \tag{8.69}$$

for a. e. $(x,t) \in Q_\infty$.

Theorem 8.4.2. *Let the assumptions of Theorem 8.4.1 be satisfied. Let $u_0 \in H_0^1(\Omega)$, and let (8.2) hold with $p = 2$ if $N \geq 2$. Suppose that $S = \{y \in \mathbb{R}^N \mid \phi(y) = 0\}$. Then every ω-limit point $(\tilde{u},\tilde{\nu})$ of (u,ν) is a Young measure equilibrium solution of problem (P).*

To prove Theorem 8.4.1, we need some preliminary results.

Lemma 8.4.3. *Let the assumptions of Theorem 8.4.1 be satisfied. Then there exist a subsequence of the sequence $\{u_m\}$ defined in (8.68) (not relabeled for simplicity) and $\tilde{u} \in L^\infty(0,\infty;X_p) \cap C([0,\infty);L^p(\Omega)) \cap W^{1,p}(Q_T)$ $(p \in (1,2])$ with $\partial_t\tilde{u} \in L^2(Q_\infty)$ such that for all $T \in (0,\infty)$,*

$$u_m \to \tilde{u} \quad \text{in } C([0,T];L^p(\Omega)) \quad (p \in (1,2]); \tag{8.70a}$$

$$u_m \rightharpoonup \tilde{u} \quad \text{in } H^1(Q_T) \text{ if } N = 1, \quad u_m \rightharpoonup \tilde{u} \text{ in } W^{1,p}(Q_T) \text{ if } N \geq 2; \tag{8.70b}$$

$$u_m \rightharpoonup \tilde{u} \quad \text{in } L^r(0,T;X_p) \quad (r \in [1,\infty)); \tag{8.70c}$$

$$u_m \overset{*}{\rightharpoonup} \tilde{u} \quad \text{in } L^\infty(0,T;X_p); \tag{8.70d}$$

$$\nabla u_m \overset{*}{\rightharpoonup} \nabla\tilde{u} \quad \text{in } L^\infty(0,T;L^p(\Omega)) \quad (p \in (1,2]). \tag{8.70e}$$

Moreover,

$$\partial_t u_m \rightharpoonup \partial_t\tilde{u} \quad \text{in } L^2(Q_\infty). \tag{8.70f}$$

Proof. Recall that $u \in L^\infty(0,\infty;X_p)$ and $\partial_t u \in L^2(Q_\infty)$; moreover, $u \in W^{1,p}(Q_T) \cap C([0,\infty);L^p(\Omega))$ for all $T \in (0,\infty)$ ($p \in (1,2]$; see Propositions 8.2.8 and 8.2.9 and Theorem 8.3.1).

Clearly, for all $n \in \mathbb{N}$, we have

$$\|u_m\|_{L^\infty(0,\infty;X_p)} + \|\partial_t u_m\|_{L^2(Q_\infty)} \le \|u\|_{L^\infty(0,\infty;X_p)} + \|\partial_t u\|_{L^2(Q_\infty)}, \tag{8.71}$$

$$\|u_m\|_{H^1(Q_T)} \le \{T\|u\|^2_{L^\infty(\mathbb{R}_+,X_p)} + \|u_t\|^2_{L^2(Q_\infty)}\}^{\frac{1}{2}} \quad \text{if } N = 1, \tag{8.72}$$

$$\|u_m\|_{W^{1,p}(Q_T)} \le \{T\|u\|^p_{L^\infty(\mathbb{R}_+,X_p)} + (\lambda(\Omega)\,T)^{1-\frac{p}{2}}\|u_t\|^p_{L^2(Q_\infty)}\}^{\frac{1}{p}} \quad \text{if } N \ge 2 \quad (p \in (1,2]) \tag{8.73}$$

for all $T \in (0,\infty)$. As in the proof of Proposition 8.2.8, the convergence in (8.70a)–(8.70b) follows from inequalities (8.72)–(8.73), whereas the convergence in (8.70c)–(8.70f) follows from inequality (8.71). Hence the result follows. □

Lemma 8.4.4. *Let the assumptions of Theorem 8.4.1 be satisfied. Then the sequence $\{\nu^m\} \subseteq \mathfrak{Y}^+(Q_\infty;\mathbb{R}^N)$ of translated Young measures defined in (8.68) is tight in $\mathfrak{Y}^+(Q_T;\mathbb{R}^N)$ for all $T \in (0,\infty)$.*

Proof. Fix $T \in (0,\infty)$. By estimates (8.21) and (8.29) the sequence $\{|\nabla u_{\epsilon_k}|\}$ is bounded in $L^p(Q_\infty)$ with $p \in (1,\infty)$ if $N = 1$ or $p \in (1,2]$ if $N \ge 2$ and thus by Lemma 2.8.12 is q-uniformly integrable in $Q_\tau = \Omega \times (0,\tau)$ for every $\tau > 0$ and $q \in (1,p)$. Then by Proposition 5.4.10, for every $m \in \mathbb{N}$,

$$\begin{aligned} C &\ge \lim_{k\to\infty} \iint_{Q_{m,T}} |\nabla u_{\epsilon_k}|^q\,dxdt = \iint_{Q_{m,T}} dxdt \int_{\mathbb{R}^N} |y|^q\,d\nu_{(x,t)}(y) \\ &= \iint_{Q_T} dxdt \int_{\mathbb{R}^N} |y|^q\,d\nu_{(x,t+t_m)}(y) = \iint_{Q_T} dxdt \int_{\mathbb{R}^N} |y|^q\,d\nu^m_{(x,t)}(y) \end{aligned} \tag{8.74}$$

with some constant $C > 0$, where $Q_{m,T} := \Omega \times (t_m, t_m + T)$.

Let $K_j := \{y \in \mathbb{R}^N \mid |y| \le j\}$, and thus K_j is compact, and $|y| > j$ ($j \in \mathbb{N}$) on the complementary set K_j^c. Then by the Chebyshev inequality and (8.74) we have

$$j^q\,\nu^m(Q_T \times K_j^c) \le \iint_{Q_T} dxdt \int_{\mathbb{R}^N} |y|^q\,d\nu^m_{(x,t)}(y) \le C,$$

whence

$$\nu^m(Q_T \times K_j^c) \le \frac{C}{j^q} \quad \text{for all } j,m \in \mathbb{N} \tag{8.75}$$

(see (8.74)). Fix arbitrary $\epsilon > 0$, and choose $j_0 > (\frac{C}{\epsilon})^{\frac{1}{q}}$. Then by (8.75) we have that $\nu^m(Q_T \times K_{j_0}^c) < \epsilon$ for all $m \in \mathbb{N}$, and thus by Definition 5.3.4 the conclusion follows. □

Lemma 8.4.5. *Let the assumptions of Theorem 8.4.1 be satisfied. Then there exist a subsequence of the sequence $\{\nu^m\}$ defined in (8.68) (not relabeled for simplicity) and $\tilde\nu \in \mathfrak{Y}^+(Q_\infty;\mathbb{R}^N)$ such that $\nu^m \overset{n}{\rightharpoonup} \tilde\nu$ in $\mathfrak{Y}^+(Q_T;\mathbb{R}^N)$ for all $T\in(0,\infty)$.*

Proof. The result follows from Lemma 8.4.4 and Theorem 5.3.11 by a standard diagonal argument. □

Lemma 8.4.6. *Let the assumptions of Theorem 8.4.1 be satisfied. Let $\{\nu^m\}\subseteq\mathfrak{Y}^+(Q_\infty;\mathbb{R}^N)$ be the sequence defined in (8.68), and let $\rho\in C(\mathbb{R}^N)$ satisfy $|\rho(y)|\le M(1+|y|^q)$ for some $M>0$ and $q\in(0,p)$ (with $p\in(1,\infty)$ if $N=1$ or $p\in(1,2]$ if $N\ge 2$). Then for every $T\in(0,\infty)$:*
(i) *the map $(x,t)\mapsto\int_{\mathbb{R}^N}\rho(y)\,d\nu^m_{(x,t)}(y)$ belongs to $L^1(Q_T)$;*
(ii) *for all $\zeta\in L^\infty(Q_T)$,*

$$\lim_{j\to\infty}\iint_{Q_T}\zeta(x,t)\,dxdt\int_{\mathbb{R}^N}f_j(|y|)\rho(y)\,d\nu^m_{(x,t)}(y)=\iint_{Q_T}\zeta(x,t)\,dxdt\int_{\mathbb{R}^N}\rho(y)\,d\nu^m_{(x,t)}(y)\quad(8.76)$$

with $f_j\in C_c([0,\infty))$ as in (8.56), uniformly with respect to $m\in\mathbb{N}$.

Proof. (i) In view of inequalities (8.21) and (8.29), by Lemma 2.8.12, for every $q\in(0,p)$, the sequence $\{|\nabla u_{\epsilon_k}|^q\}\subseteq L^1(Q_{m,T})$ is bounded and uniformly integrable. Then by the growth assumption on ρ and Proposition 5.4.10 we have

$$\begin{aligned}\iint_{Q_T}dxdt\left|\int_{\mathbb{R}^N}\rho(y)\,d\nu^m_{(x,t)}(y)\right| &\le M\iint_{Q_T}dxdt\int_{\mathbb{R}^N}(1+|y|^q)\,d\nu^m_{(x,t)}(y)\\ &=M\iint_{Q_{m,T}}dxdt\int_{\mathbb{R}^N}(1+|y|^q)\,d\nu_{(x,t)}(y)\\ &=M\lim_{k\to\infty}\iint_{Q_{m,T}}(1+|\nabla u_{\epsilon_k}|^q)\,dxdt<\infty\end{aligned}$$

(possibly extracting a subsequence of $\{u_{\epsilon_k}\}$, not relabeled). Hence the claim follows.

(ii) Set $g_j:=1-f_j$; then $0\le g_j\le 1$ in $[0,\infty)$ and $g_j=0$ in $[0,j]$ ($j\in\mathbb{N}$). Since the sequence $\{|\nabla u_{\epsilon_k}|^q\}\subseteq L^1(Q_{m,T})$ is bounded and uniformly integrable (see (i)), by the growth assumption on ρ and Remark 2.8.3 the same holds for every sequence $\{g_j(|\nabla u_{\epsilon_k}|)|\rho(\nabla u_{\epsilon_k})|\}$ ($j\in\mathbb{N}$). Then by Proposition 5.4.10 there exists a subsequence of $\{g_j(|\nabla u_{\epsilon_k}|)|\rho(\nabla u_{\epsilon_k})|\}$ (not relabeled, possibly depending on j, m) such that for all $\zeta\in L^\infty(Q_T)$,

$$\begin{aligned}&\iint_{Q_{m,T}}\zeta(x,t)\,dxdt\int_{\mathbb{R}^N}g_j(|y|)|\rho(y)|\,d\nu_{(x,t)}(y)\\ &=\lim_{k\to\infty}\iint_{Q_{m,T}}[g_j(|\nabla u_{\epsilon_k}|)|\rho(\nabla u_{\epsilon_k})|](x,t)\,\zeta(x,t)\,dxdt\quad(j,m\in\mathbb{N}).\end{aligned}\quad(8.77)$$

It follows that for all $\zeta \in L^\infty(Q_T)$ and $j, m \in \mathbb{N}$,

$$\left| \iint_{Q_{m,T}} \zeta(x,t)\, dxdt \int_{\mathbb{R}^N} g_j(|y|)\rho(y)\, d\nu_{(x,t)}(y) \right| \\ \leq \|\zeta\|_{L^\infty(Q_T)} \sup_{k\in\mathbb{N}} \iint_{Q_{m,T}} [g_j(|\nabla u_{\epsilon_k}|)|\rho(\nabla u_{\epsilon_k})|](x,t)\, dxdt. \tag{8.78}$$

By the growth condition on ρ there exists $M_0 > 0$ such that $|\rho(y)| \leq M_0\, |y|^q$ for all $y \in \mathbb{R}^N$, $|y| \geq 1$. Set $E_{j,k,m} := \{(x,t) \in Q_{m,T} \mid |\nabla u_{\epsilon_k}|(x,t) \geq j\}$; thus $g_j(|\nabla u_{\epsilon_k}|) = 0$ on $(E_{j,k,m})^c$ and $g_j(|\nabla u_{\epsilon_k}|) \leq 1$ on $E_{j,k,m}$. Then for any fixed $j, k, m \in \mathbb{N}$, we get

$$\iint_{Q_{m,T}} [g_j(|\nabla u_{\epsilon_k}|)|\rho(\nabla u_{\epsilon_k})|](x,t)\, dxdt \\ = \iint_{E_{j,k,m}} [g_j(|\nabla u_{\epsilon_k}|)|\rho(\nabla u_{\epsilon_k})|](x,t)\, dxdt \\ \leq M_0 \iint_{E_{j,k,m}} |\nabla u_{\epsilon_k}|^q(x,t)\, dxdt \\ \leq M_0 \Big(\iint_{E_{j,k,m}} |\nabla u_{\epsilon_k}|^p(x,t)\, dxdt \Big)^{\frac{q}{p}} [\lambda_{N+1}(E_{j,k,m})]^{1-\frac{q}{p}}. \tag{8.79}$$

On the other hand, by the Chebyshev inequality we have

$$\lambda_{N+1}(E_{j,k,m}) \leq \Big(\iint_{E_{j,k,m}} |\nabla u_{\epsilon_k}|^p(x,t)\, dxdt \Big) j^{-p},$$

whence by (8.79) we have that

$$\iint_{Q_{m,T}} [g_j(|\nabla u_{\epsilon_k}|)|\rho(\nabla u_{\epsilon_k})|](x,t)\, dxdt \\ \leq M_0 \Big(\iint_{E_{j,k,m}} |\nabla u_{\epsilon_k}|^p(x,t)\, dxdt \Big) j^{-(p-q)} \\ \leq M_0 T \sup_{t\in(0,\infty)} \Big(\int_\Omega |\nabla u_{\epsilon_k}|^p(x,t)\, dx \Big) j^{-(p-q)} \leq K M_0 T\, j^{-(p-q)} \tag{8.80}$$

for some $K > 0$ only depending on u_0 and $\lambda_N(\Omega)$ (see the proof of inequalities (8.21) and (8.29)). From (8.78) and (8.80) we get

$$\left|\iint_{Q_T} \zeta(x,t)\,dxdt \int_{\mathbb{R}^N} \rho(y)\,d\nu^m_{(x,t)}(y) - \iint_{Q_T} \zeta(x,t)\,dxdt \int_{\mathbb{R}^N} f_j(|y|)\rho(y)\,d\nu^m_{(x,t)}(y)\right|$$
$$= \left|\iint_{Q_{m,T}} \zeta(x,t)\,dxdt \int_{\mathbb{R}^N} g_j(|y|)\rho(y)\,d\nu_{(x,t)}(y)\right| \le KM_0 T\, j^{-(p-q)}\, \|\zeta\|_\infty,$$

and thus claim (ii) follows. This completes the proof. □

Now we can prove Theorem 8.4.1.

Proof of Theorem 8.4.1. (i) Let $(\tilde{u}, \tilde{\nu})$ be the pair with components mentioned in Lemmas 8.4.3 and 8.4.5. It follows by these lemmas that requirement (i) of Definition 8.4.1 is satisfied by $(\tilde{u}, \tilde{\nu})$. Then claim (i) will follow if we prove that $(\tilde{u}, \tilde{\nu})$ also satisfies equalities (8.4) and (8.5) (with Cauchy data $\tilde{u}(\cdot, 0)$) of Definition 8.1.1.

Let us first show that $(\tilde{u}, \tilde{\nu})$ satisfies equality (8.4), that is,

$$\nabla\tilde{u}(x,t) = \int_{\mathbb{R}^N} y\, d\tilde{\nu}_{(x,t)}(y) \quad \text{for a. e. } (x,t) \in Q_T, \tag{8.81}$$

where $\tilde{\nu}_{(x,t)} \in \mathcal{P}\updownarrow\infty(\mathbb{R}^N)$ denotes the disintegration of $\tilde{\nu}$. Recalling definitions (8.68), from equality (8.4) we get

$$\nabla u_m(x,t) = \int_{\mathbb{R}^N} y\, d\nu^m_{(x,t)}(y) \tag{8.82}$$

for a. e. $(x,t) \in Q_T$ and all $m \in \mathbb{N}$. By the convergence in (8.70c), for all $\zeta \in L^\infty(Q_T)$, we have

$$\lim_{m\to\infty} \iint_{Q_T} \nabla u_m(x,t)\zeta(x,t)\,dxdt = \iint_{Q_T} \nabla\tilde{u}(x,t)\zeta(x,t)\,dxdt. \tag{8.83}$$

Let $\{f_j\} \subseteq C_c([0,\infty))$ $(j \in \mathbb{N})$ satisfy (8.56), and let $\zeta \in L^\infty(Q_T)$ be arbitrarily chosen. By Lemma 8.4.6(ii) we have

$$\lim_{j\to\infty} \iint_{Q_T} \zeta(x,t)\,dxdt \int_{\mathbb{R}^N} f_j(|y|)y\, d\nu^m_{(x,t)}(y) = \iint_{Q_T} \zeta(x,t)\,dxdt \int_{\mathbb{R}^N} y\, d\nu^m_{(x,t)}(y) \tag{8.84}$$

uniformly with respect to $m \in \mathbb{N}$. On the other hand, since the function $h_j(x,t,y) := f_j(|y|)\, y\, \zeta(x,t)$ belongs to $\mathscr{C}_b(Q_T \times \mathbb{R}^N)$ (see Definition 5.3.1), by Lemma 8.4.4 we get

$$\lim_{m\to\infty} \iint_{Q_T} \zeta(x,t)\,dxdt \int_{\mathbb{R}^N} f_j(|y|)\, y\, d\nu^m_{(x,t)}(y) = \iint_{Q_T} \zeta(x,t)\,dxdt \int_{\mathbb{R}^N} f_j(|y|)\, y\, d\tilde{\nu}_{(x,t)}(y). \tag{8.85}$$

Then by (8.82) and (8.84)–(8.85)

$$
\begin{aligned}
&\lim_{m\to\infty}\iint_{Q_T} \nabla u_m(x,t)\zeta(x,t)\,dxdt\\
&= \lim_{m\to\infty}\iint_{Q_T} \zeta(x,t)\,dxdt \int_{\mathbb{R}^N} y\,d\nu^m_{(x,t)}(y)\\
&= \lim_{m\to\infty}\lim_{j\to\infty}\iint_{Q_T} \zeta(x,t)\,dxdt \int_{\mathbb{R}^N} f_j(|y|)\,y\,d\nu^m_{(x,t)}(y)\\
&= \lim_{j\to\infty}\lim_{m\to\infty}\iint_{Q_T} \zeta(x,t)\,dxdt \int_{\mathbb{R}^N} f_j(|y|)\,y\,d\nu^m_{(x,t)}(y)\\
&= \lim_{j\to\infty}\iint_{Q_T} \zeta(x,t)\,dxdt \int_{\mathbb{R}^N} f_j(|y|)\,y\,d\tilde{\nu}_{(x,t)}(y)\\
&= \iint_{Q_T} \zeta(x,t)\,dxdt \int_{\mathbb{R}^N} y\,d\tilde{\nu}_{(x,t)}(y),
\end{aligned}
\tag{8.86}
$$

where the last equality follows by the dominated convergence theorem, since $\tilde{\nu}_{(x,t)}(\mathbb{R}^N) = 1$ for a. e. $(x,t) \in Q_T$, and the barycenter of $\tilde{\nu}$ belongs to $L^1(Q_T)$ (see Remark 5.4.2(ii)). By the arbitrariness of ζ from equalities (8.83)–(8.86) we obtain (8.81).

It is similarly seen that $(\tilde{u},\tilde{\nu})$ satisfies (8.5) with initial data $\tilde{u}(\cdot,0)$, that is,

$$
\begin{aligned}
&\int_0^t\int_\Omega \left\{\tilde{u}(x,s)\,\partial_s\zeta(x,s) - \sum_{i=1}^N \int_{\mathbb{R}^N} \phi_i(y)\,d\tilde{\nu}_{(x,s)}(y)\,\partial_{x_i}\zeta(x,s)\right\} dxds\\
&\quad= \int_\Omega \tilde{u}(x,t)\zeta(x,t)\,dx - \int_\Omega \tilde{u}(x,0)\,\zeta(x,0)\,dx
\end{aligned}
\tag{8.87}
$$

for all $\zeta \in C^1([0,T];C^1_c(\Omega))$ and $t \in (0,T)$ $(T \in (0,\infty))$. By abuse of notation denote by ζ also any extension $\zeta \in C^1((-\infty,T);C^1_c(\Omega))$, and thus for any sequence $\{t_m\} \subseteq (0,\infty)$ and $(x,t) \in Q_T$, the translate $\zeta_m(x,t) := \zeta(x,t-t_m)$ is well defined, and $\zeta_m \in C^1([0,T];C^1_c(\Omega))$. Then from (8.5) we get

$$
\begin{aligned}
&\int_{t_m}^{t+t_m}\int_\Omega \left\{u(x,s)\,\partial_s\zeta(x,s-t_m) - \sum_{i=1}^N \int_{\mathbb{R}^N} \phi_i(y)\,d\nu_{(x,s)}(y)\,\partial_{x_i}\zeta(x,s-t_m)\right\} dxds\\
&\quad= \int_\Omega u(x,t+t_m)\zeta(x,t)\,dx - \int_\Omega u(x,t_m)\,\zeta(x,0)\,dx
\end{aligned}
$$

for every sequence $\{t_m\} \subseteq (0,\infty)$ and all $t \in (0,T)$ and $T \in (0,\infty)$ with $0 < t_m < t + t_m < T$. The above equality reads

$$\int_0^t \int_\Omega \left\{ u_m(x,s)\,\partial_s\zeta(x,s) - \sum_{i=1}^N \int_{\mathbb{R}^N} \phi_i(y)\,d\nu^m_{(x,s)}(y)\,\partial_{x_i}\zeta(x,s) \right\} dxds$$
$$= \int_\Omega u_m(x,t)\zeta(x,t)\,dx - \int_\Omega u_m(x,0)\,\zeta(x,0)\,dx. \tag{8.88}$$

By the convergence in (8.70a) and (8.70c) we have

$$\lim_{m\to\infty} \int_0^t \int_\Omega u_m(x,s)\,\partial_s\zeta(x,s)\,dxds = \int_0^t \int_\Omega \tilde{u}(x,s)\,\partial_s\zeta(x,s)\,dxds, \tag{8.89a}$$

$$\lim_{m\to\infty} \int_\Omega u_m(x,t)\zeta(x,t)\,dx = \int_\Omega \tilde{u}(x,t)\zeta(x,t)\,dx, \tag{8.89b}$$

$$\lim_{m\to\infty} \int_\Omega u_m(x,0)\,\zeta(x,0)\,dx = \int_\Omega \tilde{u}(x,0)\,\zeta(x,0)\,dx. \tag{8.89c}$$

As before, let $\{f_j\} \subseteq C_c([0,\infty))$ satisfy (8.56). By Lemma 8.4.6 we have

$$\lim_{j\to\infty} \int_0^t \int_\Omega \partial_{x_i}\zeta(x,s)\,dxds \int_{\mathbb{R}^N} f_j(|y|)\phi_i(y)\,d\nu^m_{(x,s)}(y)$$
$$= \int_0^t \int_\Omega \partial_{x_i}\zeta(x,s)\,dxds \int_{\mathbb{R}^N} \phi_i(y)\,d\nu^m_{(x,s)}(y) \tag{8.90}$$

uniformly with respect to $m \in \mathbb{N}$. Then arguing as for (8.86) plainly shows that

$$\lim_{m\to\infty} \int_0^t \int_\Omega \partial_{x_i}\zeta(x,s)\,dxds \int_{\mathbb{R}^N} \phi_i(y)\,d\nu^m_{(x,s)}(y)$$
$$= \int_0^t \int_\Omega \partial_{x_i}\zeta(x,s)\,dxds \int_{\mathbb{R}^N} \phi_i(y)\,d\tilde{\nu}_{(x,s)}(y) \quad (i = 1,\dots,N). \tag{8.91}$$

From (8.89) and (8.91) we obtain equality (8.87). Hence claim (i) follows.

(ii) Let $f \in C([0,\infty))$ satisfy $f(z) = 1$ if $z \le 1$, $0 < f(z) < 1$ for every $z \in (1,\infty)$, and $f(z) \to 0$ as $z \to \infty$ so fast that the function $h : \mathbb{R}^N \to (0,\infty)$, $h(y) := f(|y|)\,\phi(y)\cdot y$, belongs to $C_0(\mathbb{R}^N)$. Since $f(|y|) > 0$ for every $y \in \mathbb{R}^N$, we have

$$S := \{y \in \mathbb{R}^N \mid \phi(y)\cdot y = 0\} = \{y \in \mathbb{R}^N \mid h(y) = 0\}.$$

Hence the conclusion will follow if we prove that

$$\iint_{Q_T} dxdt \int_{\mathbb{R}^N} h(y)\, d\tilde{\nu}_{(x,t)}(y) = 0 \tag{8.92}$$

(see inequalities (8.1d) and (8.2f)).

To this purpose, observe that since $h \in C_0(\mathbb{R}^N) \subseteq \mathscr{C}_b(Q_T \times \mathbb{R}^N)$, by Lemma 8.4.4 we have

$$\iint_{Q_T} dxdt \int_{\mathbb{R}^N} h(y)\, d\tilde{\nu}_{(x,t)}(y) = \lim_{m\to\infty} \iint_{Q_T} dxdt \int_{\mathbb{R}^N} h(y)\, d\nu^m_{(x,t)}(y). \tag{8.93}$$

Moreover, since h is bounded in $\mathbb{R}^N$, by Lemma 2.8.12 the sequence $\{h(\nabla u_{\epsilon_k})\} \subseteq L^1(Q_T)$ is bounded and uniformly integrable ($T \in (0,\infty)$). Hence by Proposition 5.4.10, for every $m \in \mathbb{N}$,

$$\iint_{Q_T} dxdt \int_{\mathbb{R}^N} h(y)\, d\nu^m_{(x,t)}(y) = \lim_{k\to\infty} \iint_{Q_{m,T}} h(\nabla u_{\epsilon_k})(x,t)\, dxdt. \tag{8.94}$$

From (8.93)–(8.94) it follows that

$$\iint_{Q_T} dxdt \int_{\mathbb{R}^N} h(y)\, d\tilde{\nu}_{(x,t)}(y) = \lim_{m\to\infty} \lim_{k\to\infty} \iint_{Q_{m,T}} h(\nabla u_{\epsilon_k})(x,t)\, dxdt. \tag{8.95}$$

Now observe that by Remark 8.2.4

$$\begin{aligned} 0 \le \iint_{Q_{m,T}} h(\nabla u_{\epsilon_k})(x,t)\, dxdt &\le \iint_{Q_{m,T}} [\phi(\nabla u_{\epsilon_k}) \cdot \nabla u_{\epsilon_k}](x,t)\, dxdt \\ &= -\frac{1}{2}\{\|u_{\epsilon_k}(\cdot, T + t_m)\|^2_{L^2(\Omega)} - \|u_{\epsilon_k}(\cdot, t_m)\|^2_{L^2(\Omega)} \\ &\quad + \epsilon_k(\|\nabla u_{\epsilon_k}(\cdot, T + t_m)\|^2_{L^2(\Omega)} - \|\nabla u_{\epsilon_k}(\cdot, t_m)\|^2_{L^2(\Omega)})\}. \end{aligned}$$

As $k \to \infty$ in the above inequality, using the convergence in (8.36a) and estimate (8.29) with $p = 2$ if $N \ge 2$, respectively, the convergence in (8.42a) and estimate (8.21) if $N = 1$ (see Remark 8.2.4), by equality (8.94) we obtain

$$0 \le \iint_{Q_T} dxdt \int_{\mathbb{R}^N} h(y)\, d\nu^m_{(x,t)}(y) \le -\frac{1}{2}[\|u(\cdot, T + t_m)\|^2_{L^2(\Omega)} - \|u(\cdot, t_m)\|^2_{L^2(\Omega)}]$$

for every $m \in \mathbb{N}$. Since the map $t \mapsto \|u(\cdot,t)\|^2_{L^2(\Omega)}$ is nonincreasing on $(0,\infty)$ (see Remark 8.2.4), there exists $\lim_{t\to\infty} \|u(\cdot,t)\|^2_{L^2(\Omega)} \in \mathbb{R}$. Then letting $m \to \infty$ in the above inequality and using (8.93), we obtain (8.92). This completes the proof. □

Proof of Theorem 8.4.2. By Theorem 8.4.1(ii), under the present assumptions, we have $\operatorname{supp} \tilde{\nu}_{(x,t)} \subseteq \{\phi = 0\}$ for a. e. $(x,t) \in Q_T$, and thus equality (8.7) is satisfied. Hence for all $\rho \in C_c^1(\Omega)$, from (8.87) we get

$$\int_\Omega \tilde{u}(x,t)\rho(x)dx = \int_\Omega \tilde{u}(x,0)\rho(x)dx$$

for all $t \in (0,\infty)$, whence $\tilde{u}(x,t) = \tilde{u}(x,0)$ by the arbitrariness of ρ. Therefore $\tilde{u}$ does not depend on t, and from equality (8.4) we obtain (8.8). Then the conclusion follows. □

8.5 Characterization of the limiting Young measure

Let (u,ν) be a global Young measure-valued solution of problem (P) given by Theorem 8.3.1. In this section, we give an explicit expression of the Young measure ν when $N = 1$ (in this connection, see Subsection 5.4.2).

Let us state some preliminaries. We suppose that ϕ changes the monotonicity character finitely many times; observe that this number is even, say $2n$ ($n \in \mathbb{N}$), by assumptions (8.1c)–(8.1d). Hence

$$\mathbb{R} = \left(\bigcup_{l=0}^{n} I_l\right) \cup \left(\bigcup_{m=1}^{n} \hat{I}_m\right) \quad (n \in \mathbb{N}), \tag{8.96}$$

where

$$I_0 := (-\infty, b_1],\ I_l := (a_l, b_{l+1}] \quad (l = 1,\dots,n-1), \quad I_n := (a_n, \infty),$$
$$\hat{I}_m := (b_m, a_m] \quad (m = 1,\dots,n)$$

and b_l is a local maximum point, a_l a local minimum point of the graph of ϕ. Set $J_l := \phi(I_l)$ and $\hat{J}_m := \phi(\hat{I}_m)$ ($l = 0,\dots,n$; $m = 1,\dots,n$); since ϕ is increasing on each interval I_l and decreasing on each interval $\hat{I}_m$, we have

$$J_0 := (-\infty, \phi(b_1)], \quad J_l := (\phi(a_l), \phi(b_{l+1})] \quad (l = 1,\dots,n-1), \quad J_n := (\phi(a_n), \infty),$$
$$\hat{J}_m := [\phi(a_m), \phi(b_m)) \quad (m = 1,\dots,n).$$

We also consider $n+1$ increasing functions $s_l : J_l \to I_l$ and $s_l := (\phi|_{I_l})^{-1}$ ($l = 0,\dots,n$) and n decreasing functions $t_m : \hat{J}_m \to \hat{I}_m$ and $t_m := (\phi|_{\hat{I}_m})^{-1}$ ($m = 1,\dots,n$). Following [83], we will use the following assumption:

$$\begin{cases} \text{The functions } s_0',\dots,s_n',\ t_1',\dots,t_n' \text{ are} \\ \text{linearly independent on any open subset of } \mathbb{R}. \end{cases} \tag{C}$$

Now we can state the following result.

Theorem 8.5.1. *Let $N = 1$, let $\Omega \equiv (a,b)$, and let (8.1) hold. Let ϕ change the monotonicity finitely many times. Let (u,v) be a global Young measure-valued solution of (P) given by Theorem 8.3.1. Then for a. e. $(x,t) \in Q_\infty$, we have*

$$v_{(x,t)} = \sum_{l=0}^{n} c_l(x,t)\, \delta_{s_l(w(x,t))} + \sum_{m=1}^{n} d_m(x,t)\, \delta_{t_m(w(x,t))}, \tag{8.97}$$

where $w \in L^2(0,\infty; H^1(\Omega)) \cap L^\infty(Q_\infty)$ is the limiting function in (8.45). Moreover, c_l, $d_m \in L^\infty(Q_\infty)$, and for a. e. $(x,t) \in Q_\infty$:
(i) $0 \le c_l(x,t) \le 1$, $0 \le d_m(x,t) \le 1$;
(ii) $\sum_{l=0}^{n} c_l(x,t) + \sum_{m=1}^{n} d_m(x,t) = 1$;
(iii) $c_0(x,t) = 1$ *if* $w(x,t) \le A := \min\{\phi(a_1), \dots, \phi(a_n)\}$, *and* $c_n(x,t) = 1$ *if* $w(x,t) \ge B := \max\{\phi(b_1), \dots, \phi(b_n)\}$.

Remark 8.5.1. Under the assumptions of Theorem 8.5.1, by equality (8.97) and Remark 8.3.1, for a. e. $(x,t) \in Q_\infty$, we have

$$\partial_x u(x,t) = \sum_{l=0}^{n} c_l(x,t)\, s_l(\phi^*(x,t)) + \sum_{m=1}^{n} d_m(x,t)\, t_m(\phi^*(x,t)).$$

To address Theorem 8.5.1, we argue as in the proof of Proposition 6.6.6. Let (u,v) be a global Young measure-valued solution of (P) given by Theorem 8.3.1, and let $\{u_{\epsilon_k}\}$ be the sequence used to construct it. Since the family $\{\phi(v_\epsilon)\}$, $v_\epsilon := \partial_x u_\epsilon$, is bounded in $L^\infty(Q_\infty)$ (see (8.26)), by Proposition 5.4.10 and a diagonal argument there exist a sequence $\{\phi(v_{\epsilon_k})\}$ and a Young measure $\tau \in \mathfrak{Y}^+(Q_\infty; \mathbb{R})$ such that for all $T \in (0,\infty)$ and $f \in C_c(\mathbb{R})$,

$$f(\phi(v_{\epsilon_k})) \rightharpoonup f_* \quad \text{in } L^1(Q_T), \quad f_*(x,t) := \int_{\mathbb{R}} f(y)\, d\tau_{(x,t)}(y). \tag{8.98}$$

On the other hand, for every $f \in C_c(\mathbb{R})$, the sequence $\{f(\phi(v_{\epsilon_k}))\}$ is bounded in $L^1(Q)$ and uniformly integrable, and thus (possibly, extracting a subsequence, not relabeled) by Proposition 5.4.10 we get

$$f(\phi(v_{\epsilon_k})) \rightharpoonup (f \circ \phi)^* \quad \text{in } L^1(Q_T), \quad (f \circ \phi)^*(x,t) := \int_{\mathbb{R}} f(\phi(y))\, dv_{(x,t)}(y). \tag{8.99}$$

By equalities (8.98)–(8.99) we have

$$\tau_{(x,t)} = (v_{(x,t)})_\phi \qquad \text{for a. e. } (x,t) \in Q_\infty, \tag{8.100}$$

that is, $\tau_{(x,t)}$ is the image of $(v_{(x,t)})$ under ϕ (see Definition 2.5.1).

Theorem 8.5.2. *Let the assumptions of Theorem* 8.5.1 *hold. Let* $\tau \in \mathfrak{Y}^+(Q_\infty;\mathbb{R})$ *be the Young measure in* (8.98). *Then*

$$\tau_{(x,t)} = \delta_{w(x,t)} \quad \textit{for a.e. } (x,t) \in Q_\infty. \tag{8.101}$$

For a. e. $(x,t) \in Q_\infty$ and all $l = 0,\dots,n$, $m = 1,\dots,n$, set

$$\sigma^l_{(x,t)} := (\nu_{(x,t)})_{\phi|_{I_l}}, \quad \hat{\sigma}^m_{(x,t)} := (\nu_{(x,t)})_{\phi|_{\hat{I}_m}}, \tag{8.102a}$$

$$\sigma_{(x,t)} := \sum_{l=0}^{n} \sigma^l_{(x,t)} + \sum_{m=1}^{n} \hat{\sigma}^m_{(x,t)}. \tag{8.102b}$$

Plainly, for any (x,t) and l, m as above, we have

$$\operatorname{supp} \sigma^l_{(x,t)} \subseteq \overline{J_l}, \quad \operatorname{supp} \hat{\sigma}^m_{(x,t)} \subseteq \overline{\hat{J}_m}, \quad \operatorname{supp} \sigma_{(x,t)} \subseteq \phi(\mathbb{R}) = \mathbb{R}. \tag{8.103}$$

By (8.102a), for every $f \in C(\mathbb{R})$, we have that

$$\int_{J_l} f(y)\, d\sigma^l_{(x,t)}(y) = \int_{I_l} (f \circ \phi)(y)\, d\nu_{(x,t)}(y), \tag{8.104a}$$

$$\int_{\hat{J}_m} f(y)\, d\hat{\sigma}^m_{(x,t)}(y) = \int_{\hat{I}_m} (f \circ \phi)(y)\, d\nu_{(x,t)}(y). \tag{8.104b}$$

Clearly, by (8.104) and (8.102b) we have $\sigma_{(x,t)} \in \mathfrak{P}(\mathbb{R})$.

In view of equality (8.100), Theorem 8.5.2 is an immediate consequence of the following proposition.

Proposition 8.5.3. *Let the assumptions of Theorem* 8.5.1 *be satisfied. Then for a. e.* $(x,t) \in Q_\infty$, *we have*

$$\sigma_{(x,t)} = (\nu_{(x,t)})_\phi, \quad \sigma_{(x,t)} = \delta_{w(x,t)}. \tag{8.105}$$

Remark 8.5.2. The first equality in (8.105) is easily proven. In fact, by (8.96) and (8.104), for all $f \in C(\mathbb{R})$ and a. e. $(x,t) \in Q_\infty$, we have

$$\begin{aligned}
&\int_{\mathbb{R}} (f \circ \phi)(y)\, d\nu_{(x,t)}(y) \\
&\quad = \sum_{l=0}^{n} \int_{I_l} (f \circ \phi)(y)\, d\nu_{(x,t)}(y) + \sum_{m=1}^{n} \int_{\hat{I}_m} (f \circ \phi)(y)\, d\nu_{(x,t)}(y) \\
&\quad = \sum_{l=0}^{n} \int_{J_l} f(y)\, d\sigma^l_{(x,t)}(y) + \sum_{m=1}^{n} \int_{\mathbb{R}} f(y)\, d\hat{\sigma}^m_{(x,t)}(y) = \int_{\phi(\mathbb{R})} f(y)\, d\sigma_{(x,t)}(y),
\end{aligned}$$

and hence the claim follows.

Relying on Theorem 8.5.2, we can prove Theorem 8.5.1.

Proof of Theorem 8.5.1. Let $\{\epsilon_k\}$ be a sequence of indices such that the convergence results in Proposition 8.2.9 and (8.45) hold. For $k \in \mathbb{N}$, set

$$Q_{k,l} := \{(x,t) \in Q_\infty \mid v_{\epsilon_k}(x,t) \in I_l\} \quad (l = 0, \dots, n),$$
$$Q_{k,m} := \{(x,t) \in Q_\infty \mid v_{\epsilon_k}(x,t) \in \hat{I}_m\} \quad (m = 1, \dots, n).$$

Clearly, a. e. in Q_∞ for all $f \in C(\mathbb{R})$, we have

$$f(v_{\epsilon_k}) = \sum_{l=0}^{n} \chi_{Q_{k,l}} f((s_l \circ \phi)(v_{\epsilon_k})) + \sum_{m=1}^{n} \chi_{Q_{k,m}} f((t_m \circ \phi)(v_{\epsilon_k})). \tag{8.106}$$

For any fixed $l, m \in \mathbb{N}$, the sequences $\{\chi_{Q_{k,l}}\}$ and $\{\chi_{Q_{k,m}}\}$ are bounded in $L^\infty(Q_\infty)$. Hence there exist two subsequences of $\{\chi_{Q_{k,l}}\}$ and $\{\chi_{Q_{k,m}}\}$ (not relabeled for simplicity) and $c_l, d_l \in L^\infty(Q_\infty)$ such that for all $T \in (0,\infty)$,

$$\chi_{Q_{k,l}} \overset{*}{\rightharpoonup} c_l, \quad \chi_{Q_{k,m}} \overset{*}{\rightharpoonup} d_m \quad \text{in } L^\infty(Q_T). \tag{8.107}$$

It is easily seen that the functions c_l, d_m ($l = 0, \dots, n$; $m = 1, \dots, n$) have the stated properties.

We will prove that for all $l = 0, \dots, n$ and $m = 1, \dots, n$, in $L^\infty(Q_T)$ ($T \in (0,\infty)$), we have

$$\chi_{Q_{k,l}} f((s_l \circ \phi)(v_{\epsilon_k})) \overset{*}{\rightharpoonup} c_l f(s_l(w)), \tag{8.108a}$$
$$\chi_{Q_{k,m}} f((t_m \circ \phi)(v_{\epsilon_k})) \overset{*}{\rightharpoonup} d_m f(t_m(w)). \tag{8.108b}$$

On the other hand, since by (8.25) the family $\{v_\epsilon\}$ is bounded in $L^\infty(Q_\infty)$ and thus bounded and uniformly integrable, by Proposition 5.4.10 $f(v_{\epsilon_k}) \rightharpoonup f^* := \int_{\mathbb{R}} f(y)\,d\nu(y)$ in $L^1(Q_T)$ for all $T \in (0,\infty)$. Therefore, letting $k \to \infty$ in (8.106) and using (8.108a), we obtain

$$\int_{\mathbb{R}} f(y)\,d\nu_{(x,t)}(y) = \sum_{l=0}^{n} c_l f(s_l(w)) + \sum_{m=1}^{n} d_m f(t_m(w)) \quad \text{a. e. in } Q_T,$$

whence by the arbitrariness of f equality (8.97) follows.

We only prove (8.108a), since the proof of (8.108b) is similar. For any $\zeta \in L^1(Q_T)$ ($T \in (0,\infty)$), we have that

$$\iint_{Q_T} [\chi_{Q_{k,l}} f((s_l \circ \phi)(v_{\epsilon_k})) - c_l f(s_l(w))]\,\zeta\,dxdt$$
$$= \iint_{Q_T} (\chi_{Q_{k,l}} - c_l) f(s_l(w))\,\zeta\,dxdt$$

$$+\iint_{Q_T} \chi_{Q_{k,l}}[f((s_l\circ\phi)(v_{\epsilon_k}))-f(s_l(w))]\,\zeta\,dxdt. \tag{8.109}$$

Since $w\in L^\infty(Q_T)$ and $f\circ s_l$ is continuous, we have $f(s_l(w))\in L^\infty(Q_T)$. Hence by the first convergence in (8.107)

$$\lim_{k\to\infty}\iint_{Q_T}(\chi_{Q_{k,l}}-c_l)f(s_l(w))\,\zeta\,dxdt=0. \tag{8.110}$$

As for the second integral in the right-hand side of (8.109), observe that by equality (8.101) and Proposition 5.4.1 $\phi(v_{\epsilon_k})\to w$, and thus (up to a subsequence) $f((s_l\circ\phi)(v_{\epsilon_k}))\to f(s_l(w))$ a. e. in Q_∞. Then by the dominated convergence theorem

$$\lim_{k\to\infty}\iint_{Q_T}\chi_{Q_{k,l}}[f((s_l\circ\phi)(v_{\epsilon_k}))-f(s_l(w))]\,\zeta\,dxdt=0. \tag{8.111}$$

From (8.109)–(8.111) the convergence in (8.108a) follows. This completes the proof. □

It remains to prove Proposition 8.5.3. By Remark 8.5.2 only the second equality in (8.105) must be proven. We outline the proof, referring the reader to [30] for details.

Denote by $e_1\le\cdots\le e_{2n}$ the set of the local extrema $\phi(a_l)$, $\phi(b_l)$ $(l=1,\dots,n)$ of the graph of ϕ. Let $E\equiv E_k:=[e_k,e_{k+1}]$ $(k=1,\dots,2n-1)$. Hence $e_1=A$, $e_{2n}=B$, and $\mathbb{R}=(-\infty,A]\cup(\bigcup_{k=1}^{2n-1}E_k)\cup[B,\infty)$. For every $l=0,\dots,n$, we have either $E\cap\overline{J_l}=\emptyset$ or $E\cap\overline{J_l}=E$, and similarly for $E\cap\overline{\hat{J}_m}$ $(m=1,\dots,n)$. Therefore there exist $p+1$ intervals J_{l_i} and p intervals $\hat{J}_{m_j}$ $(p=1,\dots,n;\ l_i\in\{0,\dots,p\};\ m_j\in\{1,\dots,p\})$ such that

$$E\subseteq\left(\bigcap_{i=0}^{p}\overline{J_{l_i}}\right)\cap\left(\bigcap_{j=1}^{p}\overline{\hat{J}_{m_j}}\right). \tag{8.112}$$

Without loss of generality, $l_0\le\cdots\le l_p$ and $m_1\le\cdots\le m_p$. Relying on Proposition 8.3.2, we can prove the following lemma.

Lemma 8.5.4. *Let the assumptions of Theorem 8.5.1 be satisfied. Then for a. e. $(x,t)\in Q_\infty$, we have:*

(i) *let $K\subseteq(-\infty,A]$ with $A:=\min\{\phi(a_1),\dots,\phi(a_p)\}$ be compact, and let $\sigma_{(x,t)}(K)>0$. Then for a. e. $y\in(-\infty,A]$,*

$$\sigma^0_{(x,t)}((-\infty,y])\,\sigma_{(x,t)}(K)=\sigma^0_{(x,t)}((-\infty,y]\cap K); \tag{8.113}$$

(ii) *let $K\subseteq[B,\infty)$ with $B:=\max\{\phi(b_1),\dots,\phi(b_p)\}$ be compact, and let $\sigma_{(x,t)}(K)>0$. Then for a. e. $y\in[B,\infty)$,*

$$\sigma^n_{(x,t)}([y,\infty))\,\sigma_{(x,t)}(K)=\sigma^n_{(x,t)}([y,\infty)\cap K); \tag{8.114}$$

(iii) *let* $E \equiv [e_k, e_{k+1}]$ $(k = 1, \dots, 2n-1)$, *let* $K \subseteq E$ *be compact, and let* $\sigma_{(x,t)}(K) > 0$. *Then for a. e.* $y \in E$,

$$\sum_{i=1}^{p+1} D_i(y) \left\{ \frac{\sigma^{l_i}_{(x,t)}(K)}{\sigma_{(x,t)}(K)} - \sigma^{l_i}_{(x,t)}(E) \right\} + \sum_{j=1}^{p} D_{j+1}(y) \left\{ \frac{\hat{\sigma}^{m_j}_{(x,t)}(K)}{\sigma_{(x,t)}(K)} - \hat{\sigma}^{m_j}_{(x,t)}(E) \right\}$$
$$= \sum_{i=1}^{p+1} s'_{l_i}(y) \left\{ \rho^{l_i}_{(x,t)}(y) - \frac{\rho^{l_i}_K(y)}{\sigma_{(x,t)}(K)} \right\} + \sum_{j=1}^{p} t'_{m_j}(y) \left\{ \hat{\rho}^{m_j}_{(x,t)}(y) - \frac{\hat{\rho}^{m_j}_K(y)}{\sigma_{(x,t)}(K)} \right\},$$

where

$$D_1(y) := 0, \quad D_i(y) := \sum_{k=1}^{i-1} (s'_{l_k} - t'_{m_k})(y) \quad (i = 2, \dots, p+1),$$
$$\rho^{l_i}_{(x,t)}(y) := \sigma^{l_i}_{(x,t)}([y, \phi(b_{l_i+1})]), \quad \hat{\rho}^{m_j}_{(x,t)}(y) := \hat{\sigma}^{m_j}_{(x,t)}([y, \phi(b_{m_j})]),$$
$$\rho^{l_i}_K(y) := \sigma^{l_i}_{(x,t)}([y, \phi(b_{l_i+1})] \cap K), \quad \hat{\rho}^{m_j}_K(y) := \hat{\sigma}^{m_j}_{(x,t)}([y, \phi(b_{m_j})] \cap K).$$

Proof. We only prove claims (i)–(ii), the proof of (iii) being analogous but more involved (see [30, Lemma 5.5]). Since $K \subseteq (-\infty, A]$ is compact, by Lemma A.9 there exists a sequence $\{f_h\} \subseteq C_c(\mathbb{R})$ such that $0 \le f_h \le 1$, $f_h = 1$ on K, and $\lim_{h\to\infty} f_h(y) = \chi_K(y)$ for all $y \in \mathbb{R}$. Let $a_0 \in I_0$ be the unique point such that $\phi(a_0) = A$, and choose k in (8.14) such that $G_\phi(y) = \int_{a_0}^{y} g(\phi(z))dz$ $(y \in \mathbb{R})$ with $g \in C^1_c(\mathbb{R})$ such that $\operatorname{supp} g \subseteq (-\infty, A]$. By equality (8.58) we have

$$\int_{\mathbb{R}} f_h(\phi(y))\, G_\phi(y)\, d\nu_{(x,t)}(y) = \int_{\mathbb{R}} f_h(\phi(y))\, d\nu_{(x,t)}(y) \int_{\mathbb{R}} G_\phi(y)\, d\nu_{(x,t)}(y),$$

whence, by the choice of g and Remark 8.5.2,

$$\int_{(-\infty,A]} f_h(y)\, (G_\phi \circ s_0)(y)\, d\sigma^0_{(x,t)}(y) = \int_{\phi(\mathbb{R})} f_h(y)\, d\sigma_{(x,t)}(y) \int_{(-\infty,A]} (G_\phi \circ s_0)(y)\, d\sigma^0_{(x,t)}(y).$$

Letting $h \to \infty$ in this equality, we get

$$\int_K (G_\phi \circ s_0)(y)\, d\sigma^0_{(x,t)}(y) = \sigma_{(x,t)}(K) \int_{(-\infty,A]} (G_\phi \circ s_0)(y)\, d\sigma^0_{(x,t)}(y),$$

whence plainly

$$\int_{(-\infty,A]} d\sigma^0_{(x,t)}(y) \int_A^y g(z) s'_0(z)\, dz = \frac{1}{\sigma_{(x,t)}(K)} \int_K d\sigma^0_{(x,t)}(y) \int_A^y g(z) s'_0(z)\, dz.$$

Interchanging the order of integration in this equality gives (8.113). In case (ii), it is similarly seen that

$$\int_{[B,\infty)} d\sigma^n_{(x,t)}(y) \int_B^y g(z)s_0'(z)\,dz = \frac{1}{\sigma_{(x,t)}(K)} \int_K d\sigma^n_{(x,t)}(y) \int_B^y g(z)s_0'(z)\,dz,$$

whence (8.114) follows. This completes the proof. □

Now we can prove Proposition 8.5.3.

Proof of Proposition 8.5.3. Suppose that $\operatorname{supp}\sigma_{(x,t)}|_{(-\infty,A]} \neq \emptyset$ and set

$$q_0 \equiv q_0(x,t) := \max\{y \in \operatorname{supp}\sigma_{(x,t)}|_{(-\infty,A]}\}.$$

Consider for any $\delta > 0$ the compact set $K = [q_0 - \delta, q_0]$. Then $\sigma_{(x,t)}(K) > 0$, and

$$\sigma^0_{(x,t)}((-\infty,y] \cap K) = 0 \quad \text{for all } y \in (-\infty, q_0 - \delta].$$

By equality (8.113) and the arbitrariness of δ we obtain that $\sigma^0_{(x,t)}((-\infty,q_0)$, and thus $\operatorname{supp}\sigma_{(x,t)}|_{(-\infty,A]} = \{q_0\}$. Moreover, equality (8.113) with $K = \{q_0\}$ gives $\sigma^0_{(x,t)}(\{q_0\}) = \sigma_{(x,t)}(\{q_0\}) = 1$. Using (8.114), it is similarly seen that $\operatorname{supp}\sigma_{(x,t)}|_{[B,\infty)}$ consists at most of one point, say $\{q_{2n}\}$, and if it is nonempty, then $\sigma^n_{(x,t)}(\{q_{2n}\}) = \sigma_{(x,t)}(\{q_{2n}\}) = 1$. A more involved argument relying on Lemma 8.5.4(iii) shows that the same holds for every interval $E_k := [e_k, e_{k+1}]$ $(k = 1, \dots, 2n-1)$, that is, if $\operatorname{supp}\sigma|_{E_k} \neq \emptyset$, then $\operatorname{supp}\sigma_{(x,t)}|_{E_k} = \{q_k\}$, and $\sigma_{(x,t)}(\{q_k\}) = 1$ for some $q_k \in E_k$ (see the proof of [30, Proposition 5.2] for details).

To summarize, the support of the measure $\sigma_{(x,t)}$ consists of at most $2n+1$ points $\{q_k\}$ such that $\sigma_{(x,t)}(\{q_k\}) = 1$ for each $k = 0, \dots, 2n$. However, since $\sigma_{(x,t)} \in \mathfrak{P}(\mathbb{R})$, this implies that $\operatorname{supp}\sigma_{(x,t)}$ consists of one point, say $q = q(x,t)$, and $\sigma_{(x,t)} = \delta_{q(x,t)}$. Then by the first equality in (8.105) we have

$$q(x,t) = \int_{\mathbb{R}} y\,d\sigma_{(x,t)}(y) = \int_{\mathbb{R}} \phi(y)\,d\nu_{(x,t)}(y) = \phi_*(x,t),$$

whence by Remark 8.3.1 we get $\sigma_{(x,t)} = \delta_{\phi_*(x,t)} = \delta_{w(x,t)}$ for a. e. $(x,t) \in Q_\infty$. This completes the proof. □

Bibliography

[1] R. A. Adams, *Sobolev Spaces*, Pure Appl. Math. **65** (Academic Press, 1975).

[2] D. R. Adams & L. I. Hedberg, *Function Spaces and Potential Theory*, Grundlehren Math. Wiss. **314** (Springer, 1996).

[3] D. R. Adams & N. G. Meyers, *Bessel potentials. Inclusion relations among classes of exceptional sets*, Indiana Univ. Math. J. **22** (1973), 873–905.

[4] J. J. Alibert & G. Bouchitté, *Non-uniform integrability and generalized Young measures*, J. Convex Anal. **4** (1997), 129–147.

[5] L. Ambrosio, N. Fusco & D. Pallara, *Functions of Bounded Variation and Free Discontinuity Problems*, Oxford Math. Monographs (Oxford University Press, 2000).

[6] L. Ambrosio, N. Gigli & G. Savaré, *Gradient Flows in Metric Spaces and in the Space of Probability Measures*, Lecture in Math. ETH Zürich (Birkhäuser, 2000).

[7] J. Appell, J. Banaś & N. Merentes, *Bounded Variation and Around*, De Gruyter Ser. Nonlinear Anal. Appl. **17** (De Gruyter, 2014).

[8] J. M. Ball, *A version of the fundamental theorem for Young measures*. In: *PDEs and Continuum Models of Phase Transitions*, Lecture Notes in Phys. **344**, 207–215 (Springer, 1989).

[9] J. M. Ball & F. Murat, *Remarks on Chacon's biting lemma*, Proc. Am. Math. Soc. **107** (1989), 655–663.

[10] P. Baras & M. Pierre, *Singularités éliminables pour des équations semilinéaires*, Ann. Inst. Fourier **34** (1984), 185–206.

[11] G. I. Barenblatt, M. Bertsch, R. Dal Passo & M. Ughi, *A degenerate pseudoparabolic regularization of a nonlinear forward-backward heat equation arising in the theory of heat and mass exchange in stably stratified turbulent shear flow*, SIAM J. Math. Anal. **24** (1993), 1414–1439.

[12] G. Bellettini, L. Bertini, M. Mariani & M. Novaga, *Convergence of the one-dimensional Cahn–Hilliard equation*, SIAM J. Math. Anal. **44** (2012), 3458–3480.

[13] Ph. Bénilan, L. Boccardo, T. Gallouët, R. Gariepy, M. Pierre& J. L. Vasquez, *An L^1-theory of existence and uniqueness of solutions of nonlinear elliptic equations*, Ann. Sc. Norm. Super. Pisa, Cl. Sci. **22** (1995), 241–273.

[14] Ph. Bénilan & H. Brezis, *Nonlinear problems related to the Thomas–Fermi equation. Dedicated to Philippe Bénilan*, J. Evol. Equ. **3** (2003), 673–770.

[15] Ph. Bénilan, H. Brezis & M. G. Crandall, *A semilinear equation in $L^1(\mathbb{R}^N)$*, Ann. Sc. Norm. Super. Pisa, Cl. Sci. **2** (1975), 523–555.

[16] Ph. Bénilan & M. G. Crandall, *Regularizing effects of homogeneous evolution equations*. In: *Contributions to Analysis and Geometry*, Amer. J. Math., Suppl., 23–39 (Johns Hopkins University Press, 1981).

[17] M. Bertsch, F. Smarrazzo, A. Terracina & A. Tesei, *A uniqueness criterion for measure-valued solutions of scalar hyperbolic conservation laws*, Atti Accad. Naz. Lincei, Rend. Lincei, Mat. Appl. **30** (2019), 137–168.

[18] M. Bertsch, F. Smarrazzo, A. Terracina & A. Tesei, *Radon measure-valued solutions of first order scalar conservation laws*, Adv. Nonlinear Anal. **9** (2020), 65–107.

[19] M. Bertsch, F. Smarrazzo, A. Terracina & A. Tesei, *Signed Radon-measure solutions for a class of flux-saturated scalar conservation laws*, Discrete Contin. Dyn. Syst. **40** (2020), 3143–3169.

[20] M. Bertsch, F. Smarrazzo, A. Terracina & A. Tesei, *Discontinuous solutions of Hamilton–Jacobi equations versus Radon measure-valued solutions of scalar conservation laws: Disappearance of singularities*, J. Dyn. Differ. Equ. (2021). 10.1007/s10884-021-09997-x.

[21] M. Bertsch, F. Smarrazzo, A. Terracina & A. Tesei, *Discontinuous viscosity solutions of first order Hamilton–Jacobi equations*, J. Hyperbolic Differ. Equ. **17** (2021), 1–42.

https://doi.org/10.1515/9783110556902-011

[22] M. Bertsch, F. Smarrazzo & A. Tesei, *On a class of forward-backward parabolic equations: formation of singularities*, J. Differ. Equ. **269** (2020), 6656–6698.

[23] L. Boccardo, T. Gallouët & L. Orsina, *Existence and uniqueness of entropy solutions for nonlinear elliptic equations with measure data*, Ann. Inst. Henri Poincaré **13** (1996), 539–551.

[24] N. Bourbaki, *Intégration (Chap. I–IV)*, Actualités Scientifiques et Industrielles **1175** (Hermann, 1952).

[25] H. Brezis, *Analyse Fonctionnelle* (Masson, 1983).

[26] H. Brezis & A. Friedman, *Nonlinear parabolic equations involving measures as initial conditions*, J. Math. Pures Appl. **62** (1983), 73–97.

[27] H. Brezis, M. Marcus & A. C. Ponce, *Nonlinear elliptic equations with measures revisited.* In: *Mathematical Aspects of Nonlinear Dispersive Equations*, Ann. of Math. Stud. **163**, 55–110 (Princeton University Press, 2007).

[28] H. Brezis & W. A. Strauss, *Semi-linear second-order elliptic equations in* L^1, J. Math. Soc. Jpn. **25** (1973), 565–590.

[29] M. Brokate & J. Sprekels, *Hysteresis and Phase Transitions*, Appl. Math. Sci. **121** (Springer, 1996).

[30] L. T. T. Bui, F. Smarrazzo & A. Tesei, *Sobolev regularization of a class of forward-backward parabolic equations*, J. Differ. Equ. **257** (2014), 1403–1456.

[31] H. Cartan, *Calcul Différentiel* (Hermann, 1971).

[32] G. Choquet, *Forme abstraite du théorème de capacitabilité*, Ann. Inst. Fourier **9** (1959), 83–89.

[33] M. G. Crandall & M. Pierre, *Regularizing effects for* $u_t + A\phi(y) = 0$ *in* L^1, J. Funct. Anal. **45** (1982), 194–212.

[34] C. M. Dafermos, *Hyperbolic Conservation Laws in Continuum Physics*, Grundlehren Math. Wiss. (Springer, 2000).

[35] B. E. J. Dahlberg & C. E. Kenig, *Non-negative solutions of generalized porous medium equations*, Rev. Mat. Iberoam. **2** (1986), 267–304.

[36] B. E. J. Dahlberg & C. E. Kenig, *Non-negative solutions to fast diffusions*, Rev. Mat. Iberoam. **4** (1988), 11–29.

[37] G. Dal Maso, F. Murat, L. Orsina, A. Prignet, *Renormalized solutions for elliptic equations with general measure data*, Ann. Sc. Norm. Super. Pisa, Cl. Sci. **28** (1999), 741–808.

[38] F. Demengel & J. Rauch, *Weak convergence of asymptotically homogeneous functions of measures*, Nonlinear Anal. **15** (1990), 1–16.

[39] F. Demengel & D. Serre, *Nonvanishing singular parts of measure valued solutions of scalar hyperbolic equations*, Commun. Partial Differ. Equ. **16** (1991), 221–254.

[40] E. DiBenedetto, *Real Analysis*, Birkhäuser Adv.Texts Basler Lehrbücher (Birkhäuser, 2002).

[41] J. Diestel & J. J. Uhl, *Vector Measures*, Math. Surveys Monogr. **15** (Amer. Math. Soc., 1977).

[42] J. L. Doob, *Measure Theory*, Grad. Texts in Math. (Springer, 1994).

[43] J. Droniou, *Intégration et Espaces de Sobolev à Valeurs Vectorielles*, Polycopié de l'Ecole Doctorale de Maths-Info de Marseille (unpublished notes, available on HAL).

[44] R. E. Edwards, *Functional Analysis: Theory and Applications* (Dover, 1994).

[45] J. Elstrodt, *Maß- und Integrationstheorie* (Springer, 2010).

[46] L. C. Evans, *Weak Convergence Methods for Nonlinear Partial Differential Equations*, CBMS Reg. Conf. Ser. Math. **74** (American Mathematical Society, 1990).

[47] L. C. Evans & R. F. Gariepy, *Measure Theory and Fine Properties of Functions* (CRC Press, 1992).

[48] L. C. Evans & M. Portilheiro, *Irreversibility and hysteresis for a forward-backward diffusion equation*, Math. Models Methods Appl. Sci. **14** (2004), 1599–1620.

[49] H. Federer, *Geometric Measure Theory*, Grundlehren Math. Wiss. **153** (Springer, 1969).

[50] A. Friedman, *Mathematics in Industrial Problems, Part 8*, IMA Vol. Math. Appl. **83** (Springer, 1997).

[51] M. Fukushima, K. Sato & S. Taniguchi, *On the closable part of pre-Dirichlet forms and the fine support of the underlying measures*, Osaka J. Math. **28**, (1991), 517–535.

[52] T. Gallouët & J.-M. Morel, *Resolution of a semilinear equation in L^1*, Proc. R. Soc. Edinb. A **96A** (1984), 275–288.

[53] M. Giaquinta, G. Modica & J. Souček, *Cartesian Currents in the Calculus of Variations*, Ergeb. Math. Grenzgeb. (3) **37** (Springer, 1998).

[54] B. H. Gilding & A. Tesei, *The Riemann problem for a forward-backward parabolic equation*, Physica D **239** (2010), 291–311.

[55] A. Gmira & L. Véron, *Boundary singularities of solutions of some nonlinear elliptic equations*, Duke Math. J. **64** (1991), 271–324.

[56] M. Grun-Rehomme, *Caractérisation du sous-différentiel d'intégrandes convexes dans les espaces de Sobolev*, J. Math. Pures Appl. **56** (1977), 149–156.

[57] M. Helmers & M. Herrmann, *Interface dynamics in discrete forward-backward diffusion equations*, Multiscale Model. Simul. **11** (2013), 1261–1297.

[58] E. Hewitt & K. Stromberg, *Real and Abstract Analysis* (Springer, 1965).

[59] K. Höllig, *Existence of infinitely many solutions for a forward backward heat equation*, Trans. Am. Math. Soc. **278** (1983), 299–316.

[60] J. L. Kelley, *General Topology* (van Nostrand, 1955).

[61] D. Kinderlehrer & P. Pedregal, *Weak convergence of integrands and the Young measure representation*, SIAM J. Math. Anal. **23** (1992), 1–19.

[62] J. Kinnunen & O. Martio, *The Sobolev capacity on metric spaces*, Ann. Acad. Sci. Fenn., Math. **21** (1996), 367–388.

[63] O. A. Ladyženskaja, V. A. Solonnikov & N. N. Ural'ceva, *Linear and Quasi-Linear Equations of Parabolic Type*, Transl. Math. Monogr. **23** (Amer. Math. Soc., 1991).

[64] E. H. Lieb & M. Loss, *Analysis*, Grad. Stud. Math. **14** (Amer. Math. Soc., 2001).

[65] E. H. Lieb & B. Simon, *The Thomas–Fermi theory of atoms, molecules and solids*, Adv. Math. **23** (1977), 22–116.

[66] B. V. Limaye, *Functional Analysis* (New Age Int. Publ., 1996).

[67] T.-P. Liu & M. Pierre, *Source-solutions and asymptotic behavior in conservation laws*, J. Differ. Equ. **51** (1984), 419–441.

[68] J. Málek, J. Nečas, M. Rokyta & M. Růžička, *Weak and Measure-Valued Solutions of Evolutionary PDEs* (Chapman & Hall, 1996).

[69] M. Marcus & L. Véron, *Removable singularities and boundary traces*, J. Math. Pures Appl. **80** (2001) 879–900.

[70] C. Mascia, A. Terracina & A. Tesei, *Evolution of stable phases in forward-backward parabolic equations*. In: *Asymptotic Analysis and Singularities*, Adv. Stud. Pure Math. **47-2**, 451–478 (Math. Soc. Japan, 2007).

[71] C. Mascia, A. Terracina & A. Tesei, *Two-phase entropy solutions of a forward-backward parabolic equation*, Arch. Ration. Mech. Anal. **194** (2009), 887–925.

[72] N. G. Meyers, *A theory of capacities for potentials of functions in Lebesgue classes*, Math. Scand. **26** (1970), 255–292.

[73] S. Müller, *Variational models for microstructure and phase transitions*. In: *Calculus of Variations and Geometric Evolution Problems*, Lecture Notes in Math. **1713**, 85–210 (Springer, 1999).

[74] A. Novick-Cohen & R. L. Pego, *Stable patterns in a viscous diffusion equation*, Trans. Am. Math. Soc. **324** (1991), 331–351.

[75] F. Otto, *Initial-boundary value problem for a scalar conservation law*, C. R. Acad. Sci. **322** (1996), 729–734.

[76] V. Padrón, *Sobolev regularization of a nonlinear ill-posed parabolic problem as a model for aggregating populations*, Commun. Partial Differ. Equ. **23** (1998), 457–486.

[77] V. Padròn, *Effect of aggregation on population recovery modeled by a forward-backward pseudoparabolic equation*, Trans. Am. Math. Soc. **356** (2003), 2739–2756.

[78] P. Pedregal, *Parametrized Measures and Variational Principles*, Progr. Nonlinear Differential Equations Appl. (Birkhäuser, 1991).

[79] P. Perona & J. Malik, *Scale space and edge detection using anisotropic diffusion*, IEEE Trans. Pattern Anal. Mach. Intell. **12** (1990), 629–639.

[80] M. Pierre, *Nonlinear fast diffusion with measures as data*. In: *Nonlinear Parabolic Equations: Qualitative Properties of Solutions*, Pitman Res. Notes Math. Ser. **149**, 179–188 (Longman, 1987).

[81] M. Pierre, *Uniqueness of the solutions of* $u_t - \Delta\varphi(u) = 0$ *with initial datum a measure*. Nonlinear Anal. **6** (1982), 175–187.

[82] P. I. Plotnikov, *Equations with alternating direction of parabolicity and the hysteresis effect*, Russ. Acad. Sci. Dokl. Math. **47** (1993), 604–608.

[83] P. I. Plotnikov, *Passing to the limit with respect to viscosity in an equation with variable parabolicity direction*, Differ. Equ. **30** (1994), 550–566.

[84] P. I. Plotnikov, *Forward-backward parabolic equations and hysteresis*, J. Math. Sci. **93** (1999), 747–766.

[85] M. M. Porzio, F. Smarrazzo & A. Tesei, *Radon measure-valued solutions for a class of quasilinear parabolic equations*, Arch. Ration. Mech. Anal. **210** (2013), 713–772.

[86] M. M. Porzio, F. Smarrazzo & A. Tesei, *Radon measure-valued solutions of nonlinear strongly degenerate parabolic equations*, Calc. Var. Partial Differ. Equ. **51** (2014), 401–437.

[87] M. M. Porzio, F. Smarrazzo & A. Tesei, *Noncoercive diffusion equations with Radon measures as initial data*, J. London Math. Soc. (to appear).

[88] D. S. Ross, *Two new moving boundary problems for scalar conservation laws*, Commun. Pure Appl. Math. **41** (1988), 725–737.

[89] D. S. Ross, *Ion etching: An application of the mathematical theory of hyperbolic conservation laws*, J. Electrochem. Soc. **135** (1988), 1235–1240.

[90] H. L. Royden, *Real Analysis* (MacMillan, 1968).

[91] D. Serre, *Systems of Conservation Laws, Vol. 1: Hyperbolicity, Entropies, Shock Waves* (Cambridge University Press, 1999).

[92] M. Saadoune & M. Valadier, *Extraction of a "good" subsequence from a bounded sequence of integrable functions*, J. Convex Anal. **2** (1995), 345–357.

[93] L. Schwartz, *Fonctions mesurables et ∗-scalairement mesurables, mesures banachiques majorées, martingales banachiques, et propriété de Radon–Nikodým*, Séminaire d'analyse fonctionelle (Polytechnique) (1974–1975), exp. 4, 1–17.

[94] J. Simon, *Compact sets in the space* $L^p(0,T;B)$, Ann. Mat. Pura Appl. **146** (1987), 65–96.

[95] M. Slemrod, *Dynamics of measure valued solutions to a backward-forward heat equation*, J. Dyn. Differ. Equ., **3** (1991), 1–28.

[96] F. Smarrazzo, *On a class of equations with variable parabolicity direction*, Discrete Contin. Dyn. Syst. **22** (2008), 729–758.

[97] F. Smarrazzo *On a class of quasilinear elliptic equations with degenerate coerciveness and measure data*, Adv. Nonlinear Stud. **18** (2018), 361–392.

[98] F. Smarrazzo & A. Terracina, *Sobolev approximation for two-phase solutions of forward-backward parabolic problems*, Discrete Contin. Dyn. Syst. **33** (2013), 1657–1697.

[99] F. Smarrazzo & A. Tesei, *Degenerate regularization of forward-backward parabolic equations: The regularized problem*, Arch. Ration. Mech. Anal. **204** (2012), 85–139.

[100] F. Smarrazzo & A. Tesei, *Degenerate regularization of forward-backward parabolic equations: The vanishing viscosity limit*, Math. Ann. **355** (2013), 551–584.

[101] M. Talagrand, *Pettis Integral and Measure Theory*, Mem. Amer. Math. Soc. **51** (Amer. Math. Soc., Providence, 1984).

[102] B. O. Turesson, *Nonlinear Potential Theory and Weighted Sobolev Spaces*, Lecture Notes in Math. **1736** (Springer, 2000).

[103] M. Valadier, *Young measures*. In: *Methods of Nonconvex Analysis*, Lecture Notes in Math. **1446**, 152–188 (Springer, 1990).

[104] J. L. Vazquez, *Measure-valued solutions and phenomenon of blow-down in logarithmic diffusion*, J. Math. Anal. Appl. **352** (2009), 515–547.

[105] L. Véron, *Singularities of Solutions of Second Order Quasilinear Equations*, Pitman Res. Notes Math. Ser. **353** (Longman, 1996).

[106] L. Véron, *Elliptic equations involving measures*. In: *Stationary Partial Differential Equations Vol. I*, Handb. Differ. Equ., 595–712 (North-Holland, 2004).

[107] J. Warga, *Optimal Control of Differential and Functional Equations* (Academic Press, 1972).

[108] K. Zhang, *Existence of infinitely many solutions for the one-dimensional Perona–Malik model*, Calc. Var. Partial Differ. Equ. **26** (2006), 171–199.

[109] W. P. Ziemer, *Weakly Differentiable Functions*, Grad. Texts in Math. (Springer, 1989).

Appendix A Topological spaces

Some topology concepts and results are collected for convenience of the reader (e. g., see [60]).

A.1

A family $\mathcal{T} \subseteq \mathcal{P}(X)$ is called a *topology* on X if (i) $\emptyset \in \mathcal{T}$, $X \in \mathcal{T}$; (ii) for any family $\{A_i\}_{i\in I} \subseteq \mathcal{T}$, we have $\bigcup_{i\in I} A_i \in \mathcal{T}$; and (iii) for any finite family $\{A_1,\dots,A_n\} \subset \mathcal{T}$, we have $\bigcap_{k=1}^n A_k \in \mathcal{T}$.

The couple $(X,\mathcal{T})$ is called a *topological space*. Sets $A \in \mathcal{T}$ are called *open*, whereas $C \subseteq X$ is called *closed* if $C^c \in \mathcal{T}$. If $\mathcal{T}_1$, $\mathcal{T}_2$ are topologies on X, then $\mathcal{T}_1$ is *weaker* than $\mathcal{T}_2$, and $\mathcal{T}_2$ is *stronger* than $\mathcal{T}_1$ if $\mathcal{T}_1 \subseteq \mathcal{T}_2$. The topology $\mathcal{P}(X)$ is called *discrete*. For any family $\mathcal{G} \subseteq \mathcal{P}(X)$, the intersection $\mathcal{T}_0(\mathcal{G}) := \bigcap\{\mathcal{T} \text{ topology} \mid \mathcal{T} \supseteq \mathcal{G}\}$ is the topology called the *topology generated by* $\mathcal{G}$ or the *minimal topology* containing $\mathcal{G}$. For any $F \subseteq X$, the family $\mathcal{T} \cap F := \{A \cap F \mid A \in \mathcal{T}\}$ is the topology on F called the *relative topology* or the *trace* of $\mathcal{T}$ on F. The topological space $(F,\mathcal{T} \cap F)$ is called a *topological subspace* of $(X,\mathcal{T})$.

Let (X,d) be a metric space with metric d. Consider the family $\{B(x_0,r) \mid x_0 \in X, r > 0\}$ of *open balls* $B(x_0,r) := \{x \in X \mid d(x_0,x) < r\}$ with *center* x_0 and *radius* r. The family $\{E \subseteq X \mid \forall x_0 \in E \exists r > 0 \text{ such that } B(x_0,r) \subseteq E\}$ is a topology on X called a *metric topology* (hence every metric space is a topological space).

A topological space $(X,\mathcal{T})$ is called:

(i) *metrizable* if there exists a metric d on X such that the corresponding metric topology coincides with $\mathcal{T}$;
(ii) *completely metrizable* if there exists a metric d on X such that the corresponding metric topology coincides with $\mathcal{T}$ and (X,d) is a *complete* metric space.

Let (X,d) be a metric space. For any nonempty $E,F \subseteq X$ the *distance between E and F* is the quantity $d(E,F) := \inf_{x\in E,y\in F} d(x,y)$, whereas the distance of a point $x \in X$ from a nonempty $E \subseteq X$ is $d(x,E) := d(\{x\},E)$. By definition, $d(x,\emptyset) := \infty$ for all $x \in X$. The *diameter* of a nonempty set $E \subseteq X$ is the quantity $\operatorname{diam}(E) := \sup_{x,y\in E} d(x,y) \in [0,\infty]$, and $\operatorname{diam}(\emptyset) := -\infty$. A subset $E \subseteq X$ is *bounded* if $\operatorname{diam}(E) < \infty$.

A.2

Let $(X_1,\mathcal{T}_2)$ and $(X_2,\mathcal{T}_2)$ be topological spaces. The topology generated by the family $\{A_1 \times A_2 \mid A_1 \in \mathcal{T}_1, A_2 \in \mathcal{T}_2\} \subseteq \mathcal{P}(X_1 \times X_2)$ is called the *product topology* and denoted $\mathcal{T}_1 \times \mathcal{T}_2$. The topological space $(X_1 \times X_2, \mathcal{T}_1 \times \mathcal{T}_2)$ is called the *topological product* of X_1 and X_2.

https://doi.org/10.1515/9783110556902-012

A.3

Let $(X, \mathcal{T})$ be a topological space. A *neighborhood* of a set $E \subseteq X$ is any set $F \subseteq X$ such that there exists $A \in \mathcal{T}$ with $E \subseteq A \subseteq F$. A neighborhood of a singleton $\{x\}$ is called a *neighborhood of* x.

A topological space X is called a *Hausdorff space* if for any $x_1, x_2 \in X$, $x_1 \neq x_2$, there exist a neighborhood E_1 of x_1 and a neighborhood E_2 of x_2 such that $E_1 \cap E_2 = \emptyset$. Every metric space is a Hausdorff space.

A.4

Let $(X, \mathcal{T})$ be a topological space. A family $\{A_i\}_{i\in I} \subseteq \mathcal{T}$ is called a *basis* of $\mathcal{T}$ if for every $A \in \mathcal{T}$, there exists $I_0 \subseteq I$ such that $A = \bigcup_{i\in I_0} A_i$. If X is a metric space, then the family $\{B(x_0, r) \mid x_0 \in X, r > 0\}$ is a basis of the metric topology.

Let $E \subseteq X$. The set $\mathring{E} := \bigcup\{A \in \mathcal{T} \mid A \subseteq E\}$ is called the *interior* of E. The *closure* of E is the set $\overline{E} := \bigcap\{C \supseteq E \mid C^c \in \mathcal{T}\}$. The *boundary* ∂E of E is defined as $\partial E := \overline{E} \setminus E^\circ$.

Let $E, F \subseteq X$. Then E is *dense* in F if $F \subseteq \overline{E}$; in particular, E is dense in X if $\overline{E} = X$. The space X is called *separable* if there exists a countable dense set $E \subseteq X$.

Proposition A.1. (i) *A topological space with countable basis is separable.* (ii) *A separable metric space has a countable basis.*

A.5

Let $(X, \mathcal{T})$ be a topological space. A family $\{A_i\}_{i\in I} \subseteq \mathcal{T}$ is called an *open cover* of a set $E \subseteq X$ if $E \subseteq \bigcup_{i\in I} A_i$. The space X is *compact* if every open cover of X contains a finite subcover. A subset $K \subseteq X$ is *compact* if the topological space $(K, \mathcal{T} \cap K)$ is compact and *relatively compact* if the closure $\overline{K}$ is compact. Every closed subset of a compact space is compact, and every compact subset of a Hausdorff space X is closed. If X_1 and X_2 are compact spaces, then the topological product $X_1 \times X_2$ is compact.

Concerning covers of subsets of $\mathbb{R}^n$ with *closed* balls, we have the following definition.

Definition A.1. A family $\mathcal{G}$ of closed balls $\{\overline{B}_r(x_0) \mid x_0 \in \mathbb{R}^n, r > 0\}$ is a *fine cover* of a set $U \subseteq \mathbb{R}^n$ if $\inf\{r \mid \overline{B}_r(x_0) \in \mathcal{G}\} = 0$ for all $x_0 \in U$.

Let $(X, \mathcal{T})$ be a topological space. A sequence $\{x_n\} \subseteq X$ is *convergent* to $x_0 \in X$ if for every neighborhood F of x_0, there exists $\bar{n} \in \mathbb{N}$ such that $x_n \in F$ for all $n > \bar{n}$; x_0 is an *accumulation point* of $\{x_n\}$ if every neighborhood of x_0 contains infinitely many elements of $\{x_n\}$. A metric space is called *complete* if every Cauchy sequence $\{x_n\} \subseteq X$ converges to some $x \in X$.

A topological space $(X, \mathcal{T})$ is *countably compact* if every countable open cover of X contains a finite subcover; X is *sequentially compact* if every sequence $\{x_n\} \subseteq X$ contains a converging subsequence. A subset $F \subseteq X$ is *sequentially compact* if the topological space $(F, \mathcal{T} \cap F)$ is sequentially compact and *relatively sequentially compact* if the closure $\overline{F}$ is sequentially compact; X is countably compact if and only if every $\{x_n\} \subseteq X$ has an accumulation point. If X is a metric space, then the following statements are equivalent: (i) X is compact; (ii) X is countably compact; (iii) X is sequentially compact. Every relatively sequentially compact subset of a metric space is bounded.

A.6

A topological space X is *locally compact* if every $x \in X$ has a compact neighborhood. A locally compact Hausdorff space is *σ-compact* if it is a countable union of compact sets.

Remark A.1. Every locally compact Hausdorff space X with countable basis is σ-compact with all its open subsets. In fact, let $\{A_n\} \subseteq \mathcal{T}$ be a countable basis, and let $A \in \mathcal{T}$. For every $x \in A$, there exists a compact neighborhood $K_x \subset A$ of x, and hence for some $k \in \mathbb{N}$, we have $x \in \bigcup_{n=1}^{k} A_n \subset \bigcup_{n=1}^{k} \overline{A}_n \subset K_x$. Clearly, every set $\bigcup_{n=1}^{k} \overline{A}_n$ is compact; moreover, the family of these sets is countable, and A is equal to their union.

In agreement with Remark A.1, the following result can be proven.

Proposition A.2. *Let X be a locally compact Hausdorff space. Then the following statements are equivalent:* (i) *X has a countable basis;* (ii) *X is metrizable and σ-compact.*

A completely metrizable topological space with countable basis is called a *Polish space*. Separable Banach spaces and locally compact Hausdorff spaces with countable basis are examples of Polish spaces.

A.7

Let X be a topological space, and let Y be a metric space. We denote by $C(X;Y)$ the space of continuous functions $f : X \to Y$ and by $C_b(X;Y) \subseteq C(X;Y)$ the subspace of bounded continuous functions. We set $C(X) \equiv C(X;\mathbb{R})$ and $C_b(X) \equiv C_b(X;\mathbb{R})$. If Y is a Banach space, then $C_b(X;Y)$ endowed with norm $f \mapsto \|f\|_\infty := \sup_{x\in X} \|f(x,\cdot)\|_Y$ is a Banach space. Let us recall the following result (e. g., see [107, Theorem I.5.1]).

Proposition A.3. *Let X be a compact metric space, and let Y be a complete separable metric space. Then $C_b(X;Y)$ endowed with norm $\|\cdot\|_\infty$ is separable.*

Let X be a topological space. A function $f : X \to \mathbb{R}$ is *upper semicontinuous* (respectively, *lower semicontinuous*) if for each $\alpha \in \mathbb{R}$, the set $\{f < \alpha\}$ (respectively, $\{f > \alpha\}$) is open.

Proposition A.4. *Let X be a countably compact topological space, and let $f : X \to \mathbb{R}$ be upper semicontinuous. Then f is bounded from above and has a maximum.*

Let (X, d_X) and (Y, d_Y) be metric spaces. A function $f : X \to Y$ is *uniformly continuous* if for every $\epsilon > 0$, there exists $\delta > 0$ such that for all $x, x' \in X$ with $d_X(x, x') < \delta$, we have $d_Y(f(x), f(x')) < \epsilon$. If X is compact, then every $f \in C(X; Y)$ is bounded and uniformly continuous.

We denote by $C_c(X) \subseteq C(X)$ the subspace of continuous real functions with compact support (recall that $\operatorname{supp} f := \overline{\{f \neq 0\}}$). By $C_0(X) \subseteq C(X)$ we denote the subspace of continuous functions with the following property: for every $\epsilon > 0$, there exists a compact subset $K \subseteq X$ such that $|f|_{K^c}| < \epsilon$. Clearly, $C_c(X) \subseteq C_0(X)$. Plainly, $C_0(X)$ endowed with norm $f \mapsto \|f\|_\infty := \sup_{x \in X} |f(x)|$ is a Banach space. If X is a locally compact Hausdorff space with countable basis, then $C_0(X)$ endowed with norm $\|\cdot\|_\infty$ is a separable Banach space.

Proposition A.5 (Dini). *Let X be a countably compact topological space. Let $\{f_n\}$ be a nonincreasing sequence of upper semicontinuous functions such that $\lim_{n\to\infty} f_n(x) = 0$ for all $x \in X$. Then $\lim_{n\to\infty} \|f_n\|_\infty = 0$.*

A.8

A topological space X is *completely regular* if for all $x \in X$ and all closed $C \subseteq X$ with $x \notin C$, there exists $f \in C(X)$ such that $f(X) \subseteq [0,1]$, $f(x) = 0$, and $f|_C = 1$. The space X is *normal* if for any closed $E, F \subset X$ with $E \cap F = \emptyset$, there exist a neighborhood U of E and a neighborhood V of F such that $U \cap V = \emptyset$. Every metric space is completely regular and normal.

Proposition A.6 (Urysohn lemma). *A topological space X is normal if and only if for any closed disjoint $E, F \subset X$, there exists $f \in C(X)$ such that $f(X) \subseteq [0,1]$, $f|_E = 0$, and $f|_F = 1$.*

Corollary A.7. *Every locally compact Hausdorff is completely regular.*

Proposition A.8 (Tietze's extension theorem). *Let X be a normal topological space, let F be a closed subset of X, and let $f \in C(F)$. Then there exists $g \in C(X)$ such that $g|_F = f$.*

The following lemma implies that $C_c(X)$ is dense in $C_0(X)$ with norm $\|\cdot\|_\infty$ if X is locally compact.

Lemma A.9. *Let X be a locally compact Hausdorff space, $K \in \mathcal{K}$, and $A \in \mathcal{T}$ with $A \supset K$. Then there exists $f \in C_c(X)$ such that $f(X) \subseteq [0,1]$, $f|_K = 1$, and $\operatorname{supp} f \subseteq A$.*

Proof. Since X is locally compact, there exists $B \in \mathcal{T}$ with $A \supseteq \overline{B} \supseteq K$. By Corollary A.7 X is completely regular, and hence for every $x \in K$, there exists $f_x \in C_c(X)$ such that $f_x(X) \subseteq [0,1]$, $f_x(x) = 1$, and $f_x|_{B^c} = 0$. Since K is compact and every set $\{f_x > \frac{1}{2}\}$ $(x \in K)$ is open, there exist $x_1, \dots, x_p$ such that $K \subseteq \bigcup_{k=1}^{p} \{f_{x_k} > \frac{1}{2}\}$. For every $x \in X$, set $g(x) := 2\max\{f_{x_1}(x), \dots, f_{x_p}(x)\}$, so that $g \in C_c(X)$, $g|_K > 1$, and $g|_{B^c} = 0$. Then $f := \min\{1, g\}$ has the stated properties. □

A.9

Lemma A.10. *Let Y be a separable normed vector space, and let $D \equiv \{y_k \mid k \in \mathbb{N}\} \subseteq Y$ be a dense countable subset. Then the map from Y^* to $[0, \infty)$ defined as*

$$Y^* \ni y^* \mapsto |||y^*||| := \sum_{k=1}^{\infty} \frac{|\langle y^*, y_k \rangle_{Y^*,Y}|}{2^k (1 + \|y_k\|_Y)} \tag{A.1}$$

is a norm on Y^*.

Proposition A.11. *Let Y be a separable normed vector space, and let*

$$\mathcal{V} := \{y^* \in Y^* \mid \|y^*\|_{Y^*} \leq 1\}. \tag{A.2}$$

Let $\mathcal{T}_{w^*}$ be the weak* *topology on Y^*, and let $\tilde{\mathcal{T}}$ be the metric topology on Y^* associated with the norm $|||\cdot|||$ defined in (A.1). Then the relative topologies $\mathcal{T}_{w^*} \cap \mathcal{V}$ and $\tilde{\mathcal{T}} \cap \mathcal{V}$ coincide.*

Therefore

$$\lim_{n\to\infty} \langle y_n^*, y \rangle_{Y^*,Y} = \langle y^*, y \rangle_{Y^*,Y} \quad \text{for every } y \in Y \quad \Leftrightarrow \quad \lim_{n\to\infty} |||y_n^* - y^*||| = 0.$$

List of Symbols

https://doi.org/10.1515/9783110556902-013

Index

https://doi.org/10.1515/9783110556902-014

De Gruyter Studies in Mathematics

Volume 85
Shai Dekel
Pointwise Variable Anisotropic Function Spaces on $\mathbb{R}^n$
ISBN 978-3-11-076176-4, e-ISBN 978-3-11-076179-5,
e-ISBN (ePUB) 978-3-11-076187-0

Volume 84
Marko Kostić
Selected Topics in Almost Periodicity, 2021
ISBN 978-3-11-076322-5, e-ISBN 978-3-11-076352-2,
e-ISBN (ePUB) 978-3-11-076360-7

Volume 83
Lydia Außenhofer, Dikran Dikranjan, Anna Giordano Bruno
Topological Groups and the Pontryagin-van Kampen Duality. An Introduction, 2021
ISBN 978-3-11-065349-6, e-ISBN 978-3-11-065493-6,
e-ISBN (ePUB) 978-3-11-065355-7

Volume 82
Friedrich Wagemann
Crossed Modules, 2021
ISBN 978-3-11-075076-8, e-ISBN 978-3-11-075095-9,
e-ISBN (ePUB) 978-3-11-075099-7

Volume 81
Ľubica Holá, Dušan Holý, Warren Moors
USCO and Quasicontinuous Mappings, 2021
ISBN 978-3-11-075015-7, e-ISBN 978-3-11-075018-8,
e-ISBN (ePUB) 978-3-11-075022-5

Volume 76
Steven Lord, Fedor Sukochev, Dmitriy Zanin
Singular Traces. Volume 1: Theory, 2021
ISBN 978-3-11-037779-8, e-ISBN 978-3-11-037805-4,
e-ISBN (ePUB) 978-3-11-039231-9

Printed and bound by CPI Group (UK) Ltd, Croydon, CR0 4YY
13/09/2026
14961093-0004